W9-ASZ-682

Experiments in Physical Chemistry

**McGRAW-HILL
BOOK COMPANY**

New York
St. Louis
San Francisco
Auckland
Bogotá
Hamburg
Johannesburg
London
Madrid
Mexico
Montreal
New Delhi
Panama
Paris
São Paulo
Singapore
Sydney
Tokyo
Toronto

DAVID P. SHOEMAKER

Professor of Chemistry
Oregon State University

CARL W. GARLAND

Professor of Chemistry
Massachusetts Institute of Technology

JEFFREY I. STEINFELD

Professor of Chemistry
Massachusetts Institute of Technology

JOSEPH W. NIBLER

Professor of Chemistry
Oregon State University

Experiments in Physical Chemistry

FOURTH EDITION

This book was set in Times Roman.
The editors were Marian D. Provenzano and Scott Amerman;
the production supervisor was John Mancia.
New drawings were done by J & R Services, Inc.
The cover was designed by John Hite.

Library of Congress Cataloging in Publication Data

Main entry under title:

Experiments in physical chemistry.

    First-2d ed. by D. P. Shoemaker and C. W. Garland;
3d ed. by D. P. Shoemaker, C. W. Garland, and
J. I. Steinfeld.
    Includes bibliographical references and index.
    1. Chemistry, Physical and theoretical—Laboratory
manuals. I. Shoemaker, David P. II. Shoemaker,
David P. Experiments in physical chemistry.
QD457.S56   1981   541.3'078   80-20081
ISBN 0-07-057005-1

**EXPERIMENTS
IN PHYSICAL
CHEMISTRY**

890 HDHD 89876

CONTENTS

Endpapers

Physical Constants and Conversion Factors
International Relative Atomic Weights

PREFACE

This book is intended as a textbook for a junior-level laboratory course in physical chemistry. It is assumed that the student will be taking concurrently (or has previously taken) a lecture course in the principles of physical chemistry. The book contains 50 selected experiments which have been tested by extensive use. Four of these experiments are new: two involve NMR measurements (Experiments 38 and 46), one concerns laser Raman spectroscopy (Experiment 41), and one describes X-ray precession photography of single crystals (Experiment 48). Four of the experiments from the previous edition have been dropped and numerous small changes have been made in the remaining experiments. The most extensive changes occur in Experiments 10, 12, 17, 42, and 45.

In addition to the experiments themselves there are nine chapters containing material of a general nature. These chapters should be useful not only in undergraduate laboratory courses but also in special project work, graduate thesis research, and general research in chemistry. In recognition of the increasing importance of the reduction, evaluation, and processing of experimental data in the physical sciences, a new chapter on least-squares fitting procedures (Chapter XX) has been introduced, and the chapter on the use of computers (Chapter XXI) has been completely rewritten and expanded. Finally, new sections on operational amplifiers and digital voltmeters have been added to Chapter XV.

The experiments in this book are not primarily concerned with "methods" or "techniques" per se or with the analytical applications of physical chemistry. We believe that an experimental physical chemistry course should serve a dual purpose: (1) to illustrate and test established theoretical principles and (2) to develop a research orientation by providing basic experience with physical measurements that yield quantitative results of important chemical interest.

Each experiment is accompanied by a theoretical development in sufficient detail to provide a clear understanding of the method to be used, the calculations required, and the significance of the final results. The depth of coverage is frequently greater than that which is available in introductory physical chemistry textbooks. Experimental procedures are described in considerable detail as an aid to the efficient use of laboratory time and teaching staff. Emphasis is given to the reasons behind the design and procedure for each experiment so that the student can learn the general principles of a variety of experimental techniques. Stimulation of individual resourcefulness through the use of special projects or variations on existing experiments is also desirable. We strongly urge that on some occasions the experiments presented here should be used as points of departure for work of a more independent nature.

We gratefully acknowledge the assistance of students, teaching assistants, and faculty colleagues at the Massachusetts Institute of Technology and Oregon State University, as well as many helpful comments from faculty who have used this book at other universities. We encourage and welcome feedback from those who are using the book, either as students or as instructors.

DAVID P. SHOEMAKER
CARL W. GARLAND
JEFFREY I. STEINFELD
JOSEPH W. NIBLER

I

INTRODUCTION

Physical chemistry deals with the physical principles underlying the properties of chemical substances. Like other branches of physical science, it contains a body of theory which has stood the test of experiment and which is continually growing as a result of new experiments. In order to learn physical chemistry, one must become familiar with the experimental foundations on which the theoretical principles are based. Indeed, in many cases, the ability to apply the principles usefully requires an intimate knowledge of those methods and practical arts that are called "experimental technique."

For this reason, a lecture course in physical chemistry should be accompanied by a program of laboratory work. Such experimental work not only should demonstrate established principles but should also develop research aptitudes by providing experience with the kind of measurements that yield important new results. This book attempts to achieve that goal. Its aim is to provide a clear understanding of the principles of important experimental methods, the design of basic apparatus, the planning of experimental procedures, and the significance of the final results. In short, the aim is to train not laboratory technicians but future research scientists.

Severe limitations of time and equipment must be faced in presenting a set of experiments as the basis of a laboratory course that will provide a reasonably broad coverage of the wide and varied field of physical chemistry. Although high-precision research measurements would occasionally require refinements in the methods described here and would often require more sophisticated and more elaborate equipment, each experiment in this book is designed so that meaningful results of reasonable accuracy can be obtained.

In the beginning, a student will seldom have the skill and the time to plan each experimental procedure by himself. However, the trial-and-error method is not the way to learn good experimental technique. Laboratory skills are developed slowly during a sort of apprenticeship. To enable a new student to make efficient use of available time, both the apparatus and the procedure for these experiments are described in considerable detail. The student should keep in mind the importance of understanding why the experiment is done in the way described. This understanding is a vital part of the experience necessary for planning special or advanced experiments of a research character. As the student becomes more experienced, it is desirable that he be required to plan more of his own procedure. This can be easily accomplished by introducing variations in the experiments described here. A change in the chemical system to be studied, the use of equipment which purposefully differs from that described, or the choice of a different method for studying the same system will force the student to work out modifications of the procedure. Finally, at the end of the course, it is recommended that some students do special projects which are completely independent of the experiments described in this book. A description of such special projects is given at the end of this chapter.

In addition to a general knowledge of laboratory techniques, creative research work requires the ability to apply two different kinds of theory. Many an experimental method is based on a special phenomenological theory of its own; this must be well understood in order to design the experiment properly and in order to calculate the desired physical property from the observed raw data. Once the desired result has been obtained, it is necessary to understand its significance and its interrelationship with other known facts. This requires a sound knowledge of the fundamental theories of physical chemistry (e.g., thermodynamics, statistical mechanics, and quantum theory). Considerable emphasis has been placed on both kinds of theory in this book.

In the final analysis, however, research ability cannot be learned merely by performing experiments described in a textbook; it has to be acquired through contact with inspired teachers and through the accumulation of considerable experience. The most this book can attempt to do is to provide a solid frame of reference for future growth.

PREPARATION FOR AN EXPERIMENT

Although most of the experiments in this book can be performed by a single person, they have been written with the assumption that a pair of students will work together as a team. Such teamwork is advantageous, since it provides an opportunity for valuable discussion of the experiment between partners. The amount of experimental work to be assigned will be based on the amount of laboratory time available for each experiment. Many of the experiments can be completed in full during a single four-hour laboratory period. Many others are designed for six to eight hours of laboratory work but may be abridged so that meaningful results can be obtained in a single four-hour period.

Some of the experiments require at least six hours and should not be attempted in a shorter time (in particular, Exps. 6, 28, 32, 33, 35).

Before the student arrives in the laboratory to perform a given experiment, it is essential that he study the experiment carefully with special emphasis on the method, the apparatus design, and the procedure. It will usually be necessary to make changes in the procedure whenever the apparatus to be used or the system to be studied differs from that described in this book. Planning such changes or even successfully carrying out the experiment as described requires a clear understanding of the experimental method.

APPARATUS AND CHEMICALS

A complete and very detailed list of equipment and chemicals† is given at the end of each experiment. The list is divided into two sections: Those items listed in the first paragraph are required for the exclusive use of a single team; those in the second paragraph are available for the common use of several teams performing the same experiment simultaneously. It is assumed that standardized stock solutions will be made up in advance and will be available for the student's use. The quantities indicated in parentheses‡ are for the use of the instructor and do *not* indicate the amounts of each chemical to be taken by a single team. In addition to the items included in these apparatus lists, it is assumed that the laboratory is equipped with analytical balances, a distilled-water supply, and a barometer as well as gas, water,

---

† A convenient source of information about the commerical availability and current prices of less common chemicals is "Chem Sources," published annually by Directories Publishing Co., Flemington, N.J. 08822.

‡ On the basis of the authors' experience with the laboratory course at MIT, it is necessary to make available these amounts per team that will do the experiment. They are scaled up from the amounts stated in the experiment to provide for possible wastage and to give a generous safety factor. It is hoped that they will be useful as a rough guide the first time an experiment is given.

and 110-V ac power lines. Also desirable, but not absolutely necessary, are dc power lines, gas-handling lines, and a rough vacuum line.

Experimental work in physical chemistry requires many complex and expensive pieces of apparatus; many of these have been constructed especially for the student's use and cannot be readily replaced. Each team should accept complete responsibility for its equipment and should *check it over carefully before starting an experiment.*

SAFETY

Experimental work is subject to hazards of many kinds, of which every person working in a laboratory should be aware. Once one is aware of the particular hazards involved in an experimental procedure, one's instinct for self-preservation usually provides a sufficient motivation for finding ways of avoiding them. The principal danger lies in ignorance of specific hazards and in forgetfulness.

A detailed analysis of all laboratory hazards and procedures for dealing with them would be beyond the scope of this book. Certain specific hazards are pointed out in connection with individual experiments. Some general remarks on the kinds of safety hazards which should be kept in mind are given in Appendix D. It is assumed that the instructing staff will provide specific warnings and reminders where needed.

RECORDING OF EXPERIMENTAL DATA

The laboratory notebook is the essential link between the laboratory and the outside world and the ultimate reference concerning what took place in the laboratory. As such, it is not only the source book for the production of reports and publications but also a permanent record which may be consulted even after a great many years. It is standard practice in experimental research work to record *everything relevant* (data, calculations, notes and comments, literature surveys, and even some graphs) directly in a bound notebook with numbered pages. Such notebooks are available with pages which are ruled vertically as well as horizontally to give a $\frac{1}{4}$-in. grid; this facilitates tabulation of columns of figures and permits rough plots of the data to be made directly on the notebook pages during the experiment. In an undergraduate laboratory course it is often convenient to make a carbon copy of the recorded data to include with the written report. (Notebooks with duplicate sets of numbered pages, in which alternate pages are perforated for removal, may be used; however, ordinary carbon paper and bond paper can be used with any style of research notebook.) In any case, the *original* is a part of a permanent notebook. Whatever style of notebook is used, the principle is the same: *Record all data directly in your notebook.*

Data may be copied into the notebook from a partner's notebook in those cases where it is clearly impossible for both partners to record data at the same time. Even then an extra carbon copy of a single original page is often better since it avoids copy errors and saves time when a large number of figures are involved. In particular, do *not* record on odd scraps of paper such incidental data as weights, barometer readings, and temperatures with the idea of copying them into the notebook at a later time. If anything must be copied from another source (calibration chart, reference books, etc.) identify it with an appropriate reference.

A ball-point pen is best for recording data, especially if carbon copies are required; otherwise a fountain pen with permanent ink is also satisfactory. Pencils are usually considered unsuitable for recording primary data. If a correction is necessary, draw a single line through the incorrect number so as to leave it legible, and then write the correct number directly above or beside the old one. If something happens to vitiate the data on an entire page, cross out these data and record the circumstances. (No original pages should ever be removed from a laboratory notebook.) Never skip notebook pages, or any significant amount of space on a notebook page, for readings to be filled in later; always record your data in serial fashion except where it is appropriate to record data in tabular form. Neatness and good organization are desirable, but legibility, proper labeling, and completeness are absolute necessities.

As each experiment is begun, make an entry for it in the index at the beginning of the book, indicating the date begun and the page number of the beginning of the record. Where the experiment has logical subdivisions, these too should be entered in the index with their beginning dates and page numbers.

What to record Every data page for an experiment should have a clear heading which includes your name (and the name of any lab partner), the subject of the experiment, the date, and a page number.

The notebook should contain all the information that would be needed to permit someone else to perform the same experiment in the same way. In addition to the measurements and observations that constitute the results of the experiment, all other data that are relevant to the interpretation of the results should be recorded. It is not necessary to duplicate information that is conveniently available in some permanent record such as a journal article or laboratory text, provided that complete references are given.

The record of an experiment should begin with a brief statement of the experiment to be performed, with balanced chemical equations where relevant. The procedure being followed should be described in all essential detail. When the procedure is described elsewhere, the notebook entry may be abbreviated to a reference to the published description. However, be certain to record any variations from the published procedure and to specify the apparatus which was used (make and model for commercial equipment, otherwise a sketch). For each chemical substance used, record the name, formula, source, grade, or stated purity (and concentration in the case of a solution).

Data should be entered directly in the notebook, in tabular form whenever possible. Use complete, explicit headings; do not rely on your memory for the meaning of unconventional symbols and abbreviations. Be sure that all numerical values are accompanied by the appropriate units. In many cases, it is necessary or at least wise to record such laboratory conditions as the ambient temperature, the atmospheric pressure, or the relative humidity.

Instrumental records, such as spectra, may be taped or stapled into the notebook if they are of a convenient size. This should not be done if it makes the notebook clumsy and awkward to use. Records which are not fastened in the notebook should be numbered and placed in a laboratory file provided for the purpose, and an appropriate reference placed in the notebook. In any case, they should be dated and signed by the investigator.

LITERATURE WORK

Every attempt has been made to write each experiment in sufficient detail so that it can be intelligently performed without the necessity of extensive outside reading. However, it is assumed that the student will refer frequently to a standard textbook in physical chemistry for any necessary review of elementary theory. Literature sources are explicitly cited for those topics of an advanced or special nature which are beyond the scope of a typical undergraduate textbook, and these numbered references are listed together at the end of each experiment. In addition, a selected list of reading pertinent to the general topic of each experiment is given under the heading General Reading. It is hoped that the interested student will do as much reading in these books and journal articles as his time allows, since such reading is an important aspect of broadening and deepening his scientific background.

As a matter of policy, very few of the experiments contain a direct reference to published values of the final result which is to be reported by the student. Any student who wishes to compare his result with the accepted literature value is expected to do the necessary library work. As a general principle it is best to refer directly to an original journal article rather than to some secondary source. It is commonly assumed that recent measurements are more precise than older ones; this assumption is based on the fact that methods and equipment are constantly being improved. But this does not mean that there is valid reason to reject or suspect a published result merely because it is old. The quality of research data depends strongly on the integrity, conscientious care, and patience of the research worker; much fine work done many years ago in certain areas of physical chemistry has not yet been improved upon. In evaluating results based on old but high-quality research data, one must, however, be alert to the possible need for corrections necessitated by more recent theoretical developments or by improved values of physical constants.

Literature search It is convenient to distinguish between two types of literature search—one which is concerned with information about a single chemical species, and one concerned with a broad area of research, a method, a class of reactions, or a class of compounds. In the former case, it is usually best to begin directly with abstract journals. In the latter case, it is more efficient to begin by checking specialized monographs (consult the card catalog in the library), review journals, and annual series such as "Advances in ———."

The chemical literature of the review type consists of collections of articles or chapters on specific topics written by presumed experts. The introduction to these reviews usually specifies the time period covered. Ideally, the reviews include critical evaluations of the work and progress in a specific area of chemical research together with a rather complete bibliography of the important original research articles pertinent to that area. However, it should not be assumed that every reference to a particular subject is given or that the author's critical interpretations are necessarily correct. A selection of the review literature which is of special value in physical chemistry is given below.

Review Journals:
    Accounts of Chemical Research
    Advances in Physics
    Chemical Reviews
    Quarterly Reviews (London)
    Reviews of Pure and Applied Chemistry

Review Series:
    "Advances in Chemistry Series," American Chemical Society, Washington, D.C. (1950–, issued at irregular intervals).
    "Advances in Chemical Physics," I. Prigogine and S. Rice (eds.), Interscience, New York (1958–).
    "Advances in Magnetic Resonance," J. S. Waugh (ed.), Academic, New York (1965–).
    "Advances in Photochemistry," W. A. Noyes et al. (eds.), Interscience-Wiley, New York (1963–).

"Advances in Quantum Chemistry," P.-O. Löwdin (ed.), Academic, New York (1964–).

"Annual Review of Physical Chemistry," H. Eyring, C. J. Christensen, and H. S. Johnston, Annual Reviews, Inc., Palo Alto, Calif. (1950–).

"Perspectives in Structural Chemistry," J. D. Dunitz and J. A. Ibers (eds.), Wiley, New York (1967–).

"Progress in Nuclear Magnetic Resonance Spectroscopy," J. W. Emsley, J. Feeney, and L. H. Sutcliffe (eds.), Pergamon, London (1966–).

"Progress in Reaction Kinetics," G. Porter (ed.), Pergamon, New York (1961–).

"Solid State Physics," H. Ehrenreich, F. Seitz, and D. Turnbull (eds.), Academic, New York (1955–).

"Structure and Bonding," C. K. Jorgensen et al. (eds.), Springer-Verlag, New York (1966–).

"Structure Reports," W. B. Pearson and J. Trotter (eds.), Oosthoek, Utrecht, vol. 8– (1940–) [vols. 1–7 were published in German under the title "Struktur-bericht"].

The use of *Chemical Abstracts* is the most effective way to make a detailed search of the literature published from 1907 to the present time. Some practice is necessary to acquire a rapid and thorough search technique. The considerations outlined below should be helpful as an introduction to the use of these abstracts.

In searching for a specific compound it is useful to keep two considerations in mind. First, it is usually advisable to begin this type of literature search with the most recent abstract journal indexes available. In this way one may find recent review articles or monographs which are sufficiently thorough in coverage that a complete search of the older literature may be avoided. Second, the use of subject indexes is generally preferred to formula indexes. Experience has shown that tabulation of references for the specific compound of interest is occasionally more complete when subject indexes are used. In addition, reference to the physical and chemical properties of the compound and to its simple derivatives are more easily obtained because these are listed in subheadings under the name of the compound. It is of obvious importance to search for the desired compound by using the correct name. To assure that the name of the compound is consistent with *Chemical Abstracts* usage, it is frequently convenient to find the compound in the formula index and then to use the name given there in a search of the subject indexes, or to refer to "The Naming and Indexing of Chemical Compounds from *Chemical Abstracts*" (Introduction to the Subject Index of Vol. 56), Chemical Abstracts Service, American Chemical Society, Washington, D.C. (1962).

Formula and subject indexes contain page numbers which refer directly to the indicated abstract volumes. The abstracts in these volumes are brief summaries of the important contents of the original research publications. Once the abstracts describing the research of interest have been located, the original research journals should be consulted whenever possible. If a necessary journal is not available, photographic reproductions of especially pertinent papers can be obtained through the use of interlibrary loan services. The availability of all journals abstracted in *Chemical Abstracts* is listed in "Access: Key to the Source Literature of the Chemical Sciences," American Chemical Society (1969).

It is important to realize that there is a delay of 6 to 12 months between the time a journal article appears and the time its abstract is printed in *Chemical Abstracts*. In addition, there is a delay of about 12 months after a volume is complete before the subject index becomes available. Thus it is often wise to check the most recent literature by scanning the title pages of likely journals. A list of journals of greatest interest to physical chemists is given in Appendix E.

Another search method involves the use of "Science Citation Index," which is published quarterly in three parts. The Source Index gives a listing of authors with the titles of their recent publications; the Subject Index is organized under key words in the title of publications; the Citation Index lists recent papers that have cited in a given publication.

In recent years, a number of computer systems have been devised to aid in the search of current literature. A system called "NASIC" is now in use at MIT. This system is tied into a Chemical Abstracts-Condensates data base, among others, and is supposed to provide both current awareness and retrospective literature searches based on subject headings or other cross references. Such computer-based services are at present extremely limited in coverage, however; and they should be used only to supplement, not to replace, a proper literature search.

Reference books The books listed below, a small selection from the very large number which are available, describe important experimental techniques and apparatus for a variety of physical measurements. The more advanced texts also deal with the theory of the methods described.

H. B. Jonassen and A. Weissberger (eds.), "Technique of Inorganic Chemistry," Interscience-Wiley, New York (1963–).
I. M. Kolthoff and P. J. Elving (eds.), "Treatise on Analytical Chemistry," Interscience, New York (1959–).
L. Marton (ed.), "Methods of Experimental Physics," Academic, New York (1959–).
A Weissberger (ed.), "Technique of Organic Chemistry," 3d ed., Interscience, New York (1959–).

Handbooks and compilations of physical properties   The sources listed below are mainly devoted to numerical tabulations of various physical properties. They are convenient, but some are distinctly secondary sources of information. It is often difficult to judge the quality of the data listed, since references to the original sources are sometimes inadequate. If at all possible, it is wise to confirm important information by consulting the original literature. A particularly useful reference for obtaining a wide variety of routine physical data is the "Handbook of Chemistry and Physics." Although a new edition of this handbook is issued every year, changes are introduced very slowly. Any recent copy is likely to be as useful as the newest one for almost all purposes.

"American Institute of Physics Handbook," D. E. Gray et al. (eds.), 3d ed., McGraw-Hill, New York (1972).

"Crystal Data," J. and G. Donnay (eds.), Am. Cryst. Assoc. Monogr. nos. 5 (1963) and 6 (1967); 3d ed., J. Donnay and H. Ondik (eds.), U.S. Natl. Bur. Stand. and Joint Comm. on Powder Diffraction Stand., vols. 1 (1972), 2 (1973), 3 and 4 (1979).

"Crystal Structures," R. W. G. Wyckoff (ed.), 2d ed., Interscience-Wiley, New York, vol. 1 (1963)–vol. 6 (1971).

"Handbook of Chemistry and Physics," R. C. Weast (ed.), 59th ed., Chemical Rubber Publishing Co., Cleveland, Ohio (1978/9).

"Interatomic Distances," The Chemical Society, London (1958); Supplement (1965).

"International Critical Tables of Numerical Data, Physics, Chemistry, and Technology," vols. I–VII, McGraw-Hill, New York (1926).

"Journal of Physical and Chemical Reference Data," published for the U.S. Natl. Bur. Stand. by American Chemical Society and American Institute of Physics, Washington, D.C. (1972–).

"JANAF Thermochemical Tables," Dow Chemical Co., U.S. Natl. Bur. Stand., Institute of Applied Technology, Washington, D.C. (1965–).

"Landolt-Börnstein Numerical Data and Functional Relationships in Science and Technology. New Series," K.-H. Hellwege (ed.), Springer-Verlag, Berlin, Group II: Atomic and Molecular Physics (1965–).

"National Standard Reference Data Series," U.S. Natl. Bur. Stand., U.S. Government Printing Office, Washington, D.C. (1964–).†

† This NSRDS series consists of many different volumes, edited by different persons, on specialized topics; check a library card catalog for further information. As of 1972, the *Journal of Physical and Chemical Reference Data* became the publication vehicle of NSRDS.

"Physical Properties of Chemical Compounds," nos. 15, 22, 29 of "Advances in Chemistry Series," American Chemical Society, Washington, D.C.
"Selected Values of Chemical Thermodynamic Properties," U.S. Natl. Bur. Stand. Circ. 500, U.S. Government Printing Office, Washington, D.C. (1952).
"Solubilities of Inorganic and Organic Compounds," H. Stephen and T. Stephen (eds.), vols. 1 and 2, Macmillan, New York (1963).
"Solubilities: Inorganic and Metal-organic Compounds," W. F. Linke (ed.), 4th ed., Van Nostrand, New York (1958–).
"Tables of Chemical Kinetics: Homogeneous Reactions," U.S. Natl. Bur. Stand. Circ. 510, U.S. Government Printing Office, Washington, D.C. (1951).
"Tables of Experimental Dipole Moments," A. L. McClellan, Freeman, San Francisco (1963).
"Tables of Thermal Properties of Gases," U.S. Natl. Bur. Stand. Circ. 564, U.S. Government Printing Office, Washington, D.C. (1955).
"Thermophysical Properties of Matter: The TPRC Data Series," Y. S. Touloukian and C. Y. Ho (eds.), IFI/Plenum, New York, vols. 1–7 (1970).

Chemical nomenclature In searching the chemical literature and in reading and writing research papers, a knowledge of systematic nomenclature for chemical compounds is indispensable. In 1957 the International Union of Pure and Applied Chemistry (IUPAC), Committee on Chemical Nomenclature, recommended the adoption of an international nomenclature. The more important nomenclature rules, together with recommended chemical symbols and terminology, were published in *J. Amer. Chem. Soc.*, **82**, 5517, 5523, 5545, 5575 (1960).

A more detailed set of IUPAC nomenclature rules for organic compounds is given in the following:

"Nomenclature of Organic Chemistry, Definitive Rules for Section A. Hydrocarbons. Section B. Fundamental Heterocyclic Systems," International Union of Pure and Applied Chemistry, 2d ed., Butterworths, London (1966).
"Nomenclature of Organic Chemistry, Definitive Rules for Section C. Characteristic Groups Containing Carbon, Hydrogen, Oxygen, Nitrogen, Halogen, Sulfur, Selenium, and/or Tellurium," International Union of Pure and Applied Chemistry, Butterworths, London (1965).

The most useful single source to consult in the event of problems of nomenclature for organic and inorganic compounds is:

"The Naming and Indexing of Chemical Compounds from *Chemical Abstracts*" (Introduction to the Subject Index of Vol. 56), Chemical Abstracts Service, American Chemical Society, Washington, D.C. (1962).

REPORTS

The final evaluation of any experimental work is based primarily on the examination of a written report. This report should be well organized and readable, so that anyone unfamiliar with the experiment can easily follow the presentation (with the aid of explicit references where necessary) and thereby obtain a clear idea as to what was actually done and what result was obtained.

An attempt should be made to use a scientific style comparable in quality to the literary style expected in an essay. Correct spelling and grammar should not be disregarded just because the report is to be read by a scientist instead of the editor of a literary magazine. The report should be as concise and factual as possible without sacrificing clarity. In particular, mathematical equations should be accompanied by enough verbal material to make their meaning clear.

Two general sources of information dealing with proper literary usage are:

H. W. Fowler, "A Dictionary of Modern English Usage," 2d ed., rev. by E. Gowers, Oxford, New York (1965).

W. Strunk, Jr., and E. B. White, "The Elements of Style," Macmillan, New York (1959).

Two related sources dealing specifically with the writing of technical papers are:

L. F. Feiser and M. Feiser, "Style Guide for Chemists," Reinhold, New York (1960).

D. H. Menzel, H. M. Jones, and L. G. Boyd, "Writing a Technical Paper," McGraw-Hill, New York (1961).

Technical details (such as recommended symbols, nomenclature, and abbreviations and the proper presentation of formulas, equations, tables, and literature citations) of importance in preparing journal articles are very fully discussed in:

"Handbook for Authors," American Chemical Society, Washington, D.C. (1967). (Specific advice on manuscripts is intended for those submitting a paper to one of the journals published by the American Chemical Society.)

"Style Manual for Guidance in the Preparation of Papers for Journals Published by the American Institute of Physics," 2d ed., American Institute of Physics, New York (1959, revised 1965).

The preparation of journal articles will not be discussed here. The recommendations given below concern student reports written as part of a laboratory course, although the same advice might well serve for technical reports of any kind.

Most important of all, the report must be an original piece of writing. Copying or even paraphrasing of material from textbooks, printed notes, or other reports is clearly dishonest and must be carefully avoided. Brief quotations, enclosed in quotation marks and accompanied by a complete reference, are permissible where a real advantage is to be gained. Certainly there is no point in giving more than a brief summary of the theory or the details of experimental procedure if these are adequately described in some readily available reference. In part, a report is likely to be judged on how clearly it states the essential points without oscillating between minute detail on one topic and vague generalities on another.

Except for general physical and numerical constants or well-known theoretical equations, any data or material taken from an outside source must be accompanied by a complete reference to that source.

The content and length of any given report will depend on the subject matter of the experiment and on the standards established by the instructor. It is our belief that at least in some cases the report should be quite complete and should include a quantitative analysis of the experimental uncertainties and a detailed discussion of the significance of the results (see the sample report given below). For many experiments a brief report (with only a qualitative treatment of errors and a short discussion) may be considered adequate. In either case, a clear presentation of the data, calculations, and results is essential to every experiment in physical chemistry.

Format   Unless otherwise instructed, all reports should be prepared on $8\frac{1}{2}$- by 11-in. paper with reasonable margins on all sides. The pages should be stapled together or bound in a folder with paper fasteners. Legibility is absolutely essential. Double-spaced typewritten reports are a pleasure to read, but they are time consuming to prepare unless one is a facile typist. Handwritten reports submitted in ink on wide-line ruled paper are perfectly satisfactory unless you are cursed with illegible hand-writing. Crossing out and the insertion of corrections are permissible, but try to keep the report as a whole reasonably neat.

Presentation of graphs   A general discussion of the graphical treatment of experimental data is given in Chap. II. As part of that discussion, the proper technique for plotting data points and drawing lines or curves is fully described. We shall be concerned here only with the final steps necessary for the presentation of such graphs as part of a report.

Vertical and horizontal axes should be drawn in, and the main divisions along each axis must be clearly marked and numbered. Each axis is then labeled with the appropriate symbol or words with the units indicated in parentheses [for example: t (sec), A (cm^2), density (g cm^{-3})]. The data points and the symbols surrounding them (usually small circles) are inked in so that the data will stand out prominently.

All light lines are "heavied up" with a sharp pencil, but the final line should not actually be drawn through symbols surrounding the data points. If several lines or curves lie close together, distinguish them from one another by using dashed lines as well as solid lines. When necessary, one can achieve further differentiation by varying the lengths of the dashes or alternating long and short dashes. If a curve is drawn to represent an equation, the points should not be inked or encircled and the curve should be drawn so as to conceal the points. The equation itself or the number by which it is designated in the text should be written beside the curve. It is good practice to indicate clearly on the graph the numerical values of any slopes, intercepts, areas, maxima, or other features that are important in the calculations.

Each figure must have a figure number and a short legend prominently displayed, and it should be referred to by number in the body of the report.

Sample report   Given on pages 15 to 22 is a sample report on a very simple experiment. Its purpose is to illustrate how a report should be organized and to indicate the kind of material it should contain. This example is not meant to provide a rigid outline; the content of any given report will necessarily depend upon the judgment of the individual student. General comments on the various sections of this sample report are given in a series of footnotes.

SPECIAL PROJECTS

In order to become a creative and independent research scientist, one must acquire a complex set of abilities. It is often necessary to invent new experimental methods or at least to adapt old ones to new needs. New apparatus must be designed, constructed, and fully tested. Most important of all, an intelligent procedure must be established for the use of this apparatus in making precise measurements. Performing assigned experiments which are described in detail is merely the first step in developing such research ability. Later on, individually supervised experimental work on an original thesis problem will often be undertaken in order to develop independence and experience with advanced research techniques. In preparation for thesis work in physical chemistry we have found it profitable to encourage interested students to perform a "special project" in lieu of two or three regular experiments.

These special projects are intended to provide experience in choosing an interesting topic, in designing an experiment with the aid of literature references, in building apparatus, and in planning an appropriate experimental procedure. At least 15 hours of laboratory time should be available for carrying out such an experiment. Although there are certain limitations which are imposed by the available time and equipment, challenging and feasible topics with a research flavor can be found in most branches

DETERMINATION OF THE DENSITY OF CRYSTALLINE GERMANIUM [a]

Maria Smith   Oct. 15, 1980

Partner: John Klein

I.   Introduction[b]

The purpose of this experiment is to measure the density of crystals of germanium.   Since the density ρ is defined by

$$\rho = W_S/V_S \tag{1}$$

it is desired to measure the volume V_S occupied by a known weight W_S of the metal sample.

The method involves the use of a pycnometer of known volume which is first weighted empty, then weighed containing the solid sample to be studied.   The difference gives the weight of the solid, W_S. Finally the pycnometer (containing the solid sample) is filled with a liquid of known density and reweighed; the weight, and therefore the volume, of the liquid can be found by difference.   Since the total volume of the pycnometer is known, one can then calculate the volume V_S which is occupied by the solid.

[a] In addition to the title of the report, the heading should include your name, your partner's name, and the date on which the report was submitted.

[b] The introduction should state the purpose of the experiment and give a *very brief* outline of the necessary theory, which is often accomplished by citing pertinent equations. (In this sample report, the theory is trivially simple.) A very short description of the experimental method, including mention of any special apparatus should also be included. The introduction should cover the above topics as concisely as possible; this sample contains about 135 words. More complicated experiments will require longer introductions, but the normal length should be between 100 and 300 words.

II. Experimental[c]

The experimental method was similar to that described in the textbook (Aardvark and Zebra, 3rd ed., Exp. 13). The design of the pycnometer used, which differs from that described in the textbook, is shown in the following sketch:

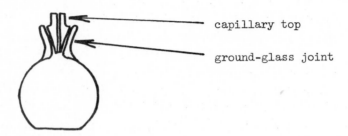

capillary top

ground-glass joint

The procedure was modified as follows: After distilled water had been added to the pycnometer containing the sample, the pycnometer (with capillary top removed) was completely immersed in a flask of distilled water and boiled under low pressure for 15 minutes to remove air trapped by the solid or dissolved by the water.[1] After this boiling, the pycnometer was equilibrated for 15 minutes in a 25°C thermostat bath before the top was inserted.

Two duplicate runs, carried out using the same procedure, were made on each of two different germanium samples. Sample I consisted of larger and somewhat more irregular pieces than did Sample II.

---

[c] This section is usually *extremely brief* and merely cites the appropriate references which describe the details of the experimental procedure. If reference is made to the textbook and/or laboratory notes assigned for the course, an abbreviated title may be cited in lieu of a complete bibliographic entry. Any references to other books or material should be assigned footnote numbers and should be properly listed at the end of the experiment in the form illustrated by the references in this book. Description of experimental procedures should be given *only* for those features not described in or differing from the reference. A simple sketch of apparatus is appropriate only when it differs from that described in the reference. NOTE: A statement of the number of runs made and the conditions under which they were carried out (concentration, temperature, etc.) should always be included at the end of this section.

III. Calculations [d]

The weight of the solid sample is given by

$$W_S = W_2 - W_1 \tag{2}$$

where W_1 is the weight of the empty pycnometer and W_2 is that of the pycnometer plus the solid sample. The weight of water contained in the pycnometer, W_L, is

$$W_L = W_3 - W_2 \tag{3}$$

where W_3 is the weight of pycnometer plus sample plus water. If the density of the liquid (water) is denoted by ρ_L, it follows from Eq. (3) that the volume of the solid sample is given by

$$V_S = V - V_L = \frac{\rho_L V + W_2 - W_3}{\rho_L} \tag{4}$$

where V is the total volume of the pycnometer. From Eqs. (1), (2), and (4), we obtain

$$\rho = \frac{W_S}{V_S} = \frac{\rho_L (W_2 - W_1)}{\rho_L V + W_2 - W_3} \tag{5}$$

Since the values of V and ρ_L are known, it is only necessary to determine W_1, W_2 and W_3 in order to calculate the density of the solid. The average values of these weights are listed below together with the values of V and ρ_L.

[d] A condensed derivation of the equations to be used may be given here. Each equation should be part of a complete sentence; number all equations consecutively throughout the entire report and refer to them by number. All symbols should be defined at the point where they first appear. A condensed tabulation of essential raw data to be used in the calculations is often useful. For a long calculation, it is very desirable to tabulate all important intermediate results. It is unnecessary and undesirable to present all computations in the report; however, a typical sample computation should be given to illustrate how the calculations were performed. A general discussion of the proper methods for handling calculations is given in Chap. II. Many reports will require graphical presentation of the data or the calculated results. Each graph should be given a figure number and a title; it should be referred to in the text by number.

$$\rho_L = 0.997048 \text{ g cm}^{-3} \text{ at } 25°C \text{ (taken from the}$$
$$\text{Handbook of Chemistry and Physics}[2])$$

$$V = 12.410 \pm 0.004 \text{ cm}^{-3} \quad \text{(given by instructor)}$$

$$W_1 = 8.6309 \text{ g}$$

Sample I: $W_2 = 42.0301 \text{ g}$

$W_3 = 48.1732 \text{ g}$

Sample II: $W_2 = 45.8479 \text{ g}$

$W_3 = 51.2944 \text{ g}$

The density of germanium can now be calculated by substitution of the above data into Eq. (5).   The computation will be shown in detail for Sample I:

$$\rho_I = \frac{(0.99705)(42.0301 - 8.6309)}{(0.99705)(12.410) + 42.0301 - 48.1732}$$

$$= \frac{(0.99705)(33.3992)}{12.3733 - 6.1431} = \frac{33.3003}{6.2302} = 5.345 \text{ g cm}^{-3}$$

The result for Sample II is 5.357 g cm^{-3}.   The weights used in these calculations have not been corrected for the effect of air buoyancy on the weighings.   Rather than correct each weight to vacuum, we may use a simple formula given by Bauer[1] for correcting the final calculated result.   This formula gives for the corrected density ρ^*,

$$\rho^* = \rho + 0.0012 \left(1 - \rho / \rho_L\right) \tag{6}$$

When Eq. (6) is applied to our results we obtain for ρ^* the following values:

$$
\begin{array}{ll}
\text{Sample I:} & 5.340 \text{ g cm}^{-3} \\
\text{Sample II:} & 5.352 \text{ g cm}^{-3} \\
\text{Average:} & 5.346 \text{ g cm}^{-3}
\end{array}
\tag{7}
$$

IV.   Uncertainties in Results[e]

According to Eq. (5) the uncertainty in ρ will depend on the uncertainty in each of five variables; however, the value of ρ_L is known to six significant figures and its uncertainty may certainly be neglected in comparison to those of the other variables. We may take as reasonable limits of error for the weighings $\lambda(W_1) = \lambda(W_2) = 0.001$ g and $\lambda(W_3) = 0.002$ g. The higher value for $\lambda(W_3)$ includes the possible failure to attain an exact filling of the pycnometer with water. For $\lambda(V)$ we take 0.004 cm^3, the value given by the instructor, although such a value seems rather high. On the basis of these uncertainty values, it is clear that the major contributors to the limit of error $\lambda(\rho)$ are the uncertainty in V and, to a lesser extent, the uncertainty in the difference W_2-W_3. The contribution to $\lambda(\rho)$ due to the uncertainty in the difference $W_2 - W_1$ is much less (since $W_2 - W_1$ is about 5.5 times larger than $W_2 - W_3$ and the uncertainty in $W_2 - W_1$ is less than that in $W_2 - W_3$) and can be neglected in obtaining an approximate value for $\lambda(\rho)$. With this simplification, the limit of error in ρ is approximately given by

$$\lambda^2(\rho) \simeq \frac{\rho^2}{(\rho_L V + W_2 - W_3)^2}[\rho_L^2 \lambda^2(V) + \lambda^2(W_2) + \lambda^2(W_3)] \quad (8)$$

[e] The type of treatment of uncertainties will depend a great deal on the nature of the experiment; see Chap. II for a detailed discussion of error analysis. The material given above is typical of a straightforward propagation-of-errors treatment. It is important to combine and simplify all expressions as much as possible in order to avoid obtaining unwieldy error equations. Since uncertainty figures need not be calculated to better than about 10 or 20 percent accuracy, one should always try to find labor-saving approximations. Where the number of runs is so small that reliable limits of error cannot be deduced from statistical considerations, limits of error must be assigned largely on the basis of experience and judgment. Make them large enough to be safe, but not ridiculously large. For a long and detailed report, a quantitative analysis of uncertainties should always be derived and a numerical value of the limit of error (or some other appropriate measure of uncertainty) should be presented. For a brief report, a qualitative discussion of the sources of error may suffice. In such a case, this section may be omitted and the error discussion included as part of the general discussion.

Thus for Sample I,

$$\lambda^2(\rho_I) = \frac{(5.34)^2}{(6.23)^2} [(.997)^2(.004)^2 + (.001)^2 + (.002)^2]$$

$$= 0.735 [15.9 + 1 + 4] \times 10^{-6} = 15.4 \times 10^{-6} \tag{9}$$

$$\lambda(\rho_I) = 0.004 \text{ g cm}^{-3}$$

Similarly, we obtain $\lambda(\rho_{II}) = 0.004$ g cm^{-3}.

V.   Discussion [f]

The values and limits of error obtained for the density of germanium at 25°C   are

Sample I: 5.340 ± 0.004 g cm^{-3}

Sample II: 5.352 ± 0.004 g cm^{-3}

The average value is 5.346 g cm^{-3}.   The value given in the Handbook of Chemistry and Physics[3] is 5.35 g cm^{-3} at 20°C; the value calculated from the volume of the crystallographic unit cell[4] and the atomic weight is 5.355 g cm^{-3}.   Only the higher of our two results is in agreement with the literature value to within our limits of error.

The values obtained for the two samples deviate from the average by slightly more than the calculated limit of error.   However, the difference is much larger than it should be considering the fact that

[f] This is the most flexible section of the entire report, and the student must depend heavily on his own judgment for the choice of topics for discussion. The final results of the experiment should be clearly presented, often in a tabular or graphical form. A comparison between these results and theoretical values or experimental values from the literature is usually appropriate. A comment should be made on any discrepancies with the accepted or expected values. In this sample discussion, comment is also made on "internal discrepancies," possible systematic errors, and the relative importance of various sources of random error; a brief suggestion is made for an improvement in the experimental method. Other possible topics include suitability of the method used compared with other methods, other applications of the method, mention of any special circumstances or difficulties which might have influenced the results, discussion of any approximations made or which could have been made, suggestions for changes or improvements in the calculations, mention of the theoretical significance of the result. At the end of several of the experiments in this book there are questions which provide topics for discussion; however, the student should usually go beyond these topics and include whatever other discussion he feels to be pertinent.

the contribution of any error in V is the same in both runs. This suggests that the material examined may be somewhat inhomogeneous, so as to yield two samples of slightly different density. We suggest the possibility that cracks or fissures inaccessible to the liquid are present in Sample I, or perhaps in both samples to different degrees. On this assumption, the greater confidence would be placed in the higher value, namely that for Sample II, although on the basis of results for only two samples there is no internal evidence that Sample II is completely free of defects. The agreement of the result for Sample II with the literature values is gratifying, but in general the best indication of reliability would be good agreement among the results for several samples.

Equations (8) and (9) show that the largest contribution to the overall error comes from the uncertainty in the pycnometer volume V. Our experimental precision indicates that by measuring the weight of the pycnometer filled with water alone, a better value of V could have been obtained. This would have reduced the uncertainty in the density but would not have improved the agreement between the two samples.

References[g]

1. N. Bauer and S. Z. Lewin, Determination of Density, in A. Weissberger (ed.), "Technique of Organic Chemistry," 3rd ed., Vol. I, Part I, Chap. IV, esp. pp. 177-180, Interscience, New York (1959).
2. "Handbook of Chemistry and Physics," 53rd ed., p. F-5, Chemical Rubber Publishing Co., Cleveland (1972).
3. *Ibid.*, p. B-92.
4. R.W.G. Wyckoff, "Crystal Structures," 2nd ed., Vol. 1, p. 26, Interscience, New York (1963).

[g] An appropriate style for referring to a book is illustrated by entry 4 above. If the publisher's name is not well known, it should be given in full (see entry 2); if the city of publication is not well known, the state or country should also be given (e.g., "Reading, Mass."). The citation style for referring to a book containing chapters by several different authors is illustrated by entry 1. The proper citation style for journal articles is shown by the many references given elsewhere in this book. For typewritten reports, it is common practice to underline only the journal volume number.

Sample Notebook Page[h]

Maria Smith   p. 1
Partner: John Klein   Oct. 8, 1980
Exp: Density of Ge
Pycnometer #7 $(V = 12.410 \pm .004 \text{cm}^3)$

	W_1 (empty)	W_2 (solid)	W_3 (solid + water)
Sample I			
Run 1	8.6313	42.0307	48.1749
Run 2	8.6308	42.0295 ~~42.0259~~	48.1715
Ave.	—	42.0301 g	48.1732 g
Sample II			
Run 3	8.6316	45.8468	51.2963
Run 4	8.6299	45.8490	51.2925
Ave.	—	45.8479 g	51.2944 g

Average W_1 = 8.6309 g
Estimated uncertainty in weighings 0.001 g
Temperature of bath = 24.96° ± 0.05 °C

[h] All data must be recorded directly in a notebook or on special data sheets. Be sure to record any identifying numbers on special apparatus and all necessary apparatus calibration data. If separate data sheets were used or if carbon copies were made of the pages in a bound research notebook, the complete data should be arranged in order at the end of the report. If a bound notebook was used, make a table of contents on the first page and list for each experiment the location of the data pages. The first page of data should have a clear and complete heading; all pages should carry the name of the student and his partner. (The instructor may also require that data sheets be checked over and initialed by a teaching assistant at the end of the experiment.)

of physical chemistry. Indeed, it is sometimes possible to make a significant start on an original research problem that will eventually lead to publishable results. The primary emphasis should, however, be placed on independent planning of the experimental work rather than on original proposals for new research.

The two (or possibly three) partners working on a given special project should plan the experiment together, starting three or four weeks in advance, and should discuss their ideas frequently with an instructor or a graduate teaching assistant. All work done in the laboratory should be supervised by an experienced research worker in order to prevent any serious safety hazards.

Some of the projects done in the laboratory course at MIT are listed below as examples of the kind of problems which might be attempted.

Heat of fusion of mercury
Spectrophotometric study of the relative stability of metal ion–EDTA complexes
Determination of the solubility of $Fe(OH)_3$ using radioactive iron
Dimerization of dye molecules in solution
Kinetics of the $H_2 + I_2 = 2HI$ reaction in the gas phase
Weak-acid catalysis of BH_4^- decomposition
Photochemistry of the cis-trans azobenzene interconversion
Isotope effect on reaction-rate constants
Susceptibility of a paramagnetic solid as a function of temperature
Dielectric constant of polypropylene glycol
X-ray study of short-range order in liquid mercury
Fluorescence and phosphorescence of complex ions in solution
Infrared study of hydrogen bonding of CH_3OD with various solvents
Raman spectra of toluene, chlorinated methanes
Light scattering near the critical point in ethane
Franck-Hertz experiment
Polanyi dilute flame reaction, e.g., $K + Br_2$
EPR study of gas-phase hydrogen and deuterium atoms
EPR spectra of methyl semiquinones
Photodissociation of NO_2
Fluorescence quenching of excited K atoms
Shock-tube kinetics: recombination of I atoms
Dielectric dispersion in high-polymer solutions

Many of these projects were quite ambitious and required hard work and enthusiasm on the part of both students and staff. Not all were completely successful in terms of precise numerical results, but each one was instructive and enjoyable. Frequently they resulted in an excellent scientific rapport between the students and the instructing staff.

Safety   It has already been emphasized that safe laboratory procedures require thoughtful awareness on the part of both students and instructors. This is especially important in the planning and execution of special projects, where new procedures need to be developed and often modified as the work progresses. Appendix D on safety hazards and safety equipment should be read before beginning a course of experimental work in physical chemistry and reviewed carefully before beginning any special project.

Chapters I, II, and XV to XXI contain general information about experimental work in physical chemistry, while Chaps. III to XIV contain the experiments which are numbered 1 to 50 consecutively throughout the book. In addition, Chaps. IV and V each contain some separate introductory material. Each figure, equation, and table is identified by a single number, and the numbers in each category run consecutively within single experiments. Outside the experiments, numbering is consecutive within single chapters. Within the experiment or chapter concerned, reference is made with the appropriate single number: e.g., Fig. 1, Eq. (8), Table 1. For cross references, double numbering is used: e.g., Fig. 38-1 refers to Fig. 1 in Exp. 38, and Eq. (V-8) refers to Eq. (8) in the introductory part of Chap. V.

II

TREATMENT OF EXPERIMENTAL DATA

The ultimate object of performing an experiment in physical chemistry is usually to obtain one or more numerical results. Between the recording of measured values and the reporting of numerical results there are processes of arithmetical calculation, some of which may involve averaging or smoothing the measured values but most of which involve the application of formulas derived from physics or physical chemistry. Part of this chapter is devoted to a discussion of general techniques for carrying out such calculations.

However, our concern with the treatment of experimental data is not ended when we have obtained the desired numerical result. An important part of the job is the determination of the degree of uncertainty to which the numerical result is subject. Every physical quantity whose a priori range of possible numerical values constitutes a continuum is subject to error in its determination. It is not possible to determine exactly what this error is: this would be equivalent to measuring the quantity without error, since correction can be made for any *known* error. But it is important to specify the highest amount by which the quantity *might* be in error. Thus one should specify the value of some parameter (such as the standard deviation) from which the probability of the existence of a random error of any given magnitude can be predicted;

and one should also give, if possible, a reasonable estimate of the possible systematic errors.

The reported value of a physical quantity, when not accompanied by a statement of its uncertainty, can be of small value. For example, suppose that the experimental value of a physical quantity is being compared with a value predicted for that quantity by a theoretical equation. If the agreement is very good, is it possibly to some degree fortuitous? If the agreement is very bad, is it outside the limits of experimental error? The significance of the degree of agreement (upon which may rest the validity of the theory) depends upon the answers to questions of this kind, and these answers require knowledge of the experimental uncertainty.

Assessment of the uncertainty involves some knowledge of the accuracy and precision of the instruments used, analysis of the experimental method and technique, determination of the degree of internal consistency in the experimental data, and, finally, a study of how errors or uncertainties in the experimental data affect the final calculated result.

ERRORS IN OBSERVATIONAL DATA

Systematic and random errors   The measurement of a physical quantity with a continuous-reading instrument is generally subject to error owing to inability of the observer to discriminate between readings differing by less than some small amount or to his inability to make the instrumental adjustments required for each reading to higher than a certain precision or to unpredictable fluctuations in the environmental conditions. Independent readings made with this instrument will generally differ by small, random amounts, and we say that the measurements are subject to *random error*. Random error may, in principle, be reduced by any arbitrary factor by taking and averaging a sufficiently large number of independent measurements. However, as the precision of the arithmetic mean of a number of measured values increases only in proportion to the square root of the number of individual values, the ultimate precision obtainable in practice is only a few times that of an individual measurement.

Even if random errors could be completely eliminated (as by taking an infinite number of measurements), the measured value may still be in error, owing to characteristics of the instrument or of the technique of using it that are the same for all measurements. Error of this kind is called *systematic error*. Examples of systematic error are calibration error in the instrument, uncompensated instrumental drift, leakage of material (e.g., gas in a pressure or vacuum system) or of electricity (as in electrometer measurements in a high-resistance circuit), incomplete fulfillment of assumed conditions for the measurement (e.g., incomplete reaction in a calorimeter, incomplete dehydration of a weighted precipitate), or some consistent operational error (parallax, uncompensated human reaction times, even personal bias). Systematic errors have been termed "determinate" or "corrigible" errors, implying that (in principle, at least) they can be eliminated or else estimated and corrected for by sufficient attention to calibration, controls and blanks, and other experimental conditions. It must be remembered, however, that systematic errors are sometimes very difficult to even identify, let alone minimize. If we define error as the difference between the observed and the true value, there is the additional philosophic difficulty that the true value is usually both unknown and unknowable. As a rule the best assurance of low systematic error is agreement between several entirely independent methods of measurement. Further discussion of systematic errors can be found in Skoog and West.[1]†

Accuracy and precision   The *precision* of a numerical result is concerned with its reproducibility when measured again with the same instrument and is therefore an expression of the uncertainty due to random error. The *accuracy* of the result is an expression of its total uncertainty including that due to systematic error. The terms accuracy and precision are also applied to instruments and methods to characterize the numerical results that can be obtained with them. Clearly it is possible to have high precision and low accuracy if there is a significant systematic error superimposed on the random error. Both are important and must be taken into consideration. Because the theory of random errors is moderately well developed, a good deal of attention is generally given to the statistical treatment of random error. Systematic error cannot be described in terms of a tidy mathematical theory, and often gets little more than passing mention. It must be emphasized, however, that systematic errors may be orders of magnitude greater than the random errors in unfavorable circumstances, and therefore the possibility of systematic error must always be kept in mind. It is the responsibility of an experimenter to make an objective estimate of possible systematic errors, but he cannot be infallible. Many published results have later been shown to be in error by amounts far greater than the claimed limits of error.

TREATMENT OF RANDOM ERRORS

Error frequency distribution[2]   Let it be supposed that a very large number of measurements x_i $(i = 1, 2, \ldots, N)$ are made of a physical quantity x and that these are subject to random errors ε_i. For simplicity we shall assume that the true value x_0 of this quantity is known and therefore that the errors are known. We are here concerned with the frequency $n(\varepsilon)$ of occurrence of errors of size ε. This can

---

† Superior numbers refer to references given at the end of chapters or experiments.

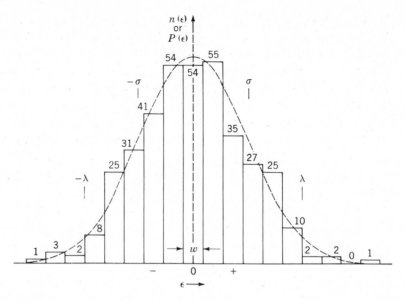

FIGURE 1

A typical distribution of errors. The bar graph represents the actual error frequency distribution $n(\varepsilon)$ for 376 measurments; the estimated normal error probability function $P(\varepsilon)$ is given by the dashed curve. Estimated values of the standard deviation σ and the 95 percent confidence limit λ are indicated in relation to the normal error curve.

be shown by means of a bar graph, like that of Fig. 1, in which the error scale is divided into ranges of equal width and the height of each bar represents the number of measurements yielding errors that fall within the respective range. The width w is chosen as a compromise between the desirability of having the numbers in each bar as large as possible and the desirability of having the number of bars as large as possible.

It is seen that even with as many as 376 measurements the graph shows irregularities, owing to the fact that the number of measurements represented by each bar is subject to statistical fluctuations that are not small in comparison with the number itself. From this graph we can make rough predictions concerning the probability that a measurement will yield an error of a given size. If we greatly increase the number of measurements represented (while perhaps decreasing the width w of the range more slowly than the rate of increase in the total number of measurements), the statistical fluctuations will become smaller in relation to the heights of the bars and our probability predictions are improved; we may even draw a smooth curve through the tops of the bars and assume it to represent an *error probability function* $P(\varepsilon)$. The vertical scale of this function should be adjusted by multiplication with an appropriate factor so that the function is normalized, i.e., so that

$$\int_{-\infty}^{\infty} P(\varepsilon) \, d\varepsilon = 1 \tag{1}$$

Its significance is that the probability that a single measurement will be in error by an amount lying in the range between ε and $\varepsilon + d\varepsilon$ is equal to $P(\varepsilon) \, d\varepsilon$.

A probability function derived in this way is approximate; the true probability function cannot be inferred from any finite number of measurements. However, it can often be assumed that the probability function is represented by a gaussian distribution called the *normal error probability function,*

$$P(\varepsilon) = \frac{1}{\sqrt{2\pi}\,\sigma} \, e^{-\varepsilon^2/2\sigma^2} \tag{2}$$

where σ is a parameter called the *standard deviation.* It is the root-mean-square error expected with this probability function:

$$\left(\overline{\varepsilon^2}\right)^{1/2} = \left(\frac{1}{\sqrt{2\pi}\,\sigma} \int_{-\infty}^{\infty} \varepsilon^2 e^{-\varepsilon^2/2\sigma^2} \, d\varepsilon \right)^{1/2} \equiv \sigma \tag{3}$$

If the true value x_0 and thus the errors ε_i themselves are known, σ can be estimated from

$$\sigma = \left(\frac{1}{N} \sum_{i=1}^{N} \varepsilon_i^2 \right)^{1/2} \tag{4}$$

The dashed curve in Fig. 1 represents a normal error probability function, with a value of σ calculated with Eq. (4) from the 376 errors ε_i.

The usual assumptions leading to the normal error probability function are that an error in a measurement is compounded of a large number of unpredictably variable contributions (which are limited in magnitude but need not conform to any particular probability function), that these contributions are mutually independent, and that on the average all of them are small in comparison with the error itself. The derivation of the normal error probability function from these assumptions is approximate unless the number of contributions is infinite and the contributions themselves are infinitesimal. The assumptions are sufficient but not altogether necessary; the normal error probability function may arise at least in part from different circumstances. The factors which, in fact, determine the distribution are seldom known in detail. Thus it is common practice to assume that the normal error probability function is applicable even in the absence of valid a priori reasons, and indeed the distributions obtained frequently conform to the normal error probability function to within ordinary statistical error, as in the case of the distribution of Fig. 1.

However, even with 376 measurements the normal error probability function is not proved to be valid in all respects; with a much larger number of measurements it may become apparent that the true probability function is somewhat skewed or flat-topped or double-peaked, etc.

So far the discussion has dealt with the errors themselves, as if we knew their magnitudes. In actual circumstances we cannot know the errors ε_i by which the measurements x_i deviate from the true value x_0, but only the deviations $(x_i - \bar{x})$ from the mean $\bar{x}$ of a given set of measurements. If the random errors follow a gaussian distribution and the systematic errors are negligible, the best estimate of the true value x_0 of an experimentally measured quantity is the arithmetic mean $\bar{x}$. If an experimenter were able to make a very large (theoretically infinite) number of measurements, he could determine the *true mean* μ exactly, and the spread of the data points about this mean would indicate the precision of the observation. Indeed, the probability function for the deviations would be

$$P(x - \mu) = \frac{1}{\sqrt{2\pi}\,\sigma} \exp\left[-\frac{(x - \mu)^2}{2\sigma^2}\right] \tag{5}$$

where μ is the mean and σ the standard deviation of the distribution of the hypothetical infinite population of all possible observations.

If a large enough number of data points are available (at least 30 and preferably 100), classical probability calculations provide a very satisfactory description of the precision of the measurements in question. As a matter of fact one very seldom has anything like 30 measurements of the same quantity; most experiments are based on averages of 6 or less. Classical probability statistics are inadequate for the treatment of small numbers of observations, and techniques developed only within recent years are necessary to avoid large errors in the estimates of error. For a finite (and usually small) number of observations of a quantity, one obtains data which show a certain amount of spread. The true mean μ and the true spread of the hypothetical infinite population of measurements are what one wishes to have. The experimental mean $\bar{x}$ and the spread of the small group of measurements are what one does have, and these include a contribution due to random error. One of the most important objectives of a statistical treatment of data is the estimation of true values from finite samples of data and the expression of the uncertainties inherent in these estimates due to random error and sample size.

The scatter of the individual values x_i about the mean is an indication of the precision of the data, and there are in common use a number of ways of expressing this scatter. These are (1) the *range*, R, which is simply the difference between largest and smallest x_i value; (2) the *average deviation*, which is the average difference between the experimental values and their mean without regard to sign; and (3) the *quartile deviation*, sometimes unfortunately called the "probable error," which is a deviation from the mean chosen so that one-half the experimental deviations are larger and one-half smaller. There are several desirable attributes which a measure of precision clearly should have. First, it should be as independent as possible of the number of observations in the experimental sample. In other words, if one takes groups of two, five, or twenty observations, the averages of the estimates of precision of these different groups should tend toward the same value. Second, it should be possible to combine different sets of measurements, including those of different size and different means to give a more accurate estimate of the precision. Third, the resulting measure of precision should be of such nature that it can be used for quantitative comparisons of different groups of experimental results, and it should provide some means of deciding whether the apparent differences are real or merely a reflection of the normal variability of the measurements from which they come.

It is found that none of the above measures of precision are ideal from these standpoints. The range is obviously a crude measure, although we shall see later how it can be used as the basis of a quick estimate of more sophisticated measures. The average deviation, although very widely used and cherished for its arithmetic simplicity, is actually useful only for qualitative comparisons. It is quite sensitive to sample size (the estimate of average deviation by taking measurements in pairs is 30 percent different from the estimate taken from large groups) and does not lend itself to use in statistical testing of significance. The quartile deviation cannot be calculated directly unless there is a large number of observations.

Variance   There is, however, a measure of precision which is unbiased by sample size, has a property of additivity such that estimates from several sources may be combined, and is directly useful in probability calculations. This is the *variance* S^2 defined by

$$S^2 \equiv \frac{1}{N-1} \sum_{i=1}^{N} (x_i - \bar{x})^2 \tag{6}$$

This form is not always the most convenient for performing actual calculations, and there are alternate forms, algebraically identical, which are often more convenient for use with a calculator.

$$S^2 = \frac{1}{N-1} \left[\sum x_i^2 - \bar{x} \sum x_i \right] = \frac{1}{N-1} \left[\sum x_i^2 - N(\bar{x})^2 \right]$$

Particular attention should be paid to the divisor $(N-1)$ which is known as the number of *degrees of freedom*, and which is the number of independent data on which the calculation of S^2 is based. Many authors, particularly in older works, use the divisor N, which is correct only if the mean value is exactly known. One degree of freedom is used up in calculating the experimental mean of a set of experimental data points. That is, one of the N deviations $(x_i - \bar{x})$ is not an independent variable since

$$\sum_{i=1}^{N} (x_i - \bar{x}) = 0 \qquad \text{or} \qquad \bar{x} = \frac{1}{N} \sum_{i=1}^{N} x_i$$

which is the equation defining $\bar{x}$. The difference between a factor of $N - 1$ or N in the denominator is trivial if there are 100 data points but becomes very important for small N.

The fundamental importance of the variance in probability calculations is suggested by the fact that S, its square root, is the standard deviation for the finite set of N data points. This square root of the variance is the best estimate of the true standard deviation σ (of the hypothetical infinite population) which can be made from a finite data set. If the error distribution is indeed normal, the probability that a measurement will be in error by an amount no larger than σ is found by integrating the probability function, Eq. (5), between the limits $\pm\sigma$. The result is 68.26 percent, or roughly two-thirds.

Confidence limits and the uncertainty of a mean When an experimenter makes several measurements, averages them, and reports a mean value, he very frequently writes his result in a form such as 21.32 ± 0.11. The implication is that the 0.11 is a measure of the uncertainty of the result. Unfortunately, many authors do not state what they mean by ± 0.11, and the reader is left with the problem of deciding. The error figure might be an average deviation, standard deviation, or quartile deviation; sometimes it is an estimate based on guesses of what the experimenter thinks his largest random error of measurement might be; sometimes it is an estimate based on an assumption about the possible systematic errors; frequently it is simply an uneducated guess. At the very least one should state clearly what is intended by the error quantity. The growing practice of giving 95 percent *confidence limits* for random errors based on small sample probability calculations is strongly recommended. These limits are based on the following reasoning.

For a population with an ideal gaussian distribution, 68 percent of all the observations fall within ± 1 standard deviation of the mean and 95 percent of all the observations fall within ± 1.96 standard deviations. If one takes from this population many groups of N observations and averages each group, the averages will likewise follow a gaussian distribution with the same grand mean but with a standard deviation equal to $S/\sqrt{N}$, where S is the standard deviation of the distribution of single measurements. Now if one takes a single group of N measurements and averages it, it can be regarded as a single observation in the frequency distribution of means of groups of N, which distribution has a standard deviation $S/\sqrt{N}$.† One would expect, therefore, to find that the means of 95 percent of the samples of N observations would fall within the range $\bar{x} \pm 1.96 \ S/\sqrt{N}$ with 95 percent confidence that the true mean actually fell within that range. This is the so-called 95 percent confidence interval. Now if one has only a small sample from which to evaluate both $\bar{x}$ and S, it is evident that the random experimental error in estimating the standard deviation from a small number of observations will introduce an additional uncertainty which must be taken into account. A rigorous solution to this problem has been obtained. For each value of N there is an appropriate correction factor, conventionally designated as t. Table 1 gives t values for various degrees of freedom, $N - 1$ in this case. More complete t tables are available in the "Handbook of Chemistry and Physics" and many of the statistics books listed under General Reading at the end of this chapter. The resulting 95 percent confidence limits $(-\lambda,\lambda)$ are given by

$$\lambda(95\% \text{ confidence}) = \frac{tS}{\sqrt{N}} \qquad (7)$$

where t is taken from the $P = 5$ percent column in Table 1. The preferred way to

† This is a quantitative statement of the intuitively obvious fact that the mean of a group of N independent measurements of equal weight has a higher precision than any single one of the measurements.

Table 1   CRITICAL VALUES OF t^a

			P			
DF^b	50	20	10	5	1	0.1
1	1.00	3.08	6.31	12.7	63.7	637.0
2	0.816	1.89	2.92	4.30	9.92	31.6
3	0.765	1.64	2.35	3.18	5.84	12.9
4	0.741	1.53	2.13	2.78	4.60	8.61
5	0.727	1.48	2.01	2.57	4.03	6.86
6	0.718	1.44	1.94	2.45	3.71	5.96
7	0.711	1.42	1.89	2.36	3.50	5.40
8	0.706	1.40	1.86	2.31	3.36	5.04
9	0.703	1.38	1.83	2.26	3.25	4.78
10	0.700	1.37	1.81	2.23	3.17	4.59
15	0.691	1.34	1.75	2.13	2.95	4.07
20	0.687	1.32	1.72	2.09	2.85	3.85
30	0.683	1.31	1.70	2.04	2.75	3.65
∞	0.674	1.28	1.64	1.96	2.58	3.29

^{a} The P values are the probabilities (in %) that the absolute value of t will exceed the tabular entry if there is no real difference between the means being tested.
^{b} DF is the number of degrees of freedom.

express a result is, for example, "21.32 ± 0.11 (95 percent confidence limits $N = 6$)." This is a thoroughly objective and unambiguous method of expressing the precision of a mean value. It must be remembered, however, that it neither gives nor suggests any information about possible systematic errors.

Short-cut methods based on the range For small samples (N from 2 to about 10) the range, $R = x(\text{largest}) - x(\text{smallest})$, can be used to obtain a reasonably accurate estimate of standard deviation. The range when multiplied by the appropriate value of K_2 from Table 2 gives an approximate but useful estimate of the standard deviation. The factor J in Table 2 is equivalent to $K_2 t / \sqrt{N}$, where t is the critical value from the $P = 5$ percent column in Table 1 for the appropriate degree of freedom. Hence, to get a quick estimate of the 95 percent confidence limits of the mean of a few observations, simply calculate $\pm JR$. In summary,

$$S \simeq K_2 R \quad \text{and} \quad \lambda(95\% \text{ confidence}) \simeq JR \tag{8}$$

These techniques, which have developed out of the statistical methodology known as quality control, are becoming widely adopted because of their convenience.

Estimation of limits of error It must be kept in mind that the above discussion of random errors is based on the notion that one has no prior information about the precision of the measurements, in which case the uncertainty must be estimated from the scatter of the data in a given sample. In practice the number of independent measurements of an experimental quantity is often very small (sometimes two or only one) owing to time limitations, and it is unrealistic or impossible to estimate the confidence limits from the range. However, the experimenter frequently has a good idea of the precision of the instrument or method he is using. In this case, and also in cases where systematic errors must be estimated as well, the 95 percent confidence

Table 2 FACTORS FOR SMALL-SAMPLE ESTIMATES BASED ON THE RANGE R

N	K_2	$J(95\%)$	$Q_c(90\%)$
2	0.89	6.4	—
3	0.59	1.3	0.94
4	0.49	0.72	0.76
5	0.43	0.51	0.64
6	0.40	0.40	0.56
7	0.37	0.33	0.51
8	0.35	0.29	0.47
9	0.34	0.26	0.44
10	0.33	0.23	0.41

limits must be assigned on the basis of the individual judgment and experience of the experimenter. The ability to assign realistic limits of error, large enough to be safe but not so large as to detract unnecessarily from the value of the measurement, is one of the marks of a good experimentalist. It would be very difficult to formulate a set of rules that would be applicable in all cases; the confidence limits assigned in any given case will depend on the characteristics and capabilities of the instrument, the reproducibility of its reading, the quality of its calibration, and the user's experience and familiarity with it. A reasonable 95 percent confidence limit in weighing a small solid object on an analytical balance with weights calibrated to 0.05 mg might be 0.3 mg; in reading a 50-ml burette it might be 0.03 ml; in determining a short time interval with a stopwatch it might be 0.4 sec; in measuring a temperature difference with a Beckmann thermometer it might be 0.01°C; etc. Estimation of standard deviations in this way is much more difficult.

SIGNIFICANCE TESTING

A frequent question which arises in quantitative experimentation is whether the observed difference between an experimentally measured quantity and the accepted, theoretical, or standard value is representative of a genuine difference or merely the natural result of random experimental error. Similarly the question may arise of whether two experimental studies differ in their results because of real differences either in the systems studied or the methods used, or whether the apparent discrepancy is no greater than might be found by chance simply because of natural variability in the data.

The basic philosophy of significance testing is simple. Two series of measurements are made and their results are observed to differ. The maximum difference which could be expected to occur if both series represent random samples of the same thing (i.e., no real difference in the quantity being measured in the two series) at a given probability level is calculated from probability theory. If the observed difference is smaller, it is concluded that there is no real evidence of difference. If it is larger, it is concluded that the difference is probably real (at whatever probability level was chosen). The conventionally accepted probability level for considering differences significant is 5 percent. It is based on the experience that if action is taken at lower levels of significance (e.g., 10 or 20 percent), there are too many embarrassing false alarms, whereas if the requirements for significance are made too stringent (e.g., at the 1 percent level), too many things get missed. Under circumstances where it is very important not to claim a significant difference when none is really there, the 1 percent or even the 0.1 percent level may be set as the basis for taking action, but in ordinary scientific practice the 5 percent level has been found to be best.

There exist many techniques for evaluating the statistical significance of various types of differences. A few of the common cases are:

1   To determine whether the variance S^2 of a set of data is, for a stated level of significance, a legitimate estimate of a *specified* value σ_0^2 of the square of the standard deviation, one uses the χ^2 test.

2   To determine whether the variances S_1^2 and S_2^2 of two sets of data are, for a stated level of significance, estimates of the *same* σ^2, one uses the F test. This test allows a judgment about possible differences in the precision of two sets of measurements, i.e., whether two variances represent genuinely different σ^2 values or whether they appear to differ only because of chance variations in the measurements.

3   To determine whether the mean $\bar{x}$ of a set of data is significantly different from a *specified* value, one uses the t test.

4   To determine whether the means $\bar{x}_1$ and $\bar{x}_2$ of two sets of data having the same variance are significantly different from each other, one also uses the t test.

The χ^2 and F distributions can also be used to determine the number of parameters needed for a satisfactory fit in the method of least squares or to test the *goodness-of-fit* which can be achieved with a specified functional form. In the latter case, the deviations between the experimental data points and the "theoretical" values calculated with the fitting function are compared with an estimate of the standard deviation in the data due to random errors. Detailed discussions of significance testing can be found in Chap. XX and in the references listed under General Reading at the end of this chapter. A brief account of the t test is given below.

The *t* test   Let us first consider whether the difference between an experimental mean $\bar{x}$ of a set of N measurements and a specified value x_0 (the theoretical, true, or accepted value) is significant. The test statistic t in this case is just the ratio of the difference $(\bar{x} - x_0)$ to the standard deviation in the mean. The uncertainty in x_0 is taken to be zero. Thus,

$$t = \frac{\bar{x} - x_0}{S/\sqrt{N}} = (\bar{x} - x_0)\left[\frac{N(N - 1)}{\sum (x_i - \bar{x})^2}\right]^{1/2} \tag{9}$$

The value of t is calculated and compared with a critical value in Table 1 corresponding to the appropriate number of degrees of freedom (DF) ($N - 1$ in this case). If the absolute value of t value exceeds the tabulated (critical) entry for a given probability level P (usually 5 percent), it is concluded that the difference $(\bar{x} - x_0)$ is significant at that level. That is, a t value as large as the critical value for DF $= N - 1$ would be expected statistically to occur in only P percent of a large number of sets of N measurements. This procedure is exactly equivalent to establishing a $(100 - P)$

percent confidence limit λ for the mean $\bar{x}$ and deciding that $\bar{x}$ and x_0 are *not* significantly different if x_0 falls within the interval $\bar{x} \pm \lambda$.

When two experimental means are to be compared, as for example two analyses of a sample by the same method, two determinations of an equilibrium constant, or the analysis of the same sample by two methods or with two instruments, and the standard deviations of the two sets of measurements are not significantly different, one can also use the t test. In the case where N_1 measurements are made in set 1, which has a variance S_1^2, and N_2 measurements are made in set 2 with a variance S_2^2, the quantity t is given by

$$t = \frac{\bar{x}_1 - \bar{x}_2}{\{[(N_1 - 1)S_1^2 + (N_2 - 1)S_2^2]/(N_1 + N_2 - 2)\}^{1/2}} \left[\frac{N_1 N_2}{N_1 + N_2}\right]^{1/2} \qquad (10)$$

The quantity in the denominator of Eq. (10) is the square root of the "pooled" variance of two sets of data;[1] if the separate variances of the two sets were identical, $S_1^2 = S_2^2 = S^2$, then this quantity would be simply S. As before, the value of t is calculated and compared with the critical values in Table 1 for $N_1 + N_2 - 2$ degrees of freedom in this case. If the value of $|t|$ exceeds the tabulated entry, then the difference in the two means is significant at that probability level.

EXAMPLES   The accepted value μ of the weight percent bromine in a standard KBr sample is 67.14. Four analyses of this sample by a new method gave a mean value of 67.07 and an estimated standard deviation of 0.04. On the basis of these data, is it likely that there is some systematic error in the new method? Using Eq. (9), we find

$$t = \frac{67.07 - 67.14}{0.04} \sqrt{4} = -3.5$$

From DF $= 3$ in Table 1, we see that there is about a 4 percent chance that the observed discrepancy is due to random errors. Thus we can say with 96 percent certainty that there is a systematic error in the new procedure.

The heat of neutralization of an amine is measured four times in a large calorimeter and six times in a small one with the following results:

Large calorimeter $N_1 = 4$, $\overline{\Delta H_1} = 7.3$ kcal/mol, $S_1 = 0.3$ kcal/mol

Small calorimeter $N_2 = 6$, $\overline{\Delta H_2} = 6.9$ kcal/mol, $S_2 = 0.4$ kcal/mol

Using Eq. (10), we find

$$t = \frac{7.3 - 6.9}{\{[3(0.09) + 5(0.16)]/8\}^{1/2}} \left[\frac{6 \times 4}{6 + 4}\right]^{1/2} = \frac{0.4}{0.366} (1.55) = 1.7$$

From DF $= 8$ in Table 1, we see that the difference in mean values is significant at the 20 percent level but is not significant at the 10 percent level. That is, there is about a 15 percent chance that the observed difference is due to random errors rather than some systematic error.

CALCULATIONS

Each directly measured experimental quantity that is to be used in the calculation of the desired final result should be measured in the laboratory to the utmost usable sensitivity of the instrument involved. This is usually greater than that corresponding to the smallest indicated scale graduation, and it is therefore usually desirable to obtain an additional digit by interpolation between scale divisions. The figure recorded for the measurement should ordinarily be one in which the last one or two digits are somewhat uncertain.

Rejection of discordant data It occasionally happens in making multiple measurements that one value differs from the rest considerably more than they differ from one another. Should the discordant value be rejected before an average is taken? This question provides one of the most severe tests to which the scientific objectivity of the experimenter is exposed, for he may (indeed, he should) reject any measurement, whether concordant or discordant, if he has valid reason to believe that it is defective. Thus, the first step would always be to check for evidence of a determinate error (used the wrong pipette, added the wrong reagent, inverted the digits in writing down the number, etc.). If none can be found, the suspect value must be included unless valid statistical arguments can be presented to show that such a large deviation on the part of a member of the hypothetical ideal population of similar measurements is highly improbable. There are a number of ways of going about this. Most of them—such as the traditional rule about discarding values which differ from the mean of the others by more than four times the average deviation of the others—are based on faulty application of large-number statistics to small-sample problems and are not to be trusted. There is a very simple approach, known as the Q test, which is both statistically sound and straightforward to use.

When one in a series of 3 to 10 measurements appears to deviate from the mean by more than seems reasonable, calculate the quantity Q, which is the ratio of the difference between the value under suspicion and the value in best agreement with it to the difference between the highest and lowest values in the series. Compare the value of Q with the critical value Q_c in Table 2 corresponding to the number of observations in the series. If Q is equal to or larger than Q_c, the suspect measurement should be rejected. If Q is less than Q_c, this measurement must be retained.

EXAMPLE   Five determinations of the baseline reading of a spectrophotometer at a standardizing wavelength were 0.32, 0.38, 0.21, 0.35, and 0.34 absorbance units. May the 0.21 be discarded?

$$Q = \frac{0.32 - 0.21}{0.38 - 0.21} = 0.65$$

From Table 2 the critical value Q_c for $N = 5$ is 0.64. The value of Q is higher, hence the 0.21 value may be discarded with 90 percent confidence.

It should be noted that only *one* value in a series may be discarded in this manner. If there is more than one divergent value in a small series, it means that the data really do show a lot of scatter. When one value in a small series is considerably off but not sufficiently to justify rejection by the Q test, there is good reason to consider reporting the median value (i.e., the middle value) instead of the mean. As an illustration of this, consider the situation if the five spectrophotometer readings had been 0.32, 0.38, 0.23, 0.35, and 0.34. The value of Q for this set (0.60) does not quite exceed the critical value Q_c, and the 0.23 value cannot be rejected. However, reporting the median value 0.34 avoids the large effect that 0.23 would have if included in calculating the mean, while still allowing it a "vote." The mean of all five values is 0.32, whereas the mean without the suspect value is 0.35.

Significant figures   It is axiomatic that experimental data contain uncertainties. Unless otherwise stated, the last digit in an estimated or interpolated reading is always assumed to be uncertain by ± 1 or more. If sufficient data are available to permit standard deviations and confidence limits to be calculated, there is no particular problem. If the data necessary for these are lacking, rules for computations using "significant figures" help to give a result which expresses to some degree the uncertainty in the experiment.[1] The main rules are summarized below:

1   Experimentally measured values are considered to be uncertain in the last digit, and perhaps slightly uncertain in the next-to-last digit. For ordinary computations, retain no digits beyond the first digit that is uncertain by more than two. This rule does not apply to the treatment of large numbers of data points by a digital computer

2   In rounding off numbers, (*a*) increase the last retained digit by one if the leftmost digit to be dropped is greater than 5 or is 5 followed by other digits; (*b*) leave the last retained digit unchanged if the leftmost digit to be dropped is less than 5; (*c*) if the leftmost digit to be dropped is exactly 5, increase the last retained digit by one if it is odd and leave it unchanged if it is even.

EXAMPLES   Rounding to four digits

1.23742	1.237
1.23751	1.238
1.23750	1.238
1.23650	1.236
1.23749	1.237

3 In addition and subtraction retain only the number of decimal places in the result as are in the component with the fewest number of decimal places.

EXAMPLE $32.7 + 3.62 + 10.008 = 46.328 = 46.3$

4 In multiplication and division the result should have a relative uncertainty that is the same general magnitude as that in the least precise component. The uncertainty of the result should be in the range of 0.2 to 2.0 times that of the least precise component.

EXAMPLES

$$346 \times 121 \times 900.0 = 37,679,400 = 3.77 \times 10^7$$

The uncertainty in 121 is about 0.8 percent, hence the answer is rounded off such that a difference of 1 in the last place will correspond to a relative uncertainty between 0.16 percent and 1.6 percent. One in 377 is 0.27 percent.

$$8.32 \times 211 \times 0.029 = 50.91008 = 51$$

The uncertainty in 0.029 is about 1 in 30 (3.3 percent); one part in 51 is 2 percent.

Arithmetical calculations   Calculations may be performed by any method that does not introduce round-off errors that are significant in comparison with the experimental errors. A slide rule, which is precise to only three significant figures, will provide enough precision for many of the computations required by the experiments in this book. However, a small electronic calculator, which will completely eliminate problems with round-off errors, is a great convenience in terms of speed and accuracy. For lengthy or repetitive calculations the use of a computer is strongly recommended (see Chap. XXI). Finally, longhand arithmetic should not be regarded as a lost art on those occasions where mechanical or electronic devices fail.

The calculations should be organized so as to prevent inadvertent loss of precision, such as frequently occurs in dealing with small differences between large quantities. Experiment 11 provides a good example of a situation of this kind. Any calculation worth doing is worth checking for possible arithmetical mistakes. Even calculations performed on a calculator may suffer from errors in entering or reading numbers.

The best way to check a calculation is to repeat it, preferably in a different way, to the same precision. When an experiment has been performed by two persons as partners, the calculations can be done by each person independently, with cross-checking of intermediate and final numerical results. A less desirable but often adequate method of checking is to perform the calculation quickly at lower precision.

Numerical methods This term is commonly used in contradistinction to analytic methods for carrying out such mathematical procedures as differentiation, integration, solution of algebraic equations, etc. Analytic methods are exact, or at least capable of being carried to any arbitrary precision; numerical methods as applied to experimental data are necessarily approximate, being limited by the finite number of data employed and by their precision.

In what follows we shall suppose that a function $y = f(x)$ is represented by a number of experimental values y_i at a series of x_i *which for convenience we shall take to be evenly spaced.* If there is reason to believe that this function should conform to some analytic form, depending on a few parameters, these parameters can be estimated by the method of least squares as described later. In this case the analytic expression can be differentiated, integrated, etc., by analytic methods, and there is no further need for numerical work.

If no clear-cut analytic form exists, these manipulations must be done numerically (or graphically, as described later). For certain purposes, especially differentiation and the solving of equations, it may be necessary to "smooth" the data so that differences from point to point do not show random fluctuations. A number of methods for performing this smoothing operation are described elsewhere.[2] We shall here give one method, based on fitting a third-degree polynominal to $2n + 1$ adjacent points. For $n = 2$, the "smoothed" value of y_i is

$$y_i' = \tfrac{1}{35}[17y_i + 12(y_{i+1} + y_{i-1}) - 3(y_{i+2} + y_{i-2})] \qquad (11)$$

For $n = 3$,

$$y_i' = \tfrac{1}{21}[7y_i + 6(y_{i+1} + y_{i-1}) + 3(y_{i+2} + y_{i-2}) - 2(y_{i+3} + y_{i-3})]$$

$$(12)$$

For $n = 4$,

$$y_i' = \tfrac{1}{231}[59y_i + 54(y_{i+1} + y_{i-1}) + 39(y_{i+2} + y_{i-2})$$
$$+ 14(y_{i+3} + y_{i-3}) - 21(y_{i+4} + y_{i-4})] \qquad (13)$$

For purposes of differentiation[3] it is useful to compute a table of successive differences as follows:

$x_0 = x_0$	y_0					
		Δy_0				
$x_1 = x_0 + w$	y_1		$\Delta^2 y_0$			
		Δy_1		$\Delta^3 y_0$		
$x_2 = x_0 + 2w$	y_2		$\Delta^2 y_1$		$\Delta^4 y_0$	
		Δy_2		$\Delta^3 y_1$		$\Delta^5 y_0$
$x_3 = x_0 + 3w$	y_3		$\Delta^2 y_2$		$\Delta^4 y_1$	
		Δy_3		$\Delta^3 y_2$		
$x_4 = x_0 + 4w$	y_4		$\Delta^2 y_3$			
		Δy_4				
$x_5 = x_0 + 5w$	y_5					

$$(14)$$

where $\Delta y_0 = y_1 - y_0$, $\Delta y_1 = y_2 - y_1$, $\Delta^2 y_0 = \Delta y_1 - \Delta y_0$, etc. The degree of smoothness of the data can be seen by inspection of the differences; there is no point in continuing the table beyond the point where the differences cease to vary in a reasonably smooth and regular manner. An expression for y as a function of x in the neighborhood of x_0 is given by the Gregory-Newton interpolation expression

$$y = y_0 + r \, \Delta y_0 + \frac{r(r - 1)}{2!} \Delta^2 y_0 + \frac{r(r - 1)(r - 2)}{3!} \Delta^3 y_0 + \cdots \qquad (15)$$

where
$$r \equiv \frac{x - x_0}{w} \qquad (16)$$

By differentiation we obtain

$$\frac{dy}{dx} = \frac{1}{w}\left(\Delta y_0 + \frac{2r-1}{2!}\Delta^2 y_0 + \frac{3r^2 - 6r + 2}{3!}\Delta^3 y_0 \right.$$

$$+ \frac{4r^3 - 18r^2 + 22r - 6}{4!}\Delta^4 y_0$$

$$\left. + \frac{5r^4 - 40r^3 + 105r^2 - 100r + 24}{5!}\Delta^5 y_0 + \cdots\right) \quad (17)$$

$$\frac{d^2 y}{dx^2} = \frac{1}{w^2}\left[\Delta^2 y_0 + (r-1)\Delta^3 y_0 + \frac{6r^2 - 18r + 11}{12}\Delta^4 y_0\right.$$

$$\left. + \frac{2r^3 - 12r^2 + 21r - 10}{12}\Delta^5 y_0 + \cdots\right] \quad (18)$$

On referring to Eq. (14) we see that the successive differences entering into Eqs. (15), (17), and (18) lie on a downward-slanting line. With as much justification an upward-slanting line may be used and, indeed, must be used if x_0 is at or very near the end of the table. No changes in the above expressions other than a replacement of Δy_0, $\Delta^2 y_0$, $\Delta^3 y_0, \ldots$ by Δy_{-1}, $\Delta^2 y_{-2}$, $\Delta^3 y_{-3}$ and a change of all signs in Eqs. (15), (17), and (18) to plus are required if we continue to define r with Eq. (16). Indeed a good procedure is to employ both downward-slanting and upward-slanting Δ's, and to restrict r to the range $-\frac{1}{2} \le r \le \frac{1}{2}$.

For evaluation of the definite integral

$$Y(a,b) = \int_a^b y \, dx \quad (19)$$

it is often possible to utilize procedures that do not require smoothing of the data.[4] The simplest procedure, based on the "trapezoidal rule," makes use of the expression

$$Y(x_0, x_n) = \frac{w}{2}(y_0 + 2y_1 + 2y_2 + \cdots + 2y_{n-1} + y_n) \quad (20)$$

This is equivalent to approximating the function by linear segments between the tabular entries (data points). A better procedure makes use of "Simpson's one-third rule":

$$Y(x_0, x_n) = \frac{w}{3}(y_0 + 4y_1 + 2y_2 + 4y_3 + 2y_4 + \cdots + 4y_{n-1} + y_n) \quad (21)$$

This rule requires that n be even, i.e., that the number of values of y be odd. This procedure is exact if y is strictly a quadratic function of x. Its use in other cases is equivalent to fitting quadratic forms (parabolas) to three points at a time, joining them at the even-numbered points. Still another rule, "Simpson's three-eighth rule,"

$$Y(x_0, x_n) = \frac{3w}{8} \left[y_0 + 3(y_1 + y_2) + 2y_3 + 3(y_4 + y_5) + 2y_6 \right.$$

$$\left. + \cdots + 3(y_{n-2} + y_{n-1}) + y_n \right] \quad (22)$$

requires that n be divisible by 3 and is equivalent to fitting third-degree functions to four points at a time. More refined procedures are available, such as Weddel's rule, but their slight superiority seldom justifies the added complications.

In cases where n is divisible neither by 2 nor by 3, the range of integration may be split into two parts, one for Simpson's one-third rule, the other for Simpson's three-eighth rule. Alternatively, if the curve is approximately linear in one or two intervals, the trapezoidal rule may be used in these intervals.

Graphical methods   These are inherently more limited in precision than numerical methods but are frequently easier and convey more of a feel for the nature of the manipulations involved. They reveal, much more clearly than a table of numbers does, such features as linearity or nonlinearity, maxima and minima, points of inflection, etc. Also, graphical methods of differentiation and integration are often easier than numerical methods. Graphical methods suffer, of course, from the fact that a plotted point has only two degrees of freedom; these we shall here assume to be represented by one independent variable x and one dependent variable y.

Use a sheet of good-quality graph paper which has a reasonable margin on all sides and which is accurately printed with thin, lightweight lines. Choose your scales of abscissas (x) and ordinates (y) so as to make best use of the area available. The type of graph paper (millimeter, inch, etc.) should be selected with a view to the use of a rational scale; the smallest divisions should preferably represent multiples of the digit 1 or 2 or perhaps 5. The axes should be clearly drawn and labeled. Plot experimental points as small dots with a sharp, hard pencil, and draw small circles (or squares, etc.) around them for better visibility. For a straight line, use a good straightedge such as a draftsman's transparent triangle, and do not hesitate to erase the line and try again until you are satisfied that the best fit has been obtained. A curve may be drawn either freehand or with ships curves or other devices with as much trial and error and erasing as is required. Any needed smoothing is done in accordance with the experimenter's judgment.

The first derivative of the plotted function is given by the slope of the tangent to the curve at the point concerned. A good way to draw the tangent is to hold a small rectangular mirror with its reflecting surface perpendicular to the paper, adjust it so that the reflection of the curve is tangent to and symmetrical with the curve itself, and then use the mirror as a straightedge to draw the tangent line. Another method

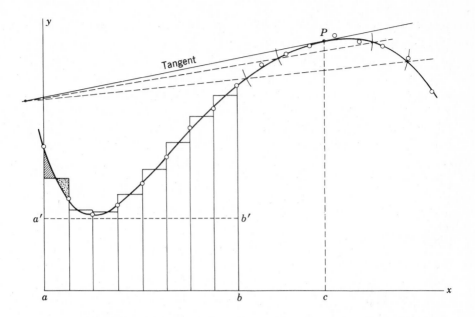

FIGURE 2
An illustration of the construction of a tangent by the method of chords and of the evaluation of a definite integral by approximating the curve by a bar graph. The top of each bar is drawn so that the two small areas thereby defined (e.g., the shaded and stippled areas on the first bar) appear equal.

is to use a compass to strike off arcs intersecting the curve on both sides and then draw a chord through the intersection points. The tangent can be drawn parallel to the chord if the curvature is uniform. If it is not, construct a second chord at a different distance in the same way and extend both lines to an intersection. Draw the (approximate) tangent line through this intersection point (see Fig. 2).

A definite integral, such as Eq. (19), can be determined by measuring the area under the curve between the desired limits (see Fig. 2). The area can be measured by use of an instrument known as a planimeter, by cutting out the area concerned with a pair of scissors and weighing it, by approximating it as well as possible by a bar graph, or by "counting squares." In the last method a count is made of the smallest squares of the graph-paper grid that lie wholly or *more than half* inside the area concerned. Whatever the method used, the area to be measured should be minimized as much as possible; there is no need to run the planimeter around the rectangular area below $a'b'$ in Fig. 2, for instance.

Whenever possible the data plotted should be made to correspond to a linear function, not only because a straight line is the easiest kind of line to draw through the points but also because deviations from a linear function are easier to detect than those from a function of any other kind. Good examples are the log p versus $1/T$ plot for the vapor pressure of a liquid (Exp. 16) and the log r versus t curve in first-order kinetics (Exp. 24). Optimum use of the graphical method is often obtained by making a *deviation plot*, in which the deviations of the experimental data from a predicted function are plotted instead of the data themselves. This is especially advantageous in certain cases where an extrapolation is to be made (Exps. 14 and 22).

PROPAGATION OF ERRORS

Once we have calculated the final result for an experiment, we are faced with the task of determining the uncertainty to which it is subject. At this point it is assumed that the limits of error in the experimental data have been estimated.† Let the desired result, for which the experiment was carried out, be designated by F, and let the directly measured quantities (e.g., weights, volumes, barometer readings, temperatures, measured emfs) be designated by $x, y, z, \ldots$. These quantities are assumed to be mutually independent. The value of F is determined by substituting the experimentally determined values of the quantities $x, y, z, \ldots$ into a formula, which will be schematically written as

$$F = f(x, y, z, \ldots)$$

Infinitesimal changes dx, dy, etc., in the experimentally determined values will produce in F the infinitesimal change

$$dF = \frac{\partial F}{\partial x} dx + \frac{\partial F}{\partial y} dy + \frac{\partial F}{\partial z} dz + \cdots \tag{23}$$

If the changes are finite rather than infinitesimal, but are small enough that the values of the partial derivatives are not appreciably affected by the changes, we have approximately

$$\Delta F = \frac{\partial F}{\partial x} \Delta x + \frac{\partial F}{\partial y} \Delta y + \frac{\partial F}{\partial z} \Delta z + \cdots \tag{24}$$

† If three or more independent measurements were made of the same experimental quantity, Eqs. (6) and (7) or Eq. (8) can be used to deduce a confidence limit for the mean value. If less than three were made (or if systematic errors must be taken into account), a confidence limit should be estimated on the basis of personal experience.

(This is equivalent to a Taylor expansion in which only the first-power terms have been retained.) Now suppose that Δx represents the experimental error $\varepsilon(x)$ in the quantity x:

$$\Delta x = \varepsilon(x) \equiv x \text{ (measured)} - x \text{ (true)} \tag{25}$$

These errors will produce an error in F,

$$\Delta F = \varepsilon(F) \equiv F \text{ (calculated from measured } x) - F \text{ (true)} \tag{26}$$

the value of which is given by

$$\varepsilon(F) = \frac{\partial F}{\partial x} \varepsilon(x) + \frac{\partial F}{\partial y} \varepsilon(y) + \frac{\partial F}{\partial z} \varepsilon(z) + \cdots \tag{27}$$

Propagation of systematic errors   Systematic or "determinate" errors are not governed by probability statistics and do not involve such concepts as the variance or standard deviation. If a systematic error in a quantity x is established, the sign and magnitude of $\varepsilon(x)$ is known, and Eq. (27) can be used for the propagation of the systematic error in F. Two important examples of this are:

$$F = x + y - z \qquad \varepsilon(F) = \varepsilon(x) + \varepsilon(y) - \varepsilon(z) \tag{28}$$

$$F = \frac{xy}{z} \qquad \frac{\varepsilon(F)}{F} = \frac{\varepsilon(x)}{x} + \frac{\varepsilon(y)}{y} - \frac{\varepsilon(z)}{z} \tag{29}$$

Propagation of random errors   In the case of random errors, we do not know the actual values of $\varepsilon(x)$, $\varepsilon(y)$, etc., and we cannot determine the actual value of $\varepsilon(F)$. However, we have already assigned to each experimental variable a standard deviation σ or a confidence limit λ. We now wish to deduce the corresponding uncertainty in the final result F, taking into consideration the high probability that the errors in the several variables will tend somewhat to cancel one another out.   Let us square both sides of Eq. (27):

$$[\varepsilon(F)]^2 = \left(\frac{\partial F}{\partial x}\right)^2 [\varepsilon(x)]^2 + \left(\frac{\partial F}{\partial y}\right)^2 [\varepsilon(y)]^2 + \cdots + 2\left(\frac{\partial F}{\partial x}\right)\left(\frac{\partial F}{\partial y}\right) \varepsilon(x)\varepsilon(y) + \cdots$$

Now let us average this expression over all values expected for $\varepsilon(x)$, $\varepsilon(y)$, ..., in accordance with the normal frequency distribution. Since the $\varepsilon(x)$ independently have average value zero, we expect the cross terms to vanish. However, the squared terms are always positive and will not vanish. If we replace the average of each squared error by the variance, we obtain

$$S^2(F) = \left(\frac{\partial F}{\partial x}\right)^2 S^2(x) + \left(\frac{\partial F}{\partial y}\right)^2 S^2(y) + \cdots \tag{30}$$

It can be seen immediately that the variance in the mean of N measurements $S^2(\bar{x}) =$

$S^2(x)/N$, which was discussed previously, can be obtained as a trivial example of Eq. (30) by taking

$$F = \bar{x} = \frac{1}{N} \sum_{i=1}^{N} x_i$$

In general, a final result F is calculated using mean values $\bar{x}, \bar{y}, \ldots$, for the observed quantities, and a confidence limit λ for each variable can be obtained from Eq. (7) or (8) or can be estimated from experience. The propagation-of-error expression in terms of confidence limits is simply

$$\lambda^2(F) = \left(\frac{\partial F}{\partial x}\right)^2 \lambda^2(\bar{x}) + \left(\frac{\partial F}{\partial y}\right)^2 \lambda^2(\bar{y}) + \cdots \tag{31}$$

Equations (30) and (31) are exact for linear functions of the variables and hold satisfactorily for other functions if the deviations of x, y, and z are less than about ± 20 percent of the mean values $\bar{x}, \bar{y}, \bar{z}$, etc. It should be pointed out that the propagation-of-variance equation is not limited to gaussian distributions.

In certain cases, the propagation of random errors can be carried out very simply:

1   For $F = ax \pm by \pm cz$,

$$\lambda^2(F) = a^2\lambda^2(x) + b^2\lambda^2(y) + c^2\lambda^2(z) \tag{32}$$

2   For $F = axyz$ (or axy/z or ax/yz or a/xyz),

$$\frac{\lambda^2(F)}{F^2} = \frac{\lambda^2(x)}{x^2} + \frac{\lambda^2(y)}{y^2} + \frac{\lambda^2(z)}{z^2} \tag{33}$$

3   For $F = ax^n$,

$$\frac{\lambda^2(F)}{F^2} = n^2 \frac{\lambda^2(x)}{x^2} \tag{34}$$

which shows that $\lambda(F)/F$, the fractional error in F, is simply n times the fractional error in x. Note the differences between Eqs. (32) and (33) for random errors and the comparable expressions in Eqs. (28) and (29) for systematic errors.

Equation (32) includes the important special case $(x - y)$ which is applicable, for example, to differences between two readings of the same variable (e.g., the weight of a liquid plus container and the weight of the empty container, the initial and final readings of a burette or a thermometer). In many cases the limits of error in the initial and final readings may be taken as equal, in which case the limit of error in the difference between the two readings is equal to $\sqrt{2}$ times the limit of error in a single reading. Equation (33) can be simplified in the special case that the fractional error in each variable is approximately the same [i.e., $\lambda(x)/x \simeq \lambda(y)/y \simeq \lambda(z)/z$]. In this case, the fractional error in F is roughly equal to the sum of the fractional errors in the individual variables divided by the square root of the number of variables.

Many propagation-of-error treatments are not so simple as those illustrated by Eqs. (32) to (34). The expression for F is often complicated enough or the functional form sufficiently awkward to justify a breakdown of the procedure into steps:

$$F = f_0(A, B, \ldots) \tag{35}$$

where $\qquad A = f_1(x_1, x_2, \ldots) \qquad B = f_2(y_1, y_2, \ldots)$, etc.

We can then write

$$\lambda^2(F) = \left(\frac{\partial F}{\partial A}\right)^2 \lambda^2(A) + \left(\frac{\partial F}{\partial B}\right)^2 \lambda^2(B) + \cdots$$

$$\lambda^2(A) = \left(\frac{\partial A}{\partial x_1}\right)^2 \lambda^2(x_1) + \left(\frac{\partial A}{\partial x_2}\right)^2 \lambda^2(x_2) + \cdots \tag{36}$$

$$\lambda^2(B) = \left(\frac{\partial B}{\partial y_1}\right)^2 \lambda^2(y_1) + \left(\frac{\partial B}{\partial y_2}\right)^2 \lambda^2(y_2) + \cdots, \text{ etc.}$$

provided that $A, B, \ldots$, are *independent*; if they are not independent (that is, in the event that some of the variables x_i are identical with some of the y_i), the calculated limit of error will be incorrect. For example, if we let

$$F = a(e^{kx} - 1) + b(y - cx) = A + B$$

it would be improper to write

$$\lambda^2(F) = \lambda^2(A) + \lambda^2(B)$$
$$\lambda^2(A) = (ake^{kx})^2 \lambda^2(x) \qquad \lambda^2(B) = b^2 \lambda^2(y) + b^2 c^2 \lambda^2(x)$$

Treatment of the function as a whole gives the correct result

$$\lambda^2(F) = (ake^{kx} - bc)^2 \lambda^2(x) + b^2 \lambda^2(y)$$

which differs from the result first given by a term $-2abcke^{kx}\lambda^2(x)$.

When the expression for F is a complicated one, it is advisable to watch for opportunities for transforming the expression or parts of the expression so that single symbols representing known quantities can be substituted for groupings of several symbols. The most advantageous simplifications are usually obtained when the quantities substituted are the end results of the calculations, such as F itself, or at least quantities which represent the later stages of the calculation of F. This is advantageous not only because it results in the greatest economy of symbols but also because it presents the most frequent opportunities for cancellation and simplification. For example, the limit of error of

$$F = C \frac{x^2}{y}$$

can be obtained from

$$\lambda^2(F) = \left(\frac{2Cx}{y}\right)^2 \lambda^2(x) + \left(\frac{Cx^2}{y^2}\right)^2 \lambda^2(y)$$

but for computational purposes it is more convenient to divide this equation by F^2, since the value of F will already have been calculated; we obtain

$$\frac{\lambda^2(F)}{F^2} = 4\frac{\lambda^2(x)}{x^2} + \frac{\lambda^2(y)}{y^2}$$

Before one begins an uncertainties treatment, it is well to examine the formula for F to determine whether any simplification can be made in it. For example, in the cryoscopic determination of the molecular weight of an unknown substance, the molecular weight is given by Eq. (12-21) as

$$M = \frac{1000g K_f}{G\,\Delta T_f}(1 - k_f\,\Delta T_f)$$

and since k_f is of the order of 0.01 deg^{-1} the term $k_f\,\Delta T_f$ is itself very small compared with unity and will contribute negligibly to the uncertainty in comparison to ΔT_f itself; therefore, the treatment of uncertainties may be based on the simpler, approximate expression

$$M = \frac{1000g K_f}{G\,\Delta T_f}$$

At various stages during the development of the derivation of the limit of error it is advisable to examine the various terms to see if any of them are negligible in comparison with other terms additive with them; if so, the negligible terms can be dropped if any material simplification of the treatment would result. A quantity can often be considered negligible for this purpose even when it could not be neglected in the calculation of F itself, for we are ordinarily interested only in obtaining a fairly rough figure for the limit of error. Thus, a term can ordinarily be dropped in the limit-of-error treatment if the effect produced in $\lambda(F)$ by dropping it is less than about 10 percent. An overgenerous (overconservative) estimate of uncertainty is to be preferred to an insufficient (overoptimistic) estimate; if a term is to be dropped, it is better that dropping it make $\lambda(F)$ a little too large than a little too small.

Uncertainties in graphically derived quantities It frequently happens that an intermediate or final result of the calculations in a given experiment is obtained from the slope or an intercept of a straight-line graph, say a plot of y against x. In such a case it is desirable to evaluate the uncertainty in the slope or in the position of the intercept. A rough procedure for doing this is based on drawing a rectangle with width $2\lambda(x_i)$ and height $2\lambda(y_i)$ around each experimental point (x_i, y_i), with the point at its center. The assignments of the limits of error $\lambda(x_i)$ and $\lambda(y_i)$ are made as described previously. The significance of the rectangle is that any point contained in it represents a possible position of the "true" point (x_i, y_i) and all points outside are ruled out as possible positions. Having already drawn the best straight line through the experimental points, and having derived from this line the slope or intercept, draw now two other (dashed) lines representing maximum and minimum values of the slope or intercept, consistent with the requirement that both lines pass through every rectangle, as shown in Fig. 3. (It should be borne in mind that the way in which the limiting lines are drawn will depend upon whether it is the slope or an interpolate intercept or an extrapolate intercept that is the quantity of interest.) Where there are a half dozen or more experimental points, it may be justifiable to neglect partially or completely one or more obviously "bad" points in drawing the original straight line and the limiting lines, provided that good judgment is exercised. The difference between the two slopes or intercepts of the limiting lines can be taken as an estimate of twice the limit of error in the slope or intercept of the best straight line.

Where the number of points is sufficiently large, the limits of error of the position of plotted points can be inferred from their scatter. Thus, an upper bound and a lower bound can be drawn, and the lines of limiting slope drawn so as to lie within these bounds. Since, as we have seen, the theory of least squares can be applied not only to yield the equation for the best straight line but also to estimate the uncertainties in the parameters entering into the equation, such graphical methods are justifiable only for rough estimates. In either case, the possibility of systematic error should be kept in mind.

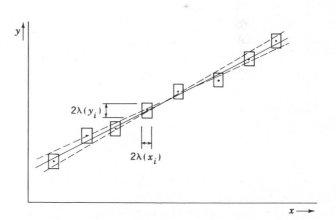

FIGURE 3
A graphical method of determining the limit of error in a slope.

Reporting the results   The final result should be reported with the estimated uncertainty and the proper units. The uncertainty may be a limit of error (confidence limits), standard error, or probable error; it is important to indicate which, in order to avoid possible confusion. A "$\pm$" sign without further explanation is generally understood to indicate a limit of error.

The number of significant figures to be retained in the reported result will depend upon the uncertainty. Normally, enough digits should be given so that the last one to the right is largely uncertain (by 3 or more) and the next to the last may be slightly uncertain (by as much as 3); that is, the number of digits given should be such that the limit of error is in the range 3 to 30 in the last two digits. The uncertainty figure should be expressed in the same units as the result to which it applies, and the last digit of the uncertainty figure should correspond to that of the digit to which it applies. Examples:

$$8.412 \pm 0.008 \text{ cal deg}^{-1} \text{ mol}^{-1}$$
$$(2.6101 \pm 0.0022) \times 10^4 \text{ cm}^{-1}$$
$$0.00522 \pm 0.00005 \text{ sec}^{-1}$$

FUNDAMENTAL LIMITATIONS ON INSTRUMENTAL PRECISION

In all experiments given in this book the precision of the measurements is governed by practical limitations of apparatus construction and operation; the precision is presumably capable of being increased by further refinement of the apparatus or the technique for using it, and it might well be imagined that there is no limit to the possible improvement. However, there are certain definite limitations imposed by physics, and although these limitations are not likely to be encountered by the beginning laboratory student, it is well that he should be aware of them.[5]

Limitations due to thermal agitation   The classical law of equipartition of energy states that every degree of freedom of a system has an average kinetic energy equal to $kT/2$, where k is Boltzmann's constant. The student is accustomed to thinking of this amount of energy (about 2×10^{-14} erg at room temperature) as being of significance only for individual atoms or molecules or at most for microscopic particles undergoing Brownian motion. However, Brownian motion may be observable in systems as large as laboratory galvanometers and may even result in a practical limit in their precision. It can be shown[5] that the rms fluctuation of scale reading in a *critically damped* galvanometer is

$$x_{\text{rms}} = (\overline{x^2})^{1/2} = (\overline{V^2})^{1/2}\frac{1}{S_V} = (\overline{i^2})^{1/2}\frac{1}{S_i}$$

$$= \left(\frac{\pi k T R}{\tau}\right)^{1/2}\frac{1}{S_V} = \left(\frac{\pi k T}{R\tau}\right)^{1/2}\frac{1}{S_i} \tag{37}$$

where R is the total resistance of the circuit (system plus external critical damping), τ is the period of the *undamped* galvanometer, and S_V and S_i are, respectively, the voltage and current sensitivities of the galvanometer (expressed as voltage or current per unit deflection). If the units used are volts, amperes, ohms, and seconds, k must be given in joules per degree per molecule (1.38×10^{-23}). For the galvanometers listed in Table XV-1, as ordinarily used, the thermal fluctuations are too small to detect (being at most of the order of a few microns on the scale). For exceedingly sensitive galvanometers used with long light paths, these fluctuations are observable and must be dealt with.

This limitation applies to a single measurement and can be overcome to some degree by making many measurements and taking the mean or by observing the mean position of the galvanometer light beam over a long period of time, provided the conditions of measurement do not change during that time. However, as in cases discussed earlier, the improvement in precision varies as the square root of the number of measurements and is therefore severely limited in practice.

Analogous "thermal noise" exists in purely electrical circuits and becomes important when exceedingly weak signals (e.g., radio signals) must be amplified.

Limitations due to particulate nature An important limitation sometimes encountered is due to the particulate nature of electricity (electrons, ions) and radiation (photons, etc.). Certain noises in vacuum tube and transistor circuits result from the discrete nature of the current carrier. An example is "shot noise," which results from statistical fluctuations in the space charge around the cathode of a vacuum tube.

The measurement of radiation intensities (see Chap. XVIII) is in certain cases (e.g., x-rays) performed by counting particles or photons one at a time. The number N counted in a time interval of given magnitude is subject to statistical fluctuations; a count of N is subject to a standard error given by

$$\sigma = \sqrt{N} \tag{38}$$

The only way of overcoming limitations imposed by the particulate nature of electricity and radiation is to scale up the experiment in physical size, time span, or intensity. Doubling the source intensity in an x-ray experiment or doubling the counting time will (on the average) result in the number of counts being doubled, while the standard error is increased by a factor of $\sqrt{2}$ and the relative error is decreased in the same proportion. To improve the relative precision by a factor of 10 requires a scale-up factor of 100; again the square-root relationship imposes a severe limitation to improvement of precision.

Limitations due to the uncertainty principle of quantum mechanics   According to quantum mechanics one cannot simultaneously measure (or define) both of two conjugate variables (position and momentum, or energy and time) to infinite precision; the product of the two standard errors may not be less than $h/2\pi$, where h is Planck's constant (6.64×10^{-27} erg sec). For instance, the energy of an excited state of an atom relative to the ground state, as determined by the wavelength of an emitted photon, is somewhat uncertain because the lifetime of the excited state is limited by collisions or other perturbations, if not by the finite probability of spontaneous emission. The uncertainty principle applies to the results of any *one* experiment comprising a pair of simultaneous measurements, as if in the example cited a single photon were to enter a very high resolution spectrograph and expose a single grain in the emulsion, the position of which could be accurately measured with a microscope comparator to obtain the wavelength. As in the case of thermal fluctuations, the precision can be improved by repeating the experiment if one can be sure that the repeated experiments are made with the same system or with identical systems under the same conditions. A spectral "line" on a photographic plate normally contains the results of an extremely large number of "experiments" of the kind described above—so large, in fact, that the square-root relation between precision and number of measurements seldom provides a severe limitation. Thus, although the uncertainty principle may largely determine the width of a spectral line, its center can often be found to a small fraction of that width.

Although real examples in which the precision of a physical measurement is materially and primarily limited by the uncertainty principle are somewhat hard to find, outside particle physics at any rate, this principle should be kept in mind as a potential limiting factor in future experiments.

REFERENCES

1. D. A. Skoog and D. M. West, "Fundamentals of Analytical Chemistry," 2d ed., pp. 25–58, Holt, New York (1969).
2. E. Whittaker and G. Robinson, "The Calculus of Observations," 4th ed., chaps. VIII and XI, Blackie, Glasgow (1944).
3. A. G. Worthing and J. Geffner, "Treatment of Experimental Data," p. 17, Wiley, New York (1943).
4. *Ibid.*, pp. 94ff.
5. H. J. J. Braddick, "The Physics of the Experimental Method," chap. IX, Chapman & Hall, London (1956).

GENERAL READING

D. C. Baird, "Experimentation," Prentice-Hall, Englewood Cliffs, N.J. (1962).

N. C. Barford, "Experimental Measurements: Precision, Error and Truth," Addison-Wesley, Reading, Mass. (1967).

B. R. Martin, "Basic Statistics for Physicists," Brookhaven National Laboratory Report 12602, Upton, N.Y. (1968).

E. M. Pugh and G. H. Winslow, "The Analysis of Physical Measurements," Addison-Wesley, Reading, Mass. (1966).

W. M. Smart, "Combinations of Observations," Cambridge, New York (1958).

E. Whittaker and G. Robinson, *op. cit.*

E. B. Wilson, Jr., "An Introduction to Scientific Research," McGraw-Hill, New York (1952).

A. G. Worthing and J. Geffner, *op. cit.*

W. J. Youden, "Statistical Methods for Chemists," Wiley, New York (1951).

III

GASES

EXPERIMENTS

1. Gas thermometry
2. Joule-Thomson effect
3. Heat-capacity ratios for gases

EXPERIMENT 1.   GAS THERMOMETRY

A fundamental attribute of temperature is that for any body in a state of equilibrium the temperature may be expressed by a number on a *temperature scale*, defined without particular reference to that body. The applicability of a universal temperature scale to all physical bodies at equilibrium is a consequence of an empirical law (sometimes called the "zeroth law of thermodynamics"), which states that, if a body is in thermal equilibrium separately with each of two other bodies, these two will be also in thermal equilibrium with each other.

However, a temperature scale must be somehow defined, in order that each attainable temperature shall have a unique numerical value. A temperature scale may be defined over a certain range by the readings of a *thermometer*, which is a body possessing some easily measurable physical property that is for all practical purposes a sensitive function of the temperature alone. The specific volume of a fluid, the electrical resistivity of a metal, and the thermoelectric potential at the junction of a pair of different metals are examples of properties frequently used. Since these properties can be much more easily and precisely measured on a relative basis than on an absolute one, it is convenient to base a temperature scale, in large part, on certain reproducible "fixed points" defined by systems whose temperatures are fixed by nature. Examples of these are the triple point of water (the temperature at which ice, liquid water, and water vapor coexist in equilibrium) and the melting or boiling points of various pure substances under 1 atm pressure. Once the temperature has been fixed at one or more points, the temperatures at other points in the range of the thermometer can be defined in terms of the value of the physical property concerned. Thus, the thermometer may serve as a device for interpolating among two or more fixed points.

In the familiar centigrade scale of temperature the two fixed points were taken as the "ice point" (temperature at which ice and air-saturated water are in equilibrium under a total pressure of 1 atm), assigned a value of 0°C, and the "steam point" (temperature at which pure water and water vapor are in equilibrium under a pressure of 1 atm), assigned a value of 100°C. A mercury thermometer with a capillary of uniform bore could be marked at 0° and at 100° on the capillary stem by use of these fixed points, and the intervening range could then be marked off into 100 equal subdivisions. It is important to recognize that the scale thus defined is not identical with one similarly defined with alcohol (for example) as the thermometric fluid; in general, the two thermometers would give different readings at the same temperature.

Thus any one physical property of any arbitrarily specified substance would seem to define a temperature scale of a rather arbitrary kind. It would seem clearly preferable to define the temperature on the basis of some fundamental law. In the middle of the nineteenth century, Clausius and Kelvin stated the second law of thermodynamics and proposed the *thermodynamic temperature scale*, which is based on that law.[1] This scale is fixed at its lower end at the absolute zero of temperature. A scale factor, corresponding to the size of the degree, must be specified to complete the definition of the temperature scale; this can be accomplished by specifying the numerical value of the temperature of a reproducible fixed point or by specifying the numerical width of the interval between two fixed points. Kelvin adopted the latter procedure in order to make the degree coincide with the (mean) centigrade degree; accordingly, the interval between the ice point and the steam point was fixed at 100°, and the absolute (or Kelvin) temperature of the ice point has been found by experiment to be approximately 273.15°. However, by recent international agreement (1948, 1954) the former procedure was substituted: The triple point of pure water is now defined as exactly 273.16°K (degrees Kelvin) and as exactly 0.01°C (degrees Celsius). The relation of the Celsius (formerly centigrade) temperature t to the Kelvin (or absolute) temperature T is†

$$t \equiv T - 273.15 \tag{1}$$

Practical difficulties arise in making very precise determinations of temperature on the thermodynamic scale; the precision of the more refined thermometric techniques considerably exceeds the accuracy with which the experimental thermometer scale may be related to the thermodynamic scale. For this reason, a practical scale known as the *International Temperature Scale* has been devised, with several fixed points and with interpolation formulas based on practical thermometers (e.g., the platinum resistance thermometer between -182.97 and $630.5°C$). This scale is intended to correspond as closely as possible to the thermodynamic scale but to permit more precision in the measurement of temperatures. Further details about this scale are given in Chap. XVI.

† The differences between the original and the new Kelvin scales and between the old centigrade and the new thermodynamic Celsius scales are small (on the order of 10^{-3} deg) and of importance only for refined measurements.

METHOD

The establishment of the International Temperature Scale has required that the thermodynamic temperatures of the fixed points be determined with as much accuracy as possible. For this purpose a device was needed that measures essentially the thermodynamic temperature and does not depend on any particular thermometric substance. On the other hand, since it was needed only for a few highly accurate measurements, it did not need to have the convenience of such instruments as resistance thermometers, mercury thermometers, and thermocouples. The device that has filled this need is the *gas thermometer*. It is based on the perfect-gas law, expressed by

$$pV = NRT \tag{2}$$

where N is the number of moles, R a universal constant, and T the "perfect-gas" temperature. At ordinary gas densities there are deviations from the perfect-gas law, but it is exact in the limit of zero gas density, where it is applicable to all gases. The "perfect-gas temperature scale" defined by Eq. (2) can be shown by thermodynamics to be identical with a thermodynamic temperature scale,† defined in terms of the second law of thermodynamics.

Gas thermometry measurements must, of course, be made with a real gas at ordinary pressures. However, it is possible to estimate the deviations from perfect-gas behavior and to convert measured pV values to "perfect-gas" pV values. For this purpose a virial equation of state for a real gas is often used:

$$\frac{p\tilde{V}}{RT} = 1 + \frac{B}{\tilde{V}} + \frac{C}{\tilde{V}^2} + \cdots \tag{3}$$

where $B, C, \ldots$ are known as the second, third, $\ldots$ *virial coefficients* and $\tilde{V}$ is the molal volume of the gas at p and T. At ordinary pressures the series converges rapidly, and for many purposes terms beyond the one containing B can be neglected. Values of second (and in some cases, third and fourth) virial coefficients are known over a wide range of temperatures from gas compressibility measurements; these values of B are useful in correcting the readings of a gas thermometer.

The establishment of the International Temperature Scale has been accomplished largely with the aid of measurements made with the helium gas thermometer. Some of the most significant work was done by Beattie and coworkers.[2]

The most precise gas thermometry method is the constant-volume method, in which a definite quantity of the gas is confined in a bulb of constant volume V at the temperature T to be determined and the pressure p of the gas is measured. A

---

† The perfect-gas properties required for this identity are (1) Boyle's law is obeyed: $p\tilde{V} = f(T)$; and (2) the internal energy per mole is a function of temperature only: $\tilde{E} = g(T)$.

problem is encountered, however, in measuring the pressure, which is usually done with a mercury manometer operating at room temperature; a way must be found to communicate between the bulb and the manometer. This is usually accomplished by connecting the bulb to the room-temperature part of the system by a slender tube and allowing a portion of the gas to occupy a relatively small, constant "dead-space" volume at room temperature. The pressure in the manometer system is then equalized to the gas pressure in this dead space by a null device, represented in the apparatus of the present experiment by the null manometer.

THEORY

The number of moles of gas N in the bulb of volume V at temperature T and in the dead space of volume v at room temperature is constant. Assuming the perfect-gas law, we have

$$\frac{pV}{RT} + \frac{pv}{RT_r} = N \tag{4}$$

where T_r is room temperature. The small region of large temperature gradient actually existing between the bulb and the dead space is replaced, without appreciable error, by a sharp division between two uniform temperatures T and T_r. Since the second term is small in comparison with the first, we shall introduce only a very small error ($\sim$0.05 percent with the present apparatus) by making the substitution

$$T_r = \frac{p_r}{p} T \tag{5}$$

where p_r is the pressure measured by the manometer when the bulb is at room temperature. We then obtain

$$\frac{pV}{RT}\left(1 + \frac{pv}{p_r V}\right) = N \tag{6}$$

Since only the pressure is being measured directly, the volume V must remain constant or else vary in a known way with the temperature (and pressure). If α is the coefficient of linear thermal expansion of the material from which the bulb is constructed, the coefficient of volume expansion is 3α. If this is assumed constant with temperature, we can write

$$\frac{pV_0}{RT}\left(1 + \frac{pv}{p_r V} + 3\alpha t\right) = N \tag{7}$$

where V_0 is the bulb volume at 0° Celsius and t is the Celsius temperature of the gas thermometer bulb. For very precise work other variables, including the dependence of the bulb volume on the difference between internal and external pressure, must be taken into account. To the precision of the present experiment, this effect of pressure is inconsequential.

Finally, if we allow for gas imperfections by including the second virial term, we can write

$$\frac{pV_0}{RT}\left(1 + \frac{pv}{p_r V} + 3\alpha t - \frac{B}{\tilde{V}}\right) = N \tag{8}$$

where

$$\frac{1}{\tilde{V}} = \frac{N}{V} \cong \frac{p_0}{RT_0} \tag{9}$$

and p_0 and T_0 are the values of p and T at (say) the ice point.

The above equations contain approximations that are acceptable for the present experiment. For precise work a much more detailed treatment is required.[2]

EXPERIMENTAL

The object of the experiment described below is to set up a gas thermometer, calibrate it at the ice point (in lieu of the experimentally more difficult triple point), and use it to determine the temperatures of one or more other fixed points. These may include the steam point, the boiling point of liquid nitrogen (in lieu of liquid oxygen, which is more difficult to obtain and more hazardous to use), the sublimation temperature of solid carbon dioxide (Dry Ice), the transition temperature of sodium sulfate decahydrate to the monohydrate and saturated solution, etc. The experiment will be performed with an apparatus which resembles in principle a research gas thermometer but is very much simpler and somewhat less precise. The apparatus is shown in Fig. 1.

Gas thermometer bulb *a*. This is a Pyrex bulb with a volume V of about 100 ml, with an attached glass capillary passing through a large rubber stopper. The upper end of this capillary tube is connected, by a length of flexible stainless-steel capillary tubing, to the null manometer.

Null manometer *b*. When the mercury surfaces in the two arms are at the same level, the pressure in the bulb is equal to that in the rest of the system, including the manometer *d*; moreover, the "dead-space" volume v then has a fixed value. This dead space is the volume of gas at temperature T_r between the top of the bulb and the mercury in the left arm. The upper part of each arm consists of a long capillary tube. This design prevents loss of mercury in the event of any accidentally large pressure difference across the null manometer. The two arms are connected at the top by a capillary stopcock, which can be opened to admit or remove gas from the bulb.

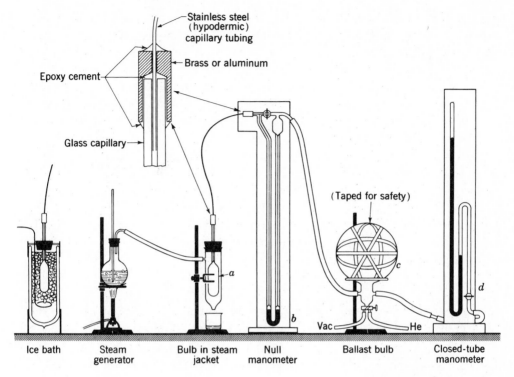

FIGURE 1
Gas thermometry apparatus.

During measurements this stopcock is kept closed, and the quantity of gas in the bulb and dead space is thereby fixed.

Ballast bulb *c*. This is a bulb of large volume (at least 5 liters) which permits small adjustments in pressure to be made by the addition or removal of small quantities of gas or air through the three-way stopcock at the bottom. One of the two lower arms of this stopcock is connected to the source of gas to be used (or to room air); the other is connected to a laboratory vacuum line, an aspirator, or a vacuum pump.

Closed-tube mercury manometer *d*. This is the instrument used for the measurement of gas pressures in this experiment. The space in the closed arm is evacuated, so the difference between the two mercury levels is a direct measure of the pressure of the gas in the other arm. For precise work, the temperature of the manometer should be determined and the manometer readings corrected (to 0°C) for the differential expansion of the mercury and the scale; however, in this experiment, if the ambient temperature is reasonably constant, the corrections should ultimately cancel out and may be omitted.

The gas thermometer bulb is first filled with the gas to be used, preferably helium or nitrogen (free of water contamination). With the stopcock on the null manometer open, the entire system is evacuated and then filled to the desired pressure with the gas. If complete evacuation is not possible (as is the case when an aspirator or a laboratory vacuum line is used), the filling must be done several times, with a delay of 1 or 2 min between each repetition to allow for diffusive mixing. When the bulb has been filled with the gas to the desired pressure, the stopcock on the null manometer is closed. After this point any gas, including air, may be allowed in the ballast bulb and manometer.

When a temperature is being measured with the thermometer, gas or air is removed from or admitted to the ballast bulb until the null manometer is balanced. When it is evident that equilibrium has been attained and the null manometer is precisely balanced, the closed-tube manometer levels are read and recorded. It is suggested that four readings be taken, alternately bringing the pressure to the null value from above and below.

Procedure   Assemble the apparatus as indicated in Fig. 1. Make sure that the stop-cock on the null manometer is properly greased. Then fill the apparatus with the gas to be used, as described above. At the final filling, adjust the pressure to about 600 Torr, and after 1 or 2 min close the null-manometer stopcock. After drift has ceased, make a reading with the bulb at room temperature. The difference between the two manometer levels is then the pressure of gas in the bulb at room temperature, p_r.

To check for possible leaks, it is advisable to restore the ballast bulb to atmospheric pressure for 10 or 15 min and then reduce the pressure so as to rebalance the null manometer. If p_r has increased significantly, a leak should be suspected.

Ice point: In the Dewar flask, prepare a "slushy" mixture of finely shaved clean ice and distilled water. This should be fluid enough to permit the gas thermometer bulb to be lowered into place and to allow a metal ring stirrer to be operated but should have sufficient ice to maintain two-phase equilibrium over the entire surface of the bulb. Mount the bulb in place, and commence stirring the mixture, moving the stirrer slowly up and down from the very top to the very bottom of the Dewar flask. Do not force the stirrer, since the bulb and its capillary stem are fragile. After equilibrium is achieved, take the ice-point pressure readings and record p_0.

Steam point: While the ice-point readings are being made, the water in the steam generator should be heated to boiling. When ready, the gas thermometer bulb should be positioned in the steam jacket so that it does not touch the wall at any point. Steam should be passed through the jacket at a rate slow enough to avoid overpressure in the jacket; steam should emerge from the bottom at low velocity. The rubber tubing connecting the steam generator to the steam jacket should run downhill all the way to permit steady drainage of any water condensed in the tubing. When drift ceases, the steam-point readings may be taken.

Repeat the ice-point measurements. The results should agree with those obtained previously to within experimental error; if the differences are larger, a leak must be suspected.

Measure the other fixed points assigned. If liquid nitrogen is used, insert the bulb into the Dewar slowly to prevent violent boiling and excessive loss of liquid.

If time permits, refill the gas thermometer bulb with another gas or reduce p_r to 300 Torr and repeat some or all of the above measurements. During the steam-point measurements, record the barometric pressure. (The barometer reading must be corrected for temperature. A discussion of the use of precision barometers is given in Chap. XVIII, and the necessary corrections are given in Appendix B.)

Also record the values of V and v for the apparatus used in the experiment.

CALCULATIONS

Equation (8) can be written in the form

$$T = Ap \tag{10}$$

where the proportionality factor

$$A = \frac{V_0}{NR}\left(1 + \frac{pv}{p_r V} + 3\alpha t - \frac{B}{\tilde{V}}\right) \tag{11}$$

is nearly constant with temperature; but for the precision attainable with this experiment, the proportionality factor should be evaluated at each temperature. Making use of the known thermodynamic temperature of the ice point, we can write

$$T = \frac{273.15}{p_0}\frac{A}{A_0}p \tag{12}$$

$$\frac{A}{A_0} = 1 + \frac{p - p_0}{p_r}\frac{v}{V} + 3\alpha t - \frac{1}{\tilde{V}}(B - B_0) \tag{13}$$

where p_0, A_0, and B_0 pertain to the ice point. Since the last two terms depend on the temperature, an approximate value of the temperature must be known before they can be evaluated; this can be obtained by setting A/A_0 equal to unity in Eq. (12). For Pyrex glass, $\alpha = 3.2 \times 10^{-6}$ deg^{-1}. Second virial coefficients for various gases are given in Table 1, and $\tilde{V}$ can be calculated from Eq. (9).

Table 1 SECOND VIRIAL COEFFICIENTS
(in cm^3 mol^{-1})

t, °C	He[a]	Ar[a]	N_2[b,c]	CO_2[b]
−250	~0			
−200	+10.4			
−150	11.4			
−100	11.7	−64.3	−51.9	
−50	11.9	−37.4	−26.4	
0	11.8	−21.5	−10.4	−154
50	11.6	−11.2	−0.4	−103
100	11.4	−4.2	+6.3	−73
150	11.0	+1.1	11.9	−51

[a] J. A. Beattie, private communication.
[b] W. Thomas, *Z. Physik*, **147**, 92 (1957).
[c] J. Otto, *Hand. d. Exp. Physik*, **8**, 2 (1929).

The student should report the temperature determined for each fixed point on both the Kelvin and Celsius scales. In cases where the temperature is a boiling point or sublimation point, calculate and report also the temperature on both scales corrected to a pressure of 1 atm. In the neighborhood of 1 atm, the boiling point of water increases with pressure by 0.037°/Torr. For liquid nitrogen, the increase is 0.013°/Torr.

DISCUSSION

What property of helium makes it particularly suitable for gas thermometry over the temperature range covered by this experiment? Is the correction for gas imperfection in this experiment of significant magnitude in relation to the experimental uncertainty? If not, by how much must the precision of the pressure measurements be improved before gas imperfection corrections become significant? How does this depend on the choice of gas to be used?

APPARATUS

Closed-tube manometer; null manometer; properly taped and mounted ballast bulb; heavy-wall pressure tubing; Dewar flask; large ring stirrer; notched cover plate for Dewar with hole for mounting gas thermometer bulb; bunsen burner; steam generator with rubber connecting tubing; steam jacket; two ring stands; ring clamp and iron gauze; two clamp holders; one large and one medium clamp.

Cylinder of helium or dry nitrogen; pure ice (3 lb); ice grinder; liquid nitrogen (1 liter); boiling chips; stopcock grease; vacuum pump or water aspirator.

REFERENCES

1. W. J. Moore, "Physical Chemistry," 4th ed., pp. 11ff., Prentice-Hall, Englewood Cliffs, N.J. (1972).
2. J. A. Beattie and coworkers, *Proc. Am. Acad. Arts Sci.*, **74**, 327 (1941); **77**, 255 (1949).

GENERAL READING

C. O. Fairchild, "Thermometry," Encyclopaedia Britannica, vol. 22, p. 110 (1959).
"Temperature: Its Measurement and Control in Science and Industry," vol. 1, Reinhold, New York (1941); vol. 2 (1955).

EXPERIMENT 2.   JOULE-THOMSON EFFECT

The Joule-Thomson effect is a measure of the deviation of the behavior of a real gas from what is defined to be ideal-gas behavior. In this experiment a simple technique for measuring this effect will be applied to a few common gases.

THEORY

An ideal gas may be defined as one for which the following two conditions apply at all temperatures for a definite quantity of the gas: (1) Boyle's law is obeyed; i.e.,

$$pV = f(T)$$

and (2) the internal energy E is independent of volume. Accordingly E is independent of pressure as well, and in the absence of other pertinent variables (due to applied fields) E is therefore a function of the temperature alone:

$$E = g(T)$$

It is apparent that the enthalpy H of an ideal gas is also a function of temperature alone:

$$H \equiv E + pV = h(T)$$

Accordingly we can write for a definite quantity of an ideal gas at all temperatures

$$\left(\frac{\partial E}{\partial V}\right)_T = \left(\frac{\partial E}{\partial p}\right)_T = \left(\frac{\partial H}{\partial V}\right)_T = \left(\frac{\partial H}{\partial p}\right)_T = 0 \tag{1}$$

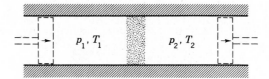

FIGURE 1
Schematic diagram of the Joule-Thomson
experiment.

The absence of any dependence of the internal energy of a gas on volume was suggested by the early experiments of Gay-Lussac and Joule. They found that when a quantity of gas in a container initially at a given temperature was allowed to expand into another previously evacuated container without work or heat flow to or from the surroundings ($\Delta E = 0$), the final temperature (after the two containers came into equilibrium with each other) was the same as the initial temperature. However, that kind of experiment (known as the Joule experiment) is of limited sensitivity, because the heat capacity of the containers is large in comparison with that of the gases studied. Subsequently Joule and Thomson[1] showed, in a different kind of experiment, that real gases do undergo small temperature changes on free expansion. This experiment utilized continuous gas flow through a porous plug under adiabatic conditions. Because of the continuous gas flow, the solid parts of the apparatus come into thermal equilibrium with the flowing gas, and their heat capacities impose a much less serious limitation than in the case of the Joule experiment.

Let it be imagined that gas is flowing slowly from left to right through the porous plug in Fig. 1. To the left of the plug the temperature and pressure of the gas are T_1 and p_1, and to the right of the plug they are T_2 and p_2. The volume of a definite quantity of gas (say 1 mol) is V_1 on the left and V_2 on the right, and the internal energy is E_1 and E_2, respectively. When 1 mol of gas flows through the plug, the work done by the system on the surroundings is

$$w = p_2 V_2 - p_1 V_1$$

Since the process is adiabatic, the change in internal energy is

$$\Delta E = E_2 - E_1 = q - w = -w$$

Combining these two equations we obtain

$$E_1 + p_1 V_1 = E_2 + p_2 V_2$$

or
$$H_1 = H_2 \tag{2}$$

Thus this process takes place at constant enthalpy.

For a process involving arbitrary infinitesimal changes in pressure and temperature, the change in enthalpy is

$$dH = \left(\frac{\partial H}{\partial p}\right)_T dp + \left(\frac{\partial H}{\partial T}\right)_p dT \tag{3}$$

In the present experiment dH is zero and dT and dp cannot be arbitrary but are related by

$$\mu \equiv \left(\frac{\partial T}{\partial p}\right)_H = -\frac{(\partial H/\partial p)_T}{(\partial H/\partial T)_p} \tag{4}$$

The quantity μ defined by this equation is known as the *Joule-Thomson coefficient*. It represents the limiting value of the experimental ratio of temperature difference to pressure difference as the pressure difference approaches zero:

$$\mu = \lim_{\Delta p \to 0} \left(\frac{\Delta T}{\Delta p}\right)_H \tag{5}$$

Experimentally ΔT is found to be very nearly linear with Δp over a considerable range; this is in accord with expectations based on the theory given below.

The denominator on the right side of Eq. (4) is the heat capacity at constant pressure C_p. The numerator is zero for an ideal gas [see Eq. (1)]. Accordingly for an ideal gas the Joule-Thomson coefficient is zero, and there should be no temperature difference across the porous plug. For a real gas, the Joule-Thomson coefficient is a measure of the quantity $(\partial H/\partial p)_T$ [which can be related thermodynamically to the quantity involved in the Joule experiment, $(\partial E/\partial V)_T$]. Using the general thermodynamic relation

$$\left(\frac{\partial H}{\partial p}\right)_T = -T\left(\frac{\partial V}{\partial T}\right)_p + V \tag{6}$$

it can be shown that, for an ideal gas satisfying the criteria already given,

$$pV = \text{const} \times T \tag{7}$$

where T is the absolute thermodynamic temperature. The coefficient $(\partial H/\partial p)_T$ is therefore a measure of the deviation from the behavior predicted by Eq. (7). On combining Eqs. (4) and (6), we obtain

$$\mu = \frac{T(\partial V/\partial T)_p - V}{C_p} \tag{8}$$

In order to predict the magnitude and behavior of the Joule-Thomson coefficient for a real gas, we can use van der Waals' equation of state,[2] which (for 1 mol) is

$$\left(p + \frac{a}{V^2}\right)(V - b) = RT \tag{9}$$

We can rearrange this equation (with neglect of the very small second-order term ab/V^2 and substitution of p/RT for $1/V$ in a first-order term) to obtain

$$pV = RT - \frac{ap}{RT} + bp$$

Thus

$$\left(\frac{\partial V}{\partial T}\right)_p = \frac{R}{p} + \frac{a}{RT^2}$$

Combination of these two equations yields

$$\left(\frac{\partial V}{\partial T}\right)_p = \frac{V - b}{T} + \frac{2a}{RT^2} \tag{10}$$

which on substitution into Eq. (8) gives the expression

$$\mu = \frac{(2a/RT) - b}{C_p} \tag{11}$$

This expression does not contain p or V explicitly, and C_p may be considered essentially independent of these variables. The temperature dependence of C_p is small, and accordingly that of μ is also small enough to be neglected over the ΔT's obtainable with a Δp of about 1 atm (namely about $1°$ or less for the gases considered here). Accordingly we may expect that μ will be approximately independent of Δp over a wide range, as stated previously.

For most gases under ordinary conditions, $2a/RT > b$ (the attractive forces predominate over the repulsive forces in determining the nonideal behavior) and the Joule-Thomson coefficient is therefore positive (gas cools on expansion). At a sufficiently high temperature the inequality is reversed, and the gas warms on expansion. The temperature at which the Joule-Thomson coefficient changes sign is called the *inversion temperature* T_I; for a van der Waals' gas

$$T_I = \frac{2a}{Rb} \tag{12}$$

This temperature is usually several hundred degrees above room temperature. However hydrogen and helium are exceptional in having inversion temperatures well below room temperatures. This results from the very small attractive forces in these gases (see Table 1 for values of van der Waals' constant a).

Table 1   VALUES OF CONSTANTS IN EQUATIONS OF STATE[a]

	He	H_2	N_2	CO_2
van der Waals:[6]				
a	0.0341	0.244	1.39	3.59
b	0.0237	0.0266	0.0391	0.0427
Beattie-Bridgeman:[3]				
A_0	0.0216	0.1975	1.3445	5.0065
a	0.05984	−0.00506	0.02617	0.07132
B_0	0.01400	0.02096	0.05046	0.10476
b	0.0	−0.04359	−0.00691	0.07235
$10^{-4}c$	0.0040	0.0504	4.20	66.00

[a] Units assumed are V in liters/mol, p in atmospheres, T in Kelvin degrees. ($R = 0.08206$ liter-atm deg^{-1}.)

Other semiempirical equations of state can be used to predict Joule-Thomson coefficients. Perhaps the best of these is the Beattie-Bridgeman equation,[3,4] which can be written (for 1 mol) as

$$p = \frac{RT(1 - \varepsilon)}{V^2}(V + B) - \frac{A}{V^2} \qquad (13)$$

where $A = A_0(1 - a/V)$, $B = B_0(1 - b/V)$, and $\varepsilon = c/VT^3$. In this equation of state there are five constants which are characteristic of the particular gas: A_0, B_0, a, b, and c. In terms of these constants and the pressure and temperature, the Joule-Thomson coefficient is given[3] by

$$\mu = \frac{1}{C_p}\left\{-B_0 + \frac{2A_0}{RT} + \frac{4c}{T^3} + \left[\frac{2B_0 b}{RT} - \frac{3A_0 a}{(RT)^2} + \frac{5B_0 c}{RT^4}\right]p\right\} \qquad (14)$$

This equation predicts a small dependence on pressure, not shown by Eq. (11) which is based on van der Waals' equation.

EXPERIMENTAL

The experimental apparatus is shown in Figs. 2 and 3. The "porous plug" is a fritted glass plate (porosity F is convenient) sealed in a 30-mm glass tube. The lower end of this tube is drawn down to 6 to 8 mm and connected by a short length of pressure tubing (securely clamped or wired) to at least 100 ft of $\frac{1}{4}$-in. copper tubing wound in a coil of convenient shape. This coil and the lower end of the glass tube (up to the fritted disk) are submerged in a constant-temperature bath in order to bring the gas to the bath temperature. The other end of the coil is connected to a supply of gas at a pressure $\frac{1}{2}$ to 1 atm above atmospheric pressure. An open-tube mercury manometer is used to measure Δp. The upper end of the glass tube (above the frit) is always at atmospheric pressure but is thermally insulated with an outer jacket filled with a medium such as vermiculite or Styrofoam. The temperature difference ΔT between the gas above the plate and the water in the bath is measured with a copper-constantan thermocouple (using the finest wires available). A sensitive galvanometer (such as Leeds and Northrup 2430d, having a sensitivity of 0.0005 μA mm^{-1} and an internal resistance of 550 Ω) is required. A potentiometer of sufficient sensitivity (± 0.1 μV) can be used, but it is not needed, since the galvanometer-thermocouple system can be calibrated directly against a Beckmann thermometer. An alternative design for a simple Joule-Thomson apparatus is described elsewhere.[5] The most obvious change in the cell design would involve mounting the "reference" junction in the high-pressure gas below the frit; this would require a change in the

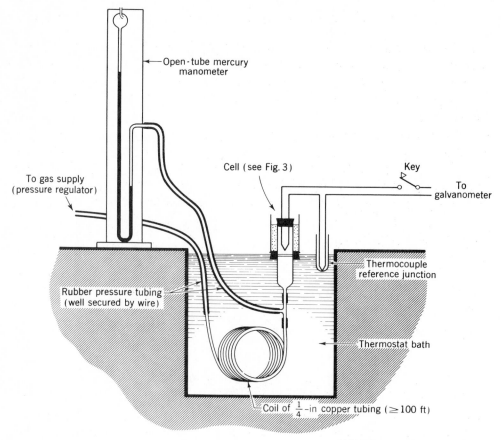

FIGURE 2
The Joule-Thomson apparatus.

calibration procedure described below. For slow flow rates through a long coil, the assumption that the high-pressure gas is at bath temperature is well justified and the present design is adequate.

Procedure   Set up the apparatus as shown in Figs. 2 and 3. The gas supply should be a cylinder or supply line equipped with a pressure regulator and a needle valve. Be sure that the supply pressure is constant; if there are other demands on the gas supply, careful planning is required. All gas connections should be securely wired or clamped. **Warning**: All changes in pressure must be carried out **very slowly**.

For the *initial* equilibration of the system, close the needle valve and adjust the main gas supply valve until the regulator gauge reads about 40 psi. Then **very slowly**

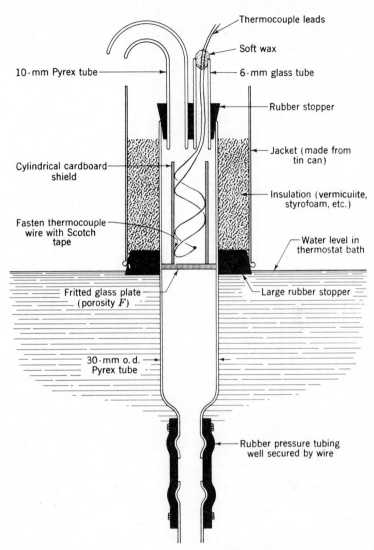

FIGURE 3
Detail of Joule-Thomson cell.

open the needle valve until the pressure in the system (as indicated by the manometer) is increasing at about the rate of 50 Torr min^{-1}. Continue to adjust the needle valve until Δp is about 750 Torr (this should take *at least* 15 min). If the gas pressure is increased rapidly at the beginning, the coil will not be able to bring the initial surge of gas to bath temperature, and the porous frit will be cooled to below its steady-state value; this will cause a very slow (2 to 3 hr) attainment of the steady-state value of

ΔT. With care in making all pressure changes, a steady state should be achieved in about 40 min.

While waiting for the system to come to a steady state, you should calibrate the thermocouple. The reference junction of the thermocouple should be embedded in a small amount of wax or oil at the bottom of a test tube which is mounted in the constant-temperature bath. The measuring junction should be removed from the Joule-Thomson cell and taped to the bulb of a Beckmann thermometer which is placed first in the bath and then in a Dewar containing water approximately 1° below the bath temperature. The temperature of the water in the Dewar can be raised or lowered by the addition of small amounts of hot or cold water, and enough thermometer and galvanometer readings can be obtained to yield a calibration curve. This curve should be obtained with the thermocouple leads connected directly across the galvanometer terminals. Another calibration curve should be obtained with an appropriate resistance in series to reduce the sensitivity by a factor of 4 or more. (With the L & N 2430d galvanometer, a 1500-Ω resistor is recommended.)

After calibration, the thermocouple wires are very carefully dried and installed in the cell. It is recommended that a cylindrical cardboard tube be placed in the cell so as to rest on the fritted disk. It may be roughly centered with a few tufts of glass wool. The thermocouple wires should be loosely coiled inside the tube and fastened near the bottom with a small piece of Scotch tape in such a way that the junction is in the center of the tube and about 5 to 10 mm above the fritted disk.

Record values of the galvanometer deflection, the pressure differential Δp, and the time until no significant change in ΔT occurs over a 10- to 15-min interval. Then *very slowly* close the needle valve so as to reduce the pressure drop Δp to approximately 600 Torr. This change should require *at least* 5 min! Record Δp and ΔT values until a steady state is achieved at this new setting (about 20 min). Repeat the procedure to obtain data at $\Delta p = 450$, 300, and 150 Torr. If time is available, make measurements on both nitrogen and carbon dioxide.

CALCULATIONS

For each gas studied, plot Δp against ΔT and draw the best straight line through these points; this line should, of course, pass through the origin. From the slope evaluate the Joule-Thomson coefficient in degrees per atmosphere.

For comparison, calculate Joule-Thomson coefficients for these gases at this temperature from the van der Waals' and/or the Beattie-Bridgeman constants given in Table 1. The molar heat capacities at constant pressure C_p of N_2 and CO_2 are 0.286 and 0.362 liter-atm deg^{-1}, respectively.

DISCUSSION

For the Joule experiment we can write

$$\eta \equiv -\left(\frac{\partial T}{\partial V}\right)_E = \frac{(\partial E/\partial V)_T}{(\partial E/\partial T)_V} = \frac{T(\partial p/\partial T)_V - p}{C_v} \qquad (15)$$

This quantity is called the Joule coefficient. It is the limit of $-(\Delta T/\Delta V)_E$, corrected for the heat capacity of the containers, as ΔV approaches zero. With van der Waals' equation of state we obtain, for 1 mol, $\eta = a/V^2 C_v$. The corrected temperature change when the two containers are of equal volume is found by integration to be $\Delta T = -a/2VC_v$, where V is the initial molar volume and C_v is the molar constant-volume heat capacity.

It will be instructive for the student to calculate this ΔT for a gas such as CO_2. In addition, he may consider the relative heat capacities of 10 liters of the gas at a pressure of 1 atm and that of the quantity of copper required to construct two spheres of this volume with walls (say) 1 mm thick and then calculate the ΔT expected to be observed with such an experimental arrangement.

APPARATUS

Joule-Thomson cell comprising glass tube with fritted disk, large rubber stopper with outer jacket, insulating medium such as vermiculite, rubber stopper with bent glass outlet tube and tube for thermocouple wires, and cardboard cylinder; clamp to hold cell in bath; thermocouple (insulated copper-constantan, No. 30) with test tube for reference junction; clamp for same; sensitive galvanometer (such as L & N 2430d); auxiliary resistance (e.g., 1500 Ω); tapping key; coil of $\frac{1}{4}$-in. copper tubing, at least 100 ft; open-tube manometer (80 cm) containing mercury; T tube; two 4- and two 36-in. (or longer) pieces of rubber pressure tubing for making connections; Beckmann thermometer; thermometer clamp; 500-ml beaker.

Constant-temperature bath regulated at 25°C; cylinders or supply lines for gases (N_2, CO_2) equipped with pressure regulators and needle valves; copper wire for securing connections of rubber tubing; Scotch tape; soft wax; paraffin wax or oil; glass wool.

REFERENCES

1. J. P. Joule and W. Thomson (Lord Kelvin), *Phil. Trans.*, **143**, 357 (1853); **144**, 321 (1854). Reprinted in "Harper's Scientific Memoirs I, The Free Expansion of Gases," Harper, New York (1898).
2. W. J. Moore, "Physical Chemistry," 4th ed., pp. 48ff., Prentice-Hall, Englewood Cliffs, N.J. (1972).

3. J. A. Beattie and W. H. Stockmayer, The Thermodynamics and Statistical Mechanics of Real Gases, in H. S. Taylor and S. Glasstone (eds.), "A Treatise on Physical Chemistry," vol. II, pp. 187ff., esp. pp. 206, 234, Van Nostrand, Princeton, N.J. (1951).
4. J. A. Beattie and O. C. Bridgeman, *J. Amer. Chem. Soc.*, **49**, 1665 (1927); *Proc. Am. Acad. Arts Sci.*, **63**, 229 (1928).
5. C. E. Hecht and G. Zimmerman, *J. Chem. Educ.*, **31**, 530 (1954).
6. "Landolt-Börnstein Physikalisch-chemische Tabellen," 5th ed., p. 254, Springer, Berlin (1923) [reprinted by Edwards, Ann Arbor, Mich. (1943)].

EXPERIMENT 3.   HEAT-CAPACITY RATIOS FOR GASES

The ratio C_p/C_v of the heat capacity of a gas at constant pressure to that at constant volume will be determined by either the method of adiabatic expansion or the sound velocity method. Several gases will be studied, and the results interpreted in terms of the contribution made to the specific heat by various molecular degrees of freedom.

THEORY

In considering the theoretical calculation of the heat capacities of gases, we shall only be concerned with perfect gases. Since $\tilde{C}_p = \tilde{C}_v + R$ for an ideal gas (where $\tilde{C}_p$ and $\tilde{C}_v$ are the molar quantities C_p/N and C_v/N), our discussion can be restricted to C_v.

The number of "degrees of freedom" for a molecule is the number of independent coordinates needed to specify its position and configuration. Hence a molecule of n atoms has $3n$ degrees of freedom. These could be taken as the three cartesian coordinates of the n individual atoms, but it is more convenient to classify them as follows:

1   Translational degrees of freedom: Three independent coordinates are needed to specify the position of the center of mass of the molecule.

2   Rotational degrees of freedom: All molecules containing more than one atom require a specification of their orientation in space. As an example, consider a rigid diatomic molecule; such a model consists of two point masses (the atoms) connected by a rigid massless bar (the chemical bond). Through the center of mass, which lies on the rigid bar, independent rotation can take place about two axes mutually perpendicular to each other and to the rigid bar. (The rigid bar itself does not constitute a third axis of rotation under ordinary circumstances for reasons based on quantum theory, there being no appreciable moment of inertia about this axis.) Rotation of a *diatomic* molecule or any linear molecule can thus be described in terms of *two* rotational degrees of

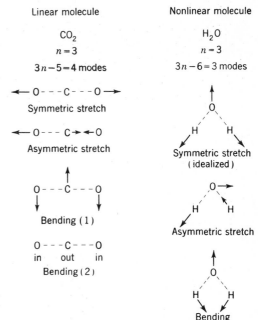

FIGURE 1
Schematic diagrams of the vibrational
normal modes for CO_2, a linear molecule, and H_2O, a bent molecule.

freedom. *Nonlinear* molecules for which the third axis has a moment of inertia
of appreciable magnitude and constitutes another axis of rotation require *three*
rotational degrees of freedom.

3 Vibrational degrees of freedom: One must also specify the displacements of
the atoms from their equilibrium positions (vibrations). The number of vibrational degrees of freedom is $3n - 5$ for linear molecules and $3n - 6$ for nonlinear molecules. These values are determined by the fact that the total number
of degrees of freedom must be $3n$. For each vibrational degree of freedom
there is a "normal mode" of vibration of the molecule, with characteristic
symmetry properties and a characteristic harmonic frequency. The vibrational
normal modes for CO_2 and H_2O are illustrated schematically in Fig. 1.

Using classical statistical mechanics one can derive the theorem of the equipartition of energy. According to this theorem $kT/2$ of energy is associated with each
quadratic term in the expression for the energy.[1] Thus there is associated with each
translational or rotational degree of freedom for a molecule a contribution to the
energy of $kT/2$ of kinetic energy and for each vibrational degree of freedom a contribution of $kT/2$ of kinetic energy and $kT/2$ of potential energy. (The corresponding
contributions to the energy per *mole* of gas are $RT/2$.)

Clearly a monatomic gas has no rotational or vibrational energy but does have a translational energy of $\frac{3}{2}RT$ per mole. The constant-volume heat capacity of a *monatomic* perfect gas is thus

$$\tilde{C}_v = \left(\frac{\partial \tilde{E}}{\partial T}\right)_V = \tfrac{3}{2}R \tag{1}$$

For diatomic or polyatomic molecules, we can write

$$\tilde{E} = \tilde{E}(\text{trans}) + \tilde{E}(\text{rot}) + \tilde{E}(\text{vib}) \tag{2}$$

In Eq. (2), contributions to the energy from electronic states have been neglected, since they are not significant at room temperature for most molecules. Also any small intermolecular energies which occur for imperfect gases are not considered.

The equipartition theorem is based on classical mechanics. Its application to translational motion is in accord with quantum mechanics as well. At ordinary temperatures the rotational results are also in accord with quantum mechanics. (The greatest deviation from the classical result is in the case of hydrogen, H_2; at temperatures below 100°K the rotational energy of H_2 is significantly below the equipartition value, as predicted by quantum mechanics.)

The vibrational energy is, however, highly quantized and depends strongly on temperature; the various vibrational modes are at ordinary temperatures only partially "active," and the degree of activity depends strongly on the temperature. As a general rule, the heavier the atoms or the smaller the force constant of the bond (i.e., the lower the vibrational frequency), the more "active" is a given degree of freedom at a given temperature and the greater is the contribution to the heat capacity. Moreover, the frequencies of modes that are predominantly bending of bonds tend to be much lower than those that are predominantly stretching of bonds. In the case of most gaseous *diatomic* molecules (where the one vibrational mode is a pure stretch) the vibrational contribution to $\tilde{C}_v$ is very small; for example, N_2 would have its classical equipartition value for $\tilde{C}_v$ only above about 4000°K. Many polyatomic molecules, especially those containing heavy atoms, will at room temperature have significant *partial* vibrational contributions to $\tilde{C}_v$.

We are now in a position to calculate for polyatomic molecules approximate or at least limiting values for $\tilde{C}_v$ and for the ratio $\tilde{C}_p/\tilde{C}_v$. For monatomic gases and all ordinary diatomic molecules (where vibration is not important at room temperature) definite values can be calculated. For a brief discussion of the calculation of $\tilde{C}_v$ (vib) for polyatomic molecules, see Exp. 40.

A.   ADIABATIC EXPANSION METHOD

For the reversible adiabatic expansion of a perfect gas, the change in energy content is related to the change in volume by

$$dE = -p \, dV = -\frac{NRT}{V} \, dV = -NRT \, d \ln V \tag{3}$$

Moreover, since E for a perfect gas is a function of temperature only, we can also write $dE = C_v \, dT$. Substituting this expression into Eq. (3) and integrating, we find that

$$\tilde{C}_v \ln \frac{T_2}{T_1} = -R \ln \frac{\tilde{V}_2}{\tilde{V}_1} \tag{4}$$

where $\tilde{C}_v$ and $\tilde{V}$ are molar quantities (that is, C_v/N, V/N). It has been assumed that C_v is constant over the temperature range involved. This equation predicts the decrease in temperature resulting from a reversible adiabatic expansion of a perfect gas.

Consider the following two-step process involving a perfect gas denoted by A:

Step I: Allow the gas to expand adiabatically and reversibly until the pressure has dropped from p_1 to p_2.

$$A(p_1, \tilde{V}_1, T_1) \rightarrow A(p_2, \tilde{V}_2, T_2) \tag{5}$$

Step II: At constant volume, restore the temperature of the gas to T_1.

$$A(p_2, \tilde{V}_2, T_2) \rightarrow A(p_3, \tilde{V}_2, T_1) \tag{6}$$

For step I, we can use the perfect-gas law to obtain

$$\frac{T_2}{T_1} = \frac{p_2 \tilde{V}_2}{p_1 \tilde{V}_1} \tag{7}$$

Substituting Eq. (7) into Eq. (4) and combining terms in $\tilde{V}_2/\tilde{V}_1$, we write

$$\ln \frac{p_2}{p_1} = \frac{-(\tilde{C}_v + R)}{\tilde{C}_v} \ln \frac{\tilde{V}_2}{\tilde{V}_1} = -\frac{\tilde{C}_p}{\tilde{C}_v} \ln \frac{\tilde{V}_2}{\tilde{V}_1} \tag{8}$$

since for a perfect gas

$$\tilde{C}_p = \tilde{C}_v + R \tag{9}$$

For step II, which restores the temperature to T_1,

$$\frac{\tilde{V}_2}{\tilde{V}_1} = \frac{p_1}{p_3} \tag{10}$$

Thus

$$\ln \frac{p_1}{p_2} = \frac{\tilde{C}_p}{\tilde{C}_v} \ln \frac{p_1}{p_3} \tag{11}$$

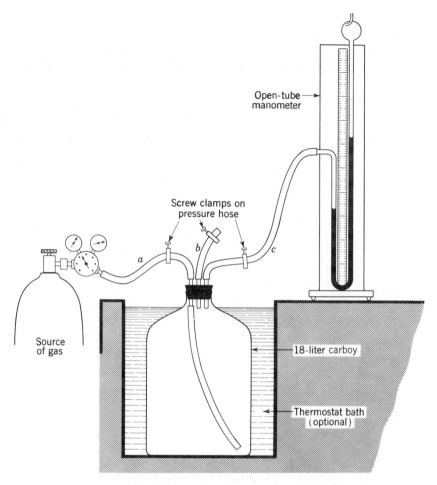

Open-tube → manometer

Screw clamps on
pressure hose

a

b

c

Source
of gas

18-liter carboy

Thermostat bath
(optional)

FIGURE 2

Apparatus for the adiabatic expansion of gases.

This can be rewritten in the form

$$\frac{\tilde{C}_p}{\tilde{C}_v} = \frac{\log p_1 - \log p_2}{\log p_1 - \log p_3} \tag{12}$$

This method, due to Clement and Desormes,[2] uses the very simple apparatus shown in Fig. 2. The change in state (5) is carried out by quickly removing and replacing the stopper of a large carboy containing the desired gas at a pressure initially somewhat higher than 1 atm pressure, so that the pressure of gas in the carboy momentarily drops to atmospheric pressure p_2. The change in state (6) consists of allowing the gas remaining in the carboy to return to its initial temperature. The initial pressure p_1 and the final pressure p_3 are read from an open-tube manometer.

The thermodynamic equations (3) to (12) apply only to that part of the gas that remains in the carboy after the stopper is replaced. We may imagine the gas initially in the carboy to be divided into two parts by an imaginary surface; the part above the surface leaves the carboy when the stopper is removed and presumably interacts irreversibly with the surroundings, but the part below the surface expands *reversibly* against this imaginary surface, doing work in pushing the upper gas out. The process is approximately adiabatic only because it is rapid; within a few seconds the gas near the walls will have received an appreciable quantity of heat by direct conduction from the walls, and the pressure can be seen to rise almost as soon as the stopper is replaced.

The experiment will presumably give a somewhat *low* result (low p_3 and therefore low $\tilde{C}_p/\tilde{C}_v$ ratio) if the expansion is to an appreciable degree irreversible, a somewhat *low* result if the stopper is left open so long that the conditions are not sufficiently adiabatic, and a somewhat *high* result if the stopper has not been removed for a long enough time to permit the pressure to drop momentarily to atmospheric. There should be no significant irreversibility if during the expansion there are no significant pressure gradients in the gas below the imaginary surface mentioned above, and such pressure gradients should not be expected if the throat area is small in comparison with the effective area of this imaginary surface. With an 18-liter carboy and a p_1 about 50 Torr above 1 atm, the volume of air forced out should be about 1 liter, which should provide a large enough surface area to fulfill this condition approximately if the throat diameter is not more than about 2 or 3 cm. The effect of the length of time the stopper is removed is much more difficult to estimate by calculation; some idea can be obtained from the duration of the sound produced when the stopper is removed and from the rate of rise of the manometer reading immediately after the stopper is replaced. For the purpose of this experiment the student may assume that, if the stopper is removed completely from the carboy to a distance of 2 or 3 in. away and replaced tightly as soon as physically possible, the desired experimental conditions are approximately fulfilled. Some additional assurance may be gained from the reproducibility obtained in duplicate runs.

The adiabatic expansion method is not the best method of determining the heat-capacity ratio. Much better methods are based on measurements of the velocity of sound in gases. One such method, described in Part B of this experiment, consists of measuring the wavelength of sound of an accurately known frequency by measuring the distance between nodes in a sonic resonance set up in a Kundt's tube. Methods also exist for determining the heat capacities directly,[3] although the measurements are not easy.

EXPERIMENTAL

The apparatus should be assembled as shown in Fig. 2. If desired, the carboy may be mounted in a thermostat bath; if so, it must be clamped securely to overcome buoyancy. The manometer is an *open-tube* manometer, one side of which is open to the atmosphere; the pressure that it measures, therefore, is the difference of pressure from atmospheric pressure. A suitable liquid for the manometer is dibutyl phthalate, which has a density of 1.046 g cm^{-3} at room temperature (20°C). To convert manometer readings (millimeters of dibutyl phthalate) to equivalent readings in millimeters of mercury, multiply by the ratio of this density to the density of mercury, which is 13.55 g cm^{-3} at 20°C. To find the total pressure in the carboy, the converted manometer readings should be added to atmospheric pressure as given by a barometer. It is unnecessary to correct all readings to 0°C, as all pressures enter the calculations as ratios.

Seat the rubber stopper firmly in the carboy and open the clamps on tubes a and b. Clamp off the tube c. The connections shown in Fig. 2 are based on the assumption that the gas to be studied is heavier than air (or the previous gas in the carboy) and therefore should be introduced at the bottom in order to force the lighter gas out at the top; in the event that the gas to be studied is lighter, the connections a and b should be reversed.

Allow the gas to be studied to sweep through the carboy for 15 min. The rate of gas flow should be about 6 liters min^{-1}, or 100 ml sec^{-1} (measure roughly in an inverted beaker held under water in the thermostat bath). Thus, 5 volumes of gas (90 liters) will pass through the carboy.

Retard the gas flow to a fraction of the flushing rate by partly closing the clamp on tube a. **Carefully** open the clamp on tube c, and then cautiously (to avoid blowing liquid out of the manometer) clamp off the exit tube b, keeping a close watch on the manometer. When the manometer has attained a reading of about 600 mm of oil, clamp off tube a. Allow the gas to come to the temperature of the thermostat bath (about 15 min), as shown by a constant manometer reading. Record this reading; when it is converted to an equivalent mercury reading and added to the barometer reading, p_1 is obtained.

Remove the stopper *entirely* (a distance of 2 or 3 in.) from the carboy, and replace it *in the shortest possible time*, making sure that it is tight. As the gas warms back up to the bath temperature, the pressure will increase and finally (in about 15 min) reach a new constant value p_3, which can be determined from the manometer reading and the barometer reading. At some point in the procedure, a barometer reading (p_2) should be taken.

Repeat the steps above to obtain two more determinations with the same gas.

For these repeat runs, long flushing is not necessary; one additional volume of gas should suffice to check the effectiveness of the original flushing.

Measurements are to be made on both helium and nitrogen. If there is sufficient time, study carbon dioxide also.

CALCULATIONS

For each of the three runs on He, N_2 (and CO_2), calculate $\tilde{C}_p/\tilde{C}_v$ using Eq. (12). Also calculate the theoretical value of $\tilde{C}_p/\tilde{C}_v$ predicted by the equipartition theorem. In the case of N_2 and CO_2, calculate the ratio both with and without vibrational contribution to $\tilde{C}_v$.

DISCUSSION

Compare your experimental ratios with those calculated theoretically, and make any deductions you can about the presence or absence of rotational and vibrational contributions, taking due account of the uncertainties in the experimental values. For CO_2, how would the theoretical ratio be affected if the molecule were nonlinear (like SO_2) instead of linear? Could you decide between these two structures from the $\tilde{C}_p/\tilde{C}_v$ ratio alone?

B.   SOUND VELOCITY METHOD

The heat-capacity ratio $\gamma \equiv C_p/C_v$ of a gas can be determined with good accuracy by measuring the speed of sound c. For an ideal gas,

$$\gamma = \frac{Mc^2}{RT} \tag{13}$$

where M is the molecular weight. A brief derivation[4] of this equation will be sketched below.

For longitudinal plane waves propagating in the x direction through a homogeneous medium of density ρ, the wave equation is

$$\frac{\partial^2 \xi}{\partial t^2} = c^2 \frac{\partial^2 \xi}{\partial x^2} \tag{14}$$

where ξ is the particle displacement. Consider a layer AB of the fluid, of thickness δx and unit cross-sectional area, which is oriented normal to the direction of propagation. The mechanical strain on AB due to the sound wave is $\partial \xi / \partial x$:

$$\frac{\partial \xi}{\partial x} = \frac{\Delta V}{V} = \frac{-\Delta \rho}{\rho} \tag{15}$$

where ΔV and $\Delta \rho$ are, respectively, the volume and density changes caused by the acoustic pressure (mechanical stress) change Δp in the ambient pressure p. From Hooke's law of elasticity,

$$-\Delta p = B_S \frac{\partial \xi}{\partial x} \tag{16}$$

where B_S is the adiabatic elastic modulus.† The net acoustic force acting on AB due to the sound wave is $-(\partial \Delta p / \partial x)\, \delta x$. According to Newton's second law of motion, this force is also equal to $(\rho\, \delta x)\, \partial^2 \xi / \partial t^2$. Thus

$$\frac{\partial\, \Delta p}{\partial x} = -\rho\, \frac{\partial^2 \xi}{\partial t^2} \tag{17}$$

or

$$\frac{\partial^2 \xi}{\partial t^2} = \frac{B_S}{\rho}\, \frac{\partial^2 \xi}{\partial x^2} \tag{18}$$

Comparison of Eqs. (14) and (18) leads to

$$\rho c^2 = B_S \tag{19}$$

and one can now rewrite Eq. (16) as $\Delta p = \rho c^2 (\Delta \rho / \rho) = c^2\, \Delta \rho$ which is equivalent to

$$c^2 = \left(\frac{\partial p}{\partial \rho}\right)_S \tag{20}$$

If sound propagation is considered as a reversible adiabatic process, we can use the expression $pV^\gamma \propto p\rho^{-\gamma} = \text{constant}$ [see Eq. (8)] to obtain

$$\left(\frac{\partial p}{\partial \rho}\right)_S = \frac{\gamma p}{\rho} \tag{21}$$

Combining Eqs. (20) and (21), one finds

$$\gamma = \frac{\rho c^2}{p} \tag{22}$$

which can be written as Eq. (13) for an ideal gas.

---

† Since the strain variations occur so rapidly that there is not time for heat flow from adjacent regions of compression and rarefaction, the adiabatic modulus is the appropriate one for sound propagation.[4]

METHOD

This experiment is based on a modified version of Kundt's tube, in which the wave-length λ of standing waves of frequency f are determined electronically. Figure 3 shows the experimental apparatus, which utilizes an audio oscillator driving a miniature speaker as the source and a microphone as the detector.

It is possible to connect the microphone directly to a sensitive voltmeter and watch the variation in the intensity of the received signal as the detector is moved along the tube. The intensity will decrease to a minimum when an antinode is reached and then increase to a maximum at a node. However, it is more precise and more convenient to use an oscilloscope to indicate the phase relationship between the speaker input and the microphone signal. The output of the audio oscillator should drive the horizontal sweep of the oscilloscope. The signal received by the microphone after propagation through the gas is applied (with amplification if necessary) to the vertical sweep of the oscilloscope. The phase difference is determined from the Lissajous figures which appear on the screen (see Chap. XVIII). When the pattern changes from a straight line tilted at 45° to the right (in phase, 0° phase angle) to another straight line tilted at 45° to the left (out of phase, 180° phase angle), the piston has moved one-half the wavelength of sound.

EXPERIMENTAL

The apparatus should be assembled as shown in Fig. 3. Details of the procedure will depend on the particular electronic components used, especially the type of oscilloscope. Additional instructions may be provided by the instructor. In most cases, the audio oscillator should be allowed to warm up for at least 1 hr to achieve stable, drift-free operation.

If the oscillator is not accurately calibrated, this can be done by (1) displaying the wave form on the screen if the oscilloscope has an accurate time base; (2) by obtaining a Lissajous figure with a standard frequency source (internal or external); or (3) holding a vibrating tuning fork near the microphone and obtaining Lissajous figures at the fundamental and several harmonics (disconnect the speaker unit during this calibration). The oscillator should be calibrated at several frequencies between 1 and 2 kHz.†

Move the piston to a position near the far end of its travel. Allow the gas to be studied to sweep through the Kundt's tube for about 10 min in order to displace all the air. Then reduce the flow rate until there is a very slow stream of gas through the tube; this will prevent air from diffusing in during the run. Adjust the gain

† The frequency unit cycles per second is called a Hertz. Thus 1 kHz = 1000 Hz = 1000 cps.

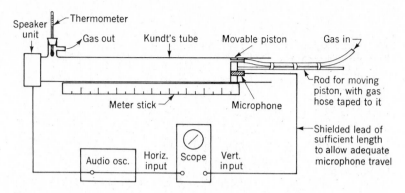

FIGURE 3

Apparatus for measuring the speed of sound in a gas. The Kundt's tube should be a glass tube ~150 cm long and ~5 cm in diameter. The close-fitting Teflon piston should move over a range of at least 100 cm. If a constant-temperature jacket is provided for the Kundt's tube, the position of the piston can be determined from the position of a fiducial mark on the rod.

controls so that the two signals are about equal in magnitude (i.e., a circle is seen when the signals differ in phase by 90°).

Set the oscillator at the desired frequency (1 kHz for N_2 and CO_2, 2 kHz for He), and slowly move the piston toward the speaker until a straight-line pattern is observed on the screen. Record the position of the piston and note whether the phase shift is 0° or 180°. Then move the piston inward again until the next straight-line pattern is observed, and again record the position. Continue to record such readings as long as a satisfactory pattern can be observed. Record the temperature of the outlet gas several times during the run.

If necessary, repeat the entire procedure for a given gas until consistent results are obtained. If time permits, repeat the procedure using a different frequency. Measurements are to be made on helium (or argon), nitrogen, and carbon dioxide.

CALCULATIONS

From the average value of the spacing $\lambda/2$ between adjacent nodes and the known frequency f, calculate the speed of sound c in each gas. Use Eq. (13) to calculate γ, and compare these experimental values with the theoretical values predicted by the equipartition theorem. In the case of N_2 and CO_2, calculate the theoretical ratio both with and without vibrational contribution to $\tilde{C}_v$.

DISCUSSION

Compare your experimental ratios with those calculated theoretically, and make any deductions you can about the presence or absence of rotational and vibrational contributions, taking due account of the uncertainties in the experimental values. For CO_2, how would the theoretical ratio be affected if the molecule were nonlinear (like SO_2) instead of linear? Could you decide between these two structures from the $\tilde{C}_p/\tilde{C}_v$ ratio alone?

Compare the speed of sound in each gas with the average speed of the gas molecules and with the average velocity component in a given direction. Explain why the speed of sound is independent of the pressure.

APPARATUS

Adiabatic expansion method   Large-volume vessel (such as 18-liter glass carboy); three-hole stopper fitted with three glass tubes; open-tube manometer, with dibutyl phthalate as indicating fluid (may contain a small amount of dye for ease of reading); three long and one short length of rubber pressure tubing; three screw clamps; cork ring and brackets, needed if vessel is to be mounted in a water bath; thermometer; 500-ml beaker.

Thermostat bath, set at 25°C (unless each vessel has an insulating jacket); cylinders of helium, nitrogen, and carbon dioxide.

Sound velocity method   Kundt's tube, complete with speaker unit (miniature cone or horn driver) and movable piston holding microphone; stable audio oscillator, preferably with calibrated dial; small crystal microphone; oscilloscope; audio-frequency amplifier, if scope gain is inadequate; electrical leads; rubber tubing; stopper; thermometer.

Cylinders of helium (or argon), nitrogen, and carbon dioxide.

REFERENCES

1. W. J. Moore, "Physical Chemistry," 4th ed., pp. 46ff., Prentice-Hall, Englewood Cliffs, N.J. (1972).
2. Lord Rayleigh, "The Theory of Sound," 2d ed., vol. II, pp. 15–23, Dover, New York (1945).
3. K. Schell and W. Heuse, *Ann. Physik*, **37,** 79 (1912); **40,** 473 (1913); **59,** 86 (1919).
4. J. Blitz, "Fundamentals of Ultrasonics," 2d ed., pp. 10–13, Butterworth, London (1967); S. M. Blinder, "Advanced Physical Chemistry," pp. 356–361, Macmillan, London (1969).

GENERAL READING

J. Blitz, *op. cit.*
J. R. Partington, "An Advanced Treatise on Physical Chemistry," vol. I, pp. 792ff., Longmans, London (1949).

IV

TRANSPORT PROPERTIES OF GASES

EXPERIMENTS

4. Viscosity of gases
5. Diffusion of gases
6. Low-pressure effusion of gases

KINETIC THEORY OF TRANSPORT PHENOMENA

In this section we shall be concerned with a molecular theory of the transport properties of gases. The molecules of a gas collide with each other frequently, and the velocity of a given molecule is usually changed by each collision that the molecule undergoes. However, when a one-component gas is in thermal and statistical equilibrium, there is a definite distribution of molecular velocities—the well-known Maxwellian distribution.[1] Figure 1 shows how molecular velocities are distributed in such a gas. This distribution is isotropic (the same in all directions) and can be characterized by a *root-mean-square velocity u*, which is given by

$$u = \left(\frac{3kT}{m}\right)^{1/2} = \left(\frac{3RT}{M}\right)^{1/2} \tag{1}$$

where k is Boltzmann's constant, R is the gas constant, m is the mass of one molecule, and M is the molecular weight. Sometimes it is more convenient to use the *mean speed $\bar{c}$*:

$$\bar{c} = \left(\frac{8kT}{\pi m}\right)^{1/2} = \left(\frac{8}{3\pi}\right)^{1/2} u \tag{2}$$

Another very important concept in kinetic theory is the average distance a molecule travels between collisions—the so-called *mean free path*. On the basis of a very simple conception of molecular collisions, the following equation for the mean free path can be derived:

$$\bar{l} = \frac{1}{\pi \bar{n} d^2}$$

where d is the "molecular diameter," or center-to-center collision distance, and $\bar{n}$ is the number of molecules per unit volume. When the fact is properly taken into

86

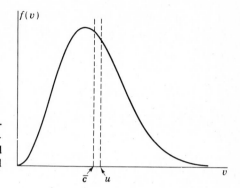

FIGURE 1
Schematic Maxwellian velocity distribution: $f(v)\ dv$ is the fraction of the molecules with velocities between v and
$v + dv$. Values of the rms velocity u and
the mean speed $\bar{c}$ are shown.

account that molecules do not all have the same velocities, a different numerical
coefficient is obtained:[1]

$$\bar{l} = \frac{1}{\sqrt{2}\ \pi \bar{n} d^2} = 3.72 \times 10^{-25} \frac{RT}{pd^2} \tag{3}$$

In this equation the perfect-gas law has been used, and the numerical coefficient
assumes that d, the molecular diameter, is in centimeters. Some typical values of $\bar{l}$
are given in Table 1.

We shall here consider in detail the viscosity, thermal conductivity, and diffusion
of gases; indeed, the kinetic theory treatment of these three transport properties is
very similar. But first let us consider the simpler problem of molecular effusion.

Effusion of gases   The theory of effusion (Knudsen flow) is quite straightforward,
since only molecular flow is involved; i.e., in the process of effusing, the molecules

Table 1   **MEAN FREE PATHS OF VARIOUS
GASES[a] AT 25°C**

Gas	Mean free path, cm	
	At 1 atm	At 0.1 Torr
He	19.4×10^{-6}	14.7×10^{-2}
Ar	7.0×10^{-6}	5.3×10^{-2}
O_2	7.1×10^{-6}	5.4×10^{-2}
N_2	6.6×10^{-6}	5.0×10^{-2}
Air	6.7×10^{-6}	5.1×10^{-2}
CO_2	4.4×10^{-6}	3.3×10^{-2}

[a] E. H. Kennard, "Kinetic Theory of Gases," McGraw-
Hill, New York (1938).

act independently of one another. For a Maxwellian distribution of velocities it can be shown[1, 2] that the number z of molecular impacts on a unit area of wall surface in unit time is

$$z = \frac{1}{\sqrt{6\pi}} u\bar{n} \tag{4}$$

where u is the rms velocity and $\bar{n}$ is the molecular concentration. Combining Eqs. (1) and (4) we obtain

$$z = \left(\frac{RT}{2\pi M}\right)^{1/2} \bar{n}$$

The number Z of *moles* colliding with 1 cm² of wall surface per second is

$$Z = \left(\frac{RT}{2\pi M}\right)^{1/2} \frac{N}{V} \tag{5}$$

where N/V is the number of moles per unit volume. For a perfect gas Eq. (5) becomes

$$Z = \frac{p}{\sqrt{2\pi MRT}} \tag{6}$$

Let two bodies of the same gas, at the same temperature but at two different pressures (p and p'), communicate through a small hole of area A, all dimensions of which are *small* in comparison with the mean free path of the gas molecules. It can then be assumed that the Maxwellian distribution of velocities is essentially un-disturbed and that the same number of molecules will enter the hole as would otherwise have collided with the corresponding area of wall surface. It will also be assumed that all molecules entering the hole will pass through to the other side. The *net* rate of effusion through the hole will be the difference between the numbers of moles flowing through the hole in unit time in the two directions:

$$\frac{dN_{\text{eff}}}{dt} = \frac{A(p - p')}{\sqrt{2\pi MRT}} \tag{7}$$

This equation assumes that the gas is perfect, but it should be emphasized that the rate of effusion calculated from it is independent of the size and shape of the molecules provided only that these are such as to give a sufficiently long mean free path at the given temperature and pressure.

Gas viscosity   Consider a body of gas under shear as shown in Fig. 4-1 and dis-cussed in the introduction to Exp. 4. Assume that all the molecules in the gas are of one kind and that the gas undergoes laminar flow. By the mass velocity of the gas at each point, we understand the (vectorial) average velocity of all molecules passing through an infinitesimal region at that point. If the velocity gradient is substantially constant within one or a few mean free paths, we may also say that the mass velocity of the gas at a given point is the (vectorial) average velocity of all molecules that have suffered their most recent collision in an infinitesimal region at that point. We shall assume that, with respect to a system of coordinates moving at the mass velocity at a given point, the distribution of velocities of gas molecules which underwent their most recent collisions at that point is the same as it would be if the gas were not under shear (i.e., that the distribution is Maxwellian and therefore isotropic). We shall also assume, for simplicity, that all molecules travel the same distance $\bar{l}$ between collisions.

Let us focus our attention on a laminar plane $x = x'$, with $v = v'$. Molecules coming from below and colliding with molecules on this layer may be assumed (in a rough approximation) to have suffered their most recent previous collision on a plane at a distance $\bar{l}$ below the plane in question. The average momentum component in the shear direction parallel to the lamina (call it the y direction) for one of these molecules is therefore

$$mv(x - \bar{l}) = mv' - m\frac{dv}{dx}\bar{l}$$

where m is the mass of a molecule. Since by hypothesis the molecules have average momentum mv' after collision, an average amount of momentum

$$-m\frac{dv}{dx}\bar{l}$$

is transferred to the laminar plane x' from below, per collision. The number of molecules reaching a unit area of the plane from below per unit time is

$$\tfrac{1}{6}\bar{n}u$$

where $\bar{n}$ is the number of molecules per unit volume and u is their rms molecular velocity. Therefore, the rate of upward transfer of momentum to a unit area of the plane from below is

$$-\tfrac{1}{6}\bar{n}um\bar{l}\frac{dv}{dx}$$

Similarly it can be shown that the rate of downward transfer of momentum from a unit area of the plane x' to the plane at $(x' - \bar{l})$ is

$$+\tfrac{1}{6}\bar{n}um\bar{l}\frac{dv}{dx}$$

A downward flow of momentum is equivalent to an upward flow of momentum of the opposite sign. Thus the net rate of *upward* flow of momentum through any given plane, per unit area, is

$$-\tfrac{1}{3}\bar{n}um\bar{l}\frac{dv}{dx}$$

This rate of change of momentum must be balanced by a force in accordance with Newton's second law:

$$\frac{f}{A} = \tfrac{1}{3}\bar{n}um\bar{\ell}\,\frac{dv}{dx} \tag{8}$$

We can write Eq. (4-1), the defining equation for the viscosity coefficient η, as

$$\frac{f}{A} = \eta\,\frac{dv}{dx} \tag{9}$$

Combining Eqs. (8) and (9) we obtain an expression for η:

$$\eta = \tfrac{1}{3}m\bar{n}u\bar{\ell} \tag{10}$$

Since $m\bar{n}$ is the density of the gas, we have on the perfect-gas assumption

$$m\bar{n} = \rho = \frac{NM}{V} = \frac{pM}{RT}$$

and we obtain

$$\eta = \tfrac{1}{3}\sqrt{3}\,p\bar{\ell}\,\sqrt{\frac{M}{RT}} \tag{11}$$

where Eq. (1) has been used for u. Equation (11) can be rearranged to give

$$\bar{\ell} = \sqrt{3}\,\frac{\eta}{p}\left(\frac{RT}{M}\right)^{1/2} \tag{12}$$

This "mean-free-path treatment" involves several rough approximations; a more sophisticated derivation[1] gives a result identical with Eq. (12) except for the numerical factor:

$$\bar{\ell} = 1.256\,\frac{\eta}{p}\left(\frac{RT}{M}\right)^{1/2} \tag{13}$$

This equation can be used to calculate the mean free path in a gas from the experimentally observed viscosity.

If we combine Eqs. (3) and (13), a theoretical expression for the coefficient of gas viscosity is obtained:

$$\eta = 2.96 \times 10^{-25}\,\frac{\sqrt{MRT}}{d^2} \tag{14}$$

According to this equation, the gas viscosity coefficient should be independent of pressure and should increase with the square root of the absolute temperature. The viscosities of gases are, in fact, found to be substantially independent of pressure over a wide range. The temperature dependence generally differs to some extent from $T^{1/2}$ because the effective molecular diameter may be dependent on how hard the molecules collide and therefore may depend somewhat on temperature. Deviation from ideal behavior in the case of air (diatomic molecules, N_2 and O_2) is demonstrated by Eq. (4-19).

Equation (14) can also be used for the calculation of molecular diameters from experimental viscosity data. These can be compared with molecular diameters as determined by other experimental methods (molecular beams, diffusion, thermal conductivity, x-ray crystallographic determination of molecular packing in the solid state, etc.).

Thermal conductivity The mechanism of thermal conduction is analogous to that of viscous resistance to fluid flow. In the case of fluid flow, where a velocity gradient exists momentum is transported from point to point by gas molecules; in thermal conduction, where a temperature gradient exists, it is kinetic energy that is transported from point to point by gas molecules. If in the mean-free-path treatment of viscosity the momentum is replaced by the average kinetic energy $\bar{\varepsilon}$ of a molecule, we obtain for one-dimensional heat flow in the x direction an equation analogous to Eq. (8):

$$\dot{q} = \tfrac{1}{3}\bar{n}u\bar{\ell}\left(\frac{-d\bar{\varepsilon}}{dx}\right) \tag{15}$$

where $\dot{q}$ is rate of heat flow per unit area, $\bar{n}$ is the number of molecules per unit volume (in cubic centimeters), u is the rms molecular velocity, and $\bar{\ell}$ is the mean free path. Now

$$\frac{d\bar{\varepsilon}}{dx} = \frac{d\bar{\varepsilon}}{dT}\frac{dT}{dx} = m\bar{C}_v\frac{dT}{dx}$$

where m is the mass of a molecule and $\bar{C}_v$ is the constant-volume specific-heat capacity (i.e., per gram) of the gas; thus we obtain

$$\dot{q} = \tfrac{1}{3}m\bar{n}u\bar{\ell}\,\bar{C}_v\left(\frac{-dT}{dx}\right) \tag{16}$$

By comparison with $\dot{q} = -K(dT/dx)$, the defining equation for the coefficient of thermal conductivity K, we find that

$$K = \tfrac{1}{3}m\bar{n}u\bar{\ell}\,\bar{C}_v \tag{17}$$

Using the expression [Eq. (10)] derived with the mean-free-path treatment for the coefficient of viscosity of a gas, we obtain the following relationship between coefficients of thermal conductivity and of viscosity:

$$\frac{K}{\eta} = \bar{C}_v \tag{18}$$

This simple expression is correct in form but requires a numerical coefficient different from unity; it must be remembered that the mean-free-path treatment is a simple treatment with several rough approximations. More advanced treatments[1] yield a modified relation

$$\frac{K}{\eta} = \frac{5}{2}\bar{C}_v \tag{19}$$

This result holds well only for monatomic gases. Interactions between diatomic and polyatomic molecules, involving transfer of rotational and vibrational energy, are difficult to treat theoretically and will not be discussed here. For monatomic gases, we can write

$$\frac{K}{\eta} = \frac{15}{4}\frac{R}{M} \tag{20}$$

where R is the gas constant and M is the molecular weight.

Diffusion   The theory of the diffusion of a gas is related to the theory of viscosity and of thermal conductivity in the sense that all three are concerned with a transport of some quantity from one point to another by the thermal motion of gas molecules. For diffusion, where a concentration gradient exists, chemical identity is transported.

Let there be two kinds of perfect gases, designated 1 and 2, present in a closed system of fixed total volume at concentrations of $\bar{n}_1$ and $\bar{n}_2$ molecules per cubic centimeter or $\bar{N}_1$ and $\bar{N}_2$ moles per cubic centimeter, respectively. These gases will have mean free paths of $\bar{\ell}_1$ and $\bar{\ell}_2$, respectively.

Let it be assumed that there is a concentration gradient in the x direction only. Consider a plane x_0, parallel to the y and z axes, and consider for a moment those molecules of kind 1 that pass through a small area $dA = dy\,dz$ on that plane. Assume for simplicity that these molecules suffered their most recent collision at or near the surface of an imaginary sphere of radius $\bar{\ell}_1$ with its center at dA. Now the mean distance from the plane x_0 up or down to the sphere is $2\bar{\ell}_1/3$. Therefore the mean concentration of these molecules, in the regions below the plane where they suffered their most recent collisions, is

$$\bar{n}_- = \bar{n}_1(x_0) - \frac{2\bar{\ell}_1}{3}\frac{d\bar{n}_1}{dx}$$

and the mean concentration similarly defined above the plane is

$$\bar{n}_+ = \bar{n}_1(x_0) + \frac{2\bar{\ell}_1}{3}\frac{d\bar{n}_1}{dx}$$

The number of molecules passing through a unit area of a given plane surface depends on the concentration of the molecules;

$$z = \tfrac{1}{4}\bar{c}\bar{n} \tag{21}$$

where $\bar{c}$ is the mean speed [see Eqs. (2) and (4)]. The *net* number of molecules of kind 1 crossing the x_0 plane in the upward direction is therefore

$$z_1 = \tfrac{1}{4}\bar{c}_1(\bar{n}_- - \bar{n}_+) = -\tfrac{1}{3}\bar{c}_1\bar{\ell}_1 \frac{d\bar{n}_1}{dx} \tag{22}$$

So far we have neglected the mass flow, if any, that is needed for the maintenance of constant and uniform pressure. If there is mass flow with a given velocity w_0 (which will be taken as positive when the mass flow is upward), the molecular velocities will be Maxwellian only with respect to a set of coordinate axes moving with that velocity, since the diffusing molecules suffered their recent collisions with molecules moving upward at that *mean* velocity. Equation (22) therefore represents the number of molecules per second crossing a unit area on a plane (parallel to the y and z axes) that is moving upward at velocity w_0. The number of molecules per second crossing a unit area on a *fixed* plane is then

$$z_1 = \bar{n}_1 w_0 - \tfrac{1}{3}\bar{c}_1\bar{\ell}_1 \frac{d\bar{n}_1}{dx}$$

$$z_2 = \bar{n}_2 w_0 - \tfrac{1}{3}\bar{c}_2\bar{\ell}_2 \frac{d\bar{n}_2}{dx} \tag{23}$$

Now for maintenance of constant pressure we must have $z_1 = -z_2$; thus we obtain

$$w_0 = \frac{1}{3(\bar{n}_1 + \bar{n}_2)}\left(\bar{c}_1\bar{\ell}_1 \frac{d\bar{n}_1}{dx} + \bar{c}_2\bar{\ell}_2 \frac{d\bar{n}_2}{dx}\right)$$

$$= \frac{1}{3\bar{n}}\frac{d\bar{n}_1}{dx}(\bar{c}_1\bar{\ell}_1 - \bar{c}_2\bar{\ell}_2) \tag{24}$$

and

$$z_1 = -\frac{1}{3\bar{n}}(\bar{n}_1\bar{\ell}_2\bar{c}_2 + \bar{n}_2\bar{\ell}_1\bar{c}_1)\frac{d\bar{n}_1}{dx} = -z_2 \tag{25}$$

where, since the pressure is constant, we have used the relation

$$-\frac{d\bar{n}_1}{dx} = \frac{d\bar{n}_2}{dx} \tag{26}$$

We can also write, in terms of moles,

$$Z_1 = -\frac{1}{3\bar{N}}(\bar{N}_1\bar{\ell}_2\bar{c}_2 + \bar{N}_2\bar{\ell}_1\bar{c}_1)\frac{d\bar{N}_1}{dx} = -Z_2$$

$$= -\tfrac{1}{3}(X_1\bar{\ell}_2\bar{c}_2 + X_2\bar{\ell}_1\bar{c}_1)\frac{d\bar{N}_1}{dx} \tag{27}$$

where X_1 and X_2 are mole fractions ($X_1 = \bar{N}_1/\bar{N}$, $X_2 = \bar{N}_2/\bar{N}$; $\bar{N} = \bar{N}_1 + \bar{N}_2$). By comparison with Eq. (5-2), the defining equation for the diffusion constant D_{12}, we find that

$$D_{12} = \tfrac{1}{3}(X_1 \bar{l}_2 \bar{c}_2 + X_2 \bar{l}_1 \bar{c}_1) \tag{28}$$

This equation appears to predict that D_{12} will be a function of the composition of the gas, although experimentally the diffusion constant is almost independent of composition. However, we must be careful in our definition of $\bar{l}_1$ and $\bar{l}_2$. We must take account of the fact that collisions of molecules of one species with one another can have no significant effect on the diffusion; such collisions do not affect the total momentum possessed by all the molecules of that species and thus do not affect the mean mass velocity of the species in its diffusion. Thus the total number of molecules of that species crossing the reference plane in a given period of time is not affected by such collisions and is the same as if such collisions did not take place at all. We should define $\bar{l}_1$ for our present purpose as the mean free path of molecules of species 1 between successive collisions with molecules of species 2 and *vice versa*. Accordingly we write [see Eq. (3)]

$$\bar{l}_1 = \frac{1}{\sqrt{2\pi \bar{n}_2 d_{12}{}^2}} \tag{29}$$

where by d_{12} we mean the center-to-center collision distance (assuming the molecules to be "hard spheres"). Equation (28) then becomes

$$D_{12} = \frac{1}{3\sqrt{2\pi d_{12}{}^2 \bar{n}}} (\bar{c}_1 + \bar{c}_2) \tag{30}$$

which is independent of the composition. If we now introduce the kinetic theory expressions for $\bar{c}_1$ and $\bar{c}_2$ as given by Eq. (2) and make use of the perfect-gas expression for $\bar{n}$, namely,

$$\bar{n} = N_0 \bar{N} = N_0 \frac{p}{RT} \tag{31}$$

where N_0 is Avogadro's number, we obtain for the diffusion constant

$$D_{12} = \frac{2}{3\pi^{3/2} N_0} \frac{(RT)^{3/2}}{p} \frac{\sqrt{1/M_1} + \sqrt{1/M_2}}{d_{12}{}^2} \tag{32}$$

and for the *self-diffusion constant* (see Exp. 5)

$$D_1 = \frac{4}{3\pi^{3/2} N_0} \frac{(RT)^{3/2}}{p} \frac{\sqrt{1/M_1}}{d_1{}^2} \tag{33}$$

More advanced treatments[1] lead to a slightly different form of expression for D_{12} and different numerical coefficients for both D_{12} and D_1:

$$D_{12} = \frac{3}{8\sqrt{2\pi} N_0} \frac{(RT)^{3/2}}{p} \frac{\sqrt{1/M_1 + 1/M_2}}{d_{12}{}^2} \tag{34}$$

and
$$D_1 = \frac{3}{8\sqrt{\pi}\,N_0}\,\frac{(RT)^{3/2}}{p}\,\frac{\sqrt{1/M_1}}{d_1{}^2} \tag{35}$$

Equations (32) and (34), which are based on a "hard-sphere" model, are in agreement in predicting no dependence of the diffusion constant on gas composition. However, for real molecules a slight composition dependence should exist, which depends upon the form of the intermolecular potential.[1]

That Eqs. (32) and (34) differ in the dependence of D_{12} on the molecular weights should not be altogether surprising, since in our simple treatment we have assumed that the velocities of molecules after collisions are uncorrelated with their velocities before collisions, while, in fact, such correlations in general exist and are functions of the ratio of the masses of the colliding particles. To take this and other remaining factors properly into account would require a treatment that is beyond the scope of this book. It may suffice to point out here that the square-root quantity in Eq. (34) is equivalent to $\sqrt{1/\mu_{12}N_0}$, where

$$\mu_{12} = \frac{m_1 m_2}{m_1 + m_2}$$

is the "reduced mass" of a system comprising a molecule of kind 1 and a molecule of kind 2.

The diffusion constant is predicted by Eqs. (32) to (35) to be inversely proportional to the total pressure. Experimentally this is the case to roughly the degree to which the perfect-gas law applies. The equations appear to predict that the diffusion constant will be proportional to the three-halves power of the temperature; however, as in the case of viscosity, significant deviations from this behavior occur, as actual molecules are not truly "hard spheres" and have collision diameters that depend on the relative speeds with which molecules collide with one another.

By means of Eq. (34) it is possible to calculate center-to-center collision distances d_{12} from a measured diffusion constant D_{12}. If three diffusion constants D_{12}, D_{13}, and D_{23} can be measured, individual molecular diameters d_1, d_2, and d_3 can be obtained if it can be assumed that the collision distances are the arithmetic averages of the molecular diameters involved, as would be the case with hard spheres:

$$d_{ij} = \tfrac{1}{2}(d_i + d_j) \tag{36}$$

If a self-diffusion constant D_1 can be measured approximately through one of the approaches discussed in Exp. 5, a molecular diameter d_1 can be obtained directly from Eq. (35). Conversely, Eqs. (34) and (35) can be used to estimate a diffusion constant, if molecular diameters are known from some other source, or to calculate self-diffusion constants from binary diffusion constants (or vice versa) by calculating first the molecular diameters. In the latter case it can be argued that any remaining approximations in the treatment will largely cancel out.

As expected, the self-diffusion constant can be related to other transport coefficients. From Eq. (28) we can write for self-diffusion

$$D_1 = \frac{2}{3}\left(\frac{8}{3\pi}\right)^{1/2} u_1 \bar{\ell}_1 \tag{37}$$

where we now interpret $\bar{\ell}_1$ as the mean free path as customarily defined. The simple mean-free-path treatment of gas viscosity gave Eq. (10), which can be written as

$$\eta = \tfrac{1}{3}\rho u \bar{\ell} \tag{38}$$

where ρ is the mass density of the gas. Thus we obtain

$$\frac{D}{\eta} = 1.8 \frac{1}{\rho} \tag{39}$$

More exact treatments[1] have shown that the coefficient of $1/\rho$ in this expression should be 1.20 for hard spheres and about 1.5 for spherical molecules with an inverse fifth-power repulsion. Clearly D may also be related to K; for monatomic gases, we can use Eq. (20) and obtain

$$\frac{D}{K} \cong \frac{0.4M}{\rho R} \tag{40}$$

REFERENCES

1. E. H. Kennard, "Kinetic Theory of Gases," McGraw-Hill, New York (1938); R. D. Present, "Kinetic Theory of Gases," McGraw-Hill, New York (1958).
2. W. J. Moore, "Physical Chemistry," 4th ed., Prentice-Hall, Englewood Cliffs, N.J. (1972).

EXPERIMENT 4.   VISCOSITY OF GASES

It is a general property of fluids (liquids and gases) that an applied shearing force that produces flow in the fluid is resisted by a force that is proportional to the gradient of flow velocity in the fluid. This is the phenomenon known as *viscosity*.

Consider two parallel plates of area A, a distance D apart, as in Fig. 1. It is convenient to imagine that D is small in comparison with any dimension of the plates in order to avoid edge effects. Let there be a uniform fluid substance between the two plates. If one of the plates is held at rest while the other moves with uniform velocity v_0 in a direction parallel to its own plane, under ideal conditions the fluid undergoes a pure shearing motion and a flow velocity gradient of magnitude v_0/D exists throughout the fluid. This is the simplest example of *laminar flow*, or pure viscous flow, which takes place under such conditions that the inertia of the fluid plays no significant

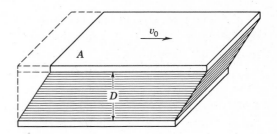

FIGURE 1
Ideal plane-parallel laminar flow.

role in determining the nature of the fluid motion. The most important of these conditions is that the flow velocities be small. In laminar flow in a system with stationary solid boundaries, the paths of infinitesimal mass elements of the fluid do not cross any of an infinite family of stationary laminar surfaces that may be defined in the system. In the simple example above, these laminar surfaces are planes parallel to the plates. When fluid velocities become high, the flow becomes *turbulent* and the momentum of the fluid carries it across such laminar surfaces so that eddies or vortices form.

In the above example, with laminar flow, the force f resisting the relative motion of the plates is proportional to the area A and to the velocity gradient v_0/D:

$$f = \eta A \frac{v_0}{D} \tag{1}$$

The constant of proportionality η is called the *coefficient of viscosity* of the fluid, or simply the *viscosity* of the fluid. The cgs unit of viscosity is the *poise*, which is equivalent to a dyne-second per square centimeter.

The viscosities of common liquids are of the order of a centipoise (cP); at 20.20°C, the viscosity of water is 1.000 cP, while that of ethyl ether is 0.23 cP, and that of glycerin is 830 cP (8.3 P). The viscosity of pitch or asphalt at room temperature is of the order of 10^{10} P, and the viscosity of glass (a supercooled liquid) is many powers of 10 higher. The viscosities of gases are of the order of 100 or 200 μP. The viscosities of liquids and soft solids decrease with increasing temperature, while those of gases increase with increasing temperature. Viscosities of all substances are substantially independent of pressure at ordinary pressures but show change at very high pressures and, apparently in the case of gases alone, at very low pressures also.

METHOD

Among the various methods[1,2] that have been used for determining the viscosities of liquids and gases are several that involve the measurement of viscous drag on a rotating disk or cylinder immersed in the fluid or a sphere falling through it. The simplest and most commonly used methods, however, depend on measurement of the rate of viscous flow through a cylindrical capillary tube, in relation to the pressure gradient along the capillary.

For absolute viscosity measurements the pressure should be constant at each end of the capillary. A simple but excellent method for gases has been described by A. O. Rankine[2] in which a constant-pressure difference is maintained across the ends of a capillary by a pellet of mercury falling in a parallel tube of considerably larger diameter (up to 3.5 mm). However, it is rarely necessary to make absolute determinations. Most viscosity measurements are made in "relative viscosimeters," in which the viscosity is proportional to the time required for the flow of a fixed quantity of fluid through the capillary, under a pressure differential that varies during the experiment *but always in the same manner*. If the pressure differential as a function of the volume of fluid passed through the capillary is accurately known, the viscosimeter can be used for absolute determinations. More commonly the unknown factors are lumped into a single apparatus constant, the value of which is determined by time-of-flow measurements on a reference substance of known viscosity. This is the principle of the Ostwald viscosimeter routinely used for measurements on liquids and solutions, in which capillary flow is induced by gravity (and accordingly the time of flow is proportional to the viscosity divided by the liquid density). It is also the principle of the method used for measurements on gases in this experiment, in which gas is caused to flow through the capillary by mercury displacement.

THEORY

Let us imagine a long cylindrical capillary tube, of length L and radius r. Let x denote the distance outward radially from the axis of the tube and z denote the distance along the axis. Under conditions of laminar flow, the laminar surfaces are a family of right circular cylinders coaxial with the tube.

Consider two such cylindrical surfaces, of radius x and $x + dx$ and length dz as shown in Fig. 2a. What is the difference in velocity dv between the two cylinders when the flow rate is such as to produce a pressure gradient dp/dz along the tube?

The force tending to cause the inner cylinder to slide past the outer one is the same as it would be for a *solid* inner piston of radius x sliding with the same relative velocity in a *solid* cylinder of radius $x + dx$, with the fluid playing the role of a "lubricant" between. Since the spacing dx is infinitesimal in comparison with the radius of curvature, we may imagine that the conditions existing between the two surfaces resemble those of Fig. 1. The force acting on an element of length dz of the tube is

$$df = \frac{dp}{dz} \pi x^2 \, dz$$

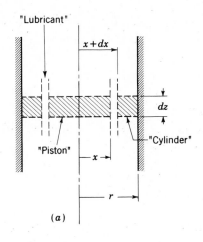

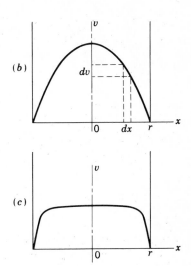

FIGURE 2

(a) Definition of variables. (b) Parabolic velocity distribution. (c) Abnormal (approximately uniform) velocity distribution existing near inlet.

The corresponding cross-sectional area is

$$dA = 2\pi x \, dz$$

and the spacing D is dx. In order to determine the corresponding velocity difference dv, we substitute for f, A, D, and v_0 in Eq. (1) and rearrange to obtain

$$dv = \frac{\pi x^2}{\eta 2\pi x} \frac{dp}{dz} \, dx = -\frac{1}{2\eta}\left(-\frac{dp}{dz}\right) x \, dx \tag{2}$$

To find the velocity at any value of x, we integrate from $x = r$, where the velocity vanishes, to $x = x$:

$$v(x) = \int_r^x dv = \frac{1}{4\eta}\left(-\frac{dp}{dz}\right)(r^2 - x^2) \tag{3}$$

This equation predicts that the distribution of velocities in the tube can be represented by a parabola, as in Fig. 2b. The volume rate of flow past a given point (volume per unit time) is

$$\phi_V = \int_0^r v(x)2\pi x \, dx = \frac{\pi r^4}{8\eta}\left(-\frac{dp}{dz}\right) \tag{4}$$

and the mean flow velocity $\bar{v}$ is†

$$\bar{v} = \frac{\phi_V}{\pi r^2} = \frac{1}{8\eta}\left(-\frac{dp}{dz}\right) r^2 = \tfrac{1}{2}v_{\max} \tag{5}$$

† It should be emphasized that $\bar{v}$ is *not* the mean molecular speed $\bar{c}$ defined in Eq. (IV-2).

If the fluid is incompressible, the volume rate of fluid flow must be constant along the tube and we find, where p_1 and p_2 are the pressures at the inlet and outlet of the tube, respectively, that

$$\phi_V = \frac{\pi r^4 (p_1 - p_2)}{8\eta L}$$

This equation, known as *Poiseuille's law*, can be applied under ordinary laminar-flow conditions to liquids, and also to gases in the limiting case where $(p_1 - p_2)$ is negligible in comparison with p_1 or p_2. For a more general treatment of gases, their compressibility must be taken into account. If we assume the perfect-gas law, we can write for the *molar* rate of flow (moles per unit time)

$$\phi_N = \frac{p}{RT} \phi_V = \frac{\pi r^4}{8\eta RT} p\left(-\frac{dp}{dz}\right) = \frac{\pi r^4}{16\eta RT}\left[-\frac{d(p^2)}{dz}\right] \tag{6}$$

Since conservation of matter requires that the mole rate of flow be constant along the tube, $d(p^2)/dz$ must be constant and accordingly we can write

$$-\frac{d(p^2)}{dz} = \frac{p_1{}^2 - p_2{}^2}{L} \tag{7}$$

We now obtain

$$\phi_N = \frac{\pi r^4 (p_1{}^2 - p_2{}^2)}{16\eta LRT} \tag{8}$$

In the apparatus to be used in this experiment (see Fig. 3) the gas contained in a bulb B is forced through a capillary tube by mercury flowing into the bulb from an upper reservoir R. The time during which the mercury level rises from a lower fiducial mark a to an upper fiducial mark b is denoted as t_{ab}. For any such apparatus of given design and dimensions and containing a constant amount of mercury, the inlet pressure p_1 is a definite function of the volume V occupied by the gas in the bulb and tubing below the capillary; p_2 is a constant (atmospheric pressure). If N is the number of moles of gas below the capillary,

$$\frac{dN}{dt} = \frac{1}{RT} \frac{d(p_1 V)}{dt} = \frac{1}{RT}\left(V \frac{dp_1}{dV} + p_1\right)\frac{dV}{dt}$$

Setting $(-dN/dt)$ to ϕ_N we obtain†

$$-\frac{dV}{dt} = \frac{\pi r^4 (p_1{}^2 - p_2{}^2)}{16\eta L[V(dp_1/dV) + p_1]} \tag{9}$$

† The minus sign is necessary, since dN/dt is negative while ϕ_N is positive.

On inverting and integrating, we find that the time required for the mercury meniscus to rise from mark a to mark b is

$$t_{ab} = \eta \, \frac{16L}{\pi r^4} \int_{V_b}^{V_a} \frac{V(dp_1/dV) + p_1}{p_1^2 - p_2^2} \, dV \equiv K\eta \qquad (10)$$

where the "apparatus constant" K contains a part $(16L/\pi r^4)$ determined by the capillary tube and a part (the integral over V) determined by the shape and dimensions of the bulb and the mercury pressure system. While it is possible to determine K from capillary dimensions and measurements of p_1 as a function of V (as described

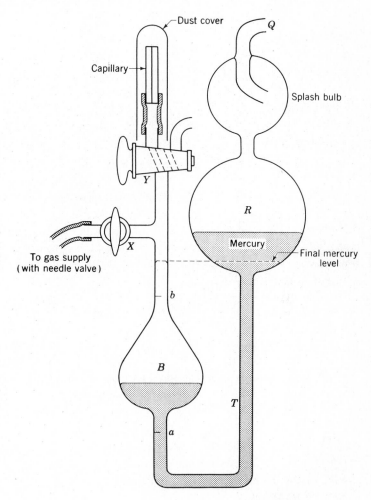

FIGURE 3
Gas viscosity apparatus.

later), more satisfactory results can be obtained by determining K from the time of flow of a gas of known viscosity.

Factors influencing apparatus design   Practical application of the theory presented above depends on the validity of a number of assumptions. One, which has already been mentioned, is the assumption that the flow is laminar. It has been shown by dimensional analysis that for any given type of hydrodynamic experiment the conditions for onset of turbulent flow depend upon the magnitude of a certain combination of pertinent experimental variables that is a pure number, called the Reynolds number R. For flow through a long, round, straight tube,

$$R = r\bar{v}\,\frac{\rho}{\eta} \tag{11}$$

where ρ is the density of the fluid. It is found empirically that laminar flow is always obtained in such a tube when R is less than 1000 regardless of the magnitude of any of the individual variables r, $\bar{v}$, ρ, η. Laminar flow can be obtained with a Reynolds number as high as 10,000 or even 25,000 if sufficiently careful attention is given to the smoothness of the walls and to the shape of the inlet, but ordinarily turbulent flow is obtained with Reynolds numbers in excess of a few thousand.[3]

Another important factor concerns the transition from a uniform velocity distribution at the inlet of the tube to the parabolic velocity distribution inside the tube (see Fig. 2b and c). The region in which this transition takes place may be of considerable length, and Eq. (8) will not apply very well over this region. Thus it is important that the length of the region be small in comparison with the length of the tube. Experiments have shown[3] that the length L' of the transition region is given approximately by the equation

$$L' = \tfrac{1}{4}Rr \tag{12}$$

where r is the radius of the tube and R is the Reynolds number.

It is important that the tube be very straight; the critical Reynolds number for onset of turbulence will be greatly reduced if the tube is appreciably curved. As a suitable criterion it might be specified that deviations from straightness should be negligible in comparison with r over distances of the order of L'.

Kinetic-energy correction   In deriving Eq. (10) it was assumed that "end effects" are negligible. Such end effects are most likely to be important at the inlet of the capillary, where a pressure drop occurs as a result of the necessity of imparting kinetic energy to the fluid.

By the law of Bernoulli, the pressure drop is equal to the "flow-average" *kinetic-energy density* of the fluid (that is, the rate of flow of kinetic energy past a given fixed point divided by the volume rate of flow of fluid past this point). For flow through a pipe or capillary,

$$\Delta p = \frac{1}{\pi r^2 \bar{v}} \int_0^r \rho \frac{v^2}{2} v 2\pi x \, dx \tag{13}$$

Beyond the transitional region of length L' we may assume the parabolic velocity distribution given by Eqs. (3) and (5):

$$v = 2\bar{v} \left(1 - \frac{x^2}{r^2} \right) \tag{14}$$

The overall kinetic-energy pressure drop from the inlet to the point where the distribution has become parabolic is obtained by substituting Eq. (14) into Eq. (13) and is

$$\Delta p = \rho(\bar{v})^2 = \frac{\rho}{\pi^2 r^4} \phi_V^2 \tag{15}$$

This quantity is called the *kinetic-energy correction*. In practice, it is subtracted from the overall pressure difference $p_1 - p_2$, since no comparable pressure change need ordinarily be considered at the outlet end. This results from the fact that the kinetic energy of the fluid carries it in a jet stream far out into the body of fluid beyond the outlet, so that the kinetic energy is dissipated (as heat, rather than as potential energy) at a considerable distance from where it could otherwise be effective in producing a pressure change at the outlet.

To take into account the effect of kinetic energy on the time of flow t_{ab}, we can subtract Δp, as given by Eq. (15), from p_1 in Eqs. (8) and (9), then invert and integrate as before. We obtain the integral appearing in Eq. (10), plus additional terms which are small. To simplify the latter we assume that $(p_1 - p_2)$ and $V \, dp_1/dV$ are negligible in comparison with p_1. We obtain

$$t_{ab} = K\eta + \frac{\rho V_{ab}}{8\pi\eta L} \tag{16}$$

where V_{ab} is the volume of the bulb between the two fiducial marks and K is the same apparatus constant as previously defined.†

> † One may ask whether the viscous drag and kinetic energy of the mercury in tube T of the apparatus shown in Fig. 3 will influence the experimental results. Simple calculations are likely to show that kinetic-energy effects (including possible turbulence) will be more important than viscous drag. A reasonable design criterion (for 0.1 percent maximum error) might be based on the application of Eq. (15) to the mercury in tube T:
>
> $$3 \frac{\rho_{Hg}}{\pi^2 r_T^4} \phi_{V\,max}^2 \leq 0.001(p_{1\,max} - p_2) \tag{17}$$
>
> where the factor 3 appears reasonable, since there are three straight portions of the tube. If the Reynolds number is high enough to indicate turbulent flow of the mercury, the factor 3 may be too small. This formula assumes a parabolic velocity distribution; a nearly uniform distribution would appear more likely in view of Eq. (12) but is a less conservative assumption, since it would imply an additional factor of $\frac{1}{2}$ on the left side of Eq. (17).

Slip correction   It has been assumed throughout the above discussion that Eq. (1) is valid down to the smallest dimensions that are of any significant importance in the experiment. This assumption appears well founded for laminar *liquid* flow, but for gases it breaks down when the mean free path is not negligibly small in comparison with the apparatus dimensions. Thus, at very low pressures or at ordinary pressures in capillaries of very small diameter, gas viscosities appear to be lower than when measured under ordinary conditions. When the mean free path of a gas is small but not negligible in comparison with the radius of the capillary, the gas behaves as if it were "slipping" at the capillary walls, rather than having zero velocity at the walls as shown in Fig. 2b and given in Eq. (3). Indeed, the mean velocity of the gas infinitesimally close to the wall is not zero, for although half the molecules in this region have suffered their most recent collision at the wall the other half have suffered their most recent collision at some distance from the wall, of the order of a mean free path $\bar{l}$. Therefore the mass velocity, instead of being zero at the wall, is a fraction approximating $4\bar{l}/r$ of the average mass velocity $\bar{v}$. Thus, if Eq. (16) yields the "apparent viscosity" η_{app}, the true viscosity can be obtained from the equation

$$\eta = \eta_{app}\left(1 + \frac{4\bar{l}}{r}\right) \tag{18}$$

Dependence on flow velocity   With regard to the tacit assumption that the viscosity coefficient is independent of flow velocities and gradients of flow velocities, it must be remembered that the molecular velocities are essentially Maxwellian in a gas under shear only when variations in mass velocity over distances of the order of a mean free path are small in comparison with the molecular velocities themselves; if these differences are considerable, it is possible that the perturbed distribution of molecular velocities will manifest itself in an altered apparent viscosity coefficient. Without going into detail, it may be stated that if $\bar{v} \ll u$ (where u is the rms molecular velocity) and if $\bar{l} \ll r$, the error in η due to perturbation of the Maxwellian distribution will be negligible.

EXPERIMENTAL

Using the apparatus shown in Fig. 3, measure the flow times t_{ab} for dry air, helium, argon, carbon dioxide, and/or any other gases provided, at the same temperature in the same apparatus. Be careful not to disturb the leveling of the apparatus; the time of flow is rather sensitive to tipping of the apparatus, and it is important to have the same leveling for all gases studied. Moisture should be removed from the air by inserting a drying tube in the rubber hose line between the compressed-air supply and the viscosity apparatus. The drying tube (Fig. 4) should be full but not too tightly packed. It should be removed from the line when other gases are studied. If CO_2 is studied, be sure it is at room temperature before starting the run, as it cools considerably on expansion at the outlet of the cylinder.

Admit one of the gases to the bulb B by slowly opening stopcock X while stopcock Y remains closed; the mercury level in the bulb should drop quite slowly. Close stopcock X when the Hg level is about 2 cm below the lower fiducial mark. Carefully vent the bulb through the open outlet from the three-way stopcock Y to flush the system. Fill and flush a second time. Refill the bulb with gas, then open stopcock Y to the capillary and measure t_{ab} with a stopwatch. Four or more runs should be made if time permits.

Repeat the above procedure for each of the gases to be studied. Also record the temperature and the corrected barometric pressure p_2. Record at least approximate values for all the pertinent apparatus factors—L, r, V_B, maximum value of $p_1 - p_2$, radius of tube T.

Optional: Make an absolute determination of the apparatus constant K. Measure the length L of the capillary tube with a millimeter scale. Fill the capillary with mercury, measure the length of the pellet of mercury contained in the capillary, expel the mercury into a weighing bottle, and weigh it. From the results calculate r for the capillary. To find p_1 as a function of V, attach to the outlet tube Q a gas burette with a parallel leveling bulb. Use water as the indicating fluid. By

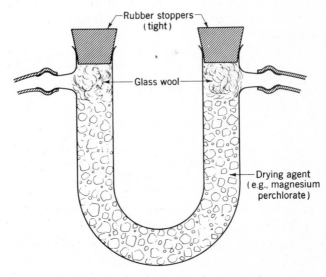

FIGURE 4
U tube for drying air.

means of one of the stopcocks (say Y) permit the mercury to rise a little at a time in bulb B. After each change measure the quantity introduced by adjusting and reading the gas burette, and measure the pressure difference $(p_1 - p_2)$ by measuring the difference in mercury levels, preferably with a cathetometer. Alternatively, connect the gas inlet tube P of the apparatus directly to an open-tube manometer to measure $(p_1 - p_2)$ directly. Plot p_1 against V and draw a curve through the points. Evaluate dp_1/dV at a number of points and calculate the integrand of Eq. (10). Evaluate the integral graphically and compute K. Then compare the value of K with the value obtained by calibration with dry air.

CALCULATIONS

For calibration purposes dry air was chosen as the standard, since it is easily available and has been carefully studied by many investigators. Data from many sources can be summarized in the empirical Sutherland expression[4]

$$\eta = \frac{(145.8 \times 10^{-7})T^{3/2}}{T + 110.4} \tag{19}$$

which gives the viscosity of dry air in poises at 1 atm as a function of absolute temperature T. At 25°C, the value is 183.7 μP (0.01837 cP). For the temperature of your experiment, calculate η for dry air from Eq. (19).

For each gas studied, average the flow times for the individual runs to obtain an average t_{ab}.

If Eq. (10) were valid, the ratio of viscosities for two gases would equal the corresponding ratio of flow times. Assuming Eq. (10) and using t_{air} and η_{air}, find approximate values of the viscosities of the other gases from their flow times.

For each gas (including air) calculate the maximum value of the Reynolds number from Eq. (11) and verify that it is below 1000. Use Eq. (12) to verify that L', the length of the transition region, is small compared with the length L of the capillary. Estimate the largest value of $\bar{v}$ from Eq. (5) and compare it with the rms molecular velocity u.

If the required conditions have been met, kinetic-energy and slip corrections may now be made in order to get more precise values of η. First calculate $\bar{l}$ for each gas from Eq. (IV-13). With Eq. (18) find η_{app} for dry air and use this value in Eq. (16) to find the value of K. Now calculate η_{app} for the other gases from Eq. (16) (one can use the approximate value of η in evaluating the kinetic-energy correction term). Finally apply the slip correction to obtain the true viscosity.

Report the final values obtained for η together with the temperature at which they were determined. Report also for each gas the calculated mean free path at 1 atm and the effective molecular diameter as calculated from the viscosity.

APPARATUS

Gas viscosity apparatus, complete with capillary tube and mercury; gas supply outlet(s) with needle valve(s); glass U tube with rubber stoppers and policemen; two lengths of gum-rubber tubing; 0 to 30°C thermometer; stopwatch; millimeter scale. If an absolute determination is to be made: additional mercury; weighing bottle; gas burette and leveling bulb; ring stands and clamps; cathetometer or open-tube manometer.

Cylinders of argon, helium, and carbon dioxide; source of compressed air; glass wool; drying agent (such as magnesium perchlorate); stopcock grease.

REFERENCES

1. J. R. Partingon, "An Advanced Treatise on Physical Chemistry," vol. I, pp. 876ff., Longmans, London (1949).
2. J. Reilly and W. N. Rae, "Physico-chemical Methods," 5th ed., vol. I, pp. 667, 690 and vol. III, p. 247, Van Nostrand, Princeton, N.J. (1953).
3. L. Prandtl, "Essentials of Fluid Dynamics," Blackie, Glasgow (1952).
4. NBS-NACA Tables of Thermal Properties of Gases, Table 2.39, Dry Air, compiled by F. C. Morey (December, 1950).

GENERAL READING

S. Pai, "Viscous Flow Theory," vol. I, Laminar Flow, Van Nostrand, Princeton, N.J. (1956).

EXPERIMENT 5.   DIFFUSION OF GASES

In this experiment, the mutual diffusion coefficients for the $Ar-CO_2$ and $He-CO_2$ systems are to be measured using a modified Loschmidt apparatus. These transport coefficients are then compared with theoretical values calculated with hard-sphere collision diameters.

When a component of a continuous fluid phase (a liquid solution or a gas mixture) is present in nonuniform concentration, at uniform and constant temperature and pressure and in the absence of external fields, that component diffuses in such a way as to tend to render its concentration uniform. For simplicity, let the concentration of a given substance $\bar{N}_1$ be a function of only one coordinate x; that is, the concentration of substance 1 varies only along the x direction, which we shall take as the upward direction. The net "flux" Z_1 of the substance passing upward past a given fixed point x_0 (i.e., amount per unit cross-sectional area per unit time) is under most conditions found to be proportional to the negative of the concentration gradient:

$$Z_1 = -D\left(\frac{d\bar{N}_1}{dx}\right)_{x_0} \tag{1}$$

where D is the *diffusion constant*, which is characteristic of the diffusing substance and usually also of the other substances present. This is called "Fick's first law of diffusion."

When only two substances are present in a closed gaseous system of fixed total volume at uniform and constant temperature, the distribution of one of the substances is completely determined when the distribution of the other is specified. The sum of the partial pressures of the two components is equal to the total pressure, and if the two components are perfect gases, the total pressure is constant and the sum of the two concentrations is also constant. Designating the two substances by the subscripts 1 and 2, we can write

$$Z_1 = -D_{12}\frac{d\bar{N}_1}{dx} = D_{12}\frac{d\bar{N}_2}{dx} \tag{2}$$

$$Z_2 = -Z_1 = -D_{21}\frac{d\bar{N}_2}{dx} = D_{21}\frac{d\bar{N}_1}{dx} \tag{3}$$

from which it can be seen that D_{12}, the diffusion constant for the diffusion of component 1 into component 2, is equal to D_{21}, the diffusion constant for the diffusion of 2 into 1.

One can also define a diffusion constant for the diffusion of a substance into itself. Suppose that we have a single molecular species (substance 1, say) present in a vessel but that the molecules initially in a certain portion of the vessel can somehow be tagged or labeled (without changing their kinetic properties) so that later they can be distinguished from the others. After a period of time the distribution of the tagged molecules could then be determined experimentally and the *self-diffusion constant* D_1 calculated. In actuality, it is not possible to tag molecules without some effect on their kinetic properties. A close approach to such tagging can be obtained, however, with the use of different isotopes that may be distinguished with a mass spectrometer or a Geiger counter. Another close approach can be obtained in a few cases by the use of molecules that are very similar in size, shape, and mass but different in chemical constitution, for example, the pair N_2 and CO or the pair CO_2 and N_2O. For each of these pairs the diffusion constant D_{12} should constitute a reasonably good estimate of the self-diffusion constants D_1 and D_2.

In the case where the two gases are distinctly different in their diffusing tendencies, the mutual diffusion constant D_{12} lies somewhere between the two self-diffusion constants D_1 and D_2 in value. It is also found that, when the two self-diffusion constants are very different, the mutual diffusion constant is largely determined by the more rapidly diffusing substance rather than by the more slowly diffusing one; thus there is a greater difference between the diffusion constants for the systems Kr–He and Kr–H_2 than for Kr–He and CO_2–He. The gas having the high self-diffusion constant (He or H_2) diffuses rapidly into the gas having the low self-diffusion constant. But there must be a *mass flow* of the resulting mixture in the opposite direction to maintain uniform pressure, so the gas with the lower diffusion constant gives the appearance of diffusing, although in reality it is mainly undergoing displacement.

THEORY

Let the concentration of gas 1 be $\bar{N}_1(x)$, a function of x only, in a vertical tube of uniform cross section A (x increasing upward). Let us calculate the rate of accumulation of gas 1 in the region from x to $x + dx$. The net flow of the gas into the region from below, per square centimeter per second, is given by Fick's first law,

$$Z_1(x) = -D_{12}\left(\frac{\partial \bar{N}_1}{\partial x}\right)_x$$

The net flow of the gas out of the region in the upward direction, per square centimeter per second, is

$$Z_1(x + dx) = -D_{12}\left(\frac{\partial \bar{N}_1}{\partial x}\right)_{x+dx}$$

$$= -D_{12}\left[\left(\frac{\partial \bar{N}_1}{\partial x}\right)_x + \left(\frac{\partial^2 \bar{N}_1}{\partial x^2}\right)_x dx\right]$$

The rate of increase of the concentration $\bar{N}_1(x)$ is given by

$$\frac{\partial \bar{N}_1(x)}{\partial t} = \frac{Z_1(x) - Z_1(x + dx)}{dx}$$

and therefore

$$\frac{\partial \bar{N}_1}{\partial t} = D_{12}\frac{\partial^2 \bar{N}_1}{\partial x^2} \tag{4}$$

This is the differential equation for diffusion in one dimension (called "Fick's second law").

Let two perfect gases be placed in a *Loschmidt apparatus*[1]—a vertical tube of length $2L$, of uniform cross section, closed at both ends, and containing a removable septum in the center. The heavier of the two gases (gas 1, say) is initially confined to the lower half and the lighter (gas 2) to the upper half in order to avoid convective mixing. Both gases are at initial concentration $\bar{N}_0$. At time $t = 0$ the septum is removed, so that the gases are free to interdiffuse. Let us define a quantity $\Delta \bar{N}$ as follows:

$$\Delta \bar{N} = \bar{N}_1 - \bar{N}_2 \equiv 2\bar{N}_1 - \bar{N}_0 \tag{5}$$

The differential equation to be solved,

$$\frac{\partial \Delta \bar{N}}{\partial t} = D \frac{\partial^2 \Delta \bar{N}}{\partial x^2} \tag{6}$$

is the differential equation (4) in $\bar{N}_1$ minus a similar one in $\bar{N}_2$, and

$$D = D_{12} = D_{21}$$

Taking the origin at the middle of the tube, we write the boundary conditions

$$\begin{array}{llll} t = 0: & \Delta \bar{N} = \bar{N}_0 & -L \leq x < 0 \\ & \Delta \bar{N} = -\bar{N}_0 & 0 < x \leq L \end{array} \tag{7}$$

$$t = \infty: \quad \Delta \bar{N} = 0 \qquad -L \leq x \leq L \tag{8}$$

$$\text{All } t: \quad \frac{\partial \Delta \bar{N}}{\partial x} = 0 \qquad x = -L$$

$$\frac{\partial \Delta \bar{N}}{\partial x} = 0 \qquad x = L \tag{9}$$

The last boundary condition follows from the fact that there is no flow of gas through the two ends of the tube; see Eq. (1). Boundary condition (7) can be expressed in terms of a Fourier series expansion[2] of $\Delta \bar{N}$:

$$t = 0: \quad \Delta \bar{N} = -\frac{4}{\pi} \bar{N}_0 \left(\sin \frac{\pi}{2L} x + \tfrac{1}{3} \sin \frac{3\pi}{2L} x + \tfrac{1}{5} \sin \frac{5\pi}{2L} x + \cdots \right) \tag{7'}$$

This is the Fourier series for a "square wave" of amplitude $\bar{N}_0$ and wavelength $2L$.

We shall omit the procedure for finding the solution of Eq. (6) and simply give the result:

$$\frac{\Delta \bar{N}}{\bar{N}_0} = \frac{\bar{N}_1 - \bar{N}_2}{\bar{N}_1 + \bar{N}_2} = -\frac{4}{\pi} \left(e^{-\pi^2 Dt/4L^2} \sin \frac{\pi}{2L} x + \tfrac{1}{3} e^{-9\pi^2 Dt/4L^2} \sin \frac{3\pi}{2L} x \right.$$

$$\left. + \tfrac{1}{5} e^{-25\pi^2 Dt/4L^2} \sin \frac{5\pi}{2L} x + \cdots \right) \tag{10}$$

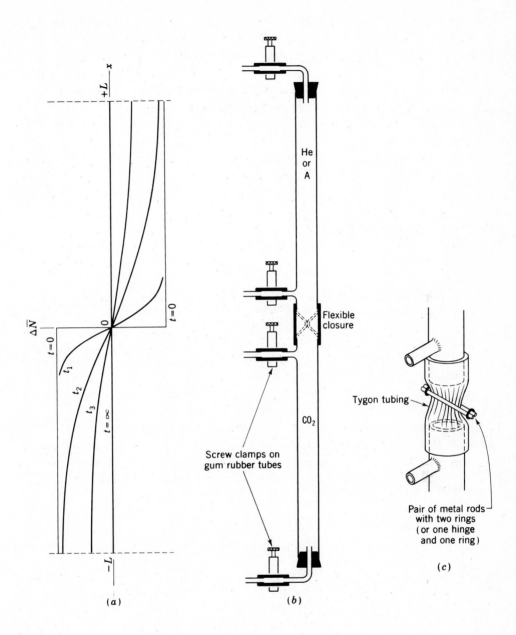

FIGURE 1

Loschmidt diffusion apparatus. (*a*) Curves showing schematically the quantity $\Delta \bar{N} = \bar{N}_1 - \bar{N}_2$. (*b*) Cell. (*c*) Detail of flexible closure.

The reader can easily verify that the solution satisfies the differential equation and the boundary conditions (7) or (7'), (8), and (9). The behavior of $\Delta\bar{N}$ with x and t is shown qualitatively in Fig. 1a.

After time t, let us replace the removable septum and analyze the two portions of the tube for total quantity of one of the gases, say gas 1. The total amount of gas 1 in the upper part will be equal to the total amount of gas 2 in the lower part. Let W_1 and N_1 be the weight and number of moles of gas 1 in the lower part and W_1' and N_1' be the weight and number of moles of gas 1 in the upper part, and let similar symbols with subscript 2 pertain to gas 2. Define a quantity f by

$$f \equiv \frac{N_1 - N_2}{N_1 + N_2} = \frac{(\bar{N}_1 - \bar{N}_2)_{av}}{\bar{N}_0} = \frac{1}{L}\int_{-L}^{0}\frac{\Delta\bar{N}}{\bar{N}_0}\,dx \tag{11}$$

Since $N_1' = N_2$, we have also

$$f = \frac{N_1 - N_1'}{N_1 + N_1'} = \frac{W_1 - W_1'}{W_1 + W_1'} \tag{12}$$

This expression permits the calculation of f from the results of analyzing separately the upper and lower parts of the tube for gas 1. From Eqs. (10) and (11), we obtain

$$f = \frac{8}{\pi^2}\left(e^{-\pi^2 Dt/4L^2} + \tfrac{1}{9}e^{-9\pi^2 Dt/4L^2} + \tfrac{1}{25}e^{-25\pi^2 Dt/4L^2} + \cdots\right) \tag{13}$$

At $t = 0$, the sum of the series in parentheses is equal to $\pi^2/8$, so the quantity f is initially unity. After a short time all terms except the first become negligible, and f therefore decays with time in a simple exponential manner.

At what time t should the run be stopped in order to obtain the most precise value of D, assuming that the only significant errors are those in determining f and that the uncertainty in f is constant (independent of the time)? Neglecting all but the first term of Eq. (13), we find, on solving for D and differentiating, that

$$\frac{\partial D}{\partial f} = -\frac{1}{t}\frac{4L^2}{\pi^2}\frac{1}{f}$$

We wish to find at what value of t the derivative $\partial D/\partial f$ is at a minimum:

$$0 = \frac{\partial^2 D}{\partial t\,\partial f} = \frac{4L^2}{\pi^2}\left(\frac{1}{t}\frac{1}{f^2}\frac{\partial f}{\partial t} + \frac{1}{t^2}\frac{1}{f}\right) = \frac{4L^2}{\pi^2 ft}\left(-\frac{\pi^2 D}{4L^2} + \frac{1}{t}\right)$$

thus the optimum time is

$$t_{opt} = \frac{4L^2}{\pi^2 D} \tag{14}$$

At this optimum time, Eq. (13) gives

$$f = 0.81057(0.367879 + 0.000014 + \cdots) = 0.2982$$

Note that neglect of the second and higher terms of Eq. (13) proves to be justified if the experiment is run for any time in the neighborhood of the optimum time or for any greater time.

METHOD

In this experiment we shall measure the diffusion constants for the systems CO_2–He and CO_2–Ar at room temperature. In each run, CO_2, the heavier gas, will be initially in the lower portion of a vertical tube. The gas in the two portions at the end of the run will be analyzed for CO_2 by sweeping it through absorption tubes (U tubes) filled with Ascarite (asbestos impregnated with NaOH, which reacts with CO_2) and with magnesium perchlorate (to absorb the water liberated in the first reaction); see Fig. 2. These tubes are weighed before and after sweeping the gas through them. The diffusion cell, a modified Loschmidt type, is shown in Fig. 1b. In place of a removable septum at the center, the two halves of this tube are joined by a flexible plastic (Tygon) sleeve of approximately the same inner diameter as the glass, which can be pinched tight to achieve a closure as shown in Fig. 1c.

A variant of the same basic method is shown in Fig. 3. In this case, a vacuum system is used for handling the gases, a large-bore stopcock is used in place of the flexible closure, and a different method is used for the quantitative analysis of CO_2. Instead of absorbing the CO_2 on Ascarite, the CO_2–Ar or CO_2–He mixture is slowly pumped through a copper coil immersed in liquid nitrogen. The CO_2 will be efficiently trapped and after all the Ar or He has been removed, the trap is allowed to warm to room temperature and the CO_2 pressure is determined.

In round numbers the length L in the apparatus to be employed is 50 cm, and the two diffusion constants are of the order of magnitude 0.6 cm^2 sec^{-1} for CO_2–He and 0.15 cm^2 sec^{-1} for CO_2–Ar. Using these figures the optimum diffusion times can be estimated from Eq. (14). The actual time should approximate this optimum time but in any case should not differ from it by more than 30 percent.

EXPERIMENTAL

Procedure A   In carrying out the experiment with the apparatus shown in Figs. 1 and 2, the student should follow a careful plan worked out with his partner, taking into account the following:

   1   Even a tiny leak is serious; all stoppers and clamps *must be tight*. Rubber and Tygon tubing must be in good condition.

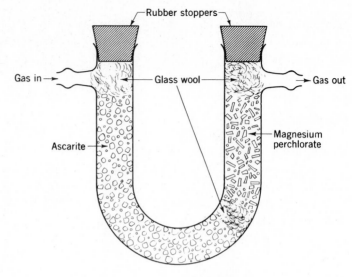

FIGURE 2
U tube for absorbing CO_2 gas.

2 When one gas is being flushed out by another, the gas being introduced should enter at the lower end of the vessel if it is the heavier and at the upper end if it is the lighter.

3 Considerable heat is evolved in the reaction of CO_2 with Ascarite; a U tube that has absorbed a considerable amount of CO_2 should be allowed to cool for about 15 min before it is weighed.

4 Carbon dioxide gas emerges rather cool from the tank valve. Time should be allowed for it to warm to room temperature before the diffusion run is started. It is wise to vent the lower chamber, containing CO_2, momentarily at the side arm before the run to restore atmospheric pressure.

5 If the cylinder is not equipped with a regulator valve, turn off the gas supply at the cylinder before closing off the system. The inlet of the system should be clamped off before the outlet.

6 The CO_2-containing gas should be swept into the drying tubes no faster than 5 bubbles per sec. Use a dibutyl-phthalate bubbler for metering gas flow. One bubble is approximately 0.1 cm^3. This rate can be obtained by synchronization with the ticking of a watch. It is well to adjust the flow rate (with the outlet valve on the compressed air line) *before* attaching to the system. The sweeping air should go *first* through the bubbler and *then* through a purifying tube (to remove CO_2 and H_2O) before it enters the system. In case the flow rate is preadjusted, it is important to make sure that the system is open at both ends at the moment the air tube is attached.

7   Drying tubes should be closed off with rubber policemen when gas is not being passed through them. Take care that the drying tube always has the same two policemen.

8   Do not neglect the volumes of gas that are contained in any significant lengths of rubber tubing. Before using a rubber tube that may previously have contained CO_2, flush it out momentarily with air. Also, do not neglect the possibility of a heavy gas spilling out of the system if it is open at two places at different levels for more than a second or two.

Prepare as shown in Fig. 2 and weigh two U tubes, A and B. It is not necessary to correct weights to vacuum readings. A third U tube, which need not be weighed, should be prepared for purification of the air to be used for sweeping.

Flush out the lower half of the cell with CO_2, using a flushing rate of at least 10 to 15 bubbles per sec through the bubbler. *Make sure the center partition is closed* before flushing. Pass at least six volumes of gas through to be sure of complete replacement.

Connect the lower half-cell to U tubes A and B in series, and sweep out the gas in the lower half-cell into the U tubes with three or four volumes of air (purified with the third U tube), at 5 bubbles per sec. Without stopping the sweeping, *quickly* pull off tube A and reconnect tube B. Immediately close the ends of tube A. After one or two more additional volumes have been swept through tube B, weigh tube B first, then tube A. Tube B should not have gained more than a few milligrams. Add the two gains in weight; this quantity is $W_1 + W_1'$, the total weight of CO_2 in the cell.

Now flush out the lower half-cell with CO_2 and the upper one with the other gas (He or Ar) to be studied, as above. This should be done during the above weighings to save time.

Begin the diffusion by *carefully* (not too suddenly) opening the center partition and simultaneously starting a stopwatch. After the estimated optimum time, close the partition.

Sweep the contents of the lower tube into U tubes A and B as before. After the weighings, repeat with the upper tube. The value of $W_1 + W_1'$ should agree with the value obtained above to within a very few milligrams; if greater disagreement is obtained, a leak must be suspected.

Repeat the procedure with the second gas combination. If time permits, repeat each of the above runs, with the diffusion cell inverted, and in each case take as your value of f the mean of the two values obtained.

Measure the length L as precisely as possible. Also record the barometric pressure and the ambient temperature.

Procedure B   In carrying out the experiment with the apparatus shown in Fig. 3, the student should review the basic operation of a vacuum system (see Chap. XVII).

The diffusion tube was designed and constructed such that the length L of the upper section is equal to the length of the lower section *plus* the length of the bore of stopcock A. Hence the stopcock bore and the lower section of the tube are both filled with CO_2.

The CO_2 and He (or Ar) gas tanks are connected to the diffusion tube as shown in Fig. 3. Stopcocks A through F should be opened and the line should be pumped on for about 20 min. Close stopcock F and check for an increase in pressure. The rate of increase should be no greater than 5 Torr/hr. Then evacuate the entire apparatus again for a few minutes.

Close stopcock D and fill the system with CO_2 to a pressure between 750 and 770 Torr. Close stopcock B and read the CO_2 pressure exactly. After closing stopcock A, pump away the remaining CO_2. Fill the upper portion of the tube with He to a pressure *as close as possible* to the CO_2 pressure (the difference should be less than 1 Torr). Close stopcock C and evacuate the remaining portion of the apparatus. Begin the diffusion by *carefully* opening stopcock A and simultaneously starting a stopwatch or timer (be sure to align the stopcock bore with the vertical axis of the upper and lower tubes). After the estimated optimum time, close stopcock A and record the elapsed time.

Place a Dewar flask of liquid nitrogen around the copper coil. After the trap has been cooled (as indicated by an absence of violent boiling of the liquid nitrogen), *first* close stopcock D and then open C. After about 30 sec, very slowly open stopcock D until the pressure (watch the manometer) just begins to decrease. If the stopcock is opened too much, the trapping of CO_2 will not be complete. Adjust stopcock D to give a pressure decrease of about 20 Torr every 15 sec. It should take not more than a few minutes to remove the helium. Then close both stopcocks D and C. To remove any He entrapped in the solid CO_2, remove the Dewar flask from the coil and allow the CO_2 to vaporize. However, do not wait for the coil to reach room temperature. Place the Dewar flask around the coil again and when the solidification is complete, fully open stopcock D for about a minute (*leave stopcock C closed*). If the above operations were performed properly, then at least 99 percent of the CO_2 will have been trapped.

With stopcock D closed, remove the Dewar flask and allow the bulk of the CO_2 to vaporize. Then place a liter beaker containing water at room temperature around the coil (this will hasten the warming up of the coil). After about a minute, remove the beaker and dry the coil with a towel. Wait a few minutes for the coil to come to thermal equilibrium with the room and then measure the CO_2 pressure. This pressure in the upper half of the tube is denoted by p_1'.

Now evacuate all of the system above stopcock A. Then close stopcock C and open A. Repeat the procedure described above in order to determine the CO_2 pressure associated with the gas in the lower half of the tube.

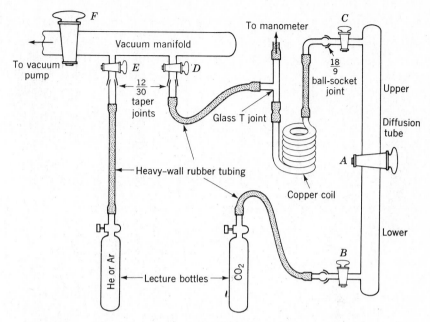

FIGURE 3
Modified Loschmidt apparatus, to be used according to procedure B. Stopcock *A* should have a very large bore which is as close as possible to the diameter of the diffusion tube. Stopcocks *B* to *E* should be high-vacuum type. As a safety precaution, each gas lecture bottle should be connected to the system via a reducing valve or a pressure-relief valve (not shown).

The quantity f in Eq. (12) is defined in terms of pressures by $f = (p_1 - p'_1)/(p_1 + p'_1)$.

Finally repeat the complete procedure with Ar as the second gas. In addition to determining the diffusion coefficients at atmospheric pressure, measure one of them at half an atmosphere.

Measure the length L as precisely as possible, and also record the ambient temperature.

CALCULATIONS

Your experimental data should be summarized in a table which includes for each run the initial pressure, the temperature, the length of the run, and the final CO_2 weights (or pressures) from the upper and lower parts of the cell.

By use of Eqs. (12) and (13) neglecting all but the first term in the series, calculate D for each gas combination. Then, with this value of D, calculate the second term in the series; if it is more than 1 percent of the first term, recalculate D. The resulting values of D are to be designated D_{12} for CO_2–He, D_{13} for CO_2–Ar.

Gas viscosity measurements yield molecular diameters of 2.58 Å (angstroms) for helium, 3.42 Å for argon, and 4.00 Å for carbon dioxide.[3] With these values, calculate d_{12} and d_{13} from Eq. (IV-36), and calculate D_{12} and D_{13} using Eq. (IV-34). Compare with your experimental values.

The molecular diameters cited above were obtained from an analysis of viscosity data in terms of the Lennard-Jones "6,12 potential":[3]

$$u(r) = 4\varepsilon \left[\left(\frac{\sigma}{r} \right)^{12} - \left(\frac{\sigma}{r} \right)^{6} \right] \tag{15}$$

where $u(r)$ is the potential energy of two molecules at a distance r. The diameters correspond to σ values (the distance at which the potential energy equals zero). Although these values are often called hard-sphere diameters (why?), one should remember that they do not result from the solution of the hard-sphere expression, Eq. (IV-14). Alternatively, if one wishes to calculate an *accurate* diffusion coefficient, it is meaningless to use such σ values in conjunction with Eq. (IV-34).

Optional: At 273°K, D_{23} (for He–Ar) is 0.653 cm^2 sec^{-1}. From D_{12}, D_{13}, and D_{23} at *room temperature* calculate d_{12}, d_{13}, and d_{23}. (To correct diffusion constants from one temperature to another, assume a $T^{3/2}$ dependence if the temperature change is small. This is only approximate, since the d's may vary in some degree with the temperature.) Then obtain d_1, d_2, and d_3 and use these to calculate D_1, D_2, and D_3 from Eq. (IV-35). Determine the ratios of the self-diffusion constants to their respective viscosities. How do these compare with the theoretical ratios discussed in the introductory section of Chap. IV?

DISCUSSION

The flexible closure or the large-bore stopcock, though simple and convenient, is admittedly less satisfactory than a removable diaphragm closure would be. Discuss its disadvantages from the points of view of (1) mixing produced in the opening and closing operations; (2) inexactness of the matching of its diameter with that of the tubes; and (3) any other defects that occur to you. Try wherever possible to employ arguments that are not wholly qualitative; estimate orders of magnitude where you can. If possible, suggest an improved closure design. What additional techniques might be used with this experimental method to enable D_{23} (for He–Ar) to be measured directly? Using your value for D_{13} estimate the time necessary for a diffusion run on CO_2–Ar in the present apparatus to achieve a state where $W_1 = 1.01 W_1'$ (i.e., where the concentration of CO_2 is only 1 percent greater in the lower half than in the upper half).

APPARATUS

Procedure A   Diffusion cell (Loschmidt type) with sturdy support; clamping device for flexible closure; two one-hole rubber stoppers, with short bent tubes, for ends of cell; four 1-in. lengths of rubber tubing; four screw clamps; four short glass tubes; three glass U tubes with rubber stoppers and policemen; bubbler, containing dibutyl phthalate; three long and two short lengths of rubber tubing; needle valve; two ring stands; four clamps with clamp holders; 0 to 30°C thermometer; stopwatch; meter stick.

Cylinders of helium, argon, and carbon dioxide; source of compressed air; glass wool; Ascarite; drying agent (such as magnesium perchlorate).

Procedure B   Vacuum system with mechanical pump and manifold having provision for two attachments via taper joints; Loschmidt diffusion cell with stopcocks attached at top and bottom; two standard taper joints to fit manifold; two ball-socket joints to fit diffusion cell; glass (or metal) T joint; five long and one short length of heavy-wall rubber pressure tubing; tightly wound copper coil; large Dewar flask; closed-tube manometer; 1-liter beaker; towel; stopwatch or electrical timer; meter stick; 0 to 30°C thermometer.

Liquid nitrogen; cylinders or lecture bottles of helium, argon, and carbon dioxide.

REFERENCES

1. J. R. Partington, "An Advanced Treatise on Physical Chemistry," vol. I, Longmans, London (1949).
2. Any text on advanced calculus, such as R. Courant, "Differential and Integral Calculus," Blackie, Glasgow (1934); or P. Franklin, "Methods of Advanced Calculus," McGraw-Hill, New York (1944).
3. J. O. Hirschfelder, C. F. Curtiss, and R. B. Bird, "Molecular Theory of Gases and Liquids," 2d ed., pp. 1035–1044, 1110, Wiley, New York (1967).

GENERAL READING

E. H. Kennard, "Kinetic Theory of Gases," McGraw-Hill, New York (1938).
R. D. Present, "Kinetic Theory of Gases," McGraw-Hill, New York (1958).

EXPERIMENT 6.   LOW-PRESSURE EFFUSION OF GASES

In Exp. 4 we were concerned with the viscous flow of gas through a capillary tube. One of the conditions for that experiment was that the mean free path of the gas must be small compared with the diameter of the tube. Actually, the flow ceases to be purely viscous and begins to assume some molecular character when the slip correction becomes significant. When the mean free path becomes very large in comparison with the diameter of a tube or hole, the flow is completely molecular in character.

At a given temperature, the mean free path $\bar{l}$ is inversely proportional to the pressure; for most gases, $\bar{l}$ is of the order of 10^{-5} cm at 1 atm and about 0.05 cm at a pressure 0.1 Torr (see Table IV-1). Clearly, it is difficult to obtain the necessary conditions for molecular flow at 1 atm pressure, since extremely small holes or pores are needed. Demonstrations of molecular flow at 1 atm with pinholes punched in foils are of no value, although they may appear to give correct flow time ratios for two gases if the choice of a pair of gases for study is fortuitously such that the molecular diameters are about the same. In general the results of such experiments show a dependence of flow time on molecular diameter, which is a property of viscous flow and not true of pure molecular flow. Valid molecular flow can be achieved, however, using holes with diameters of about 0.01 cm if the gas pressure is 0.10 Torr or less.[1] Such low-pressure effusion of helium, argon, and carbon dioxide will be studied in this experiment.

METHOD AND THEORY

Effusion will be studied for gas at low pressure flowing through a small pinhole into a vacuum. The experimental apparatus is shown in Fig. 1. The important features of this apparatus are: (1) a system for obtaining high vacuum; (2) a large bulb B (about 5 liters), to contain the gas to be studied, at an initial pressure of 0.1 to 0.2 Torr; (3) an orifice O in the form of a pinhole of the order of 0.1 mm in diameter in very thin platinum foil, through which the gas in bulb B may effuse into the high-vacuum part of the system; (4) a vacuum gauge for measuring periodically the pressure of the gas remaining in bulb B; and (5) an expansion system for filling bulb B with gas at the desired initial pressure.

For this method we can write Eq. (IV-7) as

$$-\frac{dN}{dt} = \frac{Ap}{\sqrt{2\pi MRT}} \tag{1}$$

where N is the number of moles of gas in the bulb B at pressure p and temperature T, A is the area of the pinhole, and M is the molecular weight of the gas. In Eq. (1) we have made use of the fact that the pressure outside the pinhole is essentially zero (10^{-5} Torr or less). Applying the perfect-gas law to the gas in the bulb B (volume V) we write

$$dN = \frac{V}{RT} dp \tag{2}$$

Combining Eqs. (1) and (2) gives

$$\frac{dp}{p} = -\frac{A}{V}\sqrt{\frac{RT}{2\pi M}}\,dt = -\frac{dt}{\tau} \tag{3}$$

where τ is the "relaxation time" for the system and is

$$\tau = \frac{V}{A}\sqrt{\frac{2\pi M}{RT}} \tag{4}$$

For a given apparatus and choice of gas, τ is a constant if T does not vary; therefore, Eq. (3) can be integrated to give

$$\ln\frac{p}{p_0} = -\frac{1}{\tau}t$$

or
$$\log p = \log p_0 - \frac{t}{2.303\tau} \tag{5}$$

where p is the pressure in bulb B at time t and p_0 is the initial pressure (at $t = 0$).

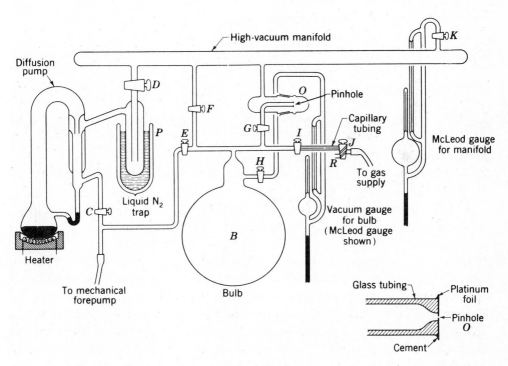

FIGURE 1
Apparatus for the low-pressure effusion of gases.

A plot of log p versus t should be a straight line. From the slope we can evaluate τ and then use Eq. (4) with known values of A, V, and T to calculate the molecular weight of the gas. Even if we do not know the value of A/V, if we obtain data for two different gases, we can determine the ratio of their molecular weights.

EXPERIMENTAL

Before the run is started, the entire system should be evacuated to a pressure of 10^{-5} Torr or less and liquid nitrogen should be placed in the Dewar flask around trap P. Either this will be done by an instructor before the period, or instructions will be issued for the proper procedure to be used. The pressure should be checked with the McLeod gauge which connects to the manifold through stopcock K. For a general discussion of high-vacuum systems and components see Chap. XVII.

The actual apparatus may differ from that shown in Fig. 1, in which case special operating instructions will be available. The procedure given here will refer to the system shown. When the system is completely evacuated, all stopcocks will be open except E and J.

The pressure gauge which is connected to the bulb by stopcock H will be either a small-volume McLeod gauge or a thermocouple gauge. If it is a McLeod gauge, readings during the run should be taken as quickly as possible, since a small amount of gas is cut off in the gauge during a pressure reading. If the gauge is a thermocouple gauge, this precaution is not necessary but the gauge must be calibrated against the manifold McLeod gauge with each gas used. This calibration can be done after the run by the following procedure: Close D and E; open H, F, and K; admit gas to the entire line at the highest required pressure through J and I. Now read both the McLeod and the thermocouple gauge. Reduce the pressure by opening D momentarily and repeat the measurements; continue until the desired range has been covered, and plot a calibration curve. (For precise calibration at very low pressures, one can insert a cold trap between the manifold and the McLeod gauge to eliminate mercury vapor.) A discussion of the thermocouple gauge is given in Chap. XVII.

Procedure   Connect the gas inlet hose to a cylinder of helium, argon, or carbon dioxide which has a regulator valve, and adjust the pressure to about 2 psi above 1 atm. Carefully turn stopcock J so that the hose can be flushed out through outlet R of this stopcock. Close stopcock I and turn J so as to fill the short length of capillary tubing with the gas. Now close J, F, and G, and open I to allow the slug of gas to expand into the bulb B and its associated pressure gauge. Close I and measure the pressure in bulb B. If the pressure is between 100 to 150 μm (micrometers) (0.10 to 0.15 Torr), the run can be started; if above 150 μm, open stopcock E *momentarily* to reduce the pressure to the range 100 to 150 μm.

To start the effusion, open stopcock G which connects the bulb B to the high-vacuum manifold through the pinhole at O. Take pressure readings on the gas in B at 5-min intervals. Continue readings until the pressure is down to about 5 percent of its initial value. Record the ambient temperature near bulb B. At least once during the run, check the manifold pressure (with the McLeod gauge at K) to verify that it is less than 10^{-4} Torr. Except for this measurement, keep stopcock K closed during the run.

To terminate a run, open F and I to allow the diffusion pump to evacuate the system to a high vacuum.

Repeat this procedure with one or both of the other gases. When this experiment has been completed, leave the apparatus evacuated unless otherwise instructed.

Record the necessary apparatus constants A and V.

CALCULATIONS

Convert all pressure-gauge readings obtained during a run to micrometers of mercury (p) and plot log p against the time in *seconds*. Alternatively, if a McLeod gauge was used, one can simplify the calculation by plotting 2 log h versus t, since the pressure is proportional to h^2, where h is the gauge reading.

From each plot, determine τ from the slope and p_0 from the intercept at $t = 0$. Compare this p_0 value with the pressure prior to opening stopcock G. Using Eq. (4) compute the molecular weight M for each gas studied. Also compute the ratio of molecular weights from the ratio of τ values. The agreement with accepted values is expected to be better for the ratio than for the individual values because the effect of uncertainties in A and V has been eliminated.

DISCUSSION

Calculate the mean free path of each gas at pressure p_0. How does this compare with the pinhole diameter?

Estimate the effect of neglecting the back effusion from the manifold into the bulb.

For the highest effusion rates attained in this experiment, what is the minimum pumping speed required for the vacuum system to maintain a manifold pressure no higher than 10^{-4} Torr?

An alternative method employs effusion through a pinhole from one bulb to another of equal volume. How would you define a relaxation time τ' for this method, and what relationship would τ' bear to τ?

Suggest how an effusion experiment could be designed and carried out to measure a very low vapor pressure for a solid.

APPARATUS

High-vacuum system; effusion apparatus such as that shown in Fig. 1; heavy-wall rubber tubing; stopwatch; Dewar flask for cold trap.

Cylinders of helium, argon, and carbon dioxide; stopcock grease; liquid nitrogen (1 liter).

REFERENCE

1. M. Knudsen, *Ann. Physik*, **28,** 75 (1909).

GENERAL READING

E. H. Kennard, "Kinetic Theory of Gases," McGraw-Hill, New York (1938).

THERMOCHEMISTRY

EXPERIMENTS

PRINCIPLES OF CALORIMETRY

We are concerned here with the problem of determining experimentally the enthalpy change ΔH or the energy change ΔE accompanying a given isothermal change in state of a system, normally one in which a chemical reaction occurs. We can write the reaction schematically in the form

$$A(T_0) + B(T_0) = C(T_0) + D(T_0) \tag{1}$$

$$\underset{\text{initial state}}{} \qquad \underset{\text{final state}}{}$$

In practice we do not actually carry out the change in state isothermally; this is not necessary because ΔH and ΔE are independent of the path. In calorimetry we usually find it convenient to use a path composed of two steps:

Step I. A change in state is carried out *adiabatically* in the calorimeter vessel to yield the desired products but in general at another temperature:

$$A(T_0) + B(T_0) + S(T_0) = C(T_1) + D(T_1) + S(T_1) \tag{2}$$

where S represents those parts of the system (e.g., inside wall of the calorimeter vessel, stirrer, thermometer, solvent) that are always at the same temperature as the reactants or products because of the experimental arrangement; these parts, plus the reactants or products, constitute the system under discussion.

Step II. The products of step I are brought to the initial temperature T_0 by adding heat to (or taking it from) the system:

$$C(T_1) + D(T_1) + S(T_1) = C(T_0) + D(T_0) + S(T_0) \tag{3}$$

As we shall see, it is often unnecessary to carry out this step in actuality.

By adding Eqs. (2) and (3) we obtain Eq. (1) and verify that these two steps describe a complete path connecting the desired initial and final states. Accordingly,

125

ΔH or ΔE for the change in state (1) is the sum of the values of this quantity pertaining to the two steps:

$$\Delta H = \Delta H_I + \Delta H_{II} \tag{4a}$$

$$\Delta E = \Delta E_I + \Delta E_{II} \tag{4b}$$

The particular convenience of the path described is that the heat q for step I is zero, while the heat q for step II can be either measured or calculated. It can be measured directly by carrying out step II (or its inverse) through the addition to the system of a measurable quantity of heat or electrical energy, or it can be calculated from the temperature change $(T_1 - T_0)$ resulting from adiabatic step I if the heat capacity of the product system is known. For step I,

$$\Delta H_I = q_p = 0 \qquad \text{constant pressure} \tag{5a}$$

$$\Delta E_I = q_v = 0 \qquad \text{constant volume} \tag{5b}$$

Thus, if *both* steps are carried out at constant pressure,

$$\Delta H = \Delta H_{II} \tag{6a}$$

and if *both* are carried out at constant volume,

$$\Delta E = \Delta E_{II} \tag{6b}$$

Whether the process is carried out at constant pressure or at constant volume is a matter of convenience. In nearly all cases it is most convenient to carry it out at constant pressure; the experiments on heats of ionic reaction and heats of solution are examples. An exception to the general rule is the determination of a heat of combustion, which is conveniently carried out at constant volume in a bomb. However, we can easily calculate ΔH from ΔE as determined from a constant-volume process (or ΔE from ΔH as determined from a constant-pressure process) by use of the equation

$$\Delta H = \Delta E + \Delta(pV) \tag{7}$$

When all reactants and products are condensed phases, the $\Delta(pV)$ term is negligible in comparison with ΔH or ΔE, and the distinction between these two quantities is unimportant. When gases are involved, as in the case of combustion, the $\Delta(pV)$ term is likely to be significant in magnitude, though still small, in comparison with ΔH or ΔE. Since it is small, we can employ the perfect-gas law and rewrite Eq. (7) in the form

$$\Delta H = \Delta E + RT \, \Delta N_{gas} \tag{8}$$

where ΔN_{gas} is the *increase* in the number of moles of gas in the system.

We must now concern ourselves with procedures for determining ΔH or ΔE for step II. We might envisage step II being carried out by adding heat to the system or taking heat away from the system and measuring q for this process.† Usually, however, it is much easier to measure work than heat. In particular, electrical work, which can be degraded to "Joule heat" by a heating coil in the system, can be used conveniently in carrying out either step II or its inverse, whichever is endothermic. Since the heat is dissipated *inside* the system, the work is negative:

$$-w_{el} = \int E \, dq = \int Ei \, dt$$

For very precise work, one should measure both E and i during the heating period. In many instances, however, it is possible to assume that the resistance of the heating coil R_H is constant and make only measurements of i by determining the potential drop E_s across a standard resistance R_s in series with the heating coil. In such a case, one can write

$$-w_{el} = \int i^2 R_H \, dt = \frac{R_H}{R_s^2} \int E_s^2 \, dt \tag{9}$$

Note that the electrical work as given by Eq. (9) is in joules when resistance is in ohms, potential in volts, and time in seconds. To convert this to calories, it must be divided by 4.184. If the electric heating is done adiabatically,

$$\Delta H_{II} = -w_{el} \qquad \text{constant pressure} \tag{10a}$$

$$\Delta E_{II} = -w_{el} \qquad \text{constant volume} \tag{10b}$$

Our discussion so far has been limited to determining ΔH_{II} or ΔE_{II} by directly carrying out step II (or its inverse). However, it is often not necessary to carry out this step in actuality. If we know or can determine the heat capacity of the system, the temperature change $(T_1 - T_0)$ resulting from step I provides all the additional information we need:

$$\Delta H_{II} = \int_{T_1}^{T_o} C_p(C + D + S) \, dT \tag{11a}$$

$$\Delta E_{II} = \int_{T_1}^{T_o} C_v(C + D + S) \, dT \tag{11b}$$

The heat capacities ordinarily vary only slightly over the small temperature ranges involved; accordingly we can write

$$\Delta H_{II} = C_p(C + D + S)(T_0 - T_1) \tag{12a}$$

$$\Delta E_{II} = C_v(C + D + S)(T_0 - T_1) \tag{12b}$$

† This could be done by placing the system in thermal contact with a heat reservoir (such as a large water bath) of known heat capacity until the desired change has been effected and calculating q from the measured change in the temperature of the reservoir.

where C_p and C_v are average values over the temperature range or to a good approximation may be regarded as constants independent of temperature.

The heat capacity, if not known, must be determined. A direct method, which depends on the assumption of the constancy of heat capacities over a small range of temperature, is to measure the temperature rise $(T_2' - T_1')$ produced by the dissipation of a measured quantity of electrical energy. We then obtain

$$\left.\begin{matrix} C_p \\ C_v \end{matrix}\right\} = \frac{-w_{el}}{T_2' - T_1'}$$

(13a)

(13b)

at constant pressure or at constant volume, respectively. This method is exemplified in the experiments on heats of ionic reaction and heats of solution.

An indirect method of determining the heat capacity is to carry out another reaction altogether, for which the heat of reaction is known, in the same calorimeter under the same conditions. This method depends on the fact that in most calorimetric measurements on chemical reactions the heat-capacity contributions of the actual product species (C and D) are very small, and often negligible, in comparison with the contribution due to the parts of the system denoted by the symbol S. In a bomb calorimeter experiment the reactants or products amount to a gram or two, while the rest of the system is equivalent to about 2500 g of water. In calorimetry involving dilute aqueous solutions, the heat capacities of such solutions can in a first approximation be taken equal to that of equivalent weights or volumes of water. Thus we may write, in place of Eqs. (12a) and (12b),

$$\left.\begin{matrix} \Delta H_{II} \\ \Delta E_{II} \end{matrix}\right\} = C(S)(T_0 - T_1)$$

(14a)

(14b)

for constant-pressure and constant-volume processes, respectively. In Eqs. (14a) and (14b) we have omitted any subscript from the heat capacity as being largely meaningless, since only solids and liquids, with volumes essentially independent of pressure, are involved. The value of $C(S)$ can be calculated from the heat of the known reaction and the temperature change $(T_2' - T_1')$ produced by it, as follows:

$$C(S) = \begin{cases} \dfrac{-\Delta H_{known}}{T_2' - T_1'} & \text{constant pressure} \\[4mm] \dfrac{-\Delta E_{known}}{T_2' - T_1'} & \text{constant volume} \end{cases}$$

(15a)

(15b)

This method is exemplified in the experiment on heats of combustion by bomb calorimetry.

Let us now consider step I, the adiabatic step, and the measurement of the temperature difference $(T_1 - T_0)$ which is the fundamental measurement of calorimetry. It is an idealization to assume that step I is truly adiabatic; as no thermal insulation is perfect, some heat will in general leak into or out of the system during the time required for the change in state to occur and for the thermometer to come into equilibrium with the product system.

In addition, we usually have a stirrer present in the calorimeter to aid in the mixing of reactants or for hastening thermal equilibration. The mechanical work done on the system by the stirrer results in the continuous addition of energy to the system at a small, approximately constant rate. During the time required for the change in state and thermal equilibration to occur, the amount of energy introduced can easily be significant.

The following are the most important experimental approaches aimed at minimizing the effect of nonadiabatic conditions and the effect of the stirrer; they can be used separately or in combination:

1 The calorimeter may be built in such a way as to minimize heat conduction in or out of the system. A vessel with an evacuated jacket (Dewar flask), often with silvered surfaces to minimize the effect of heat radiation, may be used.

2 One may interpose between system and surroundings an "adiabatic jacket," reasonably well insulated from both. This jacket is so constructed that its temperature can be adjusted at will from outside, as, for example, by supplying electrical energy to a heating circuit. In use, the temperature of the jacket is continuously adjusted so as to be as close as possible to that of the system, so that no significant quantity of heat will tend to flow between the system and the jacket.

3 The stirring system may be carefully designed to produce the least rate of work consistent with the requirement of thorough and reasonably rapid mixing.

4 We may assume that the rate of gain or loss of energy by the system resulting from heat leak and stirrer work is reasonably constant with time at any given temperature. We may therefore assume that the temperature, as a function of time, should be initially and finally linear. Thus we can estimate what the temperature T_1 would be in the event of instant equilibration by plotting temperature against time for a period long enough to obtain a curve with a long portion that is essentially linear and extrapolating this linear portion back to the time of initiation of the reaction, as shown in Fig. 1. Where the time required for the change in state is long, as in the case of determining the heat capacity by electrical heating, the rate of heat leak presumably changes continuously from the initiation of electrical heating to its termination, and it would be improper to extrapolate to the time of initiation of the heating. The appropriate procedure in this case would be to extrapolate the linear portions of the heating curves for

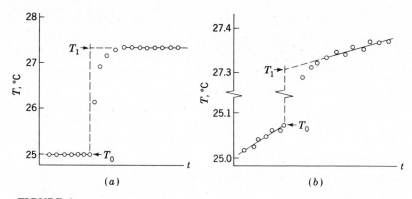

FIGURE 1

(a) Schematic plot of temperature T vs. time t showing the extrapolation for obtaining T_1. (b) Plot with interrupted temperature axis and greatly expanded temperature scale. Such a plot will eliminate the contribution of graphical errors to the uncertainty in $T_1 - T_0$.

the periods before initiation and after termination to the middle of the heating period.

In the experiments on heats of combustion we make use of approaches 2 (optionally), 3, and 4. In the experiments on heats of ionic reaction and heats of solution we make use of 1, 3, and 4.

In the above discussion, we have been concerned with determining ΔH or ΔE for a chemical change, with arbitrary amounts of reactants. The quantities in which we are interested are the molar quantities $\Delta \tilde{H}$ or $\Delta \tilde{E}$ corresponding to one formula of reaction. The number of formulas of reaction N can be calculated from the quantity of the limiting reactant. Thus we have

$$\Delta \tilde{H} = \frac{\Delta H}{N} \tag{16a}$$

or

$$\Delta \tilde{E} = \frac{\Delta E}{N} \tag{16b}$$

GENERAL READING

J. M. Sturtevant, Calorimetry, in A. Weissberger (ed.), "Technique of Organic Chemistry," 3d ed., vol. I, part I, chap. X, Interscience, New York (1959).

G. N. Lewis and M. Randall (revised by K. S. Pitzer and L. Brewer), "Thermodynamics," 2d ed., McGraw-Hill, New York (1961).

F. H. MacDougall, "Thermodynamics and Chemistry," 3d ed., Wiley, New York (1939).

EXPERIMENT 7.   HEATS OF COMBUSTION

In this experiment a bomb calorimeter is employed in the determination of the heat of combustion of an organic substance such as naphthalene, $C_{10}H_8$. The heat of combustion of naphthalene is $-\Delta\tilde{H}$ for the reaction at constant temperature and pressure

$$C_{10}H_8(s) + 12O_2(g) = 10CO_2(g) + 4H_2O(l) \tag{1}$$

THEORY

A general discussion of calorimetric measurements is presented in the section Principles of Calorimetry. That material should be reviewed as background for this experiment. It should be noted that no specification of pressure is made for the reaction in Eq. (1) other than that it is constant. (Indeed, in this experiment the reaction is not actually carried out at constant pressure, though the results are corrected to constant pressure in the calculations.) In fact energy and enthalpy changes attending physical changes are generally small in comparison with those attending chemical changes, and those involved in pressure changes on condensed phases (owing to their small molal volumes and low compressibilities) and even on gases (owing to their resemblance to perfect gases) at constant temperature are very small. Thus, energy and enthalpy changes accompanying chemical changes can be considered as being independent of pressure for nearly all practical purposes.

EXPERIMENTAL

The calorimeter (Emerson design), shown in Fig. 1, consists of a high-pressure stainless-steel bomb which sits in a calorimeter pail containing 2000 ml of water. Projecting downward into this pail through a hole in the cover is a motor-driven stirrer and a precision thermometer. All these parts constitute what has been denoted by the letter S in the Principles of Calorimetry. The bomb contains a pair of electrodes to which are attached a short length of fine iron wire in contact with the specimen. Electrical ignition of this wire provides the means of initiating combustion of the sample, which burns in a small metal pan. The bomb is constructed in two parts, which are put together by means of a large ring nut and a lead gasket. A needle valve at the top is provided for filling the bomb with oxygen to about 300 psi.

The system as shown in Fig. 1 is surrounded by the Daniels adiabatic jacket,[1] which consists of two concentric metal cans with a space in between that can be filled with water. Adjustment of the temperature of the jacket is done by passage of electric current between the two cans, by conduction through the water. The water must have a certain degree of electrical conductivity. Often ordinary tap water or a mixture of tap water and distilled water can be used; more predictable performance can be obtained by adding a small quantity of electrolyte to distilled water (0.05 g NaCl/liter of water). The electric current is controlled by a tapping key to maintain the jacket temperature as nearly as possible at the temperature registered by

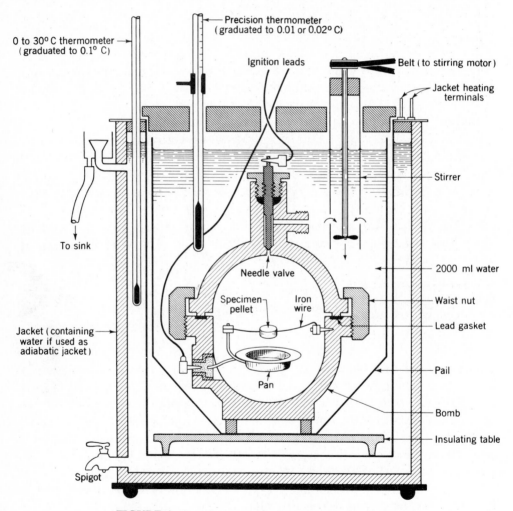

FIGURE 1

Bomb calorimeter (Emerson design), shown with Daniels adiabatic jacket which may also be used empty as an air jacket.

the calorimeter thermometer. The use of this adiabatic jacket should in principle eliminate the need to plot and extrapolate temperature values. However, there are inevitable errors (and some experimental awkwardness) in adjusting the jacket temperature during a run. It is also possible to operate the calorimeter without controlling the temperature of the Daniels adiabatic jacket. In this case the space between the two cans is left empty to provide thermal insulation. The heat leak that does occur can be compensated for very well by the extrapolation technique illustrated in Fig. V-1. This insulating air-jacket technique is strongly recommended, and a detailed procedure for the use of the adiabatic jacket will not be given.

Calorimeters of the Parr design are also widely used. They are available with insulating air jackets and with adiabatic water jackets in which water is circulated through the cover as well as the body of the jacket. The temperature of the adiabatic jacket can be adjusted manually by varying the flow through hot and cold water inlets. Detailed instructions for the operation of the Parr adiabatic jacket are available elsewhere,[2] but it is recommended that an air-jacket technique be used. The principal advantage of the Parr calorimeter lies in the design of the self-sealing, nickel-alloy bomb. This bomb, shown in Fig. 2, has a head which is precision machined to fit into the top of the cylindrical body of the bomb. Closure is achieved by compressing a rubber ring which seals against the cylinder wall. No bench socket and long-handled wrench are necessary to seal the bomb. The retaining cap is merely screwed down hand-tight; on pressurizing the bomb, a seal develops automatically. Releasing the pressure will break the seal and allow the cap to be unscrewed manually. The head shown in Fig. 2 has a single spring-loaded check valve which closes automatically when the bomb is pressurized. To vent this type of bomb, the knurled nut on the top of the valve is unscrewed one turn (to reduce the spring tension) and pressed down. A double valve head, which has a check valve for filling and a needle valve for venting the bomb, is also available. Another difference between the Emerson and Parr designs involves the electrical lead connections to the bomb. In the Parr bomb, one lead goes directly to an insulated terminal and the other lead makes contact automatically through the oval calorimeter pail which holds the bomb.

Following the procedure given below, two runs should be made with benzoic acid to determine the heat capacity of the calorimeter and two runs should be made with naphthalene or another substance (perhaps an unknown).

Procedure   The successful operation of this experiment requires close attention to detail, as there are possible sources of trouble that may prevent the experiment from working properly. It is for this reason, and because laboratory scheduling usually does not allow the student to work out his own technique by trial and error, that the procedure is presented in considerable detail. The basic procedure is very similar for both the Emerson and the Parr calorimeters. The text below is written for the Emerson

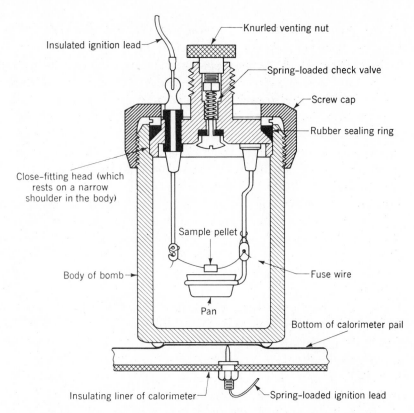

Insulated ignition lead

Knurled venting nut

Spring-loaded check valve

Screw cap

Rubber sealing ring

Close-fitting head (which
rests on a narrow
shoulder in the body)

Sample pellet

Body of bomb

Fuse wire

Pan

Bottom of calorimeter pail

Insulating liner of calorimeter

Spring-loaded ignition lead

FIGURE 2

Parr single-valve bomb, shown in contact with bottom of calorimeter pail. One electrical contact is made automatically through the pail to the body of the bomb.

calorimeter. *Changes required when using the Parr calorimeter are given in square brackets.* Further details on the operation of the Parr calorimeter are available in a manual published by the manufacturer.[2]

The student's attention is particularly drawn to the following **warning**: There is a hazard of electric shock and of short circuit from exposed terminals and metal parts of the calorimeter, especially when outer-jacket heating is employed with the Emerson calorimeter. Keep the working space dry. The bomb is expensive and should be handled carefully. In particular, be very careful not to scratch or dent the closure surfaces. When the bomb is dismantled its various parts should be placed gently on a clean, folded towel.

Condition of apparatus: The bomb must be clean and dry with no bits of iron wire in the terminals. Make sure that the jacket is completely empty and that all switches on the control box are off. Check to see that the inside of the calorimeter is dry. By testing in a beaker of water, verify that the stirrer is operating so as to impel the water *downward* [not important for the Parr].

Filling of the bomb: Cut the iron wire, free from sharp bends or kinks, to the correct length and weigh it accurately. Press pellets of the substance concerned, of weight 0.5 ± 0.1 g for naphthalene and 0.8 ± 0.1 g for benzoic acid. Shave them to the desired weight with a spatula if necessary. Fuse the wire into the pellet by heating the wire with current provided by a 1.5-V dry cell. Blow gently on the wire to prevent it from overheating. The pellet should be at the center of the wire. Weigh it accurately. The weight of the pellet alone is obtained by difference. Handle it very carefully after weighing.

Install the pellet and wire in the bomb. The pellet should be over the pan, and the wire should touch *only* the terminals. Carefully assemble the bomb and tighten the waist nut with a wrench. [Screw down the retaining cap hand-tight for the Parr; *never* use a wrench.]

Attach the bomb to the oxygen-filling apparatus. [The knurled nut on the check valve of the Parr must be removed first.] Line up the fitting carefully before tightening, as misalignment may result in damaged threads. Open the needle valve *one turn* only. *Carefully* open the supply valve, fill the bomb to 300 psi, then close the needle valve hand-tight only and remove the bomb. [The Parr has an automatic check valve for filling; admit oxygen *slowly* until the pressure reaches 400 psi.] Release the pressure to flush out most of the atmospheric nitrogen originally present in the bomb. Retighten and refill to the same pressure. If trouble is encountered in filling the bomb, check the condition of the gasket on the filling apparatus. Check the bomb for leaks by immersion in water. If a leak is found around the waist, retighten the waist nut with the wrench and try again. [Leaks are less likely to occur with the Parr. If a leak does occur, vent the bomb, loosen and rotate the head slightly, and then retighten the screw cap.] An occasional bubble—one every 5 or 10 sec—is inconsequential.

Assembly of calorimeter: Dry the bomb. Make electrical connections *tightly* to the top and side; the side connector *must not touch the waist ring nut.* Place the bomb in the dry pail. Check to see that the side connector does not touch the pail. Set the pail in the calorimeter, making sure that it is centered and does not touch the inner wall of the calorimeter. [For the Parr, the bomb and the pail must be in position in the calorimeter before the single connection to the top of the bomb can be made.]

Fill a 2000-ml volumetric flask with water at 25°C. A convenient way of doing this is to use both hot and cold water; add the two as required, swirling and checking with a thermometer until the flask is almost full, then make up to the mark. Pour the water from the flask carefully into the pail; avoid splashing. Allow the flask to drain a minute.

Put the calorimeter lid in place and introduce the stirrer as far as it will go.

Clamp the precision thermometer in place as low as it will go without obscuring the scale. The clamp should be as low as possible and *not too tight*. The jaws of the clamp should be rubber-covered to protect the thermometer. The thermometer should register within half a degree of 25°C.

If a test meter or an improvised tester consisting of a flashlight bulb and a dry cell is available, it is advisable to check the electrical continuity between the two prongs of the plug on the ignition cord. With all switches off, plug the stirrer and ignition cords into the control box (see Fig. 3). Recheck the wiring and then plug the control box into a 110-V ac outlet. Turn on the stirrer and make sure that it runs smoothly. [The Parr stirrer motor operates directly on 110-V, 60-cycle ac; no series resistor such as that on the control box is required.]

Instead of the ignition circuit shown in the control box of Fig. 3, a low-voltage transformer can be used. Transformer ignition circuits are available commerically as accessories from calorimeter manufacturers or can be easily constructed. A suitable design is shown in Fig. 4.

Making the run: Begin time-temperature readings, reading the precision thermometer once every 30 sec and recording both the time and the temperature. Estimate the temperature to thousandths of a degree if feasible. Tap the thermometer *gently* before each reading. *Do not interrupt the time-temperature readings until the run is over.* The calorimeter pail temperature should change at a very slow linear rate (of

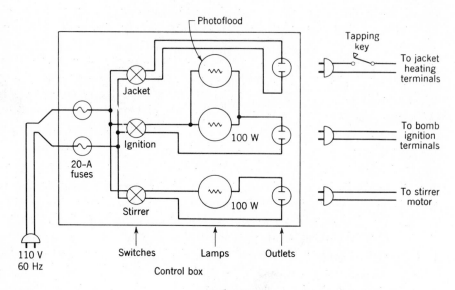

FIGURE 3
Electrical circuit for Emerson bomb calorimeter.

FIGURE 4
Transformer ignition circuit for bomb
calorimeter. The switch is a momentary-
contact push switch which is only on
when it is held down. Release the switch
if ignition does not occur after about
5 sec (the pilot light will indicate when
the fuse wire burns through). The choice
of resistor R will depend on the type of
calorimeter and the length of fuse wire
to be used (a current of 3 to 4 A is
desired).

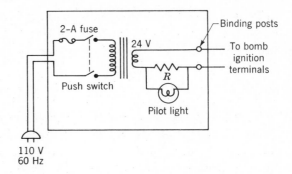

the order of 0.001° min^{-1}). After this steady rate has persisted for at least 5 min,
the bomb may be ignited.

To ignite, turn the ignition switch on and then immediately off. Record the
exact time. A dull brief flash may be observable in the light bulbs indicating the
passage of current for the instant required to burn the wire through. In nearly all
such cases the burning wire will ignite the pellet, and after a delay of 10 or 15 sec
the temperature will begin to rise. After a few minutes the pail temperature should
again show a slow, steady rate of change. Do not discontinue readings; continue
them until the time since ignition has been at least four times the period required for
attainment of this steady rate. The purpose of this is to provide a valid basis for
extrapolation. When readings are completed, turn all switches off.

Disassemble the apparatus, release the bomb pressure, and open the bomb.
Remove and weigh any unburned iron wire; ignore "globules" unless attempts to
crush them reveal that they are fused metal rather than oxide. Subtract the weight
of unburned iron wire from the initial iron-wire weight to obtain the net weight of
iron burned. If the inside of the bomb is found to be coated with soot, the amount
of oxygen present at the time of ignition was presumably insufficient to give complete
combustion and the run should be discarded. Wipe dry all bomb parts.

CALCULATIONS

For each run, plot temperature versus time using an expanded, interrupted temperature scale as shown in Fig. V-1*b*, and perform the indicated extrapolation to
determine the temperature change.

The heat capacity $C(S)$ is found by determining the temperature rise $(T'_2 - T'_1)$
obtained in the combustion of a known weight of benzoic acid (and, of course, a

known weight of iron wire) and by making use of Eq. (V-15b). For calculating the energy change produced, the specific (i.e., per gram) energies of combustion of benzoic acid (BA) and iron wire (Fe) given below can be used:[3]

$$\Delta \bar{E}_{BA} = -6316 \text{ cal g}^{-1}$$
$$\Delta \bar{E}_{Fe} = -1600 \text{ cal g}^{-1} \tag{2}$$

In this experiment ΔE for the combustion of a weighed specimen of naphthalene or other substance is determined from the rise in temperature $(T_1 - T_0)$ and the heat capacity $C(S)$ by use of Eqs. (V-6b) and (V-14b). The value of ΔE obtained includes a contribution for the combustion of the iron wire; this must be subtracted to yield the contribution from the naphthalene alone. The molar energy change $\Delta \tilde{E}$ is then obtained from the number of moles of the reactant that is present in the least equivalent amount (in this case, naphthalene) in accordance with Eq. (V-16b). The molar enthalpy change $\Delta \tilde{H}$ can then be obtained by use of Eq. (V-8).

Report the individual and average values of the heat capacity $C(S)$ and the individual and average values of the molar enthalpy change $\Delta \tilde{H}$ for the combustion of naphthalene, calculated using the average $C(S)$.

DISCUSSION

How does the order of magnitude of the error introduced into the experimental result by the assumption of the perfect-gas law in Eq. (V-8) compare with the uncertainties inherent in the measurements in this experiment? What is the magnitude of the uncertainty introduced by lack of knowledge of the specific heat of the sample? Does your ΔH value pertain to the initial or the final temperature?

APPARATUS

Bomb calorimeter complete with bomb, pail, jacket, insulating table, cover, and stirring motor with power cord; one set of electrical leads which connect to bomb; 110-V power source unit, equipped with fuse and bulbs; precision calorimeter thermometer covering range from 19 to 35°C; magnifying thermometer-reader; 0 to 30° thermometer; stopwatch; 2000-ml volumetric flask; 1.5-V dry cell; 500-ml beaker. If adiabatic jacket is to be used: tapping key and set of leads for jacket heating; water of proper conductivity for filling jacket.

Bomb-filling apparatus on table securely bolted to the floor; long-handled wrench for closing and opening bomb; cylinder of oxygen with appropriate fittings; large pail for leak-testing bomb; benzoic acid (5 g); naphthalene (5 g) or other solid to be studied; pellet press; spatula; 0.004-in.-diameter iron wire (50 cm); device for checking electrical continuity (optional).

REFERENCES

1. F. Daniels, *J. Amer. Chem. Soc.*, **38**, 1473 (1916); T. W. Richards, *J. Amer. Chem. Soc.*, **31**, 1275 (1909).
2. "Oxygen Bomb Calorimetry and Combustion Methods," Tech. Manual 130, Parr Instrument Co., Moline, Ill. (1960).
3. "Handbook of Chemistry and Physics," 53d ed., Chemical Rubber Publishing Co., Cleveland (1972).

GENERAL READING

J. M. Sturtevant, Calorimetry, in A. Weissberger (ed.), "Technique of Organic Chemistry," 3d ed., vol. I, part I, chap. X, pp. 523–654, Interscience, New York (1959).

EXPERIMENT 8.   STRAIN ENERGY OF THE CYCLOPROPANE RING

The strain energy of the cyclopropane ring is determined from thermochemical measurements on two closely related compounds: the *n*-butyl ester of cyclopropane-carboxylic acid and the methyl ester of cyclohexanecarboxylic acid. Both compounds can be synthesized from the appropriate acid chlorides, and their heats of combustion are to be measured with a bomb calorimeter.

THEORY

The molar enthalpy change $\Delta\tilde{H}_d$ for dissociation of an organic compound into separated atoms in their gaseous state

$$C_m H_n O_p N_q \cdots = mC(g) + nH(g) + pO(g) + qN(g) + \cdots \qquad (1)$$

is determined principally by the number and kinds of chemical bonds in the molecule. In many cases, the "heat of dissociation" $\Delta\tilde{H}_d$ can be closely approximated by the sum of so-called "bond energies" (more properly, bond enthalpies) B_i for all the bonds in the molecule:†

$$\Delta\tilde{H}_d = \sum_i B_i \qquad (2)$$

† Strictly speaking a heat of dissociation calculated from bond energies applies to a compound in the vapor state, but the heat of vaporization from the liquid or solid state (typically 5 to 12 kcal mol^{-1}) is ordinarily smaller in magnitude than the errors inherent in the bond-energy treatment.

For all conventional chemical bonds, the B_i values are positive; for a stable molecule, $\Delta \tilde{H}_d$ must have a positive sign.

Certain molecules are made less stable by the presence of a *strain energy* **S** corresponding to the bending or stretching of bonds from their normal state as a result of geometrical requirements, or are made more stable by the presence of a *resonance energy* **R** corresponding to aromatic character, conjugation, hyper-conjugation, etc.:

$$\Delta \tilde{H}_d = \sum_i B_i - S + R \tag{3}$$

The quantities **S** and **R** are defined so as to be positive in sign. We shall be concerned here with the strain energy of the cyclopropane ring, in which the C—C—C bond angles are constrained by geometry to be 60° rather than their normal values which are in the neighborhood of the tetrahedral angle 109.5°.

Consider the compounds *n*-butylcyclopropanecarboxylate (I) and methyl-cyclohexanecarboxylate (II), which have the same molecular composition $C_8H_{14}O_2$:

The quantity $\sum B_i$ should have the same value for both compounds, since each has 7 C—C, 1 C=O, 2 C—O, and 14 C—H bonds. However, molecule I possesses appreciable strain energy while molecule II does not. (The cyclohexyl ring is staggered, so that the angles are close to 109.5°.) Neither molecule would be expected to have appreciable resonance energy; in any case, resonance effects would be about the same for both molecules and would cancel out. Hence

$$S(I) = \Delta \tilde{H}_d(II) - \Delta \tilde{H}_d(I) \tag{4}$$

The quantity $-\Delta \tilde{H}_d$ is closely related to, but should be carefully distinguished from, the heat of formation of a compound, which is the molar enthalpy change resulting from the formation of the compound from the elements *in their standard states* at the given temperature and pressure (usually 298°K, 1 atm). The standard state of carbon at 298°K, 1 atm is graphite, C(s); those of hydrogen, oxygen, and nitrogen are the gases $H_2(g)$, $O_2(g)$, and $N_2(g)$. $\Delta \tilde{H}_f$ for the change in state

$$mC(s) + \tfrac{1}{2}nH_2(g) + \tfrac{1}{2}pO_2(g) + \tfrac{1}{2}qN_2(g) \to C_mH_nO_pN_q \quad (298°K, 1 \text{ atm}) \tag{5}$$

is related to $\Delta \tilde{H}_d$ by

$$-\Delta \tilde{H}_f = \Delta \tilde{H}_d - m \Delta \tilde{H}_d{}^0[C(s)] - \tfrac{1}{2}n \Delta \tilde{H}_d{}^0[H_2(g)] - \tfrac{1}{2}p \Delta \tilde{H}_d{}^0[O_2(g)]$$
$$- \tfrac{1}{2}q \Delta \tilde{H}_d{}^0[N_2(g)]$$

$$= \Delta \tilde{H}_d - 171.7m - 52.09n - 59.16p - 113.0q \tag{6}$$

where the numerical values are given in kcal mol^{-1}. The quantities $\Delta \tilde{H}_d^0$ are the heats of dissociation of the elements *in their standard states* into separated gaseous atoms; their values have been obtained from a variety of calorimetric and spectroscopic measurements.[1] The sign of $\Delta \tilde{H}_f$ is usually, but not always, negative.

Since the change in state (5) will seldom take place as a single reaction in a laboratory calorimeter, the heat of formation is not measured directly. However, by Hess's law, the desired enthalpy change can be obtained by a suitable algebraic combination of the enthalpy changes for other changes in state corresponding to actual reactions. As an example of the kind of reaction that is useful for this purpose, we consider combustion with high-pressure oxygen gas:

$$C_m H_n O_p(s \text{ or } l) + (m + \tfrac{1}{4}n - \tfrac{1}{2}p)O_2(g) \rightarrow mCO_2(g) + \tfrac{1}{2}nH_2O(l) \quad (298°K, 1 \text{ atm})$$

$$(7)$$

We will here restrict ourselves to the combustion of solids and liquids; combustion of gases requires a very different experimental technique. Moreover, we will not consider compounds containing elements other than C, H, and O, since the combustion products of compounds containing other elements may be somewhat variable; nitrogen, for example, may appear in the product as $N_2(g)$, oxides of nitrogen, or nitric acid. Note that in Eq. (7) the products are gaseous CO_2 and liquid water. The heat of formation for the organic compound $C_m H_n O_p$ is related to the heat of combustion $\Delta \tilde{H}_c$, which is the enthalpy change for the change in state (7), by

$$-\Delta \tilde{H}_f = \Delta \tilde{H}_c + 94.05m + 34.16n \tag{8}$$

where, again, all quantities are in kcal mol^{-1} at 298°K, 1 atm. The sign of $\Delta \tilde{H}_c$ is, for all practical purposes, invariably negative. The second and third terms on the right arise from the heats of combustion of graphite and gaseous hydrogen. The heat of dissociation of the compound is then given by

$$\Delta \tilde{H}_d = \Delta \tilde{H}_c + 265.8m + 86.25n + 59.16p \tag{9}$$

From Eqs. (4) and (9) we see that

$$S(I) = \Delta \tilde{H}_c(II) - \Delta \tilde{H}_c(I) \tag{10}$$

all other terms having canceled out.

Equation (9) may be used to calculate $\Delta \tilde{H}_d$ from the experimentally determined heat of combustion of compound II for comparison with the sum of tabulated bond energies [see Eq. (2)]. A short list of bond energies[1,2] is given in Table 1. These bond energies are mostly averages of values derived from heats of formation of several compounds. The values for double and triple bonds are subject to greater variability than those for single bonds and should be used with some caution. Values

Table 1   SELECTED BOND ENERGIES
(in kcal mol^{-1})

Bond	B	Bond	B
C—C	83.1	C=C	146
C—H	98.8	C=O	177
C—O	84.0	C=N	147
C—N	69.7	N=N	100
C—Cl	78.5		
C—Br	65.9	C≡C	200
C—I	57.4	C≡N	213
O—H	110.6		
N—H	93.4		

of $\Delta\tilde{H}_d$ calculated with bond energies are typically in error by 1 or 2 percent in favorable cases and by as much as 5 percent (or even more) in bad cases. It should be no surprise if the value obtained from Eq. (2) for compound II (which is negligibly affected by strain) differs from that calculated with Eq. (9) by an amount that exceeds the small difference between the enthalpies of I and II due to strain. However, it may be expected that since molecules I and II are very similar in all respects other than strain, the errors inherent in the application of Eq. (3) and Table 1 will be essentially the same for both molecules. Accordingly we may expect that Eq. (10) will give a fairly good measure of the strain energy **S** in molecule I.

It must be kept in mind that the strain energy **S** represents a small difference between two large quantities, $\Delta\tilde{H}_c(\text{I})$ and $\Delta\tilde{H}_c(\text{II})$, obtained from measurements on two different substances; the difference is only of the order of a few percent. Thus it is clear that all physical measurements involved in the calorimetry must be carried out with the highest possible precision (one or two parts per thousand) and also that the two different substances must be highly pure. The principal impurity to be feared in compound I is n-butanol, which boils only about 45° lower than compound I at a pressure of 90 Torr. The replacement of a small percentage of compound I by an impurity such as butanol will produce a much smaller percentage error in $\Delta\tilde{H}_c$; the error will depend on the difference between the heats of combustion per gram of the two substances. Water, which has no heat of combustion, will cause a percent error in $\Delta\tilde{H}_c$ equal to its percentage as an impurity.

EXPERIMENTAL

Since the two esters I and II are not available commercially, they will be synthesized by a procedure which is basically that of Jeffery and Vogel.[3] The starting materials for the syntheses are the free acids, cyclopropanecarboxylic acid and cyclohexanecarboxylic acid. Each acid is first converted to the acid chloride with thionyl chloride, and the acid chloride is then allowed to react with the appropriate alcohol to form the ester. Alternatively, the acid chlorides can be obtained commercially from the Aldrich Chemical Co., Inc., Milwaukee, Wisconsin, and the first part of the syntheses can be dropped. The purity of the two esters is checked by refractometry, infrared spectroscopy, and vapor-phase chromatography, and the heats of combustion are determined with a bomb calorimeter. It is suggested that each student in the team synthesize and check the purity of one of the esters and that both work together on the calorimetric measurements.

Synthesis   In following the directions given below, one should adhere closely to the amounts of material used in each step. The yields given for the two esters are those obtained routinely and correspond to a competent execution of each preparation. Substantially lower yields may require repetition of the preparation in order to obtain sufficient material for the calorimetry work. Both HCl and SO_2 are evolved in parts of this experiment and all operations where either of these gases are evolved must be carried out in a hood.†

   Preparation of acid chlorides: Cyclopropanecarboxylic acid chloride is prepared according to the reaction

$$\triangleright\!\!\!-\!\overset{\overset{\textstyle O}{\|}}{C}\!-\!OH + SOCl_2 \rightarrow \triangleright\!\!\!-\!\overset{\overset{\textstyle O}{\|}}{C}\!-\!Cl + HCl(g) + SO_2(g) \qquad (11)$$

Assemble in a hood an apparatus consisting of a single-neck 50-ml round-bottom flask, a Claissen adapter, a vertical water-cooled condenser for reflux (joined to the main stem of the adapter), and a dropping funnel (joined at the side arm of the adapter). The flask should contain a small Teflon-covered stirring bar and should sit in a heating mantle on top of a magnetic stirring unit.

   Allow 13.1 g (0.11 mol) of freshly distilled or reagent-grade thionyl chloride to run into the flask from the dropping funnel; close the stopcock as soon as the thionyl chloride has run into the flask. Start the magnetic stirrer and the flow of water through the condenser. Heat the flask until the thionyl chloride has begun to reflux. Add 6.25 g (0.073 mol) of cyclopropanecarboxylic acid dropwise from the dropping funnel to the flask. As soon as an evolution of gas more vigorous than that due to the refluxing thionyl chloride becomes apparent, turn off the external source of heat. The rate of addition of acid should be adjusted so that constant evolution of gas is

----

† Reactions and distillations carried out in a hood are subject to strong drafts of air and should be protected. In particular, distillation apparatus should be carefully wrapped in glass wool and aluminum foil or both, to prevent excessive and uneven cooling.

obtained—the addition usually requires about 15 min. After the addition of acid has been completed, close the stopcock and allow the system to stir at room temperature until the evolution of gas has slowed considerably. Reflux the mixture for 20 min. Then allow the system to cool and rearrange the apparatus to a conventional apparatus for simple distillation. In carrying out this rearrangement, close each entry to the reaction flask as each piece of equipment (dropping funnel, condenser) is removed from the flask since the acid chloride decomposes slowly in contact with moist air. The small amount of residual thionyl chloride (bp 79°/760 Torr) is then distilled from the acid chloride. At this point, one may continue and isolate the acid chloride by distillation (bp 119°/760 Torr), or one may use the undistilled residue in the next step of the synthesis. The final yields of ester are comparable from both routes.

The procedure for preparing cyclohexanecarboxylic acid chloride is exactly the same as that described above for the preparation of cyclopropanecarboxylic acid chloride. In this case 11.9 g (0.10 mol) of freshly distilled or reagent-grade thionyl chloride and 7.66 g (0.06 mol) of cyclohexanecarboxylic acid are used. Cyclohexanecarboxylic acid is a solid melting at 31°C but is easier to handle as a liquid. As an effective and safe (under normal precautions) method to melt the acid for weighing and addition to the thionyl chloride, apply steam to the exterior of the bottle or the dropping funnel. As with the previous synthesis, one may distill this acid chloride (bp 125°/760 Torr), or one may use it without distillation in the next step.

Preparation of n-butyl cyclopropanecarboxylate:

$$\triangleright\!\!-\!\!\overset{\overset{\displaystyle O}{\|}}{C}\!\!-\!\!Cl + n\text{-}C_4H_9OH \rightarrow \triangleright\!\!-\!\!\overset{\overset{\displaystyle O}{\|}}{C}\!\!-\!\!OC_4H_9 + HCl(g) \qquad (12)$$

Assemble in a hood the same apparatus as before, except that the condenser is replaced with a $CaCl_2$ drying tube, and the heating mantle is omitted. Start the magnetic stirrer after 5.40 g (0.073 mol) of anhydrous *n*-butanol has been placed in the 50-ml flask. Pour the cyclopropanecarboxylic acid chloride into the dropping funnel, and stopper the dropping funnel with another $CaCl_2$ drying tube. [If commercially prepared acid chloride is used, add 7.63 g (0.073 mol).] Since the reaction is exothermic, before allowing the acid chloride to react with the alcohol *prepare an ice bath* for use in case heat is evolved too rapidly. Add the acid chloride dropwise at a rate which maintains a vigorous evolution of HCl gas; the addition will usually take about 20 min. After the addition period, the ester reaction mixture is stirred for 40 min at room temperature.

After the solution is cool, add about 15 ml of ethyl ether, pour the mixture carefully into about 15 ml of water (**caution**: the reaction is exothermic), and mix well. Separate the aqueous layer, extract it twice with 15-ml portions of ether. Wash the combined layers twice with 15-ml portions of saturated aqueous $NaHCO_3$ solution. Extract the washings twice with 15-ml portions of ether; in each case, use the ether layer from the extraction of the first portion of $NaHCO_3$ solution to extract the second. Dry the combined organic layers for 20 to 30 min with $MgSO_4$, swirling to hasten equilibrium. Filter into a 100-ml round-bottom flask, attach a Vigreux column, and drive off the ether with a steam bath (in a hood).

The ester should preferably be distilled at reduced pressure (bp 110 to 112°/90 Torr) to minimize thermal decomposition. Simple distillation is likely to yield a product with a higher-than-acceptable contamination by butanol, so fractionation is recommended. A Vigreux column should be wrapped with glass wool and aluminum foil for thermal insulation and fitted with an Ace 9214 MSP vacuum distilling head. This head has a "cold-finger"condenser, rotatable so that the reflux ratio can be adjusted, and has stopcocks to permit "fraction cutting" (change of receiving vials without releasing the vacuum). The stopcock through which the distillate flows should be very lightly greased, preferably only at the top, so that grease does not contaminate the product. If in doubt, clean this stopcock with solvent and lightly regrease it.

Before distilling the reduced liquid, transfer it (together with a 5-ml ether washing from the previous flask) to a 25-ml round-bottom flask. The distillation should be carried out with a reflux ratio of at least 4:1. Three fractions should be collected: a low-boiling one containing most of the ether and residual butanol, one boiling above the boiling point of butanol but below the stable boiling point of the ester, and the last containing the ester boiling at constant temperature. The usual yield in the last fraction is 6 to 6.5 g.

Take refractive indexes of the last two fractions (Jeffery and Vogel[3] report $n_D = 1.42446$; at MIT, we have found $n_D^{25} = 1.4259$), and analyze both fractions by vapor-phase chromatography (VPC).† No impurities should show on the VPC trace when the peak height is about 90 percent full deflection. Decreased attenuation of the VPC may result in trace amounts of impurities being shown. To calibrate for the effect of butanol on the VPC spectrum, run also a sample consisting of 200 μl of the final ester fraction and 2 μl of n-butanol.

The infrared absorption spectrum of a 5 percent solution of the final fraction in carbon tetrachloride should be taken. Verification that the cyclopropyl group did not isomerize to propenyl-2 to any significant extent under the synthesis conditions should be found in the absence of any significant vinyl C–H bending band at 990 cm^{-1}. (Other such bands at 909 and 1320 cm^{-1} would be more or less obscured by other

---

† A silicone column (e.g., 10 percent 550 Dow-Corning Silicone fluid on Chrom P) at 100°C and a flow rate of 60 ml/sec is sufficient to show any significant impurities, although tailing of the ester peaks generally occurs. A Carbowax 20M column gives less tailing under the same conditions.

structural features of the spectrum. The vinyl C–H stretching bands at 3090 and 3030 cm^{-1} would be obscured by the cyclopropyl C–H stretching band at about $3060\ cm^{-1}$.) If there is any suspicion of such isomerization, a bromine test may be performed. Dilute about 0.5 g of the ester with 1 or 2 ml of CCl_4 and add one drop of a 5 percent solution of bromine in CCl_4; immediate fading of the bromine color would confirm that unsaturation is present. The presence of any significant quantity of butanol or free carboxylic acid would be indicated by absorption bands at $3625\ cm^{-1}$ or 2700 to $2500\ cm^{-1}$, respectively.

Preparation of methyl cyclohexanecarboxylate:

$$\underset{\displaystyle S}{\bigcirc}\!\!-\overset{\displaystyle O}{\overset{\|}{C}}-Cl + CH_3OH \rightarrow \underset{\displaystyle S}{\bigcirc}\!\!-\overset{\displaystyle O}{\overset{\|}{C}}-O-CH_3 + HCl(g) \qquad (13)$$

The procedure for this synthesis is essentially the same as that for the *n*-butyl cyclopropanecarboxylate described above. In the present case 3.8 g (0.12 mol) of methanol is used. [If commercially prepared acid chloride is used, add 8.80 g (0.06 mol).] Since methanol boils substantially lower than butanol, simple distillation may be used instead of fractionation, but it is advisable to do it at reduced pressure. A conventional setup with a "pig"-type adapter for fraction cutting may be used. As with the other ester, the liquid to be distilled should be transferred, with an ether washing, to a 25-ml round-bottom flask. Distill the ester at about 35 Torr (bp 90 to 92°/37 Torr) and collect three fractions: one low boiling, one from 50 to 90° at 35 Torr, and the last fraction after the boiling point of the ester is stable. The usual yield in the final fraction is 6.0 to 8.5 g of the ester.

Record the refractive indexes of the last two fractions ($n_D^{25} = 1.4410$), and analyze them by vapor-phase chromatography. Only minor impurity peaks should show on the VPC trace at low attenuation with the ester peak off scale. Obtain an infrared spectrum of a 5 percent solution of the final fraction in carbon tetrachloride. Isomerization of the ring to a terminally unsaturated straight chain is much less likely than with the cyclopropane ring. Check for absorption in the medium-strength $3085\ cm^{-1}$ and $3030\ cm^{-1}$ vinyl C–H stretching bands; vinyl C–H bending bands would tend to be obscured by other features.

Calorimetry The procedures for the operation of the bomb calorimeter are, with minor exceptions noted below, those described in detail in Exp. 7. That experiment and the general discussion presented in the section Principles of Calorimetry should be carefully studied.

The heat capacity of the calorimeter is to be determined with duplicate runs on solid benzoic acid, C_6H_5COOH. Two runs are then made on each liquid ester. Any run with unsatisfactory features in the time-temperature plot or indications of incomplete combustion (soot, etc.) must be rejected. If any pair of runs fails to give reasonably concordant results (agreement within 0.5 percent or better), additional runs on the same substance should be made.

For the present experiment the adiabatic jacket is to be used empty as an air jacket. The procedure is experimentally simpler and capable of high accuracy provided that the ambient temperature in the laboratory is reasonably steady.

Benzoic acid: A pellet of benzoic acid, 0.8 ± 0.1 g, is accurately weighed and placed in the sample pan in the bomb. Weigh also an 8-cm piece of iron ignition wire. Instead of fusing the wire into the pellet as described in Exp. 7, you may connect the wire to the terminals in such a way that it is held by light "spring action" against the surface of the pellet. In shaping the wire for this purpose avoid kinks and sharp bends and make certain that the wire does not touch the sample pan itself.

Seal the bomb, flush and fill with oxygen, test for leaks, and assemble the calorimeter as described in Exp. 7. Be sure to adjust the temperature of the 2000 ml of water to $25° \pm 1°C$ before introducing it into the calorimeter pail from the volumetric flask. When all is ready, turn on the stirring motor and begin time-temperature readings. Read the precision thermometer every 30 sec, and record both time and temperature. Tap the thermometer gently before each reading, and estimate the temperature to thousands of a degree. A temperature-time curve on an expanded scale (see Fig. V-1*b*) should be plotted concurrently. The readings should fall on a straight line, with a small positive slope due to energy input from the stirrer. When the readings have shown a straight-line behavior continuously for at least 5 min, ignite the bomb. Record the time of ignition and continue temperature readings until the total elapsed time following ignition is at least four times the period required to achieve a slow, steady rate of change in the upper temperature. Then turn off the control box switches, release the bomb pressure, open the bomb, remove and weigh any unburned iron wire. The presence of any soot indicates incomplete combustion, probably owing to insufficient oxygen; a leak is to be suspected. In such a case, the result of the run must be rejected. A run should also be rejected if the time-temperature plot contains any unusual features which make extrapolation uncertain.

Esters: About 0.7 to 0.9 g of one of the esters is introduced into the sample pan from a 2-ml disposable hypodermic syringe. The syringe is filled nearly half full, weighed on an analytical balance, discharged into the sample pan, and reweighed. Before each weighing, withdraw the plunger so as to pull a small quantity of air into the needle. This will prevent loss of liquid through dribbling from the needle tip. Naturally a drop hanging from the tip should not be wiped off at any time between the two weighings. Although the 8-cm iron wire described above may be used, a double wire obtained by attaching the middle of a 16-cm length to one terminal and the two ends to the other terminal may provide better insurance against incomplete combustion. The two wires should be carefully shaped to dip into the liquid without touching each other or the metal of the pan. Care should be taken not to wet the walls of the pan above the meniscus any more than necessary, as liquid clinging to the walls may possibly fail to burn. Handle the bomb very carefully; avoid bumping or tipping. The run is performed as described above for benzoic acid.

CALCULATIONS

The extrapolation of the post-ignition temperature versus time plots to time of firing may be carried out as illustrated in Fig. V-1b. However, especially if slowness in reaching the post-ignition steady state is cause for concern, a possibly more accurate although slightly less convenient procedure may be substituted: Extrapolate the linear portions of the pre-ignition and the post-ignition curves to a time approximately midway between the ignition time and the time at which the temperature again attains a slow, steady rate of change. Whether one extrapolation procedure or the other is chosen, the same should be used for both esters, and in as consistent a manner as possible so as to cause any systematic errors inherent in the chosen procedure to cancel out in the derivation of the strain energy. Ignore any "overshoot" bump in the post-ignition curve if it is small ($\sim 0.010°$, 1 to 2 min) and if the subsequent curve is linear.

Determine C, the heat capacity of the calorimeter system, from

$$C = \frac{6316w_{BA} + 1600w_{Fe}}{\Delta T'} \text{ cal deg}^{-1} \tag{14}$$

where w_{BA} and w_{Fe} are the weights in grams of benzoic acid and iron wire *burned* (be sure to correct for any unburned wire), and $\Delta T'$ is the (positive) difference between the two extrapolated temperature readings for the benzoic acid run. The value of C should be approximately 2500 cal deg^{-1}. The molar energy of combustion of each ester is calculated with the equation

$$\Delta \tilde{E}_c = -\frac{M}{w}(C \, \Delta T - 1600w_{Fe}) \tag{15}$$

where w is the weight in grams and M the molecular weight of the ester. It is the change in energy that is obtained directly in this experiment, since combustion in a bomb is a constant-volume process. The molar enthalpy change $\Delta \tilde{H}_c$ for combustion at constant pressure can be calculated from $\Delta \tilde{E}_c$ on the assumption that the gases O_2 and CO_2 adequately obey the perfect-gas law and that the enthalpies of all substances concerned are independent of pressure. Both of these assumptions are valid to an accuracy acceptable for our purposes. Thus

$$\Delta \tilde{H}_c \equiv \Delta \tilde{E}_c + \Delta(p\tilde{V}) = \Delta \tilde{E}_c + RT \, \Delta \tilde{N}_{\text{gas}} = \Delta \tilde{E}_c - (\tfrac{1}{4}n - \tfrac{1}{2}p)RT \qquad (16)$$

where $\Delta \tilde{N}_{\text{gas}}$ is the change in the number of moles of gas per mole of solid or liquid compound burned. Another necessary assumption, also valid to an accuracy sufficient for our purposes, is that the effect on $\Delta \tilde{H}_c$ of the presence of a small fraction of the product water in the vapor phase rather than the liquid phase is negligible. This may be checked by a simple calculation based on the volume of the bomb interior (about 200 ml), the vapor pressure of water at room temperature (about 25 Torr), and the molar heat of vaporization of water (about 10 kcal mol^{-1}).

The strain energy **S** may be calculated from the heats of combustion for the two esters by use of Eq. (10).

Report the yields, boiling ranges, and refractive indexes of the two esters. If available, the VPC and ir curves should be attached to the report. The identification number of the calorimeter and bomb, the individual and average values of the calorimeter heat capacity C and the heats of combustion $\Delta \tilde{H}_c$ for the two esters, and the apparent strain energy for the cyclopropane ring should be reported. An estimate of the uncertainties in C, the two $\Delta \tilde{H}_c$ values, and **S** should be given.

DISCUSSION

Calculate $\Delta \tilde{H}_d$ for compound II with Eq. (9), and compare this value with the sum of bond energies obtained with Eq. (2) and Table 1. Comment on the agreement between these values. Also compare your **S** value with a literature estimate of the strain energy.[4]

APPARATUS

Synthetic work One 25-, one 50-, and one 100-ml single-neck round-bottom flask; Claissen adapter; water-cooled condenser and rubber hoses; dropping funnel; Vigreux column; Ace 9214MSP vacuum distillation head; 0 to 150° thermometer; magnetic stirring unit; heating mantle; two $CaCl_2$ drying tubes; ice bath; glass wool; aluminum foil.

Refractometer; vapor-phase chromatograph; infrared spectrometer; fume hood; steam bath; pump for reduced pressure distillation; thionyl chloride (30 g); cyclopropane-carboxylic acid (7 g) and cyclohexanecarboxylic acid (8 g) or cyclopropanecarboxylic acid chloride (8 g) and cyclohexanecarboxylic acid chloride (9 g); n-butanol (6 g); methanol (4 g); ethyl ether (175 ml); saturated aqueous $NaHCO_3$ solution (40 ml); $MgSO_4$; CCl_4.

Calorimetry See detailed list at the end of Exp. 7. An additional item required here is a 2-ml disposable hypodermic syringe.

REFERENCES

1. L. Pauling, "Nature of the Chemical Bond," 3d ed., pp. 64–107, Cornell University Press, Ithaca, N.Y. (1960).
2. F. A. Cotton and G. Wilkinson, "Advanced Inorganic Chemistry," 2d ed., pp. 97–100, Interscience, New York (1966).
3. G. H. Jeffery and A. I. Vogel, *J. Chem. Soc.*, **1948**, 1804 (1948).
4. R. A. Nelson and R. S. Jessup, *J. Res. Natl. Bur. Stand.*, **48**, 206 (1952).

GENERAL READING

L. Pauling, *loc. cit.*
J. M. Sturtevant, Calorimetry, in A. Weissberger (ed.), "Technique of Organic Chemistry, vol. I: Physical Methods of Organic Chemistry," part I, chap. X, Interscience, New York (1959).

EXPERIMENT 9.   HEATS OF IONIC REACTION

It is desired in this experiment to determine the heat of ionization of water:

$$H_2O(l) = H^+(aq) + OH^-(aq) \quad \Delta \tilde{H}_1 \tag{1}$$

and the heat of the second ionization of malonic acid ($HOOC-CH_2-COOH$, hereinafter designated H_2R):

$$HR^-(aq) = H^+(aq) + R^{2-}(aq) \quad \Delta \tilde{H}_2 \tag{2}$$

These will be obtained by studying experimentally with a solution calorimeter the heat of reaction of an aqueous HCl solution with an aqueous NaOH solution, for which the ionic reaction is

$$H^+(aq) + OH^-(aq) = H_2O(l) \quad \Delta \tilde{H}_3 \tag{3}$$

and that of the reaction of an aqueous NaHR solution with an aqueous NaOH solution,

$$HR^-(aq) + OH^-(aq) = R^{2-}(aq) + H_2O(l) \quad \Delta \tilde{H}_4 \tag{4}$$

The last two equations can be written in the general form

$$A + B = \text{products} \tag{5}$$

where A represents the acid ion and B the basic ion.

THEORY

A general discussion of calorimetric measurements is presented in the section Principles of Calorimetry, which should be reviewed in connection with this experiment. We shall not here consider the concentration dependence of these enthalpy changes. Such concentration dependence is generally a small effect, since the heats of dilution involved are usually much smaller than the heats of chemical reaction (indeed they are zero for perfect solutions). Since we are here dealing with solutions of moderate concentration, particularly in the case of the NaOH solution, it may be useful to make parallel determinations of heats of dilution of the solutions concerned by a procedure similar to that described here if time permits.

EXPERIMENTAL

In this experiment 500 ml of solution A, with a precisely known concentration in the neighborhood of 0.25 M, is reacted with 50 ml of solution B at a concentration sufficient to provide a slight excess over the amount required to react with solution A. The reaction is carried out in the solution calorimeter shown in Fig. 1. The calorimeter is a vacuum bottle (Dewar flask) containing a thermometer, a motor-driven stirrer, a heating coil of precisely known electrical resistance, and a precision thermometer. Various methods of mixing the two solutions might be employed; here we use an inner vessel having an outlet hole plugged by stopcock grease which can be blown out by applying air pressure at the top.

For determining the heat capacity, a dc electric current of about 1.5 A is passed through the heating coil during a known time interval. The magnitude of the current is measured as a function of time by determining the potential difference developed across a standard resistance in series with the coil. The electrical circuit is shown in Fig. 2.

It is recommended that two complete runs be made with HCl and NaOH and two with NaHR and NaOH.

Procedure It is advisable to carry out a test of the electrical heating procedure in advance of the actual runs. Place roughly 600 ml of water in the calorimeter and introduce the stirrer-heater unit and thermometer. Make the electrical connections as shown in Fig. 2. Turn on the stirring motor and then turn on the current to the heater; **never** pass current through the heating coil when it is not immersed in a liquid, as it will overheat and may burn out. Standardize the potentiometer and record the potential difference across the resistor in series with the heating coil. Also note the rate of temperature rise during heating. Both of these observations will be useful during the actual runs.

Fill a 500-ml volumetric flask with solution A, the temperature of which should be within a few tenths of a degree of 25.0°C. (Adjust the temperature by swirling under running hot or cold water before the flask has been entirely filled, then make up to the mark.) Pour the solution into the clean and reasonably dry calorimeter and allow the flask to drain for a minute. Work a plug of stopcock grease into the capillary hole in the bottom of the inner vessel. The glass must be absolutely dry, or the grease

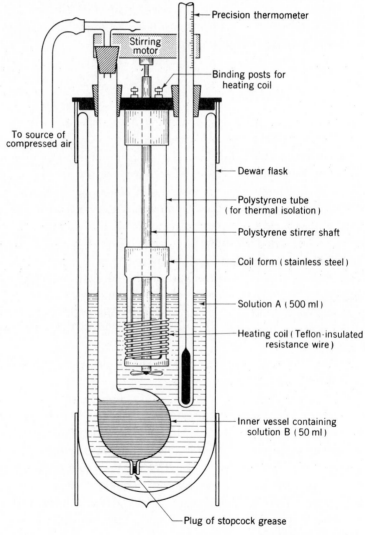

FIGURE 1
Solution calorimeter.

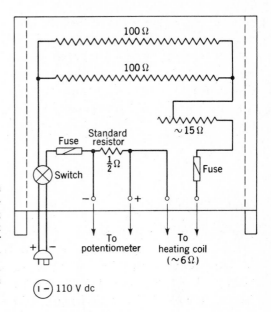

FIGURE 2
Electrical heating circuit for solution calorimeter. The unit should stand upright on the two extended ends as shown; this will permit air cooling and prevent overheating of the standard resistor. If a heating coil of higher resistance (say $\sim 60\ \Omega$) is to be used, the current can be reduced (to say ~ 0.5A) and a 1-Ω standard resistor should be used.

will not stick. Pipette in 50 ml of solution B. Place the inner vessel in the calorimeter carefully.

Introduce the stirrer-heater unit and thermometer into the calorimeter making sure that the thermometer bulb is completely immersed (level with or just below the heating coil). The inner vessel should be held in a hole in the calorimeter cover by a split stopper so that it does not rest on the bottom of the Dewar flask. Check the electrical connections to the heating coil. Connect the T tube to a compressed-air supply; turn on the air, adjust to barely audible flow, and attach the T tube to the top of the inner vessel. Turn on the stirring motor.

Start temperature-time measurements. Read the thermometer every 30 sec, estimating to thousandths of a degree if feasible. Tap the thermometer stem *gently* before each reading. After a slight but steady rate of temperature change due to stirring and heat leak has been observed for 5 min, initiate the reaction by blowing the contents of the inner vessel into the surrounding solution. This is done by placing a finger over the open end of the T tube. Release the pressure as soon as bubbling is heard. Record the time. After 15 sec, blow out the inner vessel again to ensure thermal equilibrium throughout all the solution.

After a plateau with a slight, steady rate of temperature change has prevailed for 5 min, turn on the current to the heater and *record the exact time*. Immediately measure the potential difference across the standard resistor in series with the heater. The potentiometer should be standardized and set to the expected potential (as determined above) a minute or so before the current is turned on. Repeat the potential measurement at least once every minute while the current is on.

When a temperature rise of about 1.5°C has been obtained by electric heating, turn off the current, again noting the exact time. Simultaneously blow out the inner vessel again to mix the contents with the surrounding solution. After a final plateau with a slight, steady temperature change has been achieved for 5 min, the run may be terminated. Record the resistance of the heating coil used.

CALCULATIONS

For each run, plot the temperature versus time, preferably using a greatly expanded temperature scale with an interrupted temperature axis (see Fig. V-1b). Perform the extrapolations indicated in Fig. 3 and obtain the temperature differences $(T_0 - T_1)$ and $(T_2' - T_1')$. The heat capacity is calculated by use of Eqs. (V-9) and (V-13a); the integral in Eq. (V-9) should be evaluated graphically unless the current is constant during the heating period. The enthalpy change of the ionic reaction is calculated from Eqs. (V-6a) and (V-12a). Calculate the number of formulas of reaction from the number of moles of A (the limiting reactant) present and calculate the molar enthalpy change $\Delta \tilde{H}_3$ or $\Delta \tilde{H}_4$, by use of Eq. (V-16a).

From the average values of $\Delta \tilde{H}_3$ and $\Delta \tilde{H}_4$ obtained, calculate $\Delta \tilde{H}_1$ and $\Delta \tilde{H}_2$.

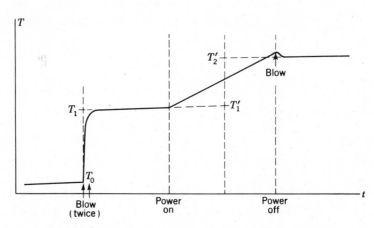

FIGURE 3
Schematic plot of temperature versus time: heat of ionic reaction.

DISCUSSION

Using literature data,[1,2] estimate the difference between the $\Delta \tilde{H}_3$ value for the concentrations employed in this experiment and the value which would apply at infinite dilution. Compare this effect with a qualitative estimate of the experimental uncertainty in the measured $\Delta \tilde{H}_3$. Indicate why there should be a large percentage error in $\Delta \tilde{H}_4$ as determined here.

APPARATUS

Solution calorimeter (1-qt Dewar); stirrer-heater unit complete with motor and power cord; inner vessel; precision calorimeter thermometer covering range from 19 to 35°C; magnifying thermometer reader; 0 to 30°C thermometer; two split rubber stoppers; glass T tube with stopper and rubber tubing; stopwatch; dc power-supply unit for heater (see Fig. 2); complete potentiometer setup (see Chap. XV); 15 electrical leads with lugs attached; 250-ml beaker; 500-ml volumetric flask; a 50- and a 10-ml pipette; 100-ml beaker; rubber pipetting bulb.

      Solutions: NaOH, concentration slightly greater than 2.5 M (0.4 liter); 0.25 M HCl (1.5 liters); 0.25 M NaHC$_3$H$_2$O$_4$ (1.5 liters; the sodium acid malonate solution can be prepared by neutralizing malonic acid with sodium hydroxide to a point just past the NaHC$_3$H$_2$O$_4$ end point); stopcock grease; source of compressed air.

REFERENCES

1. F. R. Bichowsky and F. D. Rossini, "Thermochemistry of Chemical Substances," Reinhold, New York (1936).
2. Selected Values of Chemical Thermodynamic Properties, *Natl. Bur. Stand. Circ.* 500 (1952); revised values have been published in a series of Natl. Bur. Stand. Tech. Notes (1968, 1969, 1971).

EXPERIMENT 10.   HEATS OF SOLUTION

In this experiment, techniques very similar to those used in Exp. 9 will be employed to determine the integral and differential heats of solution of a salt in aqueous solution.

THEORY

The general background for calorimetric measurements has been developed in the section Principles of Calorimetry, but it is necessary here to define and distinguish the terms *integral heat of solution* and *differential heat of solution*.[1-3]

For a given solution of specified concentration m, the integral heat of solution per mole of solute $\Delta \tilde{H}_{int}$ is the heat absorbed when one mole of solute is dissolved isothermally in enough solvent to form a final solution of concentration m. In other words, it is the enthalpy change for the change in state at constant p and T

$$A + xS = A \cdot xS \text{ (solution, conc. } m) \tag{1}$$

where A is the pure solute and S is the pure solvent. This quantity $\Delta \tilde{H}_{int}$ varies with the concentration m. As x in Eq. (1) approaches infinity (m approaches zero), $\Delta \tilde{H}_{int}$ asymptotically approaches a constant value $\Delta \tilde{H}_{\infty}$, the integral heat of solution at infinite dilution.

In many thermodynamic calculations, it is necessary to know the enthalpy change for the change in state at constant p and T

$$A = A \text{ (solution in S, conc. } m) \tag{2}$$

where A and S have the same meaning as in Eq. (1). This quantity is known as the differential heat of solution of the solute and may be thought of as the heat absorbed when 1 mol of solute is dissolved isothermally in an infinite amount of solution of concentration m. The differential heat of solution could in principle be obtained by measuring the heat absorbed per mole $(\delta Q/\delta n)$ when a very small number of moles (δn) of solute is dissolved in a finite amount of solution. In practice, however, it is much easier to calculate the differential heat of solution from the integral heat of solution determined as a function of concentration. For the change in state

$$N_1 \text{ component } 1 + N_2 \text{ component } 2 = \text{solution} \qquad \text{(const } p, T) \tag{3}$$

the enthalpy change ΔH is given by

$$\Delta H = H_{soln} - N_1 \tilde{H}_1{}^0 - N_2 \tilde{H}_2{}^0 \tag{4}$$

where $\tilde{H}_i{}^0$ is the molal enthalpy of pure component i. The value of ΔH will be a function of both the number of moles of solute (N_2) and of solvent (N_1). If Eq. (4) is differentiated with respect to N_2, while p, T, and N_1 are held constant, the result is

$$\left(\frac{\partial \Delta H}{\partial N_2}\right)_{p,T,N_1} = \left(\frac{\partial H_{soln}}{\partial N_2}\right)_{p,T,N_1} - \tilde{H}_2{}^0 = \bar{H}_2 - \tilde{H}_2{}^0 = \Delta \bar{H}_2 \tag{5}$$

where $\bar{H}_2$ is the *partial molal enthalpy* of the solute in solution and $\Delta \bar{H}_2$ is the differential heat of solution of the solute.

Now consider the solution of m moles of solute in 1000 g of solvent to give a solution of molality m. The enthalpy change ΔH will be $m \Delta \tilde{H}_{int}$; therefore,

$$\Delta \bar{H}_2 = \frac{\partial(m \, \Delta \tilde{H}_{int})}{\partial m} \tag{6}$$

and once $\Delta \tilde{H}_{int}$ is known as a function of concentration m, the differential heat of solution $\Delta \bar{H}_2$ can be calculated for any given concentration.

It is of interest to examine how the differential and integral heats of solution should vary with solution concentration as an electrolyte solution approaches infinite dilution. For this we consider the consequences of the Debye-Hückel theory of interionic attraction. From the Debye-Hückel limiting law, as expressed by Eq. (14-10), it is possible to show by straightforward thermodynamics that the limiting behavior of the differential heat of solution is expressed by

$$\bar{H}_2 - \bar{H}_2{}^0 \equiv \Delta \bar{H}_2 - \Delta \tilde{H}_\infty = 2.303 \, RT^2 \frac{d \ln A}{dT} v |z_+ z_-| \sqrt{\mu} \tag{7}$$

where $\bar{H}_2{}^0$ is the partial molal heat of solution of the solute at infinite dilution, $\Delta \tilde{H}_\infty = \bar{H}_2{}^0 - \tilde{H}_2{}^0$ is the differential *and* integral heat of solution at infinite dilution, and $v = v_+ + v_-$ is the number of ions from the salt $A_{v_+} B_{v_-}$. On combining Eq. (6) with Eq. (7), integrating, and dividing by m, we obtain for the limiting behavior of the integral heat of solution

$$\Delta \tilde{H}_{int} = \Delta \tilde{H}_\infty + \frac{2}{3} \cdot 2.303 \, RT^2 \frac{d \ln A}{dT} vA |z_+ z_-| \sqrt{\mu} \tag{8}$$

Now the Debye-Hückel "constant" A (which has the value 0.509 mol$^{-1/2}$ kg$^{1/2}$ for aqueous solutions at 298°K) is proportional to $\rho^{1/2} \varepsilon^{-3/2} T^{-3/2}$, where the solvent density ρ and dielectric constant ε both vary with temperature. If their temperature coefficients (derivatives of their natural logarithms with respect to T) are respectively α and β, then $d \ln A/dT = (\alpha/2) - (3\beta/2) - (3/2T)$. For water at 298°K, $\alpha = -0.00026$ deg^{-1}, $\beta = -0.000459$ deg^{-1}, and we obtain

$$\Delta \tilde{H}_{int} = \Delta \tilde{H}_\infty + 239v |z_+ z_-| \sqrt{\mu} \tag{9}$$

where the units of the numerical coefficient are cal mol$^{-3/2}$ kg$^{1/2}$. The final term in Eq. (9) is ordinarily quite small in comparison with the enthalpy change $\Delta \tilde{H}_\infty$. [For the most dilute solution studied in this experiment (with $\mu = m = 0.046$ mol kg^{-1}) this term corresponds to a temperature difference of only about 0.0024°K, to be compared with a measured temperature difference $T_0 - T_1$ which amounts to several tenths of a degree.] However, this term contains all the information on the concentration dependence of $\Delta \bar{H}_2$. In view of the form of Eq. (9) given by Debye-Hückel theory, it is convenient to rewrite Eq. (6) for an electrolyte solution in the form

$$\Delta \bar{H}_2 = \Delta \tilde{H}_{int} + \frac{\sqrt{m}}{2} \frac{d \, \Delta \tilde{H}_{int}}{d\sqrt{m}} \tag{10}$$

EXPERIMENTAL

The experimental method to be used is very similar to that described in Exp. 9. The apparatus is identical with that shown in Figs. 9-1 and 9-2 *except* that the inner vessel of the solution calorimeter is replaced by a short (~3-in.) length of glass tubing flared at the top to serve as a funnel for adding the solid salt. This funnel should not come in contact with the liquid below.

The experiment is carried out in five successive runs, in each of which an aliquot of solid KNO_3, weighing approximately 2.8 g, is introduced into the calorimeter. In the first run the salt is dissolved in distilled water and in other runs it is dissolved in the solution resulting from the previous run. The calorimeter should originally contain a known amount of water (about 600 g). The solution of KNO_3 in water is endothermic, and the temperature therefore drops; after each addition of KNO_3 electrical energy is dissipated inside the calorimeter by means of a heating coil. This heating accomplishes two purposes simultaneously: the temperature is returned to its initial value at the end of each run prior to the next addition of KNO_3, and the heat capacity is determined for each resulting solution.

Procedure After the equipment has been set up, a test of the electrical heating circuit should be carried out in advance of the actual runs. Follow the procedure described in Exp. 9. Be careful **never** to pass current through the heating coil when it is not immersed in a liquid.

Place approximately 2.8 g of dry KNO_3 powder (accurately weighed on an analytical balance) in each of five small well-dried and numbered weighing bottles or Erlenmeyer flasks. Stopper these flasks, and if possible, keep them thermostated at 25°C until used.

Fill a 500- and a 100-ml volumetric flask with distilled water. Adjust the temperature of the water to within a few tenths of a degree of 25.0°C by swirling the flasks under running hot or cold water. Then pour the distilled water into the clean and dry calorimeter (Dewar) and allow the flasks to drain briefly. (If desired, the flasks may be weighed before and after delivery.)

Introduce the stirrer-heater unit and thermometer into the calorimeter making sure that the thermometer bulb is completely immersed (level with or just below the heating coil). Place the funnel (*dry*) in an opening provided in the cover of the calorimeter. Make electrical connections to the heating coil and start the stirring motor.

Begin temperature-time measurements. Read the thermometer, estimating to thousandths of a degree if feasible, every 30 sec throughout the run. After a small, steady rate of temperature change due to stirring and heat leak has been observed for 5 min, one of the weighed samples of KNO_3 is emptied into the funnel and the time is recorded. The funnel should be tapped to ensure that all the salt drops into the liquid below. As the salt dissolves, the temperature will fall and then level off.

After a plateau with a steady rate of temperature change has prevailed for 5 min, turn on the current to the heating coil and *record the exact time*. Immediately measure the potential difference across the standard resistor in series with the heater. The potentiometer should be standardized and set to the expected potential a minute or so before the current is turned on. Repeat the potential measurement at least once every minute while the current is on. The current is turned off when the temperature is about 0.1°C below the initial temperature (before the salt was added) in order to allow for some "overshoot." Record the exact time. After a plateau with a small, steady rate of temperature change has been achieved for 5 min, the run is complete.

The next run can be started immediately by adding another weighed sample of KNO_3 to the solution in the calorimeter and proceeding as before. Five successive runs are to be carried out. Record the resistance of the heating coil used.

CALCULATIONS

For each run plot the temperature versus time, preferably using a greatly expanded temperature scale with an interrupted temperature axis (see Fig. V-1b). Perform the extrapolations indicated in Fig. 1 and obtain the temperature differences $(T_0 - T_1)$ and $(T_2' - T_1')$. The heat capacity is calculated by use of Eqs. (V-9) and (V-13a); the integral in Eq. (V-9) should be evaluated graphically unless the current is constant during the heating period. The C_p values for the individual runs should agree closely with each other. If an untrustworthy value of C_p is obtained for one run, discard that result and replace it with the mean of the C_p values for the preceding and following runs. The enthalpy change of the ith run ΔH_i is calculated from Eqs. (V-6a) and (V-12a). Tabulate your values of C_p and ΔH for each run.

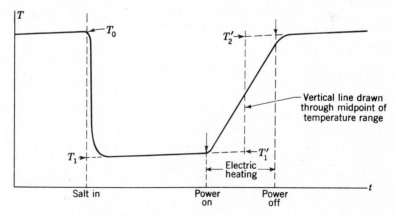

FIGURE 1
Schematic plot of temperature versus time: heat of solution.

The integral heat of solution for all the salt dissolved up to and including the nth run is given by

$$\Delta \tilde{H}_{int} = \frac{\sum\limits_{i=1}^{n} \Delta H_i}{\sum\limits_{i=1}^{n} N_i} \tag{11}$$

where N_i is the number of moles of KNO_3 added in the ith run. The molality of the solution (mol salt/1000 g water) m is calculated for each of the five solutions obtained after an addition of salt. From Eq. (11) the corresponding $\Delta \tilde{H}_{int}$ values are calculated and tabulated.

Plot $\Delta \tilde{H}_{int}$ versus $\sqrt{m}$ and extrapolate the curve to zero ionic strength to obtain $\Delta \tilde{H}_{\infty}$, the integral heat of solution at infinite dilution, taking account of the limiting behavior predicted by Eq. (9). Using Eq. (10), determine the differential heat of solution $\Delta \bar{H}_2$ at intervals of 0.05 molal over the concentration range studied.

DISCUSSION

Describe any necessary changes in the experimental equipment and/or procedure for heat of solution measurements which are exothermic.

APPARATUS

Solution calorimeter (1-qt Dewar); stirrer-heater unit complete with motor and power cord; short, wide funnel; precision calorimeter thermometer covering range from 19 to 35°C; magnifying thermometer-reader; 0 to 30°C thermometer; two split rubber stoppers; stopwatch; dc power-supply unit for heater (see Fig. 9-2); complete potentiometer setup (see Chap. XV); 15 electrical leads with lugs attached; one 500- and one 100-ml volumetric flask; five small flasks or weighing bottles to hold solid salt.

Distilled water; pure, dry KNO_3 solid (20 g); thermostat bath at 25°C (optional).

REFERENCES

1. F. H. MacDougall, "Thermodynamics and Chemistry," chap. VI, Wiley, New York (1939).
2. G. N. Lewis and M. Randall (revised by K. S. Pitzer and L. Brewer), "Thermodynamics," 2d ed., McGraw-Hill, New York (1961).
3. S. Glasstone, "Textbook of Physical Chemistry," 2d ed., p. 241, Van Nostrand, Princeton, N.J. (1946).

EXPERIMENTS

11. Partial molal volume

12. Cryoscopic determination of molecular weight

13. Freezing-point depression of strong and weak electrolytes

14. Effect of ionic strength on solubility

15. Chemical equilibrium in solution

EXPERIMENT 11. PARTIAL MOLAL VOLUME

In this experiment the partial molal volumes of sodium chloride solutions will be calculated as a function of concentration from densities measured with a pycnometer.

THEORY

Most thermodynamic variables fall into two types. Those representing *extensive* properties of a phase are proportional to the amount of the phase under consideration; they are exemplified by the thermodynamic functions V, E, H, S, A, G. Those representing *intensive* properties are independent of the amount of the phase; they include p and T. Variables of both types may be regarded as examples of homogeneous functions of degree n, that is, functions having the property

$$f(kN_1, \ldots, kN_i, \ldots) = k^n f(N_1, \ldots, N_i, \ldots) \tag{1}$$

where N_i represents for our purposes the number of moles of component i in a phase. Extensive variables are functions of degree one and intensive variables are functions of degree zero.

Among intensive variables important in thermodynamics are *partial molal quantities*, defined by the equation

$$\bar{Q}_i = \left(\frac{\partial Q}{\partial N_i} \right)_{p,T,N_{j \neq i}} \tag{2}$$

where Q may be any of the extensive quantities already mentioned. For a phase of one component, partial molal quantities are identical with so-called molal quantities, $\tilde{Q} = Q/N$. For an ideal gaseous or liquid solution, certain partial molal quantities $(\overline{V}_i, \overline{E}_i, \overline{H}_i)$ are equal to the respective molal quantities for the pure components while others $(\overline{S}_i, \overline{A}_i, \overline{G}_i)$ are not. For nonideal solutions, all partial molal quantities differ in general from the corresponding molal quantities, and the differences are frequently of interest.

A property of great usefulness possessed by partial molal quantities derives from Euler's theorem for homogeneous functions, which state that, for a homogeneous function $f(N_1, \ldots, N_i, \ldots)$ of degree n,

$$N_1 \frac{\partial f}{\partial N_1} + N_2 \frac{\partial f}{\partial N_2} + \cdots + N_i \frac{\partial f}{\partial N_i} + \cdots = nf \tag{3}$$

Applied to an extensive thermodynamic variable Q, we see that

$$N_1 \overline{Q}_1 + N_2 \overline{Q}_2 + \cdots + N_i \overline{Q}_i + \cdots = Q \tag{4}$$

Equation (4) leads to an important result. If we form the differential of Q in the usual way,

$$dQ = \frac{\partial Q}{\partial N_1} dN_1 + \cdots + \frac{\partial Q}{\partial N_i} dN_i + \cdots + \frac{\partial Q}{\partial p} dp + \frac{\partial Q}{\partial T} dT$$

and compare it with the differential derived from Eq. (4),

$$dQ = \overline{Q}_1 \, dN_1 + \cdots + \overline{Q}_i \, dN_i + \cdots + N_1 \, d\overline{Q}_1 + \cdots + N_i \, d\overline{Q}_i + \cdots$$

we obtain

$$N_1 \, d\overline{Q}_1 + \cdots + N_i \, d\overline{Q}_i + \cdots - \left(\frac{\partial Q}{\partial p}\right)_{N_i, T} dp - \left(\frac{\partial Q}{\partial T}\right)_{N_i, p} dT = 0 \tag{5}$$

For the important special case of constant pressure and temperature,

$$N_1 \, d\overline{Q}_1 + \cdots + N_i \, d\overline{Q}_i + \cdots = 0 \qquad \text{(const } p \text{ and } T) \tag{6}$$

This equation tells us that changes in partial molal quantities (resulting of necessity from changes in the N_i) are not all independent. For a binary solution one can write

$$\frac{d\overline{Q}_2}{d\overline{Q}_1} = -\frac{X_1}{X_2} \tag{7}$$

where the X_i are *mole fractions*, $X_i = N_i/\Sigma N_i$. In application to free energy this equation is commonly known as the Gibbs-Duhem equation.

We are concerned in this experiment with the partial molal volume $\overline{V}_i$, which may be thought of as the increase in the volume of an infinite amount of solution (or an amount so large that insignificant concentration change will result) when 1 mol of component i is added. This is by no means necessarily equal to the volume of 1 mol of pure i.

Partial molal volumes are of interest in part through their thermodynamic connection with other partial molal quantities such as partial molal free energy, known also as chemical potential. An important property of chemical potential is that for any given component it is equal for all phases that are in equilibrium with each other. Consider a system containing a pure solid substance (e.g., NaCl) in equilibrium with the saturated aqueous solution. The chemical potential of the solute is the same in the two phases. Imagine now that the pressure is changed isothermally. Will solute tend to go from one phase to the other, reflecting a change in solubility? For an equilibrium change at constant temperature, involving only expansion work,

$$dG = V \, dp \tag{8}$$

Differentiating with respect to N_2, the number of moles of solute, we obtain

$$d\bar{G}_2 = \bar{V}_2 \, dp \tag{9}$$

where the partial molal free energy (chemical potential) and partial molal volume appear. For the change in state

$$NaCl(s) = NaCl(aq)$$

we can write

$$d(\Delta\bar{G}_2) = \Delta\bar{V}_2 \, dp$$

or

$$\left[\frac{\partial(\Delta\bar{G}_2)}{\partial p}\right]_T = \Delta\bar{V}_2 \tag{10}$$

Thus if the partial molal volume of solute in aqueous solution is greater than the molal volume of solid solute, an increase in pressure will increase the chemical potential of solute in solution relative to that in the solid phase; solute will then leave the solution phase until a lower, equilibrium solubility is attained. Conversely, if the partial molal volume in the solution is less than that in the solid, the solubility will increase with pressure.

Partial molal volumes, and in particular their deviations from the values expected for ideal solutions, are of considerable interest in connection with the theory of solutions, especially as applied to binary mixtures of liquid components where they are related to heats of mixing and deviations from Raoult's law.

METHOD[1]

We see from Eq. (4) that the total volume of an amount of solution containing 1000 g (55.51 mol) of water and m mol of solute is given by

$$V = N_1\bar{V}_1 + N_2\bar{V}_2 = 55.51\bar{V}_1 + m\bar{V}_2 \tag{11}$$

where the subscripts 1 and 2 refer to solvent and solute, respectively. Let $\tilde{V}_1{}^0$ be the molal volume of pure water $(= 18.016/0.997044 = 18.069 \text{ cm}^3$ at $25.00°C)$. Then we define the *apparent molal volume* ϕ of the solute by the equation

$$V = N_1 \tilde{V}_1{}^0 + N_2 \phi = 55.51 \tilde{V}_1{}^0 + m\phi \tag{12}$$

which can be rearranged to give

$$\phi = \frac{1}{N_2}(V - N_1\tilde{V}_1{}^0) = \frac{1}{m}(V - 55.51\tilde{V}_1{}^0) \tag{13}$$

Now

$$V = \frac{1000 + mM_2}{d} \tag{14}$$

and

$$N_1\tilde{V}_1{}^0 = \frac{1000}{d_0} \tag{15}$$

where d is the density of the solution, d_0 is the density of pure solvent, and M_2 is the solute molecular weight. Substituting Eqs. (14) and (15) into Eq. (13), we obtain

$$\phi = \frac{1}{d}\left(M_2 - \frac{1000}{m}\frac{d - d_0}{d_0}\right) \tag{16}$$

$$= \frac{1}{d}\left(M_2 - \frac{1000}{m}\frac{W - W_0}{W_0 - W_e}\right) \tag{17}$$

In Eq. (17), the directly measured weights of the pycnometer—W_e when empty, W_0 when filled to the mark with pure water, and W when filled to the mark with solution—are used. This equation is preferable to Eq. (16) for calculation of ϕ, as it avoids the necessity of computing the densities to the high precision that would otherwise be necessary in obtaining the small difference $d - d_0$.

Now by the definition of partial molal volumes and by use of Eqs. (11) and (12),

$$\bar{V}_2 = \left(\frac{\partial V}{\partial N_2}\right)_{N_1, T, p} = \phi + N_2 \frac{\partial \phi}{\partial N_2} = \phi + m\frac{d\phi}{dm} \tag{18}$$

Also

$$\bar{V}_1 = \frac{1}{N_1}\left(N_1\tilde{V}_1{}^0 - N_2{}^2 \frac{\partial \phi}{\partial N_2}\right) = \tilde{V}_1{}^0 - \frac{m^2}{55.51}\frac{d\phi}{dm} \tag{19}$$

We might proceed by plotting ϕ versus m, drawing a smooth curve through the points, and constructing tangents to the curve at the desired concentrations in order to measure the slopes. However, for solutions of simple electrolytes, it has been found that many apparent molal quantities such as ϕ vary linearly with $\sqrt{m}$, even up to moderate concentrations.[2] This behavior is in agreement with the prediction of the Debye-Hückel theory for dilute solutions.[3] Since

$$\frac{d\phi}{dm} = \frac{d\phi}{d\sqrt{m}}\frac{d\sqrt{m}}{dm} = \frac{1}{2\sqrt{m}}\frac{d\phi}{d\sqrt{m}} \tag{20}$$

we obtain from Eqs. (18) and (19)

$$\bar{V}_2 = \phi + \frac{m}{2\sqrt{m}} \frac{d\phi}{d\sqrt{m}} = \phi + \frac{\sqrt{m}}{2} \frac{d\phi}{d\sqrt{m}} = \phi^0 + \frac{3\sqrt{m}}{2} \frac{d\phi}{d\sqrt{m}} \tag{21}$$

$$\bar{V}_1 = \bar{V}_1{}^0 - \frac{m}{55.51} \left(\frac{\sqrt{m}}{2} \frac{d\phi}{d\sqrt{m}} \right) \tag{22}$$

where ϕ^0 is the apparent molal volume extrapolated to zero concentration. Now one can plot ϕ versus $\sqrt{m}$ and draw the best *straight* line through the points. Unless the experimental data are unusually good, deviations of experimental points from the best straight line are not significant. From the slope $d\phi/d\sqrt{m}$ and the value of ϕ^0, both $\bar{V}_1$ and $\bar{V}_2$ can be obtained.

EXPERIMENTAL

Make up 200 ml of approximately 3.2 m (3.0 M) NaCl in water. Weigh the salt accurately and use a volumetric flask; then pour the solution into a dry flask. If possible, prepare this solution in advance (since the salt dissolves slowly). Solutions of $\frac{1}{2}$, $\frac{1}{4}$, $\frac{1}{8}$, and $\frac{1}{16}$ of the initial molarity are to be prepared by successive volumetric dilutions; for each dilution pipette 100 ml of solution into a 200-ml volumetric flask and make up to the mark with distilled water.

The pycnometer is rinsed with distilled water and thoroughly dried before each use. Use an aspirator, and rinse and dry by suction; use a few rinses of acetone to expedite drying. The procedure given here is for the Ostwald-Sprengel type pycnometer; the less accurate but more convenient stopper type can be used with a few obvious changes in procedure.† To fill, dip one arm of the pycnometer into the vessel containing the solution (preferably at a temperature *below* 25°C) and apply suction by mouth with a piece of rubber tubing attached to the other arm. Hang the pycnometer in the thermostat bath (25.0°C) with the main body below the surface but with the

---

† The filling of a Weld-type pycnometer must be carried out with great care. The temperature of the laboratory must be below the temperature at which the determination is to be made. Fill the pycnometer body with the liquid, and seat the capillary stopper firmly. Wipe off excess liquid around the tapered joint, cap the pycnometer, and immerse it in the thermostat bath to a level just below the cap. When temperature equilibrium has been reached, remove the cap, wipe off the excess liquid from the capillary tip, being careful not to draw liquid out of the capillary, and remove the pycnometer from the bath. As the pycnometer cools to room temperature, the liquid column in the capillary will descend; be careful not to heat the pycnometer with your hands since this will force liquid out of the capillary. Carefully clean and wipe off the whole pycnometer, including the cap, but not including the tip of the stopper. Cap the pycnometer, place it in the balance, and allow it to stand about 10 minutes before weighing.

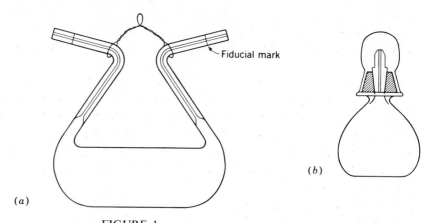

FIGURE 1
Pycnometers: (a) Ostwald-Sprengel type; (b) Weld (stopper) type.

arms emerging well above. Allow at least 15 min for equilibration. While the pycnometer is still in the bath, adjust menisci to fiducial marks with the aid of a piece of filter paper. Remove the pycnometer from the bath and quickly but thoroughly dry the outside surface with a towel and filter paper. Weigh the pycnometer, hanging it from the hook above the balance pan.

The pycnometer should be weighed empty and dry (W_e), and also with distilled water in it (W_0), as well as with each of the solutions in it (W). It is advisable to repeat W_e and W_0 as a check, inasmuch as the results of all runs depend upon them.

All weighings are to be done on an analytical balance to the highest possible precision. Record the values of all weights, then apply any weight calibration corrections. It is unnecessary to correct the weights to vacuum readings in this experiment, although for very precise work this must be done.

CALCULATIONS

The success of this experiment depends greatly upon the care with which the computations are carried out. For this reason a simple computer program for analyzing the data is strongly recommended. This computer program may either be written by the student (see Chap. XXI) or made available by the instructor. If computer facilities are unavailable, the calculations should be carried out on a small electronic calculator (or with five-place logarithm tables if necessary). All calculations should be carefully checked. If the work is done by two or more students, the partners are encouraged to work together, performing the same calculations independently and checking results with each other after each step.

Although to a great extent the weights themselves enter into the calculations, it is necessary also to have the density d of every solution to at least one part per thousand accuracy:

$$d = \frac{W_{soln}}{V} = \frac{W - W_e}{V_p} \tag{23}$$

The volume of the pycnometer V_p is obtained by use of the density of water at 25°C, d_0 (with the value 0.997044 g cm^{-3}), and $W_0 - W_e$.

The molalities m which are needed for the calculations can be obtained from the molarities M obtained from the volumetric procedures by use of the equation

$$m = \frac{1}{1 - [(M)/d](M_2/1000)} \frac{(M)}{d} = \frac{1}{[d/(M)] - (M_2/1000)} \tag{24}$$

where M_2 is the solute molecular weight (58.45) and d is the experimental density.

Calculate ϕ for each solution using Eq. (17) and plot ϕ versus $\sqrt{m}$. Draw the best straight line and obtain $d\phi/d\sqrt{m}$ from the slope. Record the value of ϕ^0, the intercept at m equal zero.

Calculate $\bar{V}_2$ and $\bar{V}_1$ for $m = 0, 0.5, 1.0, 1.5, 2.0,$ and 2.5. Plot them against m and draw a smooth curve for each of the two quantities.

In your report, present the curves (ϕ versus $\sqrt{m}$, $\bar{V}_2$ and $\bar{V}_1$ versus m) mentioned above. Present also in tabular form the quantities d, M, m, $(1000/m)(W - W_0)/(W_0 - W_e)$, and ϕ for each solution studied. Give the values obtained for the pycnometer volume V_p and for ϕ^0 and $d\phi/d\sqrt{m}$.

DISCUSSION

The density of NaCl(s) is 2.165 g cm^{-3} at 25°C. How will the solubility of NaCl in water be affected by an increase in pressure?

Discuss qualitatively whether the curves of $\bar{V}_1$ and $\bar{V}_2$ versus m behave in accord with Eq. (7).

APPARATUS

Pycnometer (approximately 70 ml) with wire loop for hanging in bath; large weighing bottle; short-stem funnel; 200-ml volumetric flask; 250-ml Erlenmeyer flask; one 250- and one 100-ml beaker; gum-rubber tube (1 to 2 ft long); one piece of filter paper; 100-ml pipette; spatula.

Constant-temperature bath set at 25°C; bath hardware for holding flasks and pycnom-eter; reagent-grade sodium chloride (30 to 40 g, or 200 ml of solution of an accurately known concentration); acetone to be used for rinsing; cleaning solution.

REFERENCES

1. F. T. Gucker, Jr., *J. Phys. Chem.*, **38,** 307 (1934).
2. D. O. Masson, *Phil. Mag.*, **8,** 218 (1929).
3. O. Redlich and P. Rosenfeld, *Z. Phys. Chem.*, **A155,** 65 (1931).

GENERAL READING

G. N. Lewis and M. Randall (revised by K. S. Pitzer and L. Brewer), "Thermodynamics," 2d ed., chap. 17, McGraw-Hill, New York (1961).

F. H. MacDougall, "Thermodynamics and Chemistry," 3d ed., chap. III, Wiley, New York (1939).

N. Bauer and S. Z. Lewin, Determination of Density, in A. Weissberger (ed.), "Technique of Organic Chemistry," 3d ed., vol. I, part I, chap. IV, Interscience, New York (1959).

EXPERIMENT 12.   CRYOSCOPIC DETERMINATION OF MOLECULAR WEIGHT

When a substance is dissolved in a given liquid solvent, the freezing point of the solvent is nearly always lowered. This phenomenon constitutes a so-called "colligative" property of the substance—a property that depends in its magnitude primarily on the number of moles of the substance that are present in relation to a given amount of solvent. In the present case, the amounts by which the freezing point is lowered, called the *freezing-point depression* ΔT_f, is approximately proportional to the number of moles of solute dissolved in a given amount of the solvent. Other colligative properties are vapor-pressure lowering, boiling-point elevation, and osmotic pressure.

THEORY

In Fig. 1, T_0 and p_0 are, respectively, the temperature and vapor pressure at which the pure solvent freezes. Strictly speaking, T_0 is the temperature at the "triple point" with a total pressure equal to p_0 rather than the freezing point at 1 atm. However, the pressure effect is small—of the order of a few thousandths of a degree—and moreover, as it is essentially the same for a dilute solution as for the pure solvent, it

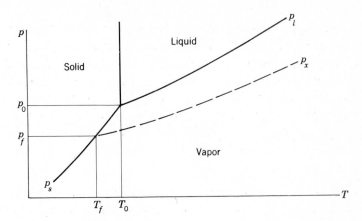

FIGURE 1

Pressure-temperature diagram for solvent and solution.

will cancel out in the calculation of ΔT_f. The curve labeled p_l is the vapor pressure of the pure liquid solvent as a function of temperature, and that labeled p_s is the vapor pressure ("sublimation pressure") of the pure solid solvent. The dashed curve labeled p_x is the vapor pressure of a solution containing solute in mole fraction X. We now introduce Assumption 1: *The solid in equilibrium with a solution at its freezing point is essentially pure solid solvent.* The validity of this assumption results primarily from the crystalline structure of the solid solvent, in which the molecules are packed in a regular manner closely dependent upon their shape and size. The substitution of one or more solute molecules for solvent molecules, to yield a solid solution, is usually accompanied by a very large increase in free energy, for the solute molecule usually has a size and shape poorly adapted to filling suitably the cavity produced by removing a solvent molecule. Exceptions occur when solute and solvent molecules are very similar, as in the toluene-chlorobenzene, benzene-pyridine, and other similar systems, and in some inorganic salt mixtures and many metal alloys.

By the second law of thermodynamics, when a solution is in equilibrium with solid solvent, the solvent vapor pressures of the two must be equal. Therefore for a solution containing solute at mole fraction X, the freezing point T_f and the corresponding equilibrium vapor pressure p_f are determined by the intersection of the sublimation-pressure curve p_s with the solution vapor-pressure curve p_x. To determine the point of intersection we write the equations for the two curves and solve them simultaneously for p and T.

The equation of the sublimation curve is conveniently obtained from the integrated Clausius-Clapeyron equation:

$$\ln \frac{p_s}{p_0} = \frac{\Delta \tilde{H}_s}{R} \left(\frac{1}{T_0} - \frac{1}{T} \right) \tag{1}$$

where $\Delta \tilde{H}_s$ is the molar heat of sublimation of the solid solvent. In order to use this equation we must introduce Assumption 2: *All assumptions required in the derivation of the integrated Clausius-Clapeyron equation may be here assumed.* These are that (a) the vapor obeys the perfect-gas law, (b) the volume of a condensed phase is negligible in comparison with that of an equivalent amount of vapor, and (c) the enthalpy change accompanying vaporization or sublimation is independent of temperature. Part (c) of this assumption may be dispensed with, as we shall see later. The vapor-pressure curve p_x for the solution may be related to that for the liquid p_l, the equation for which we may take in accordance with the above assumption as

$$\ln \frac{p_l}{p_0} = \frac{\Delta \tilde{H}_v}{R} \left(\frac{1}{T_0} - \frac{1}{T} \right) \tag{2}$$

where $\Delta \tilde{H}_v$ is the molar heat of vaporization of the liquid solvent. To obtain the equation for the p_x curve we introduce Assumption 3: *Raoult's law applies.* Raoult's law is applicable to ideal solutions and, in the limit of infinite dilution, to real solutions. It is usually a fairly good approximation for moderately dilute real solutions. According to this law,

$$\frac{p_x}{p_l} = X_0 = 1 - X \tag{3}$$

where X_0 is the mole fraction of solvent and X is the mole fraction of solute in the solution. Combining Eqs. (2) and (3) we obtain, as the equation for p_x as a function of T,

$$\ln \frac{p_x}{p_0} = \ln \frac{p_x}{p_l} + \ln \frac{p_l}{p_0} = \ln (1 - X) + \frac{\Delta \tilde{H}_v}{R} \left(\frac{1}{T_0} - \frac{1}{T} \right) \tag{4}$$

Now we are ready to solve for the coordinates of the intersection point p_f, T_f. Setting p_x equal to p_s, we obtain from Eqs. (1) and (4)

$$\ln (1 - X) = \frac{\Delta \tilde{H}_s - \Delta \tilde{H}_v}{R} \left(\frac{1}{T_0} - \frac{1}{T_f} \right)$$

or

$$\ln (1 - X) = - \frac{\Delta \tilde{H}_f}{R T_0 T_f} \Delta T_f \tag{5}$$

where

$$\Delta T_f \equiv T_0 - T_f \tag{6}$$

and $\Delta \tilde{H}_f$ is the molar heat of fusion of the liquid solvent.

It is worth noting here that Eq. (5) may be derived more elegantly on a somewhat different set of assumptions, in which those pertaining to properties of the solvent vapor do not appear. Those assumptions, together with Raoult's law, may be replaced by the single assumption that the *activity of the solvent is directly proportional to the mole fraction of the solvent in the solution.*

We wish to derive expressions for calculating the molality m of the solution (or the molecular weight M of the solute) from a measured freezing-point depression ΔT_f. The molality of the solution is given by

$$m = \frac{1000}{M_0} \frac{X}{1 - X} \tag{7}$$

where M_0 is the molecular weight of the solvent. If g is the weight of solute and G is the weight of solvent,

$$\frac{X}{X_0} = \frac{g/M}{G/M_0}$$

thus
$$M = M_0 \frac{g}{G} \frac{X_0}{X} = M_0 \frac{g}{G} \frac{1 - X}{X}$$

Combining this with Eq. (7) we obtain

$$M = \frac{g}{G} \frac{1000}{m} \tag{8}$$

The expressions commonly used for determination of molalities or molecular weights from freezing-point depressions are derived with the following approximations:

$$\ln (1 - X) \cong -X$$
$$T_0 T_f \cong T_0{}^2 \tag{9}$$
$$1 - X = X_0 \cong 1$$

With the aid of these approximations, Eq. (5) becomes

$$X = \frac{\Delta \tilde{H}_f}{R T_0{}^2} \Delta T_f \tag{10}$$

Combining Eq. (10) with Eqs. (7) and (8) we obtain

$$m = \frac{1000 \, \Delta \tilde{H}_f}{M_0 R T_0{}^2} \Delta T_f = \frac{\Delta T_f}{K_f} \tag{11}$$

and
$$M = \frac{M_0 g R T_0{}^2}{G \, \Delta \tilde{H}_f \, \Delta T_f} = 1000 \frac{g}{G} \frac{K_f}{\Delta T_f} \tag{12}$$

where
$$K_f \equiv \frac{M_0 R T_0{}^2}{1000 \, \Delta \tilde{H}_f} \tag{13}$$

K_f is called the *molal freezing-point depression constant*, since it is equal to the freezing-point depression predicted by Eq. (11) for a 1-molal solution.

It will often be the case that no further refinement of these expressions is justified. In some cases, however, the use of a higher order of approximation is worthwhile. In the treatment that follows, we shall retain terms representing no more than two orders in ΔT.

Let the right-hand side of Eq. (5) be called y,

$$y \equiv -\frac{\Delta \tilde{H}_f}{RT_0 T_f} \Delta T_f$$

and express e^y in a Maclaurin series, so that Eq. (5) becomes

$$1 - X = e^y = 1 + y + \frac{y^2}{2!} + \cdots$$

Since y is small in comparison with unity,

$$X \cong -y\left(1 + \frac{y}{2}\right) \cong -\frac{y}{1 - y/2} \tag{14}$$

The other approximations of Eq. (9) are replaced by

$$T_0 T_f = T_0^2 \frac{T_f}{T_0} = T_0^2\left(1 - \frac{\Delta T_f}{T_0}\right) \cong \frac{T_0^2}{1 + \Delta T_f/T_0} \tag{15}$$

and

$$1 - X = X_0 = 1 + y \cong \frac{1}{1 - y} \tag{16}$$

In addition, if it is not desired to retain Assumption 2c that $\Delta \tilde{H}_f$ is strictly constant, we can replace it in the expression for y by its mean value over the temperature range ΔT_f:

$$\overline{\Delta \tilde{H}_f} = \Delta \tilde{H}_f{}^0 - \Delta \tilde{C}_p \frac{\Delta T_f}{2} = \Delta \tilde{H}_f{}^0\left(1 - \frac{\Delta \tilde{C}_p}{2 \Delta \tilde{H}_f{}^0} \Delta T_f\right) \tag{17}$$

where $\Delta \tilde{H}_f{}^0$ is the molar heat of fusion of the pure solvent at its freezing point.

In combining the above equations we follow the usual rules for multiplying and dividing binominals and reject terms of second and higher order in ΔT_f in comparison with unity. We obtain

$$\frac{X}{1 - X} = \frac{\Delta \tilde{H}_f{}^0}{RT_0^2} \Delta T_f(1 + k_f \Delta T_f) \tag{18}$$

where

$$k_f = \frac{1}{T_0} + \frac{\Delta \tilde{H}_f{}^0}{2RT_0^2} - \frac{\Delta \tilde{C}_p}{2 \Delta \tilde{H}_f{}^0} \tag{19}$$

We further obtain, by use of Eqs. (7) and (8),

$$m = \frac{\Delta T_f}{K_f}(1 + k_f \Delta T_f) \tag{20}$$

$$M = 1000 \frac{g}{G} \frac{K_f}{\Delta T_f}(1 - k_f \Delta T_f) \tag{21}$$

Table 1

		Cyclohexane	Water
Molecular weight (g)	M_0	84.16	18.02
Celsius freezing point (°C)	t_0	6.68	0.00
Absolute freezing point (°K)	T_0	279.8	273.2
Molar heat of fusion at T_0 (cal mol^{-1})	$\Delta \tilde{H}_f{}^0$	640	1436
Molar heat-capacity change on fusion (cal deg^{-1} mol^{-1})	$\Delta \tilde{C}_p$	3.6	9.1
Molal freezing-point depression constant (deg molal^{-1})	K_f	20.4	1.855
Correction constant (deg^{-1})	k_f	0.003	0.005

In Table 1 are given values of the pertinent constants for two common solvents: cyclohexane and water. It will be seen that for freezing-point depressions of 1 or 2°, omission of the correction term $k_f \Delta T_f$ leads to errors of the order of 1 percent in m or M.

Where Raoult's law fails, we may expect in the case of a nondissociating and nonassociating solute that the experimental conditions will at least lie in a concentration range over which the deviation in vapor pressure can be expressed fairly well by a quadratic term in the solute mole fraction X. Within this range the main effect on the above equations will be to change k_f by a small constant amount. Thus we may expect that a plot of M [calculated either with Eq. (12) or with Eq. (21)] against ΔT_f, with a number of experimental points obtained at different concentrations, should yield an approximately straight line which, on extrapolation to $\Delta T_f = 0$, should give a good value for M. In the event of failure of the assumption of no solid solution, however, the limiting value of M itself should be expected to be in error.†

METHOD

In this experiment the freezing point of a solution containing a known weight of an "unknown" solute in a known weight of cyclohexane is determined from cooling curves. From the result at each of two concentrations, the molecular weight of the unknown is determined.

The apparatus is shown in Fig. 2. The inner test tube, containing the solution, stirrer, and thermometer, is partially insulated from a surrounding ice-salt cooling

---

† This can easily be shown by a treatment parallel to the derivation here given, taking account of the facts that the equilibrium concentration of solute in the solid phase increases (and that of solvent decreases) with increasing solute concentration in the liquid phase and that the solvent vapor pressure of the solid decreases as the solvent concentration in the solid decreases. Thus for the solid a new vapor-pressure curve should be drawn below p_s in Fig. 1, and its intersection point with p_x is to the right of that shown.

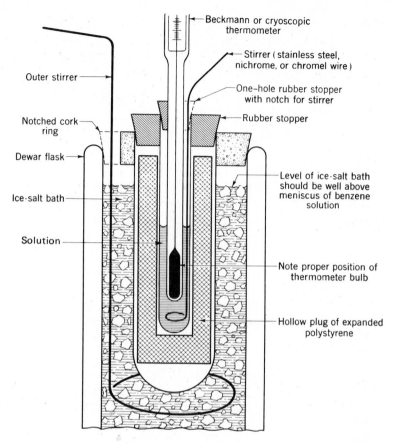

Beckmann or cryoscopic thermometer

Stirrer (stainless steel, nichrome, or chromel wire)

One–hole rubber stopper with notch for stirrer

Rubber stopper

Level of ice·salt bath should be well above meniscus of benzene solution

Note proper position of thermometer bulb

Hollow plug of expanded polystyrene

Outer stirrer

Notched cork ring

Dewar flask

Ice·salt bath

Solution

FIGURE 2
Apparatus for cryoscopic determination of molecular weight.

bath through being suspended in a larger test tube with an air space in between. To provide additional insulation the space between may be filled by a hollow plug of expanded polystyrene foam. The thermometer is either a Beckmann thermometer with a 5 or 6° range adjusted to cover the temperature range expected in the experiment or a special cryoscopic thermometer of the appropriate range. The thermometer should be graduated to 0.02 or 0.01°C.

Under the conditions of the experiment, heat flows from the inner system, at temperature T, to the ice-salt bath, at temperature T_b, at a rate which is approximately proportional to the temperature difference:

$$-\frac{dH}{dt} = A(T - T_b) \tag{22}$$

where H is the enthalpy of the inner system and A is a constant incorporating shape factors and thermal-conductivity coefficients. If $(T - T_b)$ is sufficiently large in comparison with the temperature range covered in the experiment, we can write

$$-\frac{dH}{dt} \cong \text{const} \tag{23}$$

In the absence of a phase change, the rate of change of the temperature is given by

$$-\frac{dT}{dt} = \frac{1}{C}\left(-\frac{dH}{dt}\right) \tag{24}$$

where C is the heat capacity of the inner system. When a pure liquid freezes, dT/dt vanishes as long as two phases are present and we have a "thermal arrest." When pure solid solvent separates from a liquid solution on freezing, the temperature does not remain constant because the solution becomes continually more and more concentrated and the freezing point T_f correspondingly decreases. It can be shown that in this case

$$-\frac{dT}{dt} = \frac{1}{(N_0\,\Delta\tilde{H}_f/\Delta T_f) + C}\left(-\frac{dH}{dt}\right) \tag{25}$$

where N_0 is the number of moles of solvent present in the liquid phase. Thus, when solid solvent begins to freeze out of solution on cooling, the slope changes discontinuously from that given by Eq. (24) to the much smaller slope given by Eq. (25), and we have what is called a "break" in the cooling curve.

Figure 3 shows schematically the types of cooling curves that are expected as a result of these considerations. It will be noted that the solutions may "supercool" before solidification of solvent takes place. In the present experiment, supercooling rarely exceeds about 2° and is best kept below 1° by seeding—introducing a small crystal of frozen solvent.

When supercooling occurs, the recommended procedure for estimating the true freezing point of the solution (the temperature at which freezing would have started in the absence of supercooling) is to extrapolate back that part of the curve that corresponds to freezing out of the solvent until the extrapolate intersects the cooling curve of the liquid solution. To a good approximation the extrapolation may be taken as a linear one.†

† To justify this extrapolation procedure, consider two experiments starting at the same temperature and the same time under conditions identical in all respects except that in one case supercooling is allowed to take place and in the other it is somehow prevented. By Eq. (23) the enthalpies of the two systems remain identical throughout the experiment. Except during the interval of supercooling, both systems are in equilibrium, and therefore when their enthalpies are equal, they are identical in all respects, including temperature. Therefore, except for the interval of supercooling, the two curves when superimposed are congruent throughout their entire length. The dashed line representing the extrapolation in Fig. 3c or d is therefore part of the curve for the hypothetical experiment in which supercooling was prevented.

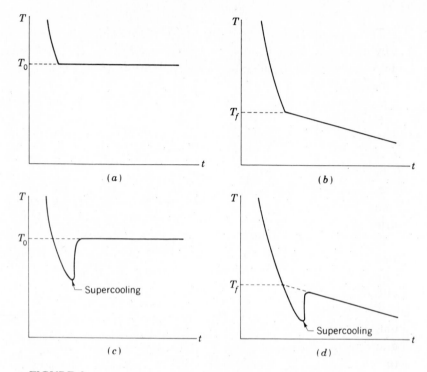

FIGURE 3

Schematic cooling curves: (a) and (c) show the cooling curves for pure solvent; (b) and (d) show the cooling curves for a solution.

This procedure is valid only when the extent of supercooling is small in comparison with the difference in temperature between the system and the ice-salt cooling bath. When this is not the case, Eq. (22) must be used in place of Eq. (23), and it is then seen that the rate of decrease of enthalpy is dependent in significant degree on whether or not supercooling takes place. Therefore the amount of supercooling must be kept small, and the bath must be kept as cold as possible.

EXPERIMENTAL

The success of this experiment depends primarily on a careful experimental technique. The solvent to be used is cyclohexane, which must be of reagent-grade purity and scrupulously dry. Pour about 25 ml of cyclohexane into a clean, dry, glass-stoppered flask for your own use. During the experiment, take care to keep the inner part of the apparatus dry and to expose the inner test tube to the air as little as possible in order to avoid condensation of moisture inside.

The ice-salt mixture should be made up freshly for each run. Mix about one part by volume of coarse rock salt with about four parts by volume of finely crushed ice in a beaker or battery jar. Allow it to stand a few minutes to become slushy, and then pour the desired amount into the Dewar flask. (This ice-salt mixture can be replaced by a mixture of 10 parts ice with 1 part denatured alcohol.)

Fill a small beaker with crushed ice and a little water, and place in it a test tube containing 1 or 2 ml of cyclohexane and a very thin glass rod. Stopper the tube with cotton or glass wool to keep out moist air from the room. This frozen cyclohexane is to be used for seeding. When it is desired to seed the system during a run, remove the glass rod, making sure it carries a small amount of frozen cyclohexane, and carefully insert it into the solution with the least possible disruption of the experiment.

Cyclohexane should be introduced into the inner test tube from a weighing bottle (weighed before and after delivery) or from a pipette. *Do not pipette cyclohexane by mouth*; use a rubber bulb. The quantity normally required is 15 ml. If a pipette is used, record the ambient temperature.

The thermometer and stirrer should be inserted with the thermometer carefully mounted in such a way that the bulb is about halfway between the bottom of the test tube and the upper surface of the liquid and concentric with the tube so that the stirrer can easily pass around it. If the thermometer is too close to the bottom, a bridge of frozen solvent can easily form which will conduct heat away from the thermometer and result in low readings.

Much depends on the technique of stirring the solution. The motion of the stirrer should carry it from the bottom of the tube up to near the surface of the liquid; before assembling the apparatus, it is well to observe carefully how high the stirrer can be raised without too frequent splashing. The stirring should be continuous throughout the run and should be at the rate of about 1 stroke per sec. The outer bath should be stirred a few times per minute.

The test tube containing cyclohexane, stopper, thermometer, and stirrer is held in the beaker of ice water, and the cyclohexane is stirred until it visibly starts to freeze. The outside of the tube is wiped dry, and the tube is allowed to warm up at least 1° above the freezing point. Then the tube is placed in the assembled apparatus and temperature readings are taken every 30 sec. Tap the thermometer gently before each reading, and estimate the temperature readings to tenths of the smallest scale division. If the temperature falls 0.5°C below the normal freezing point without evidence of freezing, seed the liquid. Once freezing has occurred, continue temperature readings for about 5 min. If the cyclohexane and apparatus are suitably dry, the temperature should remain constant or fall by no more than about 0.01 or 0.02°C during that time.

The inner test tube is removed and warmed (with stirring) in a beaker of water until the cyclohexane has completely melted. The stopper is lifted, and an *accurately weighed* pellet of the unknown compound, of about 0.6 g, is introduced. The pellet is dissolved by stirring, the inner tube is cooled with stirring in an ice bath to about the freezing temperature of pure cyclohexane and then dried on the outside, and the apparatus is reassembled. Temperature readings are taken every 30 sec throughout the run. After a depression of about 2° has been reached, it is advisable to seed the system at intervals of about 0.5° until freezing starts. After freezing has begun, temperature readings should be continued for a period at least four times as long as the estimated period of supercooling to provide data for an adequate extrapolation.

The inner tube is removed, the cyclohexane melted as before, and a second weighed pellet of the unknown is added and stirred into solution. Before this pellet is added, it is well to make a rough calculation, based on the results of the first run, to make sure that the temperatures in the second run will remain on scale and, if necessary, to modify the weight of the pellet accordingly. The second run is carried out as before, after cooling in an ice bath to about the temperature at which freezing was obtained in the first run.

If the results from these two runs are not in agreement and time allows, a repeat of both runs should be made using somewhat different pellet weights.

CALCULATIONS

If the cyclohexane was introduced with a pipette, its weight G can be calculated using the density:

$$\rho(\text{cyclohexane}) = 0.779 - 9.4 \times 10^{-4}(t - 20)$$

where t is the Celsius temperature. Plot the cooling-curve data for pure cyclohexane and for each solution studied. If supercooling took place, perform the extrapolations as shown in Fig. 3. Report the weight g of solute and the freezing-point depression ΔT_f for each run.

Calculate the molecular weight from both Eqs. (12) and (21) using the appropriate constants in Table 1. An extrapolated value of M can be obtained from the calculated values from either equation by plotting the calculated molecular weight against the depression ΔT_f. If the elementary analysis or empirical formula of the unknown is given, deduce the molecular formula and the exact molecular weight.

APPARATUS

Dewar flask; notched cork ring to fit top of Dewar; large test tube with polystyrene-foam insert; large ring stirrer; inner test tube which fits into polystyrene insert; large rubber stopper with hole for inner test tube; medium stopper with hole for thermometer and a notch for small ring stirrer; 1-qt battery jar or 1000-ml beaker; 250-ml beaker; 125-ml glass-stoppered flask for storing benzene; small test tube; glass rod (3 mm diameter and 20 cm long); 15-ml pipette; small rubber pipetting bulb; precision cryoscopic thermometer; magnifying thermometer reader; stopwatch.

Dry reagent-grade cyclohexane (50 ml); glass wool; naphthalene or an unknown solid; pellet press; acetone for rinsing; ice (3 lb); ice grinder or shaver; coarse rock salt (2 lb).

GENERAL READING

E. L. Skau, J. C. Arthur, Jr., and H. Wakeham, Determination of Melting and Freezing Temperatures, in A. Weissberger (ed.), "Technique of Organic Chemistry," 3d ed., vol. I, part I, chap. VII, Interscience, New York (1959).

F. H. MacDougall, "Thermodynamics and Chemistry," 3d ed., Wiley, New York (1939).

EXPERIMENT 13.   FREEZING-POINT DEPRESSION OF STRONG AND WEAK ELECTROLYTES

In this experiment, the freezing-point depression of aqueous solutions is used to determine the degree of dissociation of a weak electrolyte and to study the deviation from ideal behavior which occurs with a strong electrolyte.

THEORY

Use will be made of the theory developed in Exp. 12 for the freezing-point depression ΔT_f of a given solvent containing a known amount of an ideal solute; this material should be reviewed.

In the case of a dissociating (or associating) solute, the molality given by Eq. (12-11) or (12-20) is ideally the *total* effective molality—the number of moles of all solute species present, whether ionic or molecular, per 1000 g of solvent. As we shall see, ionic solute species at moderate concentrations do not form ideal solutions and, therefore, do not obey these equations. However, for a weak electrolyte the ionic concentration is often sufficiently low to permit treatment of the solution as ideal.

Weak electrolytes As an example, let us discuss a weak acid HA with nominal molality m. Due to the dissociation

$$HA = H^+ + A^-$$

the equilibrium concentrations of HA, H^+, and A^- will be $m(1 - \alpha)$, $m\alpha$, and $m\alpha$, respectively, where α is the fraction dissociated. The total molality m' of all solute species is

$$m' = m(1 + \alpha) \tag{1}$$

This molality m' can be calculated from the observed ΔT_f using Eq. (12-20). Thus, freezing-point measurements on weak electrolyte solutions of known molality m enable the determination of α.

The equilibrium constant in terms of concentrations can be calculated from

$$K_c = \frac{(H^+)(A^-)}{(HA)} = m \frac{\alpha^2}{(1 - \alpha)} \tag{2}$$

In this experiment α and K_c are to be determined at two different molalities. Since these will be obtained at two different temperatures, the values of K_c should be expected to differ slightly.

Strong electrolytes These are now generally believed to be completely dissociated in ordinary dilute solutions. However, their colligative properties when interpreted in terms of ideal solutions appear to indicate that the dissociation is a little less than complete. This fact led Arrhenius to postulate that the dissociation of strong electrolytes is indeed incomplete. Subsequently, this deviation in colligative behavior has been demonstrated to be an expected consequence of interionic attractions.

For a nonideal solution, Eq. (12-5) is replaced by

$$\ln a_0 = -\frac{\Delta \tilde{H}_f}{RT_0 T_f} \Delta T_f \cong -\frac{\Delta \tilde{H}_f}{RT_0^2} \Delta T_f \tag{3}$$

where a_0 is the *activity* of the solvent and is related to the mole fraction of solvent X_0 by

$$a_0 = \gamma_0 X_0 \tag{4}$$

The quantity γ_0 is the activity coefficient for the solvent and in electrolytic solutions differs from unity even at moderately low concentrations. Let us write $\ln a_0$ as $(\ln \gamma_0 + \ln X_0)$ in Eq. (3) and divide both sides by $\ln X_0$ to obtain

$$-\frac{\Delta \tilde{H}_f}{RT_0^2} \frac{\Delta T_f}{\ln X_0} = 1 + \frac{\ln \gamma_0}{\ln X_0} \equiv g \tag{5}$$

where g is called the *osmotic coefficient* of the solvent. Now

$$X_0 = \frac{N_0}{N_0 + \nu N_1} \tag{6}$$

where N_0 is the number of moles of solvent and vN_1 is the total number of moles of ions formed from N_1 moles of solute (for example, $v = 2$ for HCl). Thus

$$-\ln X_0 \equiv \ln\left(1 + \frac{vN_1}{N_0}\right) \cong \frac{vN_1}{N_0} \tag{7}$$

for dilute solutions. Substituting this expression for $\ln X_0$ into Eq. (5), we have

$$\frac{\Delta \tilde{H}_f}{RT_0{}^2} \frac{N_0}{vN_1} \Delta T_f = g \tag{8}$$

For a solution of molality m in a solvent with molecule weight M_0, we can replace (N_0/vN_1) by $(1000/M_0 vm)$. Equation (8) can then be written as

$$\left(\frac{\Delta \tilde{H}_f}{RT_0{}^2} \frac{1000}{M_0}\right) \frac{\Delta T_f}{vm} = \frac{\Delta T_f}{vmK_f} = g \tag{9}$$

where K_f is the molal freezing-point depression constant defined by Eq. (12-13).

For an ideal solution, $\gamma_0 = 1$ and g is unity. Then Eq. (9) is identical with Eq. (12-11), since the total molality of all solute species is vm for a completely dissociated solute of molality m. For ionic solutions, the Debye-Hückel theory predicts a value of γ_0 different from unity and therefore a deviation of g from unity. A treatment of this aspect of the Debye-Hückel theory is beyond the scope of this book, and we shall merely state the result. The osmotic coefficient g at 0°C for dilute solutions of a single strong electrolyte in water is given[1] by

$$g = 1 - 0.376\sigma|z_+z_-|\mu^{1/2} \tag{10}$$

where z_+ is the valence of the positive ion, z_- is the valence of the negative ion, and μ is the ionic strength:

$$\mu = \tfrac{1}{2} \sum_i m_i z_i{}^2 \tag{11}$$

the sum being taken over all ionic species present. The quantity σ is a function of κa where, for aqueous solutions at 0°C, κ is given by MacDougall[1] as

$$\kappa = 0.324 \times 10^8 \mu^{1/2}$$

and a is the effective ionic diameter. For a small ion, a is approximately 3×10^{-8} cm and we can take $\kappa a \simeq \mu^{1/2}$. For a uni-univalent electrolyte (such as HCl) this becomes simply

$$\kappa a \cong m^{1/2} \tag{12}$$

Therefore Eq. (10) can be written

$$g = 1 - 0.38\sigma m^{1/2} \tag{13}$$

Values of the function σ are given for several values of κa in Table 1. For a known value of m, κa for a uni-univalent electrolyte is given by Eq. (12) and one can find the appropriate value of σ by interpolating in the table. Use of this value in Eq. (13) allows one to calculate g. It should be emphasized that Eq. (13) is an approximation based on the Debye-Hückel theory and valid only in dilute solution.

EXPERIMENTAL

The apparatus used in this experiment is shown in Fig. 1. The thermometer is either a Beckmann thermometer adjusted to cover the temperature range expected in the experiment or a special cryoscopic thermometer of the appropriate range. The thermometer should be graduated to 0.01 or 0.02°C. In the experiment, an aqueous solution of a weak or strong acid is mixed with crushed ice until equilibrium is attained. The temperature is recorded, and two or more aliquots of the liquid phase are withdrawn for titration to determine the equilibrium nominal concentration m_0. The ice to be used should preferably be distilled-water ice.

The most difficult part of the experimental technique is the achievement of thorough mixing. This difficulty is aggravated by the fact that water has a maximum density near 4°C and solution at that temperature tends to settle to the bottom of the Dewar flask while colder solution tends to float near the top with the ice. The stirring must therefore be *vigorous and prolonged.* A good technique is to work the stirrer frequently above the mass of ice and with a vigorous downward thrust propel the ice all the way to the bottom of the flask. *It must not be assumed that equilibrium has been obtained until the temperature shown by the thermometer has become quite stationary and does not change when the stirring is stopped or when the manner or vigor of stirring is changed.*

Procedure   The solutions to be studied are the strong electrolyte HCl and weak electrolyte monochloroacetic acid, each at two concentrations—roughly 0.25 and 0.125 m. About 150 ml of each solution will be needed, and the 0.125 m solutions required should be prepared by diluting the 0.25 m stock solutions with distilled water. Place each solution in a clean, glass-stoppered flask packed in crushed ice. A flask of distilled water should also be packed in ice.

Wash about 300 ml of crushed, distilled-water ice with several small amounts

Table 1

κa	0.20	0.25	0.30	0.35	0.40	0.45	0.50	0.55
σ	0.7588	0.7129	0.6712	0.6325	0.5988	0.5673	0.5376	0.5108

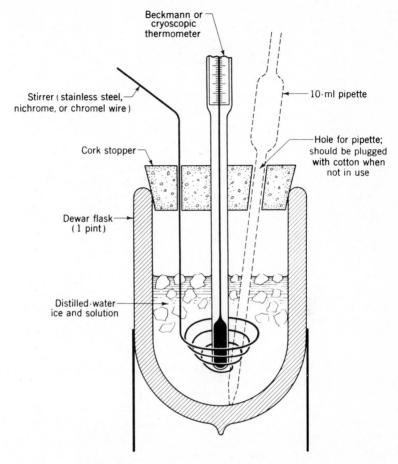

FIGURE 1
Apparatus for measuring the freezing point of aqueous solutions.

of the chilled distilled water, then fill the Dewar flask about one-third full with this washed ice. About 100 ml of the chilled distilled water is added to the Dewar, the apparatus is assembled, and the mixture well stirred to achieve equilibrium. The temperature T_0 is recorded. In reading the thermometer, tap it gently before reading and estimate to tenths of the smallest division. It should be realized that a cryoscopic thermometer may not read exactly zero at this point and may deviate by several hundredths of a degree without coming under suspicion of being defective. The important function of a thermometer of this kind is to measure temperature *differences* accurately, and it is not primarily intended to give an accurate measurement of an actual temperature.

The water is poured off and replaced by 100 ml of chilled 0.25 m HCl solution. After stirring to achieve equilibrium, as described previously, the temperature is read and a 10-ml aliquot withdrawn with a pipette. Introduce the pipette quickly, while blowing a gentle stream of air through it to prevent solution from entering it until the tip touches the bottom of the flask. This will prevent small particles of ice from being drawn into the pipette. Alternatively, attach to the tip of the pipette a filter consisting of a short length of rubber tubing containing a wad of cotton or glass wool. This aliquot is discharged into a clean weighing bottle, warmed to room temperature, and accurately weighed. It is then quantitatively transferred to a flask and titrated with 0.1 M NaOH to a methyl red or phenolphthalein end point. Stirring should be resumed vigorously for about 5 min, then another temperature reading taken and a second aliquot withdrawn. If these results are not consistent with each other, a third aliquot should be taken after further stirring.†

The solution is poured off and the experiment repeated with 0.125 m HCl. Be sure that an adequate amount of ice is present. If time permits, a run should be made with 0.0625 m HCl also.

Similar runs are carried out with the solutions of monochloroacetic acid, preferably repeating the measurement of T_0.

CALCULATIONS

Calculate the equilibrium molality m (in mol/1000 g of water) for each aliquot. If the results for the two aliquots from a given run are consistent, the average values of m and ΔT_f may be used in further calculations. For monochloroacetic acid, a weak electrolyte, calculate the effective total molality m' from Eq. (12-20) using the appropriate constants in Table 12-1. Then calculate α and K_c for each of the two concentrations studied.

For hydrochloric acid, a strong electrolyte, calculate an experimental value of g with Eq. (9) for each of the concentrations studied. In addition, use Eq. (13) to obtain a value of the osmotic coefficient g based on the Debye-Hückel theory for each concentration. Compare these experimental and theoretical values.

DISCUSSION

The colligative behavior of strong electrolytes is often expressed in terms of the Van't Hoff factor[2]

$$i = \frac{m_{\text{app}}}{m}$$

† The temperatures and the concentrations of the two aliquots may differ slightly, owing to some melting of ice, but the differences should be consistent. If they are not consistent, at least one of the aliquots was presumably not withdrawn at equilibrium.

where m_{app} is the "apparent" total molality as deduced from any colligative property when the solution is treated as ideal. Using the expressions for freezing-point depression, show the relation between i and g.

What additional data would be required in order to compare the values of K_c obtained for monochloroacetic acid at two different temperatures? Can you predict the direction of the change in K_c with T; i.e., the sign of dK_c/dT?

APPARATUS

Dewar flask (short, wide mouth, 1 pt); cork stopper with three holes; coil-type stirrer; precision cryoscopic thermometer; magnifying thermometer reader; stopwatch; two 10-ml weighing bottles; 50-ml burette; burette clamp and stand; two 100-ml volumetric flasks; one 10-, one 25-, and one 50-ml pipette; battery jar; wash bottle.

Distilled-water ice (500 g); 0.1 M sodium hydroxide solution (500 ml); 0.25 m HCl solution (400 ml) and 0.25 m monochloroacetic acid solution (400 ml); phenolphthalein indicator; stopcock grease.

REFERENCES

1. F. H. MacDougall, "Thermodynamics and Chemistry," 3d ed., Wiley, New York (1939).
2. W. J. Moore, "Physical Chemistry," 4th ed., Prentice-Hall, Englewood Cliffs, N.J. (1972).

GENERAL READING

E. L. Skau, J. C. Arthur, Jr., and H. Wakeham, Determination of Melting and Freezing Temperatures, in A. Weissberger (ed.), "Technique of Organic Chemistry," 3d ed., vol. I, part I, chap. VII, Interscience, New York (1959).

EXPERIMENT 14. EFFECT OF IONIC STRENGTH ON SOLUBILITY

The solubility of a salt in water can be influenced by the presence of other electrolytes in several ways: by a "common-ion effect," by the occurrence of a chemical reaction involving one of the ions of the salt, or by a change in the activity coefficients of the ions of the salt (which is caused by a change in the ionic strength of the solution). The first two effects are familiar ones and are often encountered in elementary treatments which make use of equilibrium constants expressed in terms of concentrations. In the present experiment we shall be concerned with the third effect and shall show

how an equilibrium constant in terms of activities can be obtained by extrapolation of solubility data to zero concentration. This constant is used in conjunction with the measured solubilities in solutions of different ionic strengths to obtain experimental values for the activity coefficient of silver acetate as a function of ionic strength. These are compared with theoretical values for activity coefficients as calculated by means of the Debye-Hückel theory.

THEORY

We are concerned here with the following equilibrium,[1] which occurs in the presence of excess solid silver acetate:

$$AgAc(s) = Ag^+ \text{ (in soln.)} + Ac^- \text{ (in soln.)} \qquad (25°C, 1 \text{ atm}) \qquad (1)$$

However, it has been shown[2] that silver acetate exists in solution not only as Ag^+ and Ac^-, but also as undissociated silver acetate AgAc, and as complex ions $AgAc_2^-$, Ag_2Ac^+ and possibly other complex ionic forms. For our purposes complex ions may be neglected, and silver acetate will be assumed to be in solution as the undissociated neutral molecules AgAc and as the free ions Ag^+ and Ac^-. The determination of the equilibrium constant K' for the change in state

$$AgAc(s) = AgAc \text{ (undiss., in soln.)} \qquad (2)$$

will not be attempted as a part of this experiment. From extensive electrochemical measurements[2] the value

$$K' = 1.05 \times 10^{-2} \text{ mol liter}^{-1} \qquad (3)$$

has been obtained at 25°C and 1 atm.† The concentration of undissociated AgAc may then be subtracted from the total concentration of silver found by titration to give the concentration c of Ag^+ ion.

For the equilibrium given in Eq. (1) the product of the *activities* of Ag^+ and Ac^- ions at the given temperature and pressure is a constant, regardless of the concentrations of these ions or of the presence of other ionic or molecular species in solution:

$$a_{Ag^+} a_{Ac^-} = K \qquad (4)$$

---

† Since the concentration of complex ions is very small, an adequate approximation to K' could be obtained from the difference between the total solubility of AgAc in pure water and the ionic concentrations of Ag^+ and Ac^- in a saturated solution as obtained from an independent measurement. For example, the concentration of dissolved ions could be determined from conductance measurements made on a saturated AgAc solution and on $AgNO_3$, $NaNO_3$, and NaAc solutions of about the same ionic concentration (see Exp. 20). Or the concentration of Ag^+ could be determined from the emf of an electrochemical cell with a silver electrode and a salt bridge to a calomel reference electrode if the activity coefficients are known.

The experimentally measurable concentrations are related to the activities by the *activity coefficients* γ:

$$a_{Ag^+} = \gamma_{Ag^+} c_{Ag^+} \qquad a_{Ac^-} = \gamma_{Ac^-} c_{Ac^-}$$

Thus,
$$K = \gamma_{Ag^+} \gamma_{Ac^-} c_{Ag^+} c_{Ac^-} = \gamma_\pm^2 c_{Ag^+} c_{Ac^-} \tag{5}$$

where $\gamma_\pm$, the *mean activity coefficient* for AgAc in the solution concerned, is the geometric mean of the two ionic activity coefficients. If there is no common ion present from another solute, and if there is no reagent present which reduces the concentration of either ion by chemical reaction, the concentrations of Ag^+ and Ac^- will be equal to each other and to the ionic solubility c, so that we can write

$$\gamma_\pm^2 c^2 = K \tag{6}$$

Measurements of the ionic solubility can therefore be used to determine the mean activity coefficient $\gamma_\pm$ for AgAc in solutions containing other ions in various concentrations, provided the constant K can be determined.

Debye-Hückel theory[3]   In the limit of zero ionic strength the activities of ions are equal to their concentrations, the activity coefficients thus becoming unity in the limit. The ionic strength μ is defined by the expression

$$\mu = \tfrac{1}{2} \sum_i c_i z_i^2 \tag{7}$$

where c_i is the concentration of the ith ionic species, z_i is the charge of the ion in units of the electronic charge, and the sum is taken over all ionic species present. The simple Debye-Hückel "limiting law"

$$\log \gamma_j = -A z_j^2 \sqrt{\mu}$$

predicts the activity coefficient of ion j at low ionic strengths but becomes seriously inexact at even moderate ionic strengths. A more complete form of the Debye-Hückel theory, which holds well for low and moderate ionic strengths, is represented by the equation[3]

$$\log \gamma_j = -z_j^2 \frac{A\sqrt{\mu}}{1 + B_j \sqrt{\mu}} \tag{8}$$

where A and B_j are constants. The value of A can be calculated theoretically; it depends on characteristics of the solvent and on the temperature. If logarithms to the base 10 are used, it has a value of about 0.509 for aqueous solutions at 25°C. B_j depends in addition on the diameter of the ion concerned and also on the diameters of other ions of opposite sign that may be present in the solution. To a good approximation, it can be shown that the *mean* activity coefficient of a given electrolyte is

$$\log \gamma_\pm = -|z_+ z_-| \frac{A\sqrt{\mu}}{1 + B\sqrt{\mu}} \tag{9}$$

where B is the mean of the B_j values of the two ions; the limiting law is, of course,

$$\log \gamma_\pm = -|z_+ z_-| A\sqrt{\mu} \tag{10}$$

When μ is calculated using concentrations in moles per *liter of solvent* (or, in the case of dilute aqueous solutions, molality which is mol/1000 g solvent), the constant B is found to be close to unity for many salts at 25°C.[4] For $\gamma_\pm$ of AgAc in the system to be studied, B has been found empirically[5] to be about 1.25 at 25°C when μ is expressed in moles per *liter of solution*.

Determination of K If we combine Eq. (9) with Eq. (6) we obtain for the case where the z's are unity

$$\log c = \tfrac{1}{2} \log K + \frac{A\sqrt{\mu}}{1 + B\sqrt{\mu}} \tag{11}$$

If $\log c$ is plotted against $\sqrt{\mu}$, a smooth curve should be obtained, and when this curve is extrapolated to $\sqrt{\mu} = 0$, the intercept will give $\tfrac{1}{2} \log K$. A more nearly linear curve can be obtained by plotting $\log c$ versus $A\sqrt{\mu}/(1 + B\sqrt{\mu})$; this facilitates extrapolation to zero concentration from data at moderate ionic strengths.

By means of this extrapolation the activity product K can be determined, and by application of Eq. (6) the activity coefficient $\gamma_\pm$ for AgAc can be calculated for each solution studied.

EXPERIMENTAL

It is desired to measure the equilibrium solubility of AgAc in several solutions of different ionic strengths. In order to achieve rapid equilibrium, one should dissolve an excess of AgAc by heating the solution; on cooling to the desired temperature this excess will readily crystallize out.

Place about 2 g of AgAc in each of five clean, dry sample bottles. The AgAc should be weighed out roughly; the sample bottles should be suitable for mounting on a tumbler or agitator in a thermostat bath. Each bottle should be indelibly numbered or tagged to provide identification. Fill each bottle three-quarters full with one of the following solutions:†

1   Distilled water
2   0.05 *M* NaNO$_3$
3   0.1 *M* NaNO$_3$
4   0.2 *M* NaNO$_3$
5   0.5 *M* NaNO$_3$

Set the sample bottles in a large beaker of hot water until the solutions have warmed up to about 55°C. Then stopper the bottles *tightly* and shake vigorously for 2 min. Place the bottles on a tumbler or agitator in a 25°C thermostat bath and allow 1 hr for equilibration. If mechanical agitation is not possible, one may clamp the bottles in the bath and shake them by hand about once every 5 min.

At the end of the hour remove the bottles from the tumbler and clamp them in the bath for 5 or 10 min to allow the solid to settle. *Without removing the bottles from the bath,* withdraw two 25-ml samples from each bottle; use a pipette fitted at the tip with a filter consisting of a short (1-in.) piece of rubber tubing packed with cotton. Remove the filter before adjusting the level of the liquid in the pipette to the mark. Use a clean, dry filter for each sample, and rinse out the pipette with a few milliliters of filtered solution before taking each sample. Discharge each sample into a 200-ml Erlenmeyer flask containing 15 ml of distilled H$_2$O and 5 ml of 6 *M* HNO$_3$. The amount of silver ion in each sample is then determined by titration with 0.1 *N* KSCN solution. Add about 1 ml of 0.1 *M* ferric alum solution to each flask as an indicator; titrate to the appearance of a faint reddish-brown color which does not disappear on shaking. While waiting for the AgAc solutions to equilibrate, make at least one practice titration using the AgNO$_3$ solution provided. Once standardized by this practice titration, the AgNO$_3$ can be used for "back-titrations" if necessary.

CALCULATIONS

Data obtained for solutions of AgAc in water and in aqueous NaNO$_3$ should be treated in the following way:

Calculate the total solubility of AgAc from your titration results. Subtract the concentration of undissociated AgAc obtained from Eq. (3) to find the ionic solubility *c* in moles per liter of solution. Also calculate the ionic strength μ of each solution

---

† If it is desired to investigate, in addition, the effect on solubility of the presence of a common ion or of a reagent which reacts with one of the ions, an additional sample may be prepared with 0.1 *M* NaAc or 0.05 *M* HNO$_3$ solution.

from concentrations in moles per liter of solution, remembering that the summation in Eq. (7) is over *all* ionic species present.

Plot $\log_{10} c$ versus $A\sqrt{\mu}/(1 + B\sqrt{\mu})$, where A is taken as 0.509, B as 1.25. Extrapolate to $\mu = 0$, taking pains to make the slope of the line as close to unity in the region of low ionic strength as the data will permit, and determine K. By means of Eq. (6) calculate the value of the mean activity coefficient $\gamma_{\pm}$ for AgAc for each solution.

The value of K determined above is in principle largely independent of the Debye-Hückel theory, in the sense that this theory has been used mainly as a convenience in extrapolating to zero ionic strength with limited data. Therefore the values of activity coefficients obtained with Eq. (6) are essentially *experimental* values. These values should be compared with *theoretical* activity coefficients calculated with the Debye-Hückel theory, using both the limiting law, Eq. (10), and the more complete expression, Eq. (9). The experimental value and both theoretical values of $\gamma_{\pm}$ for each ionic strength should be presented in tabular form.

APPARATUS

Five (or six) sample bottles (tall with narrow mouth, approximately 125 cm^3); rubber stoppers to fit; tags or labels for the sample bottles; tripod and wire gauze; bunsen burner, complete with hose; one 0 to 100°C thermometer; one 1- and one 25-ml pipette; two short ($\sim$1-in.) pieces of gum-rubber tubing; two to four 200-ml Erlenmeyer flasks; 15- or 25-ml graduated cylinder; one large (1500 to 2000 ml) beaker and one 250-ml beaker; one small (25 to 50 ml) beaker or flask; one 50-ml burette; burette clamp and stand; spatula.

Reagent-grade silver acetate (15 g); triple-beam balance; absorbent cotton or glass wool; spool of copper wire, for securing stoppers on sample bottles; constant-temperature bath at 25°C; bath clamps; tumbler or agitator for sample bottles.

The following solutions (100 ml each) should be made with precisely known concentrations: 0.05, 0.1, 0.2, 0.5 M sodium nitrate; 0.05 M silver nitrate; (0.1 M sodium acetate and 0.05 M nitric acid, if required). Also needed: standardized 0.1 N potassium thiocyanate (500 ml); approximately 6 M nitric acid (100 ml); ferric alum solution as indicator (25 ml).

REFERENCES

1. F. H. MacDougall, *J. Amer. Chem. Soc.*, **52,** 1390 (1930).
2. F. H. MacDougall and L. E. Topol, *J. Phys. Chem.*, **56,** 1090 (1952).
3. W. J. Moore, "Physical Chemistry," 4th ed., Prentice-Hall, Englewood Cliffs, N.J. (1972).
4. G. Scatchard, *Chem. Rev.*, **19,** 309 (1936).
5. F. H. MacDougall and J. Rehner, Jr., *J. Amer. Chem. Soc.*, **56,** 368 (1934).

GENERAL READING

W. J. Mader, R. D. Vold, and M. J. Vold, Determination of Solubility, in A. Weissberger (ed.), "Technique of Organic Chemistry," 3d ed., vol. I, part I, chap. XI, Interscience, New York (1959).

EXPERIMENT 15.   CHEMICAL EQUILIBRIUM IN SOLUTION

In this experiment a typical homogeneous equilibrium in aqueous solution is investigated and the validity of the law of mass action is demonstrated. The equilibrium to be studied is that which arises when iodine is dissolved in aqueous KI solutions:

$$I_2 + I^- = I_3^- \tag{1}$$

In determining the concentrations of the species present at equilibrium in the aqueous solution use is made of a heterogeneous equilibrium: the distribution of molecular iodine I_2 between two immiscible solvents, water and carbon tetrachloride.

THEORY

Homogeneous equilibrium   The thermodynamically exact equilibrium constant K_a (which must be expressed in terms of the activities of the chemical species involved) may be given to a good approximation in the case of the present reaction by the analogous expression involving concentrations.

$$K_a = \frac{a_{I_3^-}}{a_{I_2}a_{I^-}} = \frac{(I_3^-)}{(I_2)(I^-)}\frac{\gamma_{I_3^-}}{\gamma_{I_2}\gamma_{I^-}} \simeq \frac{(I_3^-)}{(I_2)(I^-)} = K_c \tag{2}$$

where the a's are activities, the γ's are activity coefficients, K_c is the equilibrium constant in terms of concentration, and the parentheses denote concentrations in moles per liter. A commonly used approximate form of the Debye-Hückel theory for the activity coefficients of ionic species[1] is

$$\log \gamma_i = -0.509 z_i^2 \sqrt{\mu}/(1 + \sqrt{\mu}) \tag{3}$$

Since both I^- and I_3^- have the same charge z and are influenced by the same ionic strength μ, the activity coefficients γ_{I^-} and $\gamma_{I_3^-}$ are equal within the accuracy of Eq. (3). It should, however, be pointed out that Eq. (3) does not take account of variations in size and shape of ions. Also the I_2 concentration in aqueous solution is very small; thus γ_{I_2} is approximately unity, since the activity coefficient of a neutral solute species in dilute solution does not deviate greatly from its limiting value at infinite dilution. Therefore, the approximation given in Eq. (2) may be expected to be good to the order of 1 percent or better in moderately dilute solutions, and K_c should be a constant independent of the concentrations of the species.

If, then, one starts with a solution of KI whose concentration C is known accurately and dissolves in it iodine, at equilibrium the iodine will be present partly as I_2 molecules and partly as I_3^- ions (a part of the I^- ions initially present having reacted to form the I_3^- ions). The total iodine concentration, $T = (I_2) + (I_3^-)$, can be found by titration with standard thiosulfate solution. If at equilibrium there is a way of measuring (I_2) separately, then K_c can be calculated, since

$$(I_3^-) = T - (I_2) \tag{4}$$

and
$$(I^-) = C - (I_3^-) \tag{5}$$

We can measure (I_2) at equilibrium by taking advantage of the fact that CCl_4, which is immiscible with aqueous solutions, dissolves molecular I_2 but not any of the ionic species involved.

Heterogeneous equilibrium   We are concerned here with the distribution of molecular iodine I_2 as the solute between two immiscible liquid phases, aqueous solution and CCl_4. At equilibrium, the concentrations of I_2 in the two phases, $(I_2)_w$ and $(I_2)_{CCl_4}$, are related by a distribution constant k:

$$k = (I_2)_w/(I_2)_{CCl_4} \tag{6}$$

This distribution law applies only to the distribution of a definite chemical species, as does Henry's law. The distribution constant k is not a true thermodynamic equilibrium constant, since it involves concentrations rather than activities. Thus it may vary slightly with the concentration of the solute (particularly because of the relatively high concentration of I_2 in the CCl_4 phase); it is therefore advantageous to determine k at a number of concentrations. It can be determined directly by titration of both phases with standard thiosulfate solution when I_2 is distributed between CCl_4 and pure water. Once k is known, (I_2) in an aqueous phase containing I_3^- can be obtained by means of a titration of the I_2 in a CCl_4 layer that has been equilibrated with this phase. The use of a distribution constant in this manner depends upon the assumption that its value is unaffected by the presence of ions in the aqueous phase

EXPERIMENTAL

Distribution ratio   Measure the distribution constant $k = (I_2)_w/(I_2)_{CCl_4}$, using CCl_4 and pure H_2O as solvents and I_2 as solute. Distilled water (200 ml) and solutions of I_2 in CCl_4 (50 ml) should be put into 500-ml glass-stoppered Erlenmeyer flasks and equilibrated with frequent shaking at 25°C. The quantities to use are given in Table 1 (runs 1 to 3); three sets of conditions are used in order to check the variation of the distribution constant with concentration. The flasks containing the solutions should be vigorously shaken for 5 min and clamped in a thermostat bath. After 10 min of thermal equilibration, remove the flasks one at a time, wrap them in a dry towel, shake them vigorously for 3 to 5 min, and then return to the thermostat bath. Repeat this procedure for at least an hour. Before removing samples for analysis let the flasks remain in the bath for 10 min after the last shaking to allow the liquid layers to separate completely. After equilibration is complete, remove one flask at a time from the thermostat bath and place in a battery jar containing water at 25°C as shown in Fig. 1. A sample of the aqueous layer and a sample of the CCl_4 layer are removed with pipettes, and the flask is stoppered and returned to the bath for an additional half hour of the above treatment, after which a second sample of each phase is removed for titration. Additional information is given in Table 1.

The purpose of taking a second sample after further equilibration is to verify that equilibrium has been achieved. If the two titrations are in substantial agreement, the average value can be used. If the results of the titrations indicate that equilibrium had not been reached when the first samples were taken, the results of the second titrations should be used although proof of equilibrium is in this case lacking.

In removing the samples for titration, it is important to avoid contamination of the sample by drops of the other phase, especially when the other phase is very much more concentrated as is the CCl_4 layer in this case. When pipetting samples of the CCl_4 solutions, blow a slow stream of air through the pipette while it is being introduced into the solution in order to minimize the contamination by the aqueous layer. It is necessary to use a rubber bulb with the pipette for these solutions; **do not**

Table 1[a]

Run no.	CCl_4 (50 ml), molarity I_2	Aqueous layer (200 ml), molarity KI	CCl_4 layer (use 20-ml pipette)		Aqueous layer (use 50-ml pipette)	
			Burette size, ml	Molarity $S_2O_3^{2-}$	Burette size, ml	Molarity $S_2O_3^{2-}$
1	0.080	0.0	50	0.1	10	0.01
2	0.040	0.0	50	0.1	10	0.01
3	0.020	0.0	10	0.1	10	0.01
4	0.080	0.15	10	0.1	50	0.1
5	0.040	0.15	10	0.1	10	0.1
6	0.080	0.03	50	0.1	10	0.1

[a] The table shows the nominal *initial* concentrations and volumes of the aqueous KI solutions and of the solutions of I_2 in CCl_4 to be used. Also given are the nominal concentrations of the thiosulfate solutions to be used for titrating and the sizes of burettes and pipettes to be used. The *actual*, precise concentrations of the solutions used should be read from the labels on the respective bottles.

pipette by mouth. Between samples rinse the pipette well with acetone and dry it before taking the next sample.

Each sample is transferred from the pipette to a 250-ml Erlenmeyer flask containing 10 ml of 0.1 M KI and titrated with the appropriate thiosulfate solution (see Table 1). The iodide is added to reduce loss of I_2 from the aqueous solution by evaporation during the titration, by forming the nonvolatile I_3^-. The reaction taking place during the titration is

$$2S_2O_3^{2-} + I_3^- \rightarrow S_4O_6^{2-} + 3I^-$$

Near the end point, as indicated by a very light yellow color of aqueous iodine solution, Thyodene or 1 ml of soluble starch solution is added; this acts as an indicator by adsorbing iodine and giving a deep blue color. At the end point this

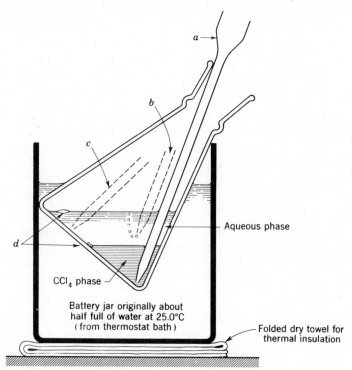

FIGURE 1

Arrangement for withdrawing samples. Use a rubber bulb for drawing up liquid in the pipette. (*a*) Position of pipette for withdrawing sample of CCl₄ phase. Tip should touch bottom at deepest point and should not be moved around. (*b*) On inserting pipette, blow *small* stream of air to keep aqueous phase from entering tip. (*c*) Position of pipette for withdrawing aqueous sample. (*d*) On inserting pipette, avoid contact with drops of CCl₄ phase (spheres on bottom or lenses floating on top).

blue color disappears sharply. If the starch has been added too soon, the color may redevelop after an apparent end point owing to diffusion of iodine from the interior of the colloidal starch particles; the titration should be continued until no blue color reappears. However, the blue color may reappear on prolonged standing because of air oxidation of the iodide ion, and this effect should be disregarded. In the two-phase titration, shake the flask vigorously after adding each portion of thiosulfate solution.

If desired, an excellent end point for the two-phase titration can be taken as the disappearance of the reddish-violet iodine color from the CCl_4 layer. This may even be done in the titration of the aqueous sample by adding 1 or 2 ml of pure CCl_4. In this case, do not add the starch indicator.

It is recommended that a practice titration be performed with one of the stock solutions of I_2 in CCl_4.

Since a thiosulfate solution is susceptible to attack by sulfur-metabolizing bacteria, it may be wise to check the standardization of the stock solution with a standard solution of KIO_3.[2] Place 1.3 to 1.4 g of KIO_3 in a weighing bottle, dry it for several hours at 110°C, cool in a desiccator, and weigh it exactly on an analytical balance. Transfer this salt to a clean 100-ml volumetric flask and make up to the mark with distilled water. Rinse a clean 5-ml pipette with several small portions of the iodate solution and then carefully deliver a 5-ml sample into a 250-ml Erlenmeyer flask. Add about 20 ml of distilled water, 5 ml of a 0.5 g/ml KI solution, and 10 ml of 1 M HCl. Titrate at once with the thiosulfate solution until the reddish color turns orange and then yellow and becomes pale. At this point add about 0.5 g of Thyodene† indicator, mix well, and titrate until the blue color disappears.

Equilibrium constant   To determine the concentrations of I_2 and I_3^- in equilibrium in aqueous solutions we equilibrate aqueous KI solutions (200 ml) with solutions of I_2 in CCl_4 (50 ml) as shown in Table 1 (runs 4 to 6). The equilibration and titration procedures are identical with those described above; additional information is given in Table 1. With the distribution constant determined above and the initial I^- concentration, the equilibrium concentrations of I_2, I_3^-, and I^- can be calculated.

Adequate time for equilibration is of great importance; the flasks should be placed in the thermostat bath as soon as possible. The temperature of the thermostat bath should be checked several times throughout the equilibration.

---

† Thyodene is a proprietary starch derivative which gives very sharp and reliable end points. It may be added directly as a powder and is therefore much more convenient for occasional use than the traditional "soluble starch" suspension. Should Thyodene be unavailable, a starch indicator solution may be prepared as follows: Mull 1 g of soluble starch powder in a mortar with several ml of boiling water. Pour the paste into 200 ml of boiling water, boil for 2 to 3 min, and allow to cool. When cool, the indicator is ready to use (about 2 ml/titration).

CALCULATIONS

Calculate the distribution constant k from the results of runs 1 to 3. If more than one run was performed, plot k versus the iodine concentration in the CCl_4 solution, $(I_2)_{CCl_4}$. If k is not constant, discuss its variation with concentration.

Using Eqs. (4), (5), and (6) and the results of runs 4 to 6, calculate the equilibrium concentrations of I_2, I_3^-, and I^- in each of the aqueous solutions. In each case use the appropriate value of k as read from the smooth curve of k versus $(I_2)_{CCl_4}$. Calculate the equilibrium constant K_c for each run. If any variation of K_c with concentration is found, do you regard it as experimentally significant?

APPARATUS

Three to six 500-ml glass-stoppered Erlenmeyer flasks; 1-qt battery jar or 1500-ml beaker; three pipettes—20, 50, 100 (or 200) ml; one 10- and one 50-ml burette; six 250-ml Erlenmeyer flasks; 10-ml graduated cylinder; three 250-ml beakers; burette clamp and stand; large rubber bulb for pipetting; wash bottle.

Constant-temperature bath set at 25°C; bath clamps for holding 500-ml Erlenmeyer flasks; pure CCl_4; acetone for rinsing; large carboy for waste liquids. Solutions: 0.08 M solution of I_2 in CCl_4 (300 ml); 0.04 M I_2 in CCl_4 (250 ml); 0.02 M I_2 in CCl_4 (100 ml); 0.15 M KI solution (800 ml); 0.03 M KI solution (400 ml); 0.1 M $Na_2S_2O_3$ solution (500 ml); 0.01 M $Na_2S_2O_3$ solution (100 ml); approximately 0.1 M KI solution (500 ml); Thyodene or 0.2 percent soluble starch solution, containing a trace of HgI_2 as preservative (50 ml).

REFERENCE

1. W. J. Moore, "Physical Chemistry," 4th ed., Prentice-Hall, Englewood Cliffs, N.J. (1972).
2. D. A. Skoog and D. M. West, "Fundamentals of Analytical Chemistry," pp. 485–492, Holt, New York (1963).

GENERAL READING

G. N. Lewis and M. Randall (revised by K. S. Pitzer and L. Brewer), "Thermodynamics," 2d ed., McGraw-Hill, New York (1961).

L. C. Craig and D. Craig, Laboratory Extraction and Countercurrent Distribution, in A. Weissberger (ed.), "Technique of Organic Chemistry," 2d ed., vol. III, part I, chap II, Interscience, New York (1956).

VII

PHASE EQUILIBRIA

EXPERIMENTS

16. Vapor pressure of a pure liquid
17. Binary liquid-vapor phase diagram
18. Binary solid-liquid phase diagram
19. Liquid-vapor coexistence curve and the critical point

EXPERIMENT 16. VAPOR PRESSURE OF A PURE LIQUID

When a pure liquid is placed in an evacuated bulb, molecules will leave the liquid phase and enter the gas phase until the pressure of the vapor in the bulb reaches a definite value which is determined by the nature of the liquid and its temperature. This pressure is called the vapor pressure of the liquid at a given temperature. The equilibrium vapor pressure is independent of the quantity of liquid and vapor present as long as both phases exist in equilibrium with each other at the specified temperature. As the temperature is increased, the vapor pressure also increases up to the critical point, at which the two-phase system becomes a homogeneous, one-phase fluid.

If the pressure above the liquid is maintained at a fixed value (say by admitting air to the bulb containing the liquid), then the liquid may be heated up to a temperature at which the vapor pressure is equal to the external pressure. At this point, vaporization will occur by the formation of bubbles in the interior of the liquid as well as at the surface; this is the boiling point of the liquid at the specified external pressure. Clearly the temperature of the boiling point is a function of the external pressure; in fact, the variation of the boiling point with external pressure is seen to be identical with the variation of the vapor pressure with temperature.

In this experiment the variation of vapor pressure with temperature will be measured and used to determine the molar heat of vaporization.

THEORY

We are concerned here with the equilibrium between a pure liquid and its vapor:

$$X(l) = X(g) \qquad (p,T) \tag{1}$$

It can be shown thermodynamically[1] that a definite relationship exists between the values of p and T at equilibrium as given by

$$\frac{dp}{dT} = \frac{\Delta S}{\Delta V} \tag{2}$$

In Eq. (2) dp and dT refer to infinitesimal changes in p and T for an equilibrium system composed of a pure substance with both phases always present; ΔS and ΔV refer to the change in S and V when one phase transforms to the other at constant p and T. Since the change in state (1) is isothermal and ΔG is zero, ΔS may be replaced by $\Delta H/T$. The result is

$$\frac{dp}{dT} = \frac{\Delta H}{T \Delta V} \tag{3}$$

Equation (2) or (3) is known as the Clapeyron equation. It is an exact expression which may be applied to phase equilibria of all kinds although it has been presented here in terms of the one-component liquid-vapor case. Since the heat of vaporization ΔH_v is positive and ΔV is positive for vaporization, it is seen immediately that the vapor pressure must increase with increasing temperature.

For the case of vapor-liquid equilibria in the range of vapor pressures less than 1 atm, one may assume that the molal volume of the liquid $\tilde{V}_l$ is negligible in comparison with that of the gas $\tilde{V}_g$, so that $\Delta \tilde{V} = \tilde{V}_g$. This assumption is very good in the low-pressure region, since $\tilde{V}_l$ is usually only a few tenths of a percent of $\tilde{V}_g$. Thus we obtain

$$\frac{dp}{dT} = \frac{\Delta \tilde{H}_v}{T \tilde{V}_g} \tag{4}$$

Since $d \ln p = dp/p$ and $d(1/T) = -dT/T^2$, we can rewrite Eq. (4) in the form

$$\frac{d \ln p}{d(1/T)} = -\frac{\Delta \tilde{H}_v}{R} \frac{RT}{p \tilde{V}_g} = -\frac{\Delta \tilde{H}_v}{Rz} \tag{5}$$

where we have introduced a *compressibility factor* z for the vapor:

$$z = \frac{p \tilde{V}_g}{RT} \tag{6}$$

Equation (5) is a convenient form of the Clapeyron equation. We can see that *if* the vapor were a perfect gas ($z \equiv 1$) and $\Delta \tilde{H}_v$ were independent of temperature, then a plot of $\ln p$ versus $1/T$ would be a straight line the slope of which would determine $\Delta \tilde{H}_v$. Indeed, for many liquids $\ln p$ is almost a linear function of $1/T$, which implies at least that $\Delta \tilde{H}_v/z$ is almost constant.

Let us now consider the question of gas imperfections, i.e., the behavior of z as a function of temperature for the saturated vapor. It is difficult to carry out p-V-T measurements on gases close to condensation and such data are scarce, but data are available for water,[2] and theoretical extrapolations[3] have been made for the vapor of "normal" liquids based on data obtained at higher temperatures. Figure 1 shows the variation of the compressibility factor z for a saturated vapor as a function of temperature in the case of water and two normal liquids, benzene and n-heptane. For the temperature axis, a "reduced" temperature T_r is used; $T_r = T/T_c$, where T_c is the critical temperature. This has the effect of almost superimposing the curves of many different substances; indeed, by the law of corresponding states such curves would be exactly superimposed. In general, it is clear that z decreases as the temperature increases. Water, due to its high critical temperature, is a reasonably ideal gas even at 100°C where z equals 0.986. But n-heptane at its 1-atm boiling point of 98°C has a value of z equal to 0.95 and is relatively nonideal. For many substances, sizable gas imperfections are present even at pressures below 1 atm.

Next we must consider the variation of $\Delta\tilde{H}_v$ with temperature. For a change in state such as Eq. (1),

$$\Delta H_{T_2} = \Delta H_{T_1} + \int_{T_1}^{T_2} \Delta C_p \, dT + \int_{p_1}^{p_2} \left(\frac{\partial \Delta H}{\partial p}\right)_T dp \qquad (7)$$

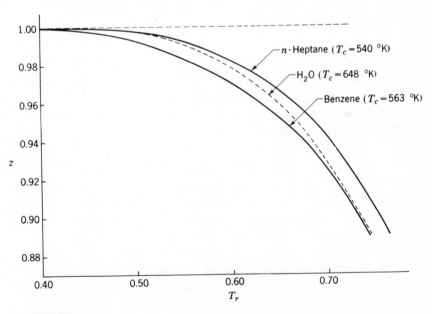

FIGURE 1
The compressibility factor z of saturated vapor as a function of reduced temperature T_r for water, benzene, and n-heptane.

Since the final term is zero for a perfect gas and small for most real gases, it is possible to approximate Eq. (7) by

$$\Delta H_{T_2} \simeq \Delta H_{T_1} + \overline{\Delta C_p}(T_2 - T_1) \tag{8}$$

where the average value over the temperature interval, $\overline{\Delta C_p}$, is used. For $\Delta \tilde{H}_v$ to be independent of temperature, $\overline{\Delta \tilde{C}_p}$ must be very close to zero, which is generally not true. Heat capacities for water, benzene, and n-heptane are given below as typical examples.[4,5] For n-heptane, the specific heat of both gas and liquid changes rapidly with temperature; use of average values will give only an order-of-magnitude result. In general, the value of $\Delta \tilde{H}_v$ will decrease as the temperature increases.

Compound	Temp. range, °C	Average values, cal deg^{-1} mol^{-1}		
		$\tilde{C}_p(g)$	$\tilde{C}_p(l)$	$\Delta \tilde{C}_p$
Water	25–100	8	18	-10
Benzene	25–80	22	35	-13
n-Heptane	25–100	~ 47	~ 58	-11

Since both $\Delta \tilde{H}_v$ and z decrease with increasing temperature, it is possible to see why $\Delta \tilde{H}_v/z$ might be almost constant and give a nearly linear plot of $\ln p$ versus $1/T$.

METHODS

There are several experimental methods of measuring the vapor pressure as a function of temperature.[6] In the *gas-saturation method* a known volume of an inert gas is bubbled slowly through the liquid, which is kept at constant temperature in a thermostat. The vapor pressure is calculated from a determination of the amount of vapor contained in the outcoming gas or from the loss in weight of the liquid. A common *static method* makes use of an *isoteniscope*, a bulb with a short U tube attached. The liquid is placed in the bulb, and some liquid is placed in the U tube. When the liquid is boiled under reduced pressure, all air is swept out of the bulb. The isoteniscope is then placed in a thermostat. At a given temperature the external pressure is adjusted so that both arms of the U tube are at the same height. At this setting, the external pressure, which is equal to the pressure of the vapor in the isoteniscope, is measured with a mercury manometer. A common *dynamic method* is one in which the variation of the boiling point with external applied pressure is measured. The total pressure above the liquid can be varied and maintained at a given value by use of a large-volume ballast bulb; this pressure is then measured with a mercury manometer. The liquid to be studied is heated until boiling occurs, and the temperature of the refluxing vapor is measured in order to avoid any effects of superheating. The experimental procedure for both the isoteniscope method and a boiling-point method is given below.

EXPERIMENTAL

1 Boiling-point method The apparatus should be assembled as shown in Fig. 2. A Claissen distilling flask can be used in this experiment by closing off the side arm with a rubber policeman. Use pressure tubing for connecting the condenser and the manometer to the ballast bulb. Fill the flask about one-third full (just above the level of the baffle) with the liquid to be studied. A few Carborundum boiling chips should be added to reduce "bumping." Be sure that the thermometer scale is visible over a

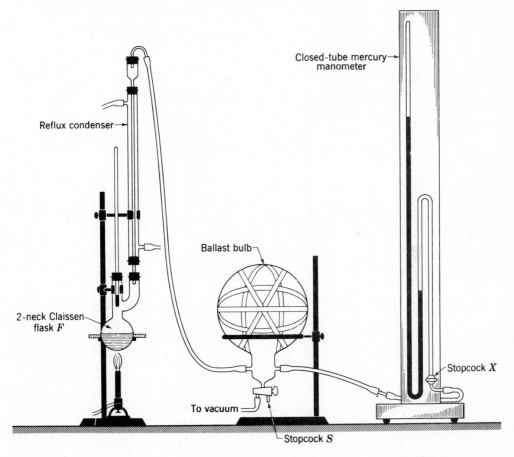

FIGURE 2
Boiling-point apparatus.

range of at least 50°C below the boiling point at 1 atm. Heating should be accomplished with either a bunsen burner or an electrical heating mantle placed directly below the flask F. The baffle is used to prevent superheating of the vapor.

Use the manometer with the greatest care; be careful not to allow liquid to condense in it. Stopcock X should be open only when the manometer is to be read and closed immediately after each reading. Stopcock X should not be open at the same time as stopcock S is open, either to vacuum or to the air.

To make a reading, adjust the heating so as to attain steady boiling of the liquid, but avoid heating too strongly. Preferably, heat the flask at a point an inch or so to one side of center. When conditions appear to be steady, open stopcock X carefully and watch the manometer. When manometer and thermometer appear to be as steady as they can be maintained, read them as nearly simultaneously as possible. The thermometer scale should be read to the nearest 0.1°C, and vapor should be condensing on and dripping from the thermometer bulb to ensure that the equilibrium temperature is obtained. In reading the manometer, record the position of each meniscus (h_1 and h_2), keeping your line of sight level with the meniscus to avoid parallax error. Estimate your readings to the nearest 0.1 mm. Record the ambient air temperature at the manometer several times during the run.

To change the pressure in the system between measurements, first remove the heater, then check to make sure that stopcock X is closed. After a short time admit some air or remove some air by opening stopcock S for a few seconds. Be especially careful in removing air from the ballast bulb to avoid strong bumping. Then open stopcock X carefully to measure the pressure on the manometer. Repeat as often as necessary to attain the desired pressure. Close stopcock X and restore the heater to its position under the flask.

Take readings at approximately the following pressures:

Pressure descending: 760, 600, 450, 350, 260, 200, 160, 130, 100, 80 Torr
Pressure ascending: 90, 110, 140, 180, 230, 300, 400, 520, 670 Torr

2   Isoteniscope method   In this technique much of the equipment shown in Fig. 2 is used, but the distilling flask and reflux condenser are replaced by an isoteniscope mounted in a glass thermostat as shown in Fig. 3. The heater is operated from a Variac voltage control, and the water bath should be vigorously stirred to ensure thermal equilibrium. The liquid to be studied is placed in the isoteniscope so that the bulb is about two-thirds full and there is no liquid in the U tube. Then the isoteniscope is placed in the thermostat (which should be at room temperature) and is connected to the ballast bulb and manometer (which are assembled and connected as in Fig. 2).

Air is swept out of the bulb by **cautiously** and slowly reducing the pressure in the ballast bulb until the liquid boils very gently. Continue pumping for about 4 min, but be careful to avoid evaporating too much of the liquid. Then tilt the isoteniscope

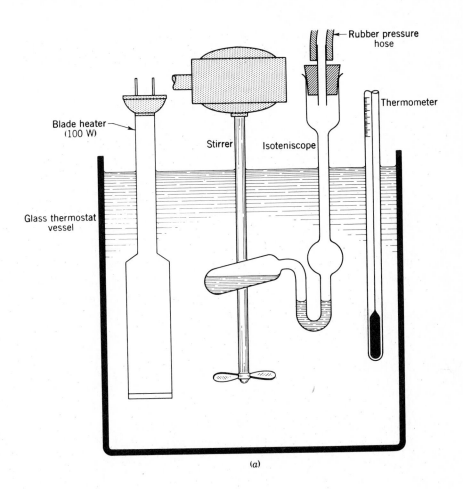

(a)

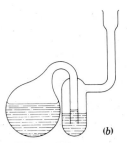

(b)

FIGURE 3
Isoteniscope: (a) Schematic diagram of the apparatus which must be connected to the ballast bulb and manometer shown in Fig. 2; (b) an alternate design for the isoteniscope. (An even more elaborate type of isoteniscope is described by Arm et al.[7])

so that some liquid from the bulb is transferred into the U tube. Carefully admit air to the ballast bulb through stopcock S until the levels of the liquid in both arms of the U tube are the same. Read and record the temperature and the pressure on the mercury manometer. (The procedure for operating and reading this manometer is described in Section 1.) To establish that all the air has been removed from the isoteniscope, **cautiously** reduce the pressure in the ballast bulb until a few bubbles are observed passing through the U-tube liquid. Then determine the equilibrium pressure reading. Repeat the above procedure, if necessary, until successive vapor-pressure readings are in good agreement.

Once the air is removed and a good pressure reading at room temperature is obtained, heat the thermostat bath to a new temperature about 5°C above room temperature. Keep the liquid levels in the U tube approximately equal at all times. When the bath temperature is steady at its new value, adjust the pressure in the ballast bulb until the levels in the U tube are equal and record both temperature and pressure.

Take readings at approximately 5°C intervals until the bath is at about 75°C, and then take readings at decreasing temperatures which are between the values obtained on heating.

CALCULATIONS

Correct all manometer pressure readings ($h_2 - h_1$) for the fact that the mercury is not at 0°C by multiplying by $(1 - 1.8 \times 10^{-4}t)$, where t is the Celsius temperature of the manometer. If t has been reasonably constant during the experiment, use an average value and apply the same correction factor to all pressures.

Convert all Celsius temperature readings to absolute temperatures T and plot $\log_{10} p$ versus $1/T$. If there is no systematic curvature, draw the best straight line through the points. If there is noticeable curvature, draw a smooth curve through the points and also draw a straight line tangent to the curve at about the midpoint.

Determine the slope of the straight line or tangent. From Eq. (5) it follows that this slope is $-\Delta\tilde{H}_v/2.303Rz$. Estimate the value of z for the saturated vapor at the appropriate temperature from Fig. 1 and calculate $\Delta\tilde{H}_v$ in calories per mole. Report the value of the heat of vaporization and of the vapor pressure for the liquid together with the applicable temperature (corresponding to the midpoint of the range studied).

DISCUSSION

Using Fig. 1 and Eq. (8), estimate the variation in $\Delta\tilde{H}_v/z$ over the range of temperatures studied. Indicate clearly whether $\Delta\tilde{H}_v/z$ should increase or decrease with increasing temperature. Does your $\log p$ versus $1/T$ plot show a curvature of the correct sign?

Evaluate a quantitative uncertainty in your value of $\Delta \tilde{H}_v/z$ by the method of "limiting slopes." Comment on this uncertainty in relation to the variation with temperature calculated above.

Discuss possible systematic sources of error.

APPARATUS

Ballast bulb (5 liters); closed-tube manometer; double-neck distillation flask (if Claissen flask is used, a rubber policeman is needed to close off side arm); a Celsius thermometer to cover the desired range, with one-hole rubber stopper to fit flask; reflux condenser with large-hole rubber stopper to fit flask; bent glass tube in a rubber stopper to fit top of condenser; bunsen burner with hose; two long pieces of rubber tubing for circulating water through condenser; three long pieces of heavy-wall, rubber pressure tubing; two condenser clamps and clamp holders; ring stand; iron ring (and clamp holder, if necessary); transite baffle with center hole to fit around bottom of flask; 0 to 30°C thermometer.

If isteniscope is used: a glass thermostat (e.g., large battery jar); mechanical stirrer; electrical blade heater; Variac; isteniscope. (An isteniscope similar to that shown in Fig. 3a can be purchased from Kontes Glass Co., Vineland, N.J. 08360.)

Liquid, such as water, n-heptane, cyclohexane, or 2-butanone.

REFERENCES

1. Any standard text on chemical thermodynamics, such as F. T. Wall, "Chemical Thermodynamics," 2d ed., Freeman, San Francisco (1965).
2. J. H. Keenan, F. G. Keyes, P. G. Hill and J. G. Moore, "Steam Tables," Wiley, New York (1969).
3. K. S. Pitzer, D. Z. Lippmann, R. F. Curl, Jr., C. M. Huggins, and D. E. Petersen, *J. Amer. Chem. Soc.*, **77**, 3433 (1955).
4. Selected Values of Properties of Hydrocarbons, *Natl. Bur. Stand. Circ.* C461 (1947).
5. "Handbook of Chemistry and Physics," 53d ed., Chemical Rubber Publishing Co., Cleveland (1972).
6. G. W. Thomson, Determination of Vapor Pressure, in A. Weissberger (ed.), "Technique of Organic Chemistry," 3d ed., vol. I, part I, chap. IX, Interscience, New York (1959).
7. H. Arm, H. Daeniker, and R. Schaller, *Helv. Chim. Acta*, **48**, 1772 (1966).

GENERAL READING

G. W. Thomson, Determination of Vapor Pressure, in A. Weissberger (ed.), *loc. cit.*

EXPERIMENT 17. BINARY LIQUID-VAPOR PHASE DIAGRAM

This experiment is concerned with the heterogeneous equilibrium between two phases in a system of two components. The particular system to be studied is cyclohexanone-tetrachloroethane at 1 atm pressure. This system exhibits a strong negative deviation from Raoult's law, resulting in a maximum boiling point.

THEORY

For a system of two components (A and B) we have from the phase rule[1]

$$F = C - P + 2 = 4 - P \tag{1}$$

where C is the number of *components* (minimum number of chemical constituents necessary to define the composition of every phase in the system at equilibrium), P is the number of *phases* (number of physically differentiable parts of the system at equilibrium), and F is the *variance* or number of *degrees of freedom* (number of intensive variables pertaining to the system that can be independently varied at equilibrium without altering the number or kinds of phases present).

When a single phase is present the pressure p, the temperature T, and the composition X_B (mole fraction of component B) of that phase can be independently varied; thus a single-phase two-component system at equilibrium is defined, except for its size,† by a point in a three-dimensional plot in which the coordinates are the intensive variables p, T, and X_B (see Fig. 1). When two phases are present at equilibrium, e.g., liquid L and vapor V, there are four variables but only two of them can be independently varied. Thus, if p and T are specified, X_{BL} and X_{BV} (the mole fractions of B in L and V) are fixed at their *limiting* values ($X_{BL}{}^0$ and $X_{BV}{}^0$) for the respective phases at this p and T. The loci of points $X_{BL}{}^0(p,T)$ and $X_{BV}{}^0(p,T)$ constitute two surfaces, shown in Fig. 1. The shaded region between them may be interpreted as representing the coexistence of two phases L and V *if* in this region X_B is interpreted as a mole fraction of B *for the system as a whole*. Within the two-phase region X_B is *not* to be regarded as one of the intensive variables constituting the variance (although in a single-phase region it is indeed one of these variables). In a two-phase region the value of X_B determines the relative proportions of the two phases in the system; as X_B varies from $X_{BL}{}^0$ to $X_{BV}{}^0$, the molar proportion x_V of vapor phase varies from zero to unity:

$$x_V = 1 - x_L = \frac{X_B - X_{BL}{}^0}{X_{BV}{}^0 - X_{BL}{}^0} \tag{2}$$

† The complete definition of the system would, of course, include also its shape, description of surfaces, specification of fields, etc.; ordinarily these have negligible effects as far as our present discussion is concerned.

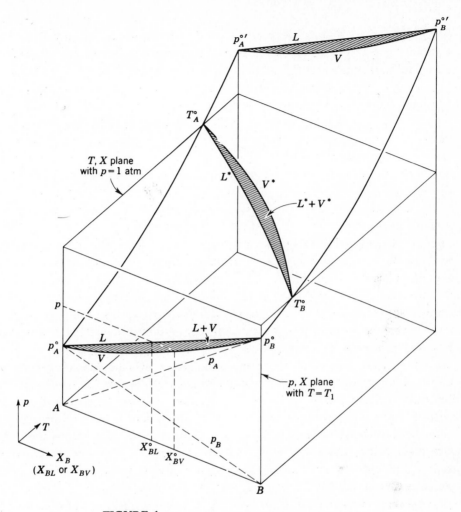

FIGURE 1
Schematic three-dimensional vapor-liquid equilibrium diagram for a two-component system obeying Raoult's law.

Figure 1 is drawn for the special case of two components which form a complete range of *ideal solutions*, i.e., solutions which obey Raoult's law with respect to both components at all compositions. According to this law the vapor pressure (or partial pressure in the vapor) of a component at a given temperature T_1 is proportional to its mole fraction in the liquid. Thus, in Fig. 1 the light dashed lines representing the partial pressures p_A and p_B and the total vapor-pressure line (L, joining $p_A^{\,0}$ and $p_B^{\,0}$) are straight lines when plotted against the liquid composition. However, the total vapor pressure as plotted against the *vapor* composition is not linear. The curved line V, joining $p_A{}^0$ and $p_B{}^0$, is convex downward, lying *below* the straight line on the constant-temperature section; its slope has everywhere the same sign as the slope of L.

The vapor pressures $p_A{}^0$ and $p_B{}^0$ of the pure liquids increase with temperature (in accord with the Clapeyron equation) as indicated by the curves joining $p_A{}^0$ with $p_A'^0$ and $p_B{}^0$ with $p_B'^0$. At a constant pressure, say 1 atm, the boiling points of the pure liquids are indicated as $T_A{}^0$ and $T_B{}^0$. The boiling point of the solution, as a function of X_{BL} or X_{BV}, is represented by the curve L^* or V^* joining these two points. Neither curve is in general a straight line. If Raoult's law is obeyed, both are convex upward in temperature, the vapor curve lying above the liquid curve in temperature and being the more convex.

In most binary liquid-vapor systems Raoult's law is a good approximation for a component only when its mole fraction is close to unity. Large deviations from this law are commonplace for the dilute component, or for both components when the mole fraction of neither is close to unity. If at a given temperature the vapor pressure of a solution is higher than that predicted by Raoult's law, the system is said to show a *positive deviation* from that law. For such a system the boiling-point curve L^* at constant pressure is usually convex downward in temperature. If at a given temperature the vapor pressure of the solution is lower than that predicted by Raoult's law, the system is said to show a *negative deviation*; in this case the curve L^* is more convex upward. These deviations from Raoult's law are often ascribed to differences between "heterogeneous" molecular attractions ($A---B$) and "homogeneous" attractions ($A---A$ and $B---B$). Thus, the existence of a positive deviation implies that homogeneous attractions are stronger than heterogeneous attractions, and a negative deviation implies the reverse. This interpretation is consistent with the fact that positive deviations are usually associated with positive heats of mixing and volume expansions on mixing while negative deviations are usually associated with negative heats and volume contractions.[2]

In many cases the deviations are large enough to result in maxima or minima in the vapor-pressure and boiling-point curves, as shown in Fig. 2. Systems for which the boiling-point curves have a maximum include acetone-chloroform and hydrogen chloride–water; systems with a minimum include methanol-chloroform, water-ethanol, and benzene-ethanol. At a maximum or a minimum the compositions of the liquid and of the vapor are the same; accordingly there is a *point of tangency* of the curves L and V and of the curves L^* and V^* at the maximum or minimum. At every value of X_B the slope of V (or V^*) has the same sign as the slope of L (or L^*); one is zero where and only where the other is zero, at the point of tangency. (A common error in curves of this kind, found even in some textbooks, is to draw a cusp—point of discontinuity of slope—in one or both curves at the point of tangency; both curves are, in fact, smooth and have continuous derivatives.)

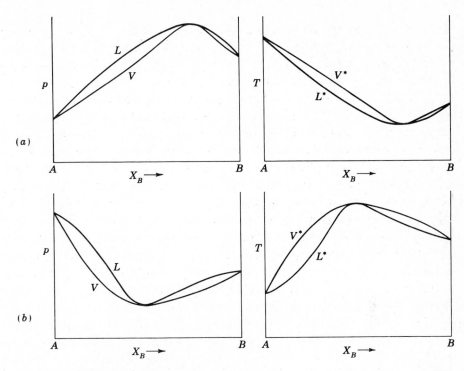

FIGURE 2
Schematic vapor-pressure and boiling-point diagrams for systems showing
(*a*) a strong positive deviation and (*b*) a strong negative deviation from Raoult's
law.

If the homogeneous attractions are very much stronger than the heterogeneous
ones, phase separation may occur in the liquid; i.e., there is a limited mutual solubility
of the two liquid components over certain pressure and temperature ranges. If the
pressures and temperatures at which two liquid phases coexist include those at which
equilibrium also exists with the vapor phase, a boiling-point diagram of a type similar
to that shown in Fig. 18-1 is found. (That figure is a *melting-point* diagram, showing
two solid phases in equilibrium with a liquid phase, but the principles are the same.
It will be noted that boundaries of fields show discontinuities in slope at points
representing the coexistence of three phases.)

Liquid-vapor phase diagrams, and boiling-point diagrams in particular, are of
importance in connection with *distillation*, which usually has as its object the partial
or complete separation of a liquid solution into its components.[3] Distillation consists

basically of boiling the solution and condensing the vapor into a separate receiver. A simple "one-plate" distillation of a binary system having no maximum or minimum in its boiling-point curve can be understood by reference to Fig. 3. Let the mole fraction of B in the initial solution be represented by X_{BL1}. When this is boiled and a small portion of the vapor condensed, a drop of distillate is obtained, with mole fraction X_{BV1}. Since this is richer in A than is the residue in the flask, the residue becomes slightly richer in B, as represented by X_{BL2}. The next drop of distillate X_{BV2} is richer in B than was the first drop. If the distillation is continued until all the residue has boiled away, the last drop to condense will be virtually pure B. To obtain a substantially complete separation of the solution into pure A and B by distillations of this kind it is necessary to separate the distillate into portions by changing the receiver during the distillation, then subsequently to distill the separate portions in the same way, and so on, a very large number of successive distillations being required. The same result can be achieved in a single distillation by use of a fractionating column containing a large number of "plates"; discussion of the operation of such a column is beyond the scope of this book. If there is a maximum in the boiling-point curve (Fig. 2b) the compositions of vapor and residue do not approach pure A or pure B, but rather the composition corresponding to the maximum. A mixture with this composition will distill without change in composition and is known as a "constant-boiling mixture" or "azeotrope." These terms are also applied to a mixture with a minimum boiling point. Azeotropes are important in chemical technology. Occasionally they are useful (as in constant-boiling aqueous hydrochloric acid, used as an analytical standard); often they are nuisances (as in the case of the azeotrope of 95 percent ethanol with 5 percent water, the existence of which prevents preparation of

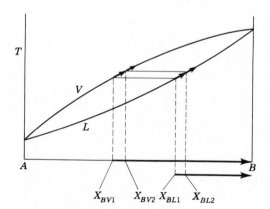

FIGURE 3
Variation of liquid and vapor compositions during distillation.

absolute ethanol by direct distillation of dilute solutions of ethanol in water). Extensive lists of azeotropes have been compiled.[4]

METHOD

A boiling-point curve can be constructed from data obtained in actual distillations in an ordinary "one-plate" distilling apparatus. Small samples of the distillate are taken directly from the condenser, after which small samples of the residue are withdrawn with a pipette. The samples of distillate and residue are analyzed, and their compositions are plotted on a boiling-point diagram against the temperatures at which they were taken. In the case of the distillate the temperature to be plotted for each sample should be an average of the initial and final values during the taking of the sample. In the case of the residue the temperature to be plotted should be that recorded at the point where the distillation is stopped to take the sample of residue.

For analysis of the samples, a physical method is often preferable to chemical methods. Chemical analysis usually is appropriate only when a simple titration of each sample is involved, as in the case of the system $HCl-H_2O$. If a physical property is chosen as the basis for an analytical method, it should be one which changes significantly and smoothly over the entire composition range to be studied. The refractive index is one property that can be used for the cyclohexanone-tetrachloroethane system. The values of n_D^{20} are 1.4507 for cyclohexanone and 1.4942 for tetrachloroethane, and log n_D^{20} is almost a linear function of the weight percent of cyclohexanone. Thus one can interpolate linearly between the values listed in Table 1, and then convert weight percents into mole fractions.

Another property that could also be used is the density, which varies in a nonlinear way between 0.9478 g/ml for cyclohexanone and 1.600 g/ml for tetrachloroethane at 20°C. In this case, a calibration curve should be constructed from known solutions or provided by the instructor. The density of each distillation

Table 1   LOGARITHM OF REFRACTIVE INDEX FOR CYCLO-
HEXANONE-TETRACHLOROETHANE MIXTURES

$\log n_D^{20}$	W% $C_6H_{10}O$	$\log n_D^{20}$	W% $C_6H_{10}O$	$\log n_{lD}^{20}$	W% $C_6H_{10}O$
0.17441	0	0.16864	40	0.16360	80
0.17298	10	0.16719	50	0.16256	90
0.17155	20	0.16582	60	0.16158	100
0.17010	30	0.16473	70		

sample can be measured by pipetting 1 ml into a small, previously weighed vial and then weighing again (to the nearest 0.1 mg).

Warning: These solutions will decompose slowly at room temperature ($\sim$1 day) and more rapidly during distillation at high temperatures. Impure solutions become yellowish, which can interfere with the refractive index measurements. Use well-purified starting materials and do not prolong the distillations unnecessarily.

EXPERIMENTAL

A simple distilling apparatus that can be used for this experiment is shown in Fig. 4. The thermometer bulb should be about level with the side arm to the condenser. Except when samples of distillate are being taken for analysis, an adequate receiving flask should be placed at the lower end of the condenser.

Before beginning the distillations, prepare twenty 5-ml shell vials for taking samples. Write on the corks the designations $1L, 1V, 2L, \ldots, 10V$ ($L = liquid$ residue; $V = $ condensed *vapor* or distillate). The samples to be taken are about 2 ml in size.

When the distillation is proceeding at a normal (not excessive) rate at about the desired temperature, quickly replace the receiver with a vial and read the thermometer. After about 2 ml has been collected, read the thermometer again, replace the receiver, and cork the vial tightly. Turn off and lower the heating mantle to halt the distillation. At the point where the temperature just begins to fall, record another thermometer reading. After the flask has cooled 10 or 20°, remove the stopper at the top of the flask and insert a 2-ml pipette equipped with a rubber bulb. Fill the pipette, discharge it into the appropriate vial, and stopper the vial.

The following procedure is recommended for economical use of materials in carrying out this experiment. The paragraph numbers correspond to sample numbers. A graduated cylinder is adequate for measuring liquids. The temperatures recommended are those appropriate for 760 Torr; at ambient pressures differing markedly from this the temperatures should be adjusted accordingly.

  1 Pure tetrachloroethane: Introduce 125 ml ($\sim$200 g) of tetrachloroethane into the flask. Distill enough to give a constant temperature (should be near 146°C at 760 Torr). Collect samples ($1V$ and $1L$) for analysis.
  2 149°C (tetrachloroethane-rich side of azeotrope): Cool the distilling flask, and return the excess distillate of paragraph 1 to the flask. Add 38 ml ($\sim$36 g) of cyclohexanone. Begin distillation. When the temperature reaches 149°C, collect about 2 ml of distillate ($2V$) and 2 ml of residue ($2L$).

3 151°C: Resume the distillation. Distill until the temperature reaches 151°C (this may take some time) and collect samples (3V, 3L).

4 154°C: Resume the distillation. When the temperature reaches 154°C, collect samples (4V, 4L).

5 157°C: Cool the flask somewhat, and add 35 ml of tetrachloroethane and 25 ml of cyclohexanone. Resume the distillation. When the temperature reaches approximately 157°C, collect samples (5V, 5L).

6 Azeotrope: Cool the flask somewhat and add 36 ml of tetrachloroethane and 54 ml of cyclohexanone. Resume the distillation until the boiling point ceases to change significantly, and take samples (6V, 6L). (If the boiling point does not become sufficiently constant, analyze the remaining residue, and make up 100 ml of solution to the composition found. Distill to constant temperature and take samples.)

7 Pure cyclohexanone: Introduce 105 ml of cyclohexanone into the clean flask and determine the boiling point as in paragraph 1. (The temperature should be near 155°C at 760 Torr.) Collect samples (7V and 7L).

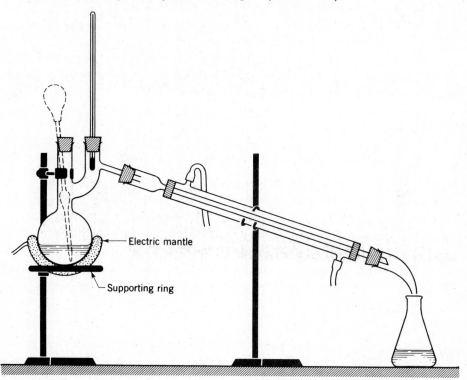

Electric mantle

Supporting ring

FIGURE 4
Distillation apparatus.

8   156.5°C(cyclohexanone-rich side of azeotrope): Cool the distilling flask, return the excess distillate of paragraph 7, and add 20 ml of tetrachloroethane. Resume the distillation, and take samples ($8V$, $8L$) at about 156.5°C.

9   157°C: Cool the flask somewhat, and add 50 ml cyclohexanone and 17 ml tetrachloroethane. Resume the distillation and collect samples ($9V$, $9L$) at about 157°C.

10   Azeotrope: Resume the distillation, continue to constant boiling temperature, and take samples ($10V$, $10L$).

As soon as possible (the samples decompose on standing), the indexes of refraction should be measured and recorded. The refractometer and the procedure for its use are described in Chap. XVIII. (If the experiment is being done by several teams using the same refractometer, it is wise to take samples to the refractometer as soon as six or eight samples are ready, or fewer if the instrument happens to be free.) If careful attention is given to the proper technique of using the refractometer, it should be possible to take readings at the rate of 1 sample per min.

At the end of the experiment all cyclohexanone-tetrachloroethane mixtures should be poured into a designated waste vessel.

At some time during the laboratory period, the barometer should be read. The ambient temperature should be recorded for the purpose of making thermometer stem corrections.

CALCULATIONS

Interpolate in Table 1 and then convert the weight percents to mole fractions. Plot the temperatures (after making any necessary stem corrections; see Chap. XVI) against the mole fractions. Draw one smooth curve through the distillate points V and another through the residue points L. Label all fields of the diagram to indicate what phases are present. Report the azeotropic composition and temperature, together with the atmospheric pressure (i.e., the properly corrected barometer reading).

APPARATUS

Claissen distilling flask; 0 to 100 or 50 to 100°C thermometer, graduated to 0.1°C; one-hole cork stopper for thermometer to fit flask; solid cork stopper; straight-tube condenser with one-hole stopper to fit distilling side arm; two lengths of rubber hose for condenser cooling water; distilling adapter with one-hole stopper to fit end of condenser; two clamps and clamp holders; two ring stands; one iron ring; electrical heating mantle (or steam bath); 20 small vials (screw-top or with corks to fit); 100-ml graduated cylinder; two wide-mouth 250-ml flasks; 2-ml pipette; pipetting bulb; two 500-ml glass-stoppered Erlenmeyer flasks.

Refractometer, thermostated at 25°C; sodium-vapor lamp (optional); eye droppers; *clean* cotton wool; acetone wash bottles; pure tetrachloroethane (300 ml) and pure cyclohexanone (350 ml); acetone for rinsing; large bottle for disposal of waste solutions.

REFERENCES

1. W. J. Moore, "Physical Chemistry," 4th ed., chap. 7, Prentice-Hall, Englewood Cliffs, N.J. (1972).
2. A. A. Noyes and M. S. Sherrill, "A Course of Study in Chemical Principles," 2d ed., p. 260, Macmillan, New York (1938).
3. A. J. Teller, *Chem. Eng.*, **61**, 168 (1954).
4. L. H. Horsley, "Azeotropic Data" (no. 6 of "Advances in Chemistry"), American Chemical Society, Washington (1952).
5. J. C. Chu, S. L. Wang, S. L. Levy, and R. Paul, "Vapor-Liquid Equilibrium Data," Edwards, Ann Arbor, Mich. (1956).

GENERAL READING

A. Findlay, "The Phase Rule and its Applications," 9th ed. (by A. N. Campbell and N. O. Smith), Dover, New York (1951).
C. S. Robinson and E. R. Gilliland, "Elements of Fractional Distillation," 4th ed., McGraw-Hill, New York (1950).
E. Hála, J. Pick, V. Fried, and O. Vitím (translated by G. Standart), "Vapour-Liquid Equilibrium," Pergamon, New York (1958).

EXPERIMENT 18. BINARY SOLID-LIQUID PHASE DIAGRAM

In this experiment we are concerned with the heterogeneous equilibrium between solid and liquid phases in a two-component system. From the many systems[1,2,3] that are suitable for study, the system naphthalene-diphenylamine has been selected for this experiment because of the simplicity of its phase diagram and the convenience of its temperature range.

THEORY

The principles underlying this experiment are identical with those discussed in Exp. 17. For our discussion here we can merely replace L (liquid) in that experiment by S (solid) and V (vapor) by L (liquid). Solid-liquid equilibria differ from liquid-vapor equilibria in being essentially independent of pressure changes of the order of a few atmospheres. This is a consequence of the Clapeyron equation, Eq. (16-3), owing to the small molar volume change associated with fusion. Accordingly we shall be concerned only with temperature-composition diagrams at 1 atm pressure.

All types of phase diagrams that have been found for liquid-vapor equilibria are also possible for solid-liquid equilibria. (The reverse of this statement, however, is not true.) In some binary systems (particularly metal systems) the two components may form *solid solutions*, sometimes covering the entire composition range from pure A to pure B. Examples of systems with solid solutions covering the entire range include copper-nickel,[4] which has a phase diagram of the type shown in Fig. 17-3, with a nearly linear dependence of melting point on composition; d- and l-carvoxime $(C_{10}H_{14}NOH)$,[5] which has a maximum melting point (analogous to the maximum boiling point shown in Fig. 17-2b); and bromobenzene-iodobenzene,[6] which has a minimum melting point (analogous to Fig. 17-2a). An important condition for the existence of a complete range of solid solutions is that A and B have the same type of crystal structure. The solid solutions must also have the same crystal structure as A and B, the atomic sites being occupied largely at random by the two kinds of atoms or molecules.

More often when solid solutions exist, they are limited in the range of their compositions. This may result from a difference between the crystal structures of pure A and pure B, or from differences in atomic or molecular size and shape resulting in "lattice incompatibility," or from factors analogous to those resulting in strong positive deviations from Raoult's law in liquids, or from any combination of these. Thus we may have two solid solutions at equilibrium, one (α) consisting predominantly of component A and the other (β) consisting predominantly of component B. These phases are often described as a solid solution of B in A and one of A in B, respectively.

Solid-liquid equilibria in a system of limited solid solubilities may yield a phase diagram of the type shown in Fig. 1. For this diagram, the compositions of the two phases (α and β, α and L, or β and L) coexisting at equilibrium at a given temperature and 1 atm pressure are found in the same way as in Exp. 17. A new feature is the possibility of the coexistence, at equilibrium, of three phases: α, β, and L. According to the phase rule, Eq. (17-1), with two components and three phases there is but one degree of freedom, and that is taken up in the arbitrary specification of the pressure as 1 atm. Thus the temperature is fixed (T_E) and the compositions of all phases are

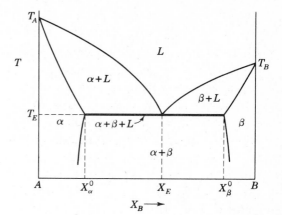

FIGURE 1
Schematic solid-liquid phase diagram at 1 atm for a binary system with limited solid solubilities and no compound formation.

fixed ($X_B = X_\alpha{}^0, X_\beta{}^0, X_E$). A system of three phases coexisting at equilibrium is represented by any point on the heavy solid horizontal line of Fig. 1. The point T_E, X_E is called the "eutectic point"; its significance will be discussed later. Diagrams of the kind shown in Fig. 1, exhibiting limited solid solubility, exist for many systems, including azobenzene-azoxybenzene,[7] bismuth-tin, and lead-tin.

Very commonly, particularly in organic systems, where a high degree of lattice incompatibility is almost always present, solid solubility is so small that it may be regarded as negligible. Here the solid-solution regions α and β shrink to the vertical lines A and B and the type of phase diagram shown in Fig. 2a is obtained. The principal features of this type of diagram can be understood at least semiquantitatively from the theory of freezing-point depression. (It will be recalled that an essential requirement for the theory of freezing-point depression developed in Exp. 12 is the absence of appreciable solid solubility.) From Eq. (12-5) we obtain the following equation for the solid-liquid curve ("liquidus curve") starting from the left at T_A, the melting point of A:

$$ T \cong T_A + \frac{RT_A{}^2}{\Delta \tilde{H}_A} \ln (1 - X_B) = T_A - \frac{RT_A{}^2}{\Delta \tilde{H}_A} \left(X_B + \frac{X_B{}^2}{2} + \cdots \right) \qquad (1) $$

where $\Delta \tilde{H}_A$ is the heat of fusion of A. Clearly this curve starts with a finite negative slope determined by the melting point and heat of fusion of pure component A; the slope increases in steepness with increasing X_B, so that the curve is convex upward. Similarly, for the liquidus curve starting from the right at T_B we obtain

$$ T \cong T_B + \frac{RT_B{}^2}{\Delta \tilde{H}_B} \ln X_B = T_B - \frac{RT_B{}^2}{\Delta \tilde{H}_B} \left[(1 - X_B) + \frac{(1 - X_B)^2}{2} + \cdots \right] \qquad (2) $$

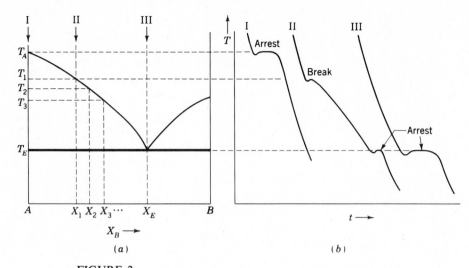

FIGURE 2

(a) Schematic solid-liquid phase diagram at 1 atm for a binary system with negligible solid solubilities and no compound formation. (b) Schematic cooling curves for this system at various overall compositions.

The eutectic composition and eutectic temperature are given by the intersection of the two liquidus curves and can be estimated by solving simultaneously Eqs. (1) and (2), on the assumption that the liquid represents an ideal solution with respect to both components over its entire composition range. This assumption is often not even roughly valid, especially in metal systems.

With the aid of Fig. 2a we can predict the general nature of the cooling curves in a system of this kind. These curves, examples of which are shown in Fig. 2b, are plots of temperature against time obtained when liquid solutions of various compositions are allowed to cool by slow leakage of heat to the surroundings; some features of such curves are discussed in Exp. 12. When a liquid consisting of pure A is cooled, the temperature falls until solid A begins to form; the temperature then remains constant until solidification is complete, whereupon it falls again. Thus, the curve shows a "thermal arrest." While two phases are present in this one-component system, there is only one degree of freedom, which is taken up in the arbitrary specification of the pressure, and the temperature is fixed. When a liquid having the eutectic composition is cooled, the behavior is similar in that a thermal arrest is obtained. Although the number of components is increased from one to two, the number of phases at the eutectic point is increased from two to three, and again we have a single degree of freedom which is taken up in the arbitrary specification of the pressure. When a liquid of some other composition—say X_1 in Fig. 2a—

is cooled, solid A begins to form at temperature T_1. This tends to deplete the liquid of component A, so that its composition passes through X_2, X_3, . . . , and the temperature falls as long as solid A alone continues to come out of solution. With two components and two phases there are two degrees of freedom; thus, at constant pressure the temperature is not fixed but varies with the composition of the liquid. However, the slope of the cooling curve is much less than that for cooling of a single phase, owing to the heat liberated by the formation of solid A; see Eqs. (12-24) and (12-25). The abrupt change in slope, which occurs when solid A begins to form, is called a "break." When the composition of the solution finally reaches X_E, solid B begins to form together with solid A, and the two solids continue to separate from solution at the temperature T_E until no liquid remains; thus we have an arrest.

It is beyond the scope of this book to consider all possible types of solid-liquid binary phase diagrams. It will suffice here to mention two other important features, which arise when A and B can combine to form a solid compound of some stoichiometric composition A_mB_n or a solid solution varying in some degree from such a composition. Usually this compound or solid solution will differ in crystal structure both from A and from B. In Fig. 3 the phase diagram for the system Mg–Ni is shown. There are two solid compounds, Mg_2Ni and $MgNi_2$, with no appreciable solid solubility. The compound $MgNi_2$ has a sharp melting point and melts to give a liquid of the same composition. The liquidus curve has a maximum (horizontal tangent, no cusp) at this composition. The compound Mg_2Ni does not melt in this way; instead it undergoes a decomposition, or "peritectic transformation," to another solid phase, namely, $MgNi_2$, and to a liquid poorer in Ni than was the original compound. During this transformation, three phases are present—liquid, Mg_2Ni, and $MgNi_2$—and the temperature (at 1 atm) is therefore fixed. Some binary phase dia-

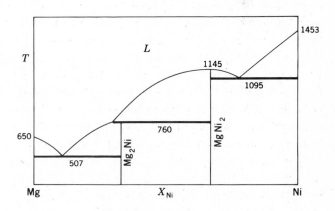

FIGURE 3
Phase diagram for the system Mg–Ni at 1 atm.

grams, particularly in metal systems, are exceedingly complicated and show many distinct compounds or phases.[1]

METHOD

The relation of cooling curves to phase diagrams (as illustrated by Fig. 2) forms the basis of "thermal analysis," an important technique for determining phase diagrams. This technique consists of cooling liquid mixtures of various compositions, making plots of temperature against time, and drawing inferences from features of these curves such as thermal arrests and breaks. A thermal arrest indicates the solidification temperature of a pure component, a compound, or a eutectic mixture (under favorable circumstances it may also indicate a peritectic transformation). A break indicates a point on the liquidus curve corresponding to the temperature at which solid first appears at the given composition.

The interpretation of such curves is often subject to experimental difficulties. A frequent difficulty is supercooling (see Fig. 12-3), which tends to obscure breaks or make them difficult to distinguish from short arrests. When the initial composition is far from the eutectic, the thermal arrest at the eutectic temperature is often hard to observe, since the amount of liquid is small relative to the amount of solid already formed and stirring becomes very difficult. When solid solubility exists, it is often very difficult to obtain the true "solidus" curve from cooling curves. This is owing to the difficulty of maintaining solid-liquid equilibrium, which requires a continuous variation in the composition of the solid. As such a variation requires either diffusion in the solid or continuous dissolution and reprecipitation of the solid, it is ordinarily very slow or nonexistent except at elevated temperatures. Owing to these difficulties, thermal analysis must often be supplemented by other techniques in elucidating a phase diagram; these include microscopic examination of solidified specimens (to determine the number of solid phases present or to distinguish regions of small crystals that form at the eutectic temperature from regions of larger crystals that form while only one solid phase is precipitating) and x-ray diffraction analysis (which enables identification of the crystalline phases that are present).

For the system naphthalene-diphenylamine, a reasonable number of cooling curves can be easily obtained using either a mercury thermometer or a thermocouple to measure the temperature. Systems with phase diagrams of greater complexity, such as phenol-α-naphthylamine, require a larger number of cooling curves; in this case, it may be necessary to assign different composition ranges to several groups who will pool their data. Work at higher temperatures on systems such as lead-tin or bismuth-tin involves no basic change in the experimental method, although some changes in the experimental arrangement may be made for convenience.[8] Low-temperature work on systems such as chlorobenzene-cyanobenzene requires a refrigerant bath (see Chap. XVI for details about several such baths). Use of a thermocouple is recommended for both high- and low-temperature systems.

EXPERIMENTAL

The materials to be used should be of reagent grade. Diphenylamine is subject to discoloration resulting from air oxidation; it should preferably be obtained in containers packed in an argon atmosphere and should be taken from a freshly opened container.

The apparatus consists of an inner test tube with a 20-ml capacity, a thermometer or thermocouple mounted in a notched one-hole rubber stopper fitting the test tube, a wire stirrer, a larger test tube to serve as an outer jacket, a large one-hole rubber stopper to support the inner test tube in the outer one, and a bath of cold water or ice and water. If a 0 to 100°C thermometer is used, it should have a stem long enough to permit all readings above 25°C to be visible above the stopper. If a thermocouple (such as copper-constantan or Chromel-Alumel) is used, the junction should be immersed in wax or oil at the bottom of a closed glass tube 6 to 8 mm in diameter. This tube can be mounted in the stopper in the same way as a thermometer. The reference junction should be in a similar tube immersed in a slushy mixture of ice and distilled water contained in a Dewar flask, and the thermoelectric potential should be measured with a potentiometer (see Chap. XV).

Make up mixtures in accordance with Table 1. Weigh materials to 0.01 g on a triple-beam balance. Note that in most instances the mixture to be studied is made by adding A or B to the previous mixture in order to minimize the quantities of materials required.

Table 1   SUGGESTED COMPOSITIONS FOR THERMAL
ANALYSIS
(A = diphenylamine; B = naphthalene)

Run no.	Wt. % B	Prepare by adding . . .		. . . to sample used in run no.
1	100	5 g	B	
2	83.3	1 g	A	1
3	66.7	1.5 g	A	2
4	50.0	2.5 g	A	3
5	33.3	5 g	A	4
6	0	5 g	A	
7	16.7	1 g	B	6
8	25.0	0.67 g	B	7
9	Eutectic	(See text)		

To obtain a cooling curve, heat the inner test tube containing the mixture in a beaker of hot water until the solid is completely melted, wipe the tube dry, place it in the outer jacket, and place the assembly in cold water or ice and water. Stir continuously, and read the temperature at regular intervals (say 30 sec). Continue readings to below 30°C if possible.†

Plot each temperature reading against time as soon as it is obtained. After each run determine break and/or arrest temperatures and plot them against weight percent *B*. From the results of runs 1 through 8 draw the liquidus curves and extrapolate them to an intersection at a point on the eutectic line. Determine the eutectic composition, and make up a mixture having this composition. Run a cooling curve on this mixture.

At the end of the experiment dispose of the mixtures and clean the glassware thoroughly. First liquefy the mixture by warming the test tube in a beaker of warm water, and then pour the contents into a designated disposal container. Remove the large rubber stopper, and wash the test tube, thermometer, and stirrer with benzene or toluene. Use several small portions in succession in order to minimize the amount of solvent required, and pour the used solvent into the designated container. Dry the apparatus.

CALCULATIONS

Convert weight-percent compositions to mole fractions. Plot the break and arrest temperatures against the overall composition X_B. Draw the eutectic line and the liquidus curves. Label all fields to show the phases present.

From the limiting slopes of the liquidus curves estimate the heats of fusion of *A* and *B*, assuming that there is no appreciable solid solubility. With these values for the heats of fusion plot ideal curves using Eqs. (1) and (2). Compare the calculated intersection temperature and composition with the eutectic temperature and composition found by experiment.

DISCUSSION

Why does the liquidus curve have a horizontal tangent at the melting point of a pure compound while it has a finite slope at the melting point of either of the two pure components?

---

† If time is limited, discontinue the cooling on all runs except 4, 8, and 9 shortly after the first definite break has occurred.

A method sometimes used for obtaining the liquidus curve in a system of sufficient optical transparency is the determination of the "clear point" (the temperature at which the last solid disappears on warming) and of the "cloud point" (the temperature at which the first solid appears on cooling) for each of a number of compositions. What would you expect to be the limitations of this procedure?

APPARATUS

Test tube with notched one-hole stopper; 0 to 100° thermometer (all the stem above 25° should be above the stopper); wire ring stirrer; outer test tube with large stopper (see Exp. 13 for above items); two 500-ml beakers; bunsen burner; tripod and wire gauze. A thermocouple with potentiometer and a small Dewar flask for the reference junction may be provided in place of the thermometer.

Diphenylamine (30 g); naphthalene (15 g); benzene or toluene (for washing glassware); ice (if needed for cold junction); waste vessel.

REFERENCES

1. M. Hansen, "Constitution of Binary Alloys," McGraw-Hill, New York (1958).
2. J. Timmermans, "Les Solutions concentrées," Masson, Paris (1936).
3. International Critical Tables (I.C.T.), vols. II (pp. 400–455) and IV (pp. 22–215), McGraw-Hill, New York (1928).
4. M. Hansen, op. cit., p. 602.
5. J. Timmermans, op. cit., p. 30.
6. Ibid., p. 88.
7. Ibid., p. 348.
8. F. Daniels, J. W. Williams, P. Bender, R. A. Alberty, C. D. Cornwell, and J. E. Harriman, "Experimental Physical Chemistry," 7th ed., pp. 123–132, McGraw-Hill, New York (1970).

EXPERIMENT 19. LIQUID-VAPOR COEXISTENCE CURVE AND THE CRITICAL POINT

The object of this experiment is to measure the densities of coexisting carbon dioxide liquid and vapor near the critical temperature. The critical density of carbon dioxide and a value of the critical temperature will also be obtained from a Cailletet-Mathias curve constructed from the data.

THEORY

At its critical point, the liquid and vapor densities of a fluid become equal. Since this point is very difficult to determine directly, an extrapolation procedure is often used. This procedure is based on the "law of rectilinear diameters,"[1,2] which states that

$$\rho_{av} \equiv \tfrac{1}{2}(\rho_l + \rho_v) = \rho_0 - cT \tag{1}$$

where ρ_{av} is the average of the densities of the liquid and vapor which are in coexistence with each other at temperature T, and ρ_0 and c are constants characteristic of the fluid. The critical point is taken to be that point at which the straight line drawn through ρ_{av} as a function of T intersects the coexistence curve (locus of the liquid and vapor density values as a function of temperature). Such a plot is known as a Cailletet-Mathias curve.

METHOD

Six to eight heavy-wall glass capillary tubes of uniform bore and known cross-sectional area are filled with different, known amounts of CO_2. The diameter of the capillary tubes can be obtained prior to filling by weighing a thread of mercury of measured length contained in them. The capillaries are then filled on a vacuum system by freezing into them CO_2 gas contained in a known volume at a known pressure and sealing them off.[3] Residue corrections for the amount of CO_2 left behind in the vacuum system after sealing off the tubes should be made if necessary. Heavy-walled capillary tubing with a nominal bore of 1 mm is convenient for use in this experiment. Lengths of 10 to 30 cm are suitable; perhaps 20 cm is the best length to use. (It may be desirable to prepare a few tubes of different lengths which are filled to the same pressure as a way of testing the reproducibility of the method.) The amounts of CO_2 to be loaded into the tubes should be chosen to yield a range of pressures between 18 and 30 atm at 25°C. *The data on the calibration and loading of the capillaries should be made available at the beginning of the laboratory period.*

The filled capillary tubes are mounted in a well-stirred, temperature-regulated water bath which can be warmed or cooled and whose temperature can be measured accurately. If a tube contains less than a critical amount of CO_2, the meniscus between the liquid and vapor will fall toward the bottom of the tube on warming. At a certain temperature, say T_0, the meniscus in a given tube will reach the bottom; thus the filling density in that tube will equal the density of the saturated vapor at the temperature T_0. This density is the total mass of CO_2 in the tube (which is known) divided by the internal volume of the tube (= cross-sectional area × length of capillary).

At temperature T_0, all of the other tubes *that still have both liquid and vapor present* will contain vapor whose density is the same as that just calculated. The volume of vapor in each of these tubes can be determined (= cross-sectional area × length of tube occupied by vapor). Therefore, the weight of CO_2 present as vapor in each of these tubes can be found (= density of vapor × volume of vapor). The weight of liquid CO_2 in each tube can then be determined (= total CO_2 − vapor). Since the volume of the liquid in each tube can also be determined (= cross-sectional area × length of tube occupied by liquid), the density of the liquid phase can be found at temperature T_0.

EXPERIMENTAL

A glass-walled constant-temperature bath with an adjustable thermoregulator and an efficient stirrer is required. This bath should also be provided with a precision thermometer and a light to illuminate the capillary tubes. A cathetometer is used to measure the height of the menisci in the capillaries, which are completely immersed in the thermostat bath. If the cathetometer is not level, it should be adjusted using the spirit levels on the base and on the telescope. Focus the telescope on one of the capillaries in the bath. The vertical cross hair should be lined up parallel to the side of the capillary. This may be achieved, if necessary, by rotating the eyepiece. When reading the height of an object, such as a meniscus, always use the same position on the horizontal cross hair—usually near the center of the field. Note that the image is inverted by the telescope.

Now check the temperature of the bath. If it is above 24°C add a little ice to cool it to that temperature; while the bath is coming to equilibrium, one can begin to measure the dimensions of the tubes (see below). If the temperature of the bath is constant at 24°C or below, record the temperature and measure the positions of the menisci in all the tubes. The readings of the cathetometer should be recorded to the nearest 0.005 cm, and one should measure one of the tubes several times to establish the reproducibility of these readings. During the period in which these cathetometer measurements of the menisci levels are being made, the bath temperature should be determined several times and recorded to the nearest 0.01°C. The temperature should not vary over a range of more than 0.04°C during the measurements, and the average of these temperature readings should be used in the subsequent calculations.

It will be convenient to number the tubes in sequence from left to right, in order to simplify the tabulation of data. Once the first set of data are recorded, begin heating the bath to the next desired temperature by making a suitable adjustment to the thermoregulator (see Chap. XVI). Recommended bath temperatures are ∼24, 26.0, 28.0, 29.2, 29.7, 30.1, 30.5°C, and higher by increments of 0.3°C (or less) until the critical temperature is reached. The bath should regulate at each temperature for at least 10 min (preferably more near T_c) before the menisci levels are measured.

Once you have begun making readings, be very careful not to jar the cathetometer or to make any more leveling adjustments. During the course of the entire experiment the only permissible movements of the cathetometer are the raising, lowering, and traversing of the telescope.

While the bath is heating up from one temperature setting to the next, carry out measurements of the dimensions of the capillary tubes. Since the top and bottom ends of these tubes are tapered or rounded, it is necessary to make two cathetometer readings at each end. As shown in Fig. 1, readings are made at the extreme tips of the inside space (a and d) and at levels where the tapering begins (b and c). The center section (bc) is a cylinder of uniform cross-sectional area; the small ends (ab and cd) can be described by cones or hemispheres.

If possible, try to observe the *critical opalescence* in carbon dioxide near the critical temperature. This milky opalescence is caused by large spatial fluctuations in the density and can only be seen if the filling density is close to the critical density of CO_2. In order to look for the opalescence, darken the room (or at least shield he bath from bright lights) and shine a flashlight on the capillary tube whose meniscus is closest to the center of the tube. Look for the scattered light at roughly 90° from the direction of the flashlight beam.

Do not be surprised if, during the experiment, you find that some menisci move towards the top of the tube with increasing temperature. Some of the menisci will become difficult to see as the critical temperature is approached, but don't give up looking too soon because only a few will disappear until just before the critical temperature.

Precaution. Do not raise the temperature of the thermostat bath appreciably above the critical temperature as the pressure increase in the capillaries may cause them to explode (the critical pressure of CO_2 is 73 atm). Although the possibility of explosion is remote, the wearing of *safety glasses*, which should be mandatory practice in all laboratory work, should be particularly emphasized in this experiment.

CALCULATIONS

First, calculate the total internal volume V^i of each of the $i = 1, 2, \ldots, N$ capillaries. The drawn-out ends can be assumed to be conical (or hemispherical) in shape. Calculate the volume of liquid V_l^i in each capillary at each temperature, and obtain the volume of the coexisting vapor V_v^i from

$$V_v^i = V^i - V_l^i \tag{2}$$

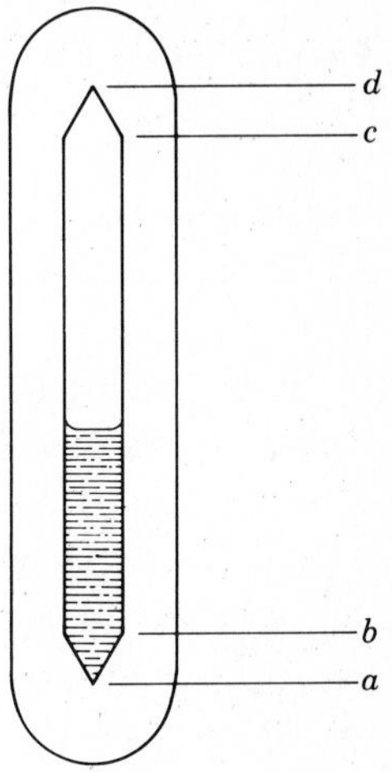

FIGURE 1
Sketch of a capillary tube with conical ends. The internal volume is given by $V = \pi r^2(h_{bc} + \frac{1}{3}h_{ab} + \frac{1}{3}h_{cd})$, where $2r$ is the inner diameter of the tube and h_{ij} represents the vertical distance between levels i and j. If the ends look more hemispherical than conical, one should find $h_{ab} \simeq h_{cd} \simeq r$; in this case, $V = \pi r^2(h_{bc} + \frac{4}{3}r)$.

The liquid and vapor densities can be determined by either a graphical method, or, preferably, by a numerical least-squares analysis as described in Chap. XX. Both of these methods are described below.

Graphical method   Plot V_l versus temperature for each capillary in which the meniscus has reached the bottom, and extrapolate to zero volume. At this temperature, call it T_0, you know the density of the vapor in each of the capillaries still containing liquid and vapor. Obtain by interpolation the volume of vapor and liquid, V_v and V_l, for each of these capillaries at T_0. From the known total weight of CO_2 in each tube, several values of the liquid density at T_0 will be obtained. Not all of these values are equally reliable. You should take this into account in arriving at the "best" value for the liquid density at T_0. If the meniscus reaches the top of any of the capillaries, a similar extrapolation of vapor volume V_v versus temperature to zero volume gives a temperature at which the liquid density in all capillaries containing both liquid and vapor is known. Several values for the vapor density can be obtained from capillaries containing liquid and vapor at this temperature, and a "best" value of the vapor density can be chosen.

Least-squares analysis   In this experiment there are two phases, and only one component is present in the tubes. Thus the phase rule tells us that there is only one degree of freedom:

$$F = C - P + 2 = 1$$

and by fixing the temperature, the vapor pressure and the densities of the liquid and vapor are also fixed. At a given temperature T, there will be N equations of the form

$$m_i = \rho_l V_l^i + \rho_v V_v^i \tag{3}$$

where m_i is the total mass of CO_2 in the ith tube. V_l^i and V_v^i are the volumes occupied by liquid and vapor in the ith tube at T. In this experiment, the number of tubes N is six or more, while only two are needed to find the two unknowns $\rho_l(T)$ and $\rho_v(T)$ in Eq. (3). Due to experimental errors no pair of density values will satisfy all N equations exactly. Thus we will rewrite Eq. (3) in terms of the residual r_i for the ith tube:

$$r_i \equiv \rho_l V_l^i + \rho_v V_v^i - m_i \tag{4}$$

The method of least squares tells us that the best possible fit to all the data is obtained when the sum of the squares of the residuals is a minimum. Thus

$$S = \sum r_i^2 = \sum_{i=1}^{N} (\rho_l V_l^i + \rho_v V_v^i - m_i)^2 \tag{5}$$

should be a minimum. This quantity can be minimized with respect to the values of the two parameters ρ_l and ρ_v by using the equations

$$\frac{\partial S}{\partial \rho_l} = 0 = 2 \sum (\rho_l V_l^i + \rho_v V_v^i - m_i) V_l^i \tag{6a}$$

$$\frac{\partial S}{\partial \rho_v} = 0 = 2 \sum (\rho_l V_l^i + \rho_v V_v^i - m_i) V_v^i \tag{6b}$$

These equations reduce to two equations in two unknowns:

$$\rho_l [\sum (V_l^i)^2] + \rho_v [\sum (V_v^i V_l^i)] = \sum m_i V_l^i \tag{7a}$$

$$\rho_l [\sum (V_l^i V_v^i)] + \rho_v [\sum (V_v^i)^2] = \sum m_i V_v^i \tag{7b}$$

A least-squares analysis can then be used to solve these equations for the best values of ρ_l and ρ_v at each temperature.

Using the resulting pairs of density values, construct a Cailletet-Mathias plot (liquid and vapor densities versus temperature). From this plot, obtain the critical temperature T_c, test the "law" of rectilinear diameters [see Eq. (1)], and obtain the critical density ρ_c.

DISCUSSION

The critical temperature and molar volume can be related to the constants of the van der Waals' equation of state:

$$\left(p + \frac{n^2 a}{V^2}\right)(V - nb) = nRT \tag{8}$$

The resulting expressions are[2]

$$\tilde{V}_c = 3b \qquad T_c = \frac{8a}{27bR} \qquad (9)$$

Using these equations and your values of T_c and $\tilde{V}_c$, calculate a and b. How well do your values agree with values tabulated in the literature? For a van der Waals' fluid, the critical pressure is given by

$$p_c = \frac{a}{27b^2} = \frac{3RT_c}{8\tilde{V}_c} \qquad (10)$$

Calculate p_c from Eq. (10) and your T_c and $\tilde{V}_c$ values, and compare it with the accepted literature value. From the empirical corresponding-states behavior of many fluids, one would expect that $p_c\tilde{V}_c/RT_c \simeq 0.28$. Will this yield a better p_c value than Eq. (10)?

Explain why the meniscus in some tubes moved down and disappeared at the bottom while the meniscus in other tubes may have moved up and disappeared at the top. Is it possible for the meniscus to remain stationary as the temperature is increased?

In recent years, it has been established experimentally and theoretically that the variation of many thermodynamic properties near a critical point can be described by the use of *critical exponents*.[4] In the case of fluids, the density difference $\rho_l(T) - \rho_v(T)$ is well represented by

$$\frac{\rho_l(T) - \rho_v(T)}{\rho_c} = B \left(\frac{T_c - T}{T_c} \right)^\beta \qquad (11)$$

where the exponent β has a value near 0.34. Thus a log-log plot of $(\rho_l - \rho_v)$ versus $(T_c - T)$ should yield a straight line of slope β. Make such a plot and determine the value of β; or, as an alternative, one can test the value 0.333 for β by making a direct plot of $(\rho_l - \rho_v)^3$ versus T. If $\beta = \frac{1}{3}$ this plot should yield a straight line.

APPARATUS

Thermostat bath provided with a heater, adjustable thermoregulator, accurate thermometer graduated to 0.02°C, a mechanical stirrer, and illuminating light; a number of sealed capillaries of uniform bore containing varying known amounts of carbon dioxide, prepared in advance; cathetometer and stand for reading the vertical position of the menisci.

REFERENCES

1. J. S. Rowlinson, "Liquids and Liquid Mixtures," 2d ed., pp. 90–94, Plenum, New York (1969).
2. W. J. Moore, "Physical Chemistry," 4th ed., pp. 20–26, 919–922, Prentice-Hall, Englewood

Cliffs, N.J. (1972); M. W. Zemansky, "Heat and Thermodynamics," 5th ed., pp. 368–371, McGraw-Hill, New York (1968).
3. M. S. Banna and R. D. Mathews, *J. Chem. Educ.*, **56**, 838 (1979).
4. H. E. Stanley, "Introduction to Phase Transitions and Critical Phenomena," pp. 9–12, 39–49, Oxford, London (1971); M. R. Moldover, *J. Chem. Phys.*, **61**, 1766 (1974).

GENERAL READING

J. S. Rowlinson, *op. cit.*, chap. 3.
H. E. Stanley, *op. cit.*, chaps. 1–3.
H. N. V. Temperley, J. S. Rowlinson, and G. S. Rushbrooke (eds.), "Physics of Simple Liquids," chap. 7, North-Holland, Amsterdam (1968).

ELECTROCHEMISTRY

EXPERIMENTS

EXPERIMENT 20. CONDUCTANCE OF SOLUTIONS

In this experiment we shall be concerned with electrical conduction through aqueous solutions. Although water is itself a very poor conductor of electricity, the presence of ionic species in solution increases the conductance considerably. The conductance of such electrolytic solutions depends on the concentration of the ions and also on the nature of the ions present (through their charges and mobilities), and conductance behavior as a function of concentration is different for strong and weak electrolytes. Both strong and weak electrolytes will be studied at a number of dilute concentrations, and from the data obtained the ionization constant for a weak electrolyte can be calculated.

THEORY

Electrolytic solutions obey Ohm's law just as metallic conductors do. Thus the current i passing through a given body of solution is proportional to the potential difference E; $E/i = R$, where R is the resistance of the body of solution in ohms. The *conductance L* is defined as the reciprocal of the resistance,

$$L = \frac{1}{R} \tag{1}$$

and is expressed in ohms^{-1}. The conductance of a homogeneous body of uniform cross section is proportional to the cross section A and inversely proportional to the length l:

$$L = \frac{LA}{l} \quad \text{or} \quad L = \frac{1}{R}\frac{l}{A} = \frac{k}{R} \tag{2}$$

231

where $\bar{L}$ is the *specific conductance* in $ohm^{-1}\ cm^{-1}$. The specific conductance is thus the reciprocal of the resistivity. The specific conductance of a solution in a cell of arbitrary design and dimensions can be obtained by first determining the cell constant k (the "effective" value of l/A) by measuring the resistance of a solution of known specific conductance. A standard solution that can be used for making this calibration is 0.02000 N potassium chloride, with $\bar{L}$ equal to 0.002768 $ohm^{-1}\ cm^{-1}$ at 25°C.[1] Once the cell constant has been found, specific conductances can be calculated from the experimental resistances by using Eq. (2).

The specific conductance $\bar{L}$ depends on the equivalent concentrations and mobilities of the ions present. For a single electrolyte giving ions A^+ and B^- and having fractional ionization α at a solute concentration c in equivalents per liter, one obtains

$$\bar{L} = \frac{\alpha c \mathscr{F}}{1000}\,(U_{A^+} + U_{B^-}) \tag{3}$$

where the U's are the true ionic mobilities and $\mathscr{F}$ is the faraday constant. Thus it is convenient to define a new quantity, the *equivalent conductance* Λ, by

$$\Lambda \equiv \frac{1000\bar{L}}{c} \tag{4}$$

Comparing Eqs. (3) and (4), we find that

$$\Lambda = \alpha \mathscr{F}(U_{A^+} + U_{B^-}) \tag{5}$$

This equivalent conductance is sometimes described as the actual conductance of that volume of solution which contains one equivalent weight of solute when placed between parallel electrodes 1 cm apart with a uniform electric field between them.

For a strong electrolyte, the fraction ionized is unity at all concentrations; thus Λ is roughly constant, varying to some extent owing to changes in mobilities with concentration but approaching a finite value Λ_0 at infinite dilution. From the effect of ion attraction on the mobilities, it can be shown theoretically[2] for strong electrolytes in dilute solution that

$$\Lambda = \Lambda_0(1 - \beta\sqrt{c}) \tag{6}$$

Using this relation, Λ_0 for strong electrolytes can be obtained experimentally from measurements of conductance as a function of concentration. At infinite dilution the ions act altogether independently, and it is then possible to express Λ_0 as the sum of the limiting conductances of the separate ions:

$$\Lambda_0 = \lambda_0^+ + \lambda_0^- \tag{7}$$

For a weakly ionized substance, Λ varies much more markedly with concentration because the degree of ionization α varies strongly with concentration. The equivalent conductance, however, must approach a constant finite value at infinite dilution, Λ_0, which again corresponds to the sum of the limiting ionic conductances. It is usually impractical to determine this limiting value from extrapolation of Λ values obtained with the weak electrolyte itself, since to obtain an approach to complete ionization the concentration must be made too small for effective measurement of conductance. However, Λ_0 for a weak electrolyte can be deduced from Λ_0 values obtained for *strong electrolytes* by the use of Eq. (7).

For sufficiently weak electrolytes, the ionic concentration is small and the effect of ion attraction on the mobilities is slight; thus we may assume the mobilities to be independent of concentration and obtain the approximate expression

$$\alpha \cong \frac{\Lambda}{\Lambda_0} \tag{8}$$

If one measures Λ for a weak electrolyte at a concentration c and calculates Λ_0 from the conductivity data for strong electrolytes as described above, it is possible to obtain the actual degree of ionization of the weak electrolyte at concentration c.

Equilibrium constant for weak electrolyte   Knowing the concentration c of the weak electrolyte, say HAc, and its degree of ionization α at that concentration, the concentrations of H^+ and Ac^- ions and of un-ionized HAc can be calculated. Then the equilibrium constant in terms of concentrations K_c can be calculated from

$$K_c = \frac{(H^+)(Ac^-)}{(HAc)} = c\,\frac{\alpha^2}{1-\alpha} \tag{9}$$

This equilibrium constant K_c given by Eq. (9) using α values obtained from Eq. (8) differs from K_a, the equilibrium constant in terms of activities, by virtue of the omission of activity coefficients $(\gamma_\pm{}^2)$ from the numerator of (9) and owing to the approximations inherent in (8). To an approximation which would be very rough at ordinary dilute concentrations but is fairly good at the very low ionic concentrations encountered in the dissociation of a weak electrolyte, the factors by which $\log K_c$ differs from $\log K_a$ are linear functions of the square root of the ionic strength.[3] We may here take the ionic strength to be αc. Thus, if $\log K_c$, as determined at a number of low concentrations, is plotted against $\sqrt{\alpha c}$, an extrapolation to $c = 0$ should give a fairly good value of $\log K_a$. However, reliable measurements at low concentrations are difficult to obtain.

METHOD

For determining ionic conductance by measuring the resistance of the solution in a conductivity cell, the use of dc circuitry is impractical, since the electrodes would quickly become "polarized"; that is, electrode reactions would take place which would set up an emf opposing the applied emf, leading to a spuriously high apparent cell resistance. Polarization can be prevented by (1) using a high (audio) frequency *alternating current*, so that the quantity of electricity carried during one half cycle is insufficient to produce any measurable polarization, and at the same time by (2) employing platinum electrodes covered with a colloidal deposit of "platinum black," having an extremely large surface area, to facilitate the adsorption of the tiny quantities of electrode reaction products produced in one-half cycle so that no measurable chemical emf is produced.

The resistance of a conductivity cell filled with an ionic solution can be measured accurately by use of a *Wheatstone-bridge* circuit employing high-frequency alternating current, where an audio oscillator is used as the source and where the detector is a telephone receiver or oscilloscope. A schematic diagram of an ac Wheatstone bridge and a discussion of the condition of balance are given in Chap. XV. In order for a sharply defined balance point to be observed, it is necessary that

$$\theta_1 = \theta_2 \quad \text{and} \quad \theta_3 = \theta_4 \tag{10}$$

where θ_i is the phase shift in the ith arm of the bridge. It may also be desirable to use

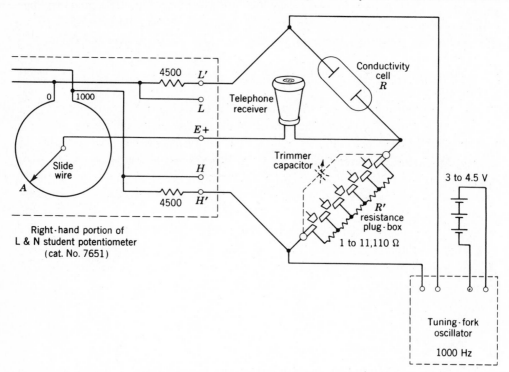

FIGURE 1
Conductance bridge (long-bridge circuit shown).

a Wagner earthing circuit (see Fig. XV-6) although this is not shown in the bridge arrangement given in Fig. 1.

The arms R_3 and R_4 are the two parts of the slide-wire of a Student Potentiometer. There are two sets of connections to this slide-wire—L and H or L' and H'— which give R_3/R_4 equal to $A/(1000 - A)$ or $(4500 + A)/(5500 - A)$, respectively. These are the so-called short and long bridges and are shown schematically in Fig. 2a and b. The reading A is the number on the 0 to 1000 scale of the slide-wire. Arm 1 contains the conductivity cell of resistance R, and arm 2 contains a precision plug box or decade resistor R'. Since the balance is always taken with A near 500, arms 3 and 4 are identical and $\theta_3 = \theta_4$ is assumed to hold. A trimmer capacitor may be placed across the plug box R' to balance the effective capacitance of the cell and make $\theta_1 = \theta_2$. Under the experimental conditions encountered here the reactances should be extremely small in comparison with the dc resistances. To a first approximation, a finite minimum signal in the detector may be taken to indicate the point of balance if a zero signal is not obtained. The capacitive reactance can then be effectively balanced with a trimmer capacitor if necessary, and a small adjustment then made to yield the final balance. In the present experiment a trimmer capacitor is seldom needed. In the case of a significant degree of polarization, very poor balance points are observed owing to large effective phase shifts resulting from severe distortion of the wave form. Although Fig. 1 shows a telephone receiver and the procedure below is written in terms of such a detector, the alternate use of an oscilloscope is strongly recommended since one can observe both the capacitive and the resistive balance (see Fig. XVIII-4).

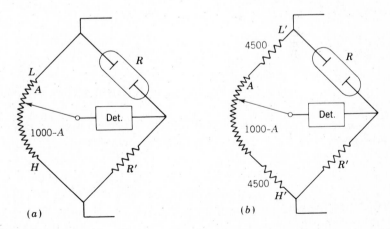

FIGURE 2
(a) Short bridge: $R = R'[A/(1000 - A)]$. (b) Long bridge: $R = R'[(4500 + A)/(5500 - A)]$.

EXPERIMENTAL

Set up the circuit as shown in Fig. 1, sandpapering all lugs before making connections. For short-bridge work, connect to terminals L and H on the potentiometer slide-wire. In this case the cell resistance R is given by

$$R = \frac{A}{1000 - A} R' \tag{11}$$

where A is the slide-wire reading and R' is the resistance of the resistance box. The short bridge is good for rapid, moderately precise work, but much greater accuracy can be obtained using the long-bridge setup. For long-bridge work, connect to terminals L' and H' on the potentiometer. In this case,

$$R = \frac{4500 + A}{5500 - A} R' \tag{12}$$

All possible measurements on solutions should be made with the long bridge. The short bridge is primarily of use in obtaining an approximate value of R. When this is known, the resistance box should be set to this value of R (or some round number within 5 percent of R) and the long bridge used for a precise determination of the cell resistance.

In making connections to the resistance plug box, be sure that all connections are *tight*. Do not use terminals both as wire clamps and as binding posts for lugs at the same time. In using the plug box, remove plugs corresponding to the desired resistance. Each plug in place shorts out a resistance inside the box. Plugs not removed must be in tight; seat them with a strong twisting motion.

The ac source for the bridge may be a battery-operated tuning-fork oscillator which gives out a signal in the audio range (1000 Hz). The power output from this oscillator can be varied by changing the voltage supplied by the batteries or by inserting a resistor in series with the bridge. The bridge should be operated with as low a power and for as short a time as possible to prevent polarization effects. Alternatively, a commercial electronic audio oscillator may be used.

Calibration check   There may be a small calibration error in the slide-wire; that is, the position of the indicator may not correspond precisely to that of the contact on the slide-wire. This should be checked by replacing the cell in the bridge circuit with a second resistance box. Set both boxes to 1000Ω and balance on the long bridge. If there is no error, the A reading should be exactly 500. To obtain a correction figure, make four readings of the balance point approaching zero or minimum hum twice from each side. Then reverse the position of the two plug boxes and take four more readings. Find the average of these values $\overline{A'}$. The correction term $(500 - \overline{A'})$ is to be *added* to all future long-bridge readings of A.

A satisfactory conductivity cell design is shown in Fig. 3. This cell is fragile

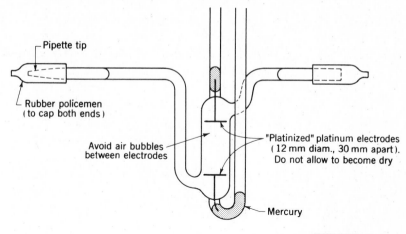

Pipette tip

Rubber policemen
(to cap both ends)

Avoid air bubbles
between electrodes

"Platinized" platinum electrodes
(12 mm diam., 30 mm apart).
Do not allow to become dry

Mercury

FIGURE 3
Conductivity cell.

and should be handled with care. The leads should be arranged to avoid placing any strain on the cell while it is mounted in the thermostat bath. The electrodes are sensitive to poisoning if very concentrated electrolytes are placed in the cell and to deterioration if allowed to become dry. (Details concerning the preparation of "platinized" platinum electrodes are given in Chap. XIX.) After each emptying, the cell must be *immediately* rinsed or filled with water or solution. While the cell is in the bath, the filling arms must be stoppered by placing rubber policemen over the ends. If the cell is allowed to stand for any long time between runs, leave it filled with conductivity water rather than solution.

Draw liquid into the cell by sucking it in through the tapered tip as you would with a pipette. Empty the cell by allowing it to drain out the other end. The procedure for filling the cell with a new solution is to empty the cell, rinse it at least *four times* with aliquots of the desired solution using about enough to fill the cell one-fourth full, shake well on each rinsing, then fill the cell, being careful to avoid bubbles on the electrodes. Be careful to avoid loss of mercury from the contact arms.

Dilution procedure   Clean thoroughly a 100-ml volumetric flask, a 250-ml wide-mouth flask, and a 25-ml pipette. Rinse these with conductivity water. Rinse the 250-ml flask with two or three *small* aliquots of the stock solution, and then take not more than 125 ml of the solution, noting the concentration given on the stock bottle.

Rinse the pipette with two or three small aliquots from the flask, and then pipette exactly 25.00 ml of the solution from the flask into the volumetric flask. Make up to the mark with conductivity water and mix thoroughly. Set this aside for the next measurement and dilution. Use the solution remaining in the 250-ml flask for rinsing and filling the conductivity cell. When the measurement of resistance of the cell containing that solution has been completed, discard the remaining solution from the 250-ml flask and rinse with two or three aliquots of the new solution from the volumetric flask. Decant the contents of the volumetric flask into the 250-ml flask, and rinse the volumetric flask with conductivity water.

Repeat the process described above as many times as necessary to obtain the dilutions needed. Care is important in this work, since dilution errors are cumulative.

Conductivity water   Ordinary distilled water is not suitable for use in this experiment, since it has too high a conductance. Much of this conductivity results from dissolved CO_2 gas from the air, which can be removed by boiling distilled water and capping a full bottle while it is still hot. Even better conductivity water can be prepared by special distillation[4] or by passing distilled water through an ion-exchange resin to "deionize" it.

Procedure   Set up the bridge circuit and make the calibration check on the slide-wire. Then rinse and fill the cell with conductivity water, and measure its resistance on the short bridge. This value should be *at least* 200,000Ω. If it is less, rinse the cell a few more times and repeat the measurement.

Record the value of R obtained. Fill the cell with 0.02000 N KCl solution and mount it in a 25°C thermostat bath. After making a short-bridge measurement to get the approximate value of R, adjust the resistance box and use the long bridge. Balance the bridge to minimum hum. If necessary, improve the balance with a trimmer capacitor placed across the resistance box; if excessive capacitance appears to be required, polarization is probably taking place and an instructor should be consulted. The final measurements on the solution being studied should consist of four readings, the balance point being approached twice from each side. The value of A on the slide-wire scale is recorded for each balance. There should be no appreciable drift in these values; if drift is observed, the cell is probably not in thermal equilibrium with the bath. Wait a few minutes and repeat the measurements. Be sure to record R' for the resistance box. The measurement on this KCl solution should be made with great care, since it determines the cell constant k and will affect all subsequent calculations.

If time permits, determine R of the cell for KCl dilutions of $\frac{1}{4}$, $\frac{1}{16}$, and $\frac{1}{64}$, and also make measurements on 0.02 N solution and three dilutions of HCl and/or KAc.

Measure R of the cell for acetic acid solutions at 0.05 N and dilutions of $\frac{1}{4}$, $\frac{1}{16}$, and $\frac{1}{64}$. For some HAc solutions (and the most dilute strong electrolyte solutions) it may be necessary to use two resistance boxes or else resort to short-bridge measurements.

At all times be sure that the thermostat bath is regulating properly; record the bath temperature.

At the end of the experiment, rinse the cell well, fill it with conductivity water, and return it with the filling arms stoppered.

CALCULATIONS

If the slide-wire calibration correction is appreciable, then all long-bridge readings should be corrected. From the result for the 0.02000 N KCl measurement, calculate the cell constant k, using $\bar{L} = 0.002768$ at 25°C.

Calculate the specific conductances $\bar{L}$ for all the solutions studied, using Eq. (2). If the conductance of the distilled water used was measurable, this should be subtracted from all the values of $\bar{L}$ obtained. Using Eq. (4), calculate equivalent conductances Λ for all solutions. The values of R, $\bar{L}$, and Λ for every solution measured should be appropriately tabulated, and the cell constant k should be given.

Those doing the strong electrolytes should plot Λ for each strong electrolyte against $\sqrt{c}$ and extrapolate to $c = 0$ to obtain Λ_0. In making the extrapolations, beware of increasingly large experimental uncertainty at the lowest concentrations and also of systematic errors due to conducting impurities or dilution errors at low concentrations. Combine these Λ_0's to obtain Λ_0 for acetic acid. For those not doing runs on strong electrolytes, the following data[6] may be used:

Solution	Λ_0 at 25°C	$d\Lambda_0/dT$
HCl	426.2	6.4
KCl	149.9	3.0
KAc	114.4	2.1

For each dilution of acetic acid, calculate α by Eq. (8) and then K_c by Eq. (9). Present in tabular form Λ, α, c, and K_c for each dilution of acetic acid. Plot log K_c against $\sqrt{\alpha c}$. If an extrapolation to zero concentration is possible, make it and obtain a value for K_a. If the data do not appear to be sufficiently good to warrant an extrapolation, report an average or best value for K_c. Report the temperature at which the measurements were made.

APPARATUS

Wheatstone bridge or potentiometer slide-wire to be used as part of a bridge; oscillator (~ 1000 Hz, with power supply if needed); detector (telephone headset, tuning eye, or oscilloscope); precision resistance plug box or decade box; decade trimmer capacitor; conductivity cell, filled with conductivity water and capped off with clean rubber policemen; holder for mounting cell in bath; two leads to connect cell to bridge; 10 electrical leads with lugs; 100-ml volumetric flask; 25-ml pipette; two 100- or 250-ml beakers; two 125-ml Erlenmeyer flasks; 500-ml glass-stoppered flask for storing conductivity water.

Constant-temperature bath set at 25°C; emery paper; conductivity water (1500 ml); *precisely* 0.02000 N KCl solution (300 ml); 0.02 N HCl solution (200 ml); 0.02 N potassium acetate solution (200 ml); 0.05 N acetic acid (250 ml).

REFERENCES

1. G. Jones and B. C. Bradshaw, *J. Amer. Chem. Soc.*, **55**, 1780 (1933).
2. L. Onsager, *Phys. Z.*, **28**, 277 (1927); W. J. Moore, "Physical Chemistry," 4th ed., chap. 10, Prentice-Hall, Englewood Cliffs, N.J. (1972).
3. D. A. MacInnes, "The Principles of Electrochemistry," chap. 18, Reinhold, New York (1939).
4. T. Shedlovsky, Conductometry, in A. Weissberger (ed.), "Technique of Organic Chemistry," 3d ed., vol. I, part IV, Interscience, New York (1960).
5. H. S. Harned and B. B. Owen, "The Physical Chemistry of Electrolytic Solutions," 2d ed., Appendix A, Reinhold, New York (1950).

GENERAL READING

R. A. Robinson and R. H. Stokes, "Electrolyte Solutions," Academic, New York (1955).
H. S. Harned and B. B. Owen, *op. cit.*
T. Shedlovsky, Conductometry, in A. Weissberger (ed.), *op. cit.*

EXPERIMENT 21. TEMPERATURE DEPENDENCE OF EMF

In this experiment the following electrochemical cell is studied:

$$Cd(s)|Cd^{2+}SO_4^{2-}(aq, c)|Cd(Hg, X_2) \qquad 1 \text{ atm}, T \qquad (1)$$

The change in state accompanying the passage of 2 faradays of *positive* electricity from left to right through the cell is given by

$$
\begin{aligned}
\text{Anode: } & Cd(s) = Cd^{2+}(aq, c) + 2e^- \\
\text{Cathode: } & 2e^- + Cd^{2+}(aq, c) = Cd(Hg, X_2) \\
\hline
\text{Net: } & Cd(s) = Cd(Hg, X_2) \qquad 1 \text{ atm}, T
\end{aligned}
\qquad (2)
$$

In the above, (s) refers to the pure crystalline solid and (Hg, X_2) represents a liquid (single-phase) cadmium amalgam in which the mole fraction of Cd is X_2.

From the emf of this cell and its temperature coefficient, the changes in free energy, entropy, and enthalpy for the above change in state are to be determined.

THEORY

When the cell operates reversibly at constant pressure and temperature, with no work being done except electrical work and expansion work,

$$\Delta G = -N \mathscr{E} \mathscr{F} \tag{3}$$

where ΔG is the increase in free energy of the system attending the change in state produced by the passage of N faradays of electricity through the cell, $\mathscr{F}$ is the Faraday constant, and $\mathscr{E}$ is the emf (electromotive force) of the cell.

From the Gibbs-Helmholtz equation, we have for a change in state at constant pressure and temperature

$$\left(\frac{\partial \Delta G}{\partial T}\right)_p = -\Delta S \tag{4}$$

Combining Eqs. (3) and (4), we find that

$$\Delta S = N \mathscr{F} \left(\frac{\partial \mathscr{E}}{\partial T}\right)_p \tag{5}$$

Knowing both ΔG and ΔS, one can find ΔH by using

$$\Delta G = \Delta H - T \Delta S \tag{6}$$

METHOD

The emf of a cell is best determined by measurements with a potentiometer, since this method gives a close approach to reversible operation of the cell. The cell potential is opposed by a potential drop across the slide-wire of the potentiometer, and at balance only very small currents are drawn from the cell (depending on the sensitivity of the galvanometer used and the fineness of control possible in adjusting the slide-wire). The potentiometer circuit is described in Chap. XV.

EXPERIMENTAL

The cell consists of a small beaker with a top which will accommodate two electrodes as shown in Fig. 1. The *cadmium electrode* is made by plating cadmium onto a platinum wire which is sealed through the bottom of a small glass tube. Electrical contact

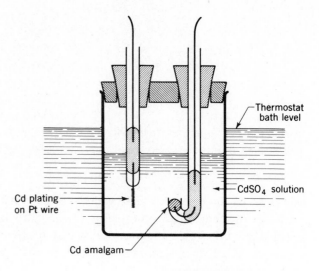

FIGURE 1
Cadmium amalgam cell.

is made to mercury contained in this tube. The *amalgam electrode* is made by placing a small quantity of the cadmium amalgam in the cup of a special J-shaped glass tube with a platinum wire sealed in to make electrical contact with mercury contained in the long arm of the tube.

Procedure   A cadmium-mercury amalgam containing 2 percent Cd by weight will be used. This amalgam can be prepared as follows: Remove the oxide coating from a thin rod of pure Cd with dilute acid, rinse the rod thoroughly in distilled water, dry, and weigh. Then dissolve the rod in the proper (weighed) amount of triple-distilled mercury. Using a medicine dropper, fill the cup of the J electrode with this amalgam. Carefully insert this electrode into the cell, which previously has been rinsed and filled with a 0.1 M $CdSO_4$ solution. The cell should be filled only half full to prevent contact of the solution with the cover.

The platinum wire of the other electrode is plated with a cadmium deposit by immersing the electrode in a beaker containing a 0.1 M solution of $CdSO_4$ and a pure cadmium rod. The electrode is connected to the negative terminal of a 1.5-V dry cell, and the cadmium rod is connected to the positive terminal. Current is passed through this plating bath until a heavy deposit of cadmium is visible on the electrode. This cadmium electrode is then transferred to the cell. Handle the electrode with care to avoid flaking off any of the cadmium deposit. It is advisable to keep the electrode wet at all times.

Mount the assembled cell in a thermostat bath. Assemble the potentiometer circuit and connect the cell. Consider the sign of ΔG for the change in state to decide on the proper connection. Have an instructor check your circuit.

After the cell has been in the thermostat bath for at least 10 min, measure the emf and repeat this measurement at least three times at 5-min intervals to verify that there is no systematic drift in the emf. The potentiometer circuit should be checked against the standard cell immediately before each reading. The emf should be determined as a function of temperature at four or more points in the range from 0 to 40°C. The same potentiometer must be used in making all these measurements.

CALCULATIONS

Plot the emf of the cell $\mathscr{E}$ versus the absolute temperature T. Report the value of $\mathscr{E}$ and the slope $(\partial\mathscr{E}/\partial T)_p$ at 25.0°C (298.2°K) from the best smooth curve through your experimental points. Using Eqs. (3), (5), and (6), calculate ΔG, ΔS, and ΔH in calories per mole, and give the change in state to which they apply.

From the known weight percent Cd in the amalgam, compute the mole fraction X_2 of cadmium. Assuming for this concentration an activity coefficient of unity (so that the activity a_2 is equal to X_2), compute the standard free-energy change $\Delta G°$ for the change in state

$$Cd(s) = Cd(Hg)$$

by means of the equation

$$\Delta G = \Delta G° + RT \ln a \tag{7}$$

DISCUSSION

Explain why $\Delta G°$ is not equal to zero, as would be expected if the standard state were an "amalgam" with $X_2 = 1$.

APPARATUS

Complete potentiometer setup (see Chap. XV); cell (50-ml beaker or weighing bottle); platinum electrode; J electrode for holding amalgam; two leads for connections to cell; two beakers; battery jar; large ring stirrer; ring stand; large clamp and clamp holder.

Provision for electroplating cadmium from 0.1 M CdSO$_4$ onto platinum electrode; dilute (3 percent) cadmium amalgam (1 ml); eye dropper; mercury for making contact to electrodes; constant-temperature baths set at 15, 25, and 35°C; approximately 0.1 M CdSO$_4$ solution (150 ml); ice (2 lb).

GENERAL READING

F. E. Smith, *Phil. Mag.* (6), **19**, 250 (1910).
C. E. Teeter, Jr., *J. Amer. Chem. Soc.*, **53**, 3927 (1931).
V. K. LaMer and W. G. Parks, *J. Amer. Chem. Soc.*, **56**, 90 (1934).

EXPERIMENT 22.   ACTIVITY COEFFICIENTS FROM CELL MEASUREMENTS

We shall be concerned with the electrochemical cell

$$Ag(s) + AgCl(s)\,|\,H^+Cl^-(aq,\ c)\,|\,H_2(g,\ p\ atm) + Pt \tag{1}$$

Measurements of emf are to be made with this cell under reversible conditions at a number of concentrations c of HCl. From these measurements, relative values of activity coefficients at different concentrations can be derived. To obtain the activity coefficients on such a scale that the activity coefficient is unity for the reference state of zero concentration, an extrapolation procedure based on the Debye-Hückel limiting law is used. By this means, the standard electrode emf of the silver–silver chloride electrode is determined and activity coefficients are determined for all concentrations studied.

THEORY

The cell (1) may be considered as a combination of the two half-cells or "electrodes" described below.

Silver–silver chloride electrode   This can be written

$$Ag(s) + AgCl(s)\,|\,Cl^-(aq,\ c) \tag{2}$$

When 1 faraday of positive electricity passes reversibly from left to right through this electrode, the change in state is

$$Ag(s) + Cl^- = AgCl(s) + e^- \tag{3}$$

and the electrode emf (the electric potential of the solution with respect to the silver metal) is accordingly

$$\mathscr{E}_1 = \mathscr{E}_1{}^0 - \frac{RT}{\mathscr{F}} \ln \frac{1}{a_{Cl^-}} \tag{4}$$

where $\mathscr{E}_1{}^0$ is the standard electrode emf for the silver–silver chloride electrode, a_{Cl^-} is the activity of the chloride ion in the aqueous solution, and $\mathscr{F}$ is the Faraday constant.

Hydrogen electrode This can be written

$$\text{Pt} + \text{H}_2(g,\ p\ \text{atm})\,|\,\text{H}^+(aq,\ c) \tag{5}$$

With the reversible passage of 1 faraday the change in state is

$$\tfrac{1}{2}\text{H}_2(g) = \text{H}^+ + e^- \tag{6}$$

and the electrode emf is accordingly

$$\mathscr{E}_2 = \mathscr{E}_2{}^0 - \frac{RT}{\mathscr{F}} \ln \frac{a_{H^+}}{f_{H_2}{}^{1/2}} \tag{7}$$

where f_{H_2} is the fugacity of the hydrogen gas and a_{H^+} the activity of the aqueous hydrogen ion. By convention, the standard electrode emf for the hydrogen electrode is zero:

$$\mathscr{E}_2{}^0 = \mathscr{E}^0_{H_2/H^+} \equiv 0 \tag{8}$$

The cell Combining the hydrogen electrode with the silver–silver chloride electrode we obtain the cell (1). The overall change in state is Eq. (3) minus Eq. (6):

$$\text{H}^+\text{Cl}^- + \text{Ag}(s) = \text{AgCl}(s) + \tfrac{1}{2}\text{H}_2(g) \tag{9}$$

The emf of the cell is given by Eq. (4) minus Eq. (7):

$$\mathscr{E} = \mathscr{E}_1 - \mathscr{E}_2 = \mathscr{E}^0 - \frac{RT}{\mathscr{F}} \ln \frac{f_{H_2}{}^{1/2}}{a_{H^+}a_{Cl^-}} \tag{10}$$

where

$$\mathscr{E}^0 = \mathscr{E}_1{}^0 - \mathscr{E}_2{}^0 = \mathscr{E}_1{}^0 \tag{11}$$

Activity coefficients Let us write

$$f_{H_2} = \gamma'p \qquad a_{H^+}a_{Cl^-} = \gamma_\pm{}^2 c^2 \tag{12}$$

where γ' is the activity coefficient for $\text{H}_2(g)$ and $\gamma_\pm$ is the mean activity coefficient for $\text{H}^+\text{Cl}^-(aq)$. Equation (10) can now be written

$$\mathscr{E} = \mathscr{E}^0 - \frac{2.303RT}{\mathscr{F}} \log \frac{p^{1/2}}{c^2} - \frac{2.303RT}{\mathscr{F}} \log \frac{\gamma'^{1/2}}{\gamma_\pm{}^2} \tag{13}$$

where p is in atmospheres and c is in moles per liter. From this equation it is clear that, if emf measurements are made on two or more cells which differ only in the concentrations c of HCl, the ratios of the corresponding activity coefficients $\gamma_\pm$ can be determined from the differences in emf. For the determination of the individual values of these activity coefficients, it is necessary to know the values of $\mathscr{E}^0$ and γ'. At 25°C, 1 atm, the activity coefficient γ' is 1.0006, which for the purpose of this experiment may be taken as unity. To determine $\mathscr{E}^0$, however, requires a procedure equivalent to determining the emf with the solute in its "reference state," at which the activity coefficient $\gamma_\pm$ is unity. However, the reference state for a solute in solution is the limiting state of zero concentration, which is inaccessible to direct experiment. However, the Debye-Hückel theory predicts the limiting behavior of $\gamma_\pm$ as the concentration approaches zero, and we can make use of this predicted behavior in devising an extrapolation procedure for the determination of $\mathscr{E}^0$. According to the Debye-Hückel limiting law,[1]

$$\log \gamma_\pm \cong -A\sqrt{\mu} \tag{14}$$

In the present case, the ionic strength μ is equal to the HCl concentration c. The value of the constant A given by the Debye-Hückel theory for a uni-univalent electrolyte in aqueous solution at 25°C is 0.509. While knowledge of this value may be helpful to the extrapolation, it is not necessary.

Let us rearrange Eq. (13) and set γ' equal to unity:

$$\mathscr{E}^0 = \mathscr{E} + \frac{2.303RT}{\mathscr{F}} \log \frac{p^{1/2}}{c^2} - \frac{2.303RT}{\mathscr{F}} \log \gamma_\pm{}^2 \tag{15}$$

Now define $\mathscr{E}^{0\prime}$ by

$$\mathscr{E}^{0\prime} \equiv \mathscr{E} + \frac{2.303RT}{\mathscr{F}} \log \frac{p^{1/2}}{c^2} + 2 \frac{2.303RT}{\mathscr{F}} A\sqrt{c} \tag{16}$$

The value of $\mathscr{E}^{0\prime}$ should be close to that of $\mathscr{E}^0$ [depending on the validity of Eq. (14) as an approximation for $\log \gamma_\pm$]; in any case $\mathscr{E}^{0\prime}$ will approach $\mathscr{E}^0$ as the concentration approaches zero. A plot of $\mathscr{E}^{0\prime}$ versus c should be approximately linear and have only a small slope, thus permitting a good extrapolation to zero concentration.[2]

Alternatively, one could define a quantity $\mathscr{E}^{0\prime\prime}$ by

$$\mathscr{E}^{0\prime\prime} \equiv \mathscr{E} + \frac{2.303RT}{\mathscr{F}} \log \frac{p^{1/2}}{c^2} \tag{17}$$

which would equal $\mathscr{E}^0$ at infinite dilution. If this quantity $\mathscr{E}^{0\prime\prime}$ is plotted against $\sqrt{c}$, we expect to obtain a curve which will give a limiting straight-line extrapolation to zero concentration.

In either case, the extrapolated intercept is the desired value of $\mathscr{E}^0$. Often it is best to make both extrapolations; with only a few points, use of $\mathscr{E}^{0'}$ is recommended.

With the value of $\mathscr{E}^0$ the activity coefficients for the various concentrations studied can be determined individually by use of Eq. (13). They can then be compared with those predicted by a more complete form of the Debye-Hückel equation, such as

$$\log \gamma_\pm = -0.509|z_+z_-| \frac{\sqrt{\mu}}{1 + B\sqrt{\mu}} \tag{18}$$

where B is approximately equal to unity.[1]

EXPERIMENTAL

The cell vessel consists of a small beaker with a stopper or cover with holes through which the hydrogen electrode assembly and the silver–silver chloride electrode assembly may be introduced. The assembled cell is shown in Fig. 1.

The hydrogen electrode consists of a mounted platinum gauze square contained within a glass sleeve having large side holes at about the level of the gauze and a side arm for admission of hydrogen near the top. The platinum gauze is "platinized," that is, coated with a deposit of platinum black by electrolytic deposition from a solution containing platinic chloride and a trace of lead acetate. This deposit should be removed with warm aqua regia and renewed if the electrode has been allowed to dry out or if there is evidence that the deposit has been "poisoned."

The silver–silver chloride electrode consists of a mounted platinum screen that has been heavily plated with silver from a cyanide bath, rinsed, aged in an acidified silver nitrate solution, rinsed, coated with a thin layer of silver chloride by anodizing in a dilute HCl solution (preferably no more than a few days before use), and kept in dilute HCl pending use. This is mounted in a glass sleeve with a small hole in the bottom to admit the cell solution and a small side hole near the top for passage of air; this sleeve protects the electrode from mechanical damage and also prevents the attainment of any significant concentration of dissolved H_2 in the solution in contact with the electrode. (Further details concerning electrode preparation can be found in Chap. XIX.)

Both electrode assemblies should be furnished ready for use. Care should be exercised to prevent the electrodes from becoming dry. The use of hydrogen in any significant quantity is attended by explosion hazard. Hydrogen usage must be kept at the minimum necessary for the experiment, and wastage and leakage must be guarded against. **Smoking and the use of flames for any purpose cannot be permitted.**

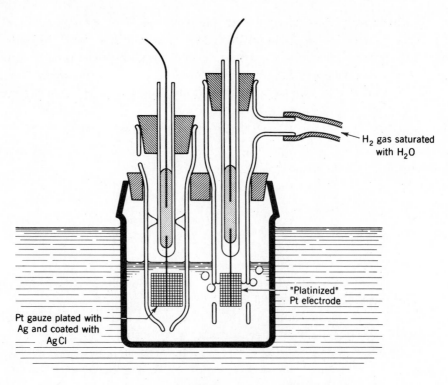

FIGURE 1
Electrochemical cell with H_2 and Ag AgCl electrodes.

The hydrogen to be used should be oxygen-free. Oxygen is most conveniently removed by passing tank hydrogen through a commercial catalytic purifier. The hydrogen should also be saturated with water vapor at room temperature (about 25°C) by bubbling it through water.

The partial pressure p of the hydrogen at the electrode should be determined from the atmospheric pressure in the laboratory by subtracting the partial pressure of water vapor at 25°C and adding the mercury equivalent of the "water head" (the average difference in liquid levels inside and outside the hydrogen electrode shell).

Procedure Fill the cell about half full with the appropriate HCl solution. Rinse the two electrodes in separate small portions of the same solution, and insert them. When the liquid level in the silver–silver chloride electrode shell has reached its equilibrium level, pass hydrogen through the cell at a moderate rate to sweep out the air and saturate the solution. After about 10 min, slow down the rate to a few bubbles per second and begin to make emf readings. Make at least four readings at 5-min intervals. These should show no significant drift.

Readings of emf are obtained with a potentiometer circuit (see Chap. XV). If a null galvanometer reading cannot be obtained with any setting of the potentiometer dial, reverse the leads and try again. The sign of the emf will depend on whether the null reading is obtained with the right-hand terminal of the cell (Pt + H_2) connected to the positive or to the negative terminal of the potentiometer.

Care should be exercised to prevent the cell from becoming polarized by accidental shorting or by passage of excessive currents during the potentiometer balancing.

Runs should be made with the following concentrations of HCl: 0.1, 0.05, 0.025, 0.0125, and 0.00625 M, obtained by successive volumetric dilution of 0.1 M HCl stock solution. Make up 200 ml of each solution.

CALCULATIONS

For each run calculate $\mathscr{E}^{0'}$ from Eq. (16) from the measured $\mathscr{E}$, the partial pressure p of H_2, and the concentration c of HCl. The value of $2.303RT/\mathscr{F}$ at 25°C is 0.05916 volt and A is 0.509.

Plot $\mathscr{E}^{0'}$ versus c and extrapolate to zero concentration to obtain $\mathscr{E}^0$ (which is equal to the standard electrode emf for the silver–silver chloride electrode). If you wish, also calculate $\mathscr{E}^{0''}$ from Eq. (17), plot it versus $\sqrt{c}$, and extrapolate to obtain another value of $\mathscr{E}^0$ as a check.

With this value of $\mathscr{E}^0$ calculate mean activity coefficients for all concentrations studied by use of Eq. (15). For each concentration, calculate also a theoretical mean activity coefficient by use of the Debye-Hückel equation in the form given in Eq. (18). Present your experimental and theoretical activity coefficients in tabular form.

APPARATUS

Complete potentiometer setup (see Chap. XV); 50-ml weighing bottle as cell; special three-hole stopper to fit cell and hold electrodes; two leads for connections to cell; hydrogen electrode, and Ag–AgCl electrode; 200-ml volumetric flask; 100-ml pipette; 250-ml flasks; two 250-ml beakers; small gas bubbler; two pieces of gum-rubber tubing; large clamp and clamp holder.

Cylinder of hydrogen gas, regulator fitted with a Deoxo purifier to remove any oxygen and a flow reducer to limit flow to 5 ft^3 hr^{-1}; large fritted-glass bubbler to saturate hydrogen with water vapor; constant-temperature bath set at 25°C; 0.1 M HCl solution (350 ml).

REFERENCES

1. W. J. Moore, "Physical Chemistry," 4th ed., chap. 12, Prentice-Hall, Englewood Cliffs, N.J. (1972); F. H. MacDougall, "Thermodynamics and Chemistry," 3d ed., Wiley, New York (1939).
2. H. S. Harned and R. W. Ehlers, *J. Amer. Chem. Soc.*, **55**, 2179 (1933).

GENERAL READING

F. Daniels and R. A. Alberty, "Physical Chemistry," 3d ed., Wiley, New York (1966).
D. A. MacInnes, "The Principles of Electrochemistry," Dover, New York (1961).

IX

CHEMICAL KINETICS

EXPERIMENTS

23. Method of initial rates: iodine clock
24. Kinetics of a hydrolysis reaction
25. Enzyme kinetics: inversion of sucrose
26. Enzyme kinetics: hydrolysis of an ester
27. Kinetics of the decomposition of benzenediazonium ion
28. Gas-phase kinetics
29. Kinetics of a fast reaction

EXPERIMENT 23.   METHOD OF INITIAL RATES: IODINE CLOCK

The homogeneous reaction in aqueous solution

$$IO_3^- + 8I^- + 6H^+ \rightarrow 3I_3^- + 3H_2O \tag{1}$$

like virtually all reactions involving more than two or three reactant molecules, takes place not in a single molecular step but in several steps. The detailed system of steps is called the *reaction mechanism*. It is one of the principal aims of chemical kinetics to obtain information to aid in the elucidation of reaction mechanisms, which are fundamental to our understanding of chemistry.

THEORY

The several steps in a reaction are usually consecutive and tend to proceed at different speeds. Usually, when the overall rate is slow enough to measure at all, it is because one of the steps tends to proceed so much more slowly than all the others that it effectively controls the overall reaction rate and can be designated the *rate-controlling* step. A steady state is quickly reached in which the concentrations of the reaction intermediates are controlled by the intrinsic speeds of the reaction steps by which they are formed and consumed. A study of the rate of the overall reaction yields information of a certain kind regarding the nature of the rate-controlling step and

251

closely associated steps. Usually, however, rate studies supply only part of the information needed to formulate uniquely and completely the correct reaction mechanism.

When the mechanism is such that the steady state is quickly attained, the rate law for Eq. (1) can be written in the form

$$-\frac{d(\text{IO}_3{}^-)}{dt} = f[(\text{IO}_3{}^-), (\text{I}^-), (\text{H}^+), (\text{I}_3{}^-), (\text{H}_2\text{O}), \ldots] \tag{2}$$

where parenthesized quantities are concentrations. In the general case the brackets might also contain concentrations of additional substances, referred to as catalysts, whose presence influences the reaction rate but which are not produced or consumed in the overall reaction. The determination of the rate law requires that the rate be determined at a sufficiently large number of different combinations of the concentrations of the various species present to enable an expression to be formulated which accounts for the observations and gives good promise of predicting the rate reliably over the concentration ranges of interest. The rate law can be written to correspond in form to that predicted by a theory based on a particular type of mechanism, but basically it is an empirical expression.

The most frequently encountered type of rate law is of the form [again using the reaction of Eq. (1) as an example]

$$-\frac{d(\text{IO}_3{}^-)}{dt} = k(\text{IO}_3{}^-)^m(\text{I}^-)^n(\text{H}^+)^p \cdots \tag{3}$$

where the exponents $m, n, p, \ldots$ are determined by experiment. Each exponent in Eq. (3) is the *order* of the reaction with respect to the corresponding species; thus, the reaction is said to be mth order with respect to $\text{IO}_3{}^-$, etc. The algebraic sum of the exponents, $m + n + p$ in this example, is the *overall order* (or, commonly, simply the *order*) of the reaction. Reaction orders are usually, but not always, positive integers within experimental error.

The order of a reaction is determined by the reaction mechanism. It is related to, and is often (but not always) equal to, the number of reactant molecules in the rate-controlling step—the "molecularity" of the reaction. Consider the following proposed mechanism for the hypothetical reaction $3\text{A} + 2\text{B} = $ products:

a.	$\text{A} + 2\text{B} = 2\text{C}$	(fast, to equil., K_a)
b.	$\text{A} + \text{C} = $ products	(slow, rate controlling, k_b)

The rate law predicted by this mechanism is

$$-\frac{d(\text{A})}{dt} = \tfrac{3}{2}k_b(\text{A})(\text{C}) = \tfrac{3}{2}k_b(\text{A})K_a{}^{1/2}(\text{A})^{1/2}(\text{B}) = k(\text{A})^{3/2}(\text{B})$$

The overall reaction involves five reactant molecules, but it is by no means necessarily of fifth order. Indeed, the rate-controlling step in this proposed mechanism is bimolecular, and the overall reaction order predicted by the mechanism is $\frac{5}{2}$. It is also important to note that this mechanism is not the only one that would predict the above $\frac{5}{2}$-order rate law for the given overall reaction; thus experimental verification of the predicted rate law would by no means constitute proof of the validity of the above proposed mechanism.

It occasionally happens that the observed exponents deviate from integers or simple rational fractions by more than experimental error. A possible explanation is that two or more simultaneous mechanisms are in competition, in which case the observed order should lie between the extremes predicted by the individual mechanisms. A possible alternative explanation is that no single reaction step is effectively rate controlling.

We now turn our attention to the experimental problem of determining the exponents in the rate law. Except in first- and second-order reactions it is usually inconvenient to determine the exponents merely by determining the time behavior of a reacting system in which many or all reactant concentrations are allowed to change simultaneously and comparing the observed behavior with integrated rate expressions. A procedure is desirable which permits the dependences of the rate on the concentrations of the different reactants to be isolated from one another and determined one at a time. In one such procedure, all the species but the one to be studied are present at such high initial concentrations relative to that of the reactant studied that their concentrations may be assumed to remain approximately constant during the reaction; the apparent reaction order with respect to the species of interest is then obtained by comparing the progress of the reaction with that predicted by rate laws for first order, second order, and so on. This procedure would often have the disadvantage of placing the system outside the concentration range of interest and thus possibly complicating the reaction mechanism.

In another procedure, which we shall call the *initial rate method*, the reaction is run for a time small in comparison with the "half-life" of the reaction but large in comparison with the time required to attain a steady state, so that the actual value of the initial rate [the initial value of the derivative on the left side of Eq. (3)] can be estimated approximately. Enough different combinations of initial concentrations of the several reactants are employed to enable the exponents to be determined separately. For example, the exponent m is determined from two experiments which differ only in the IO_3^- concentration.

In the present experiment the rate law for the reaction shown in Eq. (1) will be studied by the initial rate method, at 25°C and at a pH of about 5. The initial concentrations of iodate ion, iodide ion, and hydrogen ion will be varied independently in separate experiments, and the time required for the consumption of a definite small amount of the iodate will be measured.

METHOD

The time required for a definite small amount of iodate to be consumed will here be measured by determining the time required for the iodine produced by the reaction (as I_3^-) to oxidize a definite amount of a reducing agent, arsenious acid, added at the beginning of the experiment. Under the conditions of the experiment arsenious acid does not react directly with iodate at a significant rate but reacts with iodine as quickly as it is formed. When the arsenious acid has been completely consumed, free iodine is liberated which produces a blue color with a small amount of soluble starch which is present. Since the blue color appears rather suddenly after a reproducible period of time, this series of reactions is commonly known as the "iodine clock reaction."

The reaction involving arsenious acid may be written, at a pH of about 5,

$$H_3AsO_3 + I_3^- + H_2O \rightarrow HAsO_4^{2-} + 3I^- + 4H^+ \tag{4}$$

The overall reaction, up to the time of the starch end point, can be written, from reactions (1) and (4),

$$IO_3^- + 3H_3AsO_3 \rightarrow I^- + 3HAsO_4^{2-} + 6H^+ \tag{5}$$

Since with ordinary concentrations of the other reactants hydrogen ions are evidently produced in quantities large in comparison to those corresponding to pH 5, it is evident that buffers must be used to maintain constant hydrogen-ion concentration. As is apparent from the method used, the rate law will be determined under conditions of essentially zero concentration of I_3^-; the dependence of the rate on triiodide, which in fact has been shown to be very small,[1] will not be measured. Under these conditions, Eq. (3) is an appropriate expression for the rate.

A constant initial concentration of H_3AsO_3 is used in a series of reacting mixtures having varying initial concentrations of IO_3^-, I^-, and H^+. Since the amount of arsenious acid is the same in each run, the amount of iodate consumed up to the color change is constant, and related to the amount of arsenious acid by the stoichiometry of Eq. (5). The initial reaction rate in mole liter^{-1} sec^{-1} is thus approximately the amount consumed (per liter) divided by the time required for the blue end point to appear. From the initial rates of two reactions in which the initial concentration of only one reactant is varied and all other concentrations kept the same, it is possible to infer the exponent in the rate expression associated with the reactant which is varied. This is most conveniently done by taking logarithms of both sides of Eq. (3) and subtracting the expressions for the two runs.

EXPERIMENTAL

Solutions   Two acetate buffers, with hydrogen-ion concentrations differing by a factor of 2, will be made up by the student from stock solutions. Use will be made of the fact that at a given ionic strength the hydrogen-ion concentration is proportional to the ratio of acetic acid concentration to acetate ion concentration:

$$(H^+) = K \frac{(HAc)}{(Ac^-)} \frac{1}{\gamma_\pm^{\,2}} \tag{6}$$

where at 25°C, $K = 1.753 \times 10^{-5}$ mol liter^{-1}. The experiments will all be carried out at about the same ionic strength (0.16 ± 0.01), and accordingly the activity coefficient is approximately the same in all experiments, by the Debye-Hückel theory. It will also be seen that within wide limits the amount of buffer solution employed in a given total volume is inconsequential, provided the ionic strength of the resultant solution is always kept about the same. The solutions required are as follows:

Buffer A. Pipette 100 ml of 0.75 M NaAc solution, 100 ml of 0.22 M HAc solution, and about 20 ml of 0.2 percent soluble starch solution into a 500-lm volumetric flask, and make up to the mark with distilled water [yields $(H^+) \cong 10^{-5}\ M$].

Buffer B. Pipette 50 ml of 0.75 M NaAc solution, 100 ml of 0.22 M HAc solution, and about 10 ml of 0.2 percent soluble starch solution into a 250-ml volumetric flask, and make up to the mark with distilled water [yields $(H^+) \cong 2 \times 10^{-5}\ M$].

H_3AsO_3, 0.03 M. Should be made up from $NaAsO_2$ and brought to a pH of about 5 by addition of HAc.

KIO_3, 0.1 M.

KI, 0.2 M.

Suggested sets of initial volumes of reactant solutions, based on a final volume of 100 ml, are given in the table below.

Solution	Pipette sizes	Initial volumes of solutions, ml			
		1	2	3	4
H_3AsO_3	5	5	5	5	5
IO_3^-	5	5	10	5	5
Buffer A	20, 25	65	60	40	
Buffer B	20, 25				65
I^-	25	25	25	50	25

Two or three runs should be made on each of the four sets. Two or more runs should also be made on a set with proportions chosen by the student in which the initial compositions of *two* reacting species differ from those in set 1. In each case, the amount of buffer required is that needed to obtain a final volume of 100 ml.

It is convenient to use each pipette only for a single solution, if possible, to minimize time spent in rinsing. The pipettes should be marked to avoid mistakes.

The buffer solutions and the iodide solution should be equilibrated to 25°C by clamping flasks containing them in a thermostat bath set at that temperature. Two vessels of convenient size (ca. 250 ml) and shape (beakers or Erlenmeyer flasks), rinsed and drained essentially dry, should also be clamped in the bath. One of them should have a white-painted bottom surface (or have a piece of white cloth taped under the bottom) to aid in observing the blue end point unless other means are available to obtain a light background.

To make a run, pipette all the solutions *except* KI into one of the vessels and the KI solution into the other. Remove both vessels from the bath, and begin the reaction by pouring the iodide rapidly but quantitatively into the other solution, simultaneously starting the stopwatch. Pour the solution back and forth once or twice to complete the mixing, and place the vessel containing the final solution back into the bath. Stop the watch at the appearance of the first faint but definite blue color.

CALCULATIONS

The student should construct a table giving the actual initial concentrations of the reactants IO_3^-, I^-, and H^+. The H^+ concentrations should be calculated from the actual concentrations of NaAc and HAc in the stock solutions employed, with an activity coefficient calculated by use of the Debye-Hückel theory for the ionic strength ($\mu = 0.16$) of the reacting mixtures.

Using the known initial concentration of H_3AsO_3, calculate the initial rate for each run. From appropriate combinations of sets 1, 2, 3, and 4, calculate the exponents in the rate expression (3). Also calculate a value of the rate constant k from each run, and obtain an average value of k from all runs.

Write the rate expression, with the numerical values of the rate constant k and the experimentally obtained values of the exponents. Beside it write the temperature and ionic strength at which this expression was obtained.

Write another rate expression, in which those exponents which appear to be reasonably close (within experimental error) to integers are replaced by the integral values. Use this expression to calculate values for the initial rates of all sets studied, and compare them with the observed initial rates.

DISCUSSION

The kinetics of this reaction have been the subject of much study, and the mechanism is not yet completely elucidated with certainty. Following is an incomplete list of the mechanisms that have been proposed:

1 $IO_3^- + 2I^- + 2H^+ \xrightarrow{k} 2HIO + IO^-$ (slow)
Followed by fast reactions (Dushman[1])

2 $IO_3^- + H^+ \underset{}{\overset{K}{\rightleftharpoons}} HIO_3$ (fast, to equil.)
$I^- + H^+ \underset{}{\overset{K'}{\rightleftharpoons}} HI$ (fast, to equil.)
$HIO_3 + HI \xrightarrow{k} HIO + HIO_2$ (slow)
Followed by fast reactions (at low iodide concentrations; Abel and Hilferding[2])

3 $IO_3^- + I^- + 2H^+ \underset{}{\overset{K}{\rightleftharpoons}} H_2I_2O_3$ (fast, to equil.)
$H_2I_2O_3 \underset{}{\overset{K'}{\rightleftharpoons}} I_2O_2 + H_2O$ (fast, to equil.)
$I_2O_2 + I^- \xrightarrow{k} I_3O_2^-$ (slow)
Followed by fast reactions (Bray and Liebhafsky[3])

4 $IO_3^- + I^- + 2H^+ \underset{}{\overset{K}{\rightleftharpoons}} H_2I_2O_3$ (fast, to equil.)
$H_2I_2O_3 \xrightarrow{k} HIO + HIO_2$ (slow)
Followed by fast reactions (at low iodide concentrations; Bray and Liebhafsky[3])

5 $IO_3^- + H^+ \underset{}{\overset{K}{\rightleftharpoons}} IO_2^+ + OH^-$ (fast, to equil.)
$H^+ + OH^- \underset{}{\overset{1/K_w}{\rightleftharpoons}} H_2O$ (fast, to equil.)
$IO_2^+ + I^- \underset{}{\overset{K'}{\rightleftharpoons}} IOIO$ (fast, to equil.)
$IOIO + I^- \xrightarrow{k} I^+ + 2IO^-$ (slow)
Followed by fast reactions (Morgan, Peard, and Cullis[4])

6 $IO_3^- + H^+ \underset{}{\overset{K}{\rightleftharpoons}} IO_2^+ + OH^-$ (fast, to equil.)
$H^+ + OH^- \underset{}{\overset{1/K_w}{\rightleftharpoons}} H_2O$ (fast, to equil.)
$IO_2^+ + I^- \underset{}{\overset{K'}{\rightleftharpoons}} IOIO$ (fast, to equil.)
$IOIO \xrightarrow{k} IO^+ + IO^-$ (slow)
Followed by fast reactions (at low iodide concentrations; Morgan, Peard, and Cullis[4])

The student should discuss the above mechanisms in connection with his experimentally determined rate law.

The oxidation of iodide ion by chlorate ion ClO_3^- has also been studied.[5,6] Although the reaction appears to be attended by complications which make it difficult to study, under certain conditions it can be carried out as an "iodine clock" experiment. (For the interested student, suggested concentration ranges for 20 to 25°C are ClO_3^-, 0.05 to 0.10 M; I^-, 0.025 to 0.10 M; H^+, 0.02 to 0.04 M. A sulfate-bisulfate buffer may be used.) The resulting rate law is not identical with that for the reaction with iodate, but appears to be compatible with mechanisms analogous to several of those given above.

A different type of "clock" reaction, suitable for student investigation, is the reaction of formaldehyde with bisulfite ion.[7,8] The reaction involves a single slow step

$$HCHO + HSO_3^- \rightarrow HOCH_2SO_3^- \tag{7}$$

followed by the rapid buffer equilibrium

$$HSO_3^- \overset{K_a}{\rightleftharpoons} H^+ + SO_3^{2-} \tag{8}$$

As bisulfite ion is used up in reaction (7), the hydrogen-ion concentration adjusts itself according to the buffer equation (8). If an indicator such as phenolphthalein is added to the mixture, it will undergo a sudden color change when the pH of the solution reaches the pK_i of the indicator. The time τ required for the color change is related to the rate constant for reaction (7) by the equation[7]

$$\frac{1}{F_0 - B_0}\left[\log_{10}\frac{B_0(F_0 - B_0)K_a}{F_0 S_0 K_i'}\right] \cong 0.43k_7\tau \tag{9}$$

where B_0 is the initial bisulfite concentration, S_0 is the initial sulfite concentration, F_0 is the initial formaldehyde concentration, K_a is the bisulfite equilibrium constant (6.7×10^{-8} mol liter^{-1} at 25°C), $K_i' = RK_i$, where K_i is the indicator equilibrium constant, and R is the indicator color ratio at the end point for the indicator concentration used.

"Flowing clock" modification of the experiment   In conventional "clock" reactions, it is difficult to obtain reliable results for reaction times of less than 20 or 30 sec. Since much of modern chemical kinetics is concerned with reactions occurring on much faster time scales, it is worthwhile to introduce a modification of such experiments which makes use of a simple flow system to explore reaction times of the order of a few seconds or less.[9] Both the iodide oxidation and formaldehyde-bisulfite reactions can be investigated in this way.

The simple apparatus is shown in Fig. 1. The reactants are contained in two thermostated 250-ml graduated cylinders, each with a siphon tube connected to the T-shaped mixing chamber A, which is a 1-mm bore 3-way stopcock. The flow tube

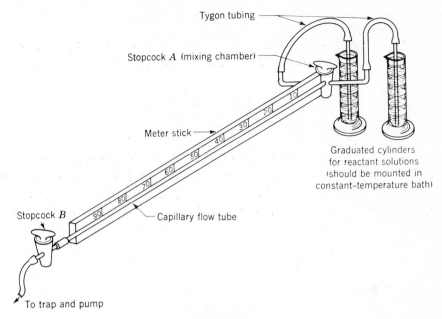

Tygon tubing

Stopcock A (mixing chamber)

Meter stick

Stopcock B

Capillary flow tube

To trap and pump

Graduated cylinders
for reactant solutions
(should be mounted in
constant-temperature bath)

FIGURE 1
Schematic diagram of the "flowing clock" apparatus.

is a 2-mm bore Pyrex capillary which lies between the mixing chamber and an inline stopcock B used to regulate the flow. A water aspirator or a house vacuum line protected with a 1-liter trap is connected to stopcock B.

The rate of flow is calibrated by filling the cylinders with tap water, applying suction, and then fully opening stopcock B. The amount of fluid V removed from the cylinders in a measured time τ_0 gives the flow rate $f = V/\tau_0$. A flow rate of 3 to 10 ml sec^{-1} is suitable for carrying out the experiments described below. Stopcock A should be carefully adjusted so that equal volumes of fluid are removed from each of the two cylinders in a given time, and then left at that setting.

The reaction time τ for a given experiment is determined by measuring the distance x along the flow tube from stopcock A at which the appropriate color change occurs, and using the relationship

$$\tau = \frac{xA}{f} \tag{10}$$

where A is the cross-sectional area of the flow tube and f is the previously determined volume flow rate.

The iodine clock experiment can be carried out using peroxide in place of iodate and sodium thiosulfate in place of arsenious acid. The rate law by analogy to Eq. (3) is

$$-\frac{d(H_2O_2)}{dt} = k(H_2O_2)^m(I^-)^n(H^+)^p \tag{11}$$

A set of experimental conditions suitable for determining the exponents in Eq. (11) is given in the table below.

| Run no. | Reservoir 1, vol in ml | | | Reservoir 2, vol in ml | | | |
	4 M H_2O_2	2 M HCl	Distilled water	0.05 M KI	0.01 M $Na_2S_2O_3$	Distilled water	Starch solution
1	125	125	0	100	100	0	50
2	62.5	125	62.5	100	100	0	50
3	125	62.5	62.5	100	100	0	50
4	125	125	0	50	100	50	50

For the formaldehyde-bisulfite reaction, a set of runs is given in the following table, where the buffer is 0.3 M in sulfite ion and 0.05 M in bisulfite ion.

| Run no. | Reservoir 1, vol in ml | | | Reservoir 2, vol in ml | |
	Buffer solution	Phenol-phthalein	Distilled water	1 M Formaldehyde solution	Distilled water
1	120	2	128	120	130
2	60	2	188	60	190
3	30	2	218	30	220
4	15	2	233	15	235

A plot of $(F_0 - B_0)^{-1}$ versus τ will test Eq. (9) since the quantity in brackets is constant. Assuming that $K_a/K_i' = 79,$[7] one can estimate k_7.

APPARATUS

Three 200-ml beakers (bottom painted white); two 250-ml and one 100-ml beakers; two 5-, one 20-, two 25-, and one 50-ml pipettes; one 250- and one 500-ml volumetric flask; four 250-ml Erlenmeyer flasks with four corks to fit; one 10-ml graduated cylinder; glass-marking pencil; stopwatch.

Constant-temperature bath (set at 25°C) with provision for mounting beakers and flasks; 0.75 M NaAc solution (300 ml); 0.22 M HAc solution (300 ml); 0.03 M H_3AsO_3 solution (150 ml); 0.1 M KIO_3 solution (150 ml); 0.2 M KI solution (500 ml); 0.2 percent soluble starch solution, with trace of HgI_2 as preservative (75 ml).

For the flowing clock modification: two 250-ml graduated cylinders; 3-way stopcock; inline stopcock; 1 m of 2-mm bore Pyrex capillary tubing; meter stick; connecting tubing;

water aspirator or other rough-vacuum source; 1-liter trap; assorted clamps and ring stands; stopwatch; thermostat bath; appropriate solutions.

REFERENCES

1. S. Dushman, *J. Phys. Chem.*, **8**, 453 (1904).
2. E. Abel and K. Hilferding, *Z. Phys. Chem.*, **136A**, 186 (1928).
3. W. C. Bray, *J. Amer. Chem. Soc.*, **52**, 3580 (1930).
4. K. J. Morgan, M. G. Peard, and C. F. Cullis, *J. Chem. Soc.*, **1951**, 1865.
5. W. C. Bray, *J. Phys. Chem.*, **7**, 92 (1903).
6. A. Skrabal and H. Schreiner, *Monatsh. Chem.*, **65**, 213 (1934).
7. P. Jones and K. B. Oldham, *J. Chem. Educ.*, **40**, 366 (1963).
8. R. L. Barrett, *J. Chem. Educ.*, **32**, 78 (1955).
9. M. L. Hoggett, P. Jones, and K. B. Oldham, *J. Chem. Educ.*, **40**, 367 (1963).

GENERAL READING

W. J. Moore, "Physical Chemistry," 4th ed., chap. 9, Prentice-Hall, Englewood Cliffs, N.J. (1972).
F. H. MacDougall, "Physical Chemistry," chap. 15, Macmillan, New York (1952).
K. J. Laidler, "Chemical Kinetics," 2d ed., McGraw-Hill, New York (1965).

EXPERIMENT 24. KINETICS OF A HYDROLYSIS REACTION

In the presence of hydrogen ions, diethyl acetal in aqueous solution hydrolyzes as follows, the reaction going substantially to completion:

$$CH_3CH(OC_2H_5)_2 + H_2O \xrightarrow{H^+} CH_3CHO + 2C_2H_5OH \qquad (1)$$

In this experiment the reaction will be shown to be first order with respect to acetal and with respect to hydrogen ion, the rate constant will be determined, and the temperature dependence of the rate constant will be used to calculate the activation energy for the reaction.

Other acetals, as well as numerous other organic compounds such as esters and epoxides, also undergo acid-catalyzed hydrolysis with similar kinetics.[1] Use has been made of kinetics measurements on reactions of this kind as a means of measuring hydrogen-ion concentration independently of hydrogen-ion activity.[2]

THEORY

A first-order reaction is ordinarily understood to be one in which the rate of disappearance of a single species is proportional to the concentration of that species in the reaction mixture. This terminology is often used even when, strictly speaking, the total reaction order is different from unity owing to the participation of additional species in the reaction which are themselves not consumed (i.e., catalysts—H^+ in the present instance) or the participation of substances which are present in such large amounts that their concentrations do not undergo significant percentage change during the reaction (H_2O in the present instance). It would be more correct in such cases to state that the reaction is first order with respect to a given disappearing species.

The rate law for a first-order reaction may be expressed by the differential equation

$$-\frac{dc_A}{dt} = kc_A \tag{2}$$

where c_A is the instantaneous concentration of the disappearing reactant (here acetal) and k is the *specific reaction rate*, or simply *rate constant*, applicable to the specified conditions (temperature and in the present case hydrogen-ion concentration) which are constant throughout the course of the reaction. If the dependence of the rate of hydrolysis of acetal on hydrogen-ion concentration is known, it may be included in the expression for the rate law. In the present experiment it will be demonstrated that this dependence is first order, and we can write accordingly

$$-\frac{dc_A}{dt} = k'c_{H^+}c_A \tag{3}$$

where k' is another rate constant at the specified temperature, independent of the hydrogen-ion concentration.

If Eq. (2) is integrated with the initial condition $c = c_0$ at $t = 0$, we obtain

$$\ln\frac{c}{c_0} = -kt \qquad c = c_0 e^{-kt} \tag{4}$$

A reaction may be shown to be first order by a variety of methods, some of which are:

1   The logarithm of the concentration c is plotted against time and the experimental points are shown to fit a straight line.

2   The time interval required for any definite fraction of the reactant to disappear is shown to be a constant, independent of when the interval is taken during the reaction. In the special case where the fraction is one-half, the time interval is called the "half-life" of the reactant.

3   The *rate* of the reaction is studied as a function of time.   A well-known example is that of radioactive decay, where the rate of decay is given by a radiation intensity as measured, say, by a Geiger counter.

4   The *initial rate* of the reaction, $-(dc/dt)_{t=0}$, is determined at a number of initial concentrations c_0, and the initial rate is shown to be directly proportional to the initial concentration (see Exp. 23).

All the above methods require that the actual concentration or its instantaneous time derivative be known at least to within a multiplicative constant.   However, there are methods useful for cases where only a linear function of the concentration

$$f = Ac + B \tag{5}$$

is available (A and B being unknown constants):

5   The reaction is allowed to go substantially to completion by waiting a very long time or by temperature change or catalysis, and a value of f_∞ is obtained. Then we can use the equation, derived from Eqs. (4) and (5),

$$\ln \frac{f - f_\infty}{f_0 - f_\infty} = -kt \tag{6}$$

in the manner of method 1 above.

6   *Guggenheim method*   In the present experiment we shall use a method[3] which eliminates the need for an f_∞ value.   From Eqs. (4) and (5) we have

$$f(t) = Ac_0 e^{-kt} + B$$

and at time $t + \Delta t$

$$f(t + \Delta t) = Ac_0 e^{-k\Delta t} e^{-kt} + B$$

Taking the difference we obtain

$$\Delta f \equiv f(t + \Delta t) - f(t) = Ac_0(e^{-k\Delta t} - 1)e^{-kt} \tag{7}$$

Thus, *for constant Δt, Δf* varies exponentially in a way similar to *c*.   If we plot its logarithm against *t* as in method 1, we should obtain a straight line, from the slope of which the rate constant can be obtained:

$$\ln \Delta f = 2.303 \log \Delta f = -kt + \text{const} \tag{8}$$

The Guggenheim treatment and its results are in principle independent of the particular constant value chosen for Δt.   The choice of a value for Δt depends on practical considerations: If it is too small, the scatter of points on the logarithmic plot will be too great, and if it is too large, the number of points that can be plotted becomes too small.   For a clear demonstration of first-order kinetics and a good determination of the rate constant, the points plotted should cover at least one natural logarithmic cycle (one power of e). A good procedure would be to run the experiment over a total time sufficient to obtain about $1\frac{1}{2}$ cycles (i.e., until increments between successive readings have decreased to about one-fifth of the initial value) and to take about one third of this total time for Δt.

This method may be seen to be very closely related to method 3 above. It uses two experimental readings separated by a constant time difference Δt to determine, not the actual rate itself, but rather a quantity which on the assumption of first-order kinetics must be directly proportional to the actual instantaneous rate. When the appropriate logarithmic plot gives a straight line, the assumption of first-order kinetics is confirmed. The method may be said to degenerate into method 3 in the limit of $\Delta t = 0$, and into method 5 in the limit $\Delta t = \infty$.

Dependence on hydrogen-ion concentration In this experiment the rate constant k will be demonstrated to be proportional to the first power of the hydrogen-ion concentration. This suggests a mechanism which can be written

$$
\begin{aligned}
&A + H_2O + H^+ = X^+ &&\text{(fast equilibrium, } K^*) \\
&X^+ \to H^+ + \text{products} &&\text{(rate determining, } k_X)
\end{aligned}
\tag{9}
$$

where A is acetal and X^+ is an intermediate complex ion. This predicts for the reaction rate

$$
-\frac{dc_A}{dt} = k_X c_{X^+} = k_X K^* c_{H^+} c_A
\tag{10}
$$

Comparison with Eqs. (2) and (3) gives

$$
k = k' c_{H^+} = k_X K^* c_{H^+}
\tag{11}
$$

It is worthy of note that the rate constant is proportional to the *concentration*, rather than to the thermodynamic *activity*, of hydrogen ion. This was explained by Brønsted in terms of the transition-state theory.[4] It arises from the fact that the equilibrium constant K^* contains an activity coefficient for the ion X^+ in the numerator and one for the ion H^+ in the denominator while all other activity coefficients present are for uncharged species. For dilute solutions of given ionic strength the Debye-Hückel theory predicts that the activity coefficients of all univalent ions are very nearly equal, and those for H^+ and X^+ accordingly should cancel out. In the same order of approximation, the activity coefficients for uncharged species are unity. (In certain other cases activity coefficients do not cancel out, and the rate constants are there found to depend on the ionic strengths of the solutions.)

As a matter of convenience, acetate buffer solutions are to be used as a means of obtaining low but stable hydrogen-ion concentrations. The hydrogen-ion concentration in the reaction mixture can be calculated directly from the buffer composition and the acetic acid dissociation constant:

$$c_{H^+} = \frac{K}{\gamma_\pm^2} \frac{c_{HAc}}{c_{Ac^-}} \tag{12}$$

At 25°C, $K = 1.753 \times 10^{-5}$ mol liter^{-1}. Alternatively, the hydrogen-ion activity of the buffer can be determined with a pH meter, and the concentration calculated from

$$c_{H^+} = \frac{a_{H^+}}{\gamma_{H^+}} \tag{13}$$

Activity coefficients can be obtained with sufficient accuracy from the Debye-Hückel expression

$$\log \gamma_{H^+} = \log \gamma_\pm = -0.509 \frac{\sqrt{\mu}}{1 + \sqrt{\mu}} \tag{14}$$

where μ is the ionic strength ($=$ NaAc concentration).

Dependence on temperature   The Arrhenius activation energy E^* is related to the temperature variation of the rate constant by the equation

$$\frac{d \ln k'}{dT} = \frac{E^*}{RT^2} \tag{15}$$

If this is integrated with the assumption of constant E^*, we obtain

$$\ln k' = 2.303 \log k' = -\frac{E^*}{RT} + \text{const} \tag{16}$$

which permits us to obtain E^* from the slope of a plot of the logarithm of the rate constant against the reciprocal of the absolute temperature. If the rate constant is available at only two temperatures, we can use the equation

$$\ln \frac{k_2'}{k_1'} = 2.303 \log \frac{k_2'}{k_1'} = \frac{E^*}{R} \frac{(T_2 - T_1)}{T_1 T_2} \tag{17}$$

Since in this experiment a buffer solution is being used at more than one temperature, we should examine the effect of possible temperature variation of the hydrogen-ion concentration in the buffer, since, in general, the dissociation constants of weak acids are functions of temperature. Failure to take this into account would have the effect of adding the heat of ionization of the weak acid to the activation energy. In the case of acetic acid, the heat of ionization happens to be very small (-0.2 kcal) in comparison with the expected energy of activation, and this small quantity is partly offset by a very small correction ($+0.09$ kcal) for the temperature variation of the activity coefficient. Therefore with an acetate buffer the effect of temperature on hydrogen-ion concentration can be neglected for the present experiment.

METHOD

As a measurable physical quantity which may be taken as a linear function of the concentration of the disappearing reactant, the *total volume V* of a given quantity of the reaction mixture might be used. To a good approximation, the volume is a linear function of the numbers of moles N_i of the component species i present:

$$V = \sum_i N_i \overline{V}_i$$

since the partial molal volumes $\overline{V}_i$ are approximately constant for both solvent and solutes in a dilute solution. In turn the numbers of moles N_i are linear functions of the concentration of the disappearing reactant. Similarly we might use the electrical conductance of the reacting mixture in the event that the species concerned are ionic, or the optical density for a given wavelength of light if the species have absorption bands.

In the present experiment we shall make use of the total volume of the solution as measured with a dilatometer, which is a vessel to which a graduated capillary of small diameter is attached so that small changes in the volume of the contained liquid can be measured by movement of the meniscus of the liquid in the capillary. The capillary need not be graduated to correspond to the total volume of solution; it is sufficient that the graduations give the volume above any fixed reference point in the capillary in any arbitrary units. The capillary reading h is then a linear function of the total volume and thus, by the above argument, is also a linear function of the concentration of the disappearing reactant.

EXPERIMENTAL

The dilatometer is shown in Fig. 1. It must be handled with great care, as it is very fragile. Special care should be exercised in cleaning it and in putting it into and taking it out of its holder.

Thermostat baths at 20.0, 25.0, and 30.0°C are required, with temperature regulation of ±0.002°C, as the volume changes produced even by small temperature changes are of the same order of magnitude as those produced by the reaction being studied. A Beckmann thermometer or its equivalent should be used for a careful and frequent monitoring of the bath regulation.

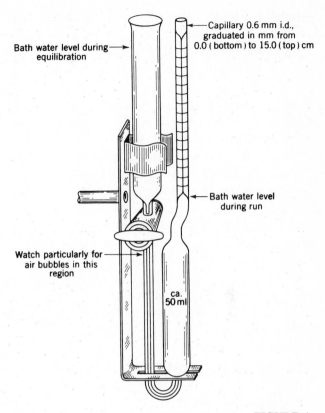

Bath water level during ⟶
equilibration

⟵ Capillary 0.6 mm i.d.,
graduated in mm from
0.0 (bottom) to 15.0 (top) cm

⟵ Bath water level
during run

Watch particularly for ⟶
air bubbles in this
region

ca.
50 ml

FIGURE 1
Dilatometer.

Materials   The acetal should be of high purity, dry, and free of acidic substances. If necessary the commercially available product may be refluxed over calcium oxide and fractionated.

Water used in the preparation of the stock solutions and water to be used for making up the final buffer solutions should be as free of dissolved gases as possible. Boiled distilled water may be used.

The following buffer solutions will be needed and should be made up volumetrically from the stock solutions provided (0.1 M NaAc, 1.5 M HAc):

1   0.02 M in NaAc and 0.3 M in HAc ($c_{H^+} \cong 3.4 \times 10^{-4}\ M$; pH $\cong$ 3.5)
2   0.02 M in NaAc and 0.6 M in HAc ($c_{H^+} \cong 6.7 \times 10^{-4}\ M$; pH $\cong$ 3.2)

To save time 250 ml of each of these solutions should be made up at the beginning of the period. At least 15 min before it will be needed, 100 ml of each solution in a volumetric flask should be placed in the thermostat bath to equilibrate. At the same time, a clean, dry 250-ml flask should be clamped in the bath for later use in preparing the reaction mixture. Flasks and solutions should be brought close to bath temperature before being placed in a bath if someone else is performing a run in the same bath.

Cleaning the dilatometer   The dilatometer should be very carefully cleaned. First, the stopcock plug should be removed and set aside for thorough cleaning. After the stopcock barrel has been scrubbed out with a test-tube brush and detergent solution and any excess stopcock grease from the stopcock leads cleaned out with a pipe cleaner, the two ends of the barrel may be plugged with corks. To facilitate further cleaning and rinsing, a one-hole rubber stopper equipped with a short length of tubing is affixed to the mouth of the dilatometer and connected to a water aspirator with pressure tubing. With the aspirator on, the end of the capillary is dipped into a battery jar or lipless beaker containing a detergent solution (see Fig. 2) until the dilatometer is partly filled. It should then be carefully shaken for a few minutes. By the same technique it is repeatedly and very thoroughly rinsed with distilled water and dried by sucking air through it. It need not be rigorously dry except in the capillary tube and in the stopcock barrel and leads. The latter can be dried with the help of a towel and pipe cleaner. The stopcock plug should be scrubbed with detergent, and its bore should be cleaned with a pipe cleaner. It should be rinsed and dried thoroughly, carefully greased as described in Chap. XVII, and reinserted into the barrel. The dilatometer can then be installed in its holder and placed in the bath to equilibrate. Between and after runs it is not necessary to clean the dilatometer so thoroughly or to regrease the stopcock. The used solution should be removed by suction, and the dilatometer then thoroughly rinsed and dried.

Procedure   Make up the reacting mixture by pouring 100 ml of buffer solution into the clean dry 250-ml flask, and then pipetting in 2 ml of acetal. The mixture should be swirled in the bath until the acetal has completely dissolved.

With the stopcock of the dilatometer closed, introduce a sufficient quantity of the reacting mixture. After all air bubbles have risen to the surface, the dilatometer can be mounted higher in the bath (so that the entire capillary is above the water surface) and the stopcock can be opened to admit the solution into the bulb. When the bulb is partly filled, the one-hole stopper with affixed tube should be attached and suction applied by mouth to reverse the flow; the purpose of this is to dislodge any air bubbles which may be trapped in the tube below the stopcock. After a few repetitions of this suction, close the stopcock and raise the dilatometer from the bath for a few seconds to permit inspection for the presence of air bubbles; if there are none, the solution may be allowed to fill the bulb completely. The rate of filling should be reduced drastically toward the end in order that it can be cut off suddenly

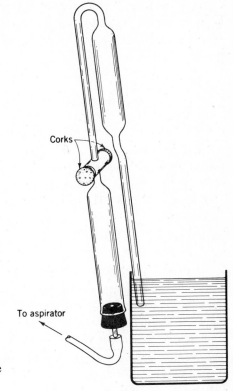

Corks

To aspirator

FIGURE 2
Method of washing and rinsing the
dilatometer.

as soon as the meniscus in on scale in the capillary. It is inadvisable to draw the meniscus down in the capillary by suction unless absolutely necessary, for drainage from the capillary wall is likely to form liquid plugs which drain slowly and unreliably during the run.

Readings of the position of the meniscus (estimated to one-tenth of the smallest scale division) should be made at 30-sec intervals, starting as soon as possible after the stopcock has been closed. They should be continued until differences between successive readings are averaging about one-fifth of the differences observed near the beginning of the run (discounting the first half dozen readings). The bath temperature should be frequently read to 0.001° and recorded.

Runs should be made under the following sets of conditions:

1   Buffer solution 1 ($c_{H^+} \cong 3.4 \times 10^{-4}\ M$) at 25.0°C
2   Buffer solution 2 ($c_{H^+} \cong 6.7 \times 10^{-4}\ M$) at 25.0°C
3   Buffer solution 2 ($c_{H^+} \cong 6.7 \times 10^{-4}\ M$) at 20.0°C
4   Buffer solution 1 ($c_{H^+} \cong 3.4 \times 10^{-4}\ M$) at 30.0°C

If time is limited, it is advisable to perform runs 2 and 3 or 1 and 4 in order to determine the activation energy E^* from data obtained at constant hydrogen-ion concentration.

CALCULATIONS

For each run a table should be constructed with the following entries: t, $h(t)$, Δh, and log Δh, where $\Delta h = h(t + \Delta t) - h(t)$; Δt should be chosen in accord with the discussion of the Guggenheim method. A plot of log Δh against t is then constructed, and the best straight line drawn through the points. Semilog paper (one decimal cycle) is convenient. The plot is likely to show some initial curvature, which should be disregarded in drawing the straight line; it is presumably due to drift toward temperature equilibrium following the filling operation. The rate constant k is calculated from the slope of the line.

If the plotted points tend to show a slight systematic oscillation about the line, a periodic variation in bath temperature due to thermoregulator cycling is probably responsible; if possible, compare the period of the oscillation with that of the thermoregulator cycling. Draw your line so it evens out these fluctuations.

The hydrogen-ion concentrations should be calculated by use of either Eq. (12) or (13). If Eq. (12) is used, the changes in Ac^- and HAc concentrations due to the small ionization of the acetic acid should not be neglected; they may be taken account of in a second approximation, the magnitude of the ionization having been estimated in a first approximation using the concentration values calculated directly from those for the stock solutions.

The results of runs 1 and 2 should be used to demonstrate that the reaction is first order with respect to hydrogen ion, and a value of k' should be calculated at 25°C.

The activation energy E^* should be calculated from the variation of k' with temperature.

APPARATUS

Dilatometer; special clamp for mounting dilatometer; one-hole rubber stopper to fit top of dilatometer, with short length of glass tubing affixed; magnifier (optional); clock or stopwatch; two clamps; three clamp holders; one 100- and two 250-ml volumetric flasks; 250-ml Erlenmeyer flask; 2-, 50-, and 100-ml pipettes; battery jar; glass-stoppered bottle for storing acetal; medium length (~2 ft) of gum-rubber tubing; corks to fit stopcock barrel.

Constant-temperature baths set at 20, 25, and 30°C with Beckmann thermometers mounted in each; water aspirator, equipped with rubber hose; stopcock grease; pipe cleaners;

acetone for rinsing; reagent-grade acetal (15 ml); 0.1 M sodium acetate (200 ml); 1.5 M acetic acid (300 ml).

REFERENCES

1. J. N. Brønsted and W. F. K. Wynne-Jones, *Trans. Faraday Soc.*, **25**, 59 (1929).
2. M. Kilpatrick, Jr., and E. F. Chase, *J. Amer. Chem. Soc.*, **53**, 1732 (1931).
3. E. A. Guggenheim, *Phil. Mag.*, **2** (7), 538 (1926).
4. W. J. Moore, "Physical Chemistry," 4th ed., chap. 9, Prentice-Hall, Englewood Cliffs, N.J. (1972).

GENERAL READING

A. A. Frost and R. G. Pearson, "Kinetics and Mechanism," 2d ed., Wiley, New York (1963).

EXPERIMENT 25.   ENZYME KINETICS: INVERSION OF SUCROSE

A very important aspect of chemical kinetics is that dealing with the rates of enzyme-catalyzed reactions. Enzymes are a class of proteins which catalyze virtually all biochemical reactions. In this experiment, we shall study the inversion of sucrose, as catalyzed by the enzyme invertase (β-fructofuranidase) derived from yeast.

THEORY

The basic mechanism postulated for enzyme-catalyzed reactions is that proposed by Michaelis and Menten in 1914. The initial step is a rapid equilibrium between the reactant (usually called the substrate S) and the enzyme E to form an enzyme-substrate complex ES:

$$E + S \rightleftharpoons ES \tag{1}$$

The equilibrium constant for this reaction is denoted by K. The next step is the decomposition of the complex to form products P and to regenerate the enzyme:

$$ES \xrightarrow{k} P + E \tag{2}$$

The rate constant of this slow step is k. Additional transformations may occur between the formation of the enzyme-substrate complex and the dissociation to products, as discussed in Exp. 26. In the present experiment, the simple mechanism given in Eqs. (1) and (2) is adequate to describe the kinetics.

The equilibrium expression for reaction (1) can be written in the form

$$(ES) = K(E)(S) \tag{3}$$

where the parentheses indicate concentrations of the species. From this equation, one can easily obtain an expression for the fraction of the total enzyme ($\sum E$) that is bound up in the enzyme-substrate complex:

$$\frac{(ES)}{(E) + (ES)} = \frac{(ES)}{(\sum E)} = \frac{K(S)}{1 + K(S)} \tag{4}$$

The rate r of formation of products is obtained by combining Eqs. (2) and (4) to give

$$r \equiv +\frac{d(P)}{dt} = k(ES) = \frac{kK (\sum E)(S)}{1 + K(S)} \tag{5}$$

For a fixed total enzyme concentration, $(\sum E)_0$, Eq. (5) can be simplified to

$$r = \frac{k'(S)}{1 + K(S)} \tag{6}$$

A plot of r versus (S), as shown in Fig. 1, starts out linearly with slope k' over the range where $K(S) \ll 1$ and the denominator in Eq. (6) is essentially unity. As (S) increases, r approaches a limiting value for $K(S) \gg 1$, namely

$$\lim_{(S) \to \infty} r = \frac{k'}{K} \tag{7}$$

The value of r reaches half its asymptotic value, namely $k'/2K$, when $(S) = 1/K$. Thus a determination of the dependence of the rate r on the substrate concentration (S) gives both k' and K. If the total enzyme concentration $(\sum E)_0$ is also known, k can be found from $k' = kK (\sum E)_0$. The most useful way of analyzing data of the type obtained in this experiment is to plot $1/r$ versus $1/(S)$ since we see from Eq. (6) that

$$\frac{1}{r} = \frac{K}{k'} + \frac{1}{k'(S)} \tag{8}$$

Such a plot is known as a Lineweaver-Burke plot. The slope determines k', and the ordinate intercept gives K/k'.

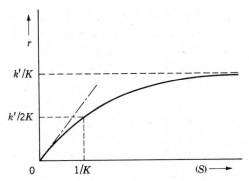

FIGURE 1
Reaction velocity versus substrate concentration.

METHOD

The enzyme yeast invertase catalyzes the conversion of sucrose to fructose and glucose according to the hydrolysis reaction

$$CH_2OH \qquad + H_2O \rightarrow \qquad CH_2OH \qquad + \qquad$$

Sucrose   Glucose   Fructose

Like a number of other enzymatic reactions, this one can also be carried out under nonphysiological conditions, with H^+ ions as a less efficient catalyst. No matter how the reaction is catalyzed, one of its most striking consequences is a large change in the optical-rotatory power of the solution as the sucrose is progressively converted to glucose and fructose.[1] This conversion actually produces a change in the direction in which the plane of polarization of light is rotated by the solution, and it is from this "inversion" that the enzyme derives its name. The inversion of sucrose in the presence of yeast was noted as early as 1832 by Persoz; and Michaelis and Menten used enzyme extracted from yeast in their classic study of the enzyme-substrate complex, in which the course of the reaction was monitored by observation of the change in optical-rotatory power accompanying it.

In the present experiment we shall use a rather simpler, although equally accurate, way of following the progress of the reaction. This method hinges on the fact that glucose and fructose, like all other monosaccharides, are active reducing agents, while the sucrose from which they are formed is not a "reducing sugar." Thus any measure of the reducing capacity of a reaction mixture becomes, in effect, a measurement of the extent to which the conversion of sucrose to glucose and fructose has proceeded in that mixture. In this experiment we shall use a 3,5-dinitrosalicylic acid anion reagent in assaying the reducing capacity of various reaction mixtures.[2] Since the oxidized and reduced form of this reagent possess significantly different visible absorption spectra, colorimetric analysis of the optical density at a suitable wavelength (5400 Å) can be used to measure the extent of hydrolysis that has occurred.

The procedure we will use is first to prepare a particular reaction system containing appropriate amounts of sucrose, enzyme, and buffer solution. This mixture is hydrolyzed for some specified time, and then "quenched" by addition of sodium hydroxide (contained in the 3,5-dinitrosalicylate reagent added at this point). The solution is heated to "develop" the color produced by the reagent in the presence of reducing activity in the solution, and the intensity of that color is determined with a spectrophotometer. After a calibration has been carried out, the spectrophotometric reading directly establishes the number of moles of fructose and glucose present in the solution. Then knowing the time during which the reaction proceeded, we can readily establish the average rate at which sucrose was converted to glucose and fructose.

EXPERIMENTAL

The reducing-sugar reagent is made up by mixing two solutions: a solution of about 25.0 g of 3,5-dinitrosalicyclic acid in 1000 ml of warm 1 M NaOH and one of 750 g of sodium potassium tartrate, $NaKC_4H_4O_4 \cdot 4H_2O$, in 750 ml of warm distilled water. These solutions are mixed slowly and diluted to a total volume of 2500 ml with warm distilled water.

Procedure   The standard assay procedure to be used (with variations as indicated) in all runs is carried out as follows:

1 Pipette into a clean test tube—in the order given—1.0 ml of enzyme solution, 0.5 ml of distilled water, 0.5 ml of buffer solution, and 1.0 ml of 0.3 M sucrose solution. Swirl the assay tube vigorously immediately before and immediately after the addition of the last reagent required to initiate the reaction.

2 Begin timing when the sucrose solution is added and the tube swirled to mix the reagents. Let the reaction proceed for *exactly* 5.0 min, or whatever other period of time is specified in a particular run.

3 At the end of the allotted period, "quench" the reaction by adding 2.0 ml of dinitrosalicylate reagent (containing NaOH as the active quenching agent), and once again vigorously swirl the tube to mix the reagents.

4 Immerse test tube in boiling water for 5 min; cover *loosely* with parafilm.

5 Cool the reaction tube by holding it in a slanted position and running cold tap water down the outside of the tube.

6   Dilute the solution with 15.0 ml of distilled water carefully dispensed from a graduated cylinder or pipette.

7   Using a Spectronic-20 or other suitable colorimeter, read the optical density of the solution at 5400 Å against a distilled-water blank.

A   Blank runs   Our first concern is to establish that the dinitrosalicylate reagent gives no appreciable test in the absence of reducing sugars. To establish this point, one should systematically examine the results of the action of the reagent on each of the other substances present in the reaction mixture. The solutions to be tested are given below.

Run	Enzyme, ml	Water, ml	Buffer, ml	0.3 M Sucrose, ml
B1	0.0	2.5	0.5	0.0
B2	0.0	1.5	0.5	1.0
B3	1.0	1.5	0.5	0.0
B4	1.0	0.5	0.5	*1.0

In the first three of these runs, mix the indicated reagents, wait 5.0 min, and then pick up the standard assay procedure at step 3. In run B4 add the enzyme, water, and buffer solutions, then add 2.0 ml of the dinitrosalicylate reagent *before adding* the 1.0 ml of sucrose solution. (This procedure is indicated in the above table and in subsequent tables by placing an asterisk before the volume of sucrose to remind you to add dinitrosalicylate reagent before adding sucrose.) Wait 5.0 min from the last addition, and pick up the standard assay procedure at step 4. Run B4 constitutes a so-called "zero-time blank"—since the sucrose is added only after the reagent that, allegedly, completely halts the enzymatic reaction—and constitutes a particularly sensitive blank run. None of the blank runs should develop an optical density greater than about 0.05 at 5400 Å.

B   Standardization runs   We must establish the optical density (at 5400 Å) produced by the action of our reagent in the presence of known amounts of glucose and fructose. The reagent offers an exceedingly delicate test, and to standardize such a reagent we use a very dilute aqueous glucose-fructose solution (hereafter denoted by GFS) containing only 0.9 g of glucose and 0.9 g of fructose per liter. Three standardization runs should be sufficient.

Run	Enzyme, ml	Water, ml	Buffer, ml	0.3 M Sucrose, ml	GFS, ml
S1	0.0	1.1	0.5	1.0	0.4
S2	0.0	0.7	0.5	1.0	0.8
S3	0.0	0.3	0.5	1.0	1.2

In each case, make up the indicated solution and then pick up the standard assay procedure at step 3. The results of blank run B2 should have shown that, in the absence of the GFS, the mixture of reagents used in the standardization runs will yield no color test. The optical densities obtained in runs S1 to S3 can thus be attributed to the action of the reagent on the GFS. These data will be used to construct a calibration curve for the colorimetric assay.

C   Enzyme activity   Carry through the standard assay procedure, described above, with the enzyme solution available in the laboratory. The observed optical density should lie in the range 0.6 to 0.9. If it is greater than 0.9, dilute the enzyme with sterile, previously boiled distilled water (see below for precautions about the preparation and storage of the enzyme) to reduce the activity to an appropriate level. If it is less than 0.6, the enzyme solution is too weak and a new solution must be made up.

D   Progress of the reaction with time   For the second of the following two sets of runs, prepare some 0.03 M sucrose solution by diluting 1 ml of the 0.3 M stock solution with 9 ml of distilled water. Runs in series C use the concentrated sucrose solution, and those in series D use the dilute solution.

Run	Enzyme, ml	Water, ml	Buffer, ml	Sucrose, ml	Time, min
C0	0.2	1.3	0.5	*1.0 (0.3 M)	0
C1	0.2	1.3	0.5	1.0 (0.3 M)	1
C3	0.2	1.3	0.5	1.0 (0.3 M)	3
C5	0.2	1.3	0.5	1.0 (0.3 M)	5
C10	0.2	1.3	0.5	1.0 (0.3 M)	10
C20	0.2	1.3	0.5	1.0 (0.3 M)	20
D0	1.0	0.5	0.5	*1.0 (0.03 M)	0
D1	1.0	0.5	0.5	1.0 (0.03 M)	1
D3	1.0	0.5	0.5	1.0 (0.03 M)	3
D5	1.0	0.5	0.5	1.0 (0.03 M)	5
D10	1.0	0.5	0.5	1.0 (0.03 M)	10
D20	1.0	0.5	0.5	1.0 (0.03 M)	20

Runs C0 and D0 are zero-time blanks in which the sucrose addition is made only *after* inactivation of the enzyme by the addition of 2.0 ml of dinitrosalicylate reagent (denoted by * in the tables). In all other runs, begin timing when the sucrose solution is added, and let the reaction proceed for *precisely* the time indicated in each case. Then pick up the standard assay procedure at step 3.

E   Dependence of initial rate on substrate concentration   Prepare, from the 0.3 M stock solution of sucrose, another 10 ml of 0.03 M sucrose solution. Taking care to use the sucrose solution called for in each case, proceed to the following runs—all of which are allowed to proceed for the standard time of 5.0 min except that run E0 is a zero-time blank.

Run	Enzyme, ml	Water, ml	Buffer, ml	Sucrose, ml
E0	1.0	0.5	0.5	*1.0 (0.3 M)
E1	1.0	0.5	0.5	1.0 (0.3 M)
E2	1.0	1.3	0.5	0.2 (0.3 M)
E3	1.0	0.5	0.5	1.0 (0.03 M)
E4	1.0	1.0	0.5	0.5 (0.03 M)
E5	1.0	1.3	0.5	0.2 (0.03 M)
E6	1.0	1.4	0.5	0.1 (0.03 M)

F Dependence of the initial rate on enzyme concentration Use the 0.3 M stock sucrose solution in all the following runs, which are to be allowed to proceed for the standard time of 5.0 min, except that run F0 is a zero-time blank.

Run	Enzyme, ml	Water, ml	Buffer, ml	0.3 M Sucrose, ml
F0	1.0	0.5	0.5	*1.0
F1	1.5	0.0	0.5	1.0
F2	1.0	0.5	0.5	1.0
F3	0.5	1.0	0.5	1.0
F4	0.2	1.3	0.5	1.0
F5	0.1	1.4	0.5	1.0

G Dependence of the rate on temperature In order to determine the temperature dependence of the enzyme-catalyzed reaction rate, the standard assay procedure is carried out at several different temperatures in a constant-temperature bath. Convenient values are 0°C (ice bath), 12, 25, 35, and 45°C. In each case, prepare the enzyme-water-buffer mixture in the assay tube, put about 2 ml of the sucrose stock solution in another test tube, and immerse both tubes in the thermostat bath for a few minutes to achieve temperature equilibrium. Next, pipette 1.0 ml of the equilibrated sucrose solution into the assay tube. Leave the assay tube in the constant-temperature bath until, 5.0 min after the addition of the sucrose solution, the reaction is terminated by the addition of 2.0 ml of the dinitrosalicylate reagent. Then complete the assay as usual.

H Non-enzymic hydrolysis of sucrose[1,2] The basic procedure for carrying out hydrolysis runs with H^+ ions as the catalyst is as follows: Pipette the indicated volume of 0.3 M sucrose solution into an assay tube. Add sufficient distilled water so that the final volume of the reaction mixture will be 4 ml (*not 3 ml, as in all preceding runs*). Add the indicated volume of 1 M HCl solution to initiate the catalytic hydrolysis, starting your timing as the acid is added. After exactly 5.0 min, stop the reaction by adding 5 ml of 1 M NaOH, which immediately neutralizes all of the catalytically active H^+ present. Add 2.0 ml of dinitrosalicylate reagent, and heat the reaction tube in a boiling-water bath for 5 min. Cool the reaction tube, and dilute by adding 9 ml of distilled water (*not 15 ml as in earlier runs*) so that the final volume is, as before, 20 ml. Measure the optical density of the solution at 5400 Å.

A set of suggested concentrations at which this procedure can be carried out are given below.

Run	0.3 *M* Sucrose, ml	Water, ml	1 *M* HCl, ml
H1	1.0	2.0	1.0
H2	1.0	1.0	2.0
H3	2.0	1.0	1.0
H4	2.0	0.0	2.0
HB	2.0	2.0	0.0

Run HB is a blank, which will allow you to see how much (if any) hydrolysis takes place in the absence of any catalyst.

Precautions In order to obtain successful results from this experiment, both careful volumetric technique and judicious handling of reagents are required. The following points should be noted.

1 When first using a pipette, rinse it with a small portion of the solution to be used and make sure that the pipette drains cleanly. If possible, always use the same pipette for the same solution; this tends to minimize the effect of systematic errors arising from discrepancies in the calibration of different pipettes. The pipettes you will need are:

> Two 1-ml transfer pipettes, used to dispense the "standard" amounts of the sucrose solution and the enzyme solution
> One 2-ml transfer pipette, used exclusively for dispensing the dinitro-salicylate reagent
> One 5-ml transfer pipette, used exclusively for dispensing the NaOH solution required in part H of the experiment
> One 1-ml graduated pipette, used for dispensing buffer solution and distilled water
> One 2-ml graduated pipette, used for dispensing "nonstandard" amounts of enzyme solution

2 The enzyme, a protein, is highly susceptible to attack by airborne micro-organisms. All glassware with which it comes in contact must be sterilized in an autoclave. In particular, a sterile Erlenmeyer flask, beaker, 1-ml pipette, spatula, and magnetic spinbar must be available. The spinbar is dropped into the finished enzyme preparation, and used to stir up the suspension just prior to withdrawing a sample. Any dilution of the enzyme stock solution must be done with water that has been sterilized by boiling prior to use; of course, the water must be cooled down before it is added to the enzyme. The enzyme solution should be covered with a thin layer of toluene, and the flask kept plugged with cotton wool. *Care must be taken not to deliver any toluene when pipetting the enzyme solution into the assay tube.*

3 The sucrose solution is also susceptible to attack by microorganisms. In this instance, it is sufficient to use a freshly prepared solution each day.

4 Fructose is very hygroscopic, and must be handled in a dry-box when making up the GFS solution.

CALCULATIONS

First, construct a calibration curve from the results of runs S1, S2, and S3. This consists of a graph showing the observed optical density versus micromoles of glucose and fructose present. The GFS, made up as previously described, will contain 5×10^{-3} mol of glucose and fructose per liter.

Using the standardization curve prepared above, the results of the standard assay in part C, and the given concentration of the enzyme stock solution in g liter^{-1}, calculate the *specific activity* of the enzyme—that is, the number of micromoles of sucrose hydrolyzed per minute per gram of enzyme present. (The specific activity of an enzyme preparation is, of course, a function of the purity of the enzyme. As inactive protein is removed from the preparation, the specific activity will rise. When the specific activity can no longer be increased by any purification method, a homogeneous enzyme preparation may have been achieved; but proof of this depends on other criteria.) The exact chemical composition of invertase is still unknown, but its molecular weight has been estimated at 100,000. Combining this datum with your calculated specific activity, calculate the *turnover number* for the enzyme—that is, the number of molecules of sucrose hydrolyzed per second per molecule of enzyme present.

From the data of runs C1 through C20 and D1 through D20, calculate the number of micromoles of sucrose hydrolyzed in each time interval. Let x denote the number of micromoles of sucrose hydrolyzed, which is equal to one-half the number of micromoles of glucose plus fructose produced. If the reaction were zero-order in sucrose then we would expect that $x = k_0 t$. Prepare a graph of the results obtained in these two series of runs, plotting x versus t, and indicate whether the data are consistent with the hypothesis that the reaction is zero-order in sucrose.

If the reaction were first-order in sucrose then (letting a stand for the number of micromoles of sucrose originally present) we should find that $a/(a - x) = k_1 t$. Prepare a graph of the results obtained in the above two series of runs, plotting $\log [a/(a - x)]$ versus t, and indicate whether your data are consistent with the hypothesis that the reaction is first-order in sucrose.

Can we properly regard the average rate measured over the first 5.0 min of reaction as an acceptable approximation to the true *initial* rate of reaction?

From the data of runs E1 through E6 and F1 through F5, determine the dependence of the initial rate on the concentrations of sucrose and of enzyme. Prepare a Lineweaver-Burke plot of $1/r$ versus $1/(S)$, and determine the constants k' and K. Are the data compatible with the Michaelis-Menten mechanism described previously?

From the data of part G, determine the initial rate r for each temperature, and plot $\log_{10} r$ versus $1/T$. From the slope of the line, determine E_{act} from

$$\frac{d \log r}{d(1/T)} = - \frac{E_{act}}{2.303 \ R} \tag{9}$$

From the data of part H, estimate the rate constant k_H for the acid-catalyzed reaction, assuming that the rate law has the form

$$\frac{-d(S)}{dt} = k_H(S)(H^+) \tag{10}$$

Do the data support this assumed rate law? Also calculate a turnover number for the H^+-catalyzed reaction; that is, the number of molecules of sucrose hydrolyzed per second per H^+ ion present. How does this figure compare with that for the invertase enzyme?

DISCUSSION

A number of additional investigations can be carried out on this same basic system. A few suggestions follow.

1 By a series of runs at different temperatures, determine E_{act} for the H^+-catalyzed reaction, and compare this figure with E_{act} for the enzyme-catalyzed reaction. Does the difference in E_{act} completely account for the ratio of turnover numbers between H^+ and the enzyme?

2 By a series of runs at different temperatures, determine the temperature at which the enzymatic reaction reaches a maximum velocity—just below that temperature at which the enzyme begins to denature. In such studies take care that the solutions have really reached temperature equilibrium before the sucrose is added to initiate reaction.

3 By a series of runs at different pH's, determine the pH at which invertase functions most effectively as a catalyst. Is the reaction rate highly dependent on pH? Why should pH have anything to do with the catalytic activity of invertase?

4 Investigate the inhibition of the enzymatic reaction by *p*-mercuribenzoate at concentrations of about 10^{-5} *M*. Show how an investigation of inhibition or "poisoning" of the catalyst might be used to establish the number of catalytically active sites on an invertase molecule.

5 Using space-filling models, construct models of the sucrose substrate and the product molecules. Try to infer what the active site on the enzyme must look like to be able to catalyze the hydrolysis.

APPARATUS

Sterile preparation of yeast invertase; pH 4.8–5 acetate buffer; 0.3 *M* sucrose; 1 *M* sodium hydroxide; 3,5-dinitrosalicylate reagent solution; 1 *M* HCl; glucose; fructose (levulose); toluene.

Two 1-ml transfer pipettes (one sterilized); one 2-ml transfer pipette; one 5-ml transfer pipette; one 1-ml graduated pipette; one 2-ml graduated pipette; one 15-ml graduated cylinder or pipette; volumetric flasks; test tubes; spinbar.

Boiling-water bath; stopwatch; colorimeter (Spectronic-20 or equivalent); 0 to 50°C constant-temperature bath; magnetic stirring unit.

Enzyme solution: Prepare with care, using apparatus that is sterilized. More than 1 liter of distilled water is boiled for 10 min, covered with Al foil, and chilled in an ice bath. Then add a precisely weighted amount of powered invertase to *cold* water in a 1-liter volumetric flask and make up to the mark. The enzyme solution should be kept tightly stoppered and chilled at all times. Before being dispensed, the solution should be stirred to homogenize it. (If a previously sterilized magnetic stirring bar is placed in each bottle, the enzyme solution can be stirred easily without introducing contamination.)

REFERENCES

1. F. A. Bettelheim, "Experimental Physical Chemistry," pp. 269–277, Saunders, Philadelphia (1971).
2. J. G. Dawber, D. R. Brown, and R. A. Reed, *J. Chem. Educ.*, **43**, 34 (1966).

EXPERIMENT 26.   ENZYME KINETICS: HYDROLYSIS OF AN ESTER

In this experiment, *p*-nitrophenylpivalate, a substrate which yields a chromophore upon reaction, is prepared and purified, and the molar extinction coefficient of the product of its hydrolysis is determined. The concentration of the absorbing product as a function of time is monitored using a recording single-beam spectrophotometer with automatic cuvette positioner. Data obtained under varying conditions of enzyme and substrate concentration are analyzed graphically. This experiment is an adaptation of one devised by Bender,[1] and the derivation of the kinetic expressions is also based on Bender's work.[2]

THEORY

Chymotrypsin is perhaps the most thoroughly studied of the many crystalline enzymes whose properties have been investigated by kinetic techniques. This enzyme (which, to be more specific, should be referred to as bovine pancreatic α-chymotrypsin A) is a protein of molecular weight 24,800. Its biological role is as a digestive enzyme; it hydrolyzes peptide or amide bonds in proteins and polypeptides. Although the "natural" substrates towards which chymotrypsin shows great specificity are amides or peptides of *N*-acylated aromatic *l*-amino acids, the enzyme is also reactive towards esters, and can hydrolyze esters of most carboxylic acids if the leaving group is good enough (i.e., if the alcohol from which the ester is derived is sufficiently acidic, as are phenols, for instance).

On treatment of chymotrypsin with diisopropylfluorophosphate (the nerve gas Sarin) a totally inactive phosphorylated enzyme is produced. The phosphorylated enzyme was shown[3] to be the diisopropylphosphoryl ester of the hydroxyl group of serine 195 in the enzyme sequence. The acylation of a specific serine in the protein is characteristic of a group of enzymes called *serine proteases*, as well as of cholinesterase. DP-chymotrypsin can be slowly dephosphorylated (and reactivated), especially in the presence of added nucleophiles. On the basis of a considerable amount of evidence, it is now felt that all the reactions catalyzed by chymotrypsin proceed via intermediate "acyl enzymes," which are esters of serine 195 of the enzyme.[4]

Chymotrypsin-catalyzed reactions, like essentially all other enzymic reactions, involve association between enzyme and substrate prior to any steps involving bond making or breaking. The overall reaction can be described by

$$E + S \xrightleftharpoons{K_s} ES \xrightarrow{k_2} ES' + P_1 \xrightarrow{k_3} E + P_1 + P_2 \tag{1}$$

where E is the free enzyme, S is the substrate, ES is the noncovalently bound enzyme-substrate complex, ES' is the acyl enzyme, and the two products P_1 and P_2 are the *leaving group* and the free acid, respectively. We can readily derive an expression for the variation of the concentration (P_1) with time (to be observed spectrophotometric-ally) by making two assumptions: (1) the association equilibrium governed by K_s is so fast that we can assume that (ES) is always at its equilibrium value, and (2) the total substrate concentration is much greater than the total enzyme concentration [i.e., $(S)_0 \gg (E)_0$]. Equation (2) is based on these assumptions, and Eq. (3) is simply a conservation equation.

$$K_s = \frac{(E)(S)}{(ES)} \cong \frac{(E)(S)_0}{(ES)} \tag{2}$$

$$(E)_0 = (E) + (ES) + (ES') \tag{3}$$

Combining these two equations, we find

$$(ES) = \frac{(E)_0 - (ES')}{1 + K_s/(S)_0} \tag{4}$$

The differential rate equations for (P_1) and (ES') follow from Eq. (1):

$$\frac{d(P_1)}{dt} = k_2(ES) \tag{5}$$

$$\frac{d(ES')}{dt} = k_2(ES) - k_3(ES') \tag{6}$$

Inserting the value of (ES) from Eq. (4) into Eq. (6),

$$\frac{d(ES')}{dt} = \frac{k_2[(E)_0 - (ES')]}{1 + K_s/(S)_0} - k_3(ES') \tag{7}$$

which can be rearranged to give

$$\frac{d(ES')}{dt} = \frac{k_2(E)_0}{1 + K_s/(S)_0} - \left[k_3 + \frac{k_2}{1 + K_s/(S)_0} \right](ES') \tag{8}$$

Equation (8) is of the form

$$\frac{d(ES')}{dt} = a - b(ES') \tag{9}$$

where $\qquad a = \dfrac{k_2(E)_0}{1 + K_s/(S)_0} \qquad$ and $\qquad b = k_3 + \dfrac{k_2}{1 + K_s/(S)_0} \qquad$ (10)

and since a and b are constants independent of time, the equation can be integrated to give

$$\frac{-\ln\left[a - b(ES')\right]}{b} = t + C \tag{11}$$

Since we know that $(ES') = 0$ at $t = 0$, we conclude that $C = (\ln a)/b$. Inserting this value into Eq. (11) and rearranging, we obtain

$$\ln\left[\frac{a - b(ES')}{a}\right] = -bt \tag{12}$$

or
$$(ES') = \frac{a}{b}(1 - e^{-bt}) \tag{13}$$

Equation (13) implies that at sufficiently long times (such that $bt \gg 1$), the concentration (ES') reaches a steady-state value of a/b, and that it approaches that steady state in an exponential manner with a rate characterized by the constant b. We can insert this expression for (ES') into Eq. (4) to determine the dependence of (ES) on time:

$$(ES) = \frac{(E)_0 - (a/b)(1 - e^{-bt})}{1 + K_s/(S)_0} \tag{14}$$

This expression can be substituted into Eq. (5), the rate equation for (P_1), to yield

$$\frac{d(P_1)}{dt} = \frac{k_2(E)_0 - k_2(a/b)(1 - e^{-bt})}{1 + K_s/(S)_0} \tag{15}$$

which may be integrated to obtain

$$(P_1) = \frac{k_2[(E)_0 - a/b]}{1 + K_s/(S)_0} t + \frac{k_2 a}{b^2[1 + K_s/(S)_0]}(1 - e^{-bt}) \tag{16}$$

Equation (16) has the form

$$(P_1) = Qt + R(1 - e^{-bt}) \tag{17}$$

and the predicted variation in (P_1) is shown in Fig. 1. At long times ($bt \gg 1$), Eq. (17) becomes $(P_1) = Qt + R$, a straight line with intercept R and slope Q. At short times, however, (P_1) is less than the value extrapolated back from the linear part of the curve, the difference being Re^{-bt}. By plotting the natural logarithm of this difference against time (i.e.. plotting the difference as a first-order rate process), one can determine a value for b.

It turns out that the experimental data for chymotrypsin reactions do indeed resemble Fig. 1. For most substrates, however, the exponential decay is so fast (i.e., b is so large) that special very fast kinetic techniques are necessary to observe it. We have chosen the p-nitrophenyl ester of pivalic acid (trimethylacetic acid) because it reacts comparatively slowly with chymotrypsin despite its good leaving group. If one can obtain values of Q, R, and b over an adequate range of total enzyme and substrate concentrations, one can test the kinetic expression assumed in Eq. (1), and

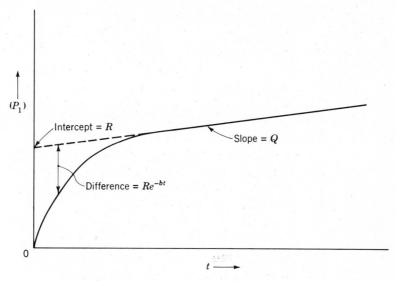

FIGURE 1
Variation of (P_1), the concentration of the leaving group, with time.

derive values for all the constants involved. Let us consider the detailed expressions for Q, R, and b by inserting a and b from Eq. (10) into Eq. (16):

$$Q = \frac{k_2 k_3 (E)_0 (S)_0}{(k_2 + k_3)(S)_0 + k_3 K_s} \tag{18}$$

$$R = \frac{k_2{}^2 (E)_0}{[k_2 + k_3 + k_3 K_s/(S)_0]^2} \tag{19}$$

$$b = \frac{(k_2 + k_3)(S)_0 + k_3 K_s}{K_s + (S)_0} \tag{20}$$

We shall now define two new constants in terms of combinations of the constants appearing in Eq. (1):

$$k_{cat} = \frac{k_2 k_3}{k_2 + k_3} \quad \text{and} \quad K_m \text{ (apparent)} = \frac{k_3 K_s}{k_2 + k_3} \tag{21}$$

Equation (18) can then be transformed into

$$Q = \frac{k_{cat}(E)_0 (S)_0}{(S)_0 + K_m} \tag{22}$$

which has the form of the more familiar Michaelis-Menten equation for enzyme kinetics, since the slope Q of the experimental curve is the "initial velocity" in a steady-state treatment. Note that $k_2/K_s = k_{cat}/K_m$.

In the particular case of p-nitrophenylpivalate hydrolysis, as in the hydrolysis of most reactive esters by chymotrypsin, the acylation step k_2 is much faster than the deacylation step k_3, since the former reaction involves a much better leaving group than the latter. We can treat the experimental data on this assumption, and check back to see if the assumption is supported by the results. The following "reciprocal equations" can be derived from Eqs. (18), (19), and (20), by assuming that $k_2 \gg k_3$ [and in the case of $1/b$ that $k_2(S)_0 \gg k_3K_s$ also]:

$$\frac{1}{Q} = \frac{k_2 + k_3}{k_2 k_3 (E)_0} + \frac{K_s}{k_2(E)_0} \frac{1}{(S)_0} \cong \frac{1}{k_3(E)_0} + \frac{K_s}{k_2(E)_0} \frac{1}{(S)_0} \tag{23}$$

$$\frac{1}{R^{1/2}} = \frac{k_2 + k_3}{k_2(E)_0^{1/2}} + \frac{k_3 K_s}{k_2(E)_0^{1/2}} \frac{1}{(S)_0} \cong \frac{1}{(E)_0^{1/2}} + \frac{k_3 K_s}{k_2(E)_0^{1/2}} \frac{1}{(S)_0} \tag{24}$$

$$\frac{1}{b} = \frac{(S)_0 + K_s}{(k_2 + k_3)(S)_0 + k_3 K_s} \cong \frac{1}{k_2} + \frac{K_s}{k_2} \frac{1}{(S)_0} \tag{25}$$

Experimental difficulties may limit the usefulness of these equations. For the hydrolysis of p-nitrophenylpivalate, the maximum substrate concentration is limited by the solubility of the ester, about 1.5×10^{-4} M. On the other hand, in order to observe the formation of product with any accuracy, $(S)_0$ must be at least 5×10^{-6} M. On plotting $1/Q$ against $1/(S)_0$ at constant $(E)_0$, you will find that $K_s/k_2(E)_0(S)_0$ is much smaller than $1/k_3(E)_0$ for the lowest $(S)_0$ you can use, so much smaller as to be comparable to the experimental error. It will thus be nearly impossible to determine the slope of this line with any accuracy. A similar circumstance will occur for a plot of $1/R^{1/2}$ against $1/(S)_0$. Thus, Q and R are effectively only useful for deriving $(E)_0$ and $k_3(E)_0$, and hence k_3. The value of $(E)_0$ is not known from the weight of enzyme used since the sample is of unknown purity, usually between 60 and 95 percent.

In the plot of $1/b$ against $1/(S)_0$, the slope is large and the intercept is small since $K_s > (S)_0$. However, if particular care is taken to obtain accurate values of b at the higher $(S)_0$ concentrations, a reasonable, if approximate, extrapolation to $1/k_2$ can be made. Thus, k_2/K_s will be known accurately, and k_2, less accurately. Then all the parameters, K_s, k_2, k_3, k_{cat}, K_m, and $(E)_0$ can at least be estimated, and the assumptions made in the derivation above can be checked.

A set of experiments at constant $(S)_0$ and varying $(E)_0$ serves as a further check on the mechanism. Equations (18), (19), and (20) predict that for constant $(S)_0$, Q and R should be directly proportional to $(E)_0$, and b should be independent of $(E)_0$.

The reactions of chymotrypsin are pH dependent, showing a dependence on deprotonation of a group with $pK_a \sim 7$. Between pH 8 and 9, the rate constants are essentially pH independent. For this experiment we use a buffer at pH 8.5 to ensure that the p-nitrophenol product ($pK_a = 7.1$) is completely ionized.

EXPERIMENTAL

If p-nitrophenylpivalate is not available, it can be prepared in the following manner. Place a solution of 4.3 g of p-nitrophenol (0.030 mol) in 5 ml dry pyridine† in a 25- or 50-ml round-bottom flask with a ground joint. Stir with a magnetic stirrer, and slowly add 3.0 g pivalyl chloride (0.024 mol) with a disposable pipette. Fit the reaction flask with a drying tube. After an initial liberation of heat, the precipitation of pyridine hydrochloride causes the reaction mixture to turn into a thick slurry. This mixture is allowed to stand for about 3 hr and then poured into 50 ml of water, followed by extraction with three 50-ml portions of ether. The combined ether fractions are washed four times with 40 ml of 5 percent aqueous $NaHCO_3$ and then four times with 40 ml of water. The organic layer is dried over $MgSO_4$ and evaporated, to yield a crude product. Recrystallization from 95 percent ethanol affords about 4 g (75 percent) of white, crystalline p-nitrophenyltrimethylacetate. This material should be recrystallized again from 95 percent ethanol to give the pure product: mp 94.5 to 95°C; λ_{max}^{KBr} 1760 cm^{-1} (ester carbonyl), 1350 cm^{-1} ($C-NO_2$). Purification can also be achieved by sublimation at 100°C and 0.15 Torr.

A pH 8.5 tris(hydroxymethyl)aminomethane buffer, 0.01 M, is prepared by weighing out 0.606 g (0.005 mol) of "Trizma Base" [a purified grade of tris(hydroxymethyl)aminomethane], quantitatively transferring it to a 500-ml volumetric flask, and adding 1.20 ml of 1 N HCl. Add distilled water to bring up to 500 ml, and check the pH of the solution with a pH meter.

A pH 4.6 acetate buffer is prepared by weighing out 0.205 g (2.5 mmol) of sodium acetate (analytical reagent anhydrous), transferring it to a 250-ml volumetric flask, and adding 0.210 ml (0.220 g = 2.73 mmol) of glacial acetic acid. Dilute to 250 ml with distilled water.

Stock p-nitrophenylpivalate solution is prepared by weighing out accurately about 19.5 mg (about 0.87 mmol) of the recrystallized, dried ester, and dissolving it in acetonitrile (spectral grade, or reagent grade distilled from P_2O_5) in a 25-ml volumetric flask. Bring to volume with acetonitrile. Calculate the molarity of the solution to three significant figures (M.W. of p-nitrophenylpivalate = 223).

Stock enzyme solution is prepared by weighing out accurately about 80 mg of chymotrypsin and adding pH 4.6 acetate buffer solution to bring up to 2 ml in a

† Reagent-grade pyridine contains water as its major impurity which must be removed by drying over KOH followed by distillation from BaO.

volumetric flask. The enzyme solution is stable for a period of a week or more at this pH if it is kept refrigerated.

The experiment is carried out using a single-beam spectrophotometer with an automatic sample positioner, an "optical-density converter," and a servo chart recorder as readout device. The sample cell is thermostated at 25.0°C by means of a constant-temperature circulator. Specific instructions will be given here for the Gilford-modified Beckman DU spectrophotometer, with Heath servorecorder, and Haake temperature unit.

At the beginning of the experiment, first turn on the constant-temperature circulator and adjust it to 25.0°C. Also turn on the spectrophotometer lamp power supply and the optical-density converter unit. Set the wavelength on the mono-chromator at 400 nm, close the slit, and turn on the tungsten lamp (visible light source). The instrument should be allowed to warm up for about 30 min.

The extinction coefficient of p-nitrophenylate anion at 400 nm is determined by hydrolyzing p-nitrophenylpivalate in 0.1 N sodium hydroxide solution. If one wanted to know the apparent extinction coefficient of p-nitrophenol at a lower pH, at which it is not completely dissociated, one would have to carry out the hydrolysis in a small volume of sodium hydroxide, neutralize, and dilute with the appropriate buffer.

Pipette 3.0 ml of 0.1 N sodium hydroxide solution into each of four cuvettes (1-cm path). Place one cuvette in the cell compartment of the spectrophotometer. To the other cuvettes, add 25, 50, and 75 μl (0.025, 0.050, and 0.075 ml) of the stock p-nitrophenylpivalate solution, respectively. These quantities should be carefully measured out, using micropipettes or a microsyringe. Cover each cell with Parafilm, shake vigorously for several seconds, and then place in the cell compartment. After about 10 min, absorbance readings can be made on the samples, using the sodium hydroxide solution as the blank against which the instrument is balanced. Note the absorbance of each sample at 3-min intervals until any two consecutive readings for a sample agree to within 0.003 absorbance units.

Calculate the apparent molar extinction coefficient ε for ionized p-nitrophenol for each sample using Beer's law:

$$A = \varepsilon c l \qquad (26)$$

where A is the absorbance (optical density), c is the concentration of the species, and l is the length of the light path in cm. Check that Beer's law holds for this compound (i.e., that ε is independent of c).

For every kinetic run with enzyme and ester, a blank containing only ester at the same concentration must be run simultaneously. The ester hydrolyzes at a detectable rate in this buffer, and the use of the absorbance of the "ester only" solution as the "blank" allows one to correct for this spontaneous rate automatically. Since the sample positioner has four positions, it is possible to follow two reactions at once, or three reactions if all three use the same initial ester concentration and thus require the same blank.

The following kinetic runs should be carried out: (1) Six or more runs using a constant aliquot of stock enzyme solution (25 μl) and different aliquots of stock ester solution (over the range from about 10 to 100 μl) added to 3 ml of buffer. (2) Three or more runs using a constant amount of ester (say 50 μl) and a range of enzyme concentrations (say 15 to 75 μl).

Procedure   A typical kinetic run is conducted as described below.

1   The wavelength selector is set at 400 nm, and the full-scale absorbance range of the chart recorder is carefully set according to the instructions provided for the particular instrument. In order to provide for the possibility of drift during the experiment, the baseline (zero of absorbance) should be set at 5 units on the chart. The full-scale absorbance range should be about twice the absorbance predicted for R, the initial "burst" of p-nitrophenol. According to the previous discussion, R should be approximately equal to $(E)_0$ for all these experiments. For experiments using a 25-μl aliquot of the stock enzyme solution, for instance, a full scale of 0.400 absorbance units is appropriate.

2   Pipette (or measure with a microsyringe) the appropriate amount of stock ester solution into each cuvette, and then add 3.0 ml of the pH 8.5 "Tris" buffer solution. Shake each cuvette to mix the solution, and place it in the thermostated cell compartment.

3   Balance the spectrophotometer for the sample in position 1 of the sample holder (this will be the "blank" for the first reaction), using the slit-width control. The blank-read switch must be in the blank position.

4   Set the timing interval selector on the sample positioner control at 10 sec dwell time, and set the recorder chart speed at the appropriate speed. The chart should move at about 120 sec/in. for the runs at the lowest values of $(S)_0$ and about 30 sec/in. for those at the highest $(S)_0$. Make a note of the chart speed on the chart and identify the kinetic run. Set the absorbance control at 0.000 and the blank-read switch on the read position.

5   Add the aliquot of stock enzyme solution to the second cuvette, and mix *vigorously but carefully* with a thin, fire-polished glass rod (2–3 mm diameter). Alternatively, remove the cuvette from the cell holder, make the addition, shake the cuvette vigorously (with the top covered with Parafilm), and quickly return it to the cell compartment. Just before mixing is begun, activate the recorder (turn to "chart" position), and then note the point on the chart corresponding to the time at which mixing was started ($t = 0$). Close the cell compartment and depress the auto, 1, and 2 buttons on the positioner control.

The instrument will now record alternately the absorption in the blank (position 1) and the reacting sample (position 2).

6   After a few minutes, the absorbance change in the reacting sample will become nearly linear. If desired, a new reaction can be started without stopping the positioner or the recorder. The cuvette in position 3 will be used as the "blank" for this run and must have the same ester concentration as that in cuvette 4. Now prepare to deliver the enzyme solution into the cuvette in position 4. When the positioner is stationary, add the enzyme to cuvette 4, mix, and close the cell compartment again. Note the time of mixing on the chart. Hold down the buttons auto, 1, and 2, and depress buttons 3 and 4 on the positioner control. (If the same blank 1 were to be used for the second run, the enzyme-containing reaction mixture would be in position 3 and only button 3 would be depressed.) Note that two reactions cannot be run simultaneously if they require very different chart speeds or full-scale absorbance settings. Also, the very fastest reactions should be run alone.

7   The progress of the reaction must be followed long enough so that the slope of the linear part of the curve can be accurately determined. This requires that the reaction be run for times long compared to the period of the initial exponential decay. Estimate the half-life (the time required for the difference Re^{-bt} in Fig. 1 to decrease to half its value) for the curve after the reaction has been running for about 5 min, and let the reaction run about 8 to 10 times the half-life. In no case should a time greater than 2000 sec be necessary.

8   When all of the runs being carried out together have been studied long enough, recheck the full-scale absorbance range of the recorder. Leaving the chart running, press only button 1 of the cuvette positioner. Wait about 10 sec, and mark the resulting line the pen has drawn on the chart. Then null the instrument using the absorbance control, and note the absorbance of cuvette 1. Return the absorbance control to 0.000, press button 2, and repeat this sequence for cuvette 2. After all the cuvettes have been calibrated, the recorder can be stopped. Check that the position of the line on the chart for each cuvette corresponds to the absorbance determined for it. If it does not, recalculate the position of zero absorbance on the chart and the full-scale absorbance range.

CALCULATIONS

Many of the calculations required in this experiment can be handled well by using a computer program either written by the student (see Chap. XXI) or made available by the instructor. The necessary calculations are described below in detail, both to explain the features of such a computer program and for the use of those who do not wish to use it.

For each kinetic run, draw a straight line through the segments of the chart tracing corresponding to the linear part of the run, and extrapolate it back to $t = 0$. Also draw a straight line through the segments corresponding to the appropriate blank. Determine the difference in chart units between the intercepts of these two lines with a vertical line drawn at $t = 0$, and convert this value into absorbance units using the known full-scale absorbance range. This yields a value of R in absorbance units. Determine the slope of the linear part of the kinetic curve, in absorbance units per second. The value of Q is this slope minus the corresponding slope of the blank.

Draw a smooth curve through the exponential part of the kinetic curve. Determine, at appropriate time intervals, the difference between the actual value and the value extrapolated from the linear portion of the kinetic curve. Plot this difference as a first-order process, and determine the first-order rate constant b. That is, plot the logarithm of the difference as a function of time, and determine the slope of the best straight line through the points. The value of b is -2.303 times this slope.

Prepare a table listing the following data for each kinetic run: the size of the aliquots of enzyme and of substrate used, $(S)_0$ in mol liter^{-1}, Q in absorbance units and in mol liter^{-1} (use the ε value you determined for this conversion), R in absorbance units sec^{-1} and in mol liter^{-1} sec^{-1}, and b in sec^{-1}. At this point you are in a position to test the kinetic expressions derived from the assumed mechanism and to derive some of the constants involved.

Consider the data obtained at a single value of $(S)_0$ with varying values of $(E)_0$. Do Eqs. (18), (19), and (20) hold for the dependence of Q, R, and b on $(E)_0$?

For the runs with constant $(E)_0$ and varying $(S)_0$, calculate $1/(S)_0$, $1/Q$, $1/R^{1/2}$, and $1/b$, and analyze these data in the following manner.

1   Plot $1/R^{1/2}$ as a function of $1/(S)_0$, and determine the intercept at $1/(S)_0 = 0$. According to Eq. (24), this intercept should be equal to $1/(E)_0^{1/2}$. Calculate $(E)_0$ and from this calculate the concentration of enzyme in milligrams per milliliter in the stock enzyme solution, using the known molecular weight of chymotrypsin (24,800). Compare this with the amount of commercial enzyme actually weighed out and estimate the purity of the sample you used.

2   Plot $1/Q$ as a function of $1/(S)_0$. The intercept of this plot should be $1/k_3(E)_0$ according to Eq. (23). Calculate k_3 using the value of $(E)_0$ determined above.

3   Carefully plot $1/b$ as a function of $1/(S)_0$. Fit the best straight line to these data, and determine the slope and intercept of the line. It is recommended that this fitting procedure be carried out analytically rather than graphically.

A least-squares computer program (see Chaps. XX–XXI) can be used to obtain both slope and intercept, as well as the standard deviation in each. From Eq. (25), we see that the slope is K_s/k_2 and the intercept (which will be determined with much less accuracy) is $1/k_2$. Determine K_s and k_2.

4 Determine k_{cat} and K_m from Eq. (21).

5 Check the following: Was the assumption $k_2 \gg k_3$ used in obtaining the simplified forms of Eqs. (23) to (25) justified? Using the constants calculated above and Eqs. (23) and (24), respectively, what would you predict for the slopes of the plots of $1/Q$ and $1/R^{1/2}$ against $1/(S)_0$? Can the data be fit with such slopes, within experimental error? How does the range of substrate concentration used compare with K_s and with K_m?

DISCUSSION

If *p*-nitrophenylpivalate has been synthesized, your report should include the yield, the melting point, and the uv and ir spectral curves of the recrystallized ester. Report your value of ε for the *p*-nitrophenoxide ion (the literature value is $1.8 \times 10^4\ M^{-1}$ cm^{-1}). Include all the kinetic curves as well as the plots of $1/Q$, $1/R^{1/2}$, and $1/b$ against $1/(S)_0$. State the conclusions that can be drawn from these data.

List all the constants obtained for the enzymic reaction, and compare them to the literature values.[2] Discuss whether the data support the kinetic scheme of Eq. (1), and whether the approximations made in obtaining the equations used in analyzing the data are justified.

You may also wish to compare the reaction at pH 8.5 with that at a different pH, say 7.0. Remember that an apparent ε value must be determined for *p*-nitrophenol at this pH.

APPARATUS

Single-beam spectrophotometer modified for kinetic measurements; automatic sample positioner; optical-density converter; chart recorder; constant-temperature circulator; four cuvettes; 500-, 250-, 25-, and 2-ml volumetric flasks; 5-ml measuring pipette; micropipettes or microsyringes; Parafilm.

p-Nitrophenylpivalate or the apparatus and reagents needed to synthesize it (*p*-nitrophenol, pivalyl chloride, pyridine, ether, 5 percent $NaHCO_3$ solution, $MgSO_4$, KOH, BaO, 95 percent ethanol); tris(hydroxymethyl)aminomethane; 1 N HCl solution; sodium acetate; glacial acetic acid; acetonitrile; chymotrypsin (Worthington Biochemical Corp., "thrice recrystallized"); 0.1 N NaOH solution; distilled or deionized water.

REFERENCES

1. M. L. Bender, F. J. Kézdy, and F. C. Wedler, *J. Chem. Educ.*, **44**, 84 (1967).
2. F. J. Kézdy and M. L. Bender, *Biochemistry*, **1**, 1097 (1962).
3. N. K. Schaffer, S. C. May, and W. H. Summerson, *J. Biol. Chem.*, **206**, 201 (1954); B. S. Hartley, *Nature*, **201**, 1284 (1964).
4. M. L. Bender and F. Kézdy, *J. Amer. Chem. Soc.*, **86**, 3704 (1964).

EXPERIMENT 27.   KINETICS OF THE DECOMPOSITION OF BENZENEDIAZONIUM ION

In an acidic aqueous solution, benzenediazonium ion ($C_6H_5N_2^+$) will decompose[1] to form nitrogen and phenol:

$$C_6H_5N_2^+ + H_2O \rightarrow C_6H_5OH + N_2(g) + H^+ \tag{1}$$

In this experiment you are to follow the course of the above reaction by a spectrophotometric measurement of the unreacted diazonium ion.[2] The reaction order with respect to $C_6H_5N_2^+$, the rate constant k, and the activation energy for the reaction are to be determined.

THEORY

The rate of reaction (1), $-d(C_6H_5N_2^+)/dt$, can be written in terms of the concentrations of the reacting species as

$$\frac{-d(C_6H_5N_2^+)}{dt} = k'(C_6H_5N_2^+)^n(H_2O)^m \tag{2}$$

As written above, the reaction would be nth order with respect to $C_6H_5N_2^+$, mth order with respect to water, and would have an overall order of $n + m$. The reaction is not acid-catalyzed, although very high acid concentrations (say 12 N HCl) seem to slightly increase the rate.[1] Thus the effect of changes in the H^+ concentration due to reaction (1) can be completely neglected. Since the present experiment will be performed in a dilute aqueous solution, the concentration of water, (H_2O), will be very nearly constant throughout the reaction. Thus the factor $(H_2O)^m$ can be absorbed into the rate constant and Eq. (2) can be rewritten as

$$\frac{-dc}{dt} = kc^n \tag{3}$$

where c represents the instantaneous concentration ($C_6H_5N_2^+$).

Equation (3) can readily be integrated to give

$$c = c_0 e^{-kt} \qquad \text{for } n = 1 \tag{4a}$$
$$c^{1-n} - c_0^{1-n} = (n-1)kt \qquad \text{for } n \neq 1 \tag{4b}$$

where c_0 is the concentration at $t = 0$. In other words, if the reaction is first order, a plot of $\log c$ versus t should give a straight line of slope $-k/2.303$ and intercept $\log c_0$. If the reaction is nth order where n (which need not be an integer) is not equal to unity, a plot of c^{1-n} versus t should give a straight line of slope $(n-1)k$ and intercept c_0^{1-n}. By making such plots for various trial values of n and determining which gives the best straight-line dependence over a wide range of concentration, one can obtain the order n of the reaction† and then determine the appropriate rate constant k_n.

If the value of a rate constant is measured at several different temperatures, it is almost always found that the temperature dependence can be represented by

$$k = \mathscr{A} e^{-E^*/RT} \tag{5}$$

where the factor $\mathscr{A}$ is independent of temperature. The Arrhenius activation energy E^* can be easily determined by plotting $\log k$ versus $1/T$. This should give a straight line of slope $-E^*/2.303R$.

METHOD

Any physical variable giving an accurate measure of the extent to which a reaction has gone toward completion can be used to obtain rate data. In this experiment, we shall monitor the concentration of unreacted diazonium ion by measuring the absorption of ultraviolet light by the solution.

According to the Beer-Lambert law,[3] the intensity of light I transmitted by an absorbing medium is given by

$$I = I_0 e^{-c\varepsilon'd} \tag{6}$$

where c is the concentration of absorbing molecules, d is the path length, I_0 is the intensity of incident light, and ε' is an extinction coefficient. The absorbance A, defined as $\log(I_0/I)$, thus gives a direct measure of the concentration:

$$A \equiv \log \frac{I_0}{I} = \frac{c\varepsilon'd}{2.303} = c\varepsilon d \tag{7}$$

where ε is the *absorptivity*. When the concentration is expressed in moles per liter, ε is called the molar absorptivity. For first-order reactions, the quantity εd will cancel

† Other methods of determining the reaction order are discussed in Exps. 23 and 24.

out of Eq. (4a) and k can be obtained directly from a plot of log A versus t. However, for orders different from first order, εd does not cancel out of Eq. (4b). In such cases, εd must be evaluated by measuring the absorbancy of a solution of known concentration in order to permit k to be calculated in concentration units.

For a brief description of spectrophotometers and their use in determining the absorbance, see Exp. 39.

EXPERIMENTAL

Kinetic runs are to be made at 25, 30, and 40°C. Well-regulated thermostat baths operating at about these temperatures will be required. The actual temperatures of each of these baths should be measured with a good thermometer and recorded.

The concentration of benzenediazonium ion can be determined by the absorbancy at wavelengths between 295 and 325 nm. Below 295 nm products of the reaction produce interfering absorption, and above 325 nm the extinction coefficient is too small to permit effective measurement of changes in the benzenediazonium ion concentration.[4] The absorbance will be measured using a spectrophotometer such as a Beckman Model DU; detailed instructions for operating this instrument will be provided in the laboratory. A wavelength of 305 nm should be used, with a slit width of 0.3 mm. For $A < 1$, use the "1" scale; when $A > 1$, use the "0.1" scale. Make two absorbancy readings on each sample.

The diazonium salt which should be used in this experiment is benzenediazonium fluoborate ($C_6H_5N_2BF_4$, M.W. 191.9). The great majority of diazonium salts are notoriously unstable solids and can decompose with explosive violence. The fluoborates are by far the safest to use and are not known to explode; however, reasonable caution should be used in preparing the compound. Since even benzenediazonium fluoborate will decompose slowly, it should not be prepared too far in advance, and it must be stored in a refrigerator. A simple high-yield procedure for its preparation has been given by Dunker, Starkey, and Jenkins.[5] Recrystallization of the product from 5 percent fluoboric acid yields white needle-like crystals which can be dried by vacuum pumping at 1 mm for several hours.†

---

† After storage at 0°C in a vacuum desiccator for several weeks, these crystals tend to stick together. After six months there is clear evidence of decomposition; fortunately, the main impurity is phenol (2 to 3 percent) which does not interfere much with rate studies in aqueous solution. By a very careful purification, it is possible to obtain a benzenediazonium fluoborate sample which can be stored at 0°C for several months without any signs of decomposition. The sample is dissolved in acetone, and then chloroform is added until a few crystals are formed. When the solution is then chilled to −20°C for 30 min, the compound will crystallize out in the form of tiny white needles. After three such recrystallizations, the sample can again be dried by pumping, and can be stored in a vacuum desiccator.

The diazonium salt should be available at the beginning of the experiment. Remove a *small* quantity from the refrigerator and warm it rapidly to room temperature. Prepare approximately 10^{-3} M solutions by placing an accurately weighed 15- to 20-mg sample of the salt into each of three 100-ml volumetric flasks, and then make up to the mark with a 0.2 M HCl solution. Label the flasks and be sure to record the exact weight of salt placed in each flask. Return the unused diazonium salt to the refrigerator at once. Be careful not to waste it.

Using a hypodermic syringe, withdraw a sample of about 5 ml of each solution *as soon as possible* after it is made up. Chill these samples by placing them in test tubes which are set in crushed ice, then put them aside for absorbancy measurements (to be made as soon as conveniently possible). These measurements will provide a value for εd in case it is needed later on.

Suspend one of the volumetric flasks in each of the three constant-temperature baths, and record the time. Allow about 20 min for the solutions to achieve thermal equilibrium. After thermal equilibrium has been attained, you may begin taking samples for spectrophotometric analysis.

When you take a sample, use the following procedure: Withdraw about 1 ml of the solution with the hypodermic syringe and use this to rinse out the syringe. Then quickly withdraw a 5-ml sample and place it in a labeled test tube set in crushed ice for chilling. Record the time of discharge into the test tube. After the sample is chilled (3 to 5 min) use about 1 ml of it to rinse out the spectrophotometer cell. Fill the cell with the remaining sample and measure its absorbancy, using the 0.2 M HCl solution as a blank. These cells are fragile and expensive; handle them with care.

The reason for chilling the sample is to slow down the reaction rate so that a negligible amount of reactant will decompose between the time you chill the sample and the time you make the absorbancy measurement. Since it is impossible to eliminate completely errors due to reaction after withdrawal and cooling, it is important to try to perform the sample removal and chilling procedure in as reproducible a fashion as possible. This will lead to a partial cancellation of such errors.†

Take as many measurements as you reasonably can without rushing. Since

† If a spectrophotometer with a cell compartment which can be temperature controlled is available, the procedure can be greatly improved and simplified. Adjust the bath temperature to 30°C, turn on the circulating pump, and wait until the cell compartment (with cell holder in place) has become stable. Then prepare a solution as described above and fill two spectrophotometer cells as soon as possible. (A third cell should already have been filled with 0.2 M HCl solution, for use as a blank.) Place the cells in the cell holder, and record the time at which the cell holder is returned to the cell compartment. Obtain absorbancy readings on both samples as soon as possible; these initial readings will provide values of εd. Allow about 20 min for the solutions to achieve thermal equilibrium, and then begin measuring the absorbancy every 15 min. Since the runs are made sequentially rather than simultaneously, it is advisable to replace the slow run at 25°C with a run at 45°C (use 5-min intervals). Fresh solutions are needed for each temperature studied.

the runs at the higher temperatures will proceed more rapidly, it will be necessary to take samples from them at more frequent intervals. A suggested interval might be once every 25 min for the 25° bath, every 15 min for the 30° bath, and every 10 min for the 40° bath.

At the end of the experiment, wash the cells thoroughly with distilled water and dry them very carefully. Store in a safe place.

CALCULATIONS

Using all the data points from any single run, prepare rough plots of the following: (1) $A^{1/2}$ versus t (order $\frac{1}{2}$), (2) $\log A$ versus t (order 1), (3) A^{-1} versus t (order 2), and (4) any additional plots you feel to be necessary in order to establish the order of the reaction.

Having determined the order of the reaction, make an accurate graph of the appropriate plot for each of the runs and determine the value of k (in concentration units) for each temperature. These plots may be of A values rather than c values. The value of the slope can then be corrected, if necessary, to give k in concentration units by utilizing the known εd value. Finally, plot $\log k$ versus $1/T$ and determine E^*.

Report your results for the order of reaction, $k(T_1)$, $k(T_2)$, $k(T_3)$, and E^*. Give correct *units* for each quantity. Use second as the unit of time, mole per liter as the unit of concentration, and kilocalorie as the unit of energy.

DISCUSSION

From your results, calculate the extent of reaction that would occur in a sample of the initial solution after 30 min at 0°C. Does this indicate that varying lengths of time between the chilling of a sample and its absorbancy measurement would introduce serious or negligible errors in your data?

APPARATUS

Spectrophotometer, such as a Beckman Model DU; two or more Corex sample cells; constant-temperature baths set at 25, 30, and 40°C; lens tissue; three 100-ml volumetric flasks; clock or stopwatch; hypodermic syringe; about 20 test tubes; plastic pail or large battery jar·beaker.

Benzenediazonium fluoborate (75 mg), stored in a refrigerator; 0.2 M hydrochloric .acid (400 ml); crushed ice.

REFERENCES

1. E. A. Moelwyn-Hughes and P. Johnson, *Trans. Faraday Soc.*, **36**, 948 (1940); M. L. Crossley, R. H. Kienle, and C. H. Benbrook, *J. Amer. Chem. Soc.*, **62**, 1400 (1940).
2. J. E. Sheats, Ph.D. Thesis, Dept. of Chemistry, M.I.T. (1965).
3. M. G. Mellon, "Analytical Absorption Spectroscopy," Wiley, New York (1950); W. J. Moore, "Physical Chemistry," 4th ed., p. 792, Prentice-Hall, Englewood Cliffs, N.J. (1972).
4. A. Wohl, *Bull. Soc. Chim. Fr.*, **6**, 1319 (1939).
5. M. F. W. Dunker, E. B. Starkey, and G. L. Jenkins, *J. Amer. Chem. Soc.*, **58**, 2308 (1936).

GENERAL READING

W. West, Spectroscopy and Spectrophotometry, in A. Weissberger (ed.), "Technique of Organic Chemistry," 3d ed., vol. I, part III, chap. XXVIII, Interscience, New York (1960).

R. Livingston, Fundamental Operations and Measurements in Obtaining Rate Data, in A. Weissberger (ed.), "Technique of Organic Chemistry," vol. VIII, part I, chap. III, Interscience, New York (1961).

EXPERIMENT 28.   GAS-PHASE KINETICS

Although a vast majority of important chemical reactions occur primarily in liquid solution, the study of simple gas-phase reactions is very important in developing a theoretical understanding of chemical kinetics. A detailed molecular explanation of rate processes in liquid solution is extremely difficult. At the present time reaction mechanisms are much better understood for gas-phase reactions; even so, this problem is by no means simple. This experiment will deal with the unimolecular decomposition of an organic compound in the vapor state. The compound suggested for study is cyclopentene or di-*t*-butyl peroxide, but several other compounds are also suitable.[1]

THEORY

A discussion of bimolecular reactions (which usually give rise to second-order kinetics) can be based quite naturally on collision theory or on transition-state theory.[1,2] We shall assume a general knowledge of these treatments as background for the discussion of unimolecular reactions (which usually give rise to first-order kinetics). The crucial question is: How does a molecule acquire the necessary energy to undergo a spontaneous unimolecular decomposition? If the "activation" occurs via a collision, one could normally expect second-order kinetics. However, Lindemann pointed out that this would not be true if the time between collisions is short compared with the average lifetime of an activated molecule.

Let us consider a simple unimolecular decomposition† in the gas phase

$$A \rightarrow B + C$$

The proposed reaction mechanism will consist of three steps: (1) collisional activation, (2) collisional deactivation, and (3) spontaneous decomposition of the activated molecule. Thus,

$$A + A \rightarrow A + A^* \qquad k_1 \tag{1}$$

$$A^* + A \rightarrow A + A \qquad k_2 \tag{2}$$

$$A^* \rightarrow B + C \qquad k_A \tag{3}$$

where A^* is an activated A molecule with a definite internal energy ε which is greater than a certain critical value ε^* necessary for reaction to occur, k_1 and k_2 are second-order rate constants, and k_A is a first-order rate constant. The overall rate of reaction is given by

$$+\frac{d(B)}{dt} = +\frac{d(C)}{dt} = -\frac{d(A)}{dt} = k_A(A^*) \tag{4}$$

where parentheses indicate concentrations of the species. As discussed below, the value of k_A will depend on the particular value of ε involved. If the concentration of A^* is small, one can make the steady-state approximation, $d(A^*)/dt = 0$ and obtain

$$+\frac{d(A^*)}{dt} = k_1(A)^2 - k_2(A^*)(A) - k_A(A^*) = 0 \tag{5}$$

Solving Eq. (5) for (A^*) and substituting the resulting expression into Eq. (4) gives

$$-\frac{d(A)}{dt} = \frac{k_1 k_A(A)^2}{k_A + k_2(A)} \tag{6}$$

High-pressure limit   At high concentrations deactivation is much more probable than decomposition, since the collision frequency is high, and deactivation should

---

† In the following treatment several details and difficulties will be overlooked; more advanced developments are given by Benson[1] and Trotman-Dickenson.[3]

occur at the first collision suffered by A*. Thus at high pressure, $k_2(A) \gg k_A$ and the reaction becomes first order:

$$-\frac{1}{(A)}\frac{d(A)}{dt} = \frac{k_1}{k_2} k_A \tag{7}$$

Also

$$\frac{k_1}{k_2} = \frac{(A^*)}{(A)} \cong f_A \tag{8}$$

where f_A is the fraction of A molecules in an activated state, with some internal energy ε. Obviously f_A depends on the value of ε, as does k_A. Thus,

$$-\frac{1}{(A)}\frac{d(A)}{dt} = \int_{\varepsilon^*}^{\infty} k_A(\varepsilon) f_A(\varepsilon) \, d\varepsilon = k_\infty \tag{9}$$

where k_∞, the value of the integral, corresponds to the experimental first-order rate constant k_{exp} determined at high pressures.

Rice and Ramsperger[4] and Kassel[5] have given an expression for k_A as a function of ε:

$$k_A(\varepsilon) = 0 \qquad \text{for } \varepsilon < \varepsilon^*$$
$$= v\left(\frac{\varepsilon - \varepsilon^*}{\varepsilon}\right)^{n-1} \qquad \text{for } \varepsilon \geq \varepsilon^* \tag{10}$$

Equation (10) is based on a model in which the molecule contains a total internal energy ε distributed among n weakly coupled harmonic oscillators. One of these oscillators is localized in a weak bond which will break when energy ε^* is concentrated in it; the other $n - 1$ oscillators are assumed to act as a reservoir of freely available energy. The probability that an energy at least as large as ε^* is localized in one oscillator is given by $(1 - \varepsilon^*/\varepsilon)^{n-1}$, and v is the frequency of energy transfer among the various oscillators in the molecule. For such a model, one would expect v to be of about the same order of magnitude as molecular vibrational frequencies—namely, 10^{13} sec^{-1}. [A more realistic and sophisticated molecular model has been proposed by N. B. Slater[1,3] which gives results very similar to those based on Eq. (10) and which also predicts a value for v of about 10^{13} sec^{-1}.]

To get a rough approximation for $f_A(\varepsilon)$ we can use a classical model in which the normal modes of vibration of the molecule A are represented by n classical harmonic oscillators. With this drastic assumption, one can use Boltzmann statistics to obtain[6]

$$f_A(\varepsilon) = \frac{1}{(n-1)! \, kT}\left(\frac{\varepsilon}{kT}\right)^{n-1} e^{-\varepsilon/kT} \tag{11}$$

When Eqs. (10) and (11) are substituted into Eq. (9) we find that

$$k_\infty = \frac{v}{(n-1)!} e^{-\varepsilon^*/kT} \int_0^{\infty} x^{n-1} e^{-x} \, dx \tag{12}$$

where $x = (\varepsilon - \varepsilon^*)/kT$. The definite integral is the gamma function $\Gamma(n) = (n - 1)!$ Thus

$$k_\infty = ve^{-\varepsilon^*/kT} = ve^{-E^*/RT} \tag{13}$$

where $E^* = N_0\varepsilon^*$. Equation (13) can be compared directly with the empirical Arrhenius equation which is used to obtain the experimental activation energy:

$$k = \mathscr{A}e^{-E_{act}/RT} \tag{14}$$

The temperature-independent factor $\mathscr{A}$ has now been given a physical significance and an estimated magnitude.

The theory presented above can be improved by using a quantum-mechanical model for the oscillations of the activated complex; this will permit a consideration of the effect of possible structural changes in the activated complex. The final result from this more sophisticated model[7] is

$$k = v^*e^{\Delta S^*/R}e^{-\Delta H^*/RT} \tag{15}$$

where v^* is an average frequency of the vibrational modes in the activated complex and ΔS^* and ΔH^* are the entropy and enthalpy changes for forming the activated complex (transition state) from a normal species. Comparing Eqs. (14) and (15) one can identify $\mathscr{A}$ with $v^*e^{\Delta S^*/R}$, but v^* is not well known, since it depends on the properties of an activated molecule (which are themselves not well known). Perhaps the best thing that can be done at present is to assign v^* a value of about 10^{13} sec^{-1} (the usual value of $\mathscr{A}$ in many reactions). Then, when the experimental value of $\mathscr{A}$ is less than 10^{12} or greater than 10^{14}, we can look for a possible explanation in terms of a value of ΔS^* which differs appreciably from zero.†

Low-pressure limit We now wish to comment on the behavior of Eq. (6) at low concentrations where the collision frequency is so low that deactivation will be very slow compared with decomposition. At low pressures, $k_2(A) \ll k_A$ and we find second-order kinetics:

$$-\frac{1}{(A)}\frac{d(A)}{dt} = k_1(A) \tag{16}$$

A detailed treatment of the change with pressure of the apparent first-order rate constant is quite complicated,[2,3] but we can easily see that

$$k_1(A) \ll \frac{k_1 k_A}{k_2} \cong k_\infty \tag{17}$$

---

† Transition-state theory gives the same result as Eq. (15) except that v^* is replaced by a universal factor kT/h ($= 6 \times 10^{12}$ sec^{-1} at $300°K$). Although this would appear to resolve the uncertainty as to the value of v^* (and thus allow better calculation of ΔS^* from $\mathscr{A}$ values), the theory involves several assumptions which are open to question on physical grounds.[7]

Thus at low initial pressures the "first-order rate constant" k_{exp} should decrease as the pressure is decreased (i.e., the half time for the reaction at constant temperature is independent of pressure at high pressure but rises as the pressure approaches zero). For reactions where $\mathscr{A} \gg 10^{13}$ sec^{-1}, this change has been observed to occur, but normally (i.e., where $\mathscr{A} \sim 10^{13}$) the change will occur at very low pressure and is not observed experimentally.

Chain mechanisms[8]   The fact that first-order kinetics are observed for a gas-phase reaction does not prove that the unimolecular mechanism described above must be involved. Indeed, very many organic decompositions which experimentally are first order have complicated free-radical chain mechanisms.

As an example, let us consider a simple *hypothetical* chain mechanism for the dehydrogenation of a hydrocarbon:

$$\begin{align}
M_a &\to H + R & k_1 \\
H + M_a &\to H_2 + R & k_2 \\
R &\to H + M_b & k_3 \\
H + R &\to M_c & k_4
\end{align} \tag{18}$$

where the M's are stable molecules and R is a free radical. The chain is initiated by the first step, propagated by many repetitions of the second and third steps (to form the stable products H_2 and M_b), and terminated by the last step. By writing the equations for $d(H)/dt$ and $d(R)/dt$ and making the steady-state assumption that (H) and (R) are constant, one can obtain

$$(H) = \left(\frac{k_1 k_3}{k_2 k_4}\right)^{1/2} \qquad (R) = \left(\frac{k_1 k_2}{k_3 k_4}\right)^{1/2} (M_a) \tag{19}$$

Now the rate of disappearance of M_a is given by

$$\begin{align}
-\frac{d(M_a)}{dt} &= k_1(M_a) + k_2(H)(M_a) \\
&= \left[k_1 + \left(\frac{k_1 k_2 k_3}{k_4}\right)^{1/2}\right](M_a) \tag{20}
\end{align}$$

and we see that the overall rate is first order. [When the chain is long, $k_2 \gg k_4$ and $k_3 \gg k_1$, so that k_1 is negligible compared with $k_1 k_2 k_3/k_4$ in Eq. (20).]

The presence of a chain reaction can usually be detected by adding small quantities of an inhibitor such as nitric oxide or propylene, which will markedly reduce the rate by reacting with the radicals and greatly shortening the chain length.[9] Another technique is to add small amounts of a compound which is known to provide relatively large concentrations of free radicals and see if the overall rate is increased. A detailed treatment of chain mechanisms will not be given here, but it should be pointed out that the overall rate law is often very complex and the apparent order of the reaction often changes considerably with pressure. To obtain a complete picture of the kinetics may require measurements over the range from a few millimeters of Hg to above 10 atm.

Wall effects   In the above discussion we have assumed that the reaction is homogeneous (i.e., no catalytic reaction at the walls of the reaction bulb). The fact that the data give first-order kinetics is not a proof that wall effects are absent. This point can be checked by packing a reaction bulb with glass spheres or thin-wall tubes and repeating the measurements under conditions where the surface-to-volume ratio is increased by a factor of 10 to 100. (This will not be done in this experiment, but the system chosen for study should be free from serious wall effects or it may not be possible to discuss the experimental results in terms of the theory of unimolecular reactions.)

METHOD[10]

The three basic experimental features of gas-phase kinetic studies are temperature control, time measurement, and the determination of concentrations. Of these, the principal problem is that of following the composition changes in the system. Perhaps the most generally applicable technique is the chemical analysis of aliquots; however, continuous methods are much more convenient. By far the easiest method is to follow the change in total pressure. This technique will be used in the present experiment. Obviously, the pressure method is possible only for a reaction which is accompanied by a change in the number of moles of gas. Also, the stoichiometry of the reaction should be straightforward and well understood so that pressure changes can be directly related to extent of reaction.

For the simple apparatus designs shown in Figs. 1 and 2 both the reactant and the products must be sufficiently volatile to avoid condensation in the manometer, which is much cooler than the reaction flask. (Heating wire may be wound around the connecting tubing between the furnace and the manometer to permit warming this section to about 50°C if necessary.) Also there should be no chemical reaction between the vapors and glass, stopcock grease, or mercury. The furnace in Fig. 1 should be capable of operating at temperatures up to 500°C, and the temperature should remain constant to within less than 1°C during a run. A large, well-insulated furnace has the disadvantage of heating up very slowly, but it will maintain a reasonably constant temperature without regulation. The Pyrex reaction bulb should be mounted in the center of the furnace. A large bulb (about 1000 ml) with a well to hold the thermocouple junction is preferable. However, a small bulb (as

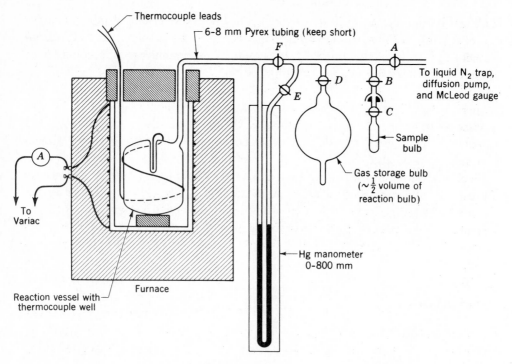

FIGURE 1
Apparatus for high-temperature gas-phase kinetics experiment.

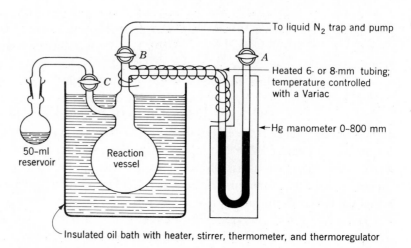

FIGURE 2
Apparatus for moderate-temperature gas-phase kinetics experiment.

small as 100 ml) can be used, and the thermocouple can be wound around the out-side of the bulb and then covered with aluminum foil. A two-junction Chromel-Alumel thermocouple, with the reference junction immersed in an ice bath, is used to measure the high temperatures achieved with this apparatus (see Chap. XVI for more information on thermocouples). The oil bath shown in Fig. 2 should be capable of operating in the range 140 to 160°C with a stability of ± 0.5°C during a run. In the case of an oil bath, there is no problem in achieving thermal contact between the reaction vessel and the thermometer. The connecting tubing between the reaction bulb and the mercury in the manometer should be short and of as small bore as feasible. The rate of the reaction should be slow enough so that the time for filling the reaction bulb and for the gas to warm up to the furnace temperature is small compared with the length of the run. Since this initial period is about 30 to 60 sec, the temperature should be chosen to give a half-life of at least 30 min.

We must finally derive an expression for the rate law in terms of the total pressure of the system. As an example, consider the gas-phase decomposition $A \to B + C$, and assume that the ideal-gas law holds for all species. Let the initial total pressure be p^0; this will also be the partial pressure of species A at zero time $(p_A{}^0)$. At some arbitrary time t, $p_B = p_C = p_A{}^0 - p_A$, and since $p = p_A + p_B + p_C$,

$$p_A = 2p^0 - p \qquad (21)$$

For a first-order reaction $\ln (A)^0/(A) = kt$; therefore

$$\ln p^0 - \ln (2p^0 - p) = \ln \frac{p^0}{2p^0 - p} = kt \qquad (22)$$

Thus a plot of $\ln (2p^0 - p)$ versus t should be a straight line. For a reaction $A \to B + C + D$ in which three product species are created, it can easily be shown that $\ln p^0 - \ln \frac{1}{2}(3p^0 - p) = kt$. Thus a plot of $\ln (3p^0 - p)$ versus t should be a straight line in this case.

For a moderately fast reaction it is often difficult to make a direct measurement of p^0. It is then best to back-extrapolate the early pressure data to zero time to obtain a value of p^0. Another possibility is to allow the reaction to go to completion (assum-ing there is no back reaction) and measure p^∞, which is equal to $2p^0$ for the particular reaction under consideration. This method is usually not so reliable, since side reac-tions which have little effect on the early stages of the reaction may influence the final pressure. (The experimental value of p^∞ is often found to be slightly less than the expected value.) Another approach is to use the Guggenheim method as in Exp. 24; this eliminates the need for a p^0 value.

In deriving Eq. (22) the dead-space volume V_D in the manometer and con-necting tube has not been considered. Since V_D changes during the reaction (change of the mercury level in the manometer), $p^0/(2p^0 - p)$ in Eq. (22) must be multiplied by a correction factor for very precise work.[11] With $V_D \sim 20$ ml and $V_B \sim 1000$ ml, this correction factor varies linearly with pressure from 1.00 to ~ 0.98 during the entire run and can be neglected.

Once the order of the reaction is established by a plot of $\ln (2p^0 - p)$ versus t, there is a rapid method of evaluating specific rate constants which utilizes the time for the reaction to proceed to a given fraction reacted. The half time and third time for a first-order reaction are given by

$$kt_{1/2} = \ln 2 = 0.691$$
$$kt_{1/3} = \ln 1.5 = 0.406$$

(23)

For the reaction $A \rightarrow B + C$, $t_{1/2}$ is the time required for the total pressure p to reach $3p^0/2$ and $t_{1/3}$ is the time required to reach $4p^0/3$. For the reaction $A \rightarrow B + C + D$, the total pressure will equal $2p^0$ at $t_{1/2}$ and $5p^0/3$ at $t_{1/3}$. Either (or both) times can be used to obtain k; it is also possible to check that the pressure data are consistent with first-order kinetics by calculating $t_{1/2}/t_{1/3}$. This ratio should be 1.70 for a first-order reaction; it is 1.50 and 2.00 for zero-order and second order, respectively.

EXPERIMENTAL

1   High-temperature reaction   The reaction suggested for study is the decomposition of cyclopentene† to cyclopentadiene and hydrogen:

(24)

This reaction is quite clean-cut, and at least 95 percent of the products are accounted for by Eq. (24). Vanas and Walters[14] have studied the reaction in the gas phase and find that p^∞/p^0 is about 1.9, which indicates the occurrence of some side reactions. However, they carried out a careful chemical analysis of the products and showed that the partial pressures of cyclopentadiene and hydrogen are equal to each other and to the increase in total pressure Δp over at least the first

---

† Cyclopentene can be obtained in very high purity as an NBS Standard Hydrocarbon Sample or obtained from the Special Products Division of the Phillip's Petroleum Co. Cyclohexene, a more readily available compound, is also suitable for study under the same conditions; however, the stoichiometry is much more complicated[12] and there are some unresolved questions about the kinetics.[13]

half of the reaction. With the exception of the last stages of the decomposition, this reaction is very well suited to the pressure method. In addition, it has been shown that the reaction is homogeneous and does not involve a chain mechanism.

The rate of reaction should be studied at a single temperature for several different initial pressures in the range 50 to 200 Torr. A furnace temperature of about 510°C will give a convenient half-life (~ 30 min). Ordinarily it is only necessary to follow the reaction until Δp exceeds $0.5p^0$. However, data should be taken on one of the runs until Δp is at least $0.75p^0$. (If possible, let this reaction mixture stand overnight and obtain a value of p^∞.) If NO gas is available, add about 2 to 3 percent of NO to the cyclopentene for one of the runs and check for any indication of inhibition; consult the instructor regarding any necessary changes in the procedure. If more than one day is available for experimental work, raise the temperature of the furnace by about 10°C and make several more runs.

Procedure The furnace should be turned on the day before measurements are to be made so that it will achieve a steady temperature. Set the power supply to a predetermined voltage which is appropriate for the desired reaction temperature. The heating current should be measured with a series ammeter and recorded periodically. At the start of the experiment, place the reference thermocouple junction in a Dewar flask filled with ice and distilled water (see Chap. XVI) and connect the thermocouple to a potentiometer circuit (see Chap. XV). Measure the thermocouple emf and check to see if the furnace is at the proper temperature. If the temperature shows appreciable long-term drift (greater than $\pm 0.5° \, hr^{-1}$) adjustment of the heater current will be necessary. On making any change in voltage setting, be careful to note the time response of the furnace as an aid in making later adjustments.

While the temperature stability of the furnace is being checked, evacuate the system (all stopcocks shown in Fig. 1 should be open *except B* and *C*) with a diffusion pump until the pressure is 10^{-4} Torr or lower. Then close stopcocks E and F and check for leads in the reaction bulb or manometer. Next close stopcock D, slowly freeze the cyclopentene with liquid nitrogen, and then open stopcocks B and C in order to pump off any air. Close C, remove the liquid nitrogen, and allow the solid to melt so that it will liberate any dissolved gases. Repeat this process at least once more, and then allow the cyclopentene to warm to room temperature. Close stopcock A and open D and C to permit the storage bulb to be filled with vapor (the vapor pressure at 25°C is about 350 Torr).

To start a run, **slowly** open stopcock F and watch the manometer. When the pressure has reached about 95 percent of the desired value for p^0, close F and start a stopwatch. Record pressure and time readings every minute during the early stages of the reaction and every few minutes thereafter until the end of the run.

Analysis of products   It is possible to design an apparatus which will permit the efficient removal of the entire reaction mixture for analysis. With the apparatus shown in Fig. 1, a qualitative analysis of the products can be made if desired. Analytical details are given by Vanas and Walters.[14]

2   Moderate-temperature reaction

The reaction suggested for study is the decomposition of *tert*-butyl peroxide (*t*BP).† Batt and Benson[15] have shown that the stoichiometry of this reaction is three (i.e., 3 mol of product for each mol of reactant) with 90 percent of the reaction accounted for by

$$(CH_3)_3COOC(CH_3)_3 \rightarrow C_2H_6 + 2(CH_3)_2CO \qquad (25)$$

The mechanism of this reaction is given by the following scheme:

$$t\text{BP} \rightarrow 2t\text{BO} \cdot \qquad (26)$$

$$t\text{BO} \cdot \rightarrow \text{Me} \cdot + (\text{Me})_2\text{CO} \qquad (27)$$

$$\text{Me} \cdot + \text{Me} \cdot \rightarrow C_2H_6 \qquad (28)$$

where the cleavage of the O—O bond in the first step is rate determining. The other 10 percent of the *t*BP decomposes according to

$$t\text{BP} \rightarrow CH_4 + (\text{Me})_2\text{CO} + \text{MeCOEt}$$
$$t\text{BP} \rightarrow 2CH_4 + \text{biacetonyl} \qquad (29)$$

both of which involve intermediate $\text{MeCOCH}_2 \cdot$ radicals. The reaction is homogeneous and chain contributions are less than 2 percent.[15] Thus, this reaction is quite well suited to the pressure method if two minor precautions are observed: (1) Fresh stopcock grease may absorb *t*BP, leading to a p^∞/p^0 ratio of ~ 2.9. Use the minimum amount of grease necessary and, if possible, do not regrease stopcocks just before a run. (2) The side arm to the manometer must be heated to avoid condensation of acetone.

The apparatus shown in Fig. 2 has a 500-ml reaction vessel which is connected to a 50-ml reservoir for liquid *t*BP via a stopcock and a standard-taper joint.‡ The connecting tubing between the reaction vessel and the mercury in the manometer must be wrapped with commercial heating tape or wound carefully with heating wire. The temperature of this section, which should be as short and of as small bore as

---

† No special handling procedures are required for *t*BP since it is one of the most stable organic peroxides known.[16] It can be distilled at 1 atm without decomposition.
‡ This apparatus is very similar in design to an apparatus described by Ellison;[17] see also the design of Trotman-Dickenson.[18]

feasible, is controlled with a variable-voltage power supply. It is not necessary to control this temperature accurately, but it should be close to the oil-bath temperature (say within 20°) and *must* be above the boiling point of acetone (56°C).

The rate of the reaction should be studied at a single temperature for several initial pressures in the range 30 to 100 Torr. A reaction temperature of about 160°C will give a convenient half-life. At least one run should be made at a lower temperature (say 150°C). In general, it is only necessary to follow the reaction until p exceeds $2p^0$, but data should be taken on one run until p is at least $2.5\,p^0$.

Procedure   If necessary, vacuum distill the tBP. Place about 10 ml of tBP in the side reservoir and close stopcock C. Open stopcocks A and B and evacuate the system while the oil bath is being heated to the desired temperature. If possible, use a diffusion pump to achieve a pressure of 10^{-4} Torr or lower. However, this is not crucial. In any case, a liquid nitrogen trap is required. Check for leaks by closing stopcocks A and B; then open them again and continue pumping.

When the reaction vessel is at constant temperature, close stopcock A and **slowly** open stopcock C. Pump down the reservoir as low as possible, but do not waste tBP by prolonged pumping. Next close stopcock C and place a beaker of hot water (60 to 65°C) around the reservoir. Stopcock B is then closed and C is opened until the desired value of p^0 is achieved. Finally, close stopcock C and start a stopwatch or timer. Record the pressure and time readings every minute during the early stages of the reaction and every few minutes thereafter.

CALCULATIONS

Plot the pressure reading for each run versus time and extrapolate to zero time to obtain a value for p^0. For each run determine $t_{1/3}$ and $t_{1/2}$, and check their ratio with that expected for first-order kinetics. Use Eqs. (23) to calculate k, in units of seconds^{-1}, for each of these times. Plot your data from the longest run in the form $\log(np^0 - p)$ versus t in seconds, where $n = 2$ or 3 depending on the stoichiometry. Draw the best straight line through the points (weighting the early points most heavily), and determine the specific rate constant k.

Tabulate T, p^0, $t_{1/3}$, and k for each run, and list your best overall value of k for each temperature studied. If data were obtained at two different temperatures, use the integrated form of Eq. (14) to calculate a value of the activation energy. Compare your result with the appropriate literature value. For cyclopentene, E_{act} is reported to be 58.8 kcal mol^{-1} on the basis of a linear plot of $\ln k$ versus $1/T$ for data over the range 485 to 545°C.[14] For *tert*-butyl peroxide, Batt and Benson[15] report 37.4 kcal mol^{-1} for the temperature range 130 to 160°C.

DISCUSSION

Calculate the frequency factor v and discuss its value in terms of the theory of unimolecular decompositions.

In the case of tBP, there is a complication due to the fact that reaction (25) is exothermic enough to cause temperature gradients in a spherical reaction vessel. Batt and Benson have shown that such gradients increase with increasing temperature, pressure, and reaction volume. At 160°C there is a gradient of roughly 0.5° during the first half of the reaction.[15] Estimate the effect that such a gradient will have on the rate constant determined in this experiment.

APPARATUS

High-temperature reaction   Gas kinetics apparatus as shown in Fig. 1, including vacuum line, sample bulb, gas storage bulb, manometer, Pyrex reaction bulb; high-temperature furnace; variable-voltage power supply; ammeter for measuring heater current; two-junction Chromel-Alumel thermocouple; 1-qt Dewar; millivolt-range potentiometer setup (see Chap. XV); thermocouple calibration table; stopwatch.

Cyclopentene (or cyclohexene); ice (3 lb); liquid nitrogen (2 liters); cylinder of NO gas, with needle valve and pressure tubing (optional).

Moderate-temperature reaction   Gas kinetics apparatus as shown in Fig. 2, including vacuum line, sample reservoir bulb, reaction vessel, heated mercury manometer; regulated oil bath; thermometer; 500-ml beaker; 1-qt Dewar; stopwatch.

tert-Butyl peroxide (15 ml); liquid nitrogen (2 liters).

REFERENCES

1. S. W. Benson, "Foundations of Chemical Kinetics," chaps. X and XI, McGraw-Hill, New York (1960).
2. Any standard physical chemistry text, such as W. J. Moore, "Physical Chemistry," 4th ed., chap. 9, Prentice-Hall, Englewood Cliffs, N. J. (1972).
3. A. F. Trotman-Dickenson, "Gas Kinetics," secs. 2.3, 2.4, 3.2, Butterworth, London (1955).
4. H. C. Ramsperger, *Chem. Revs.*, **10**, 27 (1932).
5. L. S. Kassel, "Kinetics of Homogeneous Gas Reactions," chap. 5, Reinhold (ACS Monograph), New York (1932).
6. S. W. Benson, *op. cit.*, pp. 222–223.

7. *Ibid.*, pp. 250–252.
8. E. W. R. Steacie, "Atomic and Free Radical Reactions," pp. 82–86, Reinhold (ACS Monograph), New York (1946).
9. A. F. Trotman-Dickenson, *op. cit.*, pp. 153–160.
10. A. Farkas and H. W. Melville, "Experimental Methods in Gas Reactions," Macmillan, London (1939).
11. A. O. Allen, *J. Amer. Chem. Soc.*, **56**, 2053 (1934).
12. L. Küchler, *Trans. Faraday Soc.*, **35**, 874 (1939).
13. D. Rowley and H. Steiner, *Discuss. Faraday Soc.*, **10**, 198 (1951).
14. D. W. Vanas and W. D. Walters, *J. Amer. Chem. Soc.*, **70**, 4035 (1948).
15. L. Batt and S. W. Benson, *J. Chem. Phys.*, **36**, 895 (1962).
16. N. A. Milas and D. M. Surgenov, *J. Amer. Chem. Soc.*, **68**, 205 (1946).
17. H. R. Ellison, *J. Chem. Educ.*, **48**, 205 (1971).
18. A. F. Trotman-Dickenson, *J. Chem. Educ.*, **46**, 396 (1969).

GENERAL READING

S. W. Benson, *op. cit.*
A. A. Frost and R. G. Pearson, "Kinetics and Mechanism," 2d ed., especially pp. 69–75, Wiley, New York (1963).
R. N. Pease, "Equilibrium and Kinetics of Gas Reactions," Princeton Univ. Press, Princeton, N.J. (1942).

EXPERIMENT 29.   KINETICS OF A FAST REACTION

Although conventional kinetic methods have proved very useful in studying a wide range of chemical reactions, special techniques are needed for investigating fast reactions (i.e., reactions with half-lives less than a few seconds). Such techniques may also provide a way to elucidate the rapid steps of reaction mechanisms for which only the slow rate-controlling step can be investigated by classical methods. In recent years, a wide variety of new methods have been developed for fast reactions in solution.[1,2] The earliest of these are the various flow methods (both continuous and stop-flow) which are suitable for reactions with ~ 1 sec $> t_{1/2} > 10^{-3}$ sec.[3] For half-lives below 10^{-3} sec, it is necessary to use methods which avoid the mixing of reactants, and several sophisticated techniques (such as relaxation methods) are now available for studying very fast reactions with $t_{1/2}$ values as short as 10^{-9} sec.[1,2]

In this experiment, the simplest fast-reaction technique—the continuous-flow method—will be used to study the kinetics of the formation of the ferric thiocyanate complex $FeSCN^{2+}$

THEORY

For the fast reaction between ferric and thiocyanate ions in an acid solution of constant pH, the observed behavior is consistent with the simple mechanism[4]

$$\text{Fe}^{3+} + \text{SCN}^- \underset{k_r}{\overset{k_f}{\rightleftharpoons}} \text{FeSCN}^{2+} \tag{1}$$

where k_f is the bimolecular forward rate constant and k_r is the unimolecular reverse rate constant.† The rate law resulting from mechanism (1) is

$$\frac{d(\text{FeSCN}^{2+})}{dt} = k_f(\text{Fe}^{3+})(\text{SCN}^-) - k_r(\text{FeSCN}^{2+}) \tag{2}$$

Let us recall that the equilibrium constant K is related to the rate constants by

$$K = \frac{k_f}{k_r} = \frac{(\text{FeSCN}^{2+})_\infty}{(\text{Fe}^{3+})_\infty(\text{SCN}^-)_\infty} \tag{3}$$

where the subscript ∞ denotes the equilibrium ($t = \infty$) value. Let us also note that

$$(\text{FeSCN}^{2+}) + (\text{SCN}^-) = (\text{FeSCN}^{2+})_\infty + \cdot(\text{SCN}^-)_\infty \tag{4}$$

at any time t. Using these relations, we can rewrite Eq. (2) in the form

$$\frac{d(\text{FeSCN}^{2+})}{dt} = k_f(\text{Fe}^{3+})[(\text{FeSCN}^{2+})_\infty + (\text{SCN}^-)_\infty]$$
$$- k_f[(\text{Fe}^{3+}) + K^{-1}](\text{FeSCN}^{2+}) \tag{5}$$

In order to simplify the integration of Eq. (5), let us choose the experimental conditions such that $(\text{Fe}^{3+}) \gg (\text{SCN}^-)$. This will allow us to assume that (Fe^{3+}) is essentially constant during the course of the reaction. If, in addition, the initial conditions are chosen so that $(\text{FeSCN}^{2+}) = 0$ at $t = 0$, we find

$$\ln \frac{(\text{FeSCN}^{2+})_\infty - (\text{FeSCN}^{2+})}{(\text{FeSCN}^{2+})_\infty} = -[(\text{Fe}^{3+}) + K^{-1}]k_f t \tag{6}$$

This is an approximate solution which becomes exact only when (Fe^{3+}) is constant [you can check Eq. (6) by direct differentiation]. In actual practice, $(\text{Fe}^{3+})_0$ will be chosen to be ten times larger than $(\text{SCN}^-)_0$, so that (Fe^{3+}) will vary by about 10 percent during the reaction.‡

---

† In fact, the detailed mechanism is more complex since both a direct and a base-catalyzed path exist.[4] The value of k_f as used in mechanism (1) is found to be dependent on the H^+ concentration: $k_f = k_1 + k_2/(\text{H}^+)$. Since this experiment is restricted to a single value of (H^+), the more general mechanism is omitted.

‡ With this choice of an initial ferric concentration which is much larger than the thiocyanate concentration, any formation of Fe(SCN)_2^+ species can be neglected. Also we shall work at a low pH value so that the hydrolysis of ferric ions can be neglected.

If a plot of $\ln \left[(\text{FeSCN}^{2+})_\infty - (\text{FeSCN}^{2+})\right]$ versus t is linear, then the first-order dependence on (SCN^-) and (FeSCN^{2+}) is confirmed. The rate dependence on (Fe^{3+}) has been established as first order[4] and will not be tested in this experiment.

METHOD

The continuous-flow method[3,5] is a simple and straightforward technique. Two reactant solutions are forced under pressure into a T-shaped mixing chamber, and then the reacting mixture flows down a long capillary tube. It is important that the solutions mix well and rapidly ($\sim 10^{-3}$ sec). When the mixing time is much shorter than the half-life of the reaction, a steady state is set up such that various positions along the capillary tube correspond to different reaction times. A quantitative correlation between the reaction time t and the distance x from the mixing chamber to a given point along the tube can be achieved by measuring the time interval $\Delta\tau$ needed for a known volume ΔV of solution to flow through the tube. Assuming a constant cross-sectional area $\mathscr{A}$ for the capillary, the average linear flow velocity $\bar{u}$ is given by $\bar{u} = \Delta V/(\mathscr{A}\,\Delta\tau)$. Thus, on the average, the reaction time is related to x by

$$t = x\,\frac{\mathscr{A}\,\Delta\tau}{\Delta V} \tag{7}$$

It is important that the flow of the reacting solution should be as close to "mass flow" as possible, so that all the fluid arriving at a small volume element located at some fixed point along the tube will have traveled for the same length of time after leaving the mixing chamber. Laminar flow is undesirable since it is fast near the center of the capillary and quite slow near the walls (see Exp. 4). Since turbulent flow is much closer to mass flow, we wish to use flow rates that are greater than $1000\,\eta/\rho r$ [the minimum value for turbulent flow; see Eq. (4-11)]. Deviations from mass flow will blur the time value appropriate to a given distance x. Fortunately, the resulting error in the rate constant is no more than 2 to 3 percent for turbulent flow; but it may be as high as 10 percent for streamline flow.[3,5]

In this experiment, the reaction is to be followed with a Beckman Model DU spectrophotometer. The response of this instrument is rather slow, several seconds being required to make an absorbance reading. Since the reaction to be studied is rapid (half-life of less than 1 sec), it is not possible to follow the progress of the reaction directly as in Exp. 27. However, the continuous-flow method is a very appropriate way† to study fast reactions with a slowly responding instrument. The

† The principal disadvantage of this method is the fact that large quantities of solution are required. Although the present chemicals are cheap and easily obtained, this would be a serious limitation in studying reactions involving rare or costly reactants. In those cases, one would need to use a stop-flow method with a more rapidly responding photometric system.[3]

design described in the experimental section is based directly on that first used by Dalziel; excellent detailed descriptions of his equipment are available in the literature.[3,5]

The complex $FeSCN^{2+}$ has an absorption maximum near 450 nm, whereas neither Fe^{3+} nor SCN^- absorbs in that region.[6] Therefore,

$$(A_\infty - A) = \varepsilon d[(FeSCN^{2+})_\infty - (FeSCN^{2+})] \tag{8}$$

where A is the absorbance, ε is the molar absorptivity, and d is the optical path length. From Eqs. (6) and (8) we see that a plot of $\log (A_\infty - A)$ versus t should give a straight line of slope $-k_f[(Fe^{3+}) + K^{-1}]/2.303$. Note that A_∞ values obtained at different points along the tube may not all agree with each other, since there may be variations in the capillary tube and its placement with respect to the light beam. Also note that errors in $(A_\infty - A)$ caused by a drift in the zero setting of the spectro-photometer could cause appreciable errors in the rate constant. Readings corre-sponding to a large percent reaction are most sensitive since the $(A_\infty - A)$ values are then small.

Finally, we must remember that there will be a salt effect[7] on the rate of this reaction: both the rate constant k_f and the equilibrium constant K are functions of the ionic strength μ. An (approximately constant) ionic strength of 0.40 will be chosen here to facilitate comparison with previous work.[4]

EXPERIMENTAL

A sketch of the complete assembled apparatus is shown in Fig. 1, and more de-tailed drawings of the capillary support frame are given in Fig. 2. There are two important differences between this design and that of Dalziel:[5] (1) the carboys of reactant solutions are mounted in a stationary thermostat bath and are connected to the movable capillary support frame by rubber pressure tubing, and (2) the mixing chamber is machined from a block of Lucite instead of being fabricated from glass. A detailed drawing of this mixing chamber is given in Fig. 3.

In order to assemble the apparatus, remove the phototube housing from a Beckman Model DU spectrophotometer and replace the cell holder unit (sample chamber) with the special housing in which the capillary support frame slides back and forth. Make sure that the alignment is correct (so that the capillary support frame moves smoothly) before you bolt this housing to the spectrophotometer. It is convenient to have the spectrophotometer sitting on a base with adjustable legs to facilitate this alignment. Finally, bolt the phototube housing onto the rear of this special adapter housing.

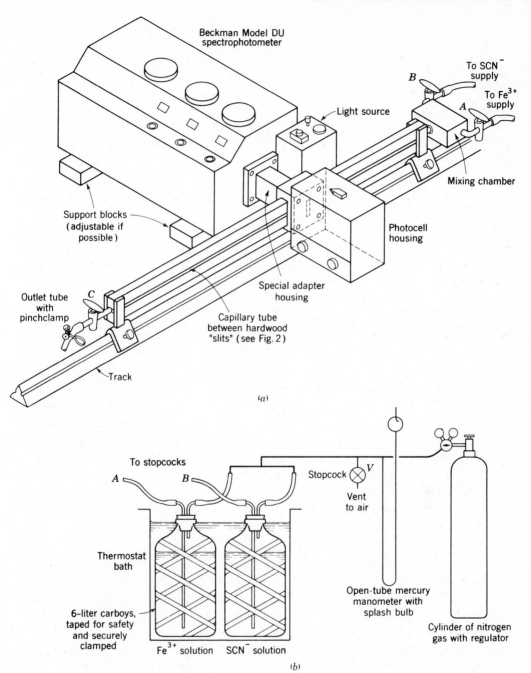

FIGURE 1

Flow-kinetics apparatus: (a) spectrophotometer setup; (b) schematic diagram of system for driving reactant solutions.

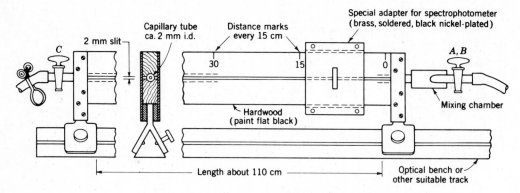

FIGURE 2

Detail of reaction capillary tube and support frame.

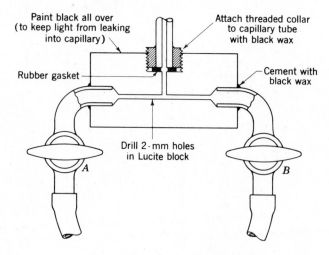

FIGURE 3

Detail of mixing chamber.

Next, fill the two carboys† with the appropriate solutions. Solution A should be 0.02 M in $Fe(NO_3)_3$, 0.2 M in $HClO_4$, and 0.14 M in $NaClO_4$; solution B should be 0.002 M in NaSCN, 0.2 M in $HClO_4$, and 0.14 M in $NaClO_4$. Be careful in handling these solutions; perchloric acid will give you dishpan hands. Now mount the carboys in the thermostat bath and connect them by a long piece of flexible pressure tubing to the appropriate stopcocks which are sealed into the mixing chamber.

---

† Before being used in this experiment, each carboy should be pressure tested to verify that there are no flaws which would make it unsafe. It is recommended that the empty carboy be placed behind a protective barrier and subjected to an internal gas pressure of 2 atm.

Again, test that the capillary frame will move smoothly back and forth through its housing. Do not force it. If there is any difficulty, ask an instructor to check the alignment. Proper alignment is crucial. Along the top of the capillary support frame, there should be a series of fiducial marks (say six or seven) which can be lined up with a reference mark on the special housing. This enables one to set the spectrophotometer at reproducible points along the capillary which are at known distances x away from the mixing chamber.

To complete the setup, connect the reagent carboys to each other via a T tube which leads to an open-tube mercury manometer and a regulated supply of a suitable driver gas (such as air or N_2).

It is important that stray light should not reach the capillary tube, since it may be reflected down the walls of the tube and be detected by the spectrophotometer. The support frame around the capillary is designed to reduce such stray light by providing a deep narrow slit-like aperture; but one must still avoid horizontal light leaks. It may be worthwhile to cover both ends of the entire support frame with blackout cloths. (Do not cover the source housing on the Model DU, as it gets extremely hot.)

Success in this experiment depends on good technique; the procedure should be followed carefully. Before beginning the measurements, you should look over Exp. 39 and read the Beckman DU spectrophotometer instruction manual for detailed information on operating this instrument.

Procedure Turn on the spectrophotometer and allow at least 20 min for it to warm up prior to use. The wavelength setting should be 455 nm throughout the entire experiment. With both reagent stopcocks A and B and the vent stopcock V closed, slowly increase the gas pressure on the reagent solutions until the manometer indicates about 500 Torr pressure above 1 atm. With the outlet stopcock C *open*, open and close the reagent stopcocks A and B (one at a time) several times to make sure that both solutions are flowing smoothly and to remove any air bubbles from the system. Use a beaker to catch the outflow from the capillary tube.† Then set the capillary frame at the first fiducial mark (one nearest to the mixing chamber) and carry out the three following steps;

  1 Open stopcock A and allow the Fe^{3+} solution to flow for a sufficient time to remove from the capillary tube any solution containing $FeSCN^{2+}$ species (until the outflow is clear). Then close stopcock A and the outlet stopcock C. In accordance with the spectrophotometer operating instructions, zero the instru-

---

† As an optional step, you can record the time required for a given volume (say 250 ml) of each separate solution to flow through the capillary tube. The times for solutions A and B should agree within a few percent.

ment with this Fe^{3+} solution (i.e., adjust the dark current, slit width, and sensitivity controls so that the instrument reads 100 percent transmission or $A = 0$).

2   To begin a run, open the outlet stopcock C and then turn both stopcocks A and B to their fully open positions. Catch the outflow of solution from the capillary in a beaker until the flow becomes stable (as indicated by a constant manometer reading, usually before 100 ml of solution is collected). Then *quickly* switch the outlet tube from the beaker to a 250-ml volumetric flask and simultaneously start a stopwatch. When this flask is full, stop the watch and record the elapsed time (~ 50 sec). Return the outlet tube to the beaker. While one student is carrying out the above flow-rate measurement, his partner should determine the absorbancy A of the reaction mixture and record that value together with the distance x from the mixing chamber. Work quickly to avoid any unnecessary waste of the reagent solutions.

3   When both the flow and absorbancy measurements are complete, close the outlet stopcock C and then as soon as possible close both stopcocks A and B. This is a crucial step in the procedure; if A and B are left open, solution may siphon from one carboy to the other. After about 2 min, determine the absorbancy again to obtain A_∞ (the infinite time value). Verify that this value does not change after one more minute.

For the next run, move the capillary support frame so as to line up the second fiducial mark and repeat steps 1 to 3 at this new distance setting. Be careful in moving the capillary support frame.

Make two runs at each of the six or seven positions along the capillary tube. Use special care in making the absorbancy readings at large values of x (corresponding to large percent reaction and a small $A_\infty - A$ value). If time permits, you should also take data at a different driving pressure. Either increase or decrease the gas pressure depending on whether you need more data at low percent reaction or at high, but it may not be safe to exceed about 700 Torr overpressure.

During the course of the experiment, more of solution A will be used up than solution B if the Fe^{3+} solution is always used in step 1 to make the zero adjustment of the spectrophotometer at each distance setting. The resulting change in the liquid level for solution A relative to that for solution B may change the relative flow rates of these solutions. This can be avoided by alternating the use of solutions A and B for making the zero adjustments (or, if necessary, corrected by merely running enough of solution B out through the capillary periodically to equalize the levels).

Before leaving the laboratory, record the concentrations of the reactant solutions, the temperature of the thermostat bath, the radius of the capillary tube, and the distances x corresponding to the various fiducial marks along the capillary support frame.

CALCULATIONS

Using Eq. (7), calculate the reaction time t corresponding to each run; tabulate these times together with the appropriate $(A_\infty - A)$ and $\log(A_\infty - A)$ values. Plot $\log(A_\infty - A)$ versus t, and determine the slope of the best straight line through the data points. From the value of this slope, calculate the rate constant k_f. The literature value[4] of K is 146 ± 5 liters mol^{-1} at 25°C and an ionic strength of 0.40, and you should take the (Fe^{3+}) value to be an average of the initial value ($t = 0$, but *after* mixing of the two solutions) and the equilibrium value ($t = \infty$, as calculated with K). Also report the back-reaction rate constant k_r. Give the proper units for each rate constant.

DISCUSSION

What fraction of the Fe^{3+} in solution A is hydrolyzed to $Fe(OH)^{2+}$ if the hydrolysis constant K_h is known[4] to be 2×10^{-3}? What are the principal sources of error in this experiment? How fast a reaction do you think could be measured with this apparatus? What is the limiting design factor in this method?

APPARATUS

Beckman Model DU spectrophotometer; 6-V battery and battery charger (or regulated dc power supply); adjustable support for spectrophotometer (optional); movable capillary support frame with mixing chamber and special housing attached; set of four retaining bolts for mounting housing on the spectrophotometer; two carboys (~ 6 liters) for reactant solutions; open-tube mercury manometer; cylinder of nitrogen gas with pressure regulator (or source of regulated compressed air); pressure tubing and hose clamps; stopcock for venting the carboys; gum-rubber tubing for outlet of capillary; a 250-ml volumetric flask; a beaker (~ 500 ml); stopwatch; blackout cloths (optional).

Constant-temperature bath (set at 25°C) with provision for mounting carboys; solution A which is 0.0200 M in $Fe(NO_3)_3$, 0.20 M in $HClO_4$, and 0.14 M in $NaClO_4$ (5 liters); solution B which is 0.00200 M in NaSCN, 0.20 M in $HClO_4$, and 0.14 M in $NaClO_4$ (5 liters).

REFERENCES

1. E. F. Caldin, "Fast Reactions in Solution," Wiley, New York (1964); *Discussions Faraday Soc.*, **17**, 133 (1954).
2. S. L. Friess, E. S. Lewis, and A. Weissberger (eds.), "Investigation of Rates and Mechanisms of Reactions," 2d ed., vol. VIII, part II of "Technique of Organic Chemistry," Interscience-Wiley, New York (1963).
3. *Ibid.*, chap. XIV, "Rapid Reactions" by F. J. Roughton and B. Chance, esp. pp. 704–748.
4. J. F. Below, Jr., R. E. Connick, and C. P. Coppel, *J. Amer. Chem. Soc.*, **80**, 2961 (1958).
5. K. Dalziel, *Biochem. J.*, **55**, 79 (1953).
6. H. S. Frank and R. L. Oswalt, *J. Amer. Chem. Soc.*, **69**, 1321 (1947).
7. W. J. Moore, "Physical Chemistry," 4th ed., pp. 464–466, Prentice-Hall, Englewood Cliffs, N.J. (1972); D. F. Eggers, Jr., N. W. Gregory, G. D. Halsey, Jr., and B. S. Rabinovitch, "Physical Chemistry," pp. 481–484, Wiley, New York (1964).

X

SURFACE PHENOMENA

EXPERIMENTS

30. Surface tension of solutions
31. Adsorption from solution
32. Physical adsorption of gases

EXPERIMENT 30.   SURFACE TENSION OF SOLUTIONS

The capillary-rise method is used to study the change in surface tension as a function of concentration for aqueous solutions of *n*-butanol and sodium chloride. The data are interpreted in terms of the surface concentration using the Gibbs isotherm.

THEORY

If a body of material is homogeneous, the value of any *extensive* property is directly proportional to the quantity of matter contained in the body:

$$Q = \bar{Q}_V V = \bar{Q}_m m = \bar{Q}_N N \tag{1}$$

where Q is the extensive quantity; V, m, and N are, respectively, the volume, mass, and number of moles of the substance involved. $\bar{Q}_V$, $\bar{Q}_m$, and $\bar{Q}_N$ are, respectively, the specific values of Q per unit volume, per unit mass, and per mole and have *intensive* magnitudes.

It is known, however, that the values of extensive properties of bodies often thought of as homogeneous are not always independent of the surface area. In fact, liquid bodies with surfaces are in general not entirely homogeneous, for the value of a given intensive quantity (say Q_V) in the region of the surface may at equilibrium deviate from the value that this quantity has in the bulk of the solution ($\bar{Q}_V$). We can write

$$Q = \int_V Q_V \, dV = \bar{Q}_V V + \int_\tau (Q_V - \bar{Q}_V) \, dV \tag{2}$$

where the second integral needs to be taken only over a region τ in the neighborhood of the surface, within which Q_V is different from $\bar{Q}_V$, the *bulk* value which prevails through the body except near the surface. We can replace the volume element dV in the second integral by $dA\,dx$, where dA is an element of surface area and dx an element of distance inward from the surface of the body. The integration over dA extends over the surface of the body, and that over dx extends to a distance τ inside the body beyond which Q_V is substantially equal to $\bar{Q}_V$. If $(Q_V - \bar{Q}_V)$ is independent of position on the surface of the body, the integration over dA can be carried out at once, and we have

$$Q = \bar{Q}_V V + \bar{Q}_A A \qquad (3)$$

where we have introduced a new quantity, called a specific surface quantity,

$$\bar{Q}_A \equiv \int_\tau (Q_V - \bar{Q}_V)\,dx \qquad (4)$$

This quantity represents the *excess*, per unit area of surface, of the quantity Q over what Q would be for a perfectly homogeneous body of the same magnitude with $Q_V = \bar{Q}_V$ throughout; see Fig. 1. In some cases $\bar{Q}_A$ may be negative, corresponding to a deficiency rather than an excess.

    If the body has surfaces of several different kinds (like a crystal with different kinds of crystal faces or a liquid with part of its surface in contact with the air and the remainder of its surface in contact with solid or other liquid phases), the parenthesized quantity in Eq. (2) is independent of surface positional coordinates only within the boundaries of each kind of surface. Each kind of surface will have in general a different $\bar{Q}_A$, and we write in such a case

$$Q = \bar{Q}_V V + \sum_i \bar{Q}_{A_i} A_i \qquad (5)$$

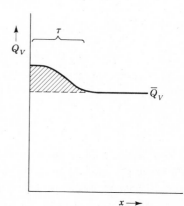

FIGURE 1
Schematic plot of Q_V vs. x. The shaded area is equal to $\bar{Q}_A$.

Surface concentration   When the extensive quantity concerned is the number N of moles of solute, the corresponding specific bulk property is $\overline{N}_V \equiv c$, the bulk concentration (concentration in the interior of the solution). The corresponding specific surface quantity $\overline{N}_A \equiv u$ is called *surface concentration* and represents excess of solute per unit area of the surface over what would be present if the internal (bulk) concentration prevailed all the way to the surface. It may be expressed in moles per square centimeter.

Surface free energy or surface tension   Under conditions of constant temperature and pressure the equilibrium state sought by any system is that of lowest free energy G, and the maximum work (other than expansion work) done by the system in any change of state under these conditions is equal to the free-energy decrease, $-\Delta G$. A body of liquid with a free surface will tend to assume the shape which gives it the lowest possible free energy at the given temperature and pressure prevailing.

The free energy of a system containing variable surface areas can be written, from Eq. (5), as follows:

$$G = G_0 + \sum_i \gamma_i A_i \tag{6}$$

where $\overline{G}_{A_i} \equiv \gamma_i$ is the specific surface free energy or *surface tension* of surface i, which may be a free surface (exposed to air or vapor or vacuum) or an interface with another liquid or solid. In the event that the surface is an interface, this quantity is called *interfacial tension*. For a stable free liquid surface it clearly must be positive. The surface tension γ_i is equal to the expenditure of work required to increase the net area of surface i by one unit of area, say 1 cm^2; if the increase in surface area is accomplished by moving a 1-cm line segment in a direction perpendicular to itself, γ_i is equal to the force, or "tension," in dynes opposing the moving of the line segment. Accordingly, it is usually expressed in units of dynes per centimeter ($=$ ergs per square centimeter).

The variation of surface tension with temperature will not be discussed here, except to remark that surface tension decreases with temperature and that the rate of decrease is large enough to require that the temperature of measurement of surface tension be kept constant, to the order of 0.1°C, by means of a thermostat.

The Gibbs isotherm   It is found that the surface tensions of solutions are in general different from those of the corresponding pure solvents. It has also been found that solutes whose addition results in a decrease in surface tension tend to concentrate slightly in the neighborhood of the surface (positive surface concentration); those whose addition results in an increase in surface tension tend to become less concentrated in the neighborhood of the surface (negative surface concentration). The migration of solute either toward or away from the surface is always such as to make the surface tension of the solution (and thus the free energy of the system) lower than it would be if the concentration of solute were uniform throughout (surface concentration equal to zero). Equilibrium is reached when the tendency for free-energy decrease due to lowering surface tension is balanced by an opposing tendency for free-energy increase due to increasing nonuniformity of solute concentration near the surface.

Let us imagine a body of solution of volume V, surface area A, bulk concentration c, and bulk osmotic pressure Π at constant temperature T and external pressure p. For arbitrary changes dA and dV in the area and volume† of the solution, the free-energy change can be written

$$dG = \gamma\, dA - \Pi\, dV \tag{7}$$

This is an exact differential; from the well-known reciprocity relation[1] we find that

$$-\left(\frac{\partial \gamma}{\partial V}\right)_A = \left(\frac{\partial \Pi}{\partial A}\right)_V \tag{8}$$

This can be rewritten

$$-\frac{d\gamma}{dc}\left(\frac{\partial c}{\partial V}\right)_A = \frac{d\Pi}{dc}\left(\frac{\partial c}{\partial A}\right)_V \tag{9}$$

where the presence of total derivatives is justified, since at constant pressure and temperature both the osmotic pressure and the surface tension are determined completely by the concentration.

Now we can write for the total number of moles of solute $N = cV + uA$ and rearrange this to give

$$c = \frac{N - uA}{V} \tag{10}$$

Differentiating, we obtain

$$\left(\frac{\partial c}{\partial V}\right)_A = -\frac{c}{V} \qquad \left(\frac{\partial c}{\partial A}\right)_V = -\frac{u}{V} \tag{11}$$

For a perfect solute

$$\Pi = \frac{N}{V} RT = cRT$$

and

$$\frac{d\Pi}{dc} = RT \tag{12}$$

We finally obtain, on combining Eqs. (9), (11), and (12),

$$\frac{u}{c} = -\frac{1}{RT}\frac{d\gamma}{dc} \tag{13}$$

† The volume change may be thought of as resulting from the motion of a piston, containing a semipermeable membrane, against the osmotic pressure Π.

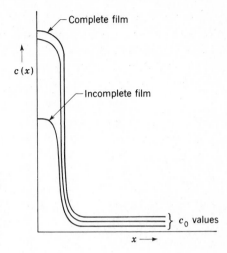

FIGURE 2
Variation of the solute concentration near a free surface for the case of positive surface adsorption. The bulk concentration is represented by c_0, and x is the distance measured inward from the surface.

This equation was first derived by Willard Gibbs and is called the *Gibbs isotherm*. We can also write it in the form†

$$u = -\frac{1}{RT}\frac{d\gamma}{d \ln c} = -\frac{1}{2.3RT}\frac{d\gamma}{d \log c} \qquad (14)$$

Surface-active substances: surface adsorption Many organic solutes in aqueous solution, particularly polar molecules and molecules containing both polar and nonpolar groupings, considerably reduce the surface tension of water. Such solutes tend to accumulate strongly at the surface where, in many cases, they form a unimolecular film of adsorbed molecules.

Usually the film is substantially complete and can accept no more solute molecules when the bulk concentration attains some small value, and hence the surface concentration undergoes little change when the bulk concentration is increased further over a wide range (see Fig. 2). From Eq. (14) we can therefore expect that over this range a plot of the surface tension of the solution against the logarithm of the bulk concentration should be linear. It will deviate from linearity both at low and at high concentrations, in both cases yielding a slope of smaller absolute magnitude.

† Equation (14) has been derived for a perfect solute and for real solutions applies only at low concentrations. In general, $d\gamma = -uRT\, d \ln a$, where a is the activity of the solute. Activity coefficients for *n*-butanol in aqueous solution at 0°C (which to fairly good approximation can be used at 25°C) are:[2] 0.1 M, 0.943; 0.2 M, 0.916; 0.4 M, 0.882; 0.6 M, 0.856; 0.8 M, 0.838; and 1.0 M, 0.823.

Solutions of electrolytes In solutions of certain electrolytes, among them NaCl and KCl, we find that the surface tension *increases* with concentration, indicating a *negative* surface concentration. This is a result of interionic electrostatic attraction, which tends to make the ions draw together and away from the surface. Measurements have shown that in dilute solution the surface tension increases linearly with concentration; thus we see from Eq. (13) that the surface concentration must be directly proportional to the bulk concentration. In actual practice the surface tension often drops initially at low concentrations and then rises linearly; the initial drop is presumably due to impurities and can be eliminated by careful work.

The ratio u/c has dimensions of length and is a measure of the "effective thickness" of the region at the surface in which the actual concentration is significantly less than the bulk concentration. Indeed, we may consider a very crude picture in which the actual concentration $c(x)$ may be supposed to be zero from the surface inward to a distance x_0 and to be equal to the bulk concentration c_0 from that distance inward. It will easily be seen from Eq. (4) that

$$x_0 = \frac{u}{c_0} \tag{15}$$

This distance, by the above argument, is independent of bulk concentration.

A model of a liquid surface in which there is no negative or positive surface concentration would have the solute uniformly distributed right out to the surface.

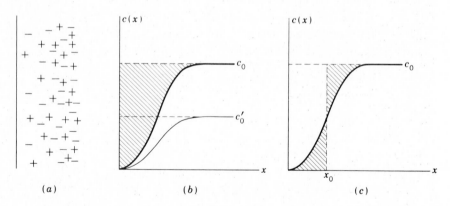

(a)   (b)   (c)

FIGURE 3
Variation of ionic solute concentration near a free surface: (*a*) schematic diagram of the distribution of ions, showing negative surface adsorption; (*b*) concentration versus distance for two different bulk concentrations (the shaded area is equal to $-u$ for a solution of bulk concentration c_0); (*c*) the "equivalent empty-layer thickness" x_0 chosen such that the two shaded areas will be equal.

In electrolyte solutions an ion tends to be surrounded by ions of the opposite charge, but at the surface there are ions which are not uniformly surrounded. There is, therefore, an unbalanced electrostatic force tending to draw these surface ions into the body of the solution. In Fig. 3a a more nearly correct model is shown, in which the ions have largely drawn away from the surface. The shaded area above the curve labeled c_0 in Fig. 3b represents the negative of the surface concentration; the lower curve corresponds to a different bulk concentration c_0' and shows the proportionality of surface concentration to bulk concentration. Figure 3c shows the significance of the "equivalent empty-layer thickness" x_0. The two shaded areas are equal.

METHOD

There are many experimental techniques for measuring the surface tension of liquids. In the *ring method* one determines the force necessary to pull a metal ring free from the surface of a liquid. The Du Nouy tensiometer, in which the ring is hung from the beam of a torsion balance, is often used, but it cannot be easily thermostated. In calculating the surface tension, it is necessary to apply a correction factor which takes into account the shape of the liquid held up by the ring. In the *drop-weight method* one determines the weight of a drop which falls from a tube of known radius. Again, a correction factor is needed because the drop that actually falls does not represent all the liquid that was supported by surface tension. In the *bubble-pressure method* one determines the maximum gas pressure obtained in forming a gas bubble at the end of a tube of known radius immersed in the liquid. Detailed descriptions of these and other methods may be found elsewhere.[3]

Capillary rise   In the absence of external forces a body of liquid tends to assume a shape of minimum area. It is normally prevented from assuming spherical shape by the force of gravity, as well as by contact with other objects. When a liquid is in contact with a solid surface, there exists a specific surface free energy for the interface, or interfacial tension γ_{12}. A solid surface itself has a surface tension γ_2, which is often large in comparison with the surface tensions of liquids. Let a liquid with surface tension γ_1 be in contact with a solid with surface tension γ_2, with which it has an interfacial tension γ_{12}. Under what circumstances will a liquid film freely spread over the solid surface and "wet" it? This will happen if, in creating liquid-solid interface and an equal area of liquid surface at the expense of an equal area of solid surface, the free energy of the entire system decreases:

$$\gamma_1 + \gamma_{12} - \gamma_2 < 0 \qquad (16)$$

If we have a vertical capillary tube which dips into a liquid, a film of the liquid will tend to run up the capillary wall if condition (16) is obeyed. Then, in order to reduce the surface of the liquid, the meniscus will tend to rise in the tube. It will rise until the force of gravity on the liquid in the capillary above the outside surface, $\pi r^2(h + r/3)\rho g$, exactly counterbalances the tension at the circumference, which is $2\pi r \gamma_1$. In these expressions ρ is the density of the liquid, g is the accelera-tion of gravity, h is the height of the liquid above the outside surface, r is the radius of the cylindrical capillary, and $r/3$ is a correction for the amount of liquid above the bottom of the meniscus, assuming it to be hemispherical (see Fig. 4). Thus we obtain

$$\gamma_1 = \frac{1}{2}\left(h + \frac{r}{3}\right)r\rho g \tag{17}$$

If Eq. (16) is not obeyed, but if instead

$$\gamma_1 \cos\theta + \gamma_{12} - \gamma_2 = 0 \tag{18}$$

for some value of θ, the liquid will not tend to spread indefinitely on the solid surface but will tend instead to give a *contact angle* θ (see Fig. 5). This may be the case with aqueous solutions or water on glass surfaces that are not entirely clean. We shall then have, instead of Eq. (17),

$$\gamma_1 \cos\theta = \frac{1}{2}\left(h + \frac{r}{3}\right)r\rho g \tag{19}$$

However, in practice, there is usually some "hysteresis"; that is, the contact angle finally attained is somewhat variable, depending on whether the liquid has been advancing over the solid surface or receding from it. Thus two different capillary

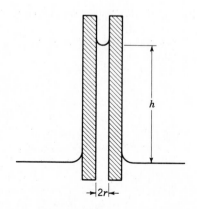

FIGURE 4

Capillary rise h in a tube of radius r.

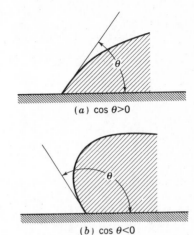

FIGURE 5
Contact angle between a liquid and
solid surface: (*a*) cos θ > 0 (e.g., water
on a glass surface which is not completely clean); (*b*) cos θ < 0 (e.g., Hg
on glass or water on paraffin).

rise heights are to be expected. If the *same* height is obtained regardless of whether
the liquid was allowed to rise from below or fall from above in the capillary, it may
be assumed that Eqs. (16) and (17) are almost certainly valid. This is nearly always
true of aqueous solutions in carefully cleaned glass capillary tubes.

EXPERIMENTAL

If the capillary tube has not been recently cleaned, it should be soaked in hot nitric
acid for several minutes and rinsed copiously with distilled water. A clean capillary
is essential to obtaining good results. When not in use, the capillary should be
stored by immersing it in a tall flask of distilled water. Assemble the apparatus as
shown in Fig. 6. Adjust the capillary tube upward or downward until the outside
liquid level is at or slightly above the zero position on the scale.

Determine the height h of capillary rise for pure water at 25.0°C. Take at
least four readings, alternately allowing the meniscus to approach its final position
from above and below. Also be sure to read the position of the outside level. If
there is not good agreement among these readings, reclean the capillary and repeat
the measurements.

Repeat the above procedure with 0.8 M n-butanol solution, dilute to precisely
three-quarters the concentration and repeat, and so on until eight concentrations
have been used (the last being 0.11 M). Rinse the apparatus and capillary with
one or two small aliquots of fresh solution at each concentration change.

If time permits, repeat the above procedure with NaCl solutions. Use solutions of concentrations approximately 4, 3, 2, and 1 M.

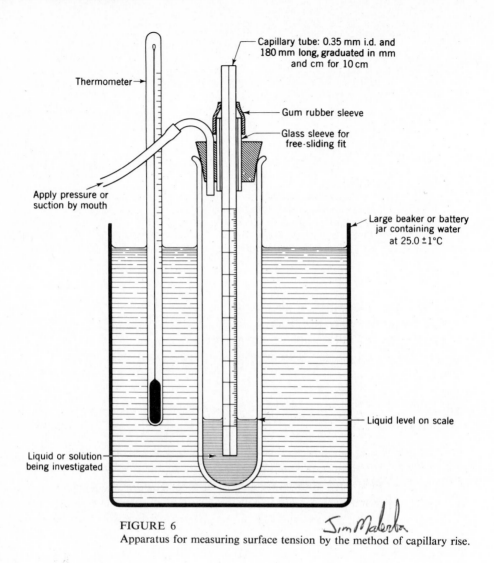

Thermometer→

Capillary tube: 0.35 mm i.d. and
180 mm long, graduated in mm
and cm for 10 cm

Gum rubber sleeve

Glass sleeve for
free-sliding fit

Apply pressure or
suction by mouth

Large beaker or battery
jar containing water
at 25.0 ±1°C

Liquid level on scale

Liquid or solution
being investigated

FIGURE 6
Apparatus for measuring surface tension by the method of capillary rise.

CALCULATIONS

From the data obtained for pure water, calculate the capillary radius r and use it in calculating the surface tension of each solution studied. Use Eq. (17) for these calculations. For the butanol solutions one may assume that the density is equal to that of pure water; however, for NaCl solutions it is necessary to use values of ρ obtained by interpolation in Table 1. At 25.0°C for pure water, the surface tension is 72.0 dynes cm^{-1} and the density is 0.9970 g cm^{-3}.

Table 1   DENSITY OF NaCl SOLUTIONS AT 25°C[a]

% by wt.	NaCl, g liter^{-1}	Moles liter^{-1}	Density, g cm^{-3}
0	0	0	0.9970
1	10.05	0.1719	1.0053
2	20.25	0.3464	1.0125
4	41.07	0.7027	1.0268
6	62.48	1.069	1.0413
8	84.47	1.445	1.0559
10	107.1	1.832	1.0707
12	130.3	2.229	1.0857
14	154.1	2.637	1.1008
16	178.6	3.056	1.1162
18	203.7	3.486	1.1319
20	229.6	3.928	1.1478
22	256.1	4.382	1.1640
24	283.3	4.848	1.1804
26	311.3	5.326	1.1972

[a] Calculated from data given in "Handbook of Chemistry and Physics," 53d ed., Chemical Rubber Publishing Co., Cleveland (1972).

For the runs on n-butanol solutions, plot the surface tension of the solution γ versus the logarithm of the bulk concentration c and determine the slope. The surface concentration (in moles per square centimeter) can be calculated from Eq. (14). Express this surface concentration in molecules per square angstrom, and obtain the "effective cross-sectional area" per molecule of adsorbed butanol in square angstroms.

For the runs on NaCl solutions, plot γ versus the bulk concentration c expressed in moles per cubic centimeter. As seen by Eq. (13) the ratio u/c can be determined from the slope of the straight line obtained. Calculate the "effective empty-layer thickness" x_0 in angstroms for NaCl. Since the laws of ideal solutions do not hold at the NaCl concentrations studied, the results obtained with NaCl have only qualitative significance and the value of x_0 obtained represents only a rough order of magnitude.

APPARATUS

One (or two) graduated capillary tubes (cleaned with concentrated HNO_3, rinsed thoroughly with distilled water, and stored in distilled water); ring stand; battery jar; test tube, with two-hole stopper assembly, to hold capillary; 0 to 30°C thermometer; one large clamp and one thermometer clamp; two clamp holders; 20-in. length of gum-rubber tubing; 200-ml volumetric flask; 50-ml pipette; 250-ml beakers.

Water aspirator with attached clean gum-rubber tube; 0.8 M aqueous solution of n-butanol (250 ml); 4 M solution of NaCl (400 ml).

REFERENCES

1. P. S. Epstein, "Textbook of Thermodynamics," Wiley, New York (1937).
2. W. D. Harkins and R. W. Wampler, *J. Amer. Chem. Soc.*, **53**, 850 (1931).
3. W. D. Harkins and A. E. Alexander, Determination of Surface and Interfacial Tension, in A. Weissberger (ed.), "Technique of Organic Chemistry," 3d ed., vol. I, part I, chap. XIV, Interscience, New York (1959).

GENERAL READING

N. K. Adams, "The Physics and Chemistry of Surfaces," 3d ed., Oxford Univ. Press, London (1941).
J. J. Bikerman, "Surface Chemistry," 2d ed., Academic, New York (1958).

EXPERIMENT 31. ADSORPTION FROM SOLUTION

In this experiment and in Exp. 35 we shall be concerned with the adsorption of molecules from the gas phase or from solution onto the surface of a solid. The term *adsorption* is used to describe the fact that there is a greater concentration of the adsorbed molecules at the surface of the solid than in the gas phase or in the bulk solution. In general one uses solid adsorbents of small particle size and often with surface imperfections such as cracks and holes which serve to increase the surface area per unit mass greatly over the apparent geometrical area. Such small, porous particles may have specific areas in the range from 10 to 1000 m^2 g^{-1}. Some examples of adsorbents commonly used in experiments of this kind are charcoal, silica gel (SiO_2), alumina (Al_2O_3), zeolites, and molecular sieves. The adsorption from aqueous solutions of acetic acid on charcoal will be investigated in the present experiment.

THEORY

The type of interaction between the adsorbed molecule and the solid surface varies over a wide range from weak nonpolar van der Waals' forces to strong chemical bonding. Examples of adsorption where ionic or covalent bonding occurs are the adsorption of chloride ions on silver chloride (ionic) or of oxygen gas on metals where oxygen-metal bonds are formed (covalent). In these cases the process is called *chemisorption*, and it is generally characterized by high heats of adsorption (from 10 to 100 kcal mol^{-1} of gas adsorbed). Chemisorption is highly specific in nature and depends on the chemical properties of both the surface molecules and the adsorbed molecules. Adsorption arising from the weaker van der Waals' and dipole forces is not so specific in character and can take place in any system at low or moderate temperatures. This type of adsorption is called *physical adsorption* and is usually associated with low heats of adsorption (less than about 10 kcal mol^{-1}). Physical adsorption forces are similar to those which cause condensation of gases into liquids or solids. When an adsorbing molecule approaches the surface of the solid, there is an interaction between that molecule and the molecule in the surface which tends to concentrate the adsorbing molecules on the surface in much the same way that a gas molecule is condensed onto the surface of bulk liquid. Another respect in which physical adsorption is similar to liquid condensation is the fact that molar heats of adsorption are of the same order of magnitude as molar heats of vaporization.

Isotherms   The amount adsorbed per gram of solid depends on the specific area of the solid, the equilibrium solute concentration in the solution (or pressure in the case of adsorption from the gas phase), the temperature, and the nature of the molecules involved. From measurements at constant temperature, one can obtain a plot of N, the number of moles adsorbed per gram of solid, versus c, the equilibrium solute concentration. This is called an *adsorption isotherm*.

In principle one can apply the Gibbs isotherm, Eq. (30-13), to the problem of adsorption at the solid-solution interface.[1] In this case, the surface concentration u will equal N/A (where A is the specific area of the solid) and $d\gamma/dc$ will refer to the change in interfacial tension with solute concentration. Although most solutes affect the solid-liquid interfacial tension in qualitatively the same way as they affect the surface tension at the air-liquid interface, very little is known quantitatively about $d\gamma/dc$ as a function of concentration. Thus, while it can be generally predicted that a solute which lowers the surface tension of water will be adsorbed on a solid surface from aqueous solution, Eq. (30-13) is not useful for even a semiquantitative calculation of the extent of such adsorption.

Often it is possible to represent experimental results over a limited range by an empirical isotherm suggested by Freundlich:[2]

$$N = Kc^a \tag{1}$$

where K and a are constants which have no physical significance but can be evaluated by a plot of log N versus log c. However, Eq. (1) fails to predict the behavior usually observed at low and at high concentrations. At low concentrations, N is often directly proportional to c; at high concentrations N usually approaches a constant limiting value which is independent of c.

Much effort has been devoted to developing a theory of adsorption which would explain the observed experimental facts. In some simple systems a theory derived by Langmuir can be applied. This theory is restricted to cases where only one layer of molecules can be adsorbed at the surface. In physical adsorption from the gas phase there is often a formation of many adsorbed layers at higher pressures, as in the case of nitrogen gas adsorbed on charcoal or silica gel at 77°K (see Exp. 32). In the case of chemisorption from the gas phase or adsorption from solution monolayer adsorption is usually observed. Monolayer adsorption is distinguished by the fact that the amount adsorbed reaches a maximum value at moderate concentrations (corresponding to complete coverage of the surface of the adsorbent by a layer one molecule thick) and remains constant with further increase in concentration. The Langmuir isotherm can be derived from either kinetic or equilibrium arguments[2,3] and is most commonly applied to the chemisorption of gases. We shall give a form appropriate to adsorption from solution:

$$\theta = \frac{kc}{1 + kc} \tag{2}$$

where θ is the fraction of the solid surface covered by adsorbed molecules and k is a constant at constant temperature. Now $\theta = N/N_m$, where N is the number of moles adsorbed per gram of solid at an equilibrium solute concentration c and N_m is the number of moles per gram required to form a monolayer. Making this substitution and rearranging Eq. (2), we obtain

$$\frac{c}{N} = \frac{c}{N_m} + \frac{1}{kN_m} \tag{3}$$

If the Langmuir isotherm is an adequate description of the adsorption process, then a plot of c/N versus c will yield a straight line with slope $1/N_m$. If the area σ occupied by an adsorbed molecule on the surface is known, the specific area A (in square meters per gram) is given by

$$A = N_m N_0 \sigma \times 10^{-20} \tag{4}$$

where N_0 is Avogadro's number and σ is given in square angstroms.

If adsorption isotherms are determined at several different temperatures, one would predict that the slopes of the c/N versus c plots should all be the same if the number of adsorption sites (that is, N_m) is independent of temperature, which is usually true. However, the intercepts should change with temperature, since k is a function of temperature. The thermodynamic theory of adsorption from solution is quite complicated, and we shall merely state that one can write

$$\left(\frac{\partial \ln c}{\partial T}\right)_{p,\theta} = \frac{\Delta H}{RT^2} \tag{5}$$

where ΔH is a differential heat for the adsorption process at a constant pressure p and a constant coverage θ. This process involves not only the adsorption of solute molecules but also the displacement of solvent molecules; this complicates the interpretation of ΔH. From Eq. (3) one can see that $(1/kN_m)$ equals $(c_{0.5}/N_m)$, where $c_{0.5}$ is the equilibrium concentration at a coverage $\theta = 0.5$ (that is, $N = \frac{1}{2}N_m$). Thus at 1 atm,

$$\frac{d \ln (1/kN_m)}{dT} = \left(\frac{\partial \ln c}{\partial T}\right)_{\theta=0.5} = \frac{\Delta H}{RT^2} \tag{6}$$

Generally ΔH in Eq. (6) is positive, which means that the extent of adsorption is greater at lower temperatures.

For the adsorption from dilute aqueous solutions of acetic acid on charcoal, the conditions for monolayer adsorption appear to be satisfied. Also, neither acetic acid nor water is appreciably soluble in charcoal, so that bulk *absorption* can be neglected.

EXPERIMENTAL

Clean and dry seven 250-ml Erlenmeyer flasks. These should either have glass-stoppered tops or be fitted with rubber stoppers. Place approximately 1 g of charcoal (weighed accurately to the nearest milligram) in six of these flasks. To each flask, add 100 ml of acetic acid solution measured accurately with a pipette. Suggested initial concentrations are 0.15, 0.12, 0.09, 0.06, 0.03, and 0.015 M, which can be made by diluting a stock solution of approximately 0.15 M acid. To the flask containing no charcoal, add 100 ml of 0.03 M acid; this sample will serve as a control. After the seven samples have been tightly stoppered, they should be shaken periodically for a period of 30 min and then allowed to stand in a thermostat bath at 25°C. Allow at least 1 hr for equilibrium and preferably several hours or overnight.

Filter all the samples through fine filter paper. Discard the first 10 ml of the filtrate as a precaution against adsorption of the acid by the filter paper. [Alternatively, remove two 25-ml aliquots with a pipette. Fit the tip of the pipette with a filter consisting of a short (1-in.) piece of rubber tubing packed with glass wool; this will prevent the transfer of any solid particles. First draw up about 10 ml in the pipette and discard this portion. Then draw up somewhat more than 25 ml, remove the filter, and adjust the liquid level to the mark.] Titrate two 25-ml aliquots with 0.1 N standardized sodium hydroxide solution using phenolphthalein as an indicator. For the titration of the 0.03 and 0.015 M samples use a 10-ml burette. Both in preparing the samples and in making the titrations use a careful analytical technique.

It is advisable to wash the charcoal adsorbent with distilled water several times to remove any impurities, then dry it in an oven at 120°C prior to use. If this has not been done, prepare an extra sample containing 1 g of charcoal and 100 ml of distilled water. Stopper and shake periodically for 30 min; then filter and titrate two 25-ml aliquots as above.

If sufficient time is available, prepare a parallel set of seven samples which are allowed to equilibrate overnight in a refrigerator at about 10°C. Filter (or withdraw aliquots) rapidly so as to prevent the temperature from changing substantially before the sample of solution is removed.

CALCULATIONS

Calculate the final concentration of acetic acid for each sample. The value for the control solution should agree with its initial value. If the sample of charcoal in pure water was studied, it should show no acidity.

From the values of the initial and final concentrations of acetic acid in 100 ml of solution calculate the number of moles present before and after adsorption and obtain the number of moles adsorbed by difference. Compute N, the number of moles of acid adsorbed per gram of charcoal. Plot an isotherm of N versus the equilibrium (final) concentration c in moles per liter.

As suggested by Eq. (3), plot c/N versus c. Draw the best straight line through these points, and calculate N_m from the slope. On the assumption that the adsorption area of acetic acid is 21 $Å^2$, calculate the area per gram of charcoal from Eq. (4).

If data were also obtained at a lower temperature, plot the isotherm (N versus c) and the Langmuir equation (c/N versus c) for this temperature on the same graphs as the 25°C data. How do the slopes and intercepts of the two Langmuir plots compare? Assuming that ΔH in Eq. (6) is independent of temperature, calculate a value for it.

APPARATUS

Seven 250-ml Erlenmeyer flasks, glass-stoppered or with rubber stoppers; three funnels; funnel holder (or three rings, with clamps and stands); fine porosity filter paper; three 250-ml beakers; stirring rod; one 10- and one 50-ml burette; burette stand and holder; several 100-ml titration flasks; a 5-, 10-, 25-, 50-, and 100-ml pipette; spatula; watch glass.

Activated charcoal (acid-free, 10 g); 0.15 M acetic acid (600 ml); 0.1 M sodium hydroxide (150 ml); phenolphthalein indicator; constant-temperature bath, set at 25°C; access to a refrigerator (optional). NOTE: If isotherms are to be measured at two temperatures, most of the equipment and chemicals should be doubled.

REFERENCES

1. S. Glasstone, "The Elements of Physical Chemistry," p. 554, Van Nostrand, Princeton, N.J. (1946).
2. W. J. Moore, "Physical Chemistry," 4th ed., pp. 484–487, Prentice-Hall, Englewood Cliffs, N.J. (1972).
3. G. S. Rushbrooke, "Introduction to Statistical Mechanics," pp. 211–214, Oxford Univ. Press, New York (1949); T. L. Hill, "Introduction to Statistical Thermodynamics," pp. 124ff., Addison-Wesley, Reading, Mass. (1960).

GENERAL READING

H. G. Cassidy, "Adsorption and Chromatography," pp. 68–79, Interscience, New York (1951).

EXPERIMENT 32.   PHYSICAL ADSORPTION OF GASES

A brief, general description of adsorption on the surface of a solid is given in Exp. 31, which deals primarily with monolayer adsorption from solution. The present experiment is concerned with the multilayer physical adsorption of a gas (the *adsorbate*) on a high-area solid (the *adsorbent*). Since such adsorption is caused by forces very similar to those which cause the condensation of a gas to a bulk liquid, appreciable adsorption only occurs at temperatures near the boiling point of the absorbate. The adsorption of N_2 gas on a high-area solid will be studied at 77.4°K (the boiling point of liquid nitrogen), and the surface area of the solid will be obtained.

THEORY

We shall be concerned here with the adsorption isotherm, i.e., the amount adsorbed as a function of the equilibrium gas pressure at a constant temperature. For physical adsorption (sometimes referred to as van der Waals' adsorption) five distinct types of isotherms have been observed;[1] we shall discuss the theory for the most common type,

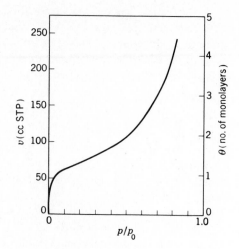

FIGURE 1
A typical isotherm for the physical adsorption of nitrogen on a high-area solid at 77°K.

a typical example of which is shown in Fig. 1. It has unfortunately become standard practice to express the amount adsorbed by v, the volume of gas in cubic centimeters at STP (standard pressure and temperature: $p = 1$ atm, $T = 273.2°$K) rather than in moles. If the adsorbed volume required to cover the entire surface with a complete *monolayer* is denoted by v_m, one can describe the isotherm in terms of the coverage θ:

$$\theta = \frac{v}{v_m} \tag{1}$$

As shown in Fig. 1, adsorption increases rapidly at high pressures and several layers of adsorbate are present even at a relative pressure $p/p_0 = 0.8$. The pressure p_0 is the saturation pressure of the gas (i.e., the vapor pressure of the liquid at that temperature). When $p/p_0 = 1.0$, bulk condensation will occur to form a liquid film on the surface ($\theta \to \infty$). One can also see from Fig. 1 that there is no clear, sharp indication of the formation of a first layer; indeed, the second and higher layers usually begin to form before the first layer is complete. Clearly we cannot obtain a good value of v_m from the adsorption isotherm without the aid of a theory which will explain the shape of the isotherm.

Brunauer, Emmett, and Teller[2] were the first to propose a theory for multilayer adsorption (BET theory). Since the behavior of adsorbed molecules is even more difficult to describe in detail than that of molecules in the liquid state, the BET theory contains some rather drastic assumptions. In spite of this, it is still the most useful theory of physical adsorption available to date. The BET theory gives a correct semiquantitative description of the shape of the isotherm and provides a good means of evaluating v_m (which is then used to obtain the surface area of the solid).

The usual form of the BET isotherm is derived for the case of a free (exposed) surface, where there is no limit on the number of adsorbed layers which may form. Such an assumption is clearly not good for a very porous adsorbent, such as one with deep cracks having a width of a few monolayer thicknesses, since the surface of these cracks can hold only a few layers even when the cracks are filled. Fortunately, the BET theory is most reliable at low relative pressures (0.05 to 0.3) where only a few complete layers have formed, and it can be applied successfully to the calculation of v_m even for porous solids.

The original derivation of the BET equation[1,2] was an extension and generalization of Langmuir's treatment of monolayer adsorption. This derivation is based on kinetic considerations—in particular on the fact that, at equilibrium, the rate of condensation of gas molecules to form each adsorbed layer is equal to the rate of evaporation of molecules from that layer. In order to obtain an expression for θ as a function of p (the isotherm) it is necessary to make several simplifying assumptions. The physical nature of these assumptions in the BET theory can be seen most clearly from a different derivation based on statistical mechanics.[3] Neither derivation will be presented here, but we shall discuss the physical model on which the BET equation is based. A detailed review of this model has been given by Hill;[4] the important assumptions are as follows.

1 The surface of the solid adsorbent is uniform; i.e., all "sites" for adsorption of a gas molecule in the first layer are equivalent.

2 Adsorbed molecules in the first layer are localized; i.e., they are confined to sites and cannot move freely over the surface.

3 Each adsorbed molecule in the first layer provides a site for adsorption of a gas molecule in a second layer, each one in the second layer provides a site for adsorption in a third, with no limitation on the number of layers.

4 There is no interaction between molecules in a given layer. Thus, the adsorbed gas is viewed as many independent stacks of molecules built up on the surface sites.

5 All molecules in the second and higher layers are assumed to be like those in the bulk liquid. In particular, the energy of these molecules is taken to be the same as the energy of a molecule in the liquid. Molecules in the first layer have a different energy owing to the direct interaction with the surface.

In brief, the statistical derivation is based on equilibrium considerations—in particular on finding that distribution of the heights of the stacks which will make the free energy a minimum.

The BET isotherm obtained from either derivation is

$$\theta = \frac{n}{S} = \frac{cx}{(1 - x)[1 + (c - 1)x]} \tag{2}$$

where S is the number of sites, n is the number of adsorbed molecules, x is the relative pressure (p/p_0), and c is a dimensionless constant greater than unity and dependent only on the temperature. Note that θ equals zero when $x = 0$ and approaches infinity as x approaches unity, in agreement with Fig. 1. Making use of Eq. (1), we can rearrange Eq. (2) to the usual form of the BET equation:

$$\frac{x}{v(1 - x)} = \frac{1}{v_m c} + \frac{(c - 1)x}{v_m c} \tag{3}$$

Thus a plot of $[x/v(1 - x)]$ versus x should be a straight line; in practice, deviations from a linear plot are often observed below $x = 0.05$ or above $x = 0.3$. From the slope s and the intercept I, both v_m and c can be evaluated:

$$v_m = \frac{1}{s + I} \qquad c = 1 + \frac{s}{I} \tag{4}$$

The volume adsorbed (in cubic centimeters STP) is related to N, the number of moles adsorbed, by

$$v = NRT_0 \tag{5}$$

where $T_0 = 273.2°K$ and $R = 82.05$ cm^3 atm. The total area of the solid is

$$A = N_0 N_m \sigma \tag{6}$$

where N_0 is Avogadro's number and σ is the cross-sectional area of an adsorbed molecule.

Although we shall not be concerned experimentally with measuring heats of adsorption, it is appropriate to comment that ΔH for the physical adsorption of a gas is always negative, since the process of adsorption results in a decrease in entropy. The *isosteric heat of adsorption* (the heat of adsorption at constant coverage θ) can be obtained by application of the Clausius-Clapeyron equation if isotherms are determined at several different temperatures; the thermodynamics of adsorption have been fully discussed by Hill.[5]

METHOD

Adsorption from the gas phase can be measured by either gravimetric or volumetric techniques. In the gravimetric method, first developed by McBain and Bakr, the weight of adsorbed gas is measured by observing the stretching of a helical spring from which the adsorbent is hung (see Fig. 2). In the volumetric method, the amount of adsorption is inferred from pV measurements which are made on the gas before and after adsorption takes place. An excellent review of the many apparatus designs in common use has been given by Joy;[6] we shall consider only a

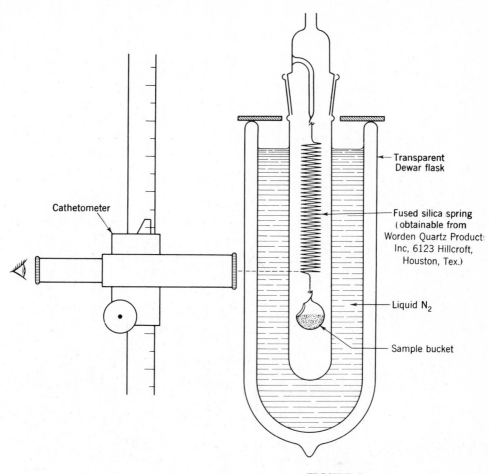

Cathetometer

Transparent
Dewar flask

Fused silica spring
(obtainable from
Worden Quartz Product
Inc, 6123 Hillcroft,
Houston, Tex.)

Liquid N$_2$

Sample bucket

FIGURE 2
Gravimetric adsorption apparatus.

conventional volumetric apparatus similar to one which has been very completely described by Barr and Anhorn.[7] This apparatus is shown in Fig. 3. The central feature in this design is a gas burette connected to a manometer which can be adjusted to maintain a constant volume in the arm containing the gas.

The gas burette shown in Fig. 3 is constructed from two 50-ml bulbs, large-bore capillary tubing, and a stopcock (F). Fiducial marks are engraved above the top bulb, between the bulbs, and below the bottom bulb. Before the burette is attached to the vacuum system, the volume between fiducial marks (V_1 and V_2) can be accurately measured by weighing the mercury required to fill each bulb. More elegant designs[6] achieve greater flexibility in operation by using five or six

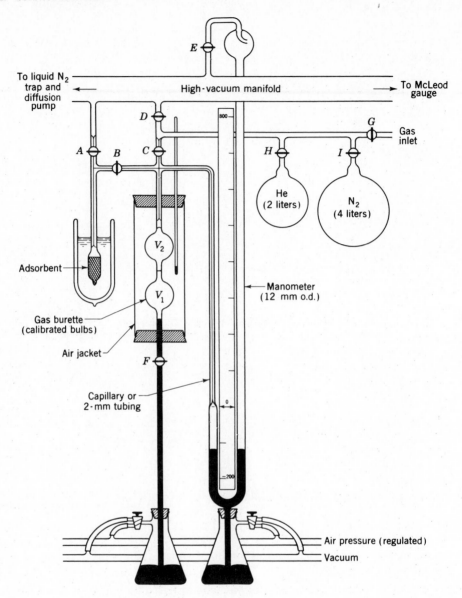

FIGURE 3
Apparatus for measuring adsorption isotherms by the volumetric method.

bulbs, but two are sufficient for the purpose of evaluating V_3, the volume of the tubing between the top of the gas burette and the zero level in the manometer. This is accomplished by measuring the pressure p of gas with both bulbs in use, then filling the lower bulb with mercury and determining the new pressure value p'. From the perfect-gas law

$$V_3 = \frac{pV_1 - (p' - p)V_2}{p' - p} \tag{7}$$

When an isotherm is being determined, the temperature of the gas in the burette system must be known and should remain constant. Although a water jacket is often used, an air jacket is sufficient for the present experiment.

Adsorbent   For the apparatus described here, a sample of a high-area solid adsorbent which has a total surface area of 150 to 250 m^2 should be used. Of the adsorbents most commonly used (charcoal, silica, alumina), silica is recommended as an excellent choice. The sample bulb should be small enough so that it is almost completely filled with the powder. Figure 4 shows a type of bulb which can easily be filled; a loose plug of Pyrex wool in the capillary will prevent loss of powder during filling and later during degassing. The exact weight of the sample should be determined before the bulb is attached to the system. After it is attached to the vacuum line, the sample must be degassed (i.e., pumped on to remove any physically adsorbed substances, principally water). This can be done by mounting a small electrical tube furnace around the sample bulb and heating to 200 to 250°C

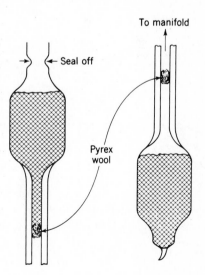

FIGURE 4
Detail drawing of the sample bulb, showing method of filling.

for several hours. Before heating, stopcock A should be opened to pump out all the air (stopcock B remains closed); this must be done cautiously to avoid forming a tight plug of powder in the capillary tube (gentle tapping will help also). After this initial degassing, a shorter degassing period will suffice between adsorption runs.

Adsorbate gas   For physical adsorption, any pure gas which does not react with the solid adsorbent can be used. However, for quantitative area measurements, a molecule of known adsorption area is required. Nitrogen gas is a very common choice and will be used in this experiment.

Liquid nitrogen is the most convenient constant-temperature bath for isotherm measurements with N_2 gas. For high-precision work, the temperature of this bath should be measured with a thermocouple so that an accurate value of p_0 can be obtained, or p_0 should be measured directly with a separate nitrogen vapor-pressure manometer.[6,7] For this experiment, it is adequate to assume that p_0 is equal to the atmospheric pressure in the room.

We can now develop the necessary equations for calculating the amount adsorbed from the pV data obtained during a run. Let us denote the known volume of the gas burette by $V_B (= V_1 + V_2 + V_3$ or $V_2 + V_3$ as the case may be) and its temperature by T_B. Since the sample bulb is not completely filled by the solid adsorbent, we must also consider the so-called dead-space volume occupied by gas. Let us denote this volume by V_s and consider it to be at a single temperature T_s. If the burette is filled with gas at an initial pressure $p_1{}^0$ and then stopcock B is opened, the pressure will drop to a new equilibrium value p_1. The number of moles adsorbed is

$$N_1 = \frac{p_1{}^0 V_B}{RT_B} - \left(\frac{p_1 V_B}{RT_B} + \frac{p_1 V_s}{RT_s} \right) \tag{8}$$

where the V's are in cubic centimeters and p's are in atmospheres. Equation (8) can be rewritten in terms of v_1 (the amount adsorbed in cubic centimeters STP) by using Eq. (5):

$$v_1 = p_1{}^0 V_B^* - p_1 (V_B^* + V_s^*) \tag{9}$$

where the p's are now in Torr and

$$V_B^* = \frac{273.2 V_B}{760 T_B} \qquad V_s^* = \frac{273.2 V_s}{760 T_s} \tag{10}$$

The value of V_B^* can be calculated from V_B and T_B, which are known. V_s^* is constant but unknown; however, it can be determined if a run is made using helium gas. Since He is not significantly adsorbed at liquid-nitrogen temperature, $v_1 = 0$ and V_s^* can be calculated from Eq. (9).

If N_2 gas is admitted to the bulb, adsorption does occur and Eq. (9) yields a value of v_1, the "volume" adsorbed at the equilibrium pressure p_1. Now stopcock B is closed and more N_2 gas is added to the burette to achieve a new initial pressure $p_2{}^0$. When B is opened, more adsorption will occur and the pressure will change to an equilibrium value p_2. The total volume adsorbed at p_2 is

$$v_2 = [p_1{}^0 V_B^* + (p_2{}^0 - p_1)V_B^*] - p_2(V_B^* + V_s^*) \tag{11}$$

The generalization of Eq. (11) for v_j, the volume adsorbed at p_j after j additions of gas to the burette, is obvious; for example,

$$v_3 = [p_1{}^0 V_B^* + (p_2{}^0 - p_1)V_B^* + (p_3{}^0 - p_2)V_B^*] - p_3(V_B^* + V_s^*)$$

Note that the bracket contains an expression for the *total* amount of gas added to the system and the second term represents the amount remaining in the gas phase at equilibrium. Since the effect of errors is cumulative in this method, care should be exercised in measuring all pressures.

Normally the entire gas burette is used ($V_B = V_1 + V_2 + V_3$), but on occasion it may be convenient to obtain two points on the isotherm from a single filling of the burette; this can be accomplished after the first point is obtained, by raising the mercury level so as to fill bulb V_1. Equation (11) is still applicable, but the value of the last term is changed, since V_B now equals $V_2 + V_3$ (in that term only) and a new equilibrium pressure must be used.

EXPERIMENTAL

Details of the operation of the vacuum system (Fig. 3) should be reviewed carefully in consultation with an instructor before beginning the experiment. In the following it is assumed that both the helium and nitrogen storage bulbs have already been filled to a pressure slightly over 1 atm, the diffusion pump is operating, and the sample has been degassed.

Stopcocks B, C, D, E, and, of course, G should be closed and the pressure determined with the McLeod gauge; it should be 10^{-5} Torr or less. Now set the mercury level exactly at the lower mark in the gas burette and close stopcock F. Also raise the mercury in the manometer to approximately the zero level. After closing stopcock A and opening C, D, and E, read the McLeod gauge again and also verify that the mercury level is the same in both arms of the manometer.

Slowly raise a narrow-mouth Dewar flask filled with liquid nitrogen into place, and clamp it so that the sample bulb is completely immersed. Then plug the mouth of the Dewar loosely with glass wool or a piece of clean towel to retard condensation of water and oxygen from the air. (Liquid O_2 dissolved in the liquid N_2 would raise the temperature of the bath.)

Close stopcocks D and E, and **very slowly** turn the stopcock on the helium storage bulb while watching the manometer. As soon as the mercury levels begin to move, cease turning the stopcock and let the gas slowly fill the gas burette to a pressure of about 300 Torr. When this pressure is reached, immediately close the stopcock on the helium storage bulb and then close C. Adjust the mercury level in the manometer so that the left arm is exactly at the zero level. Wait a few minutes for equilibrium to be achieved, readjust to the zero level if necessary, and record the pressure. Now carefully raise the mercury level in the gas burette to the center mark (i.e., fill bulb V_1 with Hg) and again adjust the manometer to the zero level. After a short wait, record the new pressure. Using the known values of V_1 and V_2 (given by the instructor), one can now calculate V_3 from Eq. (7).

Next open stopcock B slowly and allow He to enter the sample bulb. Wait for at least 3 min, then reset the manometer to the zero level and read the pressure. After several minutes more, readjust this level (if necessary) and read the pressure again. When the pressure is constant, record its value and also record T_B, the ambient temperature at the gas burette. The quantity V_s^* can be calculated from these data using Eq. (9) with $v_1 = 0$. Note that, in calculating V_B^* from Eq. (10), the appropriate value of V_B is $V_2 + V_3$.

Slowly open stopcock A and pump off the He gas. Lower the Dewar flask and allow the sample to warm up to room temperature. Open stopcocks C and D and continue pumping for at least 15 min. During this time, reset the Hg level in the gas burette to the lower mark and lower the mercury in the manometer to approximately the zero level. (Do the two arms still read the same? Open stopcock E and see if there is any change.) Now close stopcocks A and B and replace the liquid-nitrogen Dewar around the sample bulb. If time is limited and values of V_3 and V_s^* have been provided, this helium run may be omitted and the following procedure is then carried out directly after first cooling the sample.

Procedure for measuring the N_2 isotherm Close stopcocks D and E, and fill the gas burette with N_2 gas to a pressure of about 300 Torr. Follow the procedure given above in filling the burette and measuring the pressure. Record this pressure $p_1{}^0$. *Slowly* open stopcock B and allow N_2 to enter the sample bulb. Again follow the procedure given previously for obtaining the new equilibrium pressure p_1. In the case of N_2, a longer period may be necessary to achieve equilibrium, since adsorption is now taking place. The volume adsorbed v_1 can be calculated from Eq. (9). [In this case V_B in Eq. (10) is $V_1 + V_2 + V_3$.]

Now close stopcock B and add enough N_2 gas to the burette to bring the pressure up to about 100 Torr. Record this pressure $p_2{}^0$; then open B and obtain the next equilibrium pressure p_2. The volume v_2 is calculated from Eq. (11). Continue to repeat this process so as to obtain about seven points in the pressure range from 0 to about 250 Torr. Each time start with an initial pressure $p_j{}^0$ somewhat greater (say 15 to 25 percent) than the desired equilibrium pressure p_j. If time permits, obtain additional points somewhat more widely spaced over the range 250 to 650 Torr. At all times be sure that the liquid-nitrogen level completely covers the sample bulb.

Several times during the period, record the barometric pressure and T_B, the temperature at the gas burette.

CALCULATIONS

From your helium data and the known values of V_1 and V_2, calculate V_3 from Eq. (7). If the temperature at the gas burette has been fairly constant throughout the experiment, use the average value as T_B in Eq. (10) and calculate values of V_B^* when $V_B = V_1 + V_2 + V_3$ and when $V_B = V_2 + V_3$. These values can then be used in all further calculations. If T_B has varied by more than $\pm 0.5°C$, appropriate changes in V_B^* should be made where necessary. Now use Eq. (9) to calculate V_s^*.

For each equilibrium point on the isotherm, calculate a value of v, the volume adsorbed at pressure p. If the barometric pressure has been almost constant, its average value may be taken as p_0, the vapor pressure of nitrogen at the bath temperature. For each isotherm point, calculate $x = p/p_0$.

Plot the isotherm (v versus x) at 77°K and compare it qualitatively with the one shown in Fig. 1. Finally, calculate $x/v(1 - x)$ for each point and plot that quantity versus x; see Eq. (3). Draw the best straight line through the points between $x = 0.05$ and 0.3, and determine the slope and intercept of this line. From Eq. (4) calculate v_m and c for the sample studied.

Using Eqs. (5) and (6) and the known weight w of the adsorbent sample, calculate the *specific area* $\bar{A}$ in square meters per gram of solid. The cross-sectional area for an adsorbed N_2 molecule may be taken as 15.8 $Å^2$.

DISCUSSION

What factors can you think of that would tend to make the BET theory less reliable above $\theta = 0.3$, below $\theta = 0.05$?

APPARATUS

Adsorption apparatus (Fig. 3), containing high-area sample, attached to high-vacuum line equipped with McLeod gauge, diffusion pumps, and liquid-nitrogen trap (see Chap. XVII); high-purity helium and nitrogen gas; small electrical tube furnace; narrow-mouth taped

Dewar flask and clamp for mounting; glass wool or clean towel; 0 to 30°C thermometer: stopcock grease.

REFERENCES

1. J. W. Williams, R. A. Alberty, and E. O. Kraemer, The Colloidal State and Surface Chemistry, in H. S. Taylor and S. Glasstone (eds.), "A Treatise on Physical Chemistry," 3d ed., vol. II, pp. 594–611, chap. V, Van Nostrand, Princeton, N.J. (1951).
2. S. Brunauer, P. H. Emmett, and E. Teller, *J. Amer. Chem. Soc.*, **60,** 309 (1938).
3. T. L. Hill, *J. Chem. Phys.*, **14,** 263 (1946) and **17,** 772 (1949).
4. T. L. Hill, Theory of Physical Adsorption, in "Advances in Catalysis," vol. IV, pp. 225–242, Academic, New York (1952).
5. *Ibid.*, pp. 242–255.
6. A. S. Joy, *Vacuum*, **3,** 254 (1953).
7. W. E. Barr and V. J. Anhorn, *Instruments*, **20,** 454, 542 (1947).

GENERAL READING

S. Brunauer, "Physical Adsorption," chaps. I–IV, Princeton Univ. Press, Princeton, N.J. (1945).

XI

MACROMOLECULES

EXPERIMENTS

33. Osmotic pressure
34. Intrinsic viscosity: chain linkage in polyvinyl alcohol
35. Helix-coil transition in polypeptides

EXPERIMENT 33.   OSMOTIC PRESSURE

Macromolecules are very large molecules with molecular weights ranging in order of magnitude from 1000 to 1,000,000 or more. They are with a few exceptions organic molecules, and many examples such as cellulose and starch exist in nature. Macromolecules can also be made synthetically. These are often called (high) polymers because structurally they are formed by the linking together of monomer units to form chains of considerable length.

From measurements of the osmotic pressure of dilute solutions of a polymer, the number-average molecular weight of the polymer will be obtained.

THEORY

When a liquid solution of a solute B in a solvent A is separated from pure solvent A by a membrane that is permeable to A alone and both phases are at the same temperature and under the same pressure, the solvent molecules A will pass through the membrane into the solution. This can be prevented by applying a pressure to the solution which is greater, by a definite amount Π, than the pressure on the solvent. This *osmotic pressure* Π can be related to the vapor pressure of A, since at equilibrium the chemical potential of A in solution, μ_A, must equal the chemical potential of pure A, $\mu_A{}^0$. If the partial molal volume of solvent $\overline{V}_A$ is taken as being independent of pressure, the result of equating chemical potentials is[1]

$$\Pi \overline{V}_A = RT \ln \frac{p_A{}^0}{p_A} \tag{1}$$

Approximating $\overline{V}_A$ by $\tilde{V}_A$, the molal volume of pure A, and assuming Raoult's law, we obtain

$$\Pi \, \tilde{V}_A \cong -RT \ln X_A = -RT \ln (1 - X_B) \tag{2}$$

349

where X_A and X_B are the mole fractions of A and B in the solution. For *dilute* solutions, $\ln (1 - X_B)$ can be replaced by $-X_B$ to give

$$\Pi = \frac{RT}{\tilde{V}_A} X_B \cong \frac{N_B RT}{V_A} \cong \frac{N_B RT}{V} \tag{3}$$

where V_A, the volume of solvent containing N_B moles of solute, is approximately equal to the total volume of the solution V.

In general, the observed osmotic pressures do not obey the ideal-solution law of Eq. (3). Owing to the large size of high-polymer molecules, interactions are important even in dilute solutions, but these can be taken into account fairly well by including an osmotic second virial coefficient:[2]

$$\Pi = \frac{N_B RT}{V} \left(1 + \frac{B N_B}{V} \right) \tag{4}$$

This equation can be written as

$$\frac{\Pi}{C_2} = \frac{RT}{M} (1 + \beta C_2) \tag{5}$$

where M is the molecular weight of the solute, C_2 is the concentration of solute in grams per unit volume of solution $(= M N_B / V)$, and β equals B/M. Thus by plotting Π/C_2 versus C_2, it is possible to determine both M and β.

Osmotic pressure is the one colligative property of solutions which is suitable for determining very high molecular weights, since Π is of the order of several Torr even for very dilute solutions. The other colligative properties—freezing-point depression, boiling-point elevation, and vapor-pressure lowering—show effects which are too small for accurate measurement; typical values for dilute high-polymer solutions would be 5×10^{-4}°C, 10^{-4}°C, and 10^{-4} Torr, respectively.

Number-average molecular weight When a polymer sample is polydisperse (i.e., there is a distribution of molecules with different weights), a determination of molecular weight must give some kind of average value. When a colligative property is involved, the average is a *number-average* molecular weight $\overline{M}_n$, since the effect depends only on the number of molecules per unit volume and is independent of their size or shape. This is easily seen for the case of osmotic pressure by comparing Eqs. (4) and (5). Molecular-weight determinations based on properties other than the colligative properties will give different kinds of average molecular weights. One common method involves light scattering,[3] which yields a *weight-average* molecular weight $\overline{M}_w$. Both $\overline{M}_n$ and $\overline{M}_w$ are defined and discussed in Exp. 34 along with the *viscosity-average* molecular weight $\overline{M}_v$. It should be noted that $\overline{M}_w$ will always be greater than $\overline{M}_n$ for a polydisperse polymer sample. (For a monodisperse polymer, $\overline{M}_n = \overline{M}_w = \overline{M}_v$.)

METHOD

Several methods of osmotic-pressure measurement are discussed by Wagner.[4] In the dynamic-equilibrium method, flow of solvent through the membrane can be prevented by applying an external gas pressure to the solution; the equilibrium pressure which is necessary to obtain zero flow will equal the osmotic pressure. Although this method is rapid, it involves a complex cell which must be leaktight. Described below are two other methods which are frequently used. In both techniques, the osmotic pressure is determined from the internal head of solution in a capillary tube on the solution side of the membrane (see Fig. 1).

Static method   In the static method, solvent is allowed to diffuse through the membrane until there is no further change in the internal head. It is necessary to

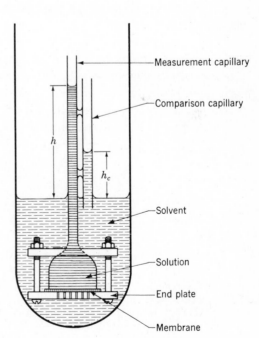

FIGURE 1
A simple osmometer; $\Delta h = h - h_c$ is the internal head corrected for capillary rise.

correct this head for the effect of surface tension (capillary rise) and to calculate the equilibrium concentration of the solution which will differ from the initial concentration due to passage of solvent through the membrane. The disadvantage of this method is that it is slow, equilibrium usually requiring several days. However, no attention is needed during this time.

Half-sum method Fuoss and Mead[4,5] have proposed a rapid method which involves constant attention for about an hour. With the solution and solvent in thermal equilibrium, the internal head is adjusted initially to be close to the expected equilibrium value. To illustrate this method let us assume that the initial head is slightly above the equilibrium value. Frequent readings are made to follow the decrease in this head with time (curve A in Fig. 2). After sufficient time has elapsed to indicate an approximate asymptotic value, the head is adjusted to be roughly as far below this asymptote as it was initially above. Now readings are made to follow the increase with time (curve B in Fig. 2). By calculating one-half the sum of the ordinates of curves A and B at several values of the time, one can obtain curve C which converges rapidly to a constant value. Since only a little solvent flows out of the solution during measurements along curve A and approximately the same amount of solvent flows into the solution during measurements along curve B, the equilibrium concentration may be assumed to be equal to the initial concentration. Correction for the effect of capillary rise is still necessary.

Osmometer Of many designs which have been used, two are especially simple to construct and operate. The Schulz-Wagner osmometer[4] is of an excellent design but is limited to static measurements. A simplified version of this cell, suitable for student use, has been described elsewhere.[6]

The osmometer suggested for this experiment is based on the design of Zimm

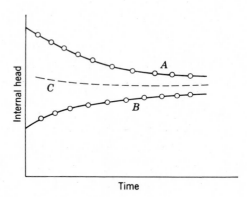

FIGURE 2
Fuoss-Mead method of half sums.
Circles indicate the experimental points;
the dashed line C indicates the average
of curves A and B.

and Myerson[7] and is shown in Fig. 3. This osmometer consists of a small, heavy-wall cylindrical cell with a filling tube and a 0.5-mm i.d. measuring capillary sealed into its side. The open ends of the cell are ground flat and smooth, so that the two membranes, which are held in place by perforated metal plates, also serve as gaskets. The end plates, machine screws, and wing nuts should be stainless steel (or brass is satisfactory if only organic liquids are to be used). The holes and grooves in the end plates may be replaced, for simplicity, by several drill holes if a disk of filter paper is used to support each membrane. Attached to the measuring capillary is a short length of capillary with the same inside diameter. This is immersed in the solvent to provide a correction for capillary rise.

A modified version of this osmometer is shown in Fig. 3d. An advantage of this "thistle-tube" design is ease of assembly since only one membrane is used and it is mounted horizontally at the bottom of the cell. Two disadvantages are the somewhat more awkward clamping arrangement and the slower diffusion of solvent. The latter point is not serious if the half-sum method is used.

Once the cell is filled with solution, a drill rod is inserted into the filling tube. By raising or lowering this steel rod, one can adjust the position of the meniscus in the measuring capillary. Thus measurements can be made using the half-sum method or using the static method where the level is set initially at the expected value in order to save time. Two designs are shown in Fig. 3 for the details of the steel rod-filling tube assembly. The simplest design, indicated in Fig. 3a, is to use 2-mm i.d. capillary tubing and a length of steel drill rod which is chosen to give a very close fit to the capillary. An alternative design, shown in Fig. 3d, utilizes a larger filling tube and a loosely fitting rod which has attached to it the neoprene tip from the plunger of a disposable syringe.

When volatile solvents are used, a small pool of mercury in the flared top of the filling tube will prevent evaporation around the metal rod. Also, a close-fitting one-hole stopper in the outer test tube which holds the pure solvent will retard evaporation of solvent and reduce absorption of water from the atmosphere if a water-soluble solvent is being used.

Membranes Of the many materials used for membranes the most convenient one is cellulose in the form of nonwaterproofed cellophane sheet (du Pont No. 300 or 600). Cellophane is a good membrane for methyl ethyl ketone, the solvent to be used in this experiment, and can be prepared in a range of porosities by treatment with strong basic solutions.[4,6] To condition a membrane in water for use with an organic solvent, it is necessary to wash it, successively, with 25, 50, 75, and 100 percent aqueous alcohol (or acetone) solutions and then displace the alcohol (or acetone) by similar washings with the desired organic solvent. Membranes should be stored in the solvent and never allowed to dry out.

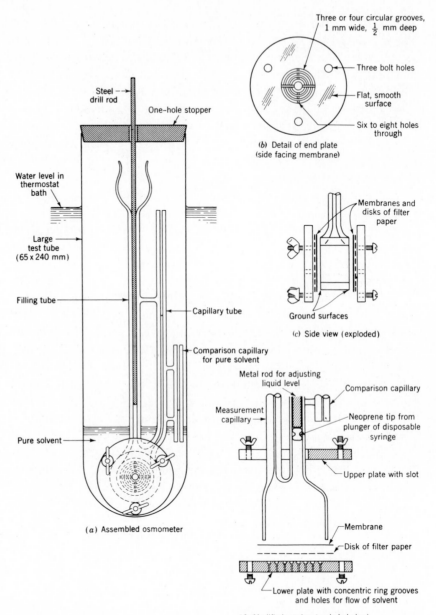

Three or four circular grooves,
1 mm wide, $\frac{1}{2}$ mm deep

Three bolt holes

Flat, smooth
surface

Six to eight holes
through

(b) Detail of end plate
(side facing membrane)

Steel
drill rod

One–hole stopper

Water level in
thermostat
bath

Large
test tube
(65 × 240 mm)

Filling tube

Capillary tube

Comparison capillary
for pure solvent

Pure solvent

(a) Assembled osmometer

Membranes and
disks of filter
paper

Ground surfaces

(c) Side view (exploded)

Metal rod for adjusting
liquid level

Comparison capillary

Measurement
capillary

Neoprene tip from
plunger of disposable
syringe

Upper plate with slot

Membrane

Disk of filter paper

Lower plate with concentric ring grooves
and holes for flow of solvent

(d) Modified version (exploded view)

FIGURE 3
Zimm-Meyerson osmometer: (*a*) the completely assembled apparatus; (*b*) detail
view of one of the end plates; (*c*) side view of the cell; (*d*) detail view of modified
version.

EXPERIMENTAL

Polystyrene is a suitable polymer for investigation in methyl ethyl ketone (MEK) solvent. It is necessary to have a reasonably high molecular weight fraction to avoid diffusion of solute molecules through the membrane. If such solute diffusion occurs, the osmotic pressure will change with time.

An initial weight concentration of polymer which will produce about an 8-cm internal head is desirable. If an approximate value of the molecular weight is known, this concentration can be estimated from Eq. (3). Make up 50 ml of this solution in a volumetric flask. Pipette 25 ml of this solution into a clean 50-ml volumetric flask, and make up to the mark with solvent. Repeat this dilution procedure at least twice to obtain solutions which are one-quarter and one-eighth of the initial concentration. If these solutions are allowed to stand for a considerable time before the osmometer is filled, be sure to seal them carefully to avoid evaporation of solvent and subsequent change of concentration.

The large test tube and the glass parts of the osmometer should be cleaned thoroughly with chromic acid cleaning solution, rinsed with distilled water, alcohol, and MEK. The osmometer is then assembled by placing a disk of smooth hard filter paper (moistened with MEK) and a round cellophane membrane (conditioned for and stored in MEK) on each of the metal end plates and bolting these plates carefully about the cell as shown in Fig. 3c or 3d. The membranes should *never* be allowed to dry out. After the osmometer is assembled, place it in the large test tube containing pure solvent.

It may be desirable to test the osmometer for leaks along the seal between the ground-glass edges of the cell and the metal end plates. This leak test can be accomplished by filling the osmometer with pure solvent and inserting the steel rod into the filling tube so as to create an internal head (liquid level in the measuring capillary above that in the comparison capillary). If the osmometer is leaktight, the level in the capillary should change only very slowly owing to diffusion through the membrane. If the level changes rapidly, a leak along the gasket is indicated and the cell should be taken apart, inspected for flaws, reassembled, and retested.

In order to fill and empty the cell it is convenient to use a large hypodermic syringe with a long, stainless-steel needle which can be inserted down the filling tube all the way to the bottom of the cell. [If the osmometer has a large filling tube, it can be filled and emptied by means of a length of Teflon "spaghetti" tubing which is inserted to the bottom of the cell through the filling tube. This spaghetti tubing is attached (by means of a tapered adapter, made by drawing down a piece of 8-mm glass tubing) to a water aspirator for emptying the osmometer or to a hypodermic syringe for filling it. When filling the cell, slowly withdraw the spaghetti tubing as the solution is introduced.]

After the osmometer has been leak-tested, rinse it several times with small quantities of solution and then fill it with the solution to about halfway up the filling tube. Be sure that there are no air bubbles in the cell; bubbles can be removed by gentle tapping or by removing some of the solution and refilling. Insert the steel drill rod into the filling tube so as to adjust the level in the capillary to approximately the expected level. Immediately rinse the syringe and needle with pure solvent to prevent any deposit of polymer from evaporation of residual solution, and introduce the osmometer into its test tube, which should contain enough solvent to produce a meniscus in the reference capillary. [If a steel rod with rubber plunger tip is used, the osmometer is *completely* filled to the top of the filling tube. Then insert the plunger tip taking care not to produce a bubble. Slowly push the plunger in, about two-thirds of the way to the bottom, allowing the excess solution to exit through the measuring capillary. Then **very slowly** pull the plunger back so as to bring the meniscus in the measuring capillary to approximately the desired position. Extreme slowness (with rotation of the plunger to avoid jerkiness) is necessary to allow the liquid in the measuring capillary to drain down the capillary walls without forming "plugs." **Caution:** Do not insert the plunger into the filling tube unless the tube contains liquid since the rubber tip may come off when the plunger is withdrawn from a dry tube.]

The osmometer should be mounted in a thermostat bath at 25°C which is regulated to within $\pm 0.01°C$ or better. A glass-walled thermostat tank is best, since it will permit the test tube to be well immersed and enable the capillary levels to be observed through the glass wall. If such a bath is not available, immerse the test tube to the level of the solvent inside and read the capillary levels over the edge of the bath. The level of the liquid in the two capillaries should be read to within 0.1 mm using a cathetometer. This instrument consists of a telescope mounted on a vertical bar engraved with a millimeter scale. The telescope is moved up or down until the liquid level is in the center of the field and the position on the bar is recorded. A more complete description is given in Chap. XVIII. The internal head Δh is the difference between the height of solution in the measuring capillary and the height of solvent in the comparison capillary $h - h_{comp}$. Determine the internal head for at least four concentrations of polymer.

Either the static method or the half-sum method described previously may be used. The method of half sums is recommended. Plot the height of the solution in the measuring capillary against time as soon as each data point is obtained. When the variation of h with time becomes quite slow, adjust the head and obtain the second set of readings. Do not wait for a constant asymptotic value of h on either set of readings. If the static method is employed, it is necessary to know the volume of the cell and the cross-sectional area of the measuring capillary in order to compute the equilibrium concentration. Since this correction is small, these dimensions need not be known to high precision. The volume of the cell can be obtained by carefully filling it with solvent using a calibrated syringe. To measure the area of the capillary, introduce a slug of mercury and measure its length. Then transfer the mercury to a weighing bottle and weigh. Using the known density of mercury, one can calculate the cross-sectional area.

When the measurements are completed, the osmometer and the test tube should be rinsed well and left filled with solvent if there is any possibility that it will be used again soon. Otherwise, the apparatus should be disassembled and its parts thoroughly rinsed and dried.

CALCULATIONS

For each solution, calculate the initial concentration in grams of solute per milliliter of solution   If the half-sum method was used, determine the equilibrium head h from the asymptotic value of the half sums as shown in Fig 2. Then calculate the internal head Δh.

If the static method was used, correct the initial concentration for diffusion of solvent to obtain the equilibrium concentration C_2.[6] Calculate the osmotic pressure Π, in dynes per square centimeter, for each solution from

$$\Pi = (\Delta h)\rho g \tag{6}$$

where the solution density ρ is taken to be the same as the density of pure solvent and g is the acceleration due to gravity. The density of pure MEK is 0.803 g cm^{-3} at 25°C.

Tabulate your values for Δh, Π, C_2, and Π/C_2. Plot Π/C_2 versus C_2, and draw the best straight line through the experimental points. (It is possible that the experimental points may show a slight curvature; in this case, draw the straight line which is tangent to the curve at zero concentration.) From the intercept and slope of this line, calculate the number-average molecular weight $\overline{M}_n$ and the osmotic second virial coefficient β using Eq. (5).

APPARATUS

One or more osmometers (complete with end plates, bolts, large test tube, drill rod, and stopper); membranes, conditioned for use in methyl ethyl ketone; four 50-ml glass-stoppered volumetric flasks; 25-ml pipette; several beakers; 25-ml hypodermic syringe with long needle;

constant-temperature bath (preferably with glass walls) set at 25°C; large clamp for mounting the osmometer; cathetometer; stopwatch; mercury (optional).

High-molecular-weight polystyrene powder; reagent-grade methyl ethyl ketone (500 ml).

REFERENCES

1. W. J. Moore, "Physical Chemistry," 4th ed., pp. 250–254, Prentice-Hall, Englewood Cliffs, N.J. (1972).
2. P. J. Flory, "Principles of Polymer Chemistry," pp. 279–282, 530–539, Cornell Univ. Press, Ithaca, N.Y. (1953).
3. *Ibid.*, pp. 283–303.
4. R. H. Wagner and L. D. Moore, Jr., Determination of Osmotic Pressure, in A. Weissberger (ed.), "Technique of Organic Chemistry," 3d ed., vol. I, part I, chap. XV, Interscience, New York (1959).
5. R. M. Fuoss and D. J. Mead, *J. Phys. Chem.*, **47,** 59 (1943).
6. F. Daniels, J. W. Williams, P. Bender, R. A. Alberty, C. D. Cornwell, and J. E. Harriman, "Experimental Physical Chemistry," 7th ed., pp. 123–132, McGraw-Hill, New York (1970).
7. B. H. Zimm and I. Myerson, *J. Amer. Chem. Soc.*, **68,** 911 (1946).

GENERAL READING

S. Glasstone, "Textbook of Physical Chemistry," 2d ed., pp. 651–673, Van Nostrand, Princeton, N.J. (1946).

EXPERIMENT 34.   INTRINSIC VISCOSITY: CHAIN LINKAGE IN POLYVINYL ALCOHOL

While the basic chemical structure of a synthetic high polymer is usually well understood, many physical properties depend on such characteristics as chain length, degree of chain branching, and molecular weight, which are not easy to specify exactly in terms of a molecular formula. Moreover, the macromolecules in a given sample are seldom uniform in chain length or molecular weight (which for a linear polymer is proportional to chain length); thus, the nature of the distribution of molecular weights is another important characteristic.

A polymer whose molecules are all of the same molecular weight is said to be *monodisperse*; a polymer in which the molecular weights vary from molecule to molecule is said to be *polydisperse*. Specimens that are approximately monodisperse can be prepared in some cases by fractionating a polydisperse polymer; this fractionation is frequently done on the basis of solubility in various solvent mixtures.

This experiment is concerned with the linear polymer polyvinyl alcohol (PVOH), $+CH_2-CHOH+_n$, which is prepared by hydrolysis of the polyvinyl acetate (PVAc) obtained from the direct polymerization of the monomer vinyl acetate, $CH_2=CH-OOCCH_3$. As ordinarily prepared polyvinyl alcohol shows a negligible amount of branching of the chains. It is somewhat unusual among synthetic high polymers in that it is soluble in water. This makes polyvinyl alcohol commercially important as a thickener, as a component in gums, and as a foaming agent in detergents.

A characteristic of interest in connection with PVOH and PVAc is the consistency of orientation of monome units along the chain. In the formula given above it is assumed that all monomer units go together "head to tail." However, occasionally a monomer unit will join onto the chain in a "head-to-head" fashion, yielding a chain of the form

$$+CH_2-CHX+_nCH_2-CHX-CHX-CH_2-CH_2-CHX+CH_2-CHX+_n\cdots$$

head-to-head linkage   reversed monomer unit

where X is Ac or OH. The frequency of head-to-head linkage depends on the relative rates of the normal growth-step reaction α

$$R\cdot + M \xrightarrow{k_\alpha} R-M\cdot \qquad (1a)$$

and the abnormal reaction β

$$R\cdot + M \xrightarrow{k_\beta} R-M\cdot \qquad (1b)$$

(where $R\cdot$ is the growing polymer radical, the arrow representing the predominant monomer orientation). The rates, in turn, must depend on the activation energies:

$$\frac{k_\beta}{k_\alpha} = \frac{A_\beta e^{-E_\beta^*/RT}}{A_\alpha e^{-E_\alpha^*/RT}} = Se^{-\Delta E^*/RT} \qquad (2)$$

where $\Delta E^* = E_\beta^* - E_\alpha^*$ is the additional thermal activation energy needed to produce abnormal addition and S is the *steric factor* representing the ratio of the probabilities that the abnormally and normally approaching monomer will not be prevented by steric or geometric obstruction from being in a position to form a bond. Presumably the activation energy for the normal reaction is the lower one, in accord with the finding of Flory and Leutner[1,2] that the frequency of head-to-head linkages in PVAc increases with increasing polymerization temperature. The quantities S

and ΔE^* can be determined from data obtained at two or more polymerization temperatures by a group of students working together. It should be of interest to see whether steric effects or thermal activation effects are the more important in determining the orientation of monomer units as they add to the polymer chain.

In this experiment, the method of Flory and Leutner will be used to determine the fraction of head-to-head attachments in a single sample of PVOH. The method depends on the fact that in PVOH a head-to-head linkage is a 1,2-glycol structure, and 1,2-glycols can be specifically and quantitatively cleaved by periodic acid or periodate ion. Treatment of PVOH with periodate should therefore break the chain into a number of fragments, bringing about a corresponding decrease in the effective molecular weight. All that is required is a measurement of the molecular weight of a specimen of PVOH before and after treatment with periodate.

METHOD[3]

For the determination of very high molecular weights, freezing-point depressions, boiling-point elevations, and vapor-pressure lowerings are too small for accurate measurement. Osmotic pressures are of a convenient order of magnitude, but measurements are time consuming. The technique to be used in this experiment depends on the determination of the intrinsic viscosity of the polymer. However, molecular-weight determinations from osmotic pressures are valuable in calibrating the viscosity method.

The coefficient of viscosity η of a fluid is defined in Exp. 4. It is conveniently measured, in the case of liquids, by determination of the time of flow of a given volume V of the liquid through a vertical capillary tube under the influence of gravity. For a virtually incompressible fluid such as a liquid, this flow is governed by Poiseuille's law in the form

$$\frac{dV}{dt} = \frac{\pi r^4 (p_1 - p_2)}{8\eta L}$$

where dV/dt is the rate of liquid flow through a cylindrical tube of radius r and length L and $(p_1 - p_2)$ is the difference in pressure between the two ends of the tube. In practice a viscosimeter of a type similar to that shown in Fig. 1 is used. Since $(p_1 - p_2)$ is proportional to the density ρ, it can be shown that for a given total volume of liquid

$$\frac{\eta}{\rho} = Bt \tag{3}$$

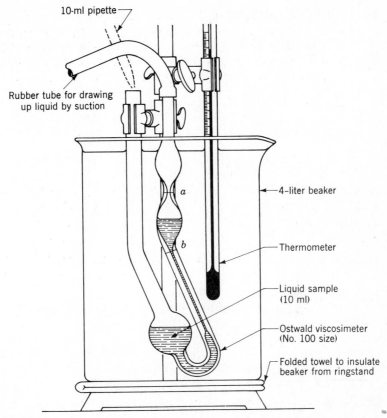

10-ml pipette

Rubber tube for drawing
up liquid by suction

a

b

4–liter beaker

Thermometer

Liquid sample
(10 ml)

Ostwald viscosimeter
(No. 100 size)

Folded towel to insulate
beaker from ringstand

FIGURE 1
Viscosimeter arrangement to be used if a glass-walled thermostat bath is not
available.

where t is the time required for the upper meniscus to fall from the upper to the
lower fiducial mark (a to b) and B is an apparatus constant which must be determined through calibration with a liquid of known viscosity (e.g., water). The
derivation of this equation is similar to that given for Eq. (4-10). For high-precision
work, it may be necessary to consider a kinetic-energy correction to Eq. (3). As in
Eq. (4-16), we can write

$$\frac{\eta}{\rho} = Bt - \frac{V}{8\pi Lt}$$

but the correction term is usually less than 1 percent of Bt and may be neglected
in this experiment.

THEORY

Einstein showed that the viscosity η of a fluid in which small rigid spheres are dilutely and uniformly suspended is related to the viscosity η_0 of the pure fluid (solvent) by the expression[4]

$$\frac{\eta}{\eta_0} - 1 = \frac{5}{2}\frac{v}{V}$$

where v is the volume occupied by all the spheres and V the total volume. The quantity $(\eta/\eta_0) - 1$ is called the *specific viscosity* η_{sp}. For nonspherical particles the numerical coefficient of v/V is greater than $\frac{5}{2}$ but should be a constant for any given shape provided the rates of shear are sufficiently low to avoid preferential orientation of the particles.

The *intrinsic viscosity*, denoted by $[\eta]$, is defined as the ratio of the specific viscosity to the weight concentration of solute, in the limit of zero concentration:

$$[\eta] \equiv \lim_{c\to 0}\frac{\eta_{sp}}{c} = \lim_{c\to 0}\left(\frac{1}{c}\ln\frac{\eta}{\eta_0}\right) \tag{4}$$

where c is usually defined as the concentration in grams of solute per 100 ml of solution. Both η_{sp}/c and $(1/c)(\ln \eta/\eta_0)$ show a reasonably linear concentration dependence at low concentrations. A plot of $(1/c)(\ln \eta/\eta_0)$ versus c usually has a small negative slope, while a plot of η_{sp}/c versus c has a positive and larger slope.[3] Careful work demands that either or both of these quantities be extrapolated to zero concentration, although $(1/c)(\ln \eta/\eta_0)$ for a single dilute solution will give a fair approximation to $[\eta]$.

If the internal density of the spherical particles (polymer molecules) is independent of their size (i.e., the volume of the molecule is proportional to its molecular weight), the intrinsic viscosity should be independent of the size of the particles, and hence of no value in indicating the molecular weight. This, however, is not the case; to see why, we must look into the nature of a polymer macromolecule as it exists in solution.

Statistically coiled molecules A high polymer such as PVOH contains many single bonds, around which rotation is possible. If the configurations around successive carbon atoms are independent and unrelated, it will be seen that two parts of the polymer chain more than a few carbon atoms apart are essentially uncorrelated in regard to direction in space. The molecule is then "statistically coiled" and resembles a loose tangle of yarn:

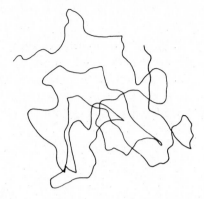

Simple statistical treatments[2] show that the mean distance between the two ends of the chain, and indeed also the effective mean diameter d of the coiled molecule regarded as a rough sphere, should be proportional to the square root of the chain length and thus to the square root of the molecular weight:

$$d \propto M^{1/2}$$

The volume v_m occupied by a molecule should then vary as $M^{3/2}$. The number of molecules in a given weight of polymer varies inversely with the molecular weight; hence the total volume of the spheres is

$$v \propto \frac{cV}{M} M^{3/2} = cVM^{1/2}$$

Therefore, $$[\eta] = KM^{1/2} \tag{5}$$

where K is a constant. This treatment is much simplified; it ignores, among other things, the problem of "excluded volume"; that is, the chain cannot coil altogether randomly because it is subject to the restriction that no two parts of the chain may be at the same point in space at the same time. This restriction becomes more and more important the higher the molecular weight. Even more serious is the effect of solvent; the above treatment tacitly assumes a "poor solvent" which would barely get the polymer into solution. A "good solvent," by solvating the polymer, makes the size of the statistical coil increase faster with chain length than it otherwise would, owing to enhancement of the excluded volume effect. Accordingly, instead of Eq. (5) we might write

$$[\eta] = KM^a \tag{6}$$

where K and a are empirical parameters characteristic both of the polymer itself and of the solvent. The exponent a varies from about 0.5, for well-coiled polymer molecules in a poor solvent, to as much as 1.7, for a rigidly extended "rod-like" polymer molecule.

Flory and Leutner,[1] working with monodisperse specimens of PVOH differing from one another in molecular weight over a wide range (obtained by fractionating polydisperse commercial PVOH), established a correlation between the molecular weight, as determined from osmotic pressure measurements, and the intrinsic viscosity. They found that for PVOH in aqueous solution at 25°C,

$$[\eta] = 2.0 \times 10^{-4} M^{0.76}$$
$$M = 7.6 \times 10^{4}[\eta]^{1.32}$$

(7)

Equation (7) also holds for a polydisperse sample of PVOH, but the molecular weight in this case is $\overline{M}_v$, the "viscosity-average" molecular weight defined below.

Number-average and viscosity-average molecular weight For a polydisperse polymer, any determination of molecular weight must yield an average of some sort. When a colligative property such as osmotic pressure is used, the average is a "number average":

$$\overline{M}_n = \frac{\displaystyle\int_0^\infty MP(M)\,dM}{\displaystyle\int_0^\infty P(M)\,dM}$$

(8)

where $P(M)$ is the molecular-weight distribution function; that is, $P(M)\,dM$ is proportional to the number of molecules with molecular weights between M and $M + dM$. However, the average obtained from intrinsic viscosity is not the same kind of average. It is called the "viscosity average" and is given by

$$(\overline{M}_v)^a = \frac{\displaystyle\int_0^\infty M^{1+a}P(M)\,dM}{\displaystyle\int_0^\infty MP(M)\,dM}$$

(9)

For a monodisperse polymer, $\overline{M}_v = \overline{M}_n = M$, but for a polydisperse polymer the two kinds of averages are not equal but are related by a constant factor which depends on the distribution function $P(M)$ and the parameter a. A commonly encountered distribution function, and one that is likely to be valid for PVOH, is one that arises if the probability of a chain-termination reaction during the polymerization is constant with time and independent of the chain length already achieved. It is also the most likely function for the product resulting from cleavage with periodate if the head-to-head structures can be assumed to be randomly distributed along the PVOH chains. This distribution function is

$$P(M) = \frac{1}{\overline{M}_n} e^{-M/\overline{M}_n} \tag{10}$$

When this function is used as the weighting function in evaluating averages, it can be shown[3,5] that

$$\frac{\overline{M}_v}{\overline{M}_n} = [(1 + a)\Gamma(1 + a)]^{1/a}$$

where Γ is the "gamma function." For $a = 0.76$ (the value given above for polyvinyl alcohol),

$$\frac{\overline{M}_v}{\overline{M}_n} = 1.89 \tag{11}$$

If a much exceeds 0.5, $\overline{M}_v$ is much closer to a "weight-average" molecular weight,

$$\overline{M}_w = \frac{\displaystyle\int_0^\infty M^2 P(M)\, dM}{\displaystyle\int_0^\infty MP(M)\, dM} \tag{12}$$

than it is to the number average $\overline{M}_n$. In fact, when $a = 1$, $\overline{M}_v$ and $\overline{M}_w$ are identical, and with the distribution assumed in Eq. (10) their ratio to $\overline{M}_n$ is 2.

Determination of frequency of head-to-head occurrences We wish to calculate the fraction of linkages which are head to head (that is, the ratio of "backward" monomer units to total monomer units), on the assumption that degradation arises exclusively from cleavage of 1,2-glycol structures and that all such structures are cleaved. Let us denote this ratio by Δ. It is equal to the *increase* in the number of molecules in the system, divided by the total number of monomer units represented by all molecules present in the system. Since these numbers are in inverse proportion to the respective molecular weights,

$$\Delta = \frac{1/\overline{M}_n' - 1/\overline{M}_n}{1/M_0} \tag{13}$$

where $\overline{M}_n$ and $\overline{M}_n'$ are number-average molecular weights before and after degradation, respectively, and M_0 is the monomer weight, equal to 44. Thus,

$$\Delta = 44 \left(\frac{1}{\overline{M}_n'} - \frac{1}{\overline{M}_n} \right) \tag{14}$$

Making use of Eq. (11) we can write

$$\Delta = 83\left(\frac{1}{M'_v} - \frac{1}{M_v}\right)$$
(15)

which permits viscosity averages to be used directly.

EXPERIMENTAL

Clean the viscosimeter thoroughly with chromic acid cleaning solution, rinse copiously with distilled water (use a water aspirator to draw large amounts of distilled water through the capillary), and dry with acetone and air. Immerse in a 25°C thermostat bath to equilibrate. Place a small flask of distilled water in a 25°C bath to equilibrate. Equilibration of water or solutions to bath temperature, in the amounts used here, should be complete in about 10 min.

If a stock solution of the polymer is not available, it should be prepared as follows: Weigh out accurately in a weighing bottle or on a watch glass 4.0 to 4.5 g of the dry polymer. Add it slowly, with stirring, to about 200 ml of hot distilled water in a beaker. To the greatest extent possible "sift" the powder onto the surface and stir gently so as not to entrain bubbles or produce foam. When all of the polymer has dissolved, let the solution cool and transfer it carefully and quantitatively into a 250-ml volumetric flask. Avoid foaming as much as possible by letting the solution run down the side of the flask. Make the solution up to the mark with distilled water and mix by *slowly* inverting a few times. If the solution appears contaminated with insoluble material that would possibly interfere with the viscosity measurements, filter it through Pyrex wool. (Since making up this solution may take considerable time, it is suggested that the calibration of the viscosimeter with water be carried out concurrently.)

In all parts of this experiment, avoid foaming as much as possible. This requires careful pouring when a solution is to be transferred from one vessel to another. It is also important to be meticulous about promptly and *very thoroughly* rinsing glassware that has been in contact with polymer solution, for once the polymer has dried on the glass surface it is quite difficult to remove. Solutions of the polymer should not be allowed to stand for many days before being used as they may become culture media for airborne bacteria.

Pipette 50 ml of the stock solution into a 100-ml volumetric flask, and make up to the mark with distilled water, observing the above precautions to prevent foaming. Mix, and place in the bath to equilibrate. In this and other dilutions, rinse the pipette *very thoroughly* with water and dry with acetone and air.

To cleave the polymer, pipette 50 ml of the stock solution into a 250-ml flask and add up to 25 ml of distilled water and 0.25 g of solid KIO_4. Warm the flask to about 70°C, and stir until all the salt is dissolved. Then clamp the flask in a thermostat bath and stir until the solution is at 25°C. Transfer quantitatively to a 100-ml volumetric flask, and make up to the mark with distilled water. Mix carefully, and place in the bath to equilibrate. (This operation can be carried out while viscosity measurements are being made on the uncleaved polymer.)

At this point, two "initial" solutions have been prepared: 100 ml of each of two aqueous polymer solutions of the same concentration ($\sim$0.9 g/100 ml), one cleaved with periodate and one uncleaved. To obtain a second concentration of each material, pipette 50 ml of the "initial" solution into a 100-ml volumetric flask and make up to the mark with distilled water. Place all solutions in the thermostat bath to equilibrate. The viscosity of both solutions of each polymer (cleaved and uncleaved) should be determined. If time permits, a third concentration (equal to one-quarter of the initial concentration) should also be prepared and measured.

The recommended procedure for measuring the viscosity is as follows:

1   The viscosimeter should be mounted vertically in a constant-temperature bath so that both fiducial marks are visible and below the water level. If a glass-walled thermostat bath which will allow readings to be made with the viscosimeter in place is not available, fill a large beaker or battery jar with water from the bath and set it on a dry towel for thermal insulation. The temperature should be maintained within $\pm0.05°$ of 25°C during a run. This will require periodic small additions of hot water.

2   Pipette the required quantity of solution (or water) into the viscosimeter. Immediately rinse the pipette copiously with water, and dry it with acetone and air before using again.

3   By mouth suction through a rubber tube, draw the solution up to a point well above the upper fiducial mark. Release the suction and measure the flow time between the upper and lower marks with a stopwatch. Obtain two or more additional runs with the same filling of the viscosimeter. Three runs agreeing within about 1 percent should suffice.

4   Each time the viscosimeter is emptied, rinse it *very thoroughly* with distilled water, then dry with acetone and air. Be sure to remove *all* polymer with water before introducing acetone.

If time permits, the densities of the solutions should be measured with a Westphal balance. Otherwise, the densities may be taken equal to that of the pure solvent without introducing appreciable error.

CALCULATIONS

At 25°C, the density of water is 0.9970 g cm^{-3} and the coefficient of viscosity η_0 is 0.8937 cP. Using your time of flow for pure water, determine the apparatus constant B in Eq. (3).

For each of the polymer solutions studied, calculate the viscosity η and the concentration c in grams of polymer per 100 ml of solution. Then calculate η_{sp}/c and $(1/c)(\ln \eta/\eta_0)$. Plot both η_{sp}/c and $(1/c)(\ln \eta/\eta_0)$ versus c and extrapolate linearly to $c = 0$ to obtain $[\eta]$ for the original and for the degraded polymer.

Calculate $\overline{M}_v$ and $\overline{M}_n$ for both the original polymer and the degraded polymer, then obtain a value for Δ. Report these figures together with the polymerization temperature for the sample studied. Discuss the relationship between Δ and the rate constants k_α and k_β.

If a group of students have studied a variety of samples polymerized at different temperatures, it will be possible to obtain estimates of the difference in thermal activation energies ΔE^* and the steric factor S [see Eq. (2)]. It may be difficult to obtain a quantitative interpretation unless the details of the polymerization (amount of initiator, polydispersity, etc.) are well known. However, you should comment qualitatively on your results in the light of what is known about the theory of chain polymerization.[2,6]

DISCUSSION

Equations (14) and (15) were derived on the assumption that a 1,2-glycol structure will result from every abnormal monomer addition. What is the result of two successive abnormal additions? Can you derive a modified expression on the assumption that the probability of an abnormal addition is independent of all previous additions? Would such an assumption be reasonable? Are your experimental results (or, indeed, those of Flory and Leutner) precise enough to make such a modification significant in experimental terms?

APPARATUS

Ostwald viscosimeter; two 100- and two 250-ml volumetric flasks; a 10- and a 50-ml pipette; 250-ml glass-stoppered flask; one 100- and two 250-ml beakers; stirring rod; bunsen burner; tripod stand and wire gauze; 0 to 100°C thermometer; length of gum-rubber tubing. If polymer stock solution is to be prepared by student: weighing bottle; funnel; Pyrex wool.

Glass-walled thermostat bath at 25°C or substitute; polyvinyl alcohol (M.W. ~ 60,000,

5 g of solid or 200 ml of 18-g liter^{-1} solution); KIO_4 (1 g); chromic acid cleaning solution (50 ml); Westphal balance, if needed.

REFERENCES

1. P. J. Flory and F. S. Leutner, *J. Polym. Sci.*, **3**, 880 (1948); **5**, 267 (1950).
2. P. J. Flory, "Principles of Polymer Chemistry," Cornell Univ. Press, Ithaca, N.Y. (1953).
3. J. F. Swindells, R. Ullman, and H. Mark, Determination of Viscosity, in A. Weissberger (ed.), "Technique of Organic Chemistry," 3d ed., vol. I, part I, chap. XII, Interscience, New York (1959).
4. A. Einstein, "Investigations on the Theory of the Brownian Movement," chap. III, Dover, New York (1956).
5. J. R. Schaefgen and P. J. Flory, *J. Amer. Chem. Soc.*, **70**, 2709 (1948).
6. R. W. Lenz, "Organic Chemistry of Synthetic High Polymers," pp. 261–271, 305–369, Interscience-Wiley, New York (1967).

GENERAL READING

F. W. Billmeyer, Jr., "Textbook of Polymer Chemistry," Interscience, New York (1957).

EXPERIMENT 35.   HELIX-COIL TRANSITION IN POLYPEPTIDES

Polymer molecules in solution can be found in many different geometric conformations, and there exist a variety of experimental methods (e.g., viscosity, light scattering, optical rotation) for obtaining information about these conformations.[1] In this experiment, the measurement of optical rotation will be used to study a special type of conformational change which occurs in many polypeptides.

The two important kinds of conformation of a polypeptide chain are the helix and the random coil. In the helical form the amide hydrogen of each "amide group"

$$
\begin{array}{c}
\text{H} \qquad \text{R} \\
\diagdown \quad \diagup \\
\text{—N—C—C—} \\
| \qquad \| \\
\text{H} \qquad \text{O}
\end{array}
$$

is internally hydrogen-bonded to the carbonyl oxygen of the third following amide group along the chain. Thus, this form involves a quite rigid, rod-like structure (see Fig. 1). Under different conditions, the polypeptide molecule may be in the form of a statistically random coil (see Exp. 34). The stable form of the polypeptide will

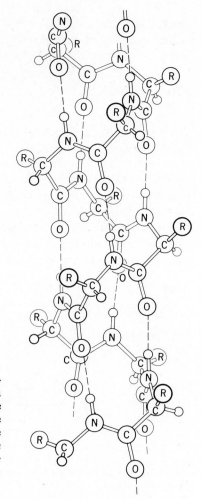

FIGURE 1
The Pauling-Corey alpha helix. In addition to the right-handed helix shown, a left-handed one is also possible (with the same *l*-amino acids). In proteins the right-handed helix is regarded as the more probable. [*Reprinted by permission from L. Pauling, "The Nature of the Chemical Bond," p. 500, Cornell University Press, Ithaca, N.Y. (1960).*]

depend on several factors—the nature of the peptide groups, the solvent, and the temperature. For example, poly-γ-benzyl-l-glutamate† (PBG)

$$\left(NH{-}CH{-}\overset{\displaystyle O}{\overset{\|}{C}} \right)_N$$
$$\underset{\underset{\displaystyle \bigcirc{-}CH_2{-}O{-}C{=}O}{\overset{\displaystyle |}{CH_2}}}{\overset{\displaystyle |}{CH_2}}$$

† Note that the usual chemical description of polypeptides is in terms of amino-acid residues, as in the PBG formula shown, rather than in terms of amide groups, which are more convenient for the present discussion.

has a helical conformation when dissolved in ethylene dichloride at 25°C, but it is in the random-coil form when dissolved in dichloroacetic acid at the same temperature. This difference is quite reasonable since a hydrogen-bonding solvent like dichloroacetic acid can form strong hydrogen bonds with the amide groups and thus disrupt the internal hydrogen bonds which are necessary for the helical form. For a mixed solvent of dichloroacetic acid and ethylene dichloride, PBG can be made to transform from the random coil to the helix by raising the temperature over a fairly narrow range. It is this rapid reversible transition which will be investigated here.

As we shall see below, the solvent plays a crucial role in determining which form is stable at low temperatures. When the helix is stable at low temperatures, the transition to the random coil at high temperatures is called a "normal" transition. For the case where the random coil is the more stable form at low temperatures, the transition is called an "inverted" transition.

THEORY

We wish to present here a very simplified and approximate statistical-mechanical theory of the helix-coil transition. The treatment is closely related to that given by Davidson.[2] Let us consider the change

$$\text{Coil} \cdot N\text{S} = \text{helix} + N\text{S} \tag{1}$$

where the polymer molecules each consist of N segments (monomer units) and are dissolved in a solvent S which can hydrogen-bond to the amide groups when the chain is in the random-coil form. We shall (for convenience) artificially simplify the physical model of internal hydrogen bonding in the helix by assuming that the hydrogen bond formed by each amide group is with the *next* amide group along the chain (see Fig. 2) rather than the third following amide group as in the actual helix. We can then assume that the chain segments are independent of each other. Let z_1 be the molecular partition function of a segment in the *random-coil form* with a solvent molecule S hydrogen-bonded to it, and z_2 be the product of the partition function of a segment in the *helical form* times the partition function of a "free" solvent molecule S. We will then define a parameter s by

$$z_2 = sz_1 \tag{2}$$

Note that s is a ratio of partition functions and many of the contributions to z_1 and z_2 (e.g., vibrational terms) will cancel out. The principal contributions to s will involve differences between the helix and random-coil form, and we may guess that s can be represented in the general form[2]

$$s = s_0 e^{-\varepsilon/kT} \tag{3}$$

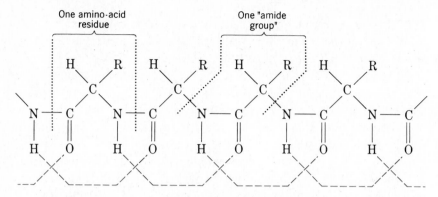

FIGURE 2
Hypothetical model of an internally hydrogen-bonded chain, with the simplification that *adjacent* "amide groups" are connected by hydrogen bonds. Note the distinction between the *amino-acid residue* and the "*amide group*"; the latter is the more convenient unit of structure for the present discussion.

where ε is the energy change per segment and s_0 is related to the entropy change per segment for the change in state (1).

First-order transition model To simplify the model even further, let us assume for the time being that a given polymer molecule is either completely in the helical form or completely in the random-coil form. This assumption will lead to a first-order transition between these two forms. The partition function for a random-coil molecule z_c is then given by

$$z_c = z_1{}^N \tag{4}$$

while that for a helical molecule z_h will be

$$z_h = z_2{}^N = s^N z_1{}^N \tag{5}$$

if one neglects end effects (e.g., at the beginning of the chain the first carbonyl oxygen is not involved in an internal hydrogen bond).

At constant pressure, the two forms will be in equilibrium with each other at some temperature T^* at which $\Delta G = 0$ for the change in state (1). Since ΔV will be quite small for change (1), we can take $\Delta A \cong 0$ and $\tilde{A}_c = \tilde{A}_h$ at T^*, where $\tilde{A}$ is the Helmholtz free energy per mole of polymer. For independent polymer molecules, we have[3]

$$\tilde{A} = -kT \ln z^{N_0} = -RT \ln z \tag{6}$$

where N_0 is Avogadro's number and z is the partition function for a single polymer molecule. Thus, one find that $z_h = z_c$ (or $s = 1$) at T^*. With the use of Eq. (3), the value of T^* can be related to the parameters ε and s_0:

$$\frac{\varepsilon}{kT^*} = \ln s_0 \qquad (7)$$

For an inert (non-hydrogen-bonding) solvent, one observes experimentally a "normal" transition: the helix, which is stable at low temperatures, is transformed at higher temperatures into the random coil. This case is represented by our model when $\varepsilon < 0$ and $s_0 < 1$. Since the solvent is inert, random-coil segments are essentially unbonded, whereas helical segments have internal hydrogen bonds and thus a lower energy ($\varepsilon_h - \varepsilon_c < 0$). In the random-coil polymer molecule, considerable rotation can occur about the single bonds in the chain skeleton, but the helical form has a rather rigid structure. Therefore, there is a decrease in the entropy per segment on changing from the relatively free rotational configurations of the flexible random coil to the more rotationally restricted helix (and thus $s_0 < 1$). From Eqs. (3) to (6), it follows that $s > 1$ at low temperatures ($T < T^*$) and thus $\tilde{A}_h < \tilde{A}_c$ and the helix is the more stable form. At high temperatures ($T > T^*$), $s \cong s_0 < 1$ and the random coil is the more stable form.

For an active (hydrogen-bonding) solvent, one observes experimentally an "inverted" transition: the random coil is stable at low temperatures and changes into the helix when the solution is heated. This case is represented by our model when $\varepsilon > 0$ and $s_0 > 1$. Here the solvent plays a dominant role since a solvent molecule is strongly hydrogen-bonded to each random-coil segment. In terms of energy, there is little difference between polymers with random-coil segments and with helical segments since there is comparable hydrogen bonding in both forms. It is now almost impossible to predict the sign of ε, but it is certainly quite reasonable to find that the energy change associated with (1) may be positive ($\varepsilon_h - \varepsilon_c > 0$). Although a helical segment will still have a lower entropy than a random-coil segment due to the rigidity of the helix, the solvated random coil is now less flexible than previously. More importantly, the free solvent molecules will have a higher entropy than solvent molecules which are bonded to the random coil (cf. "iceberg effect"[4]). Thus the entropy change associated with (1) is positive because of the release of S molecules (and therefore $s_0 > 1$). From Eqs. (3) to (6), we see that at low temperatures $s < 1$ and the random coil will be stable in this inverted case.

Recalling that ΔV is small so that $\Delta G \cong \Delta A$, we can make use of Eqs. (4) to (6) to obtain

$$\Delta \tilde{S} = -\left(\frac{\partial \Delta \tilde{G}}{\partial T} \right)_p \cong R \frac{\partial}{\partial T} \left(T \ln \frac{z_h}{z_c} \right) = NR \ln s_0 \qquad (8)$$

The entropy change *per mole of monomer*, ΔS_m, is then given by

$$\frac{\Delta \tilde{S}}{N} \equiv \Delta S_m = R \ln s_0 \qquad (9)$$

Since $\Delta G = 0$ for the first-order transition at p and T^*, we have for the enthalpy change *per mole of monomer*

$$\Delta H_m = T^* \Delta S_m = RT^* \ln s_0 = N_0 \varepsilon \qquad (10)$$

The first-order theory presented above is related to the treatment of Baur and Nosanow,[5] who used a similar but perhaps physically less realistic model to derive the same results [that is, Eqs. (7), (9), and (10)].†

Cooperative transition model It is an experimental fact that the helix-coil transition is *not* a first-order transition. In order to account for this fact, one must adopt a more realistic model by eliminating the assumption that an entire polymer molecule is all in a given form. Indeed, a polymer molecule can have some sections which are helical and others which are randomly coiled. The formation of a helical region will be a cooperative process. Since segment (i.e., amide group) n is hydrogen-bonded to segment $n + 3$, $n + 1$ to $n + 4$, $n + 2$ to $n + 5$, etc., it is difficult to initiate this ordered internal bonding; but once a single hydrogen-bond link is made, the next ones along the chain are much easier to achieve (a sort of "zipper" effect).

No attempt will be made here to develop the details of such a cooperative model. Readers with a sufficient background in statistical mechanics should refer to Davidson's presentation[2] or to the original papers in the literature.[6] The crucial idea is to introduce a "nucleation" parameter σ which is independent of temperature and very small in magnitude. It is then assumed that the partition function z_2 is changed to the value σz_2 for the *initial* segment of a helical section (i.e., the internally bonded segment directly adjacent to a solvent-bonded segment). Physically, this means that it is difficult to initiate a new helical section. For such a model, it can be shown[2,6] that X_h, the mole fraction of the segments in helical sections, will undergo a very rapid but *continuous* change from $X_h \cong 0$ when $s < 1$ to $X_h \cong 1$ when $s > 1$. Thus, a rapid cooperative transition occurs in the vicinity of a critical temperature T^* (the value of T for which $s = 1$). It is the fact that σ is very small which ensures a sharp transition region (see Fig. 3). If σ is set equal to 1, the result is $X_h = s/(1 + s)$, which corresponds to a very gradual transition.[2] If σ is set equal to 0, the transition becomes first order. For high-molecular-weight PBG, a value of

---

† In addition, Baur and Nosanow showed for the inverted case that the helical form will be stable only between T^* and another much higher transition temperature T', where the helix reverts to the random-coil form. (For PBG, it is estimated[5] that $T' \cong 700°K$.) This is physically reasonable in terms of our model: at very high temperatures, all hydrogen bonds (internal or with solvent) will be broken and the random coil will predominate owing to its higher entropy.

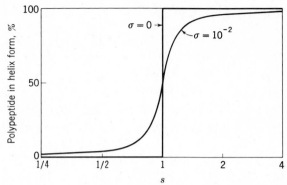

FIGURE 3

Percentage of polypeptide in helix form, as a function of the parameter s, for $\sigma = 0$ and $\sigma = 10^{-2}$.

$\sigma \cong 10^{-4}$ seems to be appropriate to obtain a fairly good quantitative fit to the experimental data.[2] It should be noted that the transition is sharpest (and the model most successful) when polypeptides with $N \sim 1000$ or more are used.

METHOD

Since the helical form of a polypeptide is rigid and rod-like, its solutions are more viscous than corresponding solutions of the random-coil form, and the transition can be detected by measuring the intrinsic viscosity (see Exp. 34). However, the highly ordered structure of the helical form causes it to have an optical-rotatory power markedly different from that of a random coil. Indeed, use of optical rotation for following the transition gives better results than does intrinsic viscosity since the optical rotation is directly related to the fraction X_h of segments in the helical form, whereas the hydrodynamic properties of the polymer molecule change substantially at the first appearance of random-coil regions in the helix (in effect, causing the otherwise straight rod to bend randomly in one or several places).

The polarimeter and its operation are described in Chap. XVIII. Although not stressed there, it is obviously necessary to use a polarimeter in which the temperature of the sample tube can be well controlled. The specific rotation $[\alpha]^t$ as determined with the sodium D doublet at $t°C$ is defined by Eq. (XVIII-9), and we note here that for a solution containing two solutes (helix and random coil) one can show that

$$[\alpha]^t = X_h[\alpha]_h^t + X_c[\alpha]_c^t \qquad (11)$$

where $[\alpha]_h^t$ and $[\alpha]_c^t$ are the specific rotations of helix and random-coil form. In a narrow range of temperature near T^*, $[\alpha]_h^t$ and $[\alpha]_c^t$ will be essentially constant. Thus measurements of $[\alpha]^t$ will determine the mole fraction X_h as a function of T.

EXPERIMENTAL

A solution of poly-γ-benzyl-l-glutamate (PBG) in a mixed solvent of dichloro-acetic acid (DCA) and ethylene dichloride, 76 volume percent in DCA, will be used. The dichloroacetic acid should be purified by vacuum distillation; the ethylene dichloride may be purified by conventional distillation in air. Mix the proper volumes of the two liquids to make about 150 to 200 ml of the desired solvent. In order to prepare approximately 100 ml of a 2.5 weight percent polymer solution, weigh accurately a sample of about 3.8 g of PBG and dissolve it in about 150 g of solvent (also accurately weighed). The resulting solution is very viscous; it must be mixed well and then allowed to stand for two or three days until any small un-dissolved particles have settled and the solution is clear. During this period, the solution should be placed in a polyethylene bottle and stored in a refrigerator. The remaining solvent should also be stored for future density determinations.

Since the solution is difficult to prepare, it is recommended that it be made available to the student. **Warning: The solvent is extremely corrosive.**

Detailed instructions for using the polarimeter and the temperature-regulating bath will be provided in the laboratory. After determining the path lengths of the cells to be used, fill a cell with pure solvent and zero-adjust the instrument. Next, rinse and fill a cell with the polymer solution.†

After mounting the solution cell in the polarimeter, it should not be necessary to move this cell during the rest of the experiment. Provisions must be made for measuring the temperature of the sample without obscuring the light path. Begin measurements with the temperature at about 10°C and read the optical rotation α at reasonably spaced temperature intervals until about 50°C is reached. The exact temperature intervals to be used are at your discretion, but should be close enough to follow the essential features of the variation of optical rotation; small intervals (perhaps 1 or 2°) should be used when the optical rotation is changing rapidly with temperature (near 30°), and larger intervals may be used when the changes with temperature are small. Be sure to allow sufficient time for thermal equilibration between readings.

In order to calculate $[\alpha]^t$, the density of the solutions must be known at each temperature. The density of the solution is close enough to that of the solvent so that they can be assumed to be identical. The density of the solvent should be determined with a pycnometer for at least three different temperatures in the range 10 to 50°C. The volume of the pycnometer should first be determined by weighing it empty and

---

† A difficulty in filling these cells is the problem of avoiding any air bubbles which might obscure part of the light path. Since the solution is viscous, this can be a tedious job and may require some time. If possible, it is wise to fill the cells about an hour before use and allow them to stand.

also filled with distilled water at a known temperature. Since the density of water as a function of temperature is well known, the volume of the pycnometer can be determined. This procedure (see Exp. 11) is then repeated using the solvent, and since the volume of the pycnometer is now known the density can be calculated. The solvent density should be plotted against temperature and the best straight line drawn through the points obtained.

Record the composition of the solution and the molecular weight (or N value) of the polymer used.

CALCULATIONS

For each data point, calculate the *specific rotation* at temperature t from

$$[\alpha]^t = 100 \frac{\alpha}{Lp\rho} \tag{12}$$

where α is the observed rotation in degrees, L is the path length of the cell in decimeters, p is the weight percent of solute in the solution, and ρ is the density of the solution. These density values are interpolated from the ρ-versus-T plot for the solvent.

Construct a plot of $[\alpha]^t$ versus temperature, and determine the limiting (asymptotic) values of the rotation: $[\alpha]_h$ from the high-temperature plateau, $[\alpha]_c$ from the low-temperature plateau. If it is assumed that these represent *temperature-independent* specific rotations for the helix and random-coil forms, respectively, Eq. (11) gives for X_h:

$$X_h = \frac{[\alpha]^t - [\alpha]_c}{[\alpha]_h - [\alpha]_c} \tag{13}$$

Using this equation, you can add to your plot a scale of X_h values so that the plot will also represent X_h versus T. Determine and report the "transition" temperature T^* at which $X_h = 0.5$.

DISCUSSION

If one *arbitrarily* defined an equilibrium constant by $K = X_h/X_c$ and used the data in this experiment together with the Van't Hoff equation $d \ln K/d(1/T) = \Delta \tilde{H}^\circ/R$ to determine the value of $\Delta \tilde{H}^\circ$, to what (if anything) would such an enthalpy change refer?

APPARATUS

Polarimeter, with one or more optical cells (complete with windows, caps, and Teflon or neoprene washers); constant-temperature control for range 10 to 50°C; pycnometer; 25-ml pipette for filling the pycnometer and a small pipette for filling the cell; rubber bulb for pipetting; 0 to 50°C thermometer; 250-ml beaker; gum-rubber tubing; lint-free tissues for wiping cell windows; 100-ml polyethylene bottles for PBG solution and excess solvent.

Poly-γ-benzyl-*l*-glutámate, M.W. 300,000 (4 g of solid or 100 ml of 2.5 weight percent solution); ethylene dichloride–dichloroacetic acid solvent (76 volume percent DCA) stored in polyethylene bottle; wash acetone. The polypeptide can be obtained from Sigma Chemical Co., P.O. Box 14508, St. Louis, Mo. 63178 or K & K Laboratories, Inc., 121 Express St., Plainview, N.Y. 11803. Check the current "Chem. Sources" for other suppliers.

REFERENCES

1. M. A. Stahmann (ed.), "Polyamino Acids, Polypeptides and Proteins," Wisconsin Univ. Press, Madison, Wis. (1962).
2. N. Davidson, "Statistical Mechanics," pp. 378–393, McGraw-Hill, New York (1962).
3. W. J. Moore, "Physical Chemistry," 4th ed., chap. 20, Prentice-Hall, Englewood Cliffs, N.J. (1972).
4. H. S. Frank and W-Y. Wen, *Discuss. Faraday Soc.*, **24**, 133 (1957); G. Nemethy and H. A. Scheraga, *J. Chem. Phys.*, **36**, 3382, 3401 (1962).
5. M. E. Baur and L. H. Nosanow, *J. Chem. Phys.*, **38**, 578 (1963).
6. B. H. Zimm and J. K. Bragg, *J. Chem. Phys.*, **31**, 526 (1959); B. Zimm, P. Doty, and K. Iso, *Proc. Natl. Acad. Sci.*, **45**, 160A (1959).

GENERAL READING

P. Doty, J. H. Bradbury, and A. M. Holtzer, *J. Amer. Chem. Soc.*, **78**, 947 (1956); P. Doty and J. T. Yang, *J. Amer. Chem. Soc.*, **78**, 498 (1956).

C. H. Bamford, A. Elliott, and W. E. Handy, "Synthetic Polypeptides," Academic, New York (1956).

XII

ELECTRIC AND MAGNETIC PROPERTIES

EXPERIMENTS

36. Dipole moment of polar molecules in solution
37. Magnetic susceptibility
38. NMR determination of paramagnetic susceptibility

EXPERIMENT 36. DIPOLE MOMENT OF POLAR MOLECULES IN SOLUTION

When a substance is placed in an electric field, such as exists between the plates of a charged condenser, it becomes to some extent electrically polarized. The polarization results at least in part from a displacement of electron clouds relative to atomic nuclei; polarization resulting from this cause is termed electronic polarization. For molecular substances atomic polarization may also be present owing to a distortion of the molecular skeleton. Taken together these two kinds of polarization are called distortion polarization. Finally, when molecules possessing permanent dipoles are present in a liquid or gas, application of an electric field produces a small preferential orientation of the dipoles in the field direction, leading to orientation polarization.

The permanent dipole moment μ of a polar solute molecule in a nonpolar solvent can be determined experimentally from measurements of the *dielectric constant* ε, given by

$$\varepsilon = \frac{C}{C_0} \tag{1}$$

where C is the capacitance of a condenser when the dielectric medium is the solution or gas in question and C_0 is the capacitance of the same condenser when the medium is a vacuum.[1]

Either of two systems can be studied in this experiment: (1) *o*- and *m*-dichlorobenzene, or (2) succinonitrile and propionitrile. In the first case, the results can be compared with the vector sum of —C—Cl bond moments obtained from the known dipole moment of monochlorobenzene. In the second case, one obtains direct information about internal rotation in a simple 1,2-disubstituted ethane

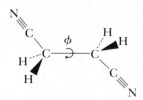

The angle ϕ is defined so that $\phi = 180°$ when the two $C \equiv N$ groups eclipse each other (on viewing the molecule along the C—C axis). Thus, $\phi = 0$ corresponds to the most stable trans configuration, and $\phi = \pm120°$ corresponds to two equivalent gauche configurations. Hindered rotation about the C—C sigma bond can be taken into account by defining temperature-dependent mole fractions in the trans, gauche plus, and gauche minus states. The dipole moment of each of these forms can be predicted from the —C—C $\equiv$ N bond moment measured in propionitrile.

THEORY

An electric dipole[2] consists of two point charges, $-q$ and $+q$, with a separation represented by a vector $\mathbf{r}$, the positive sense of which is from $-q$ to $+q$. The electric dipole moment is defined by

$$\mathbf{m} = q\mathbf{r} \qquad (2)$$

A molecule possesses a dipole moment whenever the center of gravity of negative charge does not coincide with the center of positive charge. In an electric field, all molecules have an induced dipole moment (which is aligned approximately parallel to the field direction) owing to distortion polarization. In addition, *polar molecules* have a *permanent dipole moment* (i.e., a dipole moment which exists independently of any applied field) of constant magnitude μ and a direction which is fixed relative to the molecular skeleton.

The resultant (vector sum) electric moment of the medium, per unit volume, is known as the *polarization* $\mathbf{P}$.[3] For an isotropic medium $\mathbf{P}$ is parallel to the electric field intensity $\mathbf{E}$, and to a first approximation it is proportional to $\mathbf{E}$ in magnitude. For a pure substance $\mathbf{P}$ is given by

$$\mathbf{P} = \overline{\mathbf{m}} \frac{N_0}{\tilde{V}} = \overline{\mathbf{m}} \frac{N_0\rho}{M} \qquad (3)$$

where $\overline{\mathbf{m}}$ is the average dipole moment of each molecule, N_0 is Avogadro's number, $\tilde{V}$ is the molar volume, M is the molecular weight, and ρ is the density.

The moment of a polarized dielectric is equivalent to a moment that would result from electric charges of opposite sign on opposite surfaces of the dielectric. In a condenser these "polarization charges" induce equal and opposite charges in the metal plates that are in contact with them. These induced charges are in addition to the charges that would be present at the same applied potential for the condenser with a vacuum between the plates. Accordingly, the capacitance is increased by the presence of a polarizable medium. Thus the dielectric constant ε [see Eq. (1)] is greater than unity. By electrostatic theory it can be shown[3] that

$$\varepsilon \mathbf{E} = \mathbf{E} + 4\pi \mathbf{P} \tag{4}$$

The average dipole moment $\overline{\mathbf{m}}$ for an atom or molecule in the medium is given by

$$\overline{\mathbf{m}} = \alpha \mathbf{F} \tag{5}$$

where $\mathbf{F}$ is the *local* electric field intensity and α is the polarizability (which is independent of field intensity if the field is not so intense that saturation is incipient). To a good approximation $\mathbf{F}$ may be taken as the electric field intensity at the center of a spherical cavity in the dielectric, within which the atom or molecule is contained. Thus $\mathbf{F}$ is the resultant of $\mathbf{E}$ and an additional contribution due to the polarization charges on the surface of the spherical cavity; electrostatic theory[4] gives

$$\mathbf{F} = \mathbf{E} + \frac{4\pi}{3} \mathbf{P} \tag{6}$$

Combining this with Eq. (4) we obtain

$$\mathbf{F} = \frac{\varepsilon + 2}{\varepsilon - 1} \frac{4\pi}{3} \mathbf{P} \tag{7}$$

and using Eqs. (3) and (5) we obtain for a pure substance

$$\frac{\varepsilon - 1}{\varepsilon + 2} \frac{M}{\rho} = \frac{4\pi}{3} N_0 \alpha \equiv P_M \tag{8}$$

This is the *Clausius-Mosotti* equation. The quantity P_M is called the *molar polarization* and has the dimensions of volume per mole.

If the molecules have no permanent dipole moment, only distortion polarization takes place. The corresponding polarizability is denoted by α_0. If each molecule has a permanent dipole moment of magnitude μ, there is a tendency for the moment to become oriented parallel to the field direction, but this tendency is almost completely counteracted by thermal motion which tends to make the orientation random. The component of the moment in the field direction is $\mu \cos \theta$, where θ is the angle between the dipole orientation and the field direction. The potential energy V of the dipole in a local field of intensity $\mathbf{F}$ is $-(\mu \cos \theta)F$, which is small in comparison with kT under ordinary experimental conditions. By use of the Boltzmann distribution, the average component of the permanent moment in the field direction is found to be

$$\overline{m}_\mu = [\mu \cos \theta e^{-V/kT}]_{av} = [\mu \cos \theta e^{\mu F \cos \theta/kT}]_{av}$$

$$\cong \mu \left[\cos \theta \left(1 + \frac{\mu F \cos \theta}{kT} \right) \right]_{av}$$

where the average is taken over all orientations in space. The average of $\cos \theta$ vanishes, but the average of its square is $\frac{1}{3}$; accordingly, as found by Debye,

$$\overline{m}_\mu = \frac{\mu^2}{3kT} F \tag{9}$$

Thus the total polarizability is given by

$$\alpha = \alpha_0 + \frac{\mu^2}{3kT} \tag{10}$$

and the molar polarization can be written

$$P_M = \frac{\varepsilon - 1}{\varepsilon + 2} \frac{M}{\rho} = \frac{4\pi}{3} N_0 \left(\alpha_0 + \frac{\mu^2}{3kT} \right)$$

$$= P_d + P_\mu \tag{11}$$

where P_d and P_μ are, respectively, the distortion and orientation contributions to the molar polarization:

$$P_d = \frac{4\pi}{3} N_0 \alpha_0 \quad \text{and} \quad P_\mu = \frac{4\pi}{3} N_0 \frac{\mu^2}{3kT} \tag{12}$$

Immediately it is clear that a plot of P_M versus $1/T$ utilizing measurements of ε as a function of temperature will yield both α_0 and μ. This technique is readily applicable to gases. In principle it is applicable to liquids and solutions also but is seldom convenient owing largely to the small temperature range accessible between the melting point and boiling point.

In the foregoing derivation a static (dc) electric field was assumed. The equations apply also to alternating (ac) fields, provided the frequency is low enough to enable the molecules possessing permanent dipoles to orient themselves in response to the changing electric field. Above some frequency in the upper RF or far-infrared range the permanent dipoles can no longer follow the field, and the orientation term in Eq. (11) disappears. At infrared and visible frequencies the dielectric constant cannot be measured by the use of a condenser. However, it is known from electromagnetic theory that in the absence of high magnetic polarizability (which does not exist at these frequencies for any ordinary materials)

$$\varepsilon = n^2$$

where n is the *index of refraction*. We then obtain from Eq. (11) the relation of Lorentz and Lorenz:

$$P_d = R_M = \frac{n^2 - 1}{n^2 + 2} \frac{M}{\rho} = \frac{4\pi}{3} N_0 \alpha_0 \tag{13}$$

where R_M is known as the *mole refraction*. Thus the distortion polarizability α_0 can in principle be obtained from a measurement of the refractive index at some wavelength in the far infrared where the distortion polarization is virtually complete. However, it is not experimentally convenient to measure the index of refraction in the infrared range. The index of refraction n_D measured with the visible sodium D line can usually be used instead: Although the atomic contribution to the distortion polarization is absent in the visible, and the electronic contribution is not in all cases at its dc value, these variations in the distortion polarization are small and usually negligible in comparison with the orientation polarization. The measurement of n_D can be made conveniently in an Abbe or other type of refractometer (see Chap. XVIII). Thus, P_d can be measured independently from P_M, and P_μ can be determined by difference. By this means μ can be determined from measurements made at a single temperature.

Measurements in solution   We are here concerned with a dilute solution containing a polar solute 2 in a nonpolar solvent 1. The molar polarization can be written

$$P_M = X_1 P_{1M} + X_2 P_{2M} = \frac{\varepsilon - 1}{\varepsilon + 2} \frac{(M_1 X_1 + M_2 X_2)}{\rho} \tag{14}$$

where the X's are mole fractions, M's are molecular weights, and ε and ρ (without subscripts) pertain to the solution. Since a nonpolar solvent has only distortion polarization, which is not greatly affected by interactions between molecules, we can take P_{1M} to have the same value in solution as in the pure solvent:

$$P_{1M} = \frac{\varepsilon_1 - 1}{\varepsilon_1 + 2} \frac{M_1}{\rho_1} \tag{15}$$

We can then get P_{2M} from

$$P_{2M} = \frac{1}{X_2} (P_M - P_{1M} X_1) \tag{16}$$

obtained by rearrangement of Eq. (14).

Values of P_{2M} calculated using Eq. (16) are found to vary with X_2, generally increasing as X_2 decreases. This effect arises from strong solute-solute interactions due to the permanent dipoles. This difficulty can be eliminated by extrapolating P_{2M} to infinite dilution ($X_2 = 0$) to obtain $P_{2M}{}^0$. Although this could be done by plotting P_{2M} for each of a series of solutions against X_2, it is easier and more accurate to follow the procedure of Hedestrand,[5] which is given below.

Let us assume a linear dependence of ε and ρ on the mole fraction X_2:

$$\varepsilon = \varepsilon_1 + aX_2 \tag{17}$$

$$\rho = \rho_1 + bX_2 \tag{18}$$

On writing out Eq. (16) explicitly and using Eqs. (15), (17), and (18), it is possible to rearrange terms and obtain the *limiting* expression

$$P_{2M}{}^0 = \frac{3M_1 a}{(\varepsilon_1 + 2)^2 \rho_1} + \frac{\varepsilon_1 - 1}{(\varepsilon_1 + 2)\rho_1}\left(M_2 - \frac{M_1 b}{\rho_1}\right) \tag{19}$$

Thus measurements on solutions of the *slope* of ε versus X_2 and the *slope* of ρ versus X_2 enable us to calculate the limiting molar polarization of the solute in solution. If we assume that the molar distortion polarization in an infinitely dilute solution is equal to that in the pure solute, then

$$P_{2d}{}^0 = R_{2M} = \frac{n_2{}^2 - 1}{n_2{}^2 + 2}\frac{M_2}{\rho_2} \tag{20}$$

where n_2 is the index of refraction and ρ_2 is the density measured for the solute in the pure state. From these two expressions, we can obtain the molar orientation polarization of the solute at infinite dilution:

$$P_{2\mu}{}^0 = P_{2M}{}^0 - P_{2d}{}^0 = \frac{4\pi}{3}N_0\frac{\mu^2}{3kT} \tag{21}$$

On substituting the numerical values of the physical constants, we obtain

$$\mu = 0.0128(P_{2\mu}{}^0 T)^{1/2} \times 10^{-18} \text{ esu-cm} \tag{22}$$

where $P_{2\mu}{}^0$ is given in cubic centimeters per mole and T is given in degrees Kelvin. The units 10^{-18} esu-cm in which dipole moments are conventionally given are called *debyes*.

An alternative expression for $P_{2\mu}{}^0$ has been given by Smith,[6] who improved a method of calculation first suggested by Guggenheim.[7] This expression presupposes knowledge of the index of refraction n of the *solution*, and assumes that

$$n^2 = n_1{}^2 + cX_2 \tag{23}$$

where c, like a and b, is a constant determined by experiment. It follows[6] that

$$P_{2\mu}{}^0 = \frac{3M_1}{\rho_1}\left[\frac{a}{(\varepsilon_1 + 2)^2} - \frac{c}{(n_1{}^2 + 2)^2}\right] \tag{24}$$

This expression is an approximate one which holds when $\varepsilon_1 - n_1{}^2$ is small (as in the case of benzene, where it is 0.03) and M_2 and b are not too large. It is a useful form when the index of refraction n_2 of the pure solute is inconvenient to determine; this is often the case when the pure solute is solid. Note also that Eq. (24) does not require a knowledge of the densities of the solutions.

Solvent effects   The values of $P_{2M}{}^0$ obtained from dilute solution measurements differ somewhat from P_M values obtained from the pure solute in the form of a gas. This effect is due to solvent-solute interactions in which polar solute molecules induce a local polarization in the nonpolar solvent. As a result μ as determined in solution is often smaller than μ for the same substance in the form of a gas, although it may be larger in other cases. Usually the two values agree within about 10 percent. A discussion of solvent effects is given by LeFèvre.[4]

METHOD

Heterodyne-beat method   There are several methods[8] of measuring the dielectric cell capacitance C which is needed for determining ε. Resonance methods and bridge techniques for measuring reactance are especially suitable for work on liquids or solutions having high electrical conductance. For solutions with low conductance, very accurate work can be done using the heterodyne-beat method, a block diagram of which is shown in Fig. 1.

A crystal-controlled oscillator generates a constant-frequency signal f_0, which is usually about 1 MHz. The frequency f of the variable oscillator depends on the values of L and C in the "tank" circuit:

$$f = \frac{1}{2\pi\sqrt{LC}} \tag{25}$$

The inductance L is fixed, while C is the sum of several capacities: C_X (the cell capacitance), C_P (the capacitance of a variable precision air capacitor), C_T (the capacitance of a coarse tuning capacitor), and C_S (stray capacitances from leads, etc.). The mixer combines these two RF signals to give an audio beat frequency $f - f_0$. Several circuits are available for the necessary electronic components;[9] a commercial beat-frequency oscillator, made by General Radio Company, can be

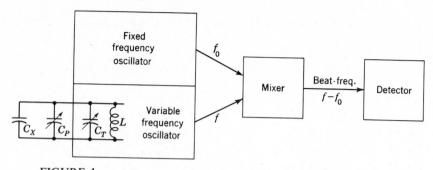

FIGURE 1

A block diagram of the heterodyne-beat apparatus for measuring the capacitance of a dielectric cell (C_X). C_P is a variable precision air capacitor, and C_T is a coarse tuning capacitor.

easily adapted by providing a means of connecting the external cell and precision air capacitors.

The detector may be a pair of earphones or "magic tuning eye" to indicate a zero beat frequency ($f = f_0$). Alternatively an oscilloscope can be used with the beat frequency $f - f_0$ applied to the vertical plates of the cathode-ray tube and a fixed audio frequency f_a applied to the horizontal plates. This fixed frequency f_a may be 60-cycle line frequency or the output of an audio oscillator (in which case the coarse tuning capacitor T may be eliminated). Either f_a or $f - f_0$ is adjusted initially to give a simple Lissajous figure as an indication of balance. In any case, it is desired to maintain a balance which indicates a constant beat frequency and thus a constant f; from Eq. (25) we see that this implies constant C. Changes in the cell capacitance C_X can be determined by reading the changes in C_P that are needed to restore balance.

If f_0 is not subject to appreciable drift, very high precision is possible, since changes in beat frequency of much less than 1 cycle are easily detected. From Eq. (25) we find that $|\Delta C/C| = 2|\Delta f/f|$ and with f about 10^6 Hz one can detect a change in capacitance of a few parts per million.

To avoid the problem of determining the stray capacitance from leads and also to reduce the influence of any drift in f_0, it is desirable to use a cell with a variable capacitance rather than one with fixed plates.† In this case, one measures the difference in capacitance between the minimum a and maximum b settings of the cell

---

† For high-precision work with a very stable oscillator, a fixed-plate capacitor cell is best, since it eliminates any error due to lack of reproducibility in setting a variable capacitor. However, in this case one must determine the lead capacitance from measurements on a material of known dielectric constant.

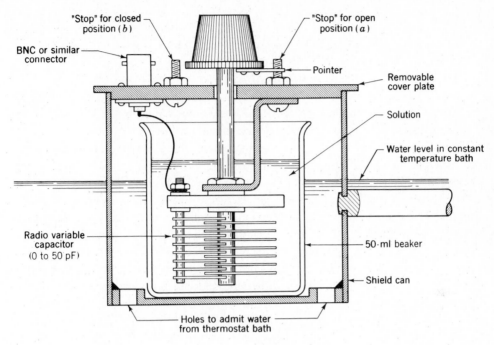

FIGURE 2
Variable-capacitance dielectric cell for solutions.

capacitor. Such a difference is not affected by the capacitance of the electrical leads and is affected only by that drift in f_0 which may occur in the short time interval between readings. The dielectric constant of a liquid or solution is now given by

$$\varepsilon = \frac{\Delta C(\text{liq})}{\Delta C(\text{air})} \tag{26}$$

where ΔC is the difference $(C_b - C_a)$ for the cell. It is possible to use air rather than vacuum readings, since the dielectric constant of air is very close to unity.

A suitable cell can be made from a small radio condenser of about 40 to 50 pF as shown in Fig. 2. The condenser, with a long shaft, is mounted on a circular insulating plate. Means should be provided for reproducible positioning in the minimum a and maximum b positions. This assembly fits into the top of a grounded brass shield, which holds a lipless beaker.

A very useful discussion of dipole moment measurements has been given by Thomson.[10]

EXPERIMENTAL

Detailed instructions for the operation of the equipment to be used should be provided by the instructor. In general, oscillators should be allowed to warm up for at least 1 hr to achieve stable, drift-free operation.

With the cell capacitor and cell beaker clean and dry, assemble the cell and mount it in the constant-temperature bath. Set the capacitor at the minimum a position and set the precision air capacitor at a high value (approximately 250 pF). If a zero-beat detector is used, adjust the coarse tuning capacitor to obtain a zero beat. When an oscilloscope and audio oscillator are used to detect the balance, adjust either the audio frequency f_a or a coarse tuning capacitor to obtain a simple, stable Lissajous figure (preferably a circle). Now record the reading of the precision air capacitor. Change the cell capacitor setting to the maximum b and restore balance (zero beat, or the same Lissajous figure) by varying the precision air capacitor. Record the new reading. Since the beat frequency is maintained constant, the sum of the cell capacitance and precision capacitance must be constant. Therefore, $\Delta C(\text{air})$ is equal to the difference between the two readings of the precision capacitor.

Rinse the cell beaker and variable capacitor with the pure solvent or the solution to be measured, fill the cell beaker to a level which will completely immerse the capacitor, and reassemble the cell. Determine $\Delta C(\text{liq})$ in the same manner as above. Handle the capacitor with care; damage to the plates can change ΔC.

Dichlorobenzene Measurements should be made on pure benzene and on dilute solutions of *o*- and *m*-dichlorobenzene. Make up 50 or 100 ml of each solution as follows. Weigh a dry, clean volumetric flask; add an appropriate amount of solute; and weigh again. Now carefully make up to the mark with benzene and reweigh. Suggested concentrations are 1, 2, 3, and 4 mole percent of solute. The densities can be calculated from these weighings. If desired, the refractive index of each solution can also be measured so that Guggenheim's method of calculation can be used.

Succinonitrile Measurements should be made on pure benzene and on dilute solutions of succinonitrile and propionitrile (or acetonitrile). Make up 50 or 100 ml of each solution in the same manner as described above. Densities can be calculated and/or refractive indexes can be measured.

CALCULATIONS

For each solution studied, calculate ε [Eq. (26)], ρ, and X_2. Plot ε and ρ versus the mole fraction of solute X_2, and draw the best straight lines through your points; see Eqs. (17) and (18). Obtain the slopes a and b; the intercepts should agree with the pure solvent results. Using Eq. (19), calculate $P_{2M}{}^0$, the molar polarization at infinite dilution. Estimate $P_{2d}{}^0$ from Eq. (20) using the literature value of the refractive index of the solute n_2, and obtain $P_{2\mu}{}^0$.

Alternatively, Guggenheim's method can be used if the indexes of refraction were measured for each solution. Plot ε and n^2 versus X_2 and draw the best straight lines through your points; see Eqs. (17) and (23). Using Eq. (24) and the appropriate slopes a and c, calculate $P_{2\mu}{}^0$. It would be instructive to follow both of these procedures and see what difference (if any) the method of extrapolation to infinite dilution has on the $P_{2\mu}{}^0$ values.

Finally, calculate the dipole moment, in debyes, from Eq. (22).

DISCUSSION

Dichlorobenzene   Compare your experimental results with the values computed from a vector addition of carbon-chlorine bond moments obtained from the dipole moment of monochlorobenzene (1.55 debyes). If they do not agree, suggest possible physical reasons for the disagreement.

Succinonitrile   An excellent discussion of this system has been presented by Braun, Stockmayer, and Orwoll,[11] who were the first to propose studying the dipole moment of succinonitrile. They show that the average square of the dipole moment is given by

$$\langle \mu^2 \rangle = \tfrac{8}{3}(1 - X_t)\mu_1{}^2 \tag{27}$$

where X_t is the mole fraction in the trans form and μ_1 is the $-C-C\equiv N$ bond moment as determined from the dipole moment of either propionitrile or acetonitrile. Thus a measurement of μ^2 for succinonitrile will determine the distribution of molecules among the three "rotational isomeric" states at the given temperature. [Note that $X_+ = X_- = \tfrac{1}{2}(1 - X_t)$ since the gauche plus and gauche minus states are equivalent.] It can also be shown from the appropriate Boltzmann populations that

$$X_t = \left(1 + 2e^{-\Delta E/RT}\right)^{-1} \tag{28}$$

where $\Delta E = E(\text{gauche}) - E(\text{trans})$. Thus, the value of X_t obtained from Eq. (27) will determine the value of ΔE and one can predict the temperature dependence of μ. Calculate a value of X_t and ΔE from your results and predict the value of μ at 280 and 350°K.

APPARATUS

Heterodyne beat-frequency oscillator (such as General Radio 1304B) comprising fixed and variable high-frequency oscillators and a mixer; beat detector (such as "magic eye" as null detector or source of stable audio frequency and oscilloscope for displaying a Lissajous figure); coarse tuning capacitor (unnecessary if a stable variable-frequency audio oscillator is employed); dielectric cell and cell holder; low-capacitance double-shielded lead for connection to cell; precision air capacitor (0 to 300 pF, such as General Radio 722D); five 50- or 100-ml volumetric flasks; a 5-ml Mohr pipette; acetone wash bottle; rubber pipette bulb.

Benzene (analytical reagent grade, 0.5 to 1 liter); o- and m-dichlorobenzene (10 to 20 ml each), or succinonitrile and propionitrile (10 to 20 ml each); acetone for rinsing.

REFERENCES

1. F. Bitter, "Currents, Fields and Particles," pp. 36ff., Technology Press, Cambridge, Mass. (1956).
2. *Ibid.*, p. 57.
3. *Ibid.*, pp. 110ff.
4. R. J. W. LeFèvre, "Dipole Moments," 3d ed., pp. 7–10 and chap. III, Methuen, London (1953).
5. G. Hedestrand, *Z. Phys. Chem.*, **B2**, 428 (1929).
6. J. W. Smith, *Trans. Faraday Soc.*, **46**, 394 (1950).
7. E. A. Guggenheim, *Trans. Faraday Soc.*, **45**, 714 (1949).
8. C. P. Smyth, Determination of Dipole Moments, in A. Weissberger (ed.), "Technique of Organic Chemistry," 3d ed., vol. I, part III, chap. XXXIX, Interscience, New York (1959).
9. J.-Y. Chien, *J. Chem. Educ.*, **24**, 494 (1947).
10. H. B. Thomson, *J. Chem. Educ.*, **43**, 66 (1966).
11. C. L. Braun, W. H. Stockmayer, and R. A. Orwoll, *J. Chem. Educ.*, **47**, 287 (1970).

GENERAL READING

P. Debye, "Polar Molecules," Reinhold, New York (1929), reprinted by Dover, New York (1945).

R. J. W. LeFèvre, *op. cit.*

H. B. Thomson, *loc. cit.*

EXPERIMENT 37.   MAGNETIC SUSCEPTIBILITY

When an object is placed in a magnetic field, in general a magnetic moment is induced in it. This phenomenon is analogous to the induction of an electric moment in an object by an electric field (see Exp. 36) but differs from it in that an induced magnetic moment may have either direction in relation to the applied field. If the induced moment is parallel to the external field (as in the electric case), the material is called *paramagnetic* or *ferromagnetic*, depending on whether the field due to the induced moment is small or large in comparison with the external field. If the moment is antiparallel to the external field, the material is called *diamagnetic*; the moment in this case is always small. This experiment will deal only with paramagnetic and diamagnetic substances in solution.

THEORY

If **I** is the magnetization (magnetic moment per unit volume, analogous to the electric polarization **P**) induced by the field **H**, the volume magnetic susceptibility χ is defined by the equation

$$\mathbf{I} = \chi\mathbf{H} \tag{1}$$

For a paramagnetic substance χ is positive, and for a diamagnetic substance it is negative; it is a dimensionless number, ordinarily very small in comparison with unity (except in the case of ferromagnetism) and essentially independent of **H** for fields readily available in the laboratory.

Magnetic susceptibilities are usually given in the literature on a weight or molar basis. Thus, while the volume susceptibility χ is induced moment per unit volume per unit applied field and is dimensionless, the weight susceptibility

$$\chi_g = \frac{\chi}{\rho} \tag{2}$$

(where ρ is the density) is induced moment per unit weight per unit applied field and typically has units of cubic centimeters per gram. The molal susceptibility

$$\chi_M = M\chi_g = \frac{M}{\rho}\chi \tag{3}$$

(where M is the molecular weight) is induced moment per mole per unit applied field and typically has units of cubic centimeters per mole.

Diamagnetism Nearly all known substances are diamagnetic. Diamagnetism results from the precession of the electronic orbits in atoms which occur when a magnetic field is present. Volume diamagnetic susceptibilities are generally very small in magnitude compared with volume paramagnetic susceptibilities for pure substances. In paramagnetic substances the observed susceptibility is the resultant of a paramagnetic contribution and a very much smaller diamagnetic contribution. In an estimation of the paramagnetism this diamagnetic contribution is often neglected, but in the case of aqueous solutions a correction should be made for the diamagnetic susceptibility of the water owing to the relatively large amount of it present.

Paramagnetism   The most important source of paramagnetism is the magnetic moment which is associated with the spin of the electron. The electron has two spin states, having spin magnetic quantum numbers $-\frac{1}{2}$ and $+\frac{1}{2}$, with the principal component of magnetic moment respectively parallel and antiparallel to the magnetic field direction. Spin paramagnetism (or in some cases ferromagnetism) exists in a substance if the atoms, molecules, or ions in it contain unequal numbers of electrons in the two possible spin states. This condition obviously exists when the atom, molecule, or ion contains an odd number of electrons [as in Fe^{3+}, Cu^{2+}, $(C_6H_5)_3C\cdot$ and other free radicals, etc.]. It may also exist when the number of electrons is even, if a degenerate electronic level (such as a d or f atomic subshell) is only partially filled. For example, the free ferrous ion Fe^{2+} has six electrons outside the argon shell. The available orbitals of lowest energy are the five degenerate (i.e., equal energy) $3d$ orbitals. Each of these may contain two electrons with their spins opposed (one with spin $+\frac{1}{2}$, the other with spin $-\frac{1}{2}$, in accord with the Pauli exclusion principle) or a single electron with either spin. No spin paramagnetism would occur if the six outer electrons of Fe^{2+} occupied three of the five $3d$ orbitals in pairs so that all spin magnetic moments cancelled. However, this would be contrary to a principle known as *Hund's first rule*, which states that, when several electronic orbitals of equal or very nearly equal energy are incompletely filled, the electrons tend to occupy as many as possible of the orbitals singly rather than in pairs, the electrons in singly occupied orbitals all having the same spin. In Fe^{2+} this rule predicts that one $3d$ orbital will contain a pair of electrons with spins opposed and the other four orbitals will each contain a single electron, the four spins being the same. Thus the free ferrous ion is paramagnetic. In molecular oxygen O_2, two molecular orbitals of equal energy each contain a single electron in accordance with Hund's first rule; consequently, oxygen gas is paramagnetic.

An atom, molecule, or ion containing one or more unpaired electrons with the same spin has a permanent magnetic moment μ. In the absence of orbital contributions to the moment (see below), μ is completely determined by the number of unpaired electrons n:

$$\mu(\text{spin only}) = \sqrt{4S(S+1)}\ \beta_B = \sqrt{n(n+2)}\ \beta_B \tag{4}$$

where S is the spin quantum number (equal to the sum of the individual electron spin quantum numbers s_i) and β_B is the "Bohr magneton" given by

$$\beta_B = \frac{eh}{4\pi mc} = 0.927 \times 10^{-20} \text{ erg gauss}^{-1} \tag{5}$$

Thus, for example, the "spin-only" magnetic moment of Fe^{2+} is

$$\mu = \sqrt{4 \times 6}\, \beta_B = 4.90 \text{ Bohr magnetons}$$

Orbital magnetic moments may also contribute to paramagnetism. An electron in an orbital with one or more units of angular momentum behaves like an electric current in a circular loop of wire and produces a magnetic moment. When all orbitals in a subshell (e.g., all five $3d$ orbitals) are equally filled (with one electron each as in Fe^{3+} or two electrons each as in Cu^+), the orbital moments cancel one another and there is no orbital contribution to the observed moment. In other cases (e.g., Fe^{2+}) an orbital contribution may arise, although usually it is "quenched" to a large extent by interactions with neighboring molecules or ions and does not contribute more than a few tenths of a Bohr magneton to the total moment. (Important exceptions are certain rare-earth ions, since quenching occurs to a much smaller extent for $4f$ orbitals than for $3d$ orbitals.) Atomic nuclei often possess spin magnetic moments, but these are so small as to have a negligible effect on magnetic susceptibility. They are important, however, in nuclear magnetic resonance spectroscopy.

In the absence of an applied field, the atomic moments in a paramagnetic substance orient themselves essentially at random owing to thermal motion and there is no net observable moment. In the presence of a magnetic field the atomic moments tend to line up with the field, but the degree of net alignment is slight because of the disorienting effect of thermal motion. It is possible to show[1] that the paramagnetic contribution to the molal susceptibility is $N_0\mu^2/3kT$, where N_0 is Avogadro's number, k is the Boltzmann constant, and T is the absolute temperature. The total molal susceptibility can be written

$$\chi_M = N_0\alpha + \frac{N_0\mu^2}{3kT} \tag{6}$$

where α is the small (negative) diamagnetism per molecule. We can write this in the form

$$\chi_M = N_0\alpha + \frac{C}{T} \tag{7}$$

where C is called the *Curie constant* for the substance concerned. If C is determined by experiment, the magnetic moment of the atom, molecule, or ion is obtained from it with the equation

$$\mu = \left(\frac{3kC}{N_0}\right)^{1/2} \tag{8}$$

Expressing this result in units of Bohr magnetons, we obtain

$$\mu = 2.824\sqrt{C} \tag{9}$$

where C is in cgs units.

Transition-metal complexes The present experiment is largely concerned with complex ions of transition-group metals, such as hexahydrated or ammoniated ferrous or ferric ions and ferro- or ferricyanides. Here each metal ion is surrounded by a number of negative or neutral groups called *ligands*. This number is six in the cases cited and in other common cases may be four or eight.

Two distinct groups of complexes can be distinguished on the basis of experimental paramagnetic susceptibilities. *High-spin complexes* are those for which the effective magnetic moment is very close to the spin-only value for the free (gaseous) transition ion. *Low-spin complexes* have much lower moments than would be predicted for the free metal ion and can even be diamagnetic.

The early theory of transition-metal complexes, due largely to Pauling,[2] involved an explanation based on distinguishing between essentially ionic and essentially covalent bonding between the metal ion and its ligands. Where the number of unpaired electrons in the complex as deduced from the measured susceptibility is the same as that expected for the free metal ion (high-spin case), the bonding with the ligands was considered to be ionic (i.e., due to Coulomb attraction as in $[Fe(III)F_6]^{3-}$ or due to electrostatic polarization of neutral ligands by the central ion as in $[Co(III)(H_2O)_6]^{3+}$). Where the number of unpaired electrons found in the complex is considerably less than the free metal ion value (low-spin case, as in most complexes with cyanides, ammonia, carbon monoxide, etc.), the bonding was considered to be covalent. It is assumed that the electrons are paired owing to the necessity of accommodating, in the atomic orbitals, some additional electrons donated by the ligands for forming electron-pair bonds. Thus, in $[Co(III)(NH_3)_6]^{3+}$ the 6 electrons outside the argon shell of Co^{3+} are augmented by 6 electron pairs from the ligands, giving 18 electrons. Of these, 12 electrons (or 6 pairs) are shared with the ligands to form 6 octahedral covalent bonds, using two $3d$, one $4s$, and three $4p$ orbitals of cobalt. The other 6 electrons are paired in the remaining three $3d$ orbitals. Since all electrons are paired with spins opposed, salts of this complex are diamagnetic.

The "crystal-field" or "ligand-field" theory of transition-metal complexes, first proposed by van Vleck,[3] has been developed extensively in recent years[4,5] and has proven to be of great value in the interpretation of a wide range of properties. This theory explains transition-metal complexes in terms of the splitting of the five-fold degenerate d level into two or more levels of different energy by perturbations due to the ligands. In the simplest version, this splitting is due purely to the electrostatic crystal field of the ligands. However, it is often necessary to consider also the metal-ligand orbital overlap (so-called "adjusted crystal-field theory"[5]). In any case, ligand-field theory is an adequate description as long as the d orbitals of the metal ion are well defined. If there is strong mixing between metal ion and ligand orbitals (as in metal carbonyls), a molecular-orbital theory is required.†

---

† As shown by van Vleck,[3] Pauling's theory is a special case of the more general **MO** theory.

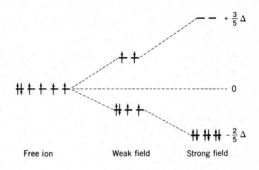

FIGURE 1
Weak field, e.g., $[Co(III)F_6]^{3-}$, and strong field, e.g., $[Co(III)(NH_3)_6]^{3+}$, splitting of the $3d$ level of Co(III) due to ligands with O_h symmetry. The splitting between the low-lying triplet and the upper doublet is defined as Δ. In the weak-field case, there are four unpaired spins; whereas all the spins are paired in the strong-field case.

We shall discuss ligand-field theory in the one-electron approximation, in which a one-electron d state is split by the ligand field and the individual states are then filled by the d electrons of the central ion (taking account of interactions). As an example, consider a d^6 complex with octahedral symmetry; Fig. 1 shows the situation for Co(III) complexes. For weak ligand fields, the splitting Δ is small and the electrons distribute themselves according to Hund's first rule. Thus a complex such as $[Co(III)F_6]^{3-}$ is a high-spin complex with four unpaired electrons. The energy of this configuration can be written as $(-\frac{2}{5}\Delta + P)$, where P is the average energy required to form an electron pair. When the ligand field is strong enough (Δ sufficiently large), the energy difference between the levels can no longer be overcome by the electron-unpairing tendency of Hund's rule. Thus a complex such as $[Co(III)(NH_3)_6]^{3+}$ is a low-spin complex: the six d electrons of the cobalt ion fill the lower triplet state with spins paired and the complex is diamagnetic. The energy of this configuration is $(-\frac{12}{5}\Delta + 3P)$. Analogous arguments can be applied to d^4, d^5, and d^7 complexes.

Thus ligand-field theory allows one to understand both high-spin and low-spin complexes. In particular, low-spin complexes can be explained without assuming a covalent electron-pair bond between the metal ion and the ligand. Indeed, there is no clear-cut distinction made between ionic and weak covalent character. According to the ligand-field theory, a low-spin complex simply means that the ligand field strength (and thus the splitting Δ) is greater than some critical value. It should be stressed that this critical value will be different for different complexes and one cannot equate increasing field strength with increasing "covalency" in any simple way.

METHOD

Most methods for the determination of a magnetic susceptibility depend upon measuring the force resulting from the interaction between a magnetic field gradient and the magnetic moment induced in the sample by the magnetic field. The x

component of this force, per unit volume of the sample, is

$$\mathbf{f}_x = \mathbf{I} \frac{\partial \mathbf{H}}{\partial x} = \chi \mathbf{H} \frac{\partial \mathbf{H}}{\partial x} \tag{10}$$

This force is such as to tend to draw the sample into the strongest part of the field if the sample is paramagnetic (χ positive) or repel it into the weakest part if the sample is diamagnetic (χ negative). The work done by the system when a volume dV of the sample is carried from a point of field strength H_1 to a point of field strength H_2 is (assuming for simplicity that H varies only as a function of x)

$$dw = dV \int f_x \, dx = dV\chi \int H \frac{dH}{dx} \, dx = dV\chi \int_{H_1}^{H_2} H \, dH$$

$$= \tfrac{1}{2}\chi(H_2{}^2 - H_1{}^2) \, dV \tag{11}$$

The Gouy balance[6]   In the Gouy balance (Fig. 2) a long tube, divided into two regions by a septum, is suspended from one side of an analytical balance so as to hang vertically in a magnetic field. The septum is in the strongest part of the field and the two ends of the tube are in regions of essentially zero field strength. The part of the tube above the septum is filled with the sample to be investigated, and the part below is empty.

In addition to the downward gravitational force acting on the tube, a force is exerted by the magnetic field. Let us calculate the work done in lowering the tube, with cross-sectional area A, by an amount Δx. This is equivalent to bringing a volume $A \, \Delta x$ of the specimen from a region of zero field strength to a region of field strength H. Thus, ignoring gravitational work,

$$w = \tfrac{1}{2}\chi H^2 A \, \Delta x = f \, \Delta x$$

Therefore the downward force on the tube due to its interaction with the magnetic field is[†]

$$f = \tfrac{1}{2}\chi H^2 A \tag{12}$$

In making a measurement, the "apparent weight" W is determined in the absence of a field and then with a field present, and the difference is equated to f:

$$(W_{\text{field}} - W_{\text{no field}}) = f = \tfrac{1}{2}\chi H^2 A \tag{13}$$

The weight W in each case is a *force* (in dynes) obtained by multiplying the mass of

---

† In case the top of the tube is not at zero field strength, we should write

$$f = \tfrac{1}{2}\chi(H_{\text{max}}^2 - H_{\text{min}}^2)A$$

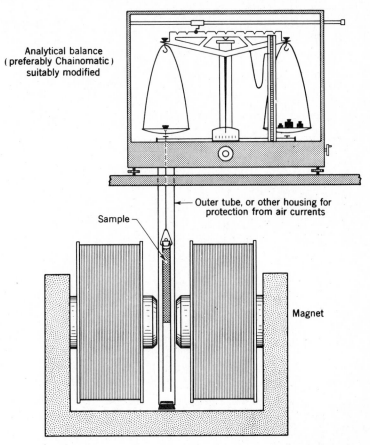

Analytical balance
(preferably Chainomatic)
suitably modified

Outer tube, or other housing for
protection from air currents

Sample

Magnet

FIGURE 2
Gouy balance.

the weights used (in grams) by the acceleration of gravity (in centimeters per second squared).

The magnet should provide a field of at least 4000 gauss and preferably 6000 or more. The field should be reasonably homogeneous over a region considerably larger than the diameter of the tube. For a sample tube up to 15 mm in diameter, a magnet with gap of 1 in. and a pole diameter of at least 3 in. is convenient.

For obtaining the weights in the presence and absence of a field it is most convenient to have an electromagnet, the field of which can easily be turned on and off. Such a magnet, with a regulated power supply, has the disadvantage of being rather expensive. A permanent magnet is usually less expensive, but special arrangements for making the no-field measurements are required. If the magnet is mounted on rails or on a pivot, it can be rolled or swung in and out of its normal

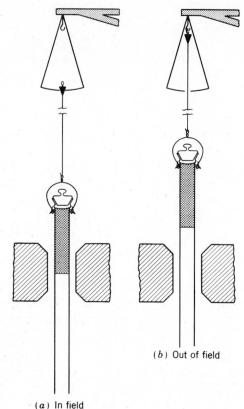

FIGURE 3
Positioning of Gouy tube for (*a*) in-field
and (*b*) out-of-field weighings when a
permanent magnet is used.

(*b*) Out of field

(*a*) In field

position. If the magnet is stationary, the Gouy tube can be hung at two different
levels (Fig. 3). If the latter method is to be used, the septum should be in the strongest part of the field in the lower position and in essentially zero field in the upper
position. The sample tube should be long enough so that in either position the
bottom end is in an essentially zero-field region *below* the strong part of the field.

A two-pan analytical balance (preferably a Chainomatic type) is mounted
on a sturdy table over the magnet as shown in Fig. 2. The Gouy tube is supported
by a nonmagnetic wire or thread which passes through a hole drilled in the base
of the balance. If an electromagnet is used, the left-hand pan may be dispensed
with and the wire attached directly to the stirrup. If, instead, a permanent magnet is used with the scheme shown in Fig. 3, a hole must be drilled in the pan so
that the wire can pass through it to a hook which is attached to the stirrup for the
no-field weighing or allowed to rest on the pan for the in-field weighing. The Gouy

tube should be protected against air currents. Provision should also be made for mounting a thermometer near the Gouy tube.

EXPERIMENTAL

The procedure to be used in operating the Gouy balance will necessarily depend on the details of construction and cannot be given here with any completeness; a set of instructions should be compiled by the instructor and posted near the apparatus. The balance should be operated in the normal manner; the beam must be off the knife-edges when the Gouy tube is being mounted, demounted, or changed in position. It is important to **remove your watch** when working near the magnet and to keep steel and iron tools or instruments out of the way.

In order to determine the "apparatus constant" $H^2A/2$ which appears in Eq. (13), measurements are made on a material of known susceptibility. For this purpose, a common standard is an aqueous solution of nickel chloride, about 30 percent $NiCl_2$ by weight. Prepare 100 ml of such a solution with an accurately known weight fraction of $NiCl_2$ in air-free distilled water. When this solution is weighed on the Gouy balance, record the ambient temperature.

Solutions of the following salts† in air-free distilled water should be studied. In each case, the concentration ($\sim 0.5\ M$) must be precisely known. If possible, prepare the solutions in advance since some of these salts dissolve quite slowly.

$KMn(VII)O_4$, potassium permanganate (*Note:* this is soluble only to about 0.4 M)

$Mn(II)SO_4$, manganous sulfate

$[Fe(II)(H_2O)_6](NH_4)_2(SO_4)_2$, ferrous ammonium sulfate (*Note:* slow air oxidation)

$K_4[Fe(II)(CN)_6]$, potassium ferrocyanide

$K_3[Fe(III)(CN)_6]$, potassium ferricyanide

---

† Magnetic measurements may be made on powdered crystalline salts rather than on aqueous solutions. The volume susceptibilities are very much larger (an attractive feature if a strong magnet is not available), but the molal susceptibilities are not so easily interpretable in terms of atomic moments, owing to interaction effects in the crystalline state. In place of Eq. (7), we write (neglecting the diamagnetic term)

$$\chi_M = \frac{C}{T + \Delta}$$

This is called the *Curie-Weiss law.* The constant Δ may be positive or negative and for most compounds is less than 75° in magnitude. To obtain a value of the Curie constant from which an atomic moment can be calculated, it is necessary to measure χ_M at more than one temperature. When the reciprocal of χ_M is plotted against absolute temperature, the reciprocal of the slope is the Curie constant C. The density ρ required in the calculations is not the crystal density, but an effective powder density determined in the manner described for solutions.

It is important to know the densities of all solutions. A satisfactory procedure is to weigh the Gouy tube empty and then filled with water to a fiducial mark near the top. From the known density of water at this temperature, the volume is calculated. The tube is then filled with the solution to be studied up to the same fiducial mark and weighed. (All these weighings are in the absence of a magnetic field.) Alternatively, the density of each solution can be determined with a Westphal balance or a good hydrometer.

CALCULATIONS

Calibration The *weight* susceptibility of an aqueous nickel chloride solution is given by Selwood[7] as

$$\chi_g = \left[\frac{10,030p}{T} - 0.720(1 - p) \right] \times 10^{-6} \qquad \text{cm}^3 \text{ g}^{-1} \qquad (14)$$

where p is the weight fraction of $NiCl_2$ and T is the absolute temperature. The second term in the brackets is the correction for the diamagnetism of the water used as solvent. This expression assumes that the solution is free of dissolved atmospheric oxygen.

From the χ_g given by this expression and the density of the $NiCl_2$ solution, the volume susceptibility χ can be calculated [Eq. (2)]. With this and the measured weight difference, determine and report the apparatus constant $H^2A/2$ appearing in Eq. (13).

Measurements on unknown solutions The weight susceptibility of a solution is related to the molal susceptibility by

$$\chi_g = \frac{\chi_M(\text{solute})}{M} p - 0.720 \times 10^{-6}(1 - p) \qquad (15)$$

where p is again the weight fraction of solute. If the concentration c of the solute in moles per liter is known, we can write the volume susceptibility directly in the form

$$\chi = \frac{c\chi_M}{1000} - 0.720 \times 10^{-6} \left(\rho - \frac{cM}{1000} \right) \qquad (16)$$

From the experimental weight difference, χ is determined by use of Eq. (13) and the known apparatus constant $H^2A/2$. From Eq. (16) the solute molal susceptibility χ_M is obtained.

If the material is paramagnetic (χ_M positive), the small negative diamagnetic term $N_0\alpha$ can be neglected in Eq. (7) and the constant C can be determined. The atomic moment μ can then be calculated from Eq. (9). With neglect of any orbital contribution, the number of unpaired electrons can be found approximately with Eq. (4). Calculate the number of unpaired electrons for each paramagnetic substance studied.

DISCUSSION

In potassium permanganate, Mn(VII) has no unpaired electrons and thus no permanent magnetic moment. However, the magnetic field induces a small, temperature-independent paramagnetism because the field couples the ground state to paramagnetic excited states.[5] The other four compounds illustrate high- and low-spin cases of octahedral d^5 and d^6 complexes. (Neutral solutions of manganese sulfate are very pale pink and contain $[Mn(II)(H_2O)_6]^{2+}$ ions.) In the case of Fe(III) and the isoelectronic Mn(II), the ^6S ground state of the free ion has no orbital angular momentum and no effective coupling to excited states. Thus the magnetic moment of their high-spin complexes should be very close to the spin-only value. Comment on the spin type and ligand field strength of each complex studied.

APPARATUS

Gouy balance (comprising a magnet, power supply if needed, suitably modified analytical balance); glass-stoppered Gouy tube; several 100-ml volumetric flasks and glass-stoppered 200-ml flasks; 0 to 30°C thermometer; Westphal balance or hydrometer (optional).

$NiCl_2$ (40 g); $MnSO_4$ (10 g); $KMnO_4$ (10 g); $K_4Fe(CN)_6$ (25 g); $K_3Fe(CN)_6$ (20 g); $Fe(NH_4)_2(SO_4)_2 \cdot 6H_2O$ (20 g); or a solution of each salt of an accurately known concentration (100 ml).

REFERENCES

1. W. J. Moore, "Physical Chemistry," 4th ed., pp. 805–810, Prentice-Hall, Englewood Cliffs, N.J. (1972).
2. L. Pauling, "The Nature of the Chemical Bond," 3d ed., pp. 161ff., Cornell Univ. Press, Ithaca, N.Y. (1960).
3. J. H. van Vleck, *J. Chem. Phys.*, **3,** 807 (1935).
4. C. J. Ballhausen, "Introduction to Ligand Field Theory," McGraw-Hill, New York (1962).
5. F. A. Cotton and G. Wilkinson, "Advanced Inorganic Chemistry," 3d ed., pp. 535–542 and chap. 20, Interscience-Wiley, New York (1972).
6. P. W. Selwood, "Magnetochemistry," 2d ed., pp. 3ff., Interscience, New York (1956).
7. *Ibid.*, p. 26.

GENERAL READING

L. E. Orgel, "An Introduction to Transition-Metal Chemistry: Ligand-Field Theory," Methuen, London (1960).

H. L. Schläfer, "Basic Principles of Ligand Field Theory," chap. 2, Interscience-Wiley, New York (1969).

E. C. Stoner, "Magnetism," Methuen, London (1948).

L. F. Bates, "Modern Magnetism," 3d ed., Cambridge, New York (1951).

J. H. van Vleck, "Electric and Magnetic Susceptibilities," Oxford Univ. Press, New York (1944).

P. W. Selwood, Determination of Magnetic Susceptibility, in A. Weissberger (ed.), "Technique of Organic Chemistry," 3d ed., vol. I, part IV, chap. XLIII, Interscience, New York (1959).

EXPERIMENT 38. NMR DETERMINATION OF PARAMAGNETIC SUSCEPTIBILITY

The energy levels of a nucleus with a magnetic moment are changed in the presence of a magnetic field. Transitions between these levels can be induced by electromagnetic radiation in the radio-frequency region, and this resonance is useful in characterizing the chemical environment of the nucleus. For this reason, nuclear magnetic resonance (NMR) spectroscopy has deveoped as one of the most powerful structural methods used in organic and inorganic chemistry. In a slightly different application, in this experiment we will examine the effect of paramagnetic "impurities" on NMR solvent resonances and will use the predicted resonance shifts to deduce the paramagnetic susceptibility and electron spin of several transition-metal complexes.

THEORY

The magnetic moment of a nucleus with nuclear spin quantum number I is

$$\mu_N = g_N \beta_N \sqrt{I(I + 1)} \tag{1}$$

where g_N is the nuclear g factor (5.5849 for a proton) and $\beta_N = eh/4\pi m_p c$ is the nuclear magneton. Substitution of the charge e and mass m_p of a proton gives a value of 5.051×10^{-24} erg/gauss for β_N. The symbol β_N is the unit of nuclear magnetic moment and is smaller than the electronic Bohr magneton β_B (defined in Exp. 37) by the electron-to-proton mass ratio.

The nuclear moment will interact with a magnetic field B to cause an energy change (Zeeman effect)

$$E_N = -g_N \beta_N M_I B \qquad (2)$$

Here M_I is the quantum number measuring the component of nuclear spin angular momentum (and magnetic moment) along the field direction and it can have values $-I, -I+1, \ldots, +I$. The effect of the field is thus to break the $2I + 1$ degeneracy and to produce energy levels whose spacing increases linearly with B (Fig. 1). Transitions among these levels can be produced by electromagnetic radiation provided that the selection rule $\Delta M_I = \pm 1$ is satisfied. In this case the resonant frequency is given by

$$\nu = \frac{\Delta E}{h} = \frac{g_N \beta_N}{h} B \qquad (3)$$

For protons, $\nu(\text{Hz}) = 4260B$ (gauss) so that, for typical fields of $\sim 10^4$ gauss, ν falls in the radio-frequency region. In practice ν is usually fixed at some convenient frequency (e.g., 60, 100, 220 MHz) and the field B is varied until resonance is achieved.

In general, the local field B at the nucleus will differ from the externally applied field H because of the magnetization I which is induced by H,

$$B = H + 4\pi I = H + 4\pi \chi H \qquad (4)$$

The volume magnetic susceptibility χ (a dimensionless quantity) is the magnetic

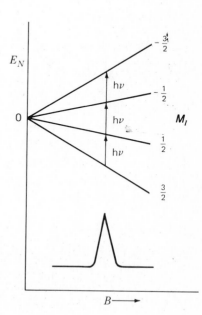

FIGURE 1
Energy levels, allowed transitions, and spectrum of a nucleus with $I = 3/2$ in a magnetic field of strength B.

analog of the electric polarizability (Exp. 36) and it can be converted to a mass susceptibility χ_g by dividing by the density ρ. Multiplication by the molecular weight M gives the molar susceptibility

$$\chi_M = M\chi_g = \frac{M\chi}{\rho} \tag{5}$$

As noted in Exp. 37, χ_M has the form

$$\chi_M = N_0\alpha + \frac{N_0\mu^2}{3kT} = N_0\alpha + \chi_M{}^P \tag{6}$$

where $N_0\alpha$ is a diamagnetic contribution while the second term is a temperature-dependent paramagnetic contribution which is dominant if the magnetic moment μ is not zero. This will be true only if the electron spin or electronic orbital angular momentum is not zero, a case which is fairly common for transition-metal compounds.

Most organic compounds are not paramagnetic, and so it is the diamagnetic susceptibility which is important in determining the resonance condition. The diamagnetic contribution arises because the orbital motion of the electrons is altered by the presence of H so that there is a net orbiting of electrons about the field lines. This circulating charge in turn generates a magnetic field H_d which is *opposed* and proportional to the applied field,

$$H_d = 4\pi\chi H = 4\pi \frac{\rho}{M} N_0\alpha H = -\sigma H \tag{7}$$

The resonant frequency is thus changed to

$$v = \frac{g_N\beta_N}{h} H(1 - \sigma) = v_0(1 - \sigma) \tag{8}$$

where v_0 is the resonance for a bare proton. The diamagnetic shielding constant σ is usually quite small ($\sim 10^{-5}$) and increases as the electron density about the nucleus is increased. Changes in B (and thus v) of a few parts per million (PPM) are typical when the chemical environment about a nucleus is changed. These chemical shifts relative to a convenient standard, such as tetramethylsilane, are easily measured with modern NMR instruments and hence serve to characterize the chemical bonding about a given nucleus. In addition, the relative intensities of, for example, proton resonances give a measure of the relative number of protons with different chemical environments (e.g., $-CH_3$ versus $-CH_2-$ groups). Finally, the coupling of magnetic moments of nearby nuclei can produce spin-spin splitting patterns which are quite useful in identifying the functional groups present in the molecule. Some further discussion of these applications is presented in Exp. 46.

If a proton of a diamagnetic molecule is present in a solution containing a paramagnetic solute, the field B at the nucleus will be increased because of the alignment of the solute magnetic moments in the applied field. Evans[1] has shown that this increase in the local field is given by

$$\Delta B = \frac{2\pi}{3} (\chi_s - \chi_0)H \tag{9}$$

where χ_s and χ_0 are the susceptibilities of the solution with and without the paramagnetic solute, respectively. According to Eqs. (3) and (9), the proton resonance will shift by an amount

$$\frac{\Delta v}{v} = \frac{\Delta B}{B} \approx \frac{\Delta B}{H} = \frac{2\pi}{3} (\chi_s - \chi_0) \tag{10}$$

with the approximation $B \approx H$ leading to negligible error.

To obtain the gram susceptibility χ_g of the pure paramagnetic material, we assume that the volume susceptibility χ_s of the solution can be written as a sum of parts

$$\chi_s = \chi_{gs}\rho_s = \chi_g m + \chi_{g0}(\rho_s - m) \tag{11}$$

Here ρ_s is the density of the solution containing m grams of paramagnetic solute per cm^3 and χ_{g0} is the gram susceptibility of the solution without the paramagnetic material. The density of the latter is ρ_0 so that $\chi_0 = \chi_{g0}\rho_0$ and we obtain from Eqs. (10) and (11) the expression

$$\chi_g = \frac{3}{2\pi m} \frac{\Delta v}{v} + \chi_{g0} + \chi_{g0} \frac{\rho_0 - \rho_s}{m} \tag{12}$$

The third term is a small correction which is unimportant for highly paramagnetic materials and is often neglected.[1] The value of χ_g for a paramagnetic material can be determined, therefore, by measuring the difference in chemical shift of a proton in the solvent and in a solution containing a known weight of the paramagnetic solute. The value of χ_{g0} can be obtained from tables such as those contained in Ref. 3 or by summing atomic susceptibilities χ_i according to Pascal's empirical relation

$$\chi_{g0} = \frac{\chi_{M_0}}{M_0} = \frac{1}{M_0} \sum_i n_i \chi_i + \lambda_i \tag{13}$$

where n_i is the number of atoms of type i in the molecule and the constitutive correction constants λ_i depend upon the nature of the multiple bonds. See Table 1 for some typical values of χ_i and λ_i.

The molar susceptibility χ_M for the solute is obtained by multiplying χ_g by the molecular weight of the paramagnetic complex. As can be seen from Eq. (6), the paramagnetic contribution χ_M^P, and hence the paramagnetic moment μ, can be extracted by a temperature-dependence study (see also Exp. 37). For accurate results, a correction should be made to m to account for solvent density changes with

Table 1 PASCAL'S CONSTANTS χ_i FOR DIAMAGNETIC SUSCEPTIBILITY[a]
(units of 10^{-6} cm^3 mol^{-1})

Co^{2+}	-12	Br$^-$	-36	B	-7.0	N open chain	-5.57
Co^{3+}	-10	CN$^-$	-18	Br	-30.6	ring	-4.61
Cr^{2+}	-15	Cl$^-$	-26	C	-6.0	monamides	-1.54
Cr^{3+}	-11	F$^-$	-11	Cl	-20.1	diamides, imides	-2.11
Cu^{2+}	-11	NO$_3^-$	-20	F	-6.3	O alcohol, ether	-4.61
Fe^{2+}	-13	OH$^-$	-12	H	-2.93	aldehyde, ketone	$+1.73$
Fe^{3+}	-10	SO$_4^{2-}$	-40	I	-44.6	carboxylic $=$ O	-3.36
K$^+$	-13			P	-26.3		
Ni^{2+}	-12			S	-15.0		

λ_i Corrections for bonds

C=C	$+5.5$	C=N	$+8.2$
C≡C	$+0.8$	C≡N	$+0.8$
C=C—C=C	$+10.6$	C in aromatic ring	-0.24

[a] See Ref. 3.

temperature.[2] Alternatively a measurement of χ_M at a single temperature can usually be combined with an adequate estimate of $N_0 \alpha$ from Pascal's constants[3] to allow one to deduce the paramagnetic contribution. If $\chi_M{}^P$ is obtained in this manner, the paramagnetic moment is given by

$$\mu = \sqrt{3kT\chi_M{}^P/N_0}$$

$$= 2.824\sqrt{T\chi_M{}^P} \text{ Bohr magnetons} \tag{14}$$

where the Bohr magneton (β_B) is 0.927×10^{-20} erg gauss^{-1}. In the absence of any orbital contributions, the value of μ is related to the number n of unpaired electrons in the d shell of the transition metal by

$$\mu(\text{spin only}) = \sqrt{n(n+2)}\, \beta_B \tag{15}$$

As noted in Exp. 37, magnetic susceptibility measurements are thus quite useful in determining the electron configuration of a paramagnetic ion in a particular ligand field.

METHOD

The basic elements of an NMR spectrometer are outlined in Fig. 2. The principal magnetic field is provided by a permanent magnet (~ 15 Kgauss), an electromagnet (25 to 50 Kgauss), or a superconducting electromagnet (50 to 75 Kgauss). Since shifts of a few PPM are to be measured, the field must be quite stable and uniform.

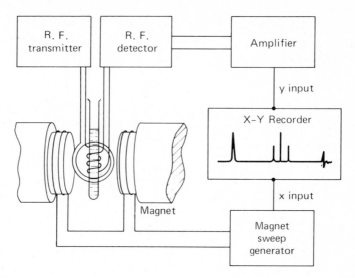

FIGURE 2
Schematic drawing of a nuclear magnetic resonance spectrometer.

This is achieved by adding small adjustment coils to the magnet and by spinning the sample tube to average out residual inhomogeneities. An additional set of Helmholtz coils is wound around the pole faces to permit small linear variation in the field in recording spectra.

The sample tube is inserted into a probe region containing two radio-frequency coils wound at 90° to each other and to the magnetic-field coils. Radiation of some fixed frequency, usually 60 MHz or higher for protons, is sent through the sample by the transmitter coil. If the magnetic field is such that Eq. (3) is satisfied, sample absorption and emission will occur. This re-radiated signal is detected by the receiver coil, which, being oriented at right angles to the transmitter, senses no signal in the absence of sample coupling. The signal from the receiver coil is amplified and displayed on a chart recorder to yield the NMR spectrum as a function of field. Since the absolute value of the magnetic field is not easily determined with high precision, field shifts are measured relative to some reference compound such as tetramethylsilane and are expressed as chemical shifts in PPM,

$$\delta_i \equiv \frac{H_r - H_i'}{H_r} \times 10^6 \qquad (16)$$

Here, H_r and H_i are field values for resonance by the reference nucleus and by nucleus i, respectively. Since v is proportional to the field, δ_i can also be written as

$$\delta_i = \frac{v_r - v_i}{v_r} \times 10^6 \qquad (17)$$

and many NMR chart displays are calibrated both in δ units and in frequency shift units appropriate to the fixed frequency source v_r (60 MHz for most small proton NMR instruments).

EXPERIMENTAL

Operation of the NMR instrument will be described by the instructor. Particular care should be taken in inserting and removing the NMR samples tubes to prevent damage to the probe. All tubes should be wiped clean prior to insertion to avoid contamination of the probe.

To provide an internal reference, the solvent is sealed in a capillary which is placed at the bottom of the NMR tube used for the sample solution. For this purpose, a melting-point capillary is closed at one end and a syringe is used to add the reference solution. For aqueous studies, a 2% solution of t-butyl alcohol in water can be used as reference and as solvent for the paramagnetic solute. The shift of the methyl resonance of the t-butyl group is then monitored. With organic ligands such as acetylacetonate (acac) groups, complexes such as $Cr(acac)_3$, $Fe(acac)_3$, and $Co(acac)_2$ are soluble in benzene and the proton resonance of the solvent is a convenient reference. The capillary is filled one-third full, the lower end of the capillary is cooled in ice, and the upper part is sealed off with a small hot flame. Alternatively, the reference solution can be placed in one compartment of a coaxial pair of cylindrical NMR tubes, which are commercially available. In either case, the spectral display should be expanded to permit an accurate measurement of the frequency shift.

The following complexes are suitable for study. Toluene is preferable to benzene as solvent (see Appendix D) but has *two* proton resonances.

d^3 $Cr(acac)_3$, chromium acetylacetonate (349.3 g/mol)
d^5 $Fe(acac)_3$, ferric acetylacetonate (353.2 g/mol)
d^8 $Co(acac)_2$, cobaltous acetylacetonate (257.2 g/mol)

The concentrations are not critical (0.02 to 0.05 M) but they must be precisely known. These salts are available commercially or they can be readily synthesized by methods described in Ref. 4.

If the NMR spectrometer is equipped with a variable temperature probe, the $Fe(acac)_3$ shifts should be measured at four or five temperatures. A 0.025 M solution in toluene can be employed over the liquid range from -95 to $+110°C$. Care should be taken to seal and test the NMR tube if the high temperature range is to be studied. Since the temperature variation of toluene density is appreciable,[2] the solute concentration should be corrected at each temperature T:

$$m_T = m_{RT} \frac{\rho_T}{\rho_{RT}} \tag{18}$$

The density of toluene at various temperatures is given in Ref. 5. Approximately 15 min should be allowed for equilibrium to be reached after changing the temperature. An accurate measurement of the probe temperature should be made in the manner described in the spectrometer manual. This usually involves a measurement of the frequency separation between OH and CH resonances in methanol or ethylene glycol since this separation changes by about 0.5 Hz per degree for a 60-MHz instrument.

Appropriate salts for alternative aqueous measurements are

d^5 $Fe(NO_3)_3 \cdot 9H_2O$ (0.015 M), ferric nitrate† (404.0 g/mol)
d^5 $K_3Fe(CN)_6$ (0.06 M), potassium ferricyanide (329.3 g/mol)
d^6 $FeSO_4 \cdot 7H_2O$ (0.02 M), ferrous sulfate (278.0 g/mol)
d^8 $NiCl_2$ (0.08 M), nickel chloride (129.6 g/mol)
d^9 $CuSO_4$ (0.08 M), cupric sulfate (159.6 g/mol)

Three or more of these solutions should be made up in 10- or 25-ml volumetric flasks using as reference solvent $\sim$2% (by volume) t-butyl alcohol in deoxygenated water. The Fe^{2+} and Fe^{3+} solutions illustrate the effect of oxidation state on the electron configuration of a single element. The Fe^{3+} and $K_3Fe(CN)_6$ solutions demonstrate the role of weak and strong ligand field splitting on the electron configuration (see discussion in Exp. 37).

The $Fe(NO_3)_3$ or $NiCl_2$ solutions can be studied from 0 to 100°C in the same manner as described for the acetylacetonate complexes. In correcting for changes in m with temperature, the density variations of pure H_2O may be used.

CALCULATIONS

Neglecting the third term of Eq. (12), the mass susceptibility of a paramagnetic solute is readily determined from m and $\Delta\nu$. The diamagnetic mass susceptibilities χ_{g0} of the solvents are -0.70×10^{-6} (benzene), -0.72×10^{-6} (toluene), and -0.72×10^{-6} (t-butyl alcohol–water solvent). For temperature-dependence studies, correct m according to Eq. (18), calculate χ_M and plot χ_M versus $1/T$. The slope of the best straight line through these points is the Curie constant C (see Exp. 37) and the magnetic moment μ is given by

---

† This compound is hydroscopic so that the weighing should be done rapidly. It is also necessary to make the solution $\sim$0.3 M in HNO_3 to prevent precipitation of $Fe(OH)_3$. The reference solvent used with this solution should have the same acid concentration.

$$\mu = \left(\frac{3kC}{N_0}\right)^{1/2} = 2.824\sqrt{C} \text{ Bohr magnetons} \tag{19}$$

The value of $N_0\alpha$ is obtained from the intercept of your plot. Compare this value of the diamagnetic susceptibility of the solute with that estimated from Pascal's constants in Table 1. For single temperature measurements, estimates of the latter type should be used to obtain χ_M^p for use in Eq. (14). Report your values of μ in Bohr magnetons and calculate the number of unpaired electrons using the spin-only formula, Eq. (15).

DISCUSSION

In an aqueous solvent, the ions Fe^{2+}, Fe^{3+}, Ni^{2+}, and Cu^{2+} are all complexed to six H_2O molecules in an octahedral arrangement. The magnetic moments can be understood in terms of the weak field splitting of the five d orbitals as described in Exp. 37. In $Fe(CN)_6^{3-}$, the CN^- ligands are strongly bound to the Fe^{3+} ion and the ligand field splitting is larger. The structures of $Cr(acac)_3$ and $Fe(acac)_3$ are also pseudo octahedral with the two oxygen atoms of each $acac^-$ ion occupying adjacent metal ligand positions. For $Co(acac)_2$, a square planar or tetrahedral arrangement of the ligands might be expected, but the energetics favoring octahedral coordination are such that there is intermolecular association to fill the two vacant ligand sites. The result is a tetramer in which a pseudo octahedral arrangement is

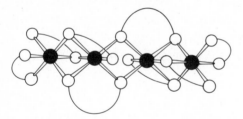

achieved by each cobalt.[6] Discuss your observed magnetic moment values in terms of the number of unpaired electrons predicted for each transition-metal complex assuming octahedral structures. Comment on any differences from the predicted spin-only moments.

APPARATUS

NMR spectrometer, with variable temperature capability if available; several 10- or 25-ml volumetric flasks; NMR tubes; melting-point capillaries; syringe; torch for sealing capillaries; ethylene glycol and methanol in NMR tubes for temperature calibrations.

Cr(acac)$_3$, Fe(acac)$_3$, Co(acac)$_3$, where acac = acetylacetonate ($\sim$1g each); benzene or toluene (250 ml); Fe(NO$_3$)$_3$ $\cdot$ 9H$_2$O, K$_3$Fe(CN)$_6$, FeSO$_4$ $\cdot$ 7H$_2$O, NiCl$_2$, CuSO$_4$ ($\sim$1 g each); 2% by volume t-butyl alcohol in deoxygenated water (250 ml); 3 N HNO$_3$ for acidifying Fe(NO$_3$)$_3$ and corresponding reference solution.

REFERENCES

1. D. F. Evans, *J. Chem. Soc.*, 2003 (1959).
2. D. Ostfield and I. A. Cohen, *J. Chem. Ed.*, **49**, 829 (1972).
3. L. N. Mulay and E. A. Boudreaux, "Theory and Applications of Molecular Diamagnetism," p. 303, Wiley-Interscience, New York (1976).
4. T. H. Crawford and J. Swanson, *J. Chem. Ed.*, **48**, 382 (1971).
5. E. W. Washburn (ed.), "International Critical Tables," vol. III, p. 28, McGraw-Hill, New York (1929).
6. F. A. Cotton and G. Wilkinson, "Advanced Inorganic Chemistry," 3d ed., p. 877, Interscience, New York (1972).

XIII

SPECTRA AND MOLECULAR STRUCTURE

EXPERIMENTS

39. Absorption spectrum of a conjugated dye
40. Infrared spectroscopy: vibrational spectrum of SO_2
41. Raman spectroscopy: vibrational spectrum of CCl_4
42. Rotation-vibration spectrum of HCl and DCl
43. Spectrum of the hydrogen atom
44. Band spectrum of nitrogen
45. Molecular flourescence of iodine
46. NMR determination of keto-enol equilibrium constants

EXPERIMENT 39.   ABSORPTION SPECTRUM OF A CONJUGATED DYE

Absorption bands in the visible region of the spectrum correspond to transitions from the ground state of a molecule to an excited electronic state which is 40 to 70 kcal above the ground state. In many substances, the lowest excited electronic state is more than 70 kcal above the ground state and no visible spectrum is observed. Those compounds which are colored (i.e., absorb in the visible) generally have some weakly bound or delocalized electrons such as the odd electron in a free radical or the π electrons in a conjugated organic molecule. In this experiment we are concerned with the determination of the visible absorption spectrum of several symmetrical polymethine dyes and with the interpretation of these spectra using the "free-electron" model.

THEORY

The visible bands for polymethine dyes arise from electronic transitions involving the π electrons along the polymethine chain. The wavelength of these bands depends on the spacing of the electronic energy levels. Bond-orbital and molecular-orbital calculations have been made for these dyes,[1] but the predicted wavelengths are in poor agreement with those observed. We shall present here the simple free-electron model first proposed by Kuhn;[2] this model contains some drastic assumptions but has proved reasonably successful for molecules like a conjugated dye.

As an example, consider a dilute solution of 1,1'-diethyl-4,4'-carbocyanine iodide (cryptocyanine):

The cation can "resonate" between the two limiting structures above, which really means that the wave function for the ion has equal contributions from both states. Thus all the bonds along this chain can be considered equivalent, with bond order 1.5 (similar to the C—C bonds in benzene). Each carbon atom in the chain and each nitrogen at the end is involved in bonding with three atoms by three localized bonds (the so-called σ bonds). The extra valence electrons on the carbon atoms in the chain and the three remaining electrons on the two nitrogens form a mobile cloud of π electrons along the chain (above and below the plane of the chain). We shall assume that the potential energy is constant along the chain and that it rises sharply to infinity at the ends; i.e., the π electron system is replaced by free electrons moving in a one-dimensional box of length L. The quantum-mechanical solution for the energy levels of this model[3] is

$$E_n = \frac{h^2 n^2}{8mL^2} \qquad n = 1, 2, 3, \ldots \tag{1}$$

where m is the mass of an electron and h is Planck's constant.

Since the Pauli exclusion principle limits the number of electrons in any given energy level to two (these two have opposite spins: $+\frac{1}{2}$, $-\frac{1}{2}$), the ground state of a molecule with N π electrons will have the $N/2$ lowest levels filled and all higher levels empty. When the molecule (or ion in this case) absorbs light, this is associated with a one electron jump from the highest filled level ($n_1 = N/2$) to the lowest empty level ($n_2 = N/2 + 1$). The energy change for this transition is

$$\Delta E = \frac{h^2}{8mL^2} (n_2{}^2 - n_1{}^2) = \frac{h^2}{8mL^2} (N + 1) \tag{2}$$

Since $\Delta E = h\nu = hc/\lambda$, where c is the speed of light and λ is the wavelength in centimeters,

$$\lambda = \frac{8mc}{h} \frac{L^2}{N + 1} \tag{3}$$

Let us denote the number of carbon atoms in a polymethine chain by p; then $N = p + 3$. Kuhn assumed that L was the length of the chain between nitrogen atoms plus one bond distance on each side; thus, $L = (p + 3)l$, where l is the bond length between atoms along the chain. Therefore,

$$\lambda = \frac{8mcl^2}{h} \frac{(p + 3)^2}{p + 4} \tag{4}$$

Putting $l = 1.39 \times 10^{-8}$ cm (the bond length in benzene, a molecule with similar bonding) and converting λ from centimeters to nanometers (1 nm $= 10^{-9}$ m), we find that

$$\lambda \text{ (in nm)} = 63.7 \frac{(p + 3)^2}{p + 4} \tag{5}$$

If there are easily polarizable groups at the ends of the chain (such as benzene rings), the potential energy of the π electrons in the chain does not rise so sharply at the ends. In effect this lengthens the path L, and we can write

$$\lambda \text{ (in nm)} = 63.7 \frac{(p + 3 + \alpha)^2}{p + 4} \tag{6}$$

where α should be a constant for a series of dyes of a given type. If such a series is studied experimentally, this empirical parameter α may be adjusted to achieve the best fit to the data; in any event, α should lie between 0 and 1.[2]

In order to compare the results of this model with the more sophisticated bond- or molecular-orbital calculations let us use Eq. (5), which assumes that $\alpha = 0$, to calculate the wavelength λ for cryptocyanine (in which $p = 9$) and compare that value with those given by Herzfeld and Sklar:[1,2]

Free electron	$\lambda = 707$ nm
Bond orbital (case 1)	$\lambda = 3900$
Bond orbital (case 2)	$\lambda = 2900$
Molecular orbital	$\lambda = 2700$

Note that only the free-electron model predicts an absorption band in the visible in agreement with observation. While the orbital calculations are poor for polymethine dyes, they are in principle a superior approach and have given excellent results for unsaturated hydrocarbons.

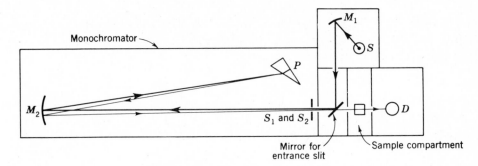

FIGURE 1
Schematic diagram of a Beckman Model DU Spectrophotometer.

METHOD

Although there are many different spectrophotometers currently available, the Beckman Model DU (shown schematically in Fig. 1) is well suited for student use. A general description of spectrophotometers will be given in terms of this instrument, but the principles involved are applicable to any other single-beam, manually operated instrument. More sophisticated dual-beam scanning instruments, such as the Cary 14, give a direct record of the absorbancy as a function of wavelength.

The source S is a battery-operated tungsten lamp which provides a "white" light for operation in the visible. (The tungsten lamp can be replaced by a hydrogen discharge tube for work in the near ultraviolet.) A condensing mirror M_1 sends a beam of light from this source through the entrance slit S_1 into the monochromator. This beam is reflected to a quartz prism by the collimating mirror M_2. The prism P is a Littrow type; the light undergoes refraction on entering the front surface, is reflected from a silvered back surface, and undergoes a second refraction on leaving the prism. (Many instruments achieve dispersion of the light by means of a diffraction grating instead of a prism.) When the prism is rotated, a narrow band which is almost monochromatic at any desired wavelength can be passed through the exit slit S_2. A sample cell,† with plane parallel windows, is mounted between the slit and the detector D. The intensity of light incident on the sample cell can be controlled by varying the slit width. The narrowest possible slit width is best, since this gives the most nearly monochromatic beam. All the light which is transmitted by the sample falls on the detector (a photomultiplier tube or some type of barrier layer cell). The photocurrent is then amplified and used to indicate the intensity of this transmitted light.

† For measurements in the visible, inexpensive Pyrex cells are very satisfactory. Corex cells can be used in the near ultraviolet, although quartz cells are necessary for most ultraviolet studies.

It is now necessary to define several terms commonly used in spectrophotometry. Absorption spectra are often characterized by the *transmittance* at a given wavelength; this is defined by

$$T = \frac{I}{I_0} \tag{7}$$

where I is the intensity of light transmitted by the sample and I_0 is the intensity of light incident on the sample. When the sample is in solution and a cell must be used, I is taken to be the intensity of light transmitted by the cell when it contains solution while I_0 is taken to be the intensity of light transmitted by the cell filled with pure solvent. Another way of describing spectra is in terms of the *absorbance* A where

$$A = \log \frac{I_0}{I} \tag{8}$$

[In the past, the quantity $\log (I_0/I)$ was often called the optical density.] A completely transparent sample would have $T = 1$ or $A = 0$, while a completely opaque sample would have $T = 0$ or $A = \infty$.

The absorbance A is related to the path length d of the sample and the concentration c of absorbing molecules by the Beer-Lambert law,[4]

$$A = \varepsilon c d \tag{9}$$

where the proportionality constant ε is called the *absorptivity* or extinction coefficient. When the concentration is expressed in moles per liter, ε is the molar absorptivity. The quantity ε is a property of the absorbing material which varies with wavelength in a characteristic manner; its value depends only slightly on the solvent used and on the temperature.

For quantitative measurements, it is important to calibrate the cells so that a correction can be made for any small difference in path length between the solution cell and the solvent cell. For analytical applications, one must check the validity of Beer's law, since slight deviations are often observed and a calibration of absorbancy versus concentration is then required. Such quantitative techniques are described elsewhere[4] and will not be necessary for conducting the present experiment.

EXPERIMENTAL

Instructions for operating the spectrophotometer will be made available in the laboratory. Turn on the instrument as instructed, and allow it to warm up for at least 20 min prior to use. For most instruments, such a warmup period is necessary to achieve stable, drift-free operation.

Several polymethine dyes should be studied, preferably a series of dyes of a given type with varying chain length. In addition to the 1,1'-diethyl-4,4'-carbocyanine iodides mentioned previously, 1,1'-diethyl-2,2'-carbocyanine iodides and 3,3'-diethyl thiacarbocyanine iodides are suitable. Other possible compounds can be found in the literature.[5]

Choose any one of the available dyes, and prepare 10 ml of a solution using methyl alcohol as the solvent. The concentration should be approximately 10^{-3} M. (If the solutions are to be kept for a long time, they should be stored in dark glass bottles to prevent slow decomposition by daylight.)

Follow the spectrophotometer operating instructions carefully, and obtain the spectrum of this initial solution. Take absorbance readings at widely spaced intervals throughout the visible region (400 to 750 nm) until the absorption band is located. Then take readings at much closer intervals throughout the band. Dilute the initial solution and redetermine the spectrum. Repeat this procedure until a spectrum is obtained with an absorbance reading of about 1 at the peak (transmittance ~ 0.1). The band shape will change with concentration, since these dyes dimerize. At the final concentration used, the monomer band (higher wavelength band) will be much more prominent than the dimer band, which will disappear or remain as a shoulder on the low wavelength side of the main peak. If the spectrophotometer cell holder will accommodate more than two cells, several concentrations can be studied simultaneously.

Make up solutions (10 ml) of the other dyes at approximately the same molar concentration that gave the best results previously. Obtain their spectra in the same way.

CALCULATIONS

Carefully plot all spectra (A versus λ) and determine λ_{max}, the wavelength at the peak, for each.

Using Eq. (6), calculate λ from the free-electron model. If a series of dyes of a single type has been studied, choose α to give the best fit for one of the series and use this value for all others in the series.

Report your experimental values of λ_{max} together with the theoretical results.

APPARATUS

Spectrophotometer, such as a Beckman Model DU; four Pyrex sample cells; lens tissue; 6-V battery and battery charger (or regulated dc power supply); several 10-ml volumetric flasks; wash bottle.

Reagent-grade methyl alcohol (150 ml); several polymethine dyes, preferably a series such as 1,1'-diethyl-4,4'-cyanine iodide, -carbocyanine iodide, and -dicarbocyanine iodide or 1,1'-diethyl-2,2'-cyanine iodide, -carbocyanine iodide, and -dicarbocyanine iodide (a few milligrams of each is sufficient). (Gallard-Schlesinger Chemical Mfg. Corp., 1001 Franklin St., Garden City, N.Y. and K & K Laboratories, Inc., 121 Express St., Plainview, N.Y. 11803, are possible but expensive suppliers of an entire series of dyes. Aldrich Chemical Co. and Eastman Organic Chemicals are more inexpensive suppliers of some of these dyes. Check the current "Chem Sources" for other suppliers.)

REFERENCES

1. K. F. Herzfeld and A. L. Sklar, *Rev. Mod. Phys.*, **14**, 294 (1942).
2. H. Kuhn, *J. Chem. Phys.*, **17**, 1198 (1949).
3. W. J. Moore, "Physical Chemistry," 4th ed., pp. 604–608, Prentice-Hall, Englewood Cliffs, N.J. (1972); or any introductory book in quantum mechanics.
4. M. G. Mellon, "Analytical Absorption Spectroscopy," Wiley, New York (1950); D. F. Boltz, "Selected Topics in Modern Instrumental Analysis," chap. 4, Prentice-Hall, Englewood Cliffs, N.J. (1952).
5. L. G. S. Brooker, *Rev. Mod. Phys.*, **14**, 275 (1942); N. I. Fisher and F. M. Hamer, *Proc. Roy. Soc.*, Ser. A, **154**, 703 (1936).

GENERAL READING

T. R. P. Gibb, "Optical Methods of Chemical Analysis," chap. II, McGraw-Hill, New York (1942).

W. West, Spectroscopy and Spectrophotometry, in A. Weissberger (ed.), "Technique of Organic Chemistry," 3d ed., vol. I, part III, chap. XXVIII, Interscience, New York (1959).

EXPERIMENT 40. INFRARED SPECTROSCOPY: VIBRATIONAL SPECTRUM OF SO_2

The infrared region of the spectrum extends from the long-wavelength end of the visible region at 1 μ out to the microwave region at about 1000 μ, 1 micron (μ) being equal to 10^{-4} cm. Although considerable work is now being done in the far infrared (wavelengths greater than 50 μ), the region between 2 and 25 μ has received the greatest attention because the vibrational frequencies of most molecules lie in this region. It is common practice to specify infrared frequencies in wavenumber units (cm^{-1}); thus the 2 to 25 μ region extends from 5000 down to 400 cm^{-1}.

THEORY

Almost all infrared work makes use of absorption techniques where radiation from a source emitting all infrared frequencies is passed through a sample of the material to be studied. When the frequency of this radiation is the same as a vibrational frequency of the molecule, the molecule may be vibrationally excited; this results in loss of energy from the radiation and gives rise to an absorption band. The spectrum of a molecule generally consists of several such bands arising from different vibrational motions of the molecule.

Associated with each vibrational energy level there are also closely spaced rotational energy levels. In general, transitions occur from a particular rotational level (quantum number J'') in a given vibrational state (quantum number v'') to a different rotational level (quantum number J') in an excited vibrational state (quantum number v'). The appearance of this rotational fine structure will depend on the resolution of the spectrometer used, as shown schematically in Fig. 1. Rotation-vibration spectra are discussed in Exp. 42 for the case of linear molecules and will not be considered here.

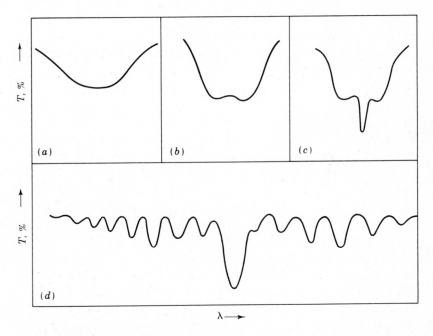

FIGURE 1
Schematic infrared band contours at (a) low, (b) and (c) medium, and (d) high resolution. For the band shown in (b) there is no Q branch ($\Delta J = 0$ transitions are forbidden); the bands shown in (c) and (d) have Q branches.

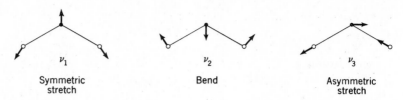

Symmetric stretch ν_1 Bend ν_2 Asymmetric stretch ν_3

FIGURE 2

Normal modes of vibration for SO_2. The lengths of the arrows which indicate displacements from the equilibrium configuration have been exaggerated for clarity.

In this experiment we shall be concerned with the infrared bands of SO_2 gas under low or medium resolution, i.e., with only the vibrational spectrum. Electron diffraction studies show that SO_2 is a symmetrical, nonlinear molecule.[1] For a nonlinear molecule containing n atoms there are $3n - 6$ vibrational degrees of freedom. Thus, SO_2 has three basic patterns of vibration called "normal modes." These are shown in Fig. 2. All vibrations of the molecule may be expressed as linear combinations of these normal modes.

The frequencies of the normal modes of vibration are called *fundamentals*. For SO_2 all three fundamentals are infrared active; they will correspond to the three most intense bands in the spectrum. The fundamentals (ν_1, ν_2, ν_3) may be assigned by analogy with the spectra of other molecules: the bending frequency ν_2 should be that of the lowest frequency band, and the asymmetric stretch ν_3 should be that of the highest frequency band.

The structure of SO_2 is confirmed by the presence of all three fundamentals in the infrared spectrum. When this is true, the molecule cannot have a center of symmetry[2] as in O—S—O. Linear unsymmetric structures, such as O—O—S or O——S—O, can be ruled out by the shape of the rotational band envelopes in the infrared spectrum; discussion of this topic is beyond the scope of this book.

In addition to the fundamentals, other much weaker bands may be observed in the spectrum at high pressure. Such weak bands are either overtones ($2\nu_i$, $3\nu_i$, ...) or combination bands ($\nu_i \pm \nu_j$, $2\nu_i \pm \nu_j$, ...) which arise from the fact that small anharmonicities tend to couple the normal vibrations. On the basis of intensity, the transitions which involve binary overtones ($2\nu_i$) and binary combinations ($\nu_i \pm \nu_j$) are the most likely to be observed. Thus, any weak bands observed for SO_2 may be assigned as overtones or combinations of the strong fundamentals. Any such assignment should take into account rules which determine whether a given combination will be active in the infrared,[3] but this subject is beyond the scope of this book and will not be discussed.

Valence force model The normal modes of vibration can be expressed in terms of the atomic masses (m_S and m_O), the O—S—O bond angle (2α), and several force constants which define the potential energy of the molecule. These force constants are analogous to the Hooke's law constant for a spring. If a force model is chosen involving fewer force constants than the number of normal modes, the force constants are overdetermined and a check on the validity of the model is then possible. The valence force model assumes that there is a large restoring force along the line of a chemical bond when the distance between the two atoms at the ends of this bond is changed. It also assumes a restoring force for any change in angle between two adjacent bonds. Adopting the notation used by Herzberg,[4] we can write the potential energy of SO_2 as

$$V = \tfrac{1}{2}[k_1(Q_1{}^2 + Q_2{}^2) + k_\delta\delta^2]$$

where Q_1 and Q_2 are changes in the S—O distances (whose equilibrium value is denoted by l), and δ is the change in the bond angle 2α. The quantities k_1 and k_δ/l^2 are force constants (usually given in dynes per centimeter) for stretching and bending motions, respectively. Using the classical mechanics of harmonic oscillators, one can obtain[4] for SO_2

$$4\pi^2 v_3{}^2 = \left(1 + \frac{2m_O}{m_S}\sin^2\alpha\right)\frac{k_1}{m_O} \tag{1}$$

$$4\pi^2(v_1{}^2 + v_2{}^2) = \left(1 + \frac{2m_O}{m_S}\cos^2\alpha\right)\frac{k_1}{m_O} + \frac{2}{m_O}\left(1 + \frac{2m_O}{m_S}\sin^2\alpha\right)\frac{k_\delta}{l^2} \tag{2}$$

$$16\pi^4 v_1{}^2 v_2{}^2 = 2\left(1 + \frac{2m_O}{m_S}\right)\frac{k_1}{m_O{}^2}\frac{k_\delta}{l^2} \tag{3}$$

If one uses v in wavenumber units and $m_O = 16$, $m_S = 32$ (atomic weight units), $4\pi^2$ should be replaced by $4\pi^2 c^2 M_1 = 5.889 \times 10^{-2}$ ($M_1 = \tfrac{1}{16}$ mass of an O^{16} atom in grams). For SO_2, $2\alpha = 120°$.

Vibrational partition function[5] The thermodynamic quantities for a perfect gas can usually be expressed as a sum of translational, rotational, and vibrational contributions (see Exp. 3). We shall consider here the heat capacity at constant volume. At room temperature and above, the translational and rotational contributions to C_v are constant; for SO_2 (a nonlinear polyatomic molecule),

$$\tilde{C}_v(\text{trans}) = \frac{3R}{2}$$

$$\tilde{C}_v(\text{rot}) = \frac{3R}{2} \tag{4}$$

The vibrational contribution to C_v varies with temperature and can be calculated from the vibrational partition function z_{vib}, which is expressed in terms of the frequencies of the normal modes. $z_{vib} = \prod_i z_i$, where for a harmonic oscillator of frequency v_i we may write [making use of Eq. (42-1)]

$$z_i \equiv \sum_j e^{-\varepsilon_j/kT} = \sum_{v=0}^{\infty} e^{-(v+1/2)hv_i/kT} = \frac{e^{-hv_i/2kT}}{1 - e^{-hv_i/kT}} \tag{5}$$

The vibrational contribution to C_v is related to z_{vib} by

$$\tilde{C}_v(\text{vib}) = R \frac{\partial}{\partial T} \left(T^2 \frac{\partial \ln z_{vib}}{\partial T} \right) \tag{6}$$

From Eqs. (5) and (6)

$$\tilde{C}_v(\text{vib}) = R \sum_i \frac{u_i^2 e^{-u_i}}{(1 - e^{-u_i})^2} \tag{7}$$

where $u_i = hv_i/kT$ and the summation is over all the normal modes.

METHOD[6]

A large variety of infrared spectrometers are in current use. In this section, a brief general description will be given for a simple, double-beam instrument, such as the Perkin-Elmer Model 337 shown schematically in Fig. 3.

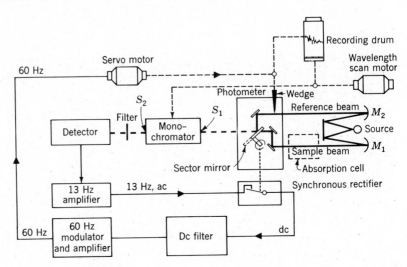

FIGURE 3
Schematic drawing of a double-beam infrared spectrophotometer.

The source provides a continuous spectrum of infrared radiation. Two common sources are the "globar" and the "Nernst glower," which are heated, respectively, to red-orange or white heat by passage of an electric current. Two separate images of the source, the reference beam and the sample beam, are collimated by spherical mirrors M_1 and M_2 and brought to a focus at the entrance slit S_1 of the monochromator. The absorption cell containing the sample is placed in the sample beam as shown. The sample absorbs certain characteristic frequencies in this beam. The other beam serves as a reference. The two beams are recombined by a rotating semicircular sector mirror into a single beam consisting of alternate pulses of sample and reference radiation, which passes into the monochromator through the entrance slit S_1. In the monochromator, the beam of radiation is dispersed so that a narrow band of almost monochromatic radiation leaves through the exit slit S_2. This dispersion can be accomplished by either a prism or a diffraction grating. Diffraction gratings give greater dispersion; however, several different orders may appear at a given grating angle, necessitating the interposition of a filter between the exit slit of the monochromator and the detector. Table 1 lists several prism materials and the region in which each has its best resolution. The monochromator shown in Fig. 4 is of a Littrow design; that is, the radiation is passed twice through the prism P to obtain increased dispersion. By rotation of the prism, it is possible to scan through a wide spectral range. The scan motor which drives the prism or grating is linked mechanically to a recording drum. Thus, a given drum position always corresponds exactly to a particular prism (or grating) setting.

Table 1

Prism material	Region of best resolution, cm^{-1}
LiF	2000–4000
CaF_2	1200–2400
NaCl	700–1400
KBr	400–800

The radiation passing through the exit slit S_2 of Fig. 4 is focused by mirror M_6 on the detector D. Infrared detectors are devices which are very sensitive to heat, such as a thermocouple or bolometer (see Chap. XVIII). If the energy in both the reference and sample beams is equal, the alternating pulses reaching the detector are equal in intensity. The result is that the detector produces a dc voltage which is not amplified by the ac amplifier of the instrument.

When the intensity of the sample beam is reduced by absorption at the charac-

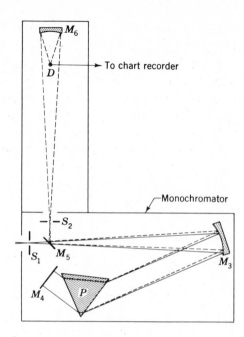

FIGURE 4
Monochromator-detector modules of a
typical spectrophotometer.

teristic wavelengths or frequencies of the sample, the two signals are unequal and the combined beam is modulated at 13 Hz (the rotational frequency of the semicircular sector mirror). The amplitude of the variation is proportional to the difference in intensity of the two beams. The ac portion of the light intensity, having been converted into an alternating voltage by the detector, is amplified by the 13-Hz tuned amplifier. This alternating signal is called the "error signal" and its amplitude is proportional to the difference between true transmittance and the transmittance indicated by the pen position. The synchronous rectifier changes the amplified 13-Hz signal into direct voltage whose polarity indicates which beam, sample or reference, is more intense. This rectifier is just a switch operated by a magnet mounted on the same shaft as the sector mirror. The switch applies the signal to the dc filter only during the half cycle of the sector mirror which admits the sample beam to the detector. The effect therefore is to transmit negative half-wave rectified voltage to the filter when the sample beam is less intense than the reference beam. The filter smooths out the 13-Hz variations before the dc voltage is applied to the 60-Hz modulator and amplifier, where the signal is turned into a 60-Hz voltage of the correct phase to make the servomotor move the optical wedge into the reference beam. When the two beams are the same, the signal disappears and the servomotor stops. The pen indicates the new transmittance because it is mechanically connected to the optical wedge. When the sample beam is more intense than the reference beam, the half-wave rectified voltage is positive, the phase of the 60-Hz voltage is reversed, and the servomotor runs the other way.

The choice of 13 Hz for the photochopper (rotating sector mirror) frequency is dictated by the response time of the detector. The necessity to filter out this variation increases the time it takes the photometer to respond to a change in intensity of the sample beam to about 1 sec. Thus in order to be able to separate closely spaced sharp spectral lines, the scan rate of the wavelength drive must be slow enough to allow the photometer to follow the changes in intensity.

In addition to infrared absorption by the sample contained in the cell, there will be absorption by water vapor and carbon dioxide in the air. With a single-beam instrument it is necessary to sweep the spectrometer with dry nitrogen gas to prevent the occurrence of interfering H_2O and CO_2 bands. In a double-beam spectrophotometer, an empty matched reference cell should be placed in the reference beam to compensate for the *absence* of these gases from the sample cell.

It is necessary to calibrate the spectrometer to know what frequency corresponds to a given prism setting. Such a calibration is performed with standard samples (H_2O vapor, CO_2, NH_3, polystyrene film, etc.) having sharp bands of known frequency. Calibrations are quite stable, and a check at one frequency may be sufficient to verify a previous calibration.

EXPERIMENTAL

Specific instructions for the operation of the infrared spectrometer to be used will be given in the laboratory. **Use this machine carefully**; if in doubt ask the instructor.

The gas cell is constructed from a short (usually 10 cm) length of large-diameter Pyrex tubing with a vacuum stopcock attached. Infrared-transparent windows are sealed on the ends with clear Glyptal resin. If a KBr prism is available for work below 700 cm^{-1}, then KBr windows are needed. If only a NaCl prism is to be used in studying the region from 5000 to 700 cm^{-1}, then NaCl windows will serve. Sodium chloride windows will fog in a moist atmosphere and should be protected when not in use.

Filling the cell   An arrangement for filling the cell is given in Fig. 5. Attach the cell at D. With stopcock C open and B closed, open stopcock A and pump out the system. Make sure that the needle valve of the SO_2 cylinder is closed. Then open B and continue pumping.

Close A and **slowly** open the valve on the SO_2 cylinder. Fill the system to about 1.2 atm with SO_2. Close the valve on the cylinder and then close B and C. Remove the cell from the vacuum line and take a spectrum.

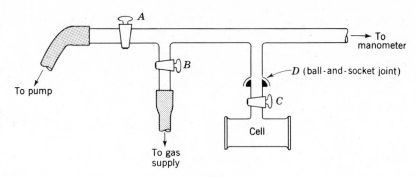

FIGURE 5
Gas-handling system for filling infrared cell.

Return the cell to the line. With B closed and C closed, open A and pump out the line for a few minutes. Now close A and open C. Open A **very slowly** and carefully to reduce the pressure to about 300 Torr. Close A and C, remove the cell, and record the spectrum. Repeat this process and record the spectrum at pressures of 100, 40, 15, and 5 Torr.

Recording of spectra   Several spectra may be recorded on a single chart, using different colored inks to distinguish them. With NaCl windows on the cell and using an NaCl prism it is not possible to observe v_2, the lowest frequency fundamental, which lies below 700 cm^{-1}. If this band is not to be studied, a chart recording in the KBr region may be made available for reference or the literature value of v_2 may be given.

CALCULATIONS AND DISCUSSION

Assign the fundamentals of SO_2 gas and report their frequencies. Then assign any other bands observed. Is it possible to infer the v_2 frequency from a combination band in the NaCl region? Is any indication of rotational structure observed?

Calculate k_1 and k_δ/l^2 from Eqs. (1) and (3) using your values of v_1, v_2, and v_3. Check the valence force model by calculating both sides of Eq. (2) and comparing them.

Using your values of v_1, v_2, and v_3, calculate $\tilde{C}_v(\text{vib})$ at 298 and 500°K from Eq. (7). The quantity u_i is equal to $v_i/0.6951T$ when v_i is in cm^{-1} units. Tables of Einstein functions $u^2 e^{-u}/(1 - e^{-u})^2$ are available.[7] Compare the spectroscopic $\tilde{C}_v = 3R + \tilde{C}_v(\text{vib})$ with the experimental $\tilde{C}_v$ value obtained from directly measured values[8] of $\tilde{C}_p$; $\tilde{C}_v = \tilde{C}_p - R = 7.3$ cal mol^{-1} deg^{-1} at 298°K and 9.0 at 500°K.

APPARATUS

Infrared spectrometer (single-beam instrument such as Perkin-Elmer Model 12 or 112 or double-beam instrument such as Baird Model AB2, Perkin-Elmer Infracord or Model 337, Beckman IR-4); gas cell with NaCl (or KBr) windows; vacuum line for filling cell; cylinder of SO_2 gas with needle valve.

REFERENCES

1. G. Herzberg, "Molecular Spectra and Molecular Structure II. Infrared and Raman Spectra of Polyatomic Molecules," p. 285, Van Nostrand, Princeton, N.J. (1945).
2. *Ibid.*, pp. 251–261.
3. *Ibid.*, pp. 261–269.
4. *Ibid.*, pp. 168–172.
5. W. J. Moore, "Physical Chemistry," 4th ed., pp. 196, 775–781, Prentice-Hall, Englewood Cliffs, N.J. (1972); or T. L. Hill, "An Introduction to Statistical Thermodynamics," pp. 147–162, Addison-Wesley, Reading, Mass. (1960).
6. W. West, Spectroscopy and Spectrophotometry, in A. Weissberger (ed.), "Technique of Organic Chemistry," 3d ed., vol. I, part III, chap. XXVIII, Interscience, New York (1959).
7. K. S. Pitzer, "Quantum Chemistry," Appendix 13, Prentice-Hall, Englewood Cliffs, N.J. (1953).
8. G. N. Lewis and M. Randall (revised by K. S. Pitzer and L. Brewer), "Thermodynamics," 2d ed., p. 419ff., McGraw-Hill, New York (1961).

GENERAL READING

G. Herzberg, *op. cit.*, chaps. II and III.
W. West, *loc. cit.*
J. I. Steinfeld, "Molecules and Radiation," chap. 8, MIT Press, Cambridge, Mass. (1974, 1978).

EXPERIMENT 41.   RAMAN SPECTROSCOPY: VIBRATIONAL SPECTRUM OF CCl₄

In Exp. 40, the vibrational frequencies of a molecule were determined by measuring the direct absorption of infrared radiation due to transitions between vibrational energy levels. Such changes in molecular state can also be caused by inelastic scattering of higher-energy visible photons. The energy (or frequency) *shifts* in the scattered radiation then give a direct measure of the vibrational frequencies of the molecule. Although the intensity of this scattered light is quite low, Sir C. V. Raman was able to observe such scattering in 1928 and the effect was later named after him. With the advent of extremely intense monochromatic laser sources, Raman spectroscopy has become much faster and more sensitive, and instrumentation for such measurements is now reasonably common. Some of the advantages and unique features of Raman spectroscopy will be demonstrated in this experiment on liquid CCl_4.

THEORY

In Exp. 36, the effect of a static or low-frequency electric field on a molecule was considered and it was seen that a dipole moment is induced because of movement of the charged electrons and nuclei. At high optical frequencies ($\sim 10^{15}$ Hz) the nuclei cannot respond rapidly enough to follow the field but polarization of the electron distribution can occur. For an isolated molecule, an oscillating radiation field of intensity $\mathbf{E}$ will induce a dipole moment of magnitude

$$\mu_{ind} = \alpha \mathbf{E} \tag{1}$$

where α is the molecular polarizability. The field $\mathbf{E}$ oscillates at the frequency v_0 of the light,

$$\mathbf{E} = \mathbf{E}_0 \cos 2\pi v_0 t \tag{2}$$

and hence the induced dipole will also change at this frequency. According to classical electromagnetic theory, any oscillating dipole will radiate energy; hence light of frequency v_0 is emitted in all directions (except that parallel to the dipole). Such elastically scattered† light is termed Rayleigh scattering and, typically, one out of a million incident photons will be so scattered.

Inelastic or Raman scattering of light can be understood classically as arising from modulation of the electron distribution, and hence the molecular polarizability, because of vibrations of the nuclei. For example, for a diatomic molecule, α can be represented adequately by the first two terms of a power series in the vibrational coordinate Q:

$$\alpha = \alpha_0 + \left(\frac{d\alpha}{dQ}\right)_0 Q \tag{3}$$

Here the subscript zero indicates that the parameters α_0 and $(d\alpha/dQ)_0$ are evaluated at the equilibrium position. In the harmonic oscillator model, Q oscillates at the

---

† Elastic scattering means no energy change in the scattering process, hence no frequency shift.

vibrational frequency v_v according to the relation

$$Q = A \cos 2\pi v_v t \qquad (4)$$

where A is the maximum amplitude of vibration. Substitution of Eqs. (2) to (4) into Eq. (1) reveals the time dependence of μ_{ind}:

$$\mu_{ind} = \alpha_0 \mathbf{E}_0 \cos 2\pi v_0 t + \left(\frac{d\alpha}{dQ}\right)_0 A\mathbf{E}_0 \cos 2\pi v_0 t \cos 2\pi v_v t \qquad (5)$$

$$= \alpha_0 \mathbf{E}_0 \cos 2\pi v_0 t + \frac{1}{2}\left(\frac{d\alpha}{dQ}\right)_0 A\mathbf{E}_0[\cos 2\pi(v_0 + v_v)t + \cos 2\pi(v_0 - v_v)t]$$

The first term is responsible for Rayleigh scattering at v_0 while the second and third terms produce inelastic Raman scattering at frequencies which are higher (anti-Stokes) and lower (Stokes), respectively, than the incident-light frequency. The frequency v_0 usually corresponds to light in the visible region (typically the 5145-Å line of the argon-ion laser), and the Raman-shifted light then occurs in the visible region but with an intensity that is 10^{-8} to 10^{-12} times that of the incident light.

The quantum mechanical treatment of light scattering[1] predicts the same general results but more explicitly involves the vibrational energy levels and wave functions of the molecule. Figure 1 shows the usual energy level representation of

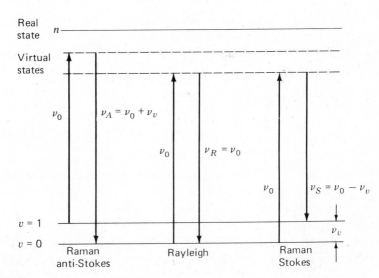

FIGURE 1
Energy levels and transition frequencies involved in Raman spectroscopy.

the scattering process. The virtual states are not real energy levels but serve as convenient intermediate levels for picturing the overall scattering process, which occurs in a time less than the period of a molecular vibration ($\sim 10^{-13}$ sec). Should the incident radiation have a frequency that causes actual absorption to a real state n, the lifetime of the upper state is much longer ($\sim 10^{-8}$ sec) and the resultant intense emission is termed fluorescence. In the absence of such a resonance, the ratio of the Raman intensity of the Raman anti-Stokes and Stokes lines is predicted[2] to be

$$\frac{I_A}{I_S} = \left(\frac{v_0 + v_v}{v_0 - v_v}\right)^4 \exp \frac{-hv_v}{kT} \tag{6}$$

Because of the v^4 dependence, blue light is scattered more efficiently than red (a fact which accounts for the blue color of the sky). However, the Boltzmann exponential factor is dominant in Eq. (6), and the anti-Stokes features are always much weaker than the corresponding Stokes lines.

Raman selection rules For polyatomic molecules, a number of Stokes Raman bands are observed, each corresponding to an allowed transition between two vibrational energy levels of the molecule. (An allowed transition is one for which the intensity is not uniquely zero.) As in the case of infrared spectroscopy (see Exp. 40), only the fundamental transitions (corresponding to frequencies v_1, v_2, v_3, ...) are usually intense enough to be observed, although weak overtone and combination Raman bands are sometimes detected. For molecules with appreciable symmetry, some fundamental transitions may be absent in the Raman and/or infrared spectra. The essential requirement is that the transition moment P (whose square determines the intensity) be nonzero; i.e.,

$$P_{\text{ind}}(0 \rightarrow 1) = \int \psi_0 \mu_{\text{ind}} \psi_1 \, dQ = E \int \psi_0 \alpha \psi_1 \, dQ \neq 0 \tag{7}$$

for the Raman case, and

$$P(0 \rightarrow 1) = \int \psi_0 \mu \psi_1 \, dQ \neq 0 \tag{8}$$

for the infrared case. Here ψ_0 and ψ_1 are wave functions for states with vibrational quantum numbers $v = 0$ and $v = 1$, and the integration extends over the full range of the coordinate Q. Since these two constraints are not the same, infrared and Raman spectra generally differ in the number, frequencies, and relative intensities of observed bands. The prediction of allowed transitions for both cases represents an elegant example of the role of symmetry and group theory in chemistry. Although a discussion of this topic is beyond the scope of this book, the application of group theory to vibrational spectroscopy is presented in several standard texts (e.g., Ref. 2

to 4). A summary of the conclusions reached for a few simple molecular structures is provided in Table 1. Here, the general form of the vibrational coordinates is shown with the mode numbering and the infrared and Raman activity for the fundamental vibrations. Each molecular type has a unique set of symmetry elements (rotation axes, reflection planes, etc.). Associated with these elements are operations (such as rotation about symmetry axes, reflections, etc.) which transform the molecule from the one configuration in space to another which is indistinguishable. This set of operations defines the point group, which is indicated in the table. The point group is associated with a unique set of symmetry species (denoted Σ_g^+, A_1, B_1, E, F_2, etc.), each of which behaves differently under the symmetry operations of the molecule. Every vibrational coordinate (and wave function) belongs to one of these species, as indicated in the table; and it is knowledge of this symmetry which permits the prediction of the infrared and Raman activity of a given fundamental transition.

Raman depolarization ratio   A more complete discussion of vibrational selection rules must take into account the vector properties of the dipole moment and the tensor character of the polarizability α. Thus, Eq. (1) should be written

$$\begin{pmatrix} \mu_x \\ \mu_y \\ \mu_z \end{pmatrix}_{\text{ind}} = \begin{pmatrix} \alpha_{xx} & \alpha_{xy} & \alpha_{xz} \\ \alpha_{yx} & \alpha_{yy} & \alpha_{yz} \\ \alpha_{zx} & \alpha_{zy} & \alpha_{zz} \end{pmatrix} \begin{pmatrix} E_x \\ E_y \\ E_z \end{pmatrix} \tag{9}$$

Such considerations are of considerable importance in the study of oriented molecules, as in a crystal, but are less important for liquids and gases where the molecular orientation is random. However, in the case of Raman spectroscopy, not all information about the components of α is lost by orientational averaging. In particular, one finds that it is possible to distinguish totally symmetric from other molecular vibrations from a measurement of the depolarization ratio

$$\rho_l = \frac{I_\perp}{I_{\parallel}} \tag{10}$$

of the Raman lines. Here, $I_\perp$ and $I_{\parallel}$ are the intensities of scattered light with polarization perpendicular ($\perp$) and parallel ($\parallel$) to the polarization of the linearly polarized exciting light. Theory shows that ρ_l is related to combinations of the tensor components of α and that $\rho_l = \frac{3}{4}$ for any vibration which is *not* totally symmetric and $0 \le \rho_l \le \frac{3}{4}$ for those that are totally symmetric.[5] Raman bands for which $\rho_l = \frac{3}{4}$ are called depolarized, and those with $\rho_l < \frac{3}{4}$ are called polarized. Totally symmetric vibrations are those in which the symmetry of the molecule does not change during the vibrational motion (e.g., the Σ_g^+, A_1', and A_1 vibrations of Table 1). For such vibrations, ρ_l values near zero are observed for highly symmetric

Table 1 NORMAL MODES OF VIBRATION FOR SEVERAL MOLECULAR TYPES

Type	Point group	Vibrational modes,[a] symmetries, infrared-Raman activities, and Raman polarizations[b]			

B—A—B $D_{\infty h}$

ν_1

Σ_g^+ (Rp)

ν_{2x} ν_{2y}

Π_u (IR)

ν_3

Σ_u^- (IR)

B—A—B C_{2v}

ν_1

A_1 (IR, Rp)

ν_2

A_1 (IR, Rp)

ν_3

B_1 (IR, Rdp)

A—B D_{3h}

ν_1

A_1' (Rp)

ν_2

A_2'' (IR)

ν_3

E' (IR, Rdp)

ν_4

E' (IR, Rdp)

C_{3v}

ν_1

A_1 (IR, Rp)

ν_2

A_1 (IR, Rp)

ν_3

E (IR, Rdp)

ν_4

E (IR, Rdp)

T_d

ν_1

A_1 (Rp)

ν_2

E (Rdp)

ν_3

F_2 (IR, Rdp)

ν_4

F_2 (IR, Rdp)

[a] Only one of two E or three F modes in the respective degenerate sets is shown for the AB$_3$ and AB$_4$ cases. Each mode of a degenerate set oscillates at the same frequency, and the description of the nuclear motion associated with each component of a degenerate mode is somewhat arbitrary.

[b] Rp indicates an allowed Raman band that is polarized; Rdp indicates an allowed depolarized Raman band.

molecules such as CO_2, SF_6, and CCl_4. Thus, the measurement of ρ_l will be used in this experiment to aid in assigning the frequency of the totally symmetric v_1 vibration of CCl_4.

Valence force model The simplest harmonic oscillator model for the potential energy of a tetrahedral molecule such as CCl_4 can be written as

$$V = \tfrac{1}{2}k(R_1^2 + R_2^2 + R_3^2 + R_4^2) + \tfrac{1}{2}k_\delta(\delta_{12}^2 + \delta_{13}^2 + \delta_{14}^2 + \delta_{23}^2 + \delta_{24}^2 + \delta_{34}^2)$$
$$(11)$$

where R_i is the change in the length of CCl bond i (whose equilibrium length is l) and δ_{ij} is the change in the angle between bonds i and j. From classical mechanics one obtains[6] the following relations between the vibrational frequencies and the force constants k and k_δ/l^2:

$$4\pi^2 v_1^2 = \frac{k}{m_{Cl}} \tag{12}$$

$$4\pi^2 v_2^2 = \left(\frac{3}{m_{Cl}}\right)\frac{k_\delta}{l^2} \tag{13}$$

$$4\pi^2(v_3^2 + v_4^2) = \left(1 + \frac{4m_{Cl}}{3m_C}\right)\frac{k}{m_{Cl}} + \left(1 + \frac{8m_{Cl}}{3m_C}\right)\left(\frac{2}{m_{Cl}}\right)\frac{k_\delta}{l^2} \tag{14}$$

$$16\pi^4 v_3^2 v_4^2 = \left(1 + \frac{4m_{Cl}}{m_C}\right)\left(\frac{2}{m_{Cl}}\right)\frac{k_\delta}{l^2}\frac{k}{m_{Cl}} \tag{15}$$

(See Exp. 40 for a discussion of the units to be used in these equations.) Since there are two force constants and four frequencies, one can test the quality of this force field by separately determining k and k_δ/l^2 from Eqs. (12) and (13) and from Eqs. (14) and (15).

METHOD

The chief components of a Raman spectrometer are a light source, optics for sample illumination and the collection of scattered light, a dispersing monochromator, a photomultiplier detector, and the necessary electronics for signal processing (Fig. 2). Since about 1965, laser sources have completely replaced mercury-arc sources because lasers provide much higher excitation intensities and greatly simplify sample illumination and scattered-light collection problems. The most common excitation source is the 5145-Å or the 4880-Å line of an argon-ion laser. Typical powers are 0.1 to 2 W in a spectral width of 0.25 cm^{-1} or less. The laser beam is normally focussed into the sample by a lens (not shown in Fig. 2) to form a

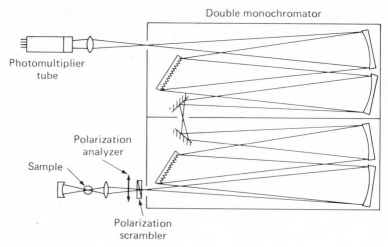

FIGURE 2

Schematic diagram of a Raman spectrometer. The exciting laser beam enters from below ($\perp$ to the plane of the drawing).

scattering cylinder approximately 0.1 mm in diameter and 10 mm in length. The scattered radiation is usually collected at an angle of 90° to the direction of the source beam by a large lens and is dispersed by a monochromator. Since the Raman light intensity is quite low, stray Rayleigh light scattered off the slits, gratings, and mirrors of the spectrometer must be minimized. For this purpose, a double (or triple) monochromator is employed. Further reduction of stray light is achieved by using holographically ruled gratings with 1200 to 2400 lines per millimeter for high dispersion and hence high resolution. The detector is a sensitive photomultiplier which can sense single photons with a quantum efficiency of about 20 percent. A photoelectron produced at the cathode of the tube is amplified by a factor of approximately 10^6 as it is accelerated through the 10- to 16-dynode stages of the photomultiplier. The resulting current pulse at the anode is easily measured; and, since the amplitude of this signal pulse is usually quite different from that of a random noise pulse, discriminator circuitry can be used to reject most of the background pulses. The signal pulses are then counted and an analog voltage proportional to the pulse rate is displayed on a chart recorder. The gratings in the spectrometer are rotated synchronously so that scattered light of various frequencies is swept across the exit slit to the detector in the course of a scan, and the recorder indicates the scattered light intensity as a function of frequency. Vibrational frequencies are deduced from this Raman spectrum by measuring the wavenumber values of each peak and finding the shift from the known exciting frequency.

The spectrometer is easily calibrated by using the emission lines of a mercury or neon lamp.

If depolarization ratios are to be measured, a polaroid analyzer and a scrambler are placed in front of the entrance slit of the spectrometer. Since the laser source is polarized, ρ_l is obtained by measuring $I_{\parallel}$ and $I_{\perp}$ when the polaroid is oriented to pass only light whose polarization is parallel or perpendicular, respectively, to the polarization of the exciting light (see Fig. 3). The scrambler serves to completely depolarize the light passed by the polaroid; this is necessary since the reflection efficiency of the gratings in the spectrometer is polarization-dependent. It should be mentioned that the theoretical value of $\frac{3}{4}$ for depolarized Raman bands applies strictly only for 90° scattering. Since the exciting source is focussed and the collection angle is usually large, measured values of ρ_l can exceed 3/4. If very precise values are desired, a collimated excitation beam and a smaller collection angle near 90° is used in conjunction with calibration by bands of known ρ_l value.

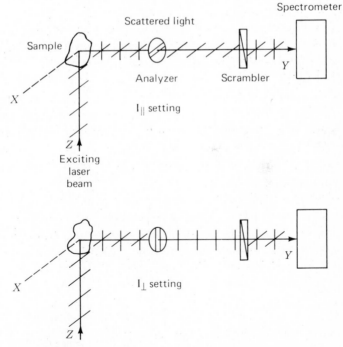

FIGURE 3
Polarization arrangements for Raman depolarization measurements.

EXPERIMENTAL

Detailed instructions for the use of the Raman spectrometer will be provided in the laboratory.† *It is particularly important to follow all safety instructions regarding the use of the excitation laser since serious eye damage can occur if the beam, or even a small reflection from it, should strike the eye.* **Safety goggles must be worn when inserting samples in the beam and when adjusting the optics for maximum signal.**

A convenient liquid cell for Raman spectroscopy is a 1-cm glass or quartz cuvette, with a clear bottom if the excitation beam enters from below. Alternatively, melting-point capillaries can serve as inexpensive cells; these have the added virtue when studying a scarce or expensive sample that very little sample volume is required. A 4-in. length of thin-wall capillary tubing is melted at the center and separated to form two 2-in. tubes. Any excess glass is removed and the end sealed while twirling to form a hemispherical bottom. This tube can be used as is, or a flat bottom can be achieved by inserting a flat stainless steel or tungsten rod and pressing the heated bottom against a flat carbon plate. A hypodermic syringe can be used to add sample to the capillary up to a height of 1 to 2 cm. The top can then be sealed if the sample is sensitive to air or volatile.

The scattering cylinder of the focussed laser beam must be aligned vertically to be parallel to the entrance slit of the spectrometer. The collection lens is used to image this cylinder onto the slit. The sample is then inserted at the laser focal point with its long axis parallel to the beam and a quick survey scan is taken until a Raman band is found. The sample, the excitation beam, and the collection lens are then carefully adjusted for maximum signal at this spectrometer setting.

Record the Stokes Raman spectrum of CCl_4 for shifts of 150 to 1000 cm^{-1} from the exciting line. Record both parallel and perpendicular polarization scans so that you can determine the depolarization ratio of all bands. Indicate the spectrometer reading on the chart at several points in the scan in order to determine the Raman shifts for all bands.

Block the laser beam or spectrometer entrance slit and adjust the spectrometer to an anti-Stokes shift of 1000 cm^{-1}. **Caution:** *exposure of the sensitive phototube to the intense Rayleigh scattering line can seriously damage the detector.* Scan the anti-Stokes spectrum from -1000 to -150 cm^{-1} in the parallel polariza-

---

† If a Raman spectrometer is not available, a suitable Stokes and anti-Stokes scan of CCl_4 is provided in the review of Raman spectroscopy in Ref. 7. A high-resolution scan of the 460 cm^{-1} band of liquid and solid CCl_4 is available in Ref. 8. Several Raman "dry lab" experiments have been described in Refs. 9 and 10.

tion configuration and, using appropriate sensitivity expansion,† measure the ratio of anti-Stokes to Stokes peak heights for each band.

You will note some structure in the Stokes band near 460 cm^{-1}, which is due to chlorine isotopic frequency shifts. (See Exp. 42 for a discussion of isotope effects in diatomic molecules.) Rescan this region at higher resolution (1 cm^{-1} or less) with an expansion of the chart display and measure the frequencies of each of the components.

CALCULATIONS AND DISCUSSION

Assign the fundamentals of CCl_4 using the fact that the v_1 and v_3 CCl_4 stretching motions are expected to occur at higher frequencies than the v_2 and v_4 CCl_4 bending motions. The depolarization ratios can be used to determine which of the two high-frequency bands should be assigned as v_1. To decide on the v_2 and v_4 assignments, use the fact that the only infrared band observed below 600 cm^{-1} is a feature near 300 cm^{-1}.

The v_3 vibration will be observed as a doublet due to Fermi-resonance splitting, which can occur if an overtone or combination energy level happens to fall near the energy level of a fundamental vibration and if both levels have the same symmetry. The v_3 frequency can be taken as the mean of the doubtlet peaks. From your values of v_1, v_2, and v_4, deduce which combination band is involved in the Fermi-resonance with v_3.

Tabulate your Stokes and anti-Stokes frequencies and also the intensity ratios. Use Eq. (6) to calculate a temperature for each mode and compare the resulting values with room temperature.

Calculate the valence force constants k and k_δ/l^2 from Eqs. (12) and (13), and then compare these values with those obtained from Eqs. (14) and (15). A more complete force field analysis[11] predicts that k in Eq. (12) should be replaced by $k_r + 3k_{rr}$ while k in Eqs. (14) and (15) should be replaced by $k_r - k_{rr}$, where k_r is the bond stretching force constant while k_{rr} is a constant due to bond-bond interaction. Is k_{rr} important in CCl_4?

---

† Known scale expansions should be chosen which give peaks that are nearly full scale so that the heights can be accurately measured. Some errors will arise in the measurement of Stokes and anti-Stokes intensities because of variation with frequency of spectrometer transmission and detector efficiency. These effects should be unimportant for bands involving small Raman shifts, and calibration corrections can be made if necessary for very accurate measurements.

From the natural abundance of ^{35}Cl (75.5 percent) and ^{37}Cl (24.5 percent), calculate the relative intensities expected for the various components of the band near 460 cm^{-1} and compare these with your observations. Suggest a reasonable modification of one of Eqs. (12) to (15) which would permit you to predict the frequency spacing of these components.

Alternative or additional liquids suitable for Raman investigation might include CS_2, PCl_3, $CHCl_3$, or benzene. Aqueous solutions can also be studied since the Raman spectrum of water is weak and quite broad; possible samples would be 1 to 3 M solutions of Na_2CO_3, $NaNO_3$, Na_2SO_3, or Na_2SO_4. (Such samples are difficult to study with infrared spectroscopy because of solvent absorption and the solubility of most infrared cell windows.)

APPARATUS

Raman spectrometer with argon- or krypton-ion laser source; liquid sample cell or melting-point capillaries; CCl_4; safety goggles (such as those available from Glendale Optical Co., 130 Crossways Park Drive, Woodbury, NY 11797).

REFERENCES

1. J. A. Koningstein, "Introduction to the Theory of the Raman Effect," Reidel, Dordrecht, Holland (1972).
2. E. B. Wilson, Jr., J. C. Decius, and P. C. Cross, "Molecular Vibrations," p. 51, McGraw-Hill, New York (1955).
3. D. C. Harris and M. D. Bertolucci, "Symmetry and Spectroscopy," Oxford, London (1978).
4. F. A. Cotton, "Chemical Applications of Group Theory," Wiley-Interscience, New York (1963).
5. E. B. Wilson, Jr., J. C. Decius, and P. C. Cross, op. cit., p. 47.
6. G. Herzberg, "Molecular Spectra and Molecular Structure II. Infrared and Raman Spectra of Polyatomic Molecules," p. 182, Van Nostrand, Princeton, N.J. (1945).
7. R. S. Tobias, J. Chem. Ed., 44, 2, 70 (1967).
8. H. J. Sloane, Appl. Spectrosc., 25, 430 (1971).
9. F. P. DeHaan, J. C. Thibeault, and D. K. Ottesen, J. Chem. Ed., 51, 263 (1974).
10. L. C. Hoskins, J. Chem. Ed., 52, 568 (1975); 54, 642 (1977).
11. E. B. Wilson, Jr., J. C. Decius, and P. C. Cross, op. cit., p. 131.

GENERAL READING

R. S. Tobias, op. cit.
G. Herzberg, op. cit., chap. II.
F. A. Cotton, op. cit., chaps. 1–5, 9.

EXPERIMENT 42.   ROTATION-VIBRATION SPECTRUM OF HCl AND DCl

This experiment is concerned with the rotational fine structure of the infrared vibrational spectrum of linear molecules such as HCl. From an interpretation of the details of this spectrum it is possible to obtain the moment of inertia of the molecule and thus the internuclear separation. In addition, the pure vibrational frequency determines a force constant which is a measure of the bond strength. By a study of DCl also, the isotope effect can be observed.

THEORY

The simplest model of a vibrating diatomic molecule is a harmonic oscillator, for which the potential energy depends quadratically on the change in internuclear distance. The allowed energy levels of a harmonic oscillator, as calculated from quantum mechanics,[1] are

$$E(v) = hv(v + \tfrac{1}{2}) \tag{1}$$

where v is the vibrational quantum number having integral values $0, 1, 2, \ldots$, v is the vibrational frequency, and h is Planck's constant.

The simplest model of a rotating diatomic molecule is a rigid rotor or "dumbbell" model in which the two atoms of mass m_1 and m_2 are considered to be joined by a rigid, weightless rod. The allowed energy levels for a rigid rotor may be shown by quantum mechanics[1] to be

$$E(J) = \frac{h^2}{8\pi^2 I} J(J + 1) \tag{2}$$

where the rotational quantum number J may take integral values $0, 1, 2, \ldots$. The quantity I is the moment of inertia, which is related to the internuclear distance r and the reduced mass $\mu = m_1 m_2/(m_1 + m_2)$ by

$$I = \mu r^2 \tag{3}$$

Since a real molecule is undergoing both rotation and vibration simultaneously, a first approximation to its energy levels $E(v,J)$ would be the sum of expressions (1) and (2). A more complete expression for the energy levels of a diatomic molecule[2,3] is given below, with the levels expressed as *term values* T in cm^{-1} units rather than as energy values E in ergs:

$$T(v, J) = \frac{E(v, J)}{hc} = \tilde{v}_e(v + \tfrac{1}{2}) - x_e \tilde{v}_e(v + \tfrac{1}{2})^2 + B_e J(J + 1)$$

$$- D_e J^2(J + 1)^2 - \alpha_e(v + \tfrac{1}{2})J(J + 1) \tag{4}$$

where $\tilde{\nu}_e$ is the frequency in cm^{-1} for the molecule vibrating about its equilibrium internuclear separation r_e and

$$B_e = \frac{h}{8\pi^2 I_e c} \qquad \text{.} \tag{5}$$

The first and third terms on the right-hand side of Eq. (4) are the harmonic-oscillator and rigid-rotor terms with r equal r_e. The second term (involving the constant x_e) takes into account the effect of anharmonicity. Since the real potential $V(r)$ for a molecule differs from a harmonic potential V_{harm} (see Fig. 1), the real vibrational levels are not quite those given by Eq. (1) and a correction term is required. The fourth term (involving the constant D_e) takes into account the effect of centrifugal stretching. Since a chemical bond is not truly rigid but more like a stiff spring, it stretches somewhat when the molecule rotates. Such an effect is important only for high J values, since the constant D_e is very small; we shall neglect this term. The last term in Eq. (4) accounts for interaction between vibration and rotation. During a vibration the internuclear distance r changes; this changes the moment of inertia and affects the rotation of the molecule. The constant α_e is also quite small, but this term should not be neglected.

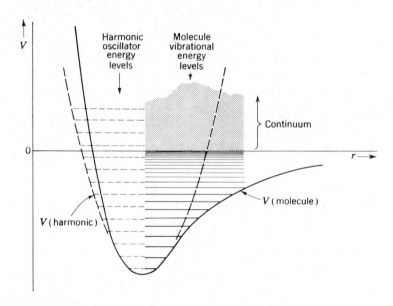

FIGURE 1

Schematic diagram showing potential energy V as a function of internuclear separation r for a diatomic molecule. The harmonic potential V_{harm} is indicated by the dashed curve. The vibrational levels are also shown.

Selection rules The harmonic-oscillator, rigid-rotor selection rules[3] are $\Delta v = \pm 1$ and $\Delta J = \pm 1$; that is, infrared emission or absorption can occur only when these "allowed" transitions take place. For an anharmonic diatomic molecule, the $\Delta J = \pm 1$ selection rule is still valid but transitions corresponding to $\Delta v = \pm 2, \pm 3$, etc. (overtones) can now be observed weakly.[3] Since we are interested in the most intense absorption band (the "fundamental"), we are concerned with transitions from various J'' levels of the vibrational ground state ($v'' = 0$) to J' levels in the first excited vibrational state ($v' = 1$). From the selection rule we know that the transition must be from J'' to $J' = J'' \pm 1$. Since $\Delta E = h v = h c \tilde{v}$, the frequency $\tilde{v}$ (in wavenumbers)

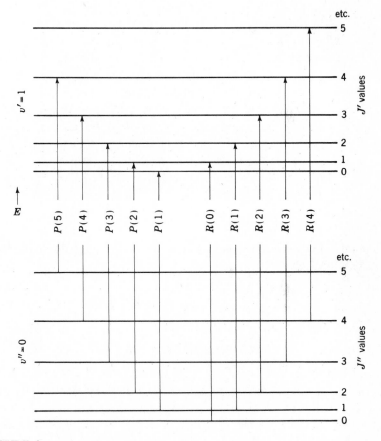

FIGURE 2
Rotational energy levels for the ground vibrational state ($v'' = 0$) and the first excited vibrational state ($v' = 1$) in a diatomic molecule. The vertical arrows indicate allowed transitions in the R and P branches; numbers in parentheses index the value J'' of the lower state. Transitions in the Q branch ($\Delta J = 0$) are not shown since they are not infrared active.

for this transition will be just $T(v',J') - T(v'',J'')$. When $\Delta J = +1$ $(J' = J'' + 1)$ and $\Delta J = -1$ $(J' = J'' - 1)$, we find, respectively, from Eq. (4) that

$$\tilde{v}_R = \tilde{v}_0 + (2B_e - 3\alpha_e) + (2B_e - 4\alpha_e)J'' - \alpha_e J''^2 \qquad J'' = 0, 1, 2, \ldots \qquad (6)$$

$$\tilde{v}_P = \tilde{v}_0 - (2B_e - 2\alpha_e)J'' - \alpha_e J''^2 \qquad\qquad\qquad J'' = 1, 2, 3, \ldots \qquad (7)$$

where $\tilde{v}_0$, the frequency of the *forbidden* transition from $v'' = 0$, $J'' = 0$ to $v' = 1$, $J' = 0$, is

$$\tilde{v}_0 = \tilde{v}_e - 2x_e\tilde{v}_e \qquad (8)$$

The two series of lines given in Eqs. (6) and (7) are called R and P branches, respectively. These allowed transitions are indicated in the energy-level diagram (Fig. 2). If α_e were negligible, Eqs. (6) and (7) would predict a series of equally spaced lines with separation $2B_e$ except for a missing line at $\tilde{v}_0$. The effect of interaction between rotation and vibration (nonzero α_e) is to draw the lines in the R branch closer together and spread the lines in the P branch farther apart as shown for a typical spectrum in Fig. 3. For convenience let us introduce a new quantity m, where $m = J'' + 1$ for the R branch and $m = -J''$ for the P branch as shown in Fig. 3. It is now possible to replace Eqs. (6) and (7) by a single equation

$$\tilde{v} = \tilde{v}_0 + (2B_e - 2\alpha_e)m - \alpha_e m^2 \qquad (9)$$

where m takes all integral values except zero. The separation between adjacent lines $\Delta\tilde{v}(m)$ will be

$$\Delta\tilde{v}(m) = \tilde{v}(m + 1) - \tilde{v}(m) = (2B_e - 3\alpha_e) - 2\alpha_e m \qquad (10)$$

Thus a plot of $\Delta\tilde{v}(m)$ versus m can be used to determine both B_e and α_e. With these known, $\tilde{v}_0$ can be obtained from any of the lines by using Eq. (9).

Isotope effect When an isotopic substitution is made in a diatomic molecule, the equilibrium bond length r_e and the force constant k are unchanged, since they depend only on the behavior of the bonding electrons. However, the reduced mass μ does change, and this will affect the vibration and rotation of the molecule. For a harmonic oscillator model, the frequency v_{harm} is given by[1]

$$v_{harm} = \frac{1}{2\pi}\left(\frac{k}{\mu}\right)^{1/2} \qquad (11)$$

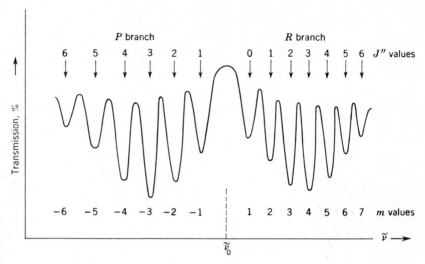

FIGURE 3
Schematic rotation-vibration infrared spectrum for a diatomic molecule.

If an asterisk is used to distinguish one isotopic molecule from another, Eq. (11) leads to the relation

$$\frac{\nu^*_{harm}}{\nu_{harm}} = \left(\frac{\mu}{\mu^*}\right)^{1/2} \tag{12}$$

For a real molecule the anharmonicity should be taken into account,[3] but since this is complicated and would require information from overtone vibrations ($\Delta v > 1$), we shall take Eq. (12) as an approximation and state that $\tilde{\nu}^*_0/\tilde{\nu}_0$ should be close to $(\mu/\mu^*)^{1/2}$. In the case of rotation, the isotope effect can be easily stated. From the definitions of B_e and I, we see that

$$\frac{B^*_e}{B_e} = \frac{\mu}{\mu^*} \tag{13}$$

Since HCl gas is a mixture of HCl^{35} and HCl^{37} molecules, an isotope effect should be present. However, HCl is predominantly HCl^{35} and the ratio of the reduced masses is only 1.0015; therefore only quite high resolution work would detect this effect. For this experiment, we shall assume that the HCl bands obtained are those of HCl^{35}.

If deuterium is substituted for hydrogen, the ratio of the reduced masses, $\mu(DCl^{35})/\mu(HCl^{35})$, is 1.946. Thus there should be a very large isotope effect.

EXPERIMENTAL

A general description of infrared experimental techniques is given in Exp. 40. For this experiment high resolution is required; an LiF (or CaF$_2$) prism or a grating must be used. In addition, a careful calibration is required; CO and CH$_4$ are suitable gases for checking the calibration, if this is necessary. Detailed instructions for operating the spectrometer will be given in the laboratory. **Use this instrument carefully.**

The cell design and procedure for filling the cell are identical with those described in Exp. 40. Hydrogen chloride gas is available commercially in small cylinders. Make several rapid scans of the HCl spectrum at various pressures to determine what pressure gives the best spectrum, and then record the spectrum at a very slow chart speed. Narrow slit widths are necessary to obtain the desired sharp bands.

Preparation of DCl   If DCl gas is not available in a commercial cylinder or bulb, it must be prepared in the laboratory. Deuterium chloride gas can be synthesized readily by the reaction between benzoyl chloride and heavy water:

$$C_6H_5COCl + D_2O \rightarrow C_6H_5COOD + DCl(g)$$
$$C_6H_5COOD + C_6H_5COCl \rightarrow (C_6H_5CO)_2O + DCl(g) \qquad (14)$$

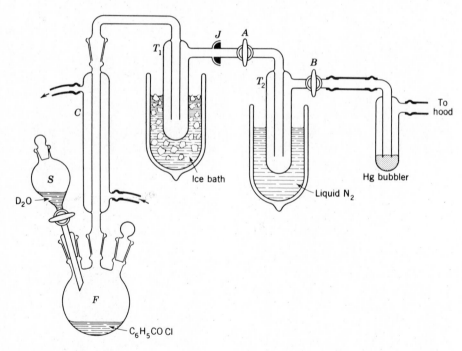

FIGURE 4
Apparatus for preparing deuterium chloride.

An arrangement for carrying out this reaction is shown in Fig. 4. Approximately 5 ml (0.3 mol) of D_2O in the separatory funnel S is added slowly to 140 g (1 mol) of benzoyl chloride in flask F. With a water-cooled reflux condenser C attached and either ice or Dry Ice and trichloroethylene in trap T_1, the flask is gently heated. Stopcocks A and B are left open, and the system is swept out by allowing some DCl gas to escape through the mercury bubbler into a hood. Stopcock B is then closed, and trap T_2 is cooled with liquid nitrogen. After 20 min remove the heat from flask F. Wait a few minutes, then close stopcock A and disconnect the ball-and-socket joint J. Keep trap T_2 in liquid nitrogen, and attach it to the system used to fill the cell (at stopcock B shown in Fig. 40-5). After the line is pumped out, the cell can be filled by allowing the DCl in the trap to warm up slowly until the desired pressure is achieved. (If a bulb of DCl gas is available, use the procedure given in Exp. 40.)

CALCULATIONS

Select your best HCl and your best DCl spectrum, and index the lines with the appropriate m values (see Fig. 3). Make a table of these m values and the corresponding frequencies $\tilde{v}(m)$. Express the frequencies in cm^{-1} units to tenths of a cm^{-1} (if possible). Then list the differences between adjacent lines $\Delta\tilde{v}(m)$, which will be roughly $2B_e$ but should vary with m. Plot $\Delta\tilde{v}(m)$ against m and draw the best straight line through the points. Compute B_e and α_e from the intercept $(2B_e - 3\alpha_e)$ at $m = 0$ and the slope $(-2\alpha_e)$. Using these B_e and α_e values and Eq. (9), calculate $\tilde{v}_0$ from several of the lines with low m values. For both HCl and DCl, present a table of $\tilde{v}_0$, B_e, and α_e values all in cm^{-1} units.

Alternatively, a least-squares computer program may be used to find B_e and α_e from the observed data.

Calculate I_e, the moment of inertia, and r_e, the internuclear distance, for both HCl and DCl. Compute the ratios $\tilde{v}_0^*/\tilde{v}$ and B_e^*/B_e and compare them with the predicted values given by Eqs. (12) and (13). Do your spectra show any indications of a Cl^{35}–Cl^{37} isotope effect? Calculate the splitting expected for this effect in HCl and in DCl. Explain the observed band intensities in terms of the Boltzmann population of the ground-state levels.

DISCUSSION

An interesting extension of this experiment is to use spectroscopic data to calculate the equilibrium constant K for the gas-phase exchange reaction

$$DCl + HBr = HCl + DBr \qquad (15)$$

From statistical thermodynamics,[4] one can show in the ideal-gas approximation that K for isotopic exchange is related to the molecular partition functions z by the equations

$$K \equiv \frac{p_{HCl}p_{DBr}}{p_{DCl}p_{HBr}} = \frac{z_{HCl}z_{DBr}}{z_{DCl}z_{HBr}} \tag{16}$$

$$= \left[\frac{M_{HCl}M_{DBr}}{M_{DCl}M_{HBr}}\right]^{3/2} \left[\frac{B_{DCl}B_{HBr}}{B_{HCl}B_{DBr}}\right] \left[\frac{z_{HCl}{}^{vib}z_{DBr}{}^{vib}}{z_{DCl}{}^{vib}z_{HBr}{}^{vib}}\right] \tag{17}$$

The term involving the molecular weights M_i comes from the translational contribution to K; the term involving B_i values represents the rotational contribution; and $z_i{}^{vib}$ is the vibrational partition function for species i defined in Eq. (40-5). The expression given in Eq. (17) corresponds to the rigid-rotor, harmonic-oscillator approximation. If the values of B and v are determined experimentally for HBr gas (or taken from Ref. 3), one can use the isotopic mass relations given in Eqs. (12) and (13) to obtain B and v for DBr. Combining these data with those on HCl and DCl, a theoretical or "spectroscopic" value of K can be determined from Eq. (17).

An experiment has been described in which infrared intensities are used to

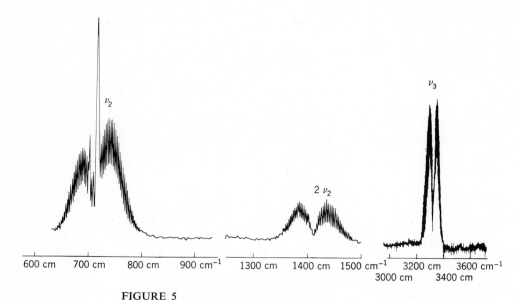

FIGURE 5

Portions of the infrared absorption spectrum of HCN. The v_2 bending vibration is a perpendicular band and therefore has allowed P, Q, and R branches. The v_3 [C—H stretching] vibration is a parallel band with P and R branches only. Note that the $2v_2$ overtone band has the appearance of a parallel band, so that no Q branch is observed.

obtain an experimental determination of K (Ref. 5), but this method is not straight-forward. The accurate measurement of intensities is complicated by pressure-broadening effects and limitations in spectrometer resolution. A preferable and much more accurate determination of K can be made with a mass spectrometer.

Several other linear molecules, such as HCN and DCN or C_2H_2 and C_2D_2, are suitable for extending the study of rotational-vibrational spectra, and these serve to demonstrate that the principles used in diatomic molecule spectroscopy are applicable to polyatomic molecules as well. The infrared absorption spectrum of HCN is reproduced in Fig. 5. The v_3 band (3240 to 3370 cm^{-1} in HCN; 2570 to 2680 cm^{-1} in DCN) is most suitable for analysis; over 20 members of the P and R branches can be resolved in each system. If a Cary 14 spectrophotometer is avail-able, the $(v_1 + v_3)$ C—H stretching combination bands of C_2H_2 at 6661 cm^{-1} (15,000 Å) and of C_2D_2 at 5127 cm^{-1} (19,500 Å) can be analyzed.

From the moments of inertia for both the ^1H and ^2H isotopic species for each pair, the student can solve for the separate C—H and C—N (in HCN) or C—C (in C_2H_2) bond distances. The acetylene spectrum shows the additional feature of a 3 : 1 line intensity alternation due to nuclear spin statistics and the Pauli principle.[2,3]

APPARATUS

High-resolution infrared spectrometer; gas cell with NaCl windows; vacuum line for filling cell; cylinder of HCl gas with needle valve; three-neck round-bottom flask; reflux condenser; glass-stoppered dropping funnel; two traps, one with a stopcock on each arm; mercury bubbler; exhaust hood; bunsen burner or heating mantle; CO and CH_4 gas for calibration check (optional).

Heavy water (5 ml at least 70 percent D_2O); benzoyl chloride (140 g); ice or Dry Ice and trichloroethylene; liquid nitrogen; or a 5-liter flask of DCl gas (available from Stohler Isotope Chemicals, 92 Beckwith Place, Rutherford, N.J.). Optional: gas samples of HBr, HCN, DCN, C_2H_2, and C_2D_2.

REFERENCES

1. W. J. Moore, "Physical Chemistry," 4th ed., pp. 767–770, Prentice-Hall, Englewood Cliffs, N.J. (1972); or L. Pauling and E. B. Wilson, "Introduction to Quantum Mechanics," McGraw-Hill, New York (1935).
2. K. S. Pitzer, "Quantum Chemistry," Prentice-Hall, Englewood Cliffs, N.J. (1953).
3. G. Herzberg, "Molecular Spectra and Molecular Structure I. Spectra of Diatomic Molecules," 2d ed., chap. III, Van Nostrand, Princeton, N.J. (1950).
4. T. L. Hill, "Introduction to Statistical Thermodynamics," chap. 10, Addison-Wesley, Reading, Mass. (1960).
5. J. E. Ruark, J. I. Ivers, and J. L. Roberts, *J. Chem. Ed.*, **51**, 758 (1974).

GENERAL READING

G. Herzberg, *loc. cit.*

W. West, Spectroscopy and Spectrophotometry, in A. Weissberger (ed.), "Technique of Organic Chemistry," 3d ed., vol. I, part III, chap. XXVIII, Interscience, New York (1959).

J. I. Steinfeld, "Molecules and Radiation," chap. 4, MIT Press, Cambridge, Mass. (1974, 1978).

EXPERIMENT 43. SPECTRUM OF THE HYDROGEN ATOM

Since the hydrogen atom has only one orbital electron surrounding a nucleus consisting of a single proton, it has a particularly simple spectrum. In this experiment, part of this spectrum will be determined. The observed frequencies of the lines can then be compared with the values predicted by quantum mechanics.

THEORY

When a high-voltage discharge takes place in H_2 gas at low pressures, many molecules are dissociated into atoms by electron impact. These atoms are in excited electronic states. Such H atoms in excited electronic states spontaneously undergo transitions to lower energy electronic states with the emission of radiation. Quantum mechanics gives an expression[1,2] for the allowed electronic energy levels of an H atom in terms of a single quantum number† n;

$$E_n = -\frac{2\pi^2\mu e^4}{h^2}\frac{1}{n^2} \qquad n = 1, 2, 3, \ldots \tag{1}$$

where e is the charge on the electron and h is Planck's constant. The reduced mass μ is given in terms of the mass of the electron m and the mass of the proton M by

$$\mu = \frac{mM}{m + M} \tag{2}$$

There is no selection rule for n; that is, transitions may occur between any of the levels given by Eq. (1). An energy-level diagram for the H atom is given in Fig. 1. Transitions from upper energy levels to lower levels are shown by the vertical lines. These transitions form several series of lines depending on the quantum

---

† Owing to spin-orbit and quantum electrodynamical effects, the energy levels have a small dependence on another quantum number l, where l can assume any integral value less than n. This energy splitting is very small and can be neglected here.

number of the lower state. Note that the energy levels converge toward a limit as $n \to \infty$. The shaded area above $n = \infty$ indicates a continuum of energy states corresponding to complete separation of the proton and the electron.

We shall be concerned with the Balmer series, the lines of which fall in the visible region of the spectrum. From Eq. (1), it is possible to predict the frequency of a transition from any upper state with quantum number n_1 to any lower state with quantum number n_2, since

$$\Delta E = E_{n_1} - E_{n_2} = h\nu = hc\tilde{\nu} \tag{3}$$

where c is the speed of light and $\tilde{\nu}$ is the frequency in units of cm^{-1} (equal to the reciprocal of the wavelength λ in centimeters). Thus,

$$\tilde{\nu} = \frac{2\pi^2 \mu e^4}{h^3 c} \left(\frac{1}{n_2^{\,2}} - \frac{1}{n_1^{\,2}} \right) = \mathcal{R} \left(\frac{1}{n_2^{\,2}} - \frac{1}{n_1^{\,2}} \right) \tag{4}$$

The quantity $\mathcal{R}$ is called the *Rydberg constant* and has the calculated value 109,678 cm^{-1}. For the Balmer series, $n_2 = 2$ and $n_1 = 3, 4, 5, \ldots$. Equation (4) predicts a series of lines which converge to a high-frequency limit at $\mathcal{R}/n_2^{\,2}$.

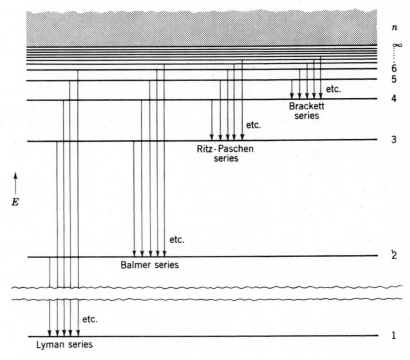

FIGURE 1
Electronic energy-level diagram for the hydrogen atom.

METHOD

The frequency (or wavelength) of these Balmer lines can be determined experimentally by comparing the hydrogen spectrum with a reference spectrum for which the wavelengths are known. The iron or mercury spectra are common choices for reference spectra. (Such reference spectral lines must, of course, have been measured absolutely, as by an interferometric technique.) In this experiment, an Hg lamp will be used because the lines of the Hg spectrum can be identified quite easily. In other work the Fe arc is preferable, since there are many more lines (see Exp. 44).

A variety of spectrographs are in current use for work of this kind; although differing in details they are basically similar. Figure 2 shows a schematic diagram of a typical spectrograph. The sources are mounted in such a way that the hydrogen discharge tube will illuminate half of the entrance slit S and the mercury reference lamp will illuminate the other half. The Hg lamp is off the optical axis of the spectrograph, and a front surfaced mirror M (or a 90° prism) is set at 45° to permit an image of this source to be focused on the slit by lens L_1. The lens L_2 renders the light from the slit parallel before it passes through the dispersing prism P. A camera lens L_3 focuses an image of the slit on the photographic plate.†

The two common prism materials for use in the visible and near-ultraviolet regions are glass and quartz. Glass is preferable in the visible, since it gives a higher dispersion (and therefore better resolution). For wavelengths shorter than about

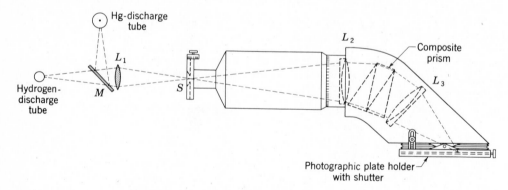

FIGURE 2
Spectrograph and source optics, viewed from above. (NOTE: the mirror M is just below the light path from the hydrogen discharge tube.)

† To align the sources properly, one may open the slit wide and look in at the camera end of the spectrograph. Adjust the source so that it appears in the center of the slit opening. Then close the slit to a narrow opening, and focus the instrument, using the Hg source until the yellow Hg line is well resolved into a doublet sharply focused in the plane of the photographic plate. A ground-glass plate may be placed in the film position and viewed with a small magnifier.

3000 to 3500 Å, most glasses absorb light strongly but quartz prisms may be used. Quartz also begins to absorb light below about 1900 Å. Other prism materials (such as fluorite) may be used below 1900 Å, but diffraction gratings are generally used in the far ultraviolet. Since the Balmer lines are more and more closely spaced and have decreasing intensity as the short-wavelength limit is approached, both high resolution and long exposures are needed to observe lines with high n_1 values.

Sources   The best Hg source is a low-pressure mercury lamp (General Electric G4T4/1, 4W germicidal lamp) which operates from a small ballast transformer (G.E. 89G435, 0.16 A, 4-6-8 W). **There is intense ultraviolet radiation from such a lamp which is dangerous to the eyes.** Therefore a Pyrex glass jacket should be placed over the Hg arc to filter out this ultraviolet radiation.

Many designs for hydrogen discharge tubes are available.[3] A convenient commercial hydrogen source is a Geissler tube made by the Tube Light Engineering Co., New York. In this tube the pressure has been adjusted to give strong Balmer lines (0.1 to 0.5 Torr). In addition, there is always a weak *band* spectrum due to molecular hydrogen; hydrogen atoms, formed by the electrode discharge, may recombine on the walls to give H_2 molecules in excited electronic states which will emit radiation. Such a band spectrum should be so weak that it will not interfere with determining the Balmer lines. The Geissler discharge tube operates from a neon-sign transformer at about 5000 V and 15 mA; it should glow steadily with a bright red color.

Hg reference spectrum   A print of the low-pressure Hg-arc spectrum is shown in Fig. 3. The upper half of the spectrum was strongly exposed in order to bring out the weak lines, and the lower half was exposed only a short time in order to show the

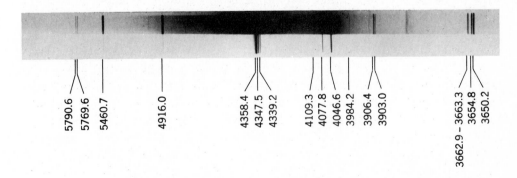

FIGURE 3
Mercury-arc reference spectrum. The upper half of this print has been strongly exposed to show the weak lines clearly. The wavelengths (in angstroms) of the Hg lines are given.

strong lines clearly. The wavelengths of these lines[4] [in air, see Eq. (5)] are given below the spectrum. The spectrum presented in Fig. 3 was made using Kodak 103a-O spectrographic plates; one can also use type 103a-F plates, which are much more sensitive in the long-wavelength end of the visible range.

EXPERIMENTAL

Instructions for the operation of the spectrograph to be used will be given in the laboratory. The placement of the sources and the focus and slit width settings of the spectrograph will be made in advance by the instructor. The sources should be arranged so that one half of the slit is illuminated by the Hg lamp and the other half by the hydrogen discharge tube. A shutter over the slit permits exposures to be made of each source independently without moving the photographic plate. This is important when wavelengths are to be determined by comparison. The photographic plate is placed in a holder which can be raised or lowered to permit several such pairs of spectra to be taken.

The photographic plates are extremely light sensitive. The plate holder must be loaded and unloaded (and developing must be done) in total darkness. The plate should be placed in the holder in such a way that the emulsion side faces the shutter. The emulsion side can be detected in the dark by moistening a finger and touching the plate lightly near an edge. The emulsion side will feel sticky. Having loaded the holder, turn on the lights and place the holder in position on the spectrograph. Set the plate-holder position close to one end of its traverse. Check to see that the slit is properly illuminated by each source. Instructions for operating the sources will be given. The hydrogen discharge tube operates on **high voltage which is possibly lethal**; be very careful. Close the shutter over the slit.

Darken the room and open the shutter on the plate holder. Take an exposure with the hydrogen discharge tube and then with the Hg lamp. Each source should be turned off when not in use. Close the shutter over the slit, move the plate holder, and take another pair of exposures with exposure times different from the first set. Repeat this process several times.

After the final pair of exposures, *close the shutter on the plate holder* and remove the holder from the spectrograph. Turn on the room lights. Set up three developing trays: one with developer, one with water, one with fixer. Set a timer for 3 min. Darken the room and place the plate in developer, *emulsion side up*; start the timer. Agitate gently and remove after 3 min. Drain over the developer tray, rinse briefly in water, and place in the fixer. Allow to remain in the fixer at least 10 min. Lights may be turned on after about 5 min.

Examine the plate. If for some reason the plate is not usable, start over again but omit any long exposures if time does not permit them. If the plate is satisfactory, rinse it under running water for 20 min and then allow it to drain and dry several hours (preferably overnight).

Measurement of spectral lines   If a comparator is available, mount the plate on the stage, focus on a suitable pair of exposures, and align the plate so that the spectral lines at both ends of the plate are parallel with the cross hair of the microscope. Starting at one end of the plate, move the stage or microscope until a line is under the cross hair and record the position. Continue this process for all lines, both Hg and H atom (disregard the H_2 molecular band spectrum). Always approach a line from the same direction to avoid errors due to backlash in the screw.

If a comparator is not available, a good enlarger may be used instead. Mount the plate in the enlarger, and project the spectra onto a large sheet of white paper mounted on a drafting board. Choose a suitable pair of exposures and mark the positions of all Hg lines and all H atom lines (disregard H_2 molecular band spectrum) with a sharp pencil. Avoid working near the edge of the image, where there may be distortion. Using a centimeter scale, measure the positions of all lines relative to some arbitrary zero position. Indicate which lines are Hg and which H.

Compare your Hg spectrum with that shown in Fig. 3, and assign wavelengths to all Hg lines. On a large piece of graph paper, plot a calibration curve of Hg wavelengths versus position in centimeters. Read the wavelengths λ of the hydrogen lines from this curve.

CALCULATIONS

Obtain the frequencies of the Balmer lines $\tilde{\nu}$ (in cm^{-1}) from the wavelength values λ (in Å). Present a table of both $\tilde{\nu}$ and λ values together with the value of n_1 for each line observed.

Plot $\tilde{\nu}$ against $1/n_1{}^2$ on a large sheet of graph paper. If the points fall on a straight line, this is a partial confirmation of Eq. (4). From the slope of this line, determine an experimental value of $\mathscr{R}$. In all this work, wavelengths in air have been used; these are related to vacuum wavelengths by

$$\lambda_{\text{air}} = \frac{\lambda_{\text{vac}}}{n_{\text{air}}} \tag{5}$$

where n_{air} is the index of refraction of air which equals 1.00027 at these wavelengths. To obtain a value of $\mathscr{R}$ referred to vacuum, one must divide the experimental value by n_{air}. Compare your vacuum value of $\mathscr{R}$ with the theoretical value.†

---

† The experimental value of $\mathscr{R}$ for hydrogen (109,677.581 cm^{-1}) using vacuum wavelengths is known with greater precision than the theoretical value, which is affected by uncertainties in e, h, c, and μ.

DISCUSSION

Show that a typical line in the Lyman series ($n_2 = 1$) lies in the ultraviolet and that a typical line in the Ritz-Paschen series ($n_2 = 3$) lies in the infrared.

Why do the Balmer lines become weaker toward shorter wavelengths? (The first Balmer line at about 6500 Å may appear weak owing to poor film sensitivity in the red; it is actually the most intense of all the lines.) What are some experimental factors which influence the line width?

APPARATUS

Spectrograph; low-pressure mercury lamp; hydrogen discharge tube; mounts and power supplies for both sources; front surface mirror; lens; spectroscopic plates (Kodak 103a-F); three developing trays; developer (Kodak D19); acid fixer; flashlight; timer; darkroom.

Comparator microscope; or enlarger, drawing board, T square and triangle, good centimeter scale.

REFERENCES

1. W. J. Moore, "Physical Chemistry," 4th ed., p. 630, Prentice-Hall, Englewood Cliffs, N.J. (1972).
2. G. Herzberg, "Atomic Spectra and Atomic Structure," 2d ed., pp. 11–38, Dover, New York (1944).
3. G. R. Harrison, R. C. Lord, and J. R. Loofbourow, "Practical Spectroscopy," pp. 188–192, Prentice-Hall, Englewood Cliffs, N.J. (1948).
4. K. W. F. Kohlrausch, "Ramanspektren," p. 34, Becker and Erler, Leipzig (1943).

GENERAL READING

H. E. White, "Introduction to Atomic Spectra," McGraw-Hill, New York (1934).

EXPERIMENT 44.   BAND SPECTRUM OF NITROGEN

Spectra which are observed in the visible and ultraviolet regions arise from transitions between electronic states. For atomic gases, such electronic spectra consist of individual sharp lines, as shown by the mercury spectrum in Fig. 43-3. For molecular gases, the transitions take place between different rotational-vibrational levels of the upper and lower electronic states and a very large number of lines occur. Under low resolution, groups of lines very close together have the appearance of broad bands in the spectrum. Therefore, such molecular spectra are referred to as *band spectra*.

In this experiment, part of the emission spectrum of nitrogen is to be determined and its vibrational structure analyzed. Homonuclear diatomic molecules, such as N_2, have no infrared spectrum[1] and are experimentally difficult to study by Raman spectroscopy. Thus electronic band spectra provide the most convenient source of vibrational and rotational constants. Since upper vibrational levels are involved, the effect of anharmonicity is easily detected.

THEORY

For a visible emission spectrum, transitions occur from excited electronic states to lower energy electronic states. There are many electronic states for N_2, and the energy-level diagram is quite complex[2] in contrast to the simple diagram for the hydrogen atom shown in Fig. 43-1. We shall discuss only the bands of the "second positive group," which are those to be observed in the present experiment. These bands arise from transitions from the $^3\Pi_u$ electronic state (denoted by C) to the $^3\Pi_g$ state (denoted by B), both of which are excited states. However, the theory presented here is typical of most electronic transitions for diatomic molecules.

Associated with each electronic state there is a characteristic potential curve which represents the potential energy of the nuclei as a function of their internuclear separation r. The potential curves for states B and C of N_2 are given schematically in Fig. 1; allowed vibrational levels are shown by the horizontal lines, while rotational levels have been omitted for clarity. In general, there is different bonding involved in two different electronic states. As a result the two curves do not have their minima at the same value of r and do not have exactly the same shape.

The total energy of a diatomic molecule may be separated into translational energy and internal energy. We are concerned here with the internal energy which can be expressed to a good approximation by

$$E = E_{el} + E_v + E_r \tag{1}$$

where E_r is the rotational energy, E_v is the vibrational energy, and E_{el} is the electronic energy. This electronic energy E_{el} refers to the minimum value of the potential curve for a given electronic state. The zero of energy is arbitrarily taken as the minimum in the potential curve for the lowest electronic state (ground state). It is convenient to divide Eq. (1) by the quantity hc to get the so-called "term value" T which has units of cm^{-1}. Thus

$$T = T_{el} + T_v + T_r \tag{2}$$

The advantage in this change is that the frequency $\tilde{v}$ (expressed in cm^{-1}) for a transition between two electronic states can be simply expressed by

$$\tilde{v} \equiv T' - T'' = (T'_{el} - T''_{el}) + (T'_v - T''_v) + (T'_r - T''_r) \tag{3}$$

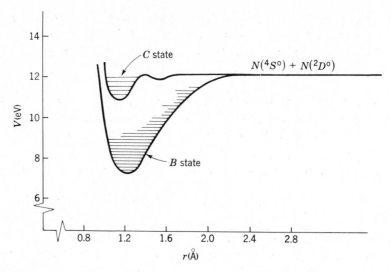

FIGURE 1
Potential energy curves for two excited electronic states of N_2.[2] B is the $^3\Pi_g$ state; C is the $^3\Pi_u$ state. (A few vibrational levels are shown.)

where the *single prime* refers to the *upper state* and the *double prime* refers to the *lower state*. Now let us denote $(T'_{el} - T''_{el})$ by $\tilde{v}_{el}$. For a given band system involving transitions from the same upper electronic state to the same lower electronic state, $\tilde{v}_{el}$ is a constant; in the case of the second positive group of N_2, $hc\tilde{v}_{el}$ corresponds to the energy separation between the minima of curves C and B in Fig. 1.

Vibrational structure   Since rotational energies are usually small compared with vibrational energies, it will be possible to neglect $(T'_r - T''_r)$ in Eq. (3) in order to discuss the vibrational structure of a band spectrum. In essence, this amounts to considering transitions among states without rotational energy. Corresponding to each spectral *line* expected on this basis, there is observed experimentally a *band* of very closely spaced lines resulting from the rotational structure which we are neglecting. From quantum mechanics,[3] one obtains for an anharmonic oscillator the approximate expression

$$T_v = \tilde{v}_e(v + \tfrac{1}{2}) - \tilde{v}_e x_e(v + \tfrac{1}{2})^2 + \tilde{v}_e y_e(v + \tfrac{1}{2})^3 + \cdots \tag{4}$$

where the vibrational quantum number v has integral values 0, 1, 2, 3, ... (Note that $\tilde{v}_e$ is not to be confused with $\tilde{v}_{el}$ introduced earlier.) The terms in $\tilde{v}_e x_e$ and $\tilde{v}_e y_e$ take into account the effect of anharmonicity. In general, the cubic and higher terms in $(v + \frac{1}{2})$ are much smaller than the quadratic term, but we shall retain the cubic term explicitly. We can now write Eq. (3) as

$$\tilde{\nu} = \tilde{\nu}_{el} + (T'_v - T''_v) \tag{5a}$$

$$= \tilde{\nu}_{el} + \tilde{\nu}'_e(v' + \tfrac{1}{2}) - \tilde{\nu}'_e x'_e(v' + \tfrac{1}{2})^2 + \tilde{\nu}'_e y'_e(v' + \tfrac{1}{2})^3$$

$$- [\tilde{\nu}''_e(v'' + \tfrac{1}{2}) - \tilde{\nu}''_e x''_e(v'' + \tfrac{1}{2})^2 + \tilde{\nu}''_e y''_e(v'' + \tfrac{1}{2})^3] \tag{5b}$$

By writing

$$\tilde{\nu}_{00} = \tilde{\nu}_{el} + \tfrac{1}{2}(\tilde{\nu}'_e - \tilde{\nu}''_e) - \tfrac{1}{4}(\tilde{\nu}'_e x'_e - \tilde{\nu}''_e x''_e) + \tfrac{1}{8}(\tilde{\nu}'_e y'_e - \tilde{\nu}''_e y''_e) \tag{6a}$$

$$\tilde{\nu}_0 = \tilde{\nu}_e - \tilde{\nu}_e x_e + \tfrac{3}{4}\tilde{\nu}_e y_e \tag{6b}$$

$$\tilde{\nu}_0 x_0 = \tilde{\nu}_e x_e - \tfrac{3}{2}\tilde{\nu}_e y_e \tag{6c}$$

$$\tilde{\nu}_0 y_0 = \tilde{\nu}_e y_e \tag{6d}$$

Eq. (5) can be simplified to

$$\tilde{\nu} = \tilde{\nu}_{00} + \tilde{\nu}'_0 v' - \tilde{\nu}'_0 x'_0 v'^2 + \tilde{\nu}'_0 y'_0 v'^3 - (\tilde{\nu}''_0 v'' - \tilde{\nu}''_0 x''_0 v''^2 + \tilde{\nu}''_0 y''_0 v''^3) \tag{7}$$

Obviously, $\tilde{\nu}_{00}$ is the frequency of the transition from $v' = 0$ (in the upper state) to $v'' = 0$ (in the lower state) which is called the 0-0 band. Since there is no selection rule for the quantum number v, any v' value can be combined with any v'' value in Eq. (7) and we should expect a large number of "lines" (bands).

When the separation between vibrational levels in the upper electronic state and in the lower electronic state is not greatly different (that is, $\tilde{\nu}'_0$ close to $\tilde{\nu}''_0$), the vibrational transitions form groups in the spectrum called sequences. For each sequence, $\Delta v \, (= v' - v'')$ is a constant. Such behavior occurs for N_2 and simplifies the analysis of the bands in the second positive group as seen from Fig. 2.

Once the vibrational transitions in the spectrum have been assigned and the frequencies measured, it is convenient to organize the values of $\tilde{\nu}_{v' \to v''}$ into a *Deslandres table*. As an example, a portion of such a table for PN vapor[4] is presented in Table 1.

Table 1   DESLANDRES TABLE OF PN BANDS[4]
(Band-head frequencies in cm^{-1}. Differences between the entries in rows 0 and 1, 1 and 2 are in parentheses.)

$v' \backslash v''$	0	1	2	3	Average differences
0	39,698.8	38,376.5	37,068.7	——	
	(1,087.4)	(1,090.7)	(1,086.8)		(1,088)
1	40,786.2	39,467.2	38,155.5	36,861.3	
	(1,072.9)	(1,069.0)		(1,071.6)	(1,071)
2	41,859.1	40,536.2	——	37,932.9	
3	——	41,597.4	40,288.3	——	

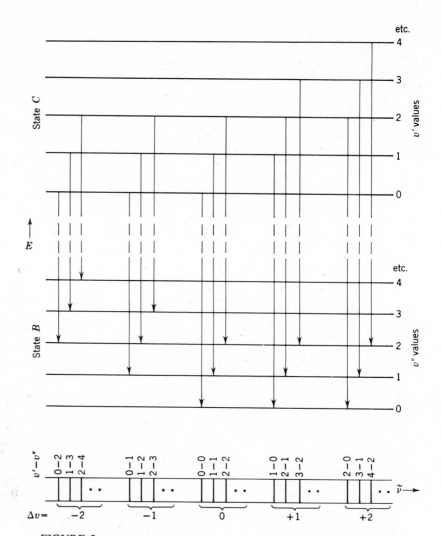

FIGURE 2

Vibrational energy levels for states B and C of N_2. The vertical arrows indicate transitions; transitions with the same Δv are grouped together to form sequences. The resulting vibrational-electronic "line" spectrum is shown at the bottom (for clarity only the first three lines in each sequence are shown).

Note that the frequencies along diagonals (which correspond to bands with the same Δv) lie close together as shown by the schematic spectrum in Fig. 2. Now consider the first two rows in the table. Since the frequencies are for transitions from two adjacent upper vibrational states ($v' = 0$ and $v' = 1$) to common lower vibrational states ($v'' = 0, 1, 2,$ or 3), there should be a constant separation between the rows— a separation corresponding to the upper state vibrational separation ($T'_{v=1} - T'_{v=0}$), as seen from Eq. (5a). The separation between the rows $v' = 1$ and $v' = 2$ will give $(T'_{v=2} - T'_{v=1})$, and so forth. The separation between successive rows should decrease as v' increases owing to the effect of anharmonicity; see Eq. (4). In exactly the same way, the separation between adjacent columns will give information about the vibrational levels of the lower electronic state. Agreement among the separations between corresponding frequencies of two rows (or two columns) is a definite check on the correctness of the entries in the Deslandres table. Inspection of the differences given in Table 1 shows a slight variation which is, however, greater than the experimental error. This is caused by the use of band-head frequencies as discussed below.

Rotational structure   There are two deficiencies with the theory as presented so far: Eq. (7) predicts a spectrum of lines rather than the observed bands, and the differences in the Deslandres table are not quite constant as they should be. Both of these failures are due to the neglect of rotational transitions. For each electronic vibrational transition discussed above, there are many rotational transitions which give rise to a large number of closely spaced lines in the vicinity of each $\tilde{v}$ as given by Eq. (7). The theory of this rotational fine structure is too complex for presentation here, especially since it involves an interaction between rotational and electronic motions in the molecule. In addition very high resolution is required to observe this fine structure experimentally. Under low resolution, the rotational lines will create the appearance of a band usually having at one end an abrupt change in intensity, called the *band head*, while the intensity falls off slowly at the other end.

   This experiment will deal with these easily observed band heads in spite of the fact that Eq. (7) is valid only for the band origins $\tilde{v}_0$. (The origin is the position in the band that corresponds to a transition between vibrational states without rotational energy.) Fortunately, the difference $\tilde{v}_{head} - \tilde{v}_0$ is small for N_2, varying from about -5 to -10 cm^{-1}. The use of band-head frequencies rather than band origins in Table 1 accounts for the slight lack of constancy in the separations, since $\tilde{v}_{head} - \tilde{v}_0$ varies somewhat.

Assignment of bands   Herzberg[5] discusses the vibrational assignment of a band system when clear sequences are observed on both sides of the 0-0 band. For this experiment, one may take one or two band heads as known from the literature and then assign the others from the sequence pattern shown in Fig. 2. Constancy of the differences in the Deslandres table will confirm the assignment.

EXPERIMENTAL

The method for determining the band spectrum of nitrogen is the same as that described in Exp. 43 for the study of the Balmer lines of hydrogen. It differs only in replacing the hydrogen discharge tube by a nitrogen discharge tube and using an iron arc for the reference spectrum instead of a mercury arc. For this experiment, the richer spectrum of lines from the iron arc is an advantage in obtaining accurate values of the wavelengths of the band heads for N_2.

Sources   A simple design for an iron arc[6] is shown in Fig. 3. An iron oxide bead is used to stabilize the arc, which may be started by shorting the gap with a carbon rod. Excellent reproductions of the Fe-arc spectrum together with the wavelengths of all the lines can be found in the literature.[7] The nitrogen source is a Geissler discharge tube (commercially available from Tube Light Engineering Co., New York) in which N_2 molecules are excited by an electrode discharge into upper electronic states. Such excited molecules will spontaneously undergo transitions to lower electronic states and in the process emit radiation. The discharge tube operates from a neon-sign transformer at about 5000 V and 15 mA; it should glow steadily with a pale greenish-blue color.

Procedure   Follow the procedure given in the experimental section of Exp. 43 to obtain several pairs of N_2 and Fe spectra. It may be convenient at the end to include one pair of exposures consisting of an Fe spectrum and an Hg spectrum. This will facilitate identification of the iron lines.

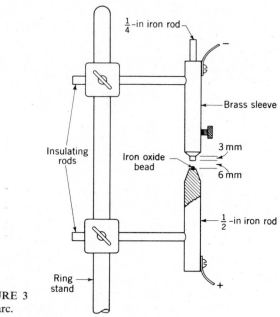

FIGURE 3
Iron arc.

The lines should be measured using a microscope comparator as described in Exp. 43. Since the iron reference spectrum contains many closely spaced lines, it is not necessary to plot a calibration curve. Record the position of every N_2 band head together with the position of one Fe line on each side of the band head. Choose Fe lines which are as close to the band-head position as possible. Determine the wavelength of the band head by linear interpolation between these iron lines. A proper identification of the iron lines is crucial; checking them against an Hg spectrum may help. Make a list of the wavelengths (in angstroms) and the frequencies (in cm^{-1}) of all band heads measured.

CALCULATIONS

Several band-head wavelengths for the second positive group of N_2 can be obtained from Fig. 8a in Herzberg[8] as an aid to assignment. Assign the v' and v'' values for all band heads observed, and construct a Deslandres table of the frequencies. Indicate in this table the differences, and check their constancy to verify the assignment. Along the bottom and right-hand side of the table, record the average values of the differences between v' and v'' levels, respectively.

Draw a vibrational energy-level diagram for both electronic states like that shown in Fig. 2, and indicate on it the separation in cm^{-1} units between those levels for which the differences $T'_{v+1} - T'_v$ or $T''_{v+1} - T''_v$ are given in the Deslandres table. Use the average values of the differences as recorded in the table.

From Eq. (4) we see that

$$T_{v+1} - T_v = \tilde{v}_e - 2\tilde{v}_e x_e(v + 1) + 3\tilde{v}_e y_e(v^2 + 2v + \tfrac{13}{12}) + \cdots \qquad (8)$$

In the case of state B, the $\tilde{v}_e y_e$ value is very small and the last term in Eq. (8) can be neglected in fitting the experimental data. In the case of state C, the $\tilde{v}_e y_e$ value is unusually large and this term must be retained. Thus the separations in the energy-level diagram should be used to determine the best values of the vibrational constants $\tilde{v}''_e$ and $\tilde{v}''_e x''_e$ for the lower electronic state B and the best values of $\tilde{v}'_e$, $\tilde{v}'_e x'_e$, and $\tilde{v}'_e y'_e$ for the upper state C. From Eqs. (6b) and (7), calculate a value of $\tilde{v}_{00}$ using several frequencies from the Deslandres table. Make a table of your values of $\tilde{v}''_e$, $\tilde{v}''_e x''_e$, $\tilde{v}'_e$, $\tilde{v}'_e x'_e$, $\tilde{v}'_e y'_e$, and $\tilde{v}_{00}$ and compare them with the literature values given by Herzberg.[9]

DISCUSSION

Is the rotational fine structure resolved in any of the bands studied? If so, compare it qualitatively with the high-resolution spectrum of the 0-2 band given in Herzberg[10] and indicate whether you feel rotational analysis would be possible from your spectrum.

APPARATUS

Medium-resolution spectrograph; nitrogen discharge tube; iron arc; low-pressure mercury lamp (optional); mounts and power supplies for all sources; front-surface mirror; lens; spectroscopic plates (Kodak 103a-F); three developing trays; developer (Kodak D19); acid fixer; flashlight; timer; darkroom.

Comparator microscope; reproduction of the iron-arc spectrum.[7]

REFERENCES

1. G. Herzberg, "Molecular Spectra and Molecular Structure I. Spectra of Diatomic Molecules," 2d ed., pp. 80, 131, Van Nostrand, Princeton, N.J. (1950).
2. F. R. Gilmore, RAND Corporation Memorandum R-4034-PR (June 1964). The potential curves are reproduced in K. E. Shuler, T. Carrington, and J. C. Light, *Appl. Optics Suppl.*, **2**, 85 (1965).
3. G. Herzberg, *op. cit.*, p. 151.
4. *Ibid.*, pp. 40–42.
5. *Ibid.*, p. 161.
6. J. Strong, "Procedures in Experimental Physics," p. 351, Prentice-Hall, Englewood Cliffs, N. J. (1939).
7. W. R. Brode, "Chemical Spectroscopy," 2d ed., pp. 619–654, Wiley, New York (1943); A. Gatterer, "Grating Spectrum of Iron," Specola Vaticana, Vatican City (1951).
8. G. Herzberg, *op. cit.*, p. 32.
9. *Ibid.*, Table 39, p. 552.
10. *Ibid.*, p. 46.

GENERAL READING

G. Herzberg, *op. cit.*, chap. IV.

J. I. Steinfeld, "Molecules and Radiation," chap. 5, MIT Press, Cambridge, Mass. (1974, 1978).

EXPERIMENT 45.   MOLECULAR FLUORESCENCE OF IODINE

When a simple diatomic molecule, such as iodine, is excited to a single vibration-rotation level of an upper electronic state by suitably monochromatic radiation, the resulting fluorescence has a particularly simple and easily interpretable intensity distribution. From the analysis of such a fluorescence spectrum, vibrational and rotational constants for the ground electronic state of the molecule can be determined.

THEORY

The phenomenon of fluorescence consists of the absorption of light by a system and its subsequent radiation, either at the same wavelength as the original exciting radiation or at different wavelengths. It should be distinguished from simple light scattering (such as the Rayleigh or Raman effects), in which the photon energy is *not* absorbed by the irradiated sample. In fluorescence, an exciting light source is used to cause a transition in a molecule from some given quantum state to some other definite state which is higher in energy. From this state, the molecule may lose energy by a nonradiative collisional process (quenching) or it can emit light as it drops to various lower vibrational-rotational energy levels. Some collisional loss of energy can also occur before emission so that several upper states can contribute to the fluorescence spectrum. An analysis of the spectrum thus can give information about the vibrational-rotational energy levels of the electronic states of a molecule. This information is important in determining the structure and other properties of molecules.

In the absence of specific *quenching* effects, fluorescence can be observed in gaseous, liquid, or solid samples. Only in the vapor phase, however, do the vibrational and rotational quantum levels of the molecule remain sufficiently unperturbed for a sharp, many-lined fluorescence spectrum to be observed. In this experiment, therefore, a gaseous sample is used. One of the substances which has been found to possess a strong vapor-phase fluorescence is molecular iodine.[1-3] Iodine has a discrete absorption spectrum in the region 6400 to 4995 Å; this absorption involves a transition from the $^1\Sigma_g^+$ ground state (denoted X) to an excited $^3\Pi_{0_u^+}$ state (denoted B).[4,5] As a consequence of the large moment of inertia and the rather weak I—I bond, the spectrum exhibits a complicated fine structure.

The fluorescence spectrum can be considerably simplified, however, if the molecule is first excited by *monochromatic* radiation to a single rotation-vibration level of the electronically excited state. The selection rules for electronic transitions in a homonuclear diatomic molecule are $\Delta J = \pm 1$ for the change in rotational quantum number and no restriction on Δv (the change in vibrational quantum number). In the absorption spectrum, allowed transitions take place from all states thermally populated at the temperature of the sample; but in monochromatically excited fluorescence, only a single level is populated, and only the emission arising from this level is observed. It is not necessary, in general, that the exciting radiation be itself produced by the substance being excited. There is always appreciable line width in emission and absorption lines, so that accidental overlapping of the spectra of different systems is to be expected fairly often. This is the case in the present experiment.

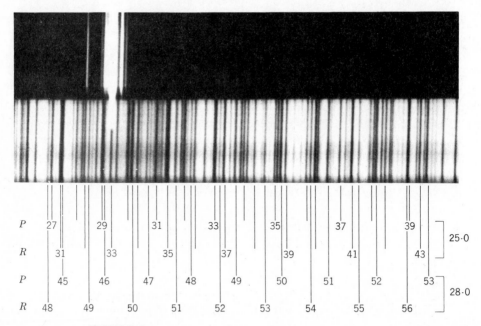

FIGURE 1

High-resolution spectrum of the visible $B \leftarrow X$ iodine absorption spectrum, showing the overlap of the mercury emission line at 5460.74 Å with a single feature of the spectrum.

Shown in Fig. 1 is a high-resolution spectrum of the "mercury green line," which can be used for exciting the iodine fluorescence. The weak satellite lines are due to nuclear hyperfine structure. Below the mercury line is shown a portion of the iodine absorption spectrum, taken on a 10-m grating spectrograph, with some of the transitions identified. The narrow central component of the mercury emission line is seen to coincide precisely with a single absorption line of the I_2 spectrum, which has been identified[6] as a transition from $v'' = 0$, $J'' = 33$ in the ground state to $v' = 25$, $J' = 34$ in the excited state. [In conventional spectroscopic notation, this is the 25–0 R(33) line.] In the mercury source described in this experiment, the exciting line is considerably broadened by the high temperature (Doppler effect) and relatively high pressure (Lorentz effect) existing in the lamp. This broadened emission line is absorbed by seven of the iodine lines, thus populating seven closely spaced rotation-vibration levels of the upper electronic state. Each of these excited states returns to the ground state by emitting a photon. This fine structure is seen only under high-resolution conditions; at moderate resolution the emission bands appear as single features corresponding to $(v' = 25) \rightarrow (v'' = 0)$, $(v' = 25) \rightarrow (v'' = 1)$,

$(v' = 25) \rightarrow (v'' = 2)$, and so on. The spacing of bands in this *progression* is thus a measure of the vibrational energy levels in the ground state of the molecule.

If an argon-ion laser is available, the 5145-Å line can be used as an alternative excitation source. Light of this wavelength excites the $v' = 43$, $J' = 12$ and 16 levels of the B state,[7] and an extended emission progression to many levels of the X state results. Under high resolution, triplet structure is observed due to overlap of the P, R doublets expected for each originating J' level; doubtlets are seen in the case of mercury excitation.

The energy of the transition (in cm^{-1}) can be represented by a polynomial analogous to that of Eq. (44-5), namely,

$$E(cm^{-1}) = \tilde{v}_0 + \tfrac{1}{2}\tilde{v}''_e - \tfrac{1}{4}\tilde{v}''_e x''_e + \tfrac{1}{8}\tilde{v}''_e y''_e + \cdots$$

$$- [\tilde{v}''_e(v'' + \tfrac{1}{2}) - \tilde{v}''_e x''_e(v'' + \tfrac{1}{2})^2 + \tilde{v}''_e y''_e(v'' + \tfrac{1}{2})^3 + \cdots] \tag{1}$$

where v'' numbers the bands, starting at the exciting frequency $\tilde{v}_0$. The higher terms in the polynomial can be fitted by means of a least-squares procedure such as described in Chap. XX.

The anharmonic terms in Eq. (1) act to reduce the spacing between adjacent levels as v'' increases. If one can observe high v'' transitions where this spacing goes to a minimum, the dissociation energy D''_0 of the ground state can be determined. D''_0 is defined as the energy difference between the $v'' = 0$ level and the *convergence limit* where the molecule dissociates and continuum states begin. D''_0 is smaller than D''_e, the dissociation energy referenced to the minimum of the potential curve, by the zero-point energy $(\tfrac{1}{2}\tilde{v}''_e - \tfrac{1}{4}\tilde{v}''_e x''_e + \tfrac{1}{8}\tilde{v}''_e y''_e + \cdots)$. Even when high v'' transitions cannot be measured, an estimate of D''_e can be obtained by extrapolating to zero the energy differences $E(v'' + 1) - E(v'')$ calculated from Eq. (1); a plot of these differences versus v'' is termed a Birge-Sponer plot.[8] If only $\tilde{v}''_e$ and x''_e terms are included, a simple analytic solution for $\tilde{D}''_e$ (in cm^{-1} units) results:

$$\tilde{D}''_e = \tilde{D}''_0 + \tfrac{1}{2}\tilde{v}''_e - \tfrac{1}{4}\tilde{v}''_e x''_e = \frac{\tilde{v}''_e}{4x_e''} - \frac{\tilde{v}_e x''_e}{4} \approx \frac{v''_e}{4x''_e} \tag{2}$$

In general, this expression gives an upper limit for $\tilde{D}''_e$; actual values for $\tilde{D}''_e$ are typically 10 to 30 percent smaller.

An accurate potential-energy curve for a molecule can be constructed by numerical methods if higher vibrational levels are known. A much simpler, but still useful, representation of the potential energy is the Morse function

$$V(cm^{-1}) = D''_e\{1 - exp[-\beta(r - r_e)]\}^2 \tag{3}$$

Here r_e is the equilibrium bond length and β is a constant which can be related to the harmonic force constant by equating a quadratic potential to an expansion of the Morse potential. The Morse potential is generally quite good near the minimum of the potential curve and it yields the correct dissociation energy at large r. It does not give $V = \infty$ at $r = 0$ as a correct potential function should, but this region of the curve is of no practical importance.

EXPERIMENTAL

The procedure for photographing the fluorescence spectrum is essentially the same as that followed in studying the hydrogen atomic emission (Exp. 43) and the nitrogen band spectrum (Exp. 44). Since the fluorescence is rather weak and in the red part of the spectrum, suitable high-speed, red-sensitive plates (Eastman Kodak Type 103a-F or 1-N) or film (Eastman Kodak Royal-X Pan Recording) should be used. Since these emulsions are very sensitive, all darkroom operations must be carried out without any safelight. [Photoelectric recording of the spectrum can, of course, be used instead if a suitable scanning spectrometer is available; in this case, a red sensitive photomultiplier (e.g., RCA 4840 or Hamamatsu $R955$) should be used.]

A suitable fluorescence cell for excitation using a mercury lamp is shown in Fig. 2. The front window should be an optical quality Pyrex flat. The cell terminates in a blackened Wood's horn to cut down on scattered light. For argon-laser excitation, a cell with three window ports should be provided. Iodine is distilled into the lower side arm, after which the cell is pumped down to 10^{-3} Torr or less residual pressure and sealed off. If the side arm is maintained at room temperature, an iodine vapor pressure of approximately 0.2 Torr will exist in the fluorescence cell, which is optimum for this experiment.

The Hg excitation source is a General Electric UA-3 medium-pressure quartz mercury arc. The lamp and the fluorescence cell should be mounted confocally in a polished elliptical reflector. A "didymium" glass filter interposed between the lamp and the cell cuts out the mercury yellow lines, which would excite a second fluorescence series and make the analysis more complicated. The fluorescence should be observable as a dull orange glow by looking down the end of the cell. If this is not

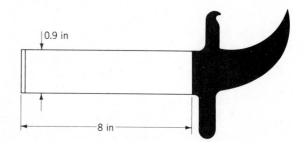

FIGURE 2
Fluorescence cell with blackened Wood's horn and side arm for iodine solid.

0.9 in

8 in

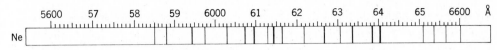

FIGURE 3
Neon emission spectrum, based on plate 11 from Pearse and Gaydon.[9]

observed, the most likely source of difficulty is leakage of air into the fluorescence cell. A half-hour exposure should produce satisfactory spectra. The emission from a neon-filled Geissler discharge tube provides a suitable calibration spectrum.

CALCULATIONS AND DISCUSSION

For photographic recording, a dispersion curve for the spectral region between 5000 and 7000 Å is obtained by carefully measuring the mercury and neon calibration lines with a traveling-microscope comparator. The neon lines can be most easily identified with the aid of the spectrum reproduced in Plate 11 of Pearse and Gaydon's "Identification of Molecular Spectra,"[9] a line drawing of which is reproduced in Fig. 3. Exact values for the wavelengths, which will be needed for a numerical least-squares analysis of the data, are given in Table 1. These same lines can be used in checking the calibration of a scanning spectrometer if such is employed in the experiment.

Measure the wavelengths of as many fluorescence bands as are observable. These should form a regular progression to the red of the exciting line. The inten-

Table 1   SOME PROMINENT LINES OF THE NEON EMISSION SPECTRUM IN THE 5400–6600 Å REGION[a]

λ_{air}(Å)	λ_{air}(Å)	λ_{air}(Å)
5400.56	6074.34	6304.79
5852.49	6096.16	6334.43
5881.89	6128.45	6382.99
5944.83	6143.06	6402.25
5974.63	6163.59	6506.53
5975.53	6217.28	6532.88
6029.99	6266.49	6598.95

[a] Taken from "Handbook of Chemistry and Physics," 41st ed., p. 2833, Chemical Rubber Publishing Co., Cleveland (1959). (Recent editions of this handbook no longer contain such wavelength tables.)

sity of each band will show an irregular variation due to the Franck-Condon principle,[10] which gives an intensity factor

$$I(v', v'') \propto |R_e(v')|^2 \cdot \left| \int \phi_{v'}(r)\phi_{v''}(r) \, dr \right|^2 \tag{4}$$

$R_e(v')$ is the effective electronic transition moment for the excited state, and the second factor is the *vibrational overlap integral* between the wave functions for the excited level v' and the ground level v''. It is possible that some of the bands may be missing because their overlap factors are too small. It is important to remember to take these into account when numbering the bands, beginning with $v'' = 0$ at the excitation line.

Use a least-squares procedure to fit the vibrational constants $\tilde{v}_e''$ and $\tilde{v}_e'' x_e''$ in Eq. (1). If your data warrant it, see if inclusion of an $\tilde{v}_e'' y_e''$ term improves your fit. Derive Eq. (2) and use your experimental results to obtain an upper limit for $\tilde{D}_e''$. Compare your constants with recent literature values.[11] From $\tilde{v}_e''$ calculate the harmonic force constant k for the I_2 bond. Derive an expression relating k to the Morse parameter β and plot the Morse potential energy curve for the ground electronic state of I_2.

APPARATUS

Medium-resolution spectrograph or spectrometer with photomultiplier; red-sensitive plates (Kodak Type 103a-F or 1-N) or film (Kodak Royal-X Pan); developing and fixing solutions; argon-ion laser or General Electric UA-3 medium-pressure mercury arc (distributed by George W. Gates Co., Franklin Square, N.Y.); didymium glass filters (Corning Glass Co.); neon discharge tube and transformer; reagent-grade iodine; vacuum system for pumping down fluorescence cell; comparator microscope.

REFERENCES

1. R. W. Wood, *Phil. Mag.*, **12**, 499 (1906).
2. F. W. Loomis, *Phys. Rev.*, **29**, 112 (1927).
3. R. W. Wood, "Physical Optics," pp. 616–647, Macmillan, New York (1934).
4. L. Mathieson and A. L. G. Rees, *J. Chem. Phys.*, **25**, 733 (1956).
5. D. H. Rank and B. S. Rao, *J. Mol. Spectrosc.*, **13**, 34 (1964).
6. J. I. Steinfeld, R. N. Zare, J. M. Lesk, and W. Klemperer, *J. Chem. Phys.*, **42**, 15 (1965).
7. R. B. Kurzel and J. I. Steinfeld, *J. Chem. Phys.*, **53**, 3293 (1970); M. Rubinson, B. Garetz, and J. I. Steinfeld, *J. Chem. Phys.*, **60**, 3082 (1974), see fn. 7.

8. G. Herzberg, "Molecular Spectra and Molecular Structure I. Spectra of Diatomic Molecules," 2d ed., p. 100, 438, Van Nostrand, Princeton, N.J. (1950).
9. R. W. R. Pearse and A. G. Gaydon, "Identification of Molecular Spectra," 4th ed., Chapman & Hall, London (1976).
10. E. U. Condon, *Am. J. Phys.*, **15**, 365 (1947).
11. K. P. Huber and G. Herzberg, "Molecular Spectra and Molecular Structure IV. Constants of Diatomic Molecules," p. 332, Van Nostrand Reinhold, New York (1979).

GENERAL READING

G. Herzberg, "Molecular Spectra and Molecular Structure I. Spectra of Diatomic Molecules," 2d ed., chaps. 2–4, 8, Van Nostrand, Princeton, N.J. (1950).

G. R. Harrison, R. C. Lord, and J. R. Loofbourow, "Practical Spectroscopy," chaps. 2, 4 (esp. 4.7), 5–7, 11, Prentice-Hall, Englewood Cliffs, N.J. (1948).

G. M. Barrow, "Introduction to Molecular Spectroscopy," chaps. 1–5, 10, McGraw-Hill, New York (1963).

P. Pringsheim, "Fluorescence and Phosphorescence," pp. 151–167, Interscience, New York (1948).

J. I. Steinfeld, *J. Chem. Educ.*, **42**, 85 (1965); "Molecules and Radiation," chap. 5, MIT Press, Cambridge, Mass. (1974, 1978).

EXPERIMENT 46. NMR DETERMINATION OF KETO-ENOL EQUILIBRIUM CONSTANTS

In this experiment, proton NMR spectroscopy is used in evaluating the equilibrium composition of various keto-enol mixtures. Chemical shifts and spin-spin splitting patterns are employed in assigning the spectral features to specific protons, and the integrated intensities yield a quantitative measure of the relative amounts of the keto and enol forms. Solvent effects on the chemical shifts and on the equilibrium constant will be investigated for one or more β-diketones and β-ketoesters.

THEORY

Chemical shifts In Exp. 38, the chemical shift in parts per million (PPM) of nucleus i relative to a reference nucleus r was defined as

$$\delta_i \equiv \frac{H_r - H_i}{H_r} \times 10^6 = \frac{v_r - v_i}{v_r} \times 10^6 \tag{1}$$

where H is the external magnetic field which produces a resonance at frequency v.

Tetramethylsilane (TMS) is usually used as the proton reference since it is chemically inert and its 12 equivalent protons give a single transition at a field H_r, higher than the field H_i found in most organic compounds. Thus δ is generally positive and increases when substituents are added which attract electrons and thereby reduce the shielding about the proton. This shielding arises because the electrons near the proton are induced to circulate by the applied field H (see Fig. 1a). This electron current produces a secondary field which *opposes* the external field so that resonance at a fixed frequency such as 60 MHz requires a higher field. This shielding effect is generally restricted to electrons localized on the nucleus of interest since random tumbling of the molecules causes the effect of secondary fields due to electrons associated with neighboring nuclei to average to zero. Nuclei such as ^{19}F, ^{13}C, and ^{11}B have more local electrons than hydrogen; hence their chemical shifts are much larger.

Long-range *deshielding* can occur in aromatic and other molecules with delo-calized π electrons. For example, when the plane of the benzene molecule is oriented perpendicular to H, circulation of the π electrons produces a ring current which induces a field at the protons, which *adds* to H (Fig. 1b). This induced field changes with benzene orientation but does not average to zero since it is not spherically symmetric. Because of this net deshielding effect, the resonance of the benzene protons occurs at $\delta = 7.27$, greatly downfield from the value $\delta = 1.43$ which is observed for cyclohexane, in which ring currents do not occur. Similar deshielding occurs for olefinic and aldehydic protons because of the π electron

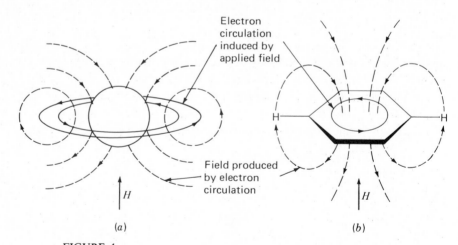

Electron circulation induced by applied field

Field produced by electron circulation

(a)

(b)

FIGURE 1

Shielding and deshielding of protons: (a) shielding of proton due to induced diamagnetic electron circulation; (b) Deshielding of protons in benzene due to aromatic ring currents.

movement. Typical values of δ for different functional groups are shown in Table 1 and additional values are available in Refs. 1 to 5. Although the resonances change somewhat for different compounds, the range for a given functional group is usually small and δ values are widely used for structural characterization in organic chemistry.

Spin-spin splitting   In Exp. 38, the energy levels of a nucleus were given as

$$E_N = -g_N \beta_N M_I B \tag{2}$$

where B is the local field at the nucleus. As a result of $\Delta M_I = \pm 1$ selection rules, a single transition would occur at

$$v = \frac{g_N \beta_N}{h} B \tag{3}$$

if all the nuclei experienced the same local field. However, high-resolution NMR spectra of most organic compounds reveal more complicated multiplet bands. These bands arise because the local magnetic field at a proton is influenced by neighboring nuclei with magnetic moments. The direction and magnitude of the induced field H_N at a proton will depend upon the orientation of the magnetic moments of the neighbors with respect to the applied field. The net field at the nucleus is given by

$$B = H(1 - \sigma) + H_N \tag{4}$$

Table 1   TYPICAL PROTON CHEMICAL SHIFTS δ

CH_3 protons		Acetylenic protons	
$(CH_3)_4Si$	0.0	$HOCH_2C{\equiv}CH$	2.33
$(CH_3)_4C$	0.92	$ClCH_2C{\equiv}CH$	2.40
CH_3CH_2OH	1.17	$CH_3COC{\equiv}CH$	3.17
CH_3COCH_3	2.07	Olefinic protons	
CH_3OH	3.38	$(CH_3)_2C{=}CH_2$	4.6
CH_3F	4.30	cyclohexene	5.57
CH_2 protons		$CH_3CH{=}CHCHO$	6.05
cyclopropane	0.22	$Cl_2C{=}CHCl$	6.45
$CH_3(CH_2)_4CH_3$	1.25	Aromatic protons	
$(CH_3CH_2)_2CO$	2.39	benzene	7.27
$CH_3COCH_2COOCH_3$	3.48	C_6H_5CN	7.54
CH_3CH_2OH	3.59	naphthalene	7.73
CH protons		α-pyridine	8.50
bicyclo[2.2.1]heptane	2.19	Aldehydic protons	
chlorocyclo-		CH_3OCHO	8.03
propane	2.95	CH_3CHO	9.72
$(CH_3)_2CHOH$	3.95	C_6H_5CHO	9.96
$(CH_3)_2CHBr$	4.17		

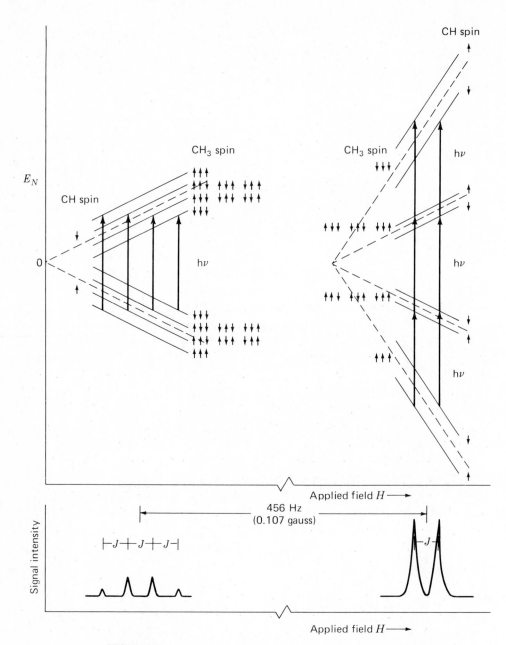

FIGURE 2

Energy levels, transitions, and 60-MHz NMR spectrum for acetaldehyde (CH_3CHO). The coupling constant $J = J_{CH_3} = J_{CH} = 2.2$ Hz.

where H is the external applied field and σ is a diamagnetic correction term [see Eq. (38-7)]. The effect of Eq. (4) is to split the energy levels in the manner illustrated for acetaldehyde in Fig. 2. It is apparent from this diagram that the external field H does not affect the small spin-spin splitting which is characterized by the coupling constant J. The quantity J is a measure of the strength of pairwise interaction of the proton spin with the spin of another nucleus. Since there are only proton-proton interactions in acetaldehyde, the same splitting occurs for both CH and CH_3 resonances. It may be noted that, since v and H are linearly related by Eqs. (3) and (4), J and chemical shifts are often reported in hertz units even though they are measured by scanning the field (in gauss) at fixed frequency.

The total integrated intensity of the CH and CH_3 multiplets follows the proton ratio of $1:3$. However, the intensity distribution within each multiplet is determined by the population of the lower level in each transition. Since the level spacing is much less than kT, the Boltzmann population factors are essentially identical for these levels. However, there is some degeneracy because rapid rotation of the CH_3 group around the $C-C$ bond makes the three protons magnetically equivalent. The number of spin orientations of the CH_3 protons which produce equivalent fields at the CH proton determine the degeneracy. The eight permutations of the CH_3 spins are shown in Fig. 2 and the intensity ratios of $1:3:3:1$ and $12:12$ are thus predicted for the CH and CH_3 multiplets. In a more general sense, it can be seen that n equivalent protons interacting with a different proton will split its resonance into $n+1$ lines whose relative intensities are given by coefficients of the terms in the binomial expansion of the expression $(\alpha + \beta)^n$. Equivalent protons also interact and produce splitting in the energy levels. However, these splittings are symmetric for upper and lower energy states so that no new NMR resonances are produced.

If a proton is coupled to more than one type of neighboring nucleus, the resultant multiplet pattern can often be understood as a simple stepwise coupling involving different J values. For example, the CH_2 octet which occurs for pure CH_3CH_2OH (Fig. 3) arises from OH doublet splitting ($J = 4.80$ Hz) of the quartet of lines caused by coupling ($J = 7.15$ Hz) with CH_3. It should be mentioned that such regular splitting and intensity patterns are only expected for two nuclei A and B if $(v_A - v_B)/J_{AB} \gtrsim 10$. The spectra for this weakly coupled case are termed *first order*. Since the difference $v_A - v_B$ (in Hz) increases with the field while J_{AB} does not, NMR spectra obtained with a high-field instrument (220 MHz) are often easier to interpret than those from a low-field spectrometer (60 MHz). However, even if the multiplets are not well separated, it is still possible to deduce accurate chemical shifts and J values using slightly more involved procedures which are outlined in most texts on NMR spectroscopy.[1-5] Such an exercise can be done as an optional

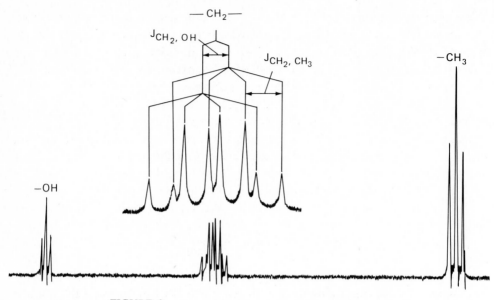

FIGURE 3

NMR spectrum of highly purified ethanol obtained at 100 MHz.

part of this experiment although it will not be necessary for the determination of equilibrium constants.

The mechanism of spin-spin coupling is known to be indirect and to involve the electrons in the bonds between interacting nuclei. The spin of the first nucleus A is preferentially coupled antiparallel to the nearest bonding electron via the so-called Fermi contact interaction, which is significant only when the electron density is non-zero at the first nucleus. (Such is the case only for electrons in s orbitals since p, d, and f orbital wave functions have zero values at the nucleus.) This electron spin align-ment information is transmitted by electron-electron interactions to the second nucleus B to produce a field which thus depends on the spin orientation of the first nucleus (Fig. 4). Since this interaction falls off rapidly with separation, only neigh-

FIGURE 4

Illustration of nuclear spin-spin interaction transmitted via polarization of bonding electrons.

boring groups produce significant splitting. A few typical spin-spin coupling constants are given in Table 2 and these, along with the chemical shifts, serve to identify proton functional groups. As mentioned above, the multiplet intensities also give useful information about neighboring groups. Thus NMR spectra can provide detailed structural information about large and complex molecules.

Keto-enol tautomerism It is well known that ketones such as acetone have an isomeric structure, which results from proton movement, called the enol tautomer, an unsaturated alcohol:

acetone (keto form)   (enol form)

For acetone, and the majority of cases where this keto-enol tautomerism is possible, the keto form is far more stable and little if any enol can be detected. However, with β-diketones and β-ketoesters, such factors as intramolecular hydrogen bonding and conjugation increase the stability of the enol form and the equilibrium can be significantly shifted to the right.

keto   enol 1   enol 2

The proton chemical environments are quite different for the keto and enol tautomers and the interconversion rate r_1 between these forms is slow enough that distinct NMR spectra are obtained for both forms. In principle, the two enols are also distinguishable when $R' \neq R''$. However the intramolecular OH proton transfer is quite rapid at normal temperatures so that a single (averaged) OH resonance is observed. In general, such averaging occurs when the conversion rate r_2 (in hertz) exceeds the frequency separation $v_1 - v_2$ (also in hertz) of the OH resonance for the two enol forms.[2] The magnetic field at the OH proton is thus averaged and resonance occurs at $(v_1 + v_2)/2$. Similarly, rapid rotation about the C—C bonds of the keto form explains why spectra due to different keto rotational conformers are not

Table 2 TYPICAL PROTON SPIN-SPIN COUPLING CONSTANTS

Coupling	J(Hz)	Coupling	J(Hz)
$\overset{\displaystyle H}{\underset{\displaystyle H}{>\!C\!<}}$	$-20 - +5$	$\overset{\displaystyle H}{\underset{\displaystyle H}{>\!C\!=\!C\!<}}$	$0 - 3.5$
$>\!CH\!-\!CH\!<$	$2 - 9$	$C\!=\!C$ with H and H	$6 - 14$
$>\!CH\!-\!(C)_n\!-\!CH\!<$	0		
benzene ring: ortho- meta- para-	6–9 1–3 1	$C\!=\!C$ with H and H	$11 - 19$

observed. Thus, distinct spectra are expected only for the two tautomers and these can be used to determine the equilibrium constant for keto-to-enol conversion:

$$K_c = \frac{(\text{enol})}{(\text{keto})} \tag{5}$$

where parentheses denote concentrations in any convenient units.

The keto arrangement shown above is the configuration which is electro-statically most favorable. It is clear that steric repulsions between R and R″ groups will be larger for this keto form than for the enol configuration, and experimental studies have confirmed that the enol concentration is larger when R and R″ are bulky.[6] This steric effect is less important in the β-ketoesters since the OR″ group increases the R ... R″ separation. For both β-ketoesters and β-diketones, α substitution of large R′ groups results in steric hindrance between R′ and R (or R″) groups, particularly for the enol tautomer, whose concentration is thereby reduced. Inductive effects have also been explored; in general, α substitution of electron-withdrawing groups such a —Cl or —CF$_3$ favor the enol form.[6]

The solvent plays an important role in determining K_c. This can occur through specific solute-solvent interactions such as hydrogen bonding or charge transfer. In addition, the solvent can reduce solute-solute interactions by dilution and thereby change the equilibrium if such interactions are different in enol-enol, enol-keto, or keto-keto dimers. Finally, the dielectric constant of the solution will depend on the solvent and one can expect the more polar tautomeric form to be favored by polar solvents. Some of these aspects are explored in this experiment.

EXPERIMENTAL

The general features of an NMR spectrometer were described briefly in Exp. 38, and more specific operating instructions will be provided by the instructor. Obtain several milliliters each of acetylacetone ($CH_3OCH_2COCH_3$, M.W. = 100.11, density = 0.98 g/ml) and ethyl acetoacetate ($CH_3CH_2OCOCH_2COCH_3$, M.W. = 130.45, density = 1.03 g/ml). Prepare small volumes of two solvents and three solutions.

> **Solvent A:** Carbon tetrachloride, spectrochemical grade (M.W.: 153.83, density: 1.58 g/ml) with 5% by volume tetramethylsilane (TMS) added. Prepare in a 5- or 10-ml volumetric flask.
>
> **Solvent B:** Methanol, spectrochemical grade (M.W.: 32.04, density: 0.791 g/ml) with 5% by volume TMS added. Prepare in a 5- or 10-ml volumetric flask.
>
> **Solution 1:** 0.20 mole fraction of acetylacetone in solvent A
>
> **Solution 2:** 0.20 mole fraction of acetylacetone in solvent B
>
> **Solution 3:** 0.20 mole fraction of ethyl acetoacetate in solvent A

Use a 1-ml pipet graduated in 0.01-ml increments to measure out 0.010 mol of solute, and then add the correct amount (0.040 mol) of solvent. You may neglect the presence of the 5% TMS when determining the necessary volumes of solvent. (Calculate these volumes before coming to lab.) All work with TMS should be carried out in a hood. All containers or samples containing TMS should be tightly sealed and stored at low temperatures because of its volatility.

Prepare an NMR tube containing about 1 in. of solvent A and another containing solvent B, and record both NMR spectra, setting the TMS signal at the chart zero. Repeat for solutions 1 to 3, taking care to scan above $\delta = 10$ PPM since the enol OH peak is shifted substantially downfield. Determine which peaks are due to solute and measure chemical shifts for all solute features. Integrate the bands carefully at least three times, expanding the vertical scale by known factors as necessary in order to obtain accurate relative intensity measurements.

CALCULATIONS

Assign all spectral features using Table 1 and other NMR reference sources.[1–7] Tabulate your results and use your integrated intensities to calculate the percent enol present in solutions 1 to 3. If possible, use the total integral corresponding to the sum of methyl (or ethyl), methylene, methyne, and enol protons. If this proves difficult because of overlap with solvent bands, indicate clearly how you used the intensities to calculate the percent enol. For both enol and keto forms, compare experimental and theoretical ratios of the different types of protons (e.g., methyl to methylene protons in the keto form).

Using Eq. (5), calculate K_c and the corresponding standard free energy difference $\Delta G°$ for the change in state keto $\rightarrow$ enol in each solution.

DISCUSSION

Discuss briefly your assignments of chemical shifts and spin-spin splitting patterns of acetylacetone and ethyl acetoacetate. Which compound has a higher concentration of enol form and what reasons can you offer to explain this result? What changes would you expect in the NMR spectra of these two compounds if the interconversion rate between enol structures were much slower?

Compare the value of K_c for acetylacetone in CCl_4 with that in CH_3OH. What does your result suggest regarding the relative polarity of the enol and keto forms? Which form is favored by hydrogen bonding and why?

Compare your values of ΔG^0 with those for the gas phase ($\Delta G^0 = -1.5$ kcal/mol for acetylacetone, $\Delta G^0 = \sim 0$ kcal/mol for ethyl acetoacetate).[7] What solvent properties might account for any differences you observe?

Additional compounds which are suitable for studies of steric effects on keto-enol equilibria include α-methylacetone ($CH_3COCHCH_3COCH_3$), diethylmalonate ($CH_3CH_2OCOCH_2COOCH_2CH_3$), ethyl benzoylacetate ($C_6H_6COCH_2$-$COOCH_2CH_3$), and t-butyl acetoacetate (CH_3COCH_2COOt-Bu). Some other possible compounds are listed in Refs. 6 and 7. Further aspects of this equilibrium that could be studied include the effects of concentration, temperature, and solvent dielectric constants on K_c.[7]

APPARATUS

NMR spectrometer with peak integrating capability; several 5- or 10-ml volumetric flasks; precision 1-ml graduated pipets; NMR tubes; spectrochemical grade CCl_4 and CH_3OH; tetramethylsilane, acetylacetone, and ethyl acetoacetate.

REFERENCES

1. J. C. Davis, Jr., "Advanced Physical Chemistry," Ronald, New York (1965).
2. J. A. Pople, W. G. Schneider, and H. J. Bernstein, "High Resolution Nuclear Magnetic Resonance," McGraw-Hill, New York (1959).

3. L. M. Jackman and S. Sternhell, "Applications of NMR Spectroscopy in Organic Chemistry," Pergamon, New York (1969).

4. J. R. Dyer, "Application of Absorption Spectroscopy of Organic Compounds," Prentice-Hall, Englewood Cliffs, N.J. (1972).

5. J. W. Akitt, "NMR and Chemistry," Chandler & Hall, London (1973).

6. J. L. Burdett and M. T. Rogers, *J. Amer. Chem. Soc.*, **86**, 2105 (1964).

7. M. T. Rogers and J. L. Burdett, *Can. J. Chem.*, **43**, 1516 (1965).

XIV

SOLIDS

EXPERIMENTS

EXPERIMENT 47.   DETERMINATION OF CRYSTAL STRUCTURE BY X-RAY DIFFRACTION

The object of this experiment is to determine the crystal structure of a solid substance from x-ray powder diffraction patterns. This involves determination of the symmetry classification (cubic, hexagonal, etc.), the type of crystal lattice (simple, body-centered, or face-centered), the dimensions of the unit cell, the number of atoms or ions of each kind in the unit cell, and the position of every atom or ion in the unit cell. Owing to inherent limitations of the powder method only substances in the cubic system should be chosen for study.

Knowledge of the crystal structure permits determination of the coordination number (the number of nearest neighbors) for each kind of atom or ion, calculation of interatomic distances, and elucidation of other structural features related to the nature of chemical binding and the understanding of physical properties in the solid state.

THEORY

A perfect crystal constitutes the repetition of a single very small unit of structure, called the *unit cell*, in a regular way so as to fill the volume occupied by the crystal (see Fig. 1).

A crystal may for some purposes be described in terms of a set of three *crystal axes* **a**, **b**, and **c**, which may or may not be of equal length and/or at right angles, depending on the symmetry of the crystal. These axes form the basis for a coordinate system with which the crystal may be described. An important property of crystals, known at least a century before the discovery of x-rays, is that for any crystal the crystal axes can be so chosen that all crystal faces can be described by equations of the form

$$hx + ky + lz = \text{positive constant} \tag{1}$$

where x, y, and z are the coordinates of any point on a given crystal face, in a co-ordinate system with axes parallel to the assigned crystal axes and with units equal to the assigned axial lengths a, b, c; and where h, k, and l are *small integers*, positive, negative, or zero. This is known as the law of rational indices. The integers h, k, and l are known as the *Miller indices*, and when used to designate a crystal face are ordin-arily taken relatively prime (i.e., with no common integral factor). Each crystal face may then be designated by three Miller indices hkl, as shown in Fig. 1. The law of rational indices historically formed the strongest part of the evidence supporting the conjecture that crystals are built up by repetition of a single unit of structure, as shown in Fig. 1.

The crystal axes for a given crystal may be chosen in many different ways; however, they are conventionally chosen to yield a coordinate system of the highest possible symmetry. It has been found that crystals can be divided into six possible systems on the basis of the highest possible symmetry that the coordinate system may possess as a result of the symmetry of the crystal. This symmetry is best described in terms of symmetry restrictions governing the values of the axial lengths a, b, and c and the interaxial angles α, β, and γ.

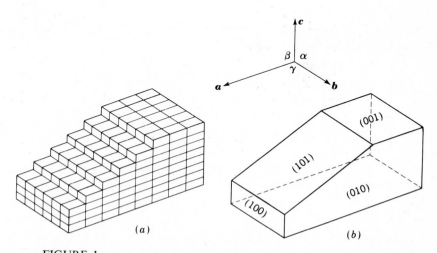

FIGURE 1
(a) Schematic diagram of a crystal, showing unit cells; (b) same crystal, showing axes and Miller indices.

The crystal systems are as follows:[1]

Triclinic system   No restrictions.

Monoclinic system   No restriction on lengths a, b, c; however, $\alpha = \gamma = 90°$; $\beta \neq 90°$.

Orthorhombic system   No restrictions on a, b, c; however, $\alpha = \beta = \gamma = 90°$.

Tetragonal system $a = b \neq c$; $\alpha = \beta = \gamma = 90°$.

Hexagonal system   *Hexagonal division:* $a = b = b' \neq c$; $\alpha = \beta = 90°$, $\gamma = \gamma' = 120°$ (there being three axes **a**, **b**, and **b'** in the basal plane, at 120° angular spacing; the **b'** axis is redundant). *Rhombohedral division:* $a = b = c$; $\alpha = \beta = \gamma \neq 90°$. (Although the coordinate systems in these two divisions are of different symmetry, hexagonal axes may be used in the description of rhombohedral crystals and vice versa.)

Cubic system $a = b = c$; $\alpha = \beta = \gamma = 90°$ (cartesian coordinates).

The symmetry of a crystal is not completely specified, however, by naming the crystal system to which it belongs. Each crystal system is further subdivided into *crystal classes*, of which there are 32. On the basis of the detailed symmetry of the atomic structure of crystals, further subdivision is possible. Symmetry will not, however, be discussed in detail here.

Since a crystal structure constitutes a regular repetition of a unit of structure, the unit cell, we may say that a crystal structure is periodic in three dimensions. The periodicity of a crystal structure may be represented by a *point lattice* in three dimensions. This is an array of points that is invariant to all the translations which leave the crystal structure invariant and to no others. We shall find the lattice useful in deriving the conditions for x-ray diffraction.

To define a crystal lattice in another way, consider the crystal structure to be divided into unit cells (parallelepipeds in shape) in such a way as to obtain the smallest unit cells possible. These unit cells are then called *primitive*. Starting with a set of crystal axes which are parallel to three edges of a unit cell and equal to them in length, a point lattice may be defined as the infinite array of points the coordinates *xyz* of which assume all possible combinations *mnp* of integral values (positive, negative, and zero), and *only* integral values. There is then one lattice point per unit cell.

It is frequently found that it is not possible to find a primitive unit cell with edges parallel to crystal axes chosen on the basis of symmetry. In such a case the crystal axes, chosen on the basis of symmetry, are proportional to the edges of a unit of structure that is larger than a primitive unit cell. Such a unit is called a *nonprimitive* unit cell, and there is more than one lattice point per nonprimitive unit cell. If the nonprimitive unit cell is chosen as small as possible consistent with the symmetry desired, it is found that the extra lattice points (those other than the corner points) lie in the center of the unit cell or at the centers of some or all of the faces of the unit cell. The coordinates of the lattice points, in such a case, are therefore either integers or half-integers.

Within a given crystal system there are in some cases several different types of crystal lattice, depending upon the type of minimum-size unit cell that corresponds to a choice of axes appropriate to the given crystal system. This unit cell may be *primitive P* or in certain cases *body-centered I, face-centered F*, or *end-centered A, B*, or *C*, depending on which pair of end faces of the unit cell is centered. The lattices are designated as primitive, body-centered, face-centered, or end-centered depending on whether the smallest possible unit cell that corresponds to the appropriate type of axes is primitive, body-centered, face-centered, or end-centered. There are in all 14 types of lattice.[2] In the cubic system there are three: primitive, body-centered, and face-centered; these are shown in Fig. 2.

Bragg reflections   Let us now determine the geometrical conditions under which diffraction of x-rays by a crystal structure would take place. These are (except for intensity considerations) the same as the conditions for diffraction from the crystal lattice, if a "point-scattering center" is placed at each point of the lattice. We shall accordingly examine the geometry of diffraction by such a crystal lattice.

The principle by which x-rays would be diffracted by such a lattice is essentially the same as that by which light is diffracted by a ruled grating. When a plane wave impinges on a point-scattering center, the scattering center radiates a spherical wave. If there are two or more such scattering centers, the spherical waves will in certain regions tend to reinforce one another and in certain other regions tend to cancel one another out, that is, interfere.

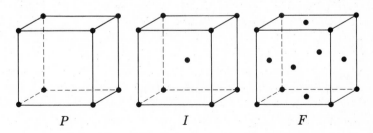

FIGURE 2
Cubic unit cells: *P*, primitive; *I*, body-centered; *F*, face-centered.

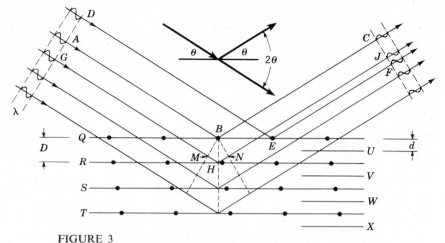

FIGURE 3

Condition for Bragg reflection from scattering centers confined to a set of equidistant, parallel planes. (Planes are perpendicular to the page; their traces are indicated by the horizontal lines.)

If there are many scattering centers, it is possible for them to be so arranged that the individual spherical waves combine in a certain region a large distance away to form a "reflected" plane wave with an amplitude which represents the sum of the amplitudes of the individual spherical waves at that distance. Such an arrangement is obtained when the scattering centers are confined to a plane surface, in which case the "reflected" plane wave is similar to that which would be obtained if the arrangement of scattering centers were replaced by a plane mirror at the same place. This is the situation represented in Fig. 3 by reflection of rays $\overline{AB}$ and $\overline{DE}$ from the plane Q. It is due to the fact that the path length, along a reflected ray of radiation, from a wave front of the incident radiation to a wave front of the reflected radiation is the same for every ray (for example, $\overline{AB} + \overline{BC} = \overline{DE} + \overline{EF}$) when the angle of incidence is equal to the angle of reflection.

It is not necessary, however, as a condition for maximum reinforcement, that all the path lengths be equal, provided that those which are not equal (for example, $\overline{AB} + \overline{BC}$ and $\overline{GH} + \overline{HJ}$ in Fig. 3) differ by a wavelength λ or by an integral number of wavelengths $n\lambda$. It can be seen from Fig. 3 that the condition for complete reinforcement is

$$n\lambda = 2D \sin \theta \tag{2}$$

(since $\overline{MH} = \overline{HN} = D \sin \theta$ and the path difference $\overline{MH} + \overline{HN}$ must be equal to $n\lambda$), where n is an integer, often called the order of the diffraction; D is the interplanar spacing; and θ is the angle which the incident ray and the reflected ray make with the reflecting planes. Equation (2) is called the *Bragg equation*, and θ is often called the

Bragg angle. The angle through which the x-rays are deflected from their original direction by reflection from the planes is twice the Bragg angle, or 2θ.

If the spacing angle θ deviates only slightly from that which satisfied the Bragg equation, reinforcement remains almost complete only if there is only a small number of reflecting planes. If, however, there is a very large (or "infinite") number of reflecting planes, the reflection from any given plane is cancelled by that from another plane a considerable distance away. Reflection should be observed, therefore, only for values of θ extremely close to those that satisfy the Bragg equation, and this is experimentally found to be the case.

We shall now modify the treatment slightly by taking note of the fact that a second-order reflection ($n = 2$) from the planes Q, R, S, T, ... corresponds to a hypothetical first-order reflection from the planes Q, U, R, V, S, W, T, X, ... only half of which contain lattice points. By inserting the required number of additional equidistant parallel planes containing no lattice points, we can dispense with the order n and write the Bragg equation in the following form, which is the form that will be used henceforth in this discussion:

$$\lambda = 2d \sin \theta \tag{3}$$

where d is now the interplanar spacing of the new set of planes. We shall now say that Eq. (3) is the necessary condition for reflection of x-rays of wavelength λ from a set of crystallographic planes with interplanar spacing d. A set of crystallographic planes is defined as an infinite array of equidistant and finitely spaced parallel planes so constructed that every point of the given crystal lattice lies on some plane of the set (though it is not necessarily true that all planes in the set contain lattice points). It is evident that sets of crystallographic planes may be constructed for a given crystal lattice in many ways (indeed, an infinite number of ways). The equations for such a set of planes are

$$
\begin{aligned}
&\vdots \\
hx + ky + lz &= -2 \\
hx + ky + lz &= -1 \\
hx + ky + lz &= 0 \\
hx + ky + lz &= 1 \\
hx + ky + lz &= 2 \\
&\vdots \\
hx + ky + lz &= N \\
&\vdots
\end{aligned}
\tag{4}
$$

where x, y, and z are coordinates as previously defined and where, when the coordinate system corresponds to a primitive unit cell, the coefficients h, k, and l may take on any integral values, positive, negative, or zero. They may be called the Miller indices of the set of crystallographic planes and are related to Miller indices as applied to crystal faces. (However, they need not be taken relatively prime.) The different integral values of N, ranging from $-\infty$ to ∞, define different planes in the set. That any given lattice point must lie on one of the planes, in the case of a coordinate system corresponding to a primitive unit cell, is easily seen from the fact that its coordinates xyz must be integers mnp; since the Miller indices are integers, the quantity $hx + ky + lz$ must be an integer, and one of the equations in the set of Eqs. (4) is satisfied.

If the coordinate system is not chosen in correspondence to a primitive unit cell, not all the lattice points have coordinates that are integers and all lattice points will lie on planes in the set only if certain restrictions are placed on the combinations of values which the Miller indices hkl may assume.

For a body-centered lattice I, some of the lattice points have coordinates that are expressible as integers mnp and some have coordinates that must be expressed as half-integers $m + \frac{1}{2}, n + \frac{1}{2}, p + \frac{1}{2}$. For the latter,

$$hx + ky + lz = hm + kn + lp + \tfrac{1}{2}(h + k + l)$$

and for this quantity to be an integer, it is necessary that $h + k + l$ be even.

For a face-centered lattice F, some of the lattice points are at $xyz = mnp$; some at $m, n + \frac{1}{2}, p + \frac{1}{2}$; some at $m + \frac{1}{2}, n, p + \frac{1}{2}$; and some at $m + \frac{1}{2}, n + \frac{1}{2}, p$, where m, n, and p are in each case any three integers. In order that $hx + ky + lz$ be an integer, it is evidently necessary that $k + l$, $h + l$, and $h + k$ simultaneously be even. An equivalent restriction is easily seen to be that h, k, and l must be either all even or all odd.

When these restrictions are not obeyed, no reflections can be obtained from the set of crystallographic planes under consideration, for there will be lattice points lying between the planes and scattering out of phase with those in the planes, resulting in complete cancellation or interference. By observing experimentally what sets of planes reflect x-rays one can deduce what the restrictions are and thereby deduce the lattice type.

The interplanar distance d is determined by the Miller indices hkl. For the cubic system it is easy to show by analytic geometry that

$$d = \frac{a_0}{\sqrt{h^2 + k^2 + l^2}} = \frac{a_0}{M} \tag{5}$$

where a_0 is the length of the edge of the unit cube and

$$M^2 \equiv h^2 + k^2 + l^2 \tag{6}$$

Lattice type   From the angles at which x-rays are diffracted by a crystal it is possible to deduce the interplanar distances d, with Eq. (3). To determine the lattice type and compute the unit-cell dimensions, it is necessary to deduce the Miller indices of the planes that show these distances. In the case of a powder specimen (where all information concerning orientations of crystal axes has been lost) the only available information regarding Miller indices is that obtainable by application of Eqs. (5) and (6).

To find which of the three types of cubic lattice is the correct one, we make use of some interesting properties of integers. From Eq. (5) we see that

$$\left(\frac{1}{d}\right)^2 = \left(\frac{1}{a_0}\right)^2 M^2 = \left(\frac{1}{a_0}\right)^2 (h^2 + k^2 + l^2) \tag{7}$$

so that, if we square our reciprocal spacings, it should be possible to find a numerical factor which will convert them into a sequence of integers, which we shall find convenient to make *relatively prime*. We shall see that the type of lattice is determined by the character of the integer sequence obtained.

For a simple cubic (primitive) lattice, all integral values are independently possible for the Miller indices h, k, and l. Now it is possible to express most, but not all, integers as the sum of the squares of three integers. In Table 1 the various possible values of M^2 are listed in the column under P, and it is seen that there are gaps where the integers 7, 15, 23, 28, 31, 39, 47, and 55 are absent. (Other gaps occur at higher values of M^2.) For those values of M^2 that are possible, the first column of the table gives the Miller indices the sum of whose squares yield the M^2 values. In some cases it is seen that there is more than one possible choice.

For a simple (primitive) cubic lattice P there are no restrictions on the Miller indices and therefore none on M^2 except as noted above. In the case of a body-centered cubic lattice, only those values of M^2 can be allowed which arise from Miller indices the sum of which is even. This has the effect of requiring M^2 to be even, as seen in the first of the two columns under I. We can then divide them by 2 and thereby reduce them to a *relatively prime* sequence, shown in the second column under I, for comparison with the sequence obtained from the $(1/d)^2$ values. We note immediately that the relatively prime sequence obtained differs from that for a primitive cubic lattice in having gaps at different places. By use of this fact it is almost always possible to distinguish between a primitive cubic lattice and a body-centered cubic lattice on the basis of a powder photograph. For the face-centered cubic lattice, application of the restriction that the indices must be all even or all odd produces the characteristic sequence 3, 4, 8, 11, 12, 16, . . . given in the column under F. If most or all of the numbers in this sequence are present, and if none of the excluded numbers is present, the lattice is evidently face-centered cubic.

If no relatively prime sequence of integers can be found to within the experimental uncertainty of the measurements, the crystalline substance presumably does not belong to the cubic system.

Table 1   POSSIBLE VALUES OF M^2 FOR CUBIC LATTICES

hkl	M	P, M^2	I M^2	I $M^2/2$	F, M^2
100	1.0000	1			
110	1.4142	2	2	1	
111	1.7321	3			3
200	2.0000	4	4	2	4
210	2.2361	5			
211	2.4495	6	6	3	
220	2.8284	8	8	4	8
300, 221	3.0000	9			
310	3.1623	10	10	5	
311	3.3166	11			11
222	3.4641	12	12	6	12
320	3.6056	13			
321	3.7417	14	14	7	
400	4.0000	16	16	8	16
410, 322	4.1231	17			
411, 330	4.2426	18	18	9	
331	4.3589	19			19
420	4.4721	20	20	10	20
421	4.5826	21			
332	4.6904	22	22	11	
422	4.8990	24	24	12	24
500, 430	5.0000	25			
510, 431	5.0990	26	26	13	
511, 333	5.1962	27		—	27
520, 432	5.3852	29			
521	5.4772	30	30	15	
440	5.6569	32	32	16	32
522, 441	5.7446	33			
530, 433	5.8310	34	34	17	
531	5.9161	35			35
600, 442	6.0000	36	36	18	36
610	6.0828	37			
611, 532	6.1644	38	38	19	
620	6.3246	40	40	20	40
621, 540, 443	6.4031	41			
541	6.4807	42	42	21	
533	6.5574	43			43
622	6.6332	44	44	22	44
630, 542	6.7082	45			
631	6.7823	46	46	23	
444	6.9282	48	48	24	48
700, 632	7.0000	49			
710, 550, 543	7.0711	50	50	25	
711, 551	7.1414	51			51
640	7.2111	52	52	26	52
720, 641	7.2801	53			
721, 633, 552	7.3485	54	54	27	
642	7.4833	56	56	28	56

When the cubic lattice type has been deduced, the unit-cell dimension a_0 can be calculated. From the unit-cell volume, the measured crystal density, and the formula weight, the number of formulas in one unit cell can be calculated.

Deduction of the structure   The arrangement of the atoms or ions in the unit cell is at least partly determined by symmetry considerations, but in most cases it is necessary to take account of the *intensities* of the Bragg reflections. The way that this is done in present crystallographic practice is far too complicated to describe here.[3] We shall here only illustrate by a simple example how it is possible to use qualitative arguments based on intensity.

If the substance is a binary compound AB, and if its unit cell is simple cubic *P* with one formula (one atom of A and one of B) per cubic cell, the relative positions of the two atoms are fixed by symmetry. This is true of the salt cesium chloride, CsCl, the structure of which is shown in Fig. 4. One of the ions, Cs^+ say, may, without loss of generality, be placed at the origin. The other ion, Cl^-, must be at the center of the unit cell; if it is in any other position, the structure will lack the threefold rotational axes of symmetry which are always present along all four body diagonals of the unit cell in the cubic system.

In many cases, including even some with one formula of a binary compound for each lattice point, the positions of the atoms or ions are not necessarily given uniquely by symmetry, though the number of possible choices may not be large. The solution of the structure in some cases may lie in a simple clue from the intensities. Let it be supposed, for example, that a certain class of powder lines are relatively weak and that omitting these lines from the calculation leads to a "pseudo lattice" with a smaller number of atoms per lattice point than in the case of the true lattice. A possible hypothesis might then be that the weak lines owe their weakness to destructive interference between two kinds of atoms or ions and that, if the chemical difference between these two kinds of atoms or ions could somehow be removed, the interference would become complete and the pseudo lattice would become a true

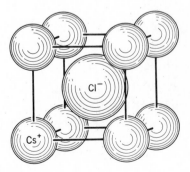

FIGURE 4
Unit cell of cesium chloride. Note that this structure is not body-centered cubic but would be if the two ions were identical.

lattice. Knowledge of the pseudo lattice (and pseudo cell) may then show where the atoms or ions must be placed (irrespective of kind), and knowledge of the true lattice will then show which atoms are of each kind.

As an example of this kind of clue, let us deduce the relative positions of the ions in cesium chloride, bromide, and iodide (which all have the same structure) from intensity considerations, without appealing to any arguments based on symmetry. Putting the $(1/d)^2$ on a relatively prime basis, we obtain the integers 1, 2, 3, 4, 5, 6, $-$, 8, 9, 10, 11, 12, 13, 14, $-$, 16, ..., which clearly indicates a primitive lattice. From the density we find that there is one formula per lattice point. However, we see that all the odd-numbered lines are somewhat weak in CsCl, much weaker in CsBr, and very faint or absent in CsI. If we neglect them, the sequence of integers obtained (on dividing by 2 to obtain relative primes) indicates a body-centered cubic "pseudo lattice" with one-half formula, or one ion (irrespective of kind), at each pseudo-lattice point. In other words, if we were unable to distinguish between the two kinds of ions, the structure would look like one with a single atom of a single kind at each point of a body-centered cubic lattice. This clearly shows that in the real structure, an ion of one kind is located at (000) and an ion of the other kind is located at $(\frac{1}{2}\frac{1}{2}\frac{1}{2})$. It may be mentioned that the virtually complete obliteration of the odd lines in CsI is due to the fact that the Cs^+ ion and the I^- ion are *isoelectronic* (that is, have the same number of electrons, namely 54, which is the number in xenon); if any odd lines are observed at all, it is owing to the fact that, because of the different nuclear charges of the two ions, the sizes of the electron clouds of the two ions are slightly different.†

When there are more than two atoms per lattice point, the structure determination will be more complicated. One frequently used procedure is that of "trial and error," in which a number of "model" structures are successively proposed and tested by calculation of intensities and comparison with experiment, until a structure is found which yields satisfactory agreement. When the number of possible structures is too large for the practical application of trial-and-error methods, methods must be used that are too advanced for description here. With their use, crystal structures have been found in which there are 500 or more atoms in a unit cell. In such cases and in most ordinary work, however, powder data would be inadequate for determining the structure, and diffraction patterns must be obtained from single crystals.

METHOD

There are several experimental techniques for realizing the diffraction conditions, the most powerful of which depends on having a single crystal of the substance to be studied;[4] see Exp. 48 following. As single-crystal methods are rather complicated,

---

† The argument here depends upon the fact that it is the electrons, rather than the nuclei, which scatter x-rays.

we shall concern ourselves here only with the so-called powder method, or Debye-Scherrer method, which makes use not of a single crystal but rather of a powder obtained by grinding up the crystalline or microcrystalline material.[5] This powder contains crystal particles of the order of magnitude of a few microns in size.

In this method, the specimen for diffraction is obtained by sticking some of the powdered material onto a fine (0.1 mm) glass fiber with a trace of Vaseline or filling a very thin-walled glass capillary tube (0.2 mm diameter) with the powder. A narrow beam of parallel monochromatic x-rays, about 0.5 mm in diameter, impinges on this specimen at right angles to its axis. The source of x-rays is usually a Coolidge-type x-ray tube with a copper (or molybdenum) target, equipped with a filter (of nickel foil, in the case of a copper target) to remove all spectral components except the desired $K\alpha$ line. The narrow beam is formed by a *collimator*, which consists basically of a conical tube 5 or 6 cm long with a pinhole or slit at each end. On the opposite side of the specimen is a conical receptacle similar to the collimator but with only an entrance pinhole or slit and no exit; this is the *beam stop*, in which the undiffracted beam is trapped. The diffracted radiation is detected by a strip of photographic film bent into a cylinder and held firmly against the inside wall of a cylindrical camera, coaxial with the specimen. The arrangement of the collimator, specimen, and photographic film is shown schematically in Fig. 5. When the beam impinges on the randomly oriented particles in a stationary powder specimen, most of the particles will not diffract the x-rays at all. Only those particles will diffract x-rays which happen by chance to be so oriented that Eq. (3) holds for Bragg reflection from some set of crystallographic planes. The direction of the diffracted rays will then deviate from the direction of the incident beam by twice the Bragg angle. Since orientation is completely random with respect to the beam axis, the rays diffracted at a given scattering angle may lie with equal probability anywhere on a right circular cone with apex angle 4θ. In practice, the specimen is usually rotated about its own axis, so that during a single rotation all or nearly all the particles present will have an opportunity to reflect x-rays from any given set of crystallographic planes, the resulting diffracted radiation being distributed rather evenly over the cone. Where the cone intersects the photographic film, the latent image of a *powder line* is formed. When the film is removed from the camera and developed, fixed, washed, and dried, it is found to have on it a number of lines, each one of which is due to reflection from one or more sets of crystallographic planes.

The positions of the lines on the film can be measured with a comparator (see Exp. 43) or microphotometer, but adequate measurements can be made with a good millimeter scale. From measurements of the positions of the lines or distances between them, the Bragg angles can be calculated. The determination of these angles is a purely geometric problem. The "effective radius" R of the film in the camera can be obtained by picking two sets of lines, A,B, and C,D as in Fig. 5, and measuring the four distances, $\overline{AC}$, $\overline{AD}$, $\overline{BC}$, and $\overline{BD}$:

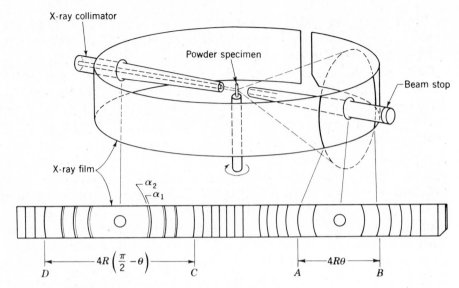

FIGURE 5
Schematic diagram illustrating x-ray powder method (Straumanis arrangement[6]).
Note splitting of back-reflection lines.

$$4R = \frac{1}{\pi} (\overline{AC} + \overline{AD} + \overline{BC} + \overline{BD}) \tag{8a}$$

If it is desired to calculate angles in degrees rather than in radians, it is convenient to calculate

$$4R° = \frac{\pi}{180} (4R) = \frac{1}{180} (\overline{AC} + \overline{AD} + \overline{BC} + \overline{BD}) \tag{8b}$$

Then

$$\theta \text{ (rad)} = \frac{\overline{AB}}{4R} \qquad \theta' = \frac{\pi}{2} - \theta \text{ (rad)} = \frac{\overline{CD}}{4R} \tag{9a}$$

$$\theta \text{ (deg)} = \frac{\overline{AB}}{4R°} \qquad \theta' = 90 - \theta \text{ (deg)} = \frac{\overline{CD}}{4R°} \tag{9b}$$

From trigonometric tables, $\sin \theta$ or $\log \sin \theta$ can then be obtained. In processing the back-reflection data, it is unnecessary to convert from $(\pi/2 - \theta)$ to θ; it is sufficient to look up $\cos (\pi/2 - \theta) = \cos \theta'$. We then calculate

$$\frac{1}{d} = \frac{2}{\lambda} \sin \theta = \frac{2}{\lambda} \cos \theta' \tag{10}$$

When the measurements are made and $(1/d)$ is computed, account should be taken of the fact that the $K\alpha$ spectral line generally used as a monochromatic source of x-rays is not truly a single line but actually a closely spaced doublet. The doublet is ordinarily not resolved in the forward-reflection part of the film, but in the back-reflection part of the film the lines are usually observably split into two components, known as $K\alpha_1$ and $K\alpha_2$ in order of increasing wavelength (and therefore in order of increasing θ), in the intensity ratio 2:1. When the two components are not resolved, the position taken for the line should be an estimate of the position of the "center of gravity," and the value of the wavelength λ used in the calculation should be the weighted mean wavelength:

$$\lambda_{\text{mean}} = \tfrac{1}{3}(2\lambda_{\alpha_1} + \lambda_{\alpha_2}) \tag{11}$$

When the components are sufficiently well resolved to make possible the measurement of the positions of the two components separately, such measurements should be made and the actual values of the wavelengths should then be used in the calculations. Each component will then yield a separate value of $1/d$; the two values obtained should be in good agreement and may be averaged for the ensuing calculations.

For copper ($K\alpha$) radiation,

$$
\begin{aligned}
\lambda_{\alpha_1} &= 1.54050 \text{ Å} \\
\lambda_{\alpha_2} &= 1.54434 \text{ Å} \\
\lambda_{\text{mean}} &= 1.5418 \text{ Å}
\end{aligned}
\tag{12}
$$

In present-day x-ray crystallography, powder photography is only rarely used for complete structure determination. It is, however, frequently used in the precise determination of unit-cell dimensions. Its most common technical use is as an analytical tool; powder patterns of thousands of crystalline substances are known.

EXPERIMENTAL

Preparation of powder specimen   A specimen of the material to be studied is finely pulverized in an agate mortar with an agate pestle. The powder is then loaded into a Pyrex or Lindemann glass capillary tube, 0.2 to 0.3 mm o.d., having a very thin wall (0.01 mm or less), so as to obtain a densely filled specimen about 1 cm in length. If the capillary tube is flared out at one end, it is easier to load; even without a flared-out end, however, the powder can be picked up by scraping one end of the capillary against the surface of the mortar and, with this end uppermost, agitating the powder by very gentle rasping with a file. One end of the capillary tube should be sealed off with a flame or with wax before filling, and the other end with wax after filling. The capillary tube is then affixed with wax to the end of a short ($\frac{3}{8}$ in. or less) length of $\frac{1}{8}$-in. brass rod; see Fig. 6.

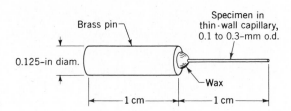

FIGURE 6
Mounted powder specimen.

Installation of the specimen   A powder camera of the Straumanis design is shown in Fig. 7. It is a valuable instrument and must be handled carefully. The camera should not be used until the procedures for using it have been demonstrated to the student by an experienced user. To install and line up the specimen, proceed as follows:†

Remove the circular cover, and lay it down, inside surface upward, in a *safe place*. Remove the two slits (collimator and beam stop), and lay them down *inside the cover*.

By means of tweezers or long-nose pliers, insert the brass pin supporting the specimen into its receptacle, where it is held by friction. Carefully replace the collimator, being sure not to force it or to drop it.

Place the lens cap over the outside collimator opening, and set the camera down with the beam-stop hole facing a well-lighted surface (beam stop *not* in place). By hand, turn the pulley wheel on the outside of the camera. As it goes around, the silhouette of the specimen should appear to go up and down. Stop it in an "up" position, and by means of the screw on top of the camera push it down to the center of the visual field. Turn the screw back up, and rotate again to see if adjustment is needed; if so, use the screw as before. Continue until the specimen appears to undergo no motion as the spindle is rotated.

Remove the lens cap, and replace the beam stop. As in the case of the collimator, do not force it or drop it. Replace the cover.

Loading the camera with film   Take the camera into the darkroom. Remove the cover and the collimator and beam stop. Turn off the room light, and under a safe-light cut, punch, and insert the film in accordance with specific instructions given. (It is advisable to rehearse the procedure ahead of time with the light on, using a piece of exposed film.) Insert the collimator and beam stop and replace the cover.

Making the x-ray exposure   This should be done under the constant supervision of a qualified person; detailed instructions cannot be given here. **Be very careful to**

† The procedure given is that applicable to the 114.6-mm camera manufactured by the North American Philips Co.[6]

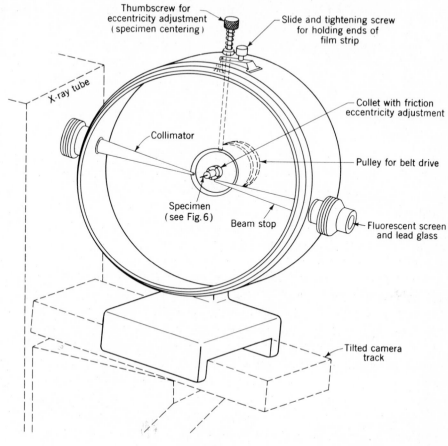

FIGURE 7
Philips powder camera (shown with cover not in place).

avoid any x-ray exposure to any part of the body; make sure that all x-ray ports are covered when the unit is in operation. The optimum exposure time depends on the composition and size of the specimen, the tube current and operating potential, and the dimensions of the collimating system. Typical exposures are 4 to 20 hr. In the absence of a recommended exposure time, make a short trial exposure and a second longer exposure if necessary.

Developing the film   The camera is taken into the darkroom and the film removed. It should be attached to a clip or other support and developed for 3 to 5 min, depending on the temperature of the developer. Agitate the film, but do not allow its surface to rub against the walls or bottom of the tank or tray. Wash the film for 30 sec and place it in the fixing bath. After the film has cleared, the room light may be turned on. The film should be in the fixing bath for at least twice as long as it takes to clear and should then be washed in running water for at least 30 min. It may then be carefully wiped with clean fingers, rinsed again with distilled water, and hung up to dry.

Measurement of crystal density   The density of the substance can be determined with a pycnometer or with a 5-ml (or smaller) volumetric flask used as a pycnometer. The liquid used should be one in which the substance is insoluble; for a water-soluble inorganic salt, medium-boiling petroleum ether is convenient. The procedure is described in Chap. I (Sample Report). Make two determinations.

CALCULATIONS

If it is not possible to take the x-ray powder pattern in the laboratory, a contact print of a powder photograph will be furnished and the x-ray equipment will be discussed and shown to the students. The instructions below should be followed with either the negative taken or the print provided.

Measure $\overline{AB}$ or $\overline{CD}$ (see Fig. 5) for as many lines as possible; estimate each distance to ± 0.1 mm. Enter the measurements, together with estimated intensity, into a table. Make a separate measurement for each component of a resolved doublet.

Pick a pair of sharp lines, A,B in the forward-reflection part of the film and another pair C,D in the back-reflection region. Measure the four distances $\overline{AC}$, $\overline{AD}$, $\overline{BC}$, $\overline{BD}$ as accurately as possible. Calculate $4R$ or $4R°$ from Eq. (8).

Calculate θ or θ' for each line or resolved component of a line [Eq. (9)] and look up $\sin \theta$ or $\cos \theta'$. Calculate $1/d$ from Eq. (10), averaging the values obtained for the two components of each resolved doublet. Calculate $(1/d)^2$.

Find a factor that will reduce the $(1/d)^2$ values to relatively prime integers within experimental error. Refer to Table 1 and identify the lattice type. Also, from Table 1, obtain the Miller indices hkl for each line.

The most precise values of a_0 are obtained from the lines in the back-reflection part of the film, where θ is close to $90°$. This can be seen by combining Eqs. (3) and (5), solving for a_0, and differentiating:

$$da_0 = -\frac{M\lambda}{2 \sin^2 \theta} \cos \theta \, d\theta$$

$$= -a_0 \cot \theta \, d\theta$$

The experimental uncertainty in measuring θ is proportional to that of measuring the distances between pairs of lines and is approximately constant over the film unless lines in the back-reflection region are unduly faint or broad. However, even if the uncertainty of measurement is a little larger in the back-reflection region, the effect is ordinarily far outweighed by the $\cot\theta$ factor. In fact, lines very close to the collimator hole should always be used if at all possible, even if they are broad and diffuse. Another good reason for using only lines in the back-reflection region is that they are much more free of shifts due to absorption of x-rays by the specimen. From a few well-chosen lines in the back-reflection region, preferably resolved doublets, calculate a_0 from Eq. (5) and average the values obtained.

From the measured density and the a_0 value determined as above, calculate the number of atoms per unit cell and per lattice point. Report the lattice type, the value of a_0, and the number of atoms per unit cell.

DISCUSSION

Determine the crystal structure, if possible, by methods similar to those described for CsCl (see Theory). Draw a diagram of the cubic unit cell showing the positions of all atoms or ions.

APPARATUS

X-ray diffraction apparatus, complete with tube stand, x-ray tube, and power supply; Debye-Scherrer powder camera, preferably Straumanis type (such as North American Philips 114.6-mm camera); film cutter and punch; thin-wall Lindemann glass capillary tubes (may be purchased from Caine Scientific Sales Co., Chicago, Ill.); agate mortar and pestle; small file; x-ray film (Eastman no-screen, 35-mm continuous strip); darkroom, equipped with x-ray developing tank or adequate trays; x-ray developer and fixer solutions; timer; thermometer; good centimeter scale; pycnometer or 5-ml volumetric flask; small pipette or eye dropper.

Small quantity of crystalline material of cubic structure for study (e.g., alkali halides; alkaline earth oxides; cuprous or silver chloride; simple metals such as aluminum or copper, finely powdered with a file); liquid (medium-boiling petroleum ether or CCl_4) for density work.

REFERENCES

1. M. J. Buerger, "Elementary Crystallography," chap. 9, Wiley, New York (1956).
2. *Ibid.*, chap. 8.
3. H. Lipson and W. Cochran, "The Determination of Crystal Structures," G. Bell, London (1953).

4. M. J. Buerger, "X-ray Crystallography," Wiley, New York (1942).
5. H. P. Klug and L. Alexander, "X-ray Diffraction Procedures for Polycrystalline and Amorphous Materials," chap. 4, Wiley, New York (1954).
6. *Ibid.*, p. 178; M. J. Buerger, *Am. Miner.*, **21**, 11 (1936); *J. Appl. Phys.*, **16,** 501 (1945); M. Straumanis and A. Ieviņš, *Z. Phys.*, **98,** 461 (1936).

GENERAL READING

W. F. deJong, "General Crystallography: A Brief Compendium," Freeman, San Francisco (1959).

W. L. Bragg, "The Crystalline State I. A General Survey," Cornell, Univ. Press Ithaca, N.Y. (1965).

R. C. Evans, "Crystal Chemistry," 2d ed., Cambridge, New York (1964).

M. J. Buerger, *op. cit.*

G. H. Stout and L. H. Jensen, "X-ray Structure Determination," Macmillan, New York (1968).

EXPERIMENT 48. SINGLE CRYSTAL X-RAY DIFFRACTION WITH THE BUERGER PRECESSION CAMERA

The introduction to X-ray crystallography given in the preceding experiment is mainly confined to the determination of a very simple cubic structure by the powder method. The present experiment is considerably more ambitious: it deals with a single crystal having a symmetry different from cubic. Analysis of this more general problem requires the use of the *reciprocal lattice*, a concept fundamental to crystallography and many aspects of solid-state physical chemistry. Another basic crystallographic concept—the *space group*—is also introduced. The specification of the space group of a crystal amounts to a complete statement of the symmetry of the infinitely periodic crystal structure.

Measurements of the reciprocal lattice by diffraction methods yield the lattice constants of the "reciprocal unit cell," from which one may calculate the lattice constants (unit-cell dimensions and angles) of the crystallographic unit cell in direct space. Once the lattice constants and the space group are established, the crystal structure can be determined (by methods beyond the scope of the present experiment) from the Bragg intensities associated with the reciprocal lattice points. A complete crystal structure is usually considered to include not only the positional coordinates of all of the atoms in the unit cell but also their average vibrational amplitudes in various directions. Structural chemistry information on molecular shapes and configurations, interatomic distances, bond angles, and intermolecular separation distances can then be easily determined.

The instrumentation used to explore the reciprocal lattice was once almost entirely photographic; nowadays most crystal structure determinations are done with four-circle diffractometers under computer control. Nevertheless photographic methods are still valuable. One such photographic method, the *precession method*, is particularly effective in illustrating the concept of the reciprocal lattice. This method uses the Buerger precession camera.[1] As we shall see, this camera effectively "photographs the reciprocal lattice" to scale, layer by layer.

This experiment is devoted to the use of precession photographs for determining the lattice constants and space group of an orthorhombic crystal.

THEORY

Some elementary concepts of crystallography—the crystal lattice, the unit cell, lattice planes, Miller indices, and the Bragg equation—have been introduced in Exp. 47. These concepts are vital prerequisites for the study of the new concepts described here.

The reciprocal lattice   In Exp. 47 we dealt with the concept of a set of lattice planes—an infinite set of equally spaced parallel planes so defined that all points of the crystal lattice lie on planes of the set (although it is not necessary that all planes of the set contain lattice points). The infinite set of equations defining the set of lattice planes is indicated in Eqs. (47-4). The set of lattice planes is designated by the *Miller indices* h, k, l, each of which can independently take on all integral values from $-\infty$ to ∞, provided the coordinates x, y, z are defined with respect to a *primitive* unit cell. If the coordinates are defined with respect to a non-primitive cell, exemplified by the body-centered or face-centered cells discussed in Exp. 47, the Miller indices are subject to certain restrictions. These lattice planes are conceptually useful in the derivation of the Bragg equation [Fig. 47-3, Eq. (47-3)].

The crystal lattice, unit cell, and lattice planes exist in ordinary three-dimensional space, referred to in crystallography as *direct space* to distinguish it from a different kind of three-dimensional space called *reciprocal space* for reasons that will soon become apparent. The *reciprocal lattice*,[2] hereinafter abbreviated r.l., may be constructed in reciprocal space as follows.

First choose an arbitrary fixed point in reciprocal space and designate it as the origin of the r.l. ($hkl = 000$), a point indicated on the illustrations by the letter O. Now choose a set of lattice planes hkl in direct space. Construct in reciprocal space, extending from the origin O, a vector $\mathbf{q}_{hkl}$ with a direction perpendicular to the set of lattice planes in direct space and a length which is the *reciprocal* of the interplanar spacing d_{hkl}:

$$\mathbf{q}_{hkl} \perp (hkl); \quad |\mathbf{q}_{hkl}| \equiv \frac{1}{d_{hkl}} \tag{1}$$

where (hkl) indicates the set of lattice planes corresponding to Miller indices hkl. The terminus of this vector is the *reciprocal lattice point hkl*, hereinafter abbreviated r.l.p. Since the set of lattice planes $\overline{hkl}$ (minus signs are customarily placed above the indices) is identical to the set hkl, there will be another vector in reciprocal space of the same length but in the opposite direction from the origin to r.l.p. $\overline{hkl}$. When this is done for all possible sets of lattice planes, the r.l. is completely defined. This procedure is perfectly general and does not depend on any particular assignment of crystal axes $\mathbf{a}$, $\mathbf{b}$, $\mathbf{c}$ to the lattice in direct space. In Fig. 1 this procedure is illustrated for the set of lattice planes (210). For the set (420), the direction of the r.l. vector $\mathbf{q}$ is the same but since d for (420) is half as large as that for (210), the vector $\mathbf{q}_{420}$ is twice as long as the vector $\mathbf{q}_{210}$. The student is urged to perform this construction for several different sets of planes on graph paper, and verify that the points generated constitute a lattice.

Corresponding to the lattice basis vectors (axes) $\mathbf{a}$, $\mathbf{b}$, $\mathbf{c}$ in direct space, there are r.l. basis vectors $\mathbf{a^*}$, $\mathbf{b^*}$, $\mathbf{c^*}$ in reciprocal space. The latter are defined in terms of the former in such a way that the following vectorial relations hold:

$$\mathbf{q}_{hkl} = h\mathbf{a^*} + k\mathbf{b^*} + l\mathbf{c^*}$$

$$\mathbf{a^*} \cdot \mathbf{a} = \mathbf{b^*} \cdot \mathbf{b} = \mathbf{c^*} \cdot \mathbf{c} = 1 \tag{2}$$

$$\mathbf{a^*} \cdot \mathbf{b} = \mathbf{b^*} \cdot \mathbf{a} = \mathbf{b^*} \cdot \mathbf{c} = \mathbf{c^*} \cdot \mathbf{b} = \mathbf{c^*} \cdot \mathbf{a} = \mathbf{a^*} \cdot \mathbf{c} = 0$$

Thus $\mathbf{a^*}$ is perpendicular to $\mathbf{b}$ and to $\mathbf{c}$, for example; it may or may not be parallel to $\mathbf{a}$ depending on the symmetry of the lattice.

For simplicity in the present experiment we shall confine ourselves to the orthorhombic system, which has orthogonal axes. In this system (as also in the tetragonal and cubic systems, where $\alpha = \beta = \gamma = 90°$)

$$a^* = \frac{1}{a} \qquad b^* = \frac{1}{b} \qquad c^* = \frac{1}{c} \tag{3}$$

It is very important to remember that these relations do *not* apply generally in the triclinic, monoclinic, and hexagonal systems, except for axes parallel to symmetry axes or normal to symmetry planes.

Any person working with diffraction will very soon appreciate the value of the r.l. in "mapping" the circumstances of diffraction. As a trivial example, consider the systematic extinctions corresponding to body-centered (I) and face-centered (F)

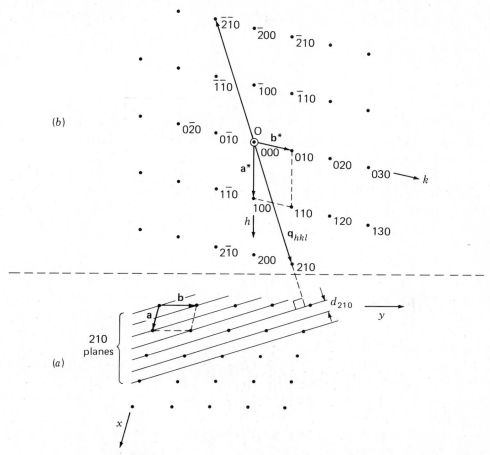

FIGURE 1

Construction of the reciprocal lattice (b) from the direct lattice (a). A two-dimensional section of each lattice is shown, corresponding to $z = 0$ in the direct lattice and $l = 0$ in the reciprocal lattice. The set of planes 210, of which the traces are shown in (a), are parallel to c (i.e., to the z axis) which comes out of the paper.

cubic lattices in Exp. 47. The condition that $h + k + l$ for an I lattice must be even means that the r.l.p. for any set of Miller indices hkl for which the sum is odd simply does not exist, and the intensity for Bragg reflection from the corresponding set of planes must be zero.†

> † The student will find it instructive to show that as a result, the r.l. corresponding to an I direct-space lattice is actually an F lattice. It can likewise be shown that the r.l. corresponding to an F direct-space lattice is an I lattice.

The Ewald construction   Let us rewrite the Bragg equation, Eq. (47-3), in the form

$$\frac{2}{\lambda} \sin \theta_{hkl} = 1/d_{hkl} \tag{4}$$

We will now convert this equation to vectorial form. Let us define a vector of unit length s_0 in the direction of the incident X-ray beam, and another of unit length s in the direction of the beam which is Bragg-reflected from the set of lattice planes hkl. These are shown in Fig. 2a together with a Bragg construction. From the vector diagram given it is clear that $|s - s_0| = 2 \sin \theta$. Because the reflection must be *specular* (i.e., like reflection from a mirror), the vector $s - s_0$ is perpendicular to the set of lattice planes. But so also is the r.l. vector q_{hkl}, which is of length $|q_{hkl}| = 1/d_{hkl}$. Therefore,

$$\frac{s}{\lambda} - \frac{s_0}{\lambda} = q_{hkl} \tag{5}$$

This is a general equation limiting the geometrical conditions of X-ray diffraction in the kinematic approximation. It is represented graphically in Fig. 2b in a diagram known as the *Ewald construction*. Let us construct a vector s_0/λ in reciprocal space (codirectional with s_0 in direct space) with its terminus at the origin O of the r.l. Describe a sphere (called the Ewald sphere) of radius $1/\lambda$ with its center S at the tail of the vector s_0/λ. Now *if* the sphere happens to pass through any r.l.p. (other than the origin), say hkl, then the geometrical conditions for diffraction are satisfied and the direction of the diffracted beam is that of a vector s/λ drawn from the center S of the Ewald sphere to the r.l.p. hkl.

The orientation of the crystal in space not only determines the orientation of the crystal axes and the direct lattice, but also determines the orientation of the r.l. If the crystal is caused to rotate in direct space, the r.l. rotates correspondingly about its origin O in reciprocal space. For a beam of strictly parallel and absolutely monochromatic X-rays (so that s_0/λ is exactly defined) and a given arbitrary orientation of the crystal, the probability that the Ewald sphere passes through any r.l.p. (other than the origin) is infinitesimal. Therefore, in *diffraction cameras* the crystal is caused to undergo some kind of rotatory, oscillatory, or precessional motion. During the corresponding motion of the r.l., some of the r.l.p.'s pass through the stationary Ewald sphere. As each r.l.p. passes through the sphere, the geometrical conditions for diffraction are satisfied for an instant and a "flash" of reflected X-rays impinges upon the photographic film and exposes some grains in the emulsion. This process is repeated at this point on the film hundreds or thousands of times because of repetition of the mechanical cycle of the instrument. When the film is developed there appears a diffraction "spot" with a size and shape determined by the size and shape of the crystal specimen and by the angle of divergence ($\gtrsim 1°$) of

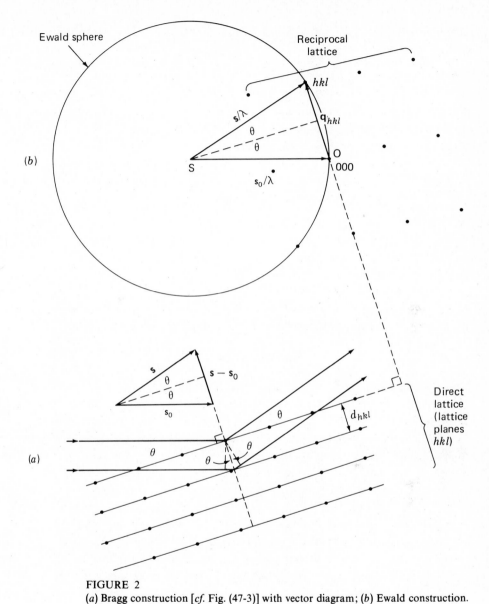

FIGURE 2
(a) Bragg construction [cf. Fig. (47-3)] with vector diagram; (b) Ewald construction.

the incident X-ray beam. The intensity (blackness) of this spot depends on con-
structive or destructive interference of the X-ray wavelets scattered by the individual atoms in the unit cell. In the Buerger precession camera the film is moved in
synchronization with the crystal motion so that the position of the spot on the

developed film reveals the position (Miller indices) of the corresponding r.l.p. in reciprocal space.

The point group and crystal class   The theory of symmetry and the development of the point groups and space groups is a complex subject.[3] We will here dispense with formal group theory; indeed we will dispense with all kinds of symmetry that are not encountered in the one crystallographic system to which the present experiment is limited: the orthorhombic system.

We must first introduce the *point group*, a collection of symmetry elements that specify the symmetry of a finite object.† The only point-symmetry elements that occur in the orthorhombic system are:

- the identity element (trivial, possessed by every object),
- the twofold rotation axis, designated by the symbol 2, representing invariance to a 180° rotation,
- the mirror plane, designated by the symbol m, representing invariance to reflection in a plane, and
- the center of inversion (often called *center of symmetry*), designated by the symbol i, representing invariance to inversion through a point.

There are 32 crystallographic point groups, and the orthorhombic system contains three (see Fig. 3):

C_{2v} or $mm2$ = two mirror planes crossing at right angles, generating a twofold axis at their line of intersection. Examples of objects conforming to this point group: an open shoe box; the molecules H_2O, SO_2, H_2CO (formaldehyde).

D_2 or 222 = three two-fold axes intersecting at right angles (actually, any two of the three suffice to generate the third). Examples: a two-bladed propeller without shaft; a molecule of 2,6,2′,6′-tetraiodobiphenyl (twisted out of the planar configuration by steric repulsion of the iodine atoms).

D_{2h} or mmm = three mirror planes intersecting at right angles, generating three two-fold axes at their lines of intersection *and* (very importantly) a center of inversion. Examples: a brick; the molecules C_2H_4 (ethylene), and B_2H_6 (bridge structure of diborane). This point group contains all of the elements of $C_{2v} - mm2$ and all of the elements of $D_2 - 222$; those two point groups are said to be *subgroups* of $D_{2h} - mmm$.

In each case, two symbols for each point group have been given. The first is the Schoenflies symbol, often used in spectroscopy. The second is the Hermann-Mauguin symbol, much more commonly used in crystallography. The three positions in the Hermann-Mauguin symbol for orthorhombic point groups correspond

---

† The point group can also be applied to an infinite object, such as a crystal lattice, a crystal structure, a reciprocal lattice, or an "intensity-weighted" reciprocal lattice.

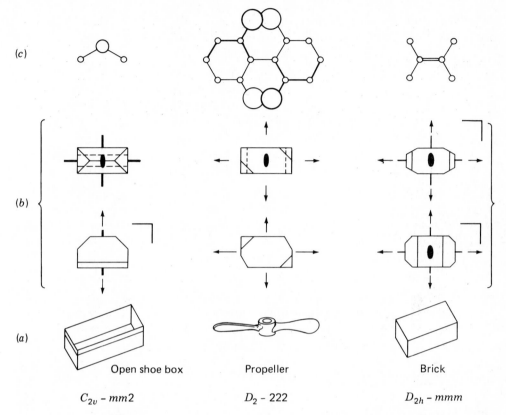

FIGURE 3
The three orthorhombic point groups with examples: (*a*) familiar objects, (*b*) hypothetical crystals showing the point group in their morphology, and (*c*) chemical molecules. In part (*b*), the conventional graphical symbols for the elements are superimposed: a right-angled line for a mirror plane parallel to the paper and a single straight line trace for one perpendicular to it; an arrow and a "lens" for a twofold rotation axis, and a small circle for a center of inversion ("center of symmetry").

to the three assigned crystal axes **a**, **b**, **c** and each is occupied by a letter or numeral designating the symmetry element operating with respect to the corresponding direction. *The direction in the case of the mirror plane is that of its normal.* Clearly, with different axial assignments, a crystal may have point group symmetry 2*mm* or *m*2*m*; these are equivalent to *mm*2 and all three correspond to the Schoenflies symbol C_{2v}.

Corresponding to each point group is a *crystal class*; all crystals having the same point-group symmetry are said to belong to the same crystal class. Thus there are three orthorhombic crystal classes, which are given the same identifying symbols as the corresponding point groups. The crystal class or point group of a given crystal is ultimately defined by its atomic structure, but the macroscopic properties of the crystal—its optical, elastic, piezoelectric, and pyroelectric properties, and also ideally its external morphology—must conform to the symmetry of the point group. However, the information required for the complete specification of the crystal point group is often not available from macroscopic properties alone.

One might then hope that all of the required information can be obtained from X-ray diffraction experiments. Unfortunately, with ordinary photographic methods such as those employed in this experiment, it is generally not possible to detect any significant difference in Bragg reflection intensity between reflections hkl and $\overline{hkl}$. That is, the X-ray experiment by itself seems to confer upon the crystal an apparent center of inversion, and thus is incapable of showing whether or not the crystal really possesses a center of inversion. (This limitation is known as Friedel's law.) The intensity-weighted r.l., i.e., the r.l. weighted with the Bragg reflection intensities, therefore possesses, for all orthorhombic crystals, the centrosymmetric $D_{2h} - mmm$ point-group symmetry. Accordingly, all orthorhombic crystals are said to have $D_{2h} - mmm$ "Laue symmetry." Without further information it is impossible to determine which of the three orthorhombic point groups is correct for a given crystal. As we shall see, the presence of *screw axes*, *glide planes*, or both can be inferred from certain systematic extinctions, and in some cases certain combinations of these can uniquely define the space group and point group. In many other cases, either physical methods must be used, or ambiguity will remain until the entire crystal structure has been successfully determined, possibly after trials with alternative space groups.

For crystals that do not conduct electricity, the final ambiguity may often be resolved by study of the piezoelectric or pyroelectric properties of the crystal. Crystals with $D_{2h} - mmm$ symmetry cannot show either the piezoelectric effect or the pyroelectric effect; those with $D_2 - 222$ can show the former but not the latter; those with $C_{2v} - mm2$ can show both. Details are given by Wooster and Breton.[4] Another method, novel and sensitive, applicable to transparent crystals (even if they have some electrical conductivity), involves determination of the capability of the crystal to generate second harmonics in a laser beam.[5]

The space group The crystal structure, idealized, is perfectly "lattice periodic" and infinite in extent in all three dimensions. A complete statement of the symmetry of the structure must take this into account. Such a statement is the specification of the space group.

An excellent introduction to space groups is given by Buerger.[6] A complete tabulation of all 230 possible space groups in three dimensions, with extensive and detailed tables and diagrams, is given in vol. I, Symmetry Groups, of "International Tables for X-Ray Crystallography,"[7] hereinafter abbreviated as *I.T.*

The same symmetry elements that we encountered in the point groups may exist in the space group, but they are repeated by the lattice and are present in an infinite number of positions. In addition, certain elements occur that do not occur in point groups because they cannot be possessed by a finite object. These include the elements of translation, corresponding to the periodic crystal lattice. In addition to these there are, in many space groups, elements that combine translations with rotations or reflections: in the orthorhombic system these are the twofold screw axis and various glide planes. The twofold screw axis corresponds to a 180° rotation *accompanied by a simultaneous translational shift of one-half a lattice translation* (lattice-point separation vector) in the direction of the rotation axis; it is designated by the symbol 2_1. The glide plane similarly corresponds to a reflection accompanied by a translational shift of one-half a lattice translation in a direction parallel to the plane of reflection; it is designated *a*, *b*, *c*, or *n* according to whether the translational shift is by $\mathbf{a}/2$, $\mathbf{b}/2$, $\mathbf{c}/2$, or a diagonal direction, e.g., $(\mathbf{b} + \mathbf{c})/2$ for an *n* glide plane normal to $\mathbf{a}$.† The twofold screw axis and the glide plane are illustrated in Fig. 4.

As in the cubic system (Exp. 47), the lattice in the orthorhombic system may be primitive *P*, body-centered *I*, or face-centered *F*. In addition, a lattice in the orthorhombic system may be end-centered—i.e., the unit cell is centered on only one pair of opposing faces instead of all three. The lattice is then designated by a capital letter corresponding to the crystal axis *not* contained in the centered face: *C* if the centering point is at $(\frac{1}{2} \frac{1}{2}, 0)$; *A* if it is at $(0, \frac{1}{2}, \frac{1}{2})$; *B* if it is at $(\frac{1}{2}, 0, \frac{1}{2})$; however, all three should be considered as examples of a single Bravais lattice type.

The structures of the space groups and their relation to the point groups may perhaps be best understood through the use of some examples. Consider the point group $C_{2v} - mm2$, which possesses a rotation axis 2 at the intersection of two perpendicular mirror planes *m*. We can immediately form five different *symmorphic* space groups by combining these elements of symmetry with the lattice translations of the different Bravais lattices: *P*, *C*, *A*(*B*), *F*, *I*. These are

$$C_{2v}^1 - Pmm2, \; C_{2v}^{11} - Cmm2, \; C_{2v}^{14} - Amm2, \; C_{2v}^{18} - Fmm2, \; C_{2v}^{20} - Imm2$$

The Hermann-Mauguin symbol is formed by placing the capital letter symbolizing the Bravais lattice in front of the Hermann-Mauguin symbol of the point group.

---

† An additional glide plane *d*, with a transitional shift of one-quarter a cell face diagonal, occurs in only two of the 59 orthorhombic space groups and will not be discussed here.

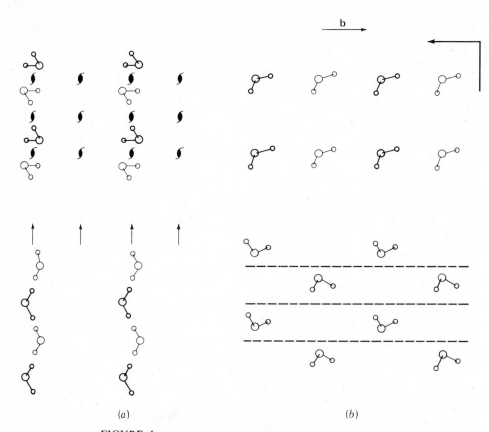

FIGURE 4

(a) 2_1 screw axis; (b) glide plane. Symbols used for these elements are shown for the cases where (top) the 2_1 axis and the *normal* to the glide plane are perpendicular to the paper, and where (bottom) they lie in the paper. A dashed line is also used for the trace of an *a* glide plane; a dotted line is used for *c*, and dash-dot-dash for *n*.

It will be noted that *Cmm2* and *Amm2* are different space groups although *C* and *A* represent the same Bravais lattice. This is because in *Cmm2* the axis 2 is perpendicular to the centered unit-cell face, while in *Amm2* it is parallel to it. It is unnecessary to consider *Bmm2* as a separate space group; it is equivalent to *Amm2* by interchange of the **a** and **b** axis labels. Indeed, the Hermann-Mauguin symbols assigned above correspond only to the conventional assignments of axis labels; the space groups may also be written in terms of other axial assignments. Thus, for example,[7]

$$C_{2v}^{14}: Amm2 \doteqdot B2mm \doteqdot Cm2m \doteqdot Am2m \doteqdot Bmm2 \doteqdot C2mm$$

where $\doteq$ means "is equivalent to". Alternative settings such as these are used when it is necessary to assign orthorhombic axes in such a way that their lengths conform to the conventional relation

$$c \leq a \leq b \tag{6}$$

which is almost universally followed in the orthorhombic system.

The symmorphic space groups are only 73 (out of 230) in number, and are rarely encountered, particularly in molecular crystals. Molecules pack better when shifted by screw axes or glide planes which permit the "bumps" on one molecule to fit into the recesses of the next, or into intermolecular spaces, and permit oppositely charged groups to come into close proximity.

In addition to the symmorphic space groups, there are additional space groups in which rotation axes are replaced (or supplemented) by screw axes, and/or mirror planes are replaced (or supplemented) by glide planes. These nonsymmorphic space groups, together with the symmorphic ones, are said to be *isomorphic* with the point group and with each other, and all are said to belong to the crystal class of the point group. In the Hermann-Mauguin symbol, the character 2 for a rotation axis is replaced where necessary by 2_1 for the screw axis, and/or the character m for a mirror plane is replaced by a, b, c, n, or d for a glide plane. For example, there are 17 nonsymmorphic space groups isomorphic to point group $C_{2v} - mm2$, some of which are

$$C_{2v}^2 - Pmc2_1, \ C_{2v}^8 - Pba2, \ C_{2v}^{10} - Pnn2, \ C_{2v}^{12} - Cmc2_1, \ C_{2v}^{15} - Abm2, \ C_{2v}^{22} - Ima2$$

These 17 plus the 5 symmorphic space groups constitute the 22 space groups in crystal class $C_{2v} - mm2$. There are 9 space groups in class $D_2 - 222$ and 28 in class $D_{2h} - mmm$, making a total of 59 space groups in the orthorhombic system.

To illustrate how a space group is presented diagrammatically in I. T., the diagrams for space group $D_{2h}^{18} - Cmca$ are reproduced in Fig. 5. The "complete" Hermann-Mauguin designation for this space group is

$$C \frac{2}{m} \frac{2}{c} \frac{2_1}{a}$$

where $2/m$ means that a twofold rotation axis and the normal to a mirror plane are both parallel to $\mathbf{a}$, etc. It will be observed in Fig. 5 that in addition to the elements explicitly given in the complete symbol, there are also b glides with normals parallel to $\mathbf{a}$ and to $\mathbf{c}$, an n glide with its normal parallel to $\mathbf{b}$, and a center of inversion. The b and n glides are generated by the interaction of other elements with the C-centering translation of the Bravais lattice.

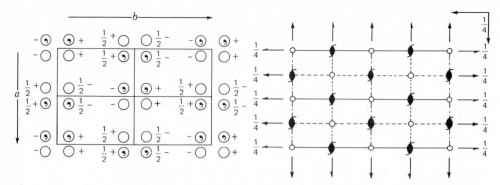

FIGURE 5

Diagrams for space group No. 64, $D_{2h}^{18} - Cmca$, reproduced by permission from I.T.[7] The right-hand diagram gives the symmetry elements and their positions in the unit cell. Each symmetry element parallel to the paper occurs at $z = 0$ and $z = \frac{1}{2}$ unless the fraction $\frac{1}{4}$ appears, in which case the element occurs at $z = \frac{1}{4}$ and $z = \frac{3}{4}$. Solid right-angled line with arrows: a and b glide planes parallel to the paper (with a single arrow this would indicate an a or a b glide depending on direction; without arrows, a mirror plane parallel to the paper). Solid, dashed, dotted, and dot-dashed lines: respectively m, b, c, and n planes perpendicular to the paper. Full arrow and half arrow: respectively 2 and 2_1 axes parallel to the paper. Lens with "tails": 2_1 axes perpendicular to the paper. Small circles: centers of inversion ("centers of symmetry"). The left-hand diagram shows a complete set of symmetry-equivalent points. The empty circles and the circles containing a comma represent points related enantiomorphically (i.e., left hand and right hand). z coordinates are indicated as follows: $+ = +z$, $- = -z$, $\frac{1}{2}^{+} = \frac{1}{2} + z$, $\frac{1}{2}^{-} = \frac{1}{2} - z$.

Systematic extinctions   We now discuss the manner in which systematic extinctions are used to determine, completely or partially, the space group in the orthorhombic system.

As in the cubic case (Exp. 47), centered lattices impose certain conditions (called extinction rules) limiting possible reflections. We have already pointed out that a lattice extinction rule corresponds to a class of r.l.p.'s that are extinguished in the reciprocal lattice. In addition to the rules given for I and F lattices in Exp. 47, the end-centered lattice gives rise to its own condition: for a C lattice it can easily be shown by use of Eq. (47-4) that reflection can only occur if $h + k$ is even. The screw axes and glide planes, in addition, impose their own conditions. For example, a 2_1 screw axis parallel to **c** permits $00l$ reflections to occur only if l is even. The extinction conditions for all lattices, screw axes, and glide planes (except d) applicable to the orthorhombic system are given in Table 1. It is important to emphasize that these are necessary, *but not sufficient*, conditions for Bragg reflection; although a reflection may be permitted by all of the conditions in Table 1, it may be accidentally extinguished by destructive interference of wavelets scattered by different atoms located in various parts of the unit cell.

One must be careful about inferring from a single extinction rule that the symmetry element corresponding to it in Table 1 is necessarily present. Suppose, for example, that all $0k0$ reflections with k odd are absent. It would be a mistake to infer from that one fact that there is necessarily a 2_1 screw axis parallel to **b**; one might only be observing a special case of $hk0$ reflections which are all absent when k is odd (as in space group $P2ab$, which does not have a 2_1 axis).

METHOD

The Buerger precession camera A sheet of photographic film is of course only two-dimensional, and for ease in identifying the diffraction spots due to Bragg reflections it is convenient to record on it the information for only a two-dimensional section of the r.l. This can be accomplished with the precession camera,[1] which produces an undistorted photograph of one selected section of the intensity-weighted r.l. This camera is shown schematically in Fig. 6.

The crystal, its center in a fixed position X in the X-ray beam, is made to undergo a precessional motion such that the normal to the r.l. section being photo-

Table 1   CONDITIONS LIMITING POSSIBLE REFLECTIONS ("EXTINCTION RULES")

Type of reflection	Lattice type, or symmetry element and orientation	Condition(s) limiting reflection
hkl	Lattice I	$h + k + l = $ even
	Lattice F	$h + k = $ even, $k + l = $ even, $l + h = $ even (i.e., h, k, l are all even or all odd)
	Lattice C	$h + k = $ even
	Lattice A	$k + l = $ even
	Lattice B	$l + h = $ even
$00l$	$2_1 \parallel \mathbf{c}$	$l = $ even
$h00$	$2_1 \parallel \mathbf{a}$	$h = $ even
$0k0$	$2_1 \parallel \mathbf{b}$	$k = $ even
$hk0$	$a \perp \mathbf{c}$	$h = $ even
	$b \perp \mathbf{c}$	$k = $ even
	$n \perp \mathbf{c}$	$h + k = $ even
$0kl$	$b \perp \mathbf{a}$	$k = $ even
	$c \perp \mathbf{a}$	$l = $ even
	$n \perp \mathbf{a}$	$k + l = $ even
$h0l$	$c \perp \mathbf{b}$	$l = $ even
	$a \perp \mathbf{b}$	$h = $ even
	$n \perp \mathbf{b}$	$l + h = $ even

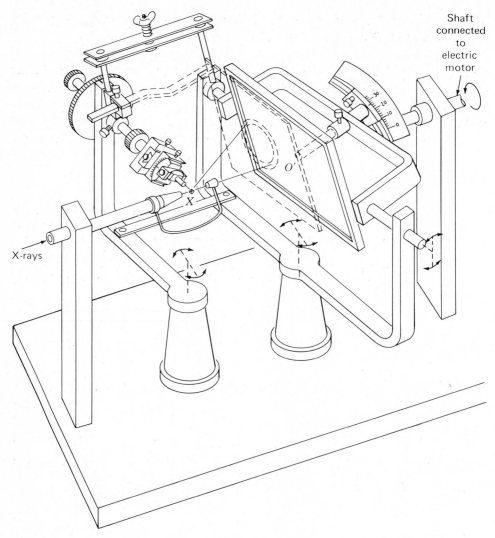

FIGURE 6
Schematic diagram of Buerger Precession Camera.

graphed rotates around the incident X-ray beam at a constant angle μ. *The photographic film is made to undergo exactly the same precessional motion* so that its plane is always parallel to the plane of the r.l. section being photographed. A "layer screen" consisting of a metal plate with an annular opening (shown with dashed lines in the figure) prevents reflections from reaching the film that do not come from the desired r.l. section. In the simplest form of operation, the r.l. section

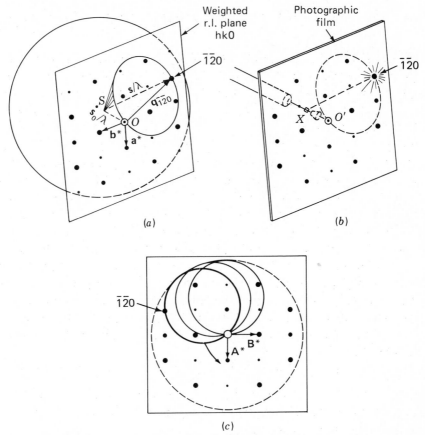

FIGURE 7

Operating principles of the Buerger precession camera. (a) Ewald sphere, with the
h k 0 section plane of the reciprocal lattice intersecting it at a circle; r.l.p. $\overline{1}\overline{2}0$ satisfies
the conditions for diffraction. (b) Parallel situation in precession camera; the
crystal occupies a position analogous to the center of the Ewald sphere, and the
film occupies a position analogous to the h k 0 section plane. Spot $\overline{1}\overline{2}0$ is getting an
instantaneous exposure. (c) The film, showing the motion of the circle of intersection.

chosen is a so-called "equatorial" section, which includes the r.l. origin O, and the
precessional motion of the film is such that the center point of the film (corresponding to the origin O of the r.l.) coincides with a point O' that remains stationary
during the precession. For "nonequatorial" sections, which do not include the
origin of the r.l., the film is displaced toward the crystal from this stationary point
by an amount and in a direction corresponding to the displacement of the desired r.l.
section from the origin.

The principles of operation are illustrated in Fig. 7. In direct space the crystal and the film are in the same relation to one another as the center of the Ewald sphere and the r.l. layer being photographed. The possible vectors s/λ are confined to the surface of a cone; the corresponding diffracted X-rays are confined to a similar cone which intersects the film in a circle. As the crystal and the film precess, the circle rotates around the center of the film, sweeping out a circular area. In general, therefore, every "spot" within that area gets exposed twice per revolution. In Fig. 7 the r.l. points and the corresponding exposed spots on the film are represented by dots the sizes of which are proportional to the intensities of the reflection.

Figure 8a shows several sections or "layers" of the r.l. intersecting the Ewald spheres; the traces shown correspond to a set of lattice planes in the r.l. The spacing between planes is designated d^*. For illustration, we here assume that axes have been assigned so that different planes of the set correspond to successive values of the Miller index l; in this case

$$d^* = \frac{1}{c} \tag{7}$$

(This relation holds, incidentally, regardless of whether the crystal axes are orthogonal or not.) We denote by ζ the distance from the equatorial r.l. plane (i.e., the one passing through the origin O) to the r.l. plane being photographed. It is equal to 0 or some multiple of d^*; with the above assumed axial assignment

$$\zeta = ld^* = \frac{l}{c} \tag{8}$$

The distance from the center of the Ewald sphere to the r.l. origin O is $1/\lambda$; the corresponding distance from the crystal X to the point O' on the cassette mechanism which remains stationary during the precessional motion is D (usually 6 cm). Therefore the factor by which distances on the film must be multiplied to convert them to distances in reciprocal space is

$$f = \frac{1/\lambda}{D} = \frac{1}{\lambda D} \tag{9}$$

Thus, if on the film (Fig. 7c) we measure the reciprocal cell dimensions A^* and B^* in some convenient units (centimeters, say), we calculate the reciprocal cell dimensions by

$$a^* = fA^* = \frac{A^*}{\lambda D}; \quad b^* = fB^* = \frac{B^*}{\lambda D} \tag{10}$$

In the interest of brevity we will not give here the equations for the various instrumental parameters required for equatorial and nonequatorial phtographs. These are given fully in Ref. 1; moreover, tables and graphs are usually supplied with the instrument.

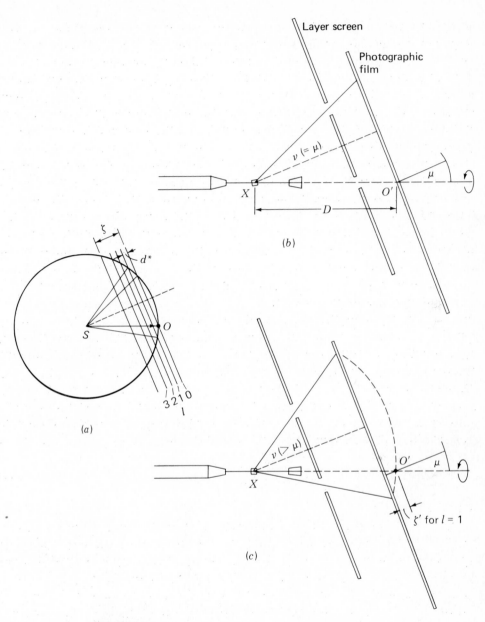

FIGURE 8
Details of geometry. Planes of reciprocal lattice sections, film, and layer screen are perpendicular to the paper; their traces are shown. (a) Ewald construction. (b) Parallel situation in the precession camera, set up to photograph $l = 0$. (c) Precession camera set up to photograph $l = 1$.

Determination of the lattice constants and space group This process is best described in terms of an illustrative hypothetical example.

A crystal in the form of a stubby needle about 0.4 mm long and 0.16 mm in diameter was mounted with the needle axis parallel to the spindle. After satisfactory alignment was achieved, an equatorial photograph was taken at $\mu = 30°$ with CuKα radiation. This is shown schematically in Fig. 9a; it showed clearly two "mirror lines" and accordingly has two-dimensional point symmetry *mm*. This did not yet prove that the crystal is orthorhombic since certain r.l. equatorial sections in monoclinic, tetragonal, hexagonal, and cubic crystals have the same apparent symmetry. The spindle was rotated 90° and another picture taken (which almost certainly would *not* show the two mirror lines if the crystal were monoclinic, but might easily show *more* symmetry if it were of higher symmetry than orthorhombic). The resulting photograph, also showing *mm* symmetry, is shown in Fig. 9b. Additional photographs at other nonequivalent spindle settings showed only a single vertical mirror line; orthorhombic symmetry of the crystal (Laue symmetry $D_{2h} - mmm$) is therefore established. Following the convention of Eq. (6), the direction in which the minimum spacing between spots in a mirror-line direction was greatest in the two films was designated the **c*** axis, which happened to be parallel with the spindle axis. The next greatest minimum separation was the vertical direction on the first film; that direction was designated the **a*** axis. The shortest was **b***, in the vertical direction in the second film. The corresponding separations A^*, B^*, C^* were measured as accurately as possible with a millimeter scale. Distances between the ends of several parallel rows of spots were measured and divided by the number of intervening spaces to give the corresponding A^* (or B^* or C^*) value; the values for the different rows were then averaged. The values obtained were

$$A^* = 0.747(2) \text{ cm} \qquad B^* = 0.611(2) \text{ cm} \qquad C^* = 0.974(3) \text{ cm}$$

where, as is conventional in crystallographic literature, the digit in parentheses represents the estimated standard deviation (e.s.d.) in the final digit of the given figure. On multiplying by the scale factor $f = 1/\lambda D = 1/[1.5418 \times 6.00(2)] = 0.1081(4)$ $\text{Å}^{-1} \text{ cm}^{-1}$ [Eq. (9)] the following tentative reciprocal lattice constants were obtained:

$$a^* = 0.0808(4) \text{ Å}^{-1} \qquad b^* = 0.0660(3) \text{ Å}^{-1} \qquad c^* = 0.1053(4) \text{ Å}^{-1}$$

These yielded the following tentative lattice constants in direct space:

$$a = 12.38(6) \text{ Å} \qquad b = 15.14(7) \text{ Å} \qquad c = 9.50(4) \text{ Å}$$

To ascertain whether glide extinctions are doubling the reciprocal cell in one or more directions, two more pictures were taken of levels tentatively identified as

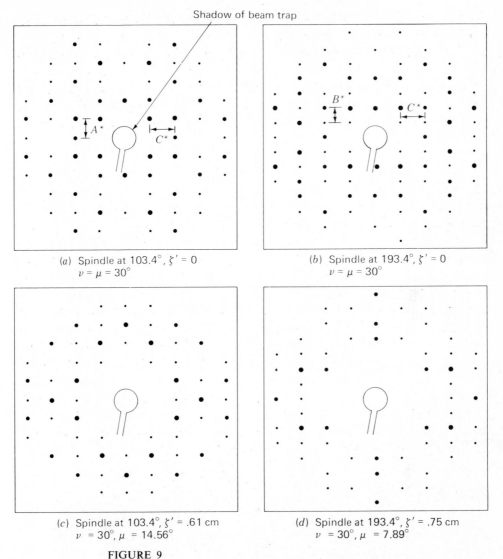

Shadow of beam trap

(a) Spindle at 103.4°, ζ' = 0
ν = μ = 30°

(b) Spindle at 193.4°, ζ' = 0
ν = μ = 30°

(c) Spindle at 103.4°, ζ' = .61 cm
ν = 30°, μ = 14.56°

(d) Spindle at 193.4°, ζ' = .75 cm
ν = 30°, μ = 7.89°

FIGURE 9
Hypothetical precession photographs (see text). These are idealized (i.e., show no idiosyncracies of spot shape, doubling, white adiation streaks, or tendency to higher intensity near the limiting circles.

$k = 1$ and $h = 1$. (See Figs. 9c and 9d.) It will be noted that there are "holes" at the centers of these two patterns, in which no diffraction spots appear. This feature of nonequatorial photographs results from the fact that the inner r.l.p.'s always remain within the rotating diffraction circle. The separations between spots, A^*, B^*, C^*, were found to be the same as on the equatorial photographs, and the mm symmetry was again apparent in both photographs. Doubling of the reciprocal cell in any direction was therefore ruled out.†

It was observed on the layer $h0l$ (Fig. 9a) that all $h00$ reflections with h odd were missing, as were all $00l$ reflections with l odd; likewise on the layer $0kl$ (Fig. 9b) the odd $0k0$ and $00l$ reflections were uniformly missing. As no other systematic absences were apparent, including absences due to centering, the space group is established as

$$D_2^4 - P2_12_12_1$$

which is isomorphic with the point group $D_2 - 222$. This is the second most common space group encountered for organic crystals; the most common is the monoclinic space group $C_{2h}^5 - P2_1/c$ (with a 2_1 screw axis parallel to the b axis, perpendicular to a c glide plane).

It is to be noted that in the case of space group $P2_12_12_1$ the reciprocal lattice systematic extinctions uniquely determine the space group; no space group giving those extinctions exists in crystal classes $C_{2v} - mm2$ and $D_{2h} - mmm$. Therefore the absence of a center of inversion is established. Another example of a space group that can be uniquely determined by extinctions is $D_{2h}^{15} - Pbca$ (or the equivalent, $Pcab$); here the three glide planes, and therefore also a center of inversion, are established by extinctions. An important and frequently encountered space group, however, is $D_{2h}^{16} - Pnma$, which by extinctions alone cannot be distinguished from $C_{2v}^9 - Pn2_1a$. If in this case the crystal shows polar external morphology, or a piezoelectric or pyroelectric effect, or second harmonic generation in a laser beam, the lower symmetry $C_{2v}^9 - Pn2_1a$ is established.

EXPERIMENTAL

The object of the present experiment is to determine by means of precession photographs the lattice constants a, b, c and the space group (or set of possible alternative space groups) for an appropriate crystalline substance. Substances with orthorhombic symmetry suitable for this experiment include $KMnO_4$, Rochelle salt (sodium potassium D-tartrate tetrahydrate), nickel dimethylglyoxime, lithium

---

† It is possible in certain cases that even the above precautions will fail to save the experimenter from being deceived by glide extinctions. For example, if the crystal has an F lattice, the body-center r.l.p.'s (h, k, l, all odd) will not be found by a procedure limited to taking the four photographs described above. Such body-center points (if they exist) will be revealed by an hhl equatorial photograph, or by a nonequatorial photograph at $k_{tent.} = \frac{1}{2}$ or at $h_{tent.} = \frac{1}{2}$. If a photograph of either kind shows the body-centered points, then the tentative a^*, b^*, c^* must be halved, and the tentative a, b, c must be doubled.

acetate dihydrate, the amino acids DL-alanine and L-threonine, and the organic compounds cyanamide, acetamide, and acetanilide. Many others may be found in Ref. 8.

Specimen preparation   If the substance chosen is soluble in water or some convenient organic solvent, diffraction specimens are usually best prepared by crystallization from the appropriate solvent. Inorganic salts and highly polar organic substances (acids, amides) are usually soluble enough in water; most others dissolve in alcohol, toluene, or heptane, or mixtures of these. For best results the solubility should not be too high (2 to 5 g per 100 ml of solvent at 25°C is convenient). Careful attention should be given to the possibility that the crystallized product may have a degree of solvation (e.g., hydration) different from that desired. A particular hydrate, for example, will differ from the anhydrous material and from other hydrates in crystal structure, unit cell, and probably symmetry. It will also differ in density, and so the measured density should be checked against the literature value for the desired material.

Usually a rather simple procedure will work for growing crystals. A saturated solution is prepared by shaking an excess of the solid material with the solvent at a temperature somewhat higher than room temperature (typically 40 to 60°C). The solution is filtered or decanted into a beaker and warmed about 5 or 10° higher than the previous temperature to insure that all particles are dissolved. The beaker is then covered and set aside overnight or for a few days. Crystals may be withdrawn with tweezers or an eye dropper and placed on filter paper to dry.

The crystal specimens should be examined under a low-power binocular microscope for morphological clues to symmetry. It may be helpful to make sketches of typical crystals in the notebook. If the crystals are transparent, further clues to the symmetry may be obtained by examining the crystals under a polarizing microscope, between crossed polarizers. An orthorhombic crystal, transmitting through parallel prismatic faces, should show optical extinction when the apparent symmetry axes are parallel to either of the two polarization directions.

The density of organic crystals and the lighter inorganic ones can often be measured by the flotation method, using mixtures of two miscible liquids having densities that bracket that of the specimen. For polar organic crystals containing no elements heavier than oxygen, light hydrocarbons such as kerosene ($\rho = 0.79$ g cm^{-3} at 25°C) and methylene iodide ($\rho = 3.32$ g cm^{-3} at 25°C) are usually satisfactory. An initial mixture of the two liquids is made in proportions corresponding roughly to an initial estimate of the crystal density, and one or a few crystals dropped in. If the crystals tend to sink, some methylene iodide is added and the mixture stirred until again homogeneous; if the crystals tend to float, some kerosene is added. When the composition has been adjusted so that the crystals neither sink nor float perceptibly, the density of the mixture is measured with a pycnometer (see Exp. 11). The method obviously fails if the crystals dissolve in or react with the solvent, or are too dense for any convenient flotation medium. If no flotation procedure is satisfactory, the density can be determined by the procedure described in the Sample Report (Chap. 1).

Taking the precession photographs   Manifestly, the necessary photographs can only be obtained in a laboratory possessing an X-ray generator, a Buerger precession camera, and personnel trained in the use of this equipment. Therefore, the myriad details of mounting the crystal, centering and aligning it on the precession camera, making the parameter settings, and taking and developing the photographs will not be given here. In the absence of the necessary equipment and qualified personnel, the exercise of determining the unit cell and space group may be performed with a set of photographs obtainable from some X-ray diffraction laboratory.

CALCULATIONS

The lattice constants and the space group (or set of alternative space groups) should be determined in the manner already described. The number Z of formula units (or molecules) of the compound in the unit cell should then be calculated with the formula

$$Z = \frac{V_c}{V_f} \tag{11}$$

V_f, the volume occupied in the crystal by one formula unit, in units of Å^3, is given by

$$V_f = \frac{F}{0.6023\rho} \tag{12}$$

where F is the formula weight in g mol^{-1} and ρ is the crystal density in g cm^{-3}; and

$$V_c = a \cdot b \cdot c \tag{13}$$

is the unit cell volume also in units of Å^3. [Eq. (13) holds only when the three cell axes are mutually orthogonal, as in the orthorhombic system.]

The number Z must be an integer within experimental error. If it is not, an error must be suspected in the X-ray measurements, in the measured density (air bubbles, etc.), or in the calculations; or perhaps the composition of the crystal is different from that assumed owing to solvation or other factors.

The student should draw as many conclusions as possible from the space group and the number of formula weights or molecules, consulting I.T. and publications on the structure in the literature for help when necessary.

REFERENCES

1. M. J. Buerger, "The Precession Method in X-ray Crystallography," Wiley, New York (1964).
2. M. J. Buerger, "Crystal Structure Analysis," chap. 15, Wiley, New York (1960).
3. M. J. Buerger, "Elementary Crystallography," chaps. 2, 4, 5, 6, 10, and 11, Wiley, New York (1956).
4. W. A. Wooster and A. Breton, "Experimental Crystal Physics," 2d ed., pp. 82–89, Clarendon Press, Oxford (1970).
5. S. C. Abrahams, *J. Appl. Cryst.*, **5**, 143 (1972); S. K. Kurtz and T. T. Perry, *J. Appl. Phys.*, **39**, 3798 (1968).
6. M. J. Buerger, "Elementary Crystallography," chaps. 12–19, Wiley, New York (1956).
7. N. F. M. Henry and K. Lonsdale (eds.), Symmetry Groups, in "International Tables for X-Ray Crystallography," vol. I, published for International Union of Crystallography by Kynoch Press, Birmingham, England (1952). A new series to be titled "International Tables for Crystallography" is planned by the International Union of Crystallography. vol. A, Symmetry in Direct Space (T. Hahn, general ed.) is in preparation and is expected to appear in 1981.
8. R. W. G. Wyckoff, "Crystal Structures," 2d ed., vols. 1–6, Wiley, New York (1963–1971). These books give excellent and well illustrated condensed descriptions of many crystal structures.

GENERAL READING

M. J. Buerger, "Crystal Structure Analysis," Wiley, New York (1960).
B. E. Warren, "X-Ray Diffraction," Addison-Wesley, Reading, Mass. (1969).
M. M. Woolfson, "An Introduction to X-Ray Crystallography," Cambridge (1970).
A. Guinier, "X-Ray Diffraction," Freeman, San Francisco (1963).

EXPERIMENT 49.   LATTICE ENERGY OF SOLID ARGON

Accurate measurements of the sublimation pressures of solid argon are to be made as a function of temperature in the range 65 to 78°K. Applying both a second- and third-law thermodynamic treatment to these equilibrium measurements, one obtains the heat of sublimation of argon. From this heat of sublimation and the Debye theory of lattice vibrations, the lattice energy of solid argon can be determined.

Representing the Ar–Ar interaction by a Lennard-Jones 6,12 potential, the lattice energy will be calculated using potential parameters obtained from the room-temperature transport properties of argon gas. This "theoretical" value is then compared with the "experimental" thermodynamic value.

THEORY

The change in state involved here is simply

$$Ar(s) = Ar(g) \qquad (\text{const } p, T) \tag{1}$$

The molar enthalpy of sublimation $\Delta \tilde{H}$ is rigorously related to the slope dp/dT of the sublimation pressure curve by the Clapeyron equation, which has been discussed in Exp. 16. The exact expression is

$$\Delta \tilde{H} = T \frac{dp}{dT} (\tilde{V}_g - \tilde{V}_s) \tag{2}$$

If $\tilde{V}_s$ is assumed to be negligible in comparison to $\tilde{V}_g$ and if the vapor is assumed to be an ideal gas, an approximate form of the Clapeyron equation is obtained

$$\Delta \tilde{H} \simeq -2.303R \frac{d \log p}{d(1/T)} \tag{3}$$

which will yield a value of $\Delta \tilde{H}$ directly from the slope of a plot of $\log p$ versus $1/T$. This approximate Clapeyron equation is very nice for examination questions or the quick analysis of crude data, but it should never be used to represent accurate experimental measurements. Fortunately, the equation of state of $Ar(g)$ has been determined experimentally down to the triple point of argon. Table 1 gives values of

Table 1 SECOND VIRIAL COEFFICIENT OF ARGON[a]
(B in units of $cm^3 \ mol^{-1}$)

T, °K	B	dB/dT	T, °K	B	dB/dT
66	−497	23.9	76	−335	11.2
68	−454	19.8	78	−314	9.9
70	−418	17.0	80	−295	8.8
72	−386	14.7	83.81	−265	7.1
74	−359	12.8	87.30	−242	5.9

[a] Extrapolated from measurements above 84°K; see M. A. Byrne, M. R. Jones, and L. A. K. Staveley, *Trans. Faraday Soc.*, **64**, 1747 (1968).

B, the second virial coefficient, and dB/dT as a function of temperature. Also, the molar volume of the solid can be taken to have the constant value 25 cm³ mol⁻¹ over the investigated temperature range. Thus, one can use the exact Clapeyron equation to determine second-law values of $\Delta\tilde{H}$.

On the basis of the third law, the thermodynamic properties of a pure substance [such as S^0, $(H^0 - H_0^0)/T$ and $(G^0 - H_0^0)/T$, where $S^0 = \int_0^T C_p^0 d\ln T$, $H^0 - H_0^0 = \int_0^T C_p^0 dT$ and $G^0 = H^0 - TS^0$] can be tabulated as functions of temperature. Table 2 gives these functions for Ar(s, cubic), Ar(l) and Ar(g). In the preparation of this table, experimental values of $\tilde{C}_p^0$ for the solid and liquid phases and the experimental enthalpy of fusion were used. For the gas phase, $\tilde{C}_p^0 = 5R/2$ (except at extremely low T) and $\tilde{S}^0$ was calculated from the Sackür-Tetrode equation. We can now proceed to evaluate third-law values of the enthalpy of sublimation. For the change in state (1), $\Delta\tilde{G}^0 = -RT\ln K$, where the equilibrium constant $K = f_g/a_s$. At low pressures, only the second virial coefficient need be considered and a satisfactory approximation for $\Delta\tilde{G}^0$ is

$$\Delta\tilde{G}^0 = -RT\ln p - Bp - \tilde{V}_s(1 - p) \tag{4}$$

Table 2   THERMODYNAMIC PROPERTIES OF ARGON[a]
(The unit of energy is the calorie.)

		Ar (s, l)			Ar (g)	
T, °K	S^0	$\dfrac{H^0 - H_0^0}{T}$	$\dfrac{G^0 - H_0^0}{T}$	S^0	$\dfrac{H^0 - H_0^0}{T}$	$\dfrac{G^0 - H_0^0}{T}$
0	0	0	0	0	0	0
10	0.261	0.198	−0.063	20.116	4.968	−15.148
20	1.496	1.056	−0.440	23.560	4.968	−18.592
30	3.014	1.968	−1.046	25.574	4.968	−20.606
40	4.435	2.710	−1.725	27.004	4.968	−22.036
50	5.705	3.309	−2.396	28.112	4.968	−23.144
60	6.848	3.803	−3.045	29.018	4.968	−24.050
70	7.893	4.230	−3.663	29.784	4.968	−24.816
72	8.095	4.311	−3.784	29.924	4.968	−24.956
74	8.296	4.394	−3.902	30.060	4.968	−25.092
76	8.496	4.475	−4.021	30.192	4.968	−25.224
78	8.694	4.557	−4.137	30.321	4.968	−25.353
80	8.893	4.638	−4.255	30.447	4.968	−25.479
83.81(s)	9.273	4.798	−4.475	30.678	4.968	−25.710
83.81(l)	12.668	8.193	−4.475			
87.30	13.101	8.289	−4.817	30.881	4.968	−25.914

[a] The entries of Ar(s) and Ar(l) were calculated from data given in refs. 1 and 2; the entries for Ar(g) are statistical thermodynamic values for a monatomic ideal gas.

Indeed, the last term in the above expression is required only for data of the highest accuracy. From experimental $\Delta\tilde{G}^0$ values and Table 2, we may now determine $\Delta\tilde{H}_0{}^0$, the enthalpy of sublimation at absolute zero:

$$\frac{\Delta\tilde{G}^0}{T} = \left(\frac{\tilde{G}^0 - \tilde{H}_0{}^0}{T}\right)_g - \left(\frac{\tilde{G}^0 - \tilde{H}_0{}^0}{T}\right)_s + \frac{\Delta\tilde{H}_0{}^0}{T} \tag{5}$$

A value of $\Delta\tilde{H}_0{}^0$ may be calculated from each experimental value of the sublimation pressure, and these $\Delta\tilde{H}_0{}^0$ values must be constant within the limits of experimental error. The value of $\Delta\tilde{H}^0$ at any temperature can be calculated from $\Delta\tilde{H}_0{}^0$ and the entries in Table 2. It is important to recognize that $\Delta\tilde{H}^0$ refers to the change in state

$$\text{Ar}(s) = \text{Ar(perfect gas)} \qquad (1 \text{ atm, const } T) \tag{6}$$

whereas $\Delta\tilde{H}$ refers to change in state (1) involving the real gas at pressure p. If the enthalpy of the solid phase is assumed to be independent of pressure for low pressures, then $\Delta\tilde{H}$ can be calculated from

$$\Delta\tilde{H} = \Delta\tilde{H}^0 + \left(B - T\frac{dB}{dT}\right)p \tag{7}$$

If all the experimental data were perfect, the second-law value of $\Delta\tilde{H}$ calculated from Eq. (2) and the third-law value calculated from Eqs. (4), (5), and (7) would be identical. However, in this case, the third-law result should be the more reliable since it is difficult to determine dp/dT with very high accuracy.

An independent value for $\Delta\tilde{H}$ has been reported by two different investigators from direct calorimetric measurements of the enthalpy of vaporization.[1-3] The calorimetric measurements, which are in good agreement with each other, give a $\Delta\tilde{H}_0{}^0$ value which is about 15 cal higher than the third-law value. This discrepancy, although small, is disturbing because it is well outside the limits of error claimed for the various experimental results. These studies of the thermodynamic properties of argon were made prior to the discovery in 1964 of a metastable hexagonal phase of argon, and the earlier investigators assumed without proof that their solid phase was the face-centered cubic phase. Until more is known concerning the hexagonal phase, it is impossible to say how the presence of this phase might have affected the earlier measurements.†

> † Face-centered cubic argon is the stable solid phase of pure argon, but upon the addition of as little as 1 percent of N_2 to Ar, the solid phase crystallizes with the close-packed hexagonal structure. A study of the sublimation pressures of solid solutions of Ar with small amounts of N_2 would give valuable information concerning the hexagonal phase, and such a study could be made with the apparatus used in this experiment.

Lattice energy If we measure the energy content of argon on an arbitrary scale which defines $\tilde{E}^0(g)$ to be zero at $0°K$, then the total energy of the crystal lattice at $0°K$ is $-\Delta\tilde{E}_0{}^0$, which for all practical purposes is the same as $-\Delta\tilde{H}_0{}^0$. There are two additive contributions to this absolute-zero energy content of solid argon: the *lattice energy* Φ_0, which is the potential energy of the argon atoms at rest in the lattice relative to a zero of energy for these atoms in the gas at infinite separation, and the zero-point vibrational energy, which arises from the quantum-mechanical vibrational motion of the atoms about their equilibrium positions.[4] In terms of the Debye characteristic temperature Θ_D, the zero-point vibrational energy equals $\frac{9}{8}R\Theta_D$. Thus, the lattice energy is given by

$$\Phi_0 = -\Delta\tilde{H}_0{}^0 - \tfrac{9}{8}R\Theta_D \tag{8}$$

where $\Theta_D = 93°K$ for argon.[2]

It is possible to calculate a theoretical value of the lattice energy for a molecular crystal if data are available on the potential energy between atoms as a function of their separation. A commonly used form for the interatomic potential (see Fig. 1) is due to Lennard-Jones;[4,5]

$$u(r) = 4\varepsilon\left[\left(\frac{\sigma}{r}\right)^{12} - \left(\frac{\sigma}{r}\right)^{6}\right] \tag{9}$$

where $u(r)$ is the potential energy for *two* atoms at a distance r. The r^{-12} term is an empirical function to describe the repulsion at short distances, and the r^{-6} term represents the r dependence of the potential energy found by London to describe the attraction at large distances due to induced dipole-dipole interaction. The Lennard-Jones potential for a given atom is characterized by the two constants ε and σ (which are shown in Fig. 1). These parameters can be evaluated from an analysis

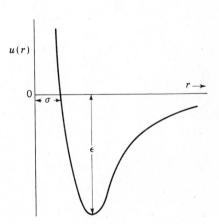

FIGURE 1
Lennard-Jones potential $u(r)$ as a function of interatomic distance r. The characteristic parameters ε and σ determine this potential curve; see Eq. (9).

of gas data (second virial coefficient, Joule-Thomson effect, or gas viscosity); the best values for argon[6] are

$$\frac{\varepsilon}{k} = 119°\text{K} \qquad \sigma = 3.41\text{Å} \tag{10}$$

For the solid it is assumed that the total potential energy (i.e., lattice energy) is the sum of all pair potentials $u_{ij}(r_{ij})$. The result of this summation for a face-centered cubic lattice (such as argon) is[7]

$$\Phi_0 = 2N_0\varepsilon \left[12.132 \left(\frac{\sigma}{d}\right)^{12} - 14.454 \left(\frac{\sigma}{d}\right)^{6} \right] \tag{11}$$

where d is the distance between nearest neighbors. Note that almost all the repulsion part of the potential comes from nearest-neighbor interactions (nearest neighbors alone would give 12 for both coefficients). Since r^{-6} falls off much more slowly than r^{-12}, the coefficient of the second term in Eq. (11) is considerably greater than 12. A knowledge of the lattice spacing for solid argon and the parameters of Eq. (10) will permit a calculation of the lattice energy for comparison with the experimental value obtained from Eq. (8).

EXPERIMENTAL

Temperatures below 77°K can be achieved with a liquid-nitrogen bath by reducing the pressure over the nitrogen by pumping, thus lowering its boiling point. It is possible to lower the bath temperature to 63°K before liquid nitrogen begins to solidify. Measurement of the temperature can be made by using a gas thermometer (see Exp. 1) or a thermocouple (Chap. XVI) or by measuring the vapor pressure of the nitrogen.

The low-temperature cell, shown in Fig. 2, consists of a copper block containing a small chamber A for solid argon and a large chamber B for use as a gas thermometer bulb or N_2 vapor-pressure bulb. If a thermocouple is used, it can be brought down the tube into chamber B and stationed near the argon chamber. The cupronickel (or stainless steel) connecting tubes are soldered through a cap with a side arm. This cap fits over the top of a tall glass Dewar flask and is attached to the Dewar with a rubber sleeve.† The side arm is connected by heavy-wall vacuum hose to a high-capacity mechanical vacuum pump.

---

† The top of a rubber surgical glove makes an excellent sleeve.

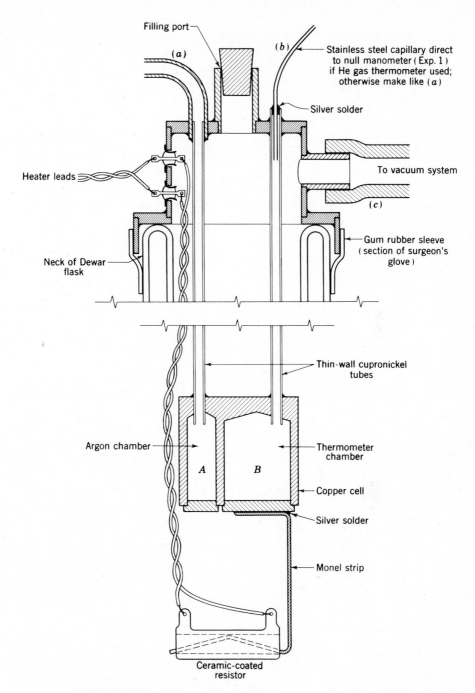

Filling port

(a)

(b)

Stainless steel capillary direct
to null manometer (Exp. 1)
if He gas thermometer used;
otherwise make like (a)

Silver solder

To vacuum system

(c)

Heater leads

Gum rubber sleeve
(section of surgeon's
glove)

Neck of Dewar
flask

Thin-wall cupronickel
tubes

Argon chamber

A   B

Thermometer
chamber

Copper cell

Silver solder

Monel strip

Ceramic-coated
resistor

FIGURE 2
Low-temperature vapor-pressure apparatus. (Metal parts to be well tinned and
soft-soldered, except where otherwise indicated.)

If the vapor pressure of liquid nitrogen is used for the temperature measurement, one allows N_2 gas to condense in chamber B and the pressures can be read directly on a regular manometer. If a copper-constantan thermocouple is used, the tube to chamber B should be sealed off at the top where the wires come out. Since chamber B is then not used, all further instructions concerning it may be disregarded.†

Procedure   The apparatus should be assembled by connecting the necessary manometers, filling lines, and pumping lines; see Fig. 3. Evacuate each chamber by closing clamp or stopcock C and stopcocks $A2$ and $B2$ and opening stopcocks $A1$ and $B1$. Then close stopcocks $A1$ and $B1$ and test for leaks in either chamber by waiting approximately 15 min to see if the manometers indicate any increase in pressure. The rate of increase in pressure should not exceed 5 Torr/hr.

Filling chambers A and B with gas should be done with care to avoid breaking the manometers. If a simple reducing valve is used, the high-pressure valve on the

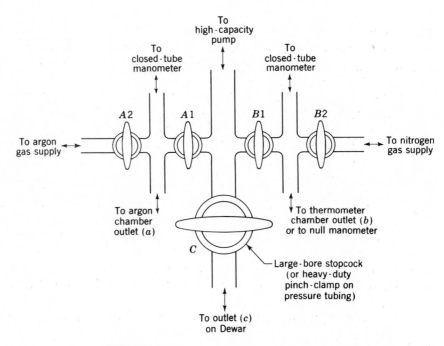

FIGURE 3
Vacuum and gas connections for the apparatus shown in Fig. 2.

---

† Alternatively, with a slight change in apparatus design, helium-gas thermometry can be used to measure the bath temperature. The procedure is described in Exp. 1.

storage cylinder should never be opened when gas is being admitted to the system. Allow the gas contained in the section between the two valves to expand into the chamber; then close the needle valve and refill this section. Repeat this procedure several times (if necessary) until the chamber is full. Flushing the chamber once or twice with the appropriate gas will reduce the possibility of contamination.

Fill chamber A and its manometer with argon at 1 atm and close stopcock $A2$. **Slowly** raise the Dewar filled with liquid nitrogen until the copper cell is almost completely immersed (do *not* immerse the connecting tubes above the cell). The argon pressure should drop to about 200 Torr. **Slowly** allow N_2 gas to enter chamber B until the pressure is constant at about 1 atm, and then close stopcock $B2$. Now raise the Dewar all the way up, and attach the cap to the top of the Dewar. The copper cell should be sufficiently far below the liquid-nitrogen level to ensure that the cell will remain submerged throughout the run. (If necessary, additional liquid nitrogen may be added through the filling port in the cap at this time.)

More argon may now be added to chamber A, although this is not necessary if the measurements are to be made on warming. Experience has shown that the argon pressures will not be reproducible unless the argon is pumped very briefly prior to any measurement made after the cell has been cooled. To avoid the possibility of pumping off too much argon, it is convenient to make the measurements when the cell is warming rather than cooling. The careful student will want to confirm the fact that reproducible results can be obtained on both warming and cooling.

Close the liquid-nitrogen filling port with a tightly fitting rubber stopper and begin pumping on the refrigerant by opening clamp or stopcock C. After the cell has cooled to approximately 65°K, the Ar pressure should be measured at some arbitrary constant temperature. Then pump chamber A briefly and measure the Ar pressure again at the same constant temperature. If the two pressures differ, repeat the procedure. Usually the pressure will be reproducible after the second pumping. No further pumping should be necessary at higher temperatures, although it is advisable to check for reproducibility. At each temperature, also record the vapor pressure of the nitrogen in chamber B. Indeed, the constancy of this pressure is your indication that the temperature is stable. The heights of the mercury columns in both manometers should be estimated to within 0.1 mm. Proper lighting conditions are most important for an accurate measurement of the position of the top of the meniscus.

If only external heat leaks served to warm up the system, the rate of warmup would be inconveniently slow. Intermittent heating with an electric heater element immersed directly in the liquid nitrogen will provide a suitable rate of heating between readings. Emphasis should be placed on obtaining good equilibrium data rather than a large number of data points. Four or five good experimental points are sufficient to establish the sublimation curve.

Warning: A closed system containing a liquid with a normal boiling point well below room temperature has been responsible for the destruction of many vacuum systems and for injuries to personnel. Make certain that the apparatus is properly pumped out or vented before it is allowed to warm up. As you should at all times in the laboratory, be sure to wear safety glasses.

CALCULATIONS

Convert all manometer readings to pressures in Torr. The correction for the density of mercury as a function of temperature must not be ignored (see p. 621). Compute the temperatures and tabulate them along with the corresponding argon sublimation pressures. If N_2 vapor pressures were used for thermometry, one can obtain T from the equation of Henning and Otto:[8]

$$\log p_{N_2} (\text{Torr}) = 7.781845 - \frac{341.619}{T} - 0.0062649T \qquad (12)$$

As a convenience, nitrogen vapor-pressure values as calculated from Eq. (12) are given at $1°$ intervals in Table 3. Either plot p_{N_2} versus T and read temperatures from this graph, or carry out a nonlinear interpolation between entries in the table. If a thermocouple was used, convert potential readings to temperatures with an appropriate conversion table. It may be wise to check the thermocouple calibration by measuring the bath temperature at atmospheric pressure. Calculate this temperature from Eq. (12) using the pressure determined with a barometer, and look up the corresponding emf. If necessary, apply a constant additive correction to all emf readings before using the tables.

Represent the argon sublimation pressures by an equation of the form

$$\log_{10} p (\text{Torr}) = -\frac{A}{T} + b \qquad (13)$$

Only two constants are used because the temperature range of the measurements is quite limited. After the constants A and b are determined, calculate the deviation

Table 3 VAPOR PRESSURE OF LIQUID NITROGEN

T, °K	p, Torr	T, °K	p, Torr	T, °K	p, Torr
64		69	250.4	74	503.2
65	131.5	70	290.4	75	571.6
66	155.7	71	335.4	76	646.8
67	183.4	72	385.5	77	729.2
68	214.8	73	441.4	78	819.3

$p_{obs} - p_{calc}$ for each of the experimental points and add this to your table of p and T values. These deviations will serve as an estimate of the precision of the measurements.

Employing the second-law approach, calculate $\Delta \tilde{H}$ at 66, 70, and 76°K from your experimental data using both the exact and approximate Clapeyron equations. Estimate the experimental error in $\Delta \tilde{H}$ and comment on any differences between the values obtained with Eqs. (2) and (3).

Utilizing the third-law approach, calculate $\Delta \tilde{H}_0{}^0$ from each of your experimental sublimation pressures. Then calculate $\Delta \tilde{H}^0$ at 66, 70, and 76°K, using what you consider to be the best value of $\Delta \tilde{H}_0{}^0$ (justify this choice). From Eq. (7), obtain $\Delta \tilde{H}$ values at the corresponding temperatures and compare these values with those obtained through the second-law treatment.

The report should show all numerical values of the various terms involved in the thermodynamic equations used and all appropriate graphs. Also compare your $\Delta \tilde{H}$ values with those reported in the literature.[2,9]

Finally, use Eq. (8) to determine the "experimental" value of the lattice energy of argon at 0°K. X-ray diffraction data[10] give 5.30 Å for the cubic unit-cell parameter of solid argon at 4°K. Find the nearest-neighbor distance d, and use Eqs. (10) and (11) to calculate a "theoretical" value of Φ_0. Compare the theoretical and experimental values.

APPARATUS

Tall Dewar flask (preferably with vertical unsilvered strip); support for Dewar; copper cell and cap assembly as shown in figure; rubber stopper to fit filling port; rubber sleeve; eight lengths of heavy-wall rubber tubing; high-capacity mechanical vacuum pump; gas-handling assembly with four stopcocks (see figure); heavy-duty hose clamp; two closed-tube manometers; Variac for heater supply. If a copper-constantan thermocouple is to be used, a Dewar flask for ice-water reference junction and a potentiometer circuit are needed.

Supplies of argon and dry nitrogen gas at 1 atm; liquid nitrogen.

REFERENCES

1. P. Flubacher, A. J. Leadbetter, and J. A. Morrison, *Proc. Phys. Soc.* (*London*), **78**, 1449 (1961).
2. R. H. Beaumont, H. Chihara, and J. A. Morrison, *Proc. Phys. Soc.* (*London*), **78**, 1462 (1961).
3. K. Clusius, *Z. Phys. Chem.*, **31B**, 459 (1936); A. Frank and K. Clusius, *Z. Phys. Chem.*, **42B**, 395 (1939).
4. J. O. Hirschfelder, C. F. Curtiss, and R. B. Bird, "Molecular Theory of Gases and Liquids," 2d ed., pp. 1035–1044, Wiley, New York (1967).

5. J. A. Beattie and W. H. Stockmayer, The Thermodynamics and Statistical Mechanics of Real Gases, in H. S. Taylor and S. Glasstone (eds.), "A Treatise on Physical Chemistry, vol. 2, States of Matter," 3d ed., chap. 2, pp. 305–306, Van Nostrand, Princeton, N.J. (1951).

6. *Ibid.*, pp. 323–328.

7. *Ibid.*, p. 309.

8. F. Henning and J. Otto, *Phys. Z.*, **37**, 634 (1936).

9. A. M. Clark, F. Dim, J. Robb, A. Michels, T. Wassenaar, and Th. Zwietering, *Physica*, **17**, 876 (1951).

10. O. G. Peterson, D. N. Batchelder, and R. O. Simmons, *Phys. Rev.*, **150**, 703 (1966).

EXPERIMENT 50.   ORDER-DISORDER TRANSITION IN AMMONIUM CHLORIDE

Solids which undergo a cooperative order-disorder transition of the lambda type exhibit anomalous variations in many of their thermodynamic properties over a small temperature range close to the transition temperature. In this experiment, the variation in heat capacity of NH_4Cl will be determined with a very simple cooling apparatus.

The order-disorder transition in ammonium chloride involves the relative orientations of the tetrahedral NH_4^+ ions in a CsCl-type structure. The most stable orientation of the NH_4^+ ion in the cubic unit cell is for the hydrogens to point toward the nearest-neighbor Cl^- ions. Thus there are two possible positions for each ammonium ion. In the completely ordered state (at low temperature), all NH_4^+ tetrahedra have the same relative orientation with respect to the crystallographic axes; i.e., the NH_4^+ ions are all "parallel" to each other. In the completely disordered state (at room temperature), the orientations are random with respect to these two positions. This simple orientational ordering is analogous to the spin ordering in a simple-cubic ferromagnet. The latter problem has been extensively studied and will be discussed below in terms of an Ising model.

THEORY

Cooperative order-disorder transitions present very challenging theoretical problems. The Ising model, which assumes that there are only two orientations possible and only nearest-neighbor interactions, has been solved analytically in two dimensions but has not yet been solved exactly in three dimensions.[1] There are now excellent numerical approximations available for a variety of lattices, but we shall present here the simplest possible approximate solution for an Ising lattice. This treatment, first applied by Bragg and Williams to the problem of ordering in alloys, makes drastic and incorrect simplifications but retains in a crude fashion many of the qualitative features which characterize a "higher-order" transition.[2,3] The Bragg-Williams approximation is equivalent to mean-field theory for ferromagnets and van der Waals' theory for fluids.

Consider a lattice of N atoms with "spins" which can be either up $(+)$ or down $(-)$. It is assumed that the behavior of the lattice vibrations is independent of the particular spin configuration; that is,

$$A = A_l + A_c \tag{1}$$

where A_l is the Helmholtz free energy due to lattice vibrations and A_c is the contribution to the free energy due to spin-spin interactions. We shall be concerned here only with A_c and the corresponding configurational partition function

$$Q_c = \sum e^{-E_c/kT} \tag{2}$$

where the summation is over all possible configurations and E_c is the configuration energy defined below. For a constant-volume system at temperature T,

$$A_c = -kT \ln Q_c \tag{3}$$

It is now assumed that only nearest-neighbor spins interact; parallel spins contribute U_p per pair to the potential energy and antiparallel spins U_a. Thus the total energy of the system contains the term $(N_{++} + N_{--})U_p + N_{+-}U_a$, where N_{++} and N_{--} denote the number of parallel nearest-neighbor spin *pairs* with spin up and spin down respectively and N_{+-} denotes the number of antiparallel *pairs*. Since $N_{++} + N_{--} + N_{+-} = zN/2$ this term can be rewritten as $\frac{1}{2}zNU_p + 2JN_{+-}$, where $J \equiv \frac{1}{2}(U_a - U_p)$ and z is the coordination number. The quantity $\frac{1}{2}zNU_p$ does not depend on the spin configuration and can be included as part of the static lattice energy. Therefore, we can write

$$E_c = 2JN_{+-} \tag{4}$$

as a general result for an Ising model. In the case of ferromagnetic ordering (all spins parallel in the ordered phase), the energy parameter $J > 0$.

Equation (2) can now be rewritten as

$$Q_c = \sum g(N_{+-}) \exp \frac{-2JN_{+-}}{kT} \tag{5}$$

where the summation is over all possible values of N_{+-}, and $g(N_{+-})$ is the number of configurations possible for a given N_{+-} value. The basic problem is to obtain expressions for N_{+-} and $g(N_{+-})$ in terms of N_+ and N_-, the number of spins oriented up and down respectively. The Bragg-Williams approximation assumes that the spins are *randomly* distributed, which leads to

$$N_{+-} = \frac{zN_+N_-}{N} \tag{6}$$

$$g(N_{+-}) = g(N_+, N_-) = \frac{N!}{N_+!\,N_-!} \tag{7}$$

These expressions are in serious error since the nearest-neighbor interactions tend to favor like pairs; however, they make it quite easy to find a solution to the Ising model. It is convenient at this point to introduce the *long-range order parameter* σ which is defined by

$$\frac{N_+}{N} \equiv \tfrac{1}{2}(1 + \sigma) \qquad \frac{N_-}{N} \equiv \tfrac{1}{2}(1 - \sigma) \tag{8}$$

When $|\sigma| = 1$ the system is completely ordered (if $\sigma = +1$, all spins are up; if $\sigma = -1$, all spins are down). When $\sigma = 0$ the system is completely disordered (random configuration of spins with half up and half down). On the basis of Eqs. (5) to (8), we can express Q_c as

$$Q_c = \sum_\sigma \frac{N!}{[(N/2)(1 + \sigma)]!\,[(N/2)(1 - \sigma)]!} \exp\left[-\frac{zNJ}{2kT}(1 - \sigma^2)\right] \tag{9}$$

As $N \to \infty$, $\ln Q_c$ is well represented by $\ln t_{max}$, where t_{max} is the largest term in the sum.[3] Using Sterling's approximation† for the logarithm of a large number, we can express $\ln t$ for arbitrary σ as

$$\ln t = -N\left[\frac{(1 + \sigma)}{2}\ln\frac{1 + \sigma}{2} + \frac{(1 - \sigma)}{2}\ln\frac{1 - \sigma}{2}\right] - \frac{zNJ}{2kT}(1 - \sigma^2) \tag{10}$$

The most probable value of σ, which gives rise to the largest term t_{max}, can be found as the solution to $d\ln t/d\sigma = 0$. The result is $zJ\sigma/kT = \tfrac{1}{2}\ln[(1 + \sigma)/(1 - \sigma)]$, which is equivalent to the more compact expression

$$\sigma = \tanh\frac{zJ}{kT}\sigma \tag{11}$$

This transcendental equation can be solved graphically by plotting both $F_1(x)$ and $F_2(x)$ versus x, where $F_1(x) = \sigma \equiv (kT/zJ)x$ and $F_2(x) = \tanh x$.[2] For all $T > T_c \equiv zJ/k$, the only root to Eq. (11) is $\sigma = 0$. For $T < T_c$, $F_1(x)$ and $F_2(x)$ cross at three points—0 and equivalent $\pm x$ values which correspond to finite σ values. The latter

† $\ln N! \simeq N\ln N - N$ for large N.

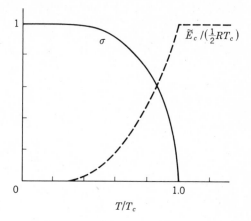

FIGURE 1
Variation of (*a*) the long-range order parameter σ and (*b*) the configurational energy E_c according to the Bragg-Williams approximation.

values can be shown to give a lower free energy than $\sigma = 0$ and thus represent the equilibrium values of σ.[2]

The resulting temperature variation of σ and $E_c = \frac{1}{2}zNJ(1 - \sigma^2)$ are shown in Fig. 1. The qualitatively important feature is the continuous but rapid variation in E_c. Thus the order-disorder transition is not first order (i.e., no latent heat). The configurational heat capacity $C_v(\text{conf}) = (\partial E_c/\partial T)_v$ in this approximation will rise rather slowly with temperature until it reaches a maximum value of $3Nk/2$ at T_c and then will drop discontinuously to zero for all $T > T_c$. Since there is a discontinuity in a second derivative of the free energy, the Bragg-Williams transition is called a *second-order transition*.

The predicted variation in the configurational heat capacity for the Bragg-Williams approximation is compared in Fig. 2 with the results of other theoretical models. Experimentally there is usually a divergence in C_v at T_c which is quite asymmetrical (anomalous variation above T_c occurs over a much narrower temperature range than that below T_c). This variation can often be empirically represented by $A(T - T_c)^{-\alpha}$ for $T > T_c$ and $A'(T_c - T)^{-\alpha'}$ for $T < T_c$, where α and α' are called *critical exponents*. Detailed treatments of such aspects of cooperative phenomena are available in an introductory monograph by Stanley.[1]

METHOD

A very simple but sensitive cooling apparatus will be used to detect order-disorder transitions and to characterize the shape of the anomalous heat-capacity variation associated with such cooperative effects. The apparatus, shown in Fig. 3, is based

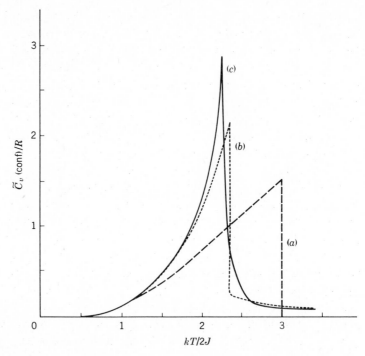

FIGURE 2

Predicted variation in the configurational heat capacity: (a) Bragg-Williams approximation, (b) quasi-chemical or Bethe approximation, (c) numerical series solution to the simple-cubic ($z = 6$) Ising model.

directly on the design of Hirakawa and Furukawa.[4] The sample and the copper shield S are imbedded in Styrofoam, which can be easily cut to shape with an Exacto knife.†
Very fine (0.05 mm) copper-constantan thermocouple wires are used, and the sample thermocouple is cemented in a hole to provide good thermal contact. This hole can be easily drilled with a hypodermic needle filled with fine Carborundum and water. Since the sensitivity is high, a small sample crystal (~ 1.5 cm^3) can be used.

A second thermocouple should be mounted on the copper shield, or a difference couple can be attached to the sample and shield (although this is awkward unless urethane is foamed in place). The inside of the inner Dewar is roughly evacuated to prevent the Styrofoam from collapsing under atmospheric pressure at low temperatures. In addition to acting as a sample holder with negligible heat capacity, Styrofoam provides very good thermal insulation at low temperatures. Thus heat transfer

---

† Urethane foam can be conveniently "foamed in place" but is not so readily removed as Styrofoam, which can be dissolved in acetone.

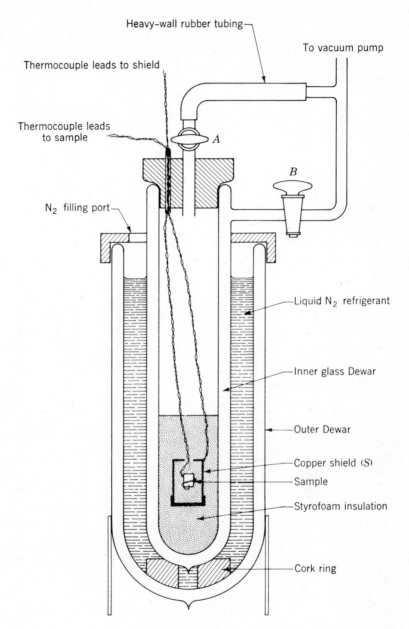

FIGURE 3
Diagram of the cooling-curve apparatus.

is slow even when there is a large temperature difference between the sample and the bath. This helps to reduce the effect of fluctuations in the bath temperature.

The temperature variation of the sample can be analyzed in terms of two expressions for the rate of heat transfer from the sample

$$\frac{dq}{dt} \equiv \frac{dH}{dt} = \frac{dH}{dT}\frac{dT}{dt} = C\frac{dT}{dt} \tag{12}$$

$$\frac{dq}{dt} = a(T_S - T) + \alpha(T_0 - T) \tag{13}$$

If we neglect the term $\alpha(T_0 - T)$ which represents the heat loss to the outside (chiefly along the thermocouple wires), Eqs. (12) and (13) can be combined to give

$$C = \frac{a(T_S - T)}{dT/dt} \tag{14}$$

where C is the heat capacity of the sample, a is the heat-transfer coefficient between the sample at temperature T and the copper shield at temperature T_S. This coefficient will depend on the thermal conductivity of the Styrofoam and the geometry of the system. In general, a will be a weak function of temperature but it can be treated as effectively constant over a limited range.

In this experiment, dT/dt is determined by recording the times when the sample temperature achieves a series of fixed evenly spaced values. The emf from the sample thermocouple is balanced with a potentiometer, and the time is recorded when zero deflection is observed on the galvanometer. The potentiometer dial is then changed by 2 or 3 μV and the time of the next null is recorded, etc. The character of the heat-capacity variation near a phase transition can be roughly seen by plotting the time difference Δt for a constant emf step (2 or 3 μV) versus the emf. However, when the anomalous heat effect is large (as in the case of NH_4Cl), the rate of cooling of the sample will become considerably less than that of the shield. This means that $(T_S - T)$ will vary with the temperature and the shape of the heat-capacity curve will differ appreciably from the shape of Δt versus T. Indeed, a large heat-capacity peak at T_c will cause a large $(T_S - T)$ value and thus a big heat flow just below T_c. This may cause a small dip in the Δt versus T curve immediately after the sample passes through T_c.

Since *differential thermal analysis* (DTA) is now a common and popular technique, it might seem attractive to use this method to study order-disorder transitions. In DTA methods, the temperature difference between the sample and a thermally similar reference material is measured. A moderately free flow of heat between the sample and the reference permits rapid equilization of the two temperatures after a transition occurs and thus enables one to detect closely spaced events. Although suitable for determining transition temperatures, DTA is not capable of sufficient quantitative accuracy to permit one to establish the shape of the heat-capacity curve or even a good value for the integrated heat effect.[5]

Another new and attractive method is *differential scanning calorimetry* (DSC). In this case, the sample temperature is maintained equal to that of a reference material by varying (and recording) the heat input required to maintain isothermal conditions. A commercial instrument, available from Perkin-Elmer Corp., Instrument Division, Norwalk, Conn., would be suitable for a sample such as NH_4Cl and could also be used above room temperature.

EXPERIMENTAL

It is assumed that the sample-shield-Styrofoam assembly, with thermocouples attached, has previously been installed in the inner Dewar flask. It is also assumed that the mass of the NH_4Cl sample pellet is known. Since single crystals of NH_4Cl are not commercially available, pressed pellets of ~ 1.5 cm^3 volume should be used.

Mount the inner Dewar in position and connect the stopcocks to a mechanical vacuum pump, as shown in Fig. 3. Then connect both sets of thermocouple leads to a good quality DPDT switch (such as those made by Leeds & Northrup). Very fine thermocouple wires are fragile; handle them with care. The DPDT switch will allow either thermocouple to be connected to a high-precision potentiometer, which must be capable of measuring potentials of a few millivolts with an accuracy of better than 1 μV.† A Leeds & Northrup K-3 potentiometer can be used but a better instrument is desirable. A high-sensitivity galvanometer ($\sim 0.2\ \mu$V mm^{-1}) is also needed; see Chap. XV. Make all necessary connections to set up the potentiometer, and standardize it against a standard cell. At the same time, prepare an ice-water mixture in a 1-qt Dewar flask as the 0°C bath for the reference junctions of both thermocouples. Fill the Dewar completely full of crushed ice, rinse the ice with a small amount of cold distilled water, and then add enough cold distilled water to fill the Dewar half full of liquid. Insert the junctions well into the liquid; a thin oil-filled tube to hold the junctions will protect them from mechanical damage.

As soon as possible, place liquid nitrogen in the outer Dewar and evacuate the inner Dewar through stopcock *A*. Stopcock *B* should be kept closed at this time, and the jacket of the inner Dewar should contain air at 1 atm to promote rapid initial cooling.‡ Allow the sample to cool until its temperature is about 10° above the transition temperature. Then open stopcock *B* and evacuate the jacket of the inner

---

† If a difference thermocouple is attached to sample and shield, its leads can be connected directly to a sensitive galvanometer capable of measuring up to 40 μV with an accuracy of a few percent.

‡ If this initial rate is too slow, replacing the air with helium will help since the thermal conductivity of He gas is about 9 times that of dry air.

Dewar, which will reduce the cooling rate. Begin recording the emf $\mathscr{E}$ of the sample thermocouple and the time t as determined in the manner described above. About every 10 min, switch to the shield thermocouple and determine $\mathscr{E}_S$ and the corresponding time. Continue to take readings until the temperature is about 40° below T_c. (A cooling rate of about 0.4° min^{-1} is convenient, although slower rates will give even better results.) Once the sample temperature falls more than 10° below T_c, some of the readings can be omitted since the heat-capacity variation is no longer as rapid. Keep the same emf interval $\Delta\mathscr{E}$ but skip a series of readings from time to time.

CALCULATIONS

If the sample temperature readings are closely spaced, dT/dt can be well approximated by $\Delta T/\Delta t$ and Eq. (14) becomes

$$C \simeq b(\mathscr{E}_S - \mathscr{E})\, \Delta t \tag{15}$$

where $\mathscr{E}_S$ and $\mathscr{E}$ represent the emf of the shield thermocouple and the sample thermocouple respectively, Δt is the time interval required for a fixed emf step $\Delta\mathscr{E}$ corresponding to an essentially constant sample temperature step ΔT, and $b = a/\Delta\mathscr{E}$. Note that the thermocouple sensitivity does not appear in Eq. (15) although a thermocouple emf-temperature calibration table is needed to determine the sample temperature.

Plot your experimental $\mathscr{E}_S$ values versus t on a large piece of graph paper, draw a smooth curve through the points, and read $\mathscr{E}_S$ values corresponding to the times at which sample $\mathscr{E}$ values are known. Calculate $(\mathscr{E}_S - \mathscr{E})\,\Delta t$ for each interval and plot this quantity versus $\mathscr{E}$. This plot is, in fact, a plot of the heat capacity in arbitrary units versus the temperature expressed in μV. A temperature scale can be added by merely noting the emf values corresponding to integral values of the absolute temperature. A heat-capacity scale in cal deg^{-1} mol^{-1} can be determined from the weight of the sample and a literature value of $\tilde{C}_p$ at some specified temperature a few degrees above T_c.[6,7] This information will allow one to specify the value of the "constant" b in Eq. (15).

The data analysis in this experiment could be easily carried out with a simple computer program. In this case, fit a polynomial to your $\mathscr{E}_S$ data and determine the value of $(\mathscr{E}_S - \mathscr{E})$ at a temperature corresponding to the midpoint of each time interval.

DISCUSSION

Analysis of C_p data for NH_4Cl in terms of critical exponents is complicated by two factors: (1) There is a large lattice heat capacity which must be properly subtracted from the total heat capacity before attempting to fit $C_p(\mathrm{conf})$. (2) There is a very small first-order discontinuity superimposed on the lambda-like transition at 1 atm (i.e., a very small latent heat and some hysteresis).[7,8] A further discussion of the detailed character of this transition is given in Ref. 9.

The present apparatus could be used to detect hysteresis in the transition temperature. Describe a possible procedure and carry it out if time permits. Does your data give any indication of a finite latent heat? What factors would contribute to obscuring such an indication? How small a latent heat could you determine with this equipment?

APPARATUS

Cooling-curve apparatus, complete with assembled inner Dewar (containing NH_4Cl sample, copper shield, and thermocouples); outer Dewar flask and cover; 1-qt Dewar for ice; high-precision potentiometer setup; good quality DPDT switch; clock or watch with sweep-second hand; mechanical vacuum pump; heavy-wall vacuum tubing.

Liquid nitrogen; distilled-water ice; supply of helium gas (optional).

REFERENCES

1. H. E. Stanley, "Introduction to Phase Transitions and Critical Phenomena," Oxford Univ. Press, London (1971).
2. D. ter Haar, "Elements of Statistical Mechanics," pp. 251–262, Rinehart, New York (1954).
3. G. S. Rushbrooke, "Introduction to Statistical Mechanics," pp. 289–296, Oxford Univ. Press, London (1949).
4. K. Hirakawa and M. Furukawa, *Jap. J. Appl. Phys.*, **9**, 971 (1970).
5. P. G. Garn, "Thermoanalytical Methods of Investigation," Academic, New York (1965); W. W. Wendlandt, "Thermal Methods of Analysis," Interscience-Wiley, New York (1964).
6. H. Chihara and M. Nakamura, *Bull. Chem. Soc. Jap.*, **45**, 133 (1972).
7. P. M. Schwartz, *Phys. Rev. B*, **4**, 920 (1971).
8. A. V. Voronel and S. R. Garber, *Sov. Phys. JETP*, **25**, 970 (1967).
9. B. B. Weiner and C. W. Garland, *J. Chem. Phys.*, **56**, 155 (1972).

XV

ELECTRICAL MEASUREMENTS

This chapter is concerned with the principles of the design and operation of several important electrical circuits.[1] Emphasis is given to galvanometers, potentiometers, Wheatstone bridges, and operational amplifiers, since they are frequently used in physical chemistry. Methods of analog-to-digital conversion for digital voltmeters are also described. At the end of the chapter, capacitance measurements are discussed briefly. The measurement of voltage with various electronic instruments such as oscilloscopes, chart recorders, and phase-sensitive detectors is discussed in Chap. XVIII.

GALVANOMETERS

The D'Arsonval (or moving-coil) galvanometer[2,3] consists of a rectangular coil of many turns of insulated copper wire mounted in the field of a magnet, usually a horseshoe-shaped permanent magnet. This coil is suspended from a fine wire or very thin metal strip, which serves as an electrical lead and also as a torsion fiber. Attached to the bottom of the coil is a loosely coiled wire, which serves as the other lead. Current flowing through the coil moves perpendicular to the magnetic lines of force and produces a torque on the coil. Therefore, the coil will rotate in the field until the magnetic torque is balanced by an opposite mechanical torque introduced by twisting the suspension. Motion of the coil can be detected in two ways: by the position of an indicating pointer attached to the coil or by the position of a light beam reflected from a small mirror attached to the suspension. The pointer-type galvanometer is cheaper, more rugged, and less sensitive than the reflecting type. For any type of D'Arsonval galvanometer the amount of deflection per unit of current through the coil will depend on certain design parameters (the size of the coil, the number of turns of wire in the coil, the magnetic field intensity, and the stiffness of the suspension).[2]

The three most significant characteristics which should be considered in choosing the proper galvanometer for a given application are sensitivity, period, and external critical damping resistance.[3] Of these, sensitivity is usually the most critical requirement. For each kind of galvanometer there is a definite *current sensitivity*, which is the current in microamperes required to produce a 1-mm scale deflection. (On reflecting-type galvanometers the ground-glass scale on which the light beam is observed is normally mounted 1 m from the mirror attached to the coil.) It is often of more direct interest to know the *voltage sensitivity* in microvolts per millimeter of scale deflection. This is equal to the current sensitivity multiplied by R, the total resistance in ohms of the entire electrical circuit through which the current is passing (i.e., R is the sum of the galvanometer coil and suspension resistance and all external resistance in series with the galvanometer). The period of a galvanometer is the time in seconds required for a complete oscillation of the undamped galvanometer. A critically damped galvanometer will almost reach its final deflection within this time, and it is therefore usually convenient to have as short a period as possible. Since most galvanometers are designed to have a period between 2 and 20 sec, this characteristic is seldom of crucial importance in choosing the proper instrument. It is, however, quite important to choose a galvanometer with the best external critical damping resistance (CDRX) to match the resistance of the circuit in which it will be used. If the galvanometer is greatly underdamped (circuit resistance much greater than the CDRX), the coil will oscillate about its final position after a deflection. If greatly overdamped (circuit resistance much less than the CDRX), the coil will move toward its final position very slowly. Either of these conditions can make the instrument difficult to use; a slight underdamping is preferable to any overdamping for most measurements.

The characteristics of several Leeds and Northrup galvanometers are given in Table 1; comparable instruments are also made by Rubicon and several other manufacturers. For student use, pointer galvanometers similar to the L & N models 2310 and 2340 are very suitable and practical types. Reflecting-type galvanometers are much more sensitive and should be used for high-precision work. The L & N models 2420, 2435, and 2430 are box galvanometers with enclosed lamp and scale.

Table 1 PERFORMANCE CHARACTERISTICS FOR SEVERAL LEEDS & NORTHRUP GALVANOMETERS[a]

L & N Model No.	Type	Current sensitivity, $\mu A\ mm^{-1}$	Voltage sensitivity for CDRX, $\mu V\ mm^{-1}$	Period, sec	Resistance, Ω CDRX	Coil + suspension
2310-d	Pointer	0.125	1380	3.5	10,000	1000
2310-e	Pointer	1.0	46	4.5	30	16
2340-a	Pointer	1.0	70	4.0	35	35
2340-b	Pointer	0.6	84	3.5	100	40
2340-c	Pointer	0.25	272.5	3.5	1000	90
2340-d	Pointer	0.1	1050	4.0	10,000	500
2420-c	Reflecting	0.025	400	3.0	15,000	1000
2420-d	Reflecting	0.04	92	3.0	2000	300
2435-a	Reflecting	0.1	7	4.0	35	35
2435-b	Reflecting	0.06	8.4	3.0	100	40
2435-c	Reflecting	0.025	27.2	3.0	1000	90
2435-d	Reflecting	0.01	105	3.5	10,000	500
2430-a	Reflecting	0.006	0.5	3.0	60	17
2430-c	Reflecting	0.005	2.1	2.5	400	25
2430-d	Reflecting	0.0005	12.8	3.0	22,500	550
2430-e	Reflecting	0.001	6.7	3.0	6500	175

[a] Models 2310 and 2420 are no longer sold; data taken from L & N Catalog ED (1956). Other information from current L & N Data Sheets.

The box contains a mirror system which folds the light beam from the moving mirror several times before it falls on the scale; this permits a 1-m light path to be confined in a box about 10 in. long.

Some pointer-type galvanometers have a clamping device to protect the coil suspension from damage while the instrument is not in use; this clamp must be released in order to use the galvanometer. At the end of an experiment, be sure to reclamp the suspension. Most reflecting-type galvanometers do not have such a clamping arrangement but may have special spring mountings to minimize the effect of shock. If a reflecting galvanometer is to be moved any appreciable distance, it is wise to attach a short length of wire across the terminals. This will provide considerable overdamping and thus protect the suspension.

POTENTIOMETER CIRCUITS

As the name implies, the potentiometer is a device for measuring electrical potential difference. Basically it is an electrical circuit containing a precision resistor employed as a variable potential divider; indeed, the name "potentiometer" is often used in electronics technology to denote any variable resistor used as a potential divider.

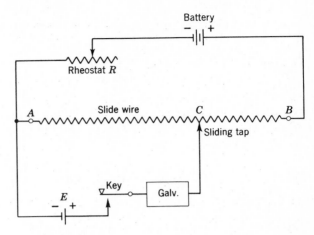

FIGURE 1
Schematic diagram of a simple slide-wire potentiometer.

Basic circuit The principle of the potentiometer can be described in terms of the simple "slide-wire" potentiometer shown schematically in Fig. 1. With a constant direct current flowing through the *uniform* resistance slide-wire AB, the potential difference between A and C should be accurately proportional to the length AC. Thus the potentiometer is a device for selecting with high precision any desired electromotive force (emf) which is less than a certain maximum value. This maximum value depends on the construction of the slide-wire and the voltage of the battery. The emf across part of the slide-wire (AC) may be compared with another emf by connecting the negative pole of the external emf E with the negative end of the slide-wire (at A) and connecting its positive pole with the positive slide-wire contact C through a galvanometer and a tapping key. When this key is depressed, current will flow through the galvanometer coil and a galvanometer scale deflection will be observed if the emf of E and that between A and C are not equal. The direction of the galvanometer deflection will obviously depend on whether the potential at C is higher or lower than that at the positive pole of E. If the position of the sliding contact is adjusted until no galvanometer deflection is observed on depressing the tapping key, the emf of E must be the same as that between A and C. Indeed, a linear scale marked off directly in volts can be placed alongside the slide-wire, and the emf between A and C can be read directly from this scale, provided that the current through the slide-wire has been properly adjusted by manipulation of the rheostat R.

The adjustment of the current through the slide-wire is accomplished by connecting a source of known emf, such as a *standard cell*, in place of the potential E. The slider C is then placed at the setting on the slide-wire scale corresponding to the known potential, and the rheostat R is adjusted until no galvanometer deflection is observed on depressing the tapping key. This "standardization" of the potentiometer against a standard cell should be repeated frequently during the experimental work because the battery voltage usually shows a tendency to drift slightly.

The accuracy of an emf measurement depends most of all on the design and the quality of the potentiometer unit used. For a precision potentiometer, the relative accuracy of measurement (ability to measure very small changes in emf) will depend a great deal on the galvanometer sensitivity while the absolute accuracy will depend more on the accuracy of the standard cell emf value.

Standard cells The electrochemical cell used as a standard is the Weston cell:

$$Cd(12.5\% \text{ Cd amalgam}) + CdSO_4 \cdot \tfrac{8}{3}H_2O(s), CdSO_4(sat), Hg_2SO_4(s) + Hg(l)$$

which has an emf of 1.0184 absolute volts at 25°C and a temperature coefficient of 4×10^{-5} V deg^{-1}. An airtight, H-shaped cell is used with platinum wires sealed through the glass to make contact with the electrodes. The anode arm (negative electrode) consists of a two-phase cadmium amalgam (overall composition 12.5 percent Cd by weight) covered with crystals of hydrated cadmium sulfate; the cathode arm (positive electrode) contains a pool of liquid mercury covered by a paste of mercurous sulfate. The rest of the cell is filled with a saturated solution of $CdSO_4$, except for a small air space to allow room for thermal expansion.

In actual practice, it is more common to use a Weston cell containing an *unsaturated* solution of $CdSO_4$ as a standard cell. This cell has the advantage that the temperature coefficient is low (1×10^{-5} V deg^{-1}) and the disadvantage that the emf may vary slowly with time. Such a standard cell has an initial emf between 1.0186 and 1.0196 V, and it should be restandardized every year or two. The best method of checking a "working" standard cell against a very reliable "reference" standard cell is to connect the two cells in opposition and measure the emf difference between them with a precision potentiometer (which is standardized against another working standard cell). In this way the effect of experimental error is almost eliminated, since it affects only a very small difference reading.

In using a standard cell, one must **never** pass more than 10^{-4} A through the cell. A protective resistance should be in series with the standard cell during the initial stages of standardizing a potentiometer. Take particular care to depress the galvanometer key only briefly, especially when the deflections are large.

Leeds and Northrup Student Potentiometer The L & N Student Potentiometer, Catalog No. 7651, is a medium-precision unit (accuracy of ± 0.5 mV on the 1.6-V scale and ± 0.01 mV on the 0.016-V scale) which is widely used for measuring emfs in the range 0.001 to 1.6 V. Indeed, many of the experiments in this book require the use of this instrument (or some other of comparable quality and similar construction).

Therefore, we shall discuss its operation in considerable detail. Figure 2 shows a complete potentiometer circuit based on the Student Potentiometer unit; note that several auxiliary components are required in addition to the potentiometer unit.

In the Student Potentiometer the potential difference is selected, not between a fixed end and a variable point, but rather between two variable points, one of which is moved in steps (by means of a step switch) and the other of which is moved continuously (a dial "slide-wire"). The stator of the step switch has 16 contact points, between each adjacent pair of which is a noninductively wound fixed resistor of 10-Ω resistance, all the resistances being precisely equal to one another. The slide-wire is in series with the 15 fixed resistances, and its resistance is precisely equal to the resistance of any one of the fixed resistors. The 16 points of the step switch are numbered in tenths of a volt from 0 to 1.5 V, and the scale of the slide-wire is so graduated (the smallest graduations being 0.0005 V) as to cover the range 0 to 0.1000 V. A potential can be estimated to 0.0001 V from the slide-wire dial reading. The desired potential difference is obtained between binding post E^-, which is connected to the rotor of the step switch, and binding post E^+, which is connected to the sliding-contact rotor of the dial slide-wire. Thus the desired potential difference or emf is the sum of a step-switch reading and a slide-wire reading. The proper current through the 15 resistances and the slide-wire is obtained from batteries through rheostats, and the current supply is normally connected to the binding posts marked BA^+ and 1. The potentiometer is standardized against a standard cell in the same way as described previously in the case of the simple slide-wire potentiometer.

If, *after standardization*, the negative lead of the battery circuit is switched from binding post 1 to that marked 0.01, the range of the Student Potentiometer changes from 0–1.6 to 0–0.016 V; that is, all readings must be multiplied by 0.01. This change of scale is accomplished by means of an internal resistive network so designed (see Fig. 2) that the current flowing through the step-switch resistors and the slide-wire is reduced to precisely 1 percent of the value established during the standardization. The total resistance in the battery circuit is, however, not affected on switching from post 1 to 0.01. This low-scale arrangement permits one to make a more precise measurement of very small emf values, such as thermocouple emfs below 16 mV or the potential drop across a 1-Ω standard resistor when less than 16 mA of direct current are flowing through it. Figure 3 shows a convenient way to rewire the DPDT switch so that it is not necessary to move lead wires when using the 0.01 scale.

The battery circuit shown in Fig. 2 consists of two 1.5-V dry cells connected in series with a battery control unit which permits both coarse and fine adjustment of the current through the potentiometer unit. It is possible to achieve better current stability by using a 2-V low-discharge wet cell. In that case, the 120-Ω fixed resistor in the control unit should be replaced by a 20-Ω resistor so that a current of 10 mA can be obtained.

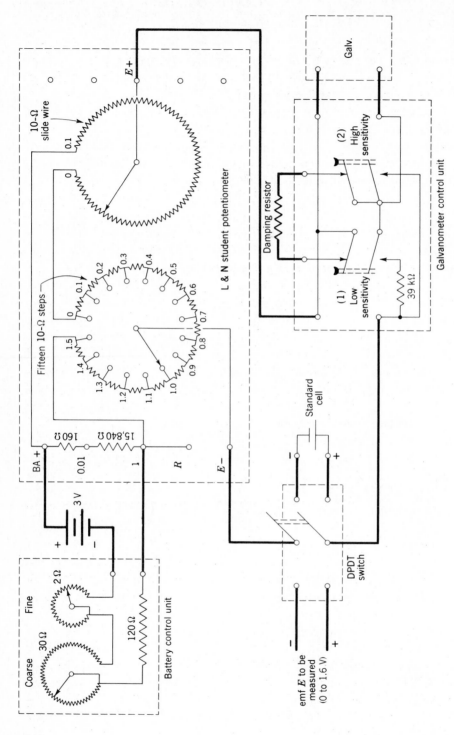

FIGURE 2

Complete wiring diagram for a potentiometer circuit based on the **L & N Student Potentiometer.** The internal-wiring diagram of the potentiometer is somewhat simplified.

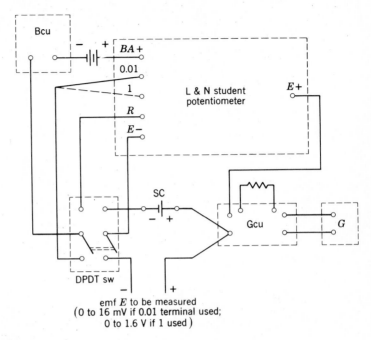

FIGURE 3
Alternative method of making connections to the Student Potentiometer. This method is of greatest value when the 0.01 scale is to be used (as shown). For internal wiring of units, see Fig. 2.

Also shown in Fig. 2 is a galvanometer control unit, which replaces a simple tapping key. There are two spring-loaded push buttons either of which, when depressed, will close the circuit through the galvanometer. When the low-sensitivity button (1) is depressed, a protective resistance of 39,000 Ω is in the galvanometer circuit. This will limit the current flow through the galvanometer and prevent damage when the potentiometer setting is far from the proper value. When the high-sensitivity button (2) is depressed, this protective resistance is *not* in the circuit. Therefore, button 1 should be used during all preliminary adjustments, and button 2 should be depressed only when the deflections are too small to detect using button 1. Either button should be depressed only briefly and not held down. There is also a position on the control unit for mounting an external damping resistor so that it will be across the galvanometer whenever *neither* button is depressed. This resistor should be chosen to provide approximately critical damping for the galvanometer used. It is also advantageous to have critical damping when button 2 is depressed; this is more difficult to arrange, as it requires the use of a galvanometer having a critical damping resistance equal to the resistance of the rest of the circuit or else the modification of the circuit to match the critical damping resistance of the galvanometer available.

Finally, a detailed outline is given below of the procedure for using the circuit shown in Fig. 2:

1 Connect the components, taking care that the batteries, standard cell, and unknown emf are connected with the correct polarities. (Until the student has become thoroughly familiar with the potentiometer circuit, it is advisable to omit the connections to the standard cell until the wiring has been checked by an instructor.)

2 Set the step switch and slide-wire dial to the value of the emf of the standard cell. By means of the DPDT switch, place the standard cell in the circuit.

3 Standardize the potentiometer against the standard cell by adjusting the coarse battery control until there is no galvanometer deflection when galvanometer button 1 is depressed. Then adjust the fine battery control until there is no deflection on depressing button 2. Do not hold button 2 down for a prolonged period.

4 Switch from the standard cell to the unknown potential E by reversing the DPDT switch. **Do not disturb the battery current setting.** Vary the potentiometer step switch and slide-wire dial setting until there is no galvanometer deflection on depressing button 1 of the galvanometer control unit; then make any necessary adjustments of the slide-wire until a null point is obtained when button 2 is depressed. NOTE: If it is impossible to achieve balance at any setting of the potentiometer, either the unknown emf E is greater than 1.6 V or E is connected into the circuit with the incorrect polarity.

5 Record the emf value for E, then restandardize and repeat the measurement.

High-precision research potentiometers Several types of potentiometers are available for high-quality research applications; each of them has certain special design features which improve the precision and the convenience of operation. Almost all research potentiometers are designed so that the instrument can be standardized against a standard cell without changing the settings of the step switches or slide-wire. This feature is incorporated in the circuit shown in Fig. 4. By means of a built-in DPDT switch, one pole of the standard cell can be connected through the galvanometer to a fixed contact A on the step switch. The other pole of the standard cell is directly connected to the movable contact on a calibration slide-wire C. Adjustment of the setting on slide-wire C allows one to obtain any potential difference between 1.018 and 1.020 V in order to match the emf of the standard cell used. When the DPDT switch is thrown to the EMF position (dashed lines in figure), the E^- terminal is connected through the galvanometer to the movable contact on the

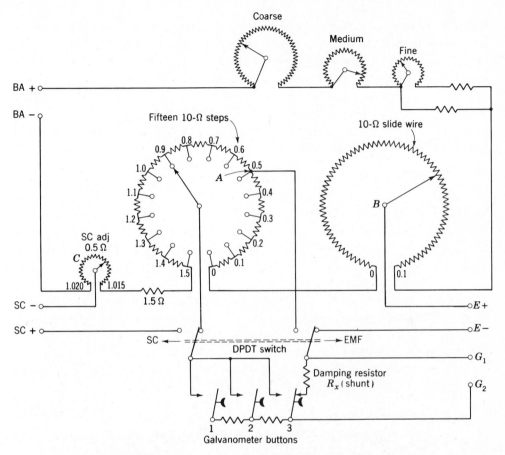

Coarse

Medium

Fine

BA +

BA −

Fifteen 10-Ω steps

0.8 0.7
0.9   0.6
1.0   0.5
              A
1.1   0.4
1.2   0.3
SC adj
0.5 Ω
1.3   0.2
C
1.020 1.015
1.4 0.1
1.5 0

10-Ω slide wire

B

0
0.1

SC −

SC +

1.5 Ω

E +

E −

SC ◄ = = = = = = = = = = = ► EMF

DPDT switch

Damping resistor
R_x (shunt)

G₁

G₂

1   2   3
Galvanometer buttons

FIGURE 4
Schematic diagram of a potentiometer unit which incorporates some of the
special features usually found in high-precision research-quality instruments.

step switch. The E^+ terminal is directly connected to the contact on the
measuring slide-wire *B*. Most research potentiometers also have built-in battery
control rheostats and galvanometer tapping keys. Figure 4 shows a typical version
of both of these features; in each case the design is more elaborate than that discussed
for the Student Potentiometer. Note that the galvanometer is always shunted by a
resistor R_x except when the highest sensitivity key (key 3) is depressed; the value of
R_x should be close to the CDRX of the galvanometer used. Some potentiometers
provide several galvanometer binding posts with different R_x values between them to
permit easy matching to the CDRX. Other important design features, which have
been omitted from Fig. 4 for the sake of clarity, are range switches to provide 1, 0.1,

and 0.01 scales in such a way that one can check against the standard cell when the measuring circuit is on any range setting; replacement of a single step switch and slide-wire by as many as five decade step switches; special arrangements to keep the resistance in the galvanometer circuit a constant independent of the settings of the step switches. This last feature is valuable for precise measurements of a thermocouple emf; since the galvanometer sensitivity is constant, deflections can be easily converted into emf values in order to follow small changes. More details on individual potentiometers are given in the manufacturers' catalogs and in the literature.[1,4]

WHEATSTONE BRIDGE CIRCUITS

Direct-current Wheatstone bridge   The dc Wheatstone-bridge circuit provides a simple means of accurately determining an unknown resistance. As shown in Fig. 5a, an arbitrary dc potential drop is established across the bridge from A to C and a galvanometer with tapping key serves as a detector of current flow from B to D. Since direct current is involved, all arms of the bridge are treated as purely resistive elements. When the bridge is balanced (i.e., no deflection on the galvanometer when the tapping key is closed), the potential at B must be the same as that at D and it follows that

$$\frac{R_1}{R_2} = \frac{R_3}{R_4} \tag{1}$$

A form of this bridge incorporating the slide-wire in an L & N Student Potentiometer is shown in Fig. 5b. The slide-wire, which has a scale A reading from 0 to 1000, constitutes two of the four arms of the bridge. When terminals L and H are used, one obtains from Eq. (1) the condition of balance

$$X = \frac{A}{1000 - A} R \tag{2}$$

Often the known resistance R is variable, for example, a decade resistance box. In this case, R can be varied until A is somewhere in the range 450 to 550. A considerably greater sensitivity can then be achieved by shifting the L lead to L' and the H lead to H', obtaining the so-called "long bridge." The two left arms of the bridge now each contain an additional resistance 4.5 times that of the slide-wire itself, and the balance condition then becomes

$$X = \frac{4500 + A}{5500 - A} R \tag{3}$$

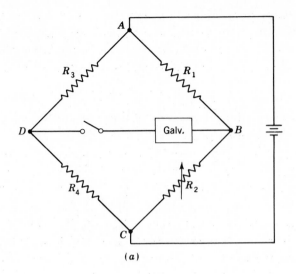

(a)

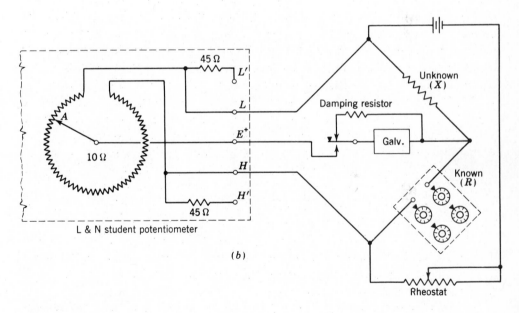

(b)

FIGURE 5

The dc Wheatstone bridge: (a) general schematic diagram of the bridge; (b) wiring diagram, based on the use of the slide-wire in an L & N Student Potentiometer. Slide-wire internal circuit slightly idealized.

In general, use of the long bridge is preferable because of the greater sensitivity provided. In some cases, particularly when X and R are small, use of the long bridge is not advantageous because the galvanometer sensitivity is the limiting factor and the galvanometer sensitivity is decreased rather than increased by going to the long bridge.

Commercial Wheatstone bridges are available in a form where R_3/R_4 can be set by a step switch to accurate decimal ratios from 10^{-3} to 10^3, and R_2 is a precision four- or five-decade resistance box. Other special features are available for certain applications; the Mueller bridge, used with four-lead platinum resistance thermometers, is shown in Fig. XVI-3.

Alternating-current Wheatstone bridge   A Wheatstone bridge can be operated with alternating as well as direct current. Indeed, the use of alternating current is necessary to prevent polarization of the electrodes in a conductance cell (see Exp. 20). The basic circuit for an ac bridge is the same as that shown in Fig. 5a except that an ac source (such as a tuning-fork oscillator or an audio oscillator operated at up to 1 or 2 kHz) is used and the galvanometer and tapping key are replaced by an audio detector (such as earphones) or an oscilloscope. A more sophisticated ac-bridge diagram for conductance measurements is shown in Fig. 6. For an ac bridge the condition of balance requires that the alternating potential at B and D be of equal amplitude and exactly in phase. This implies that

$$\frac{Z_1}{Z_2} = \frac{Z_3}{Z_4} \tag{4}$$

where Z is the impedance. For a series combination of a pure resistance R and a pure capacitance C,

$$Z(\text{series}) = R + \frac{1}{i\omega C} \tag{5}$$

where $i = \sqrt{-1}$ and $\omega = 2\pi f$; for a parallel combination,

$$\frac{1}{Z(\text{parallel})} = \frac{1}{R} + i\omega C \tag{6}$$

If the ratio arms R_3 and R_4 in Fig. 6 are identical noninductively wound resistors for which the residual capacity between turns and between each arm and external objects are equal, then the balance condition becomes simply $Z_1 = Z_2$. Since the conductance cell has an impedance Z_1 which is not purely resistive, it is necessary to place a variable capacitor C_2 across the resistor R_2. By adjusting C_2 one can compensate for the phase shift in the conductance cell, and a sharp balance of the bridge

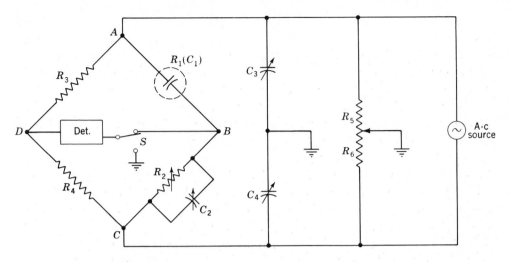

FIGURE 6
Schematic diagram of an ac Wheatstone bridge (with Wagner earthing device) suitable for conductance measurements.

can be obtained. However, the balance may be a minimum rather than a null point if B and D are not at ground potential, since some current can still flow through the detector to ground via distributed capacity. Detection of the balance point can be improved for very precise work by using a Wagner earthing circuit[5] to bring point D (and thus point B) close to ground potential. After an initial bridge balance is achieved, the detector is switched from B to ground and the Wagner earthing circuit (C_3, C_4, R_5, and R_6 in Fig. 6) is adjusted to minimize the signal through the detector. The bridge is then rebalanced, and this process is repeated until the best possible balance point is achieved. The best detector for precise measurements is an oscilloscope, since this can be used in a way which permits separate observation of both the capacitive and the resistive balance[6] (see Chap. XVIII).

Once a balance condition has been achieved, there is still the problem of relating the cell resistance R_1 to the known resistance R_2. The principal difficulty lies in the need to adopt an equivalent electrical circuit for the cell. This question is discussed in detail elsewhere,[7] but the simplest realistic model is

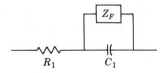

where the *series* capacitance C_1 is due to the electrical double layer at the interface between the electrode and the electrolyte and the frequency-dependent faradaic impedance Z_F is associated with the transfer of charge through the interface. Note that capacitances in parallel with R_1 have been neglected in this model. For "ideally polarized" electrodes, one could neglect the faradaic leakage (set $Z_F = \infty$) and obtain the balance condition

$$R_2 = R_1 + \frac{1}{\omega^2 C_1{}^2 R_1} \tag{7}$$

However, this expression is not likely to be valid in practice due to the depolarizing effect of dissolved oxygen. On the other hand, if C_1 is so large† that its impedance is small compared to Z_F, one finds that $R_1 = R_2$ at all frequencies. This behavior is, fortunately, that observed for platinum electrodes with a heavy deposit of platinum black such as those used in Exp. 20. The general expression which is valid for bright platinum electrodes is considerably more complicated,[7] but one can probably use the approximation $R_1 = R_2$ for moderately dilute aqueous electrolytes (where $R_1 \geq 10^3 \ \Omega$). If a frequency dependence is observed in the R_2 values at balance, then it will be necessary to measure R_2 at several frequencies and extrapolate to $\omega = \infty$.

STANDARD RESISTORS AND POTENTIAL DIVIDERS

In order to make an accurate measurement of the direct current flowing through a resistive element, a standard resistor is inserted in the circuit in series with the element and the potential drop across this standard resistor is measured with a potentiometer. Obviously, the resistance value of the standard resistor must be chosen so that the potential difference across it will be of the proper magnitude for the potentiometer used. Also, the standard resistor must be designed to have an adequate current rating (wattage), a low temperature-coefficient, and a long-term resistance stability. A small temperature variation of resistance can be achieved if the resistor consists of manganin wire wound on a large insulated spool. Such wire-wound resistors can be obtained commercially from several manufacturers, and one can easily calibrate the resistance value using a precision Wheatstone bridge. Also, resistors certified to be within 0.1 percent of the nominal value can be obtained from the Shallcross Mfg. Co. For more precise work, Leeds and Northrup offers a NBS-type standard resistor in which the resistance coil is sealed in a container filled with oil. This type of resistor has very heavy leads with two terminals on each lead—one for the current connection and one for the potentiometer connection.

---

† Since the double layer is quite thin, rather large (a few μF per cm^2 of electrode surface) capacitances are possible.

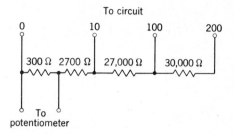

FIGURE 7
Basic circuit for a potential divider
(volt box).

It is sometimes necessary to measure the potential drop across a circuit element when the voltage is higher than the maximum possible setting on the potentiometer. The proper technique in such a case is to use a high-resistance potential divider, which can be easily constructed from two wire-wound resistors. Commercial potential dividers, called volt boxes, are available with a design similar to that shown in Fig. 7. By connecting the 0 terminal to one side of the circuit element and the 10, 100, or 200 terminal to the other side it is possible to reduce the voltage accurately by a factor of 10, 100, or 200. Since the resistance in the volt box is high, only a few milliamperes will flow through this shunt provided the voltage is kept below the rated value for each terminal (usually about 5 V/1000 Ω between the terminals used). It is preferable to have the 0 terminal near ground potential; this prevents possible leakage currents and is safer when high voltages are used.

OPERATIONAL AMPLIFIERS

In many measurement applications in physical chemistry, the current or voltage levels are very low or are derived from devices whose output impedances are poorly matched to the available recording instruments. For example, the small millivolt output of a thermocouple can be accurately measured only by using a potentiometer, as discussed earlier, or by using a very high input impedance device to approach the necessary zero-current condition. For weak fluorescence signals, the anode current of a photomultiplier is often quite small and is best measured with a current-to-voltage amplifier. Fortunately, a convenient and economical solution to such measurement requirements exists in the form of solid-state operational amplifiers. These devices are single "chips" made up of a number of transistors, resistors, capacitors, and diodes, carefully designed to produce desired characteristics of high input impedance, high gain, precise linearity, low output impedance, and broad frequency response. The function of the individual parts is described in many texts on electronics but a detailed analysis is not necessary to understand the essential features of an operational amplifier and some of its applications.

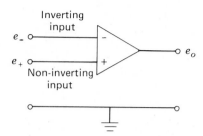

FIGURE 8
Representation of an operational
amplifier.

An operational amplifier is basically an amplifier of voltage differences, and it
is usually represented by a triangular symbol as shown in Fig. 8. The plus and
minus signs on the input terminals refer to the polarity of the output voltage
produced by a positive voltage step at either input. All voltages are referenced to
the ground line and this is often omitted from the symbol. The output voltage e_0 is
related to the input difference

$$e_0 = A(e_+ - e_- + \Delta e) \approx A(e_+ - e_-) \tag{8}$$

where A is the amplifier open-loop gain (10^4 to 10^8), which is ideally the same for
both inputs. The quantity Δe is a small voltage offset (~ 1 mV or less) which is
characteristic of the amplifier and which can be cancelled by applying a small
compensating voltage to the inverting $(-)$ input.

For stable operation, the operational amplifier is generally used with a feedback loop which can involve a resistor, a capacitor, or a diode. The essential
function of the amplifier is to produce whatever voltage is required at the output to
hold, by means of the feedback loop, the inverting input at a voltage close (within
1 mV or so) to that of the noninverting input. When the feedback is achieved with a
simple connector as shown in Fig. 9a, the output acts as a *voltage follower* with
$e_0 = e_i$. Such a circuit is commonly used when a voltage source cannot supply
sufficient current to drive a voltage measuring device or other load. The input
impedance is that of the amplifier (ranging from 10^5 to 10^{14} Ω) and the output
impedance is small (10^{-1} to 10^3 Ω).

If amplification of a small input voltage is desired, either of the circuits shown
in Figs. 9b and 9c can be used. In the latter case, the noninverting input is
grounded and the amplifier acts to hold the inverting $(-)$ terminal at a *virtual
ground*, i.e., $e_- \cong 0$, but there is no actual connection to ground. (Actually, e_-
approaches zero as the gain A approaches infinity.) The source resistance relative to
ground potential is determined by the input resistor R_i, and a current of

$$i = \frac{e_i - e_-}{R_i} \cong \frac{e_i}{R_i} \tag{9}$$

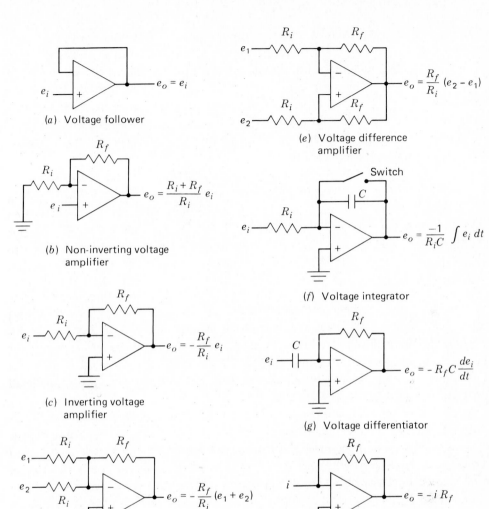

(a) Voltage follower

(b) Non-inverting voltage amplifier

(c) Inverting voltage amplifier

(d) Voltage summing amplifier

(e) Voltage difference amplifier

(f) Voltage integrator

(g) Voltage differentiator

(h) Current-to-voltage amplifier

FIGURE 9
Some circuit configurations with operational amplifiers.

flows into the inverting input terminal. Since the input resistance to the amplifier is very high, essentially all of this input current must flow out through the feedback resistor R_f, corresponding to a potential at the output of

$$e_0 = e_- - iR_f \cong -\frac{R_f e_i}{R_i} \qquad (10)$$

The polarity is thus reversed and the output voltage is amplified by the factor R_f/R_i. This gain must be less than the open-loop gain A, and the feedback resistance R_f must be small compared to the amplifier input resistance. If several voltage inputs are applied to the inverting input, as in Fig. 9d, the amplifier can be used to perform a summing operation with simultaneous amplification. Feedback arrangements to perform voltage subtraction, integration, and differentiation are shown in Figs. 9e to 9g; and other operations such as multiplication, division, and logarithmic manipulations are also possible.

For many applications in spectroscopy, electrochemistry, chromatography, and other areas, a current, instead of a voltage, is generated which is directly proportional to a property of interest. For example, electron emission from a photocathode is linearly related to variations in the incident-light intensity. In a photomultiplier, this cathode current is amplified and the anode current i (typically 10^{-9} to 10^{-3} A) can be converted to a voltage $e_0 = iR_f$ by passage through a load resistor R_f. However, if R_f is large, the potential at the anode varies significantly with anode current. As a result, the gain of the tube changes with current so that the anode current (and the voltage across R_f) is no longer linearly related to light intensity. A current-to-voltage amplifier, such as that depicted in Fig. 9h, can be used to solve this problem. This arrangement offers no impedance to current flow into the virtual ground at the inverting input and the anode potential is constant. Of course, no current actually flows into the amplifier; instead the output level drops to a negative potential to ensure that all of the photomultiplier current flows through the feedback resistor R_f. The same magnitude of voltage signal is produced as with a simple load resistor, but with opposite polarity. The amplifier in effect acts as a zero-impedance device while generating a large voltage across R_f, which can then be read with a meter or recorder of relatively low input impedance. Summation, integration, and other operations involving current sources can be done with the circuits shown in Figs. 9a to 9f if the input resistance R_i is simply eliminated.

In principle, an ideal operational amplifier produces an output which is proportional only to the differential voltage $e_+ - e_-$ and is independent of the *common mode voltage* [$CMV = \frac{1}{2}(e_+ + e_-)$]. The extent to which this is true of a real amplifier can be judged by the *common mode rejection ratio*:

$$CMRR = \frac{\text{gain for } (e_+ - e_-)}{\text{gain for CMV}}$$

Typical values of CMRR for ordinary operational amplifiers are 10^3 to 10^5. High values of this parameter are particularly important for applications where common noise signals (such as 60-Hz pickup from power lines) occur on both inputs to the amplifier. Examples include signal amplification of outputs from transducers such

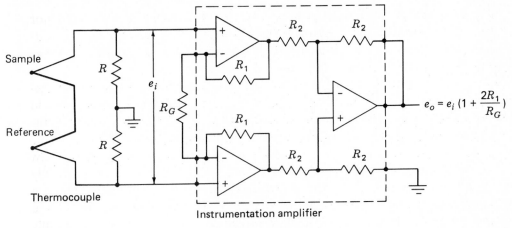

FIGURE 10

Thermocouple amplification using an instrumentation amplifier.

as thermocouples, thermistors, strain gauge bridges, and other devices. For such applications, instrumentation or isolation amplifiers are generally used.

An instrumentation amplifier consists of three or more operational amplifiers and is carefully designed to achieve gains in the range 1 to 10^4 with very high values of input impedance (10^8 to 10^{10} Ω) and CMRR (10^4 to 10^6). Figure 10 shows a simplified circuit diagram for an instrumentation amplifier used to amplify a thermocouple output. With such an arrangement, the two thermocouple leads can have common voltages as large as ± 10 V. For common mode voltages in excess of 10 V, an isolation amplifier is used in which the input is electrically isolated from the output and from the power supply. With such devices, CMRR values of 10^8 or more are possible for common mode voltages which can exceed 1000 V.

DIGITAL VOLTMETERS

Most voltage or current measurements in physical chemistry involve conversion from an analog form to a digital number at some recording stage so that subsequent numerical analysis can be performed. For example, the student acts as an analog-to-digital converter in nulling a potentiometer and then recording a Chromel-Alumel thermocouple output as, say, 10.24 mV. This digital number then might be used with a reference table to deduce that the temperature at the sensing point is 252°C. This same analog-to-digital voltage conversion (and even the conversion to temperature) can also be done with a digital voltmeter. Voltmeters are readily available which provide 3 to 6 display digits at accuracies of 0.1 to 0.003 percent of full scale. With the development of large-scale integrated circuits, the cost of such meters has dropped dramatically (ranging from $60 to $500 in 1980, depending upon the accuracy and number of digits), and it is certain that they will find increasing use in instrumentation and in instructional laboratories.

A digital voltmeter consists of an analog-to-digital conversion circuit plus a readout display. Often, lines are provided for transfer of digital results to a computer. A variety of methods are used for the digitizing operation but only the two most common techniques will be discussed here.

The dual-slope integration technique (Fig. 11) is used with most display meters where millisecond conversion times are adequate. The conversion involves two steps. In the first step the signal is applied to an operational amplifier, which provides extremely high input impedance plus gain for low-level inputs. At the start of a measurement cycle, control circuits simultaneously switch the amplifier output to the integrator input, remove the short from the integrating capacitor, and open a gate to pass clock pulses to a counter. The integrator output, which begins from an initial value of zero, produces a linear ramp with slope and direction proportional to the instantaneous amplitude and polarity of the input voltage (cf., Fig. 9f). The integration continues until a fixed number of counts have accumulated (10^4 for a 4-digit meter). At this time, in the second step of the cycle, the integrator input is switched to a reference voltage whose polarity is opposite to that of the input signal. The counter is reset to zero and again begins to count while the reference voltage drives the integrator back to zero. The slope of this second ramp is proportional to the reference voltage, and the time to reach zero is exactly proportional to the input voltage. When the integrator reaches zero, the counter gate is closed and the count displayed is numerically equal to the unknown input voltage.

The dual-slope method has a number of important features. Conversion accuracy is independent of the effect of temperature or aging on the clock frequency and integrating capacitor value since they only need remain stable during a conversion cycle. Accuracy is thus determined by the accuracy and stability of the reference source. Resolution is limited primarily by the analog resolution of the converter. Because of the integration operation the converter gives excellent high-frequency noise rejection, and acts to average the signal over the integration period. Moreover, by choosing the clock frequency to be a multiple of 60 Hz, such as 600 kHz, it is possible to achieve nearly complete rejection of 60-Hz ac fluctuations on the input signal since this ac noise averages to zero when one (or more) full 60-Hz cycle equals the 10^4 count period. The minimum period for this process is 16.67 msec so that this advantage of the dual-slope method of integration is not obtained at high digitizing rates.

For applications requiring high resolution and high speed, the successive

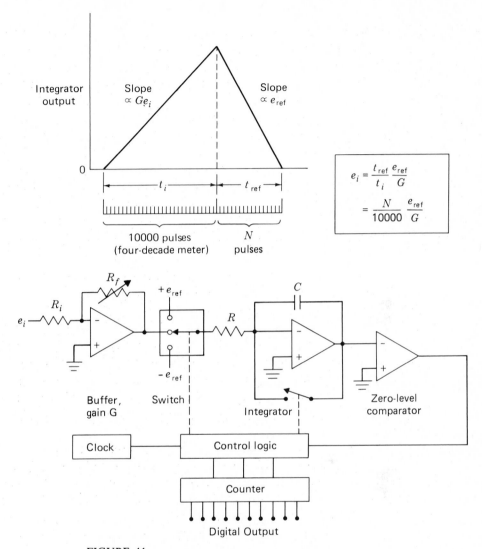

$$e_i = \frac{t_{ref}}{t_i} \frac{e_{ref}}{G}$$

$$= \frac{N}{10000} \frac{e_{ref}}{G}$$

FIGURE 11
Dual-slope type A/D converter. For calibration, either e_{ref} or the buffer gain G can be adjusted to give the correct display for a known input voltage.

approximation A/D conversion method is used. This method is illustrated in Fig. 12 and consists of comparing the unknown input against a precisely generated internal voltage from a digital-to-analog (D/A) converter. The D/A converter operates by switching in various binary fractions of a reference voltage (e_{ref} = full scale, typically 1 to 10 V) to the summing point ($-$) of an operational amplifier. The

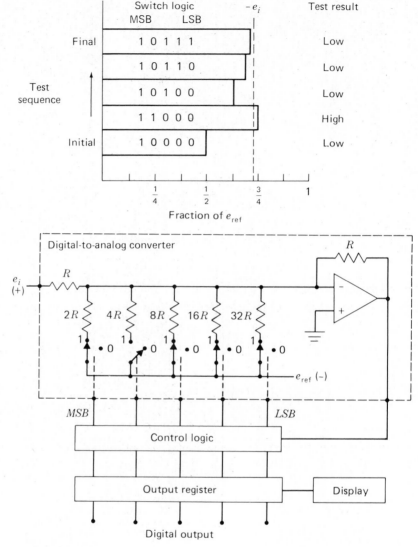

FIGURE 12

Successive approximation type A/D converter. A 5-bit converter is shown for simplicity of illustration, but commercial instruments typically use 10 to 16 bits.

polarities of e_{ref} and e_i are opposite so that subtraction occurs when e_i is also introduced into the summing point. The output polarity of the operational amplifier thus indicates the relative magnitudes of e_i and the total fraction of the reference voltage passed by the control logic. At the start of a conversion cycle, the D/A converter's most significant binary bit (MSB), which is one-half full scale, is compared with the input. (For a discussion of binary numbers, see Chap. XXI.) If smaller than the input, the MSB is left on and the next bit (which is one-fourth full scale) is added. If the MSB is larger than the input, it is turned off when the next bit is turned on. The procedure is analogous to that used in placing weights on one side of a two-pan analytical balance. The process of comparison is continued down to the least significant bit (LSB) which is 2^{-n} of full scale for a converter of n bits. At this point the output register contains the complete output digital number which can be sent to a computer or other processing device. The bit-comparison operation can be done in as little as 100 nsec so that a 10-bit conversion (with an accuracy of 1 part in $2^{10} = 1024$) can be done in 1 μsec. In practice, conversion times of 10 to 20 μsec are typical for high-resolution converters which generally have a 10- to 16-bit digitizing capability.

CAPACITANCE MEASUREMENTS

The measurement of capacitance is of importance in physical chemistry primarily for the study of dielectric properties. As discussed in Exp. 36, dipole moments can be determined from the dielectric constants of gases or dilute solutions. There is also considerable interest in the dielectric properties of liquids and solids. When there are strong dipolar interactions between molecules, as in a polar liquid, there is an appreciable relaxation time for dielectric polarization. In that case, *dielectric loss* (dissipation of electrical energy as heat) occurs and the dielectric constant is frequency dependent.[8,9] In the following discussion we shall be concerned mostly with methods of measuring the capacitance of samples with very low dielectric loss (gases or dilute solutions in nonpolar solvents).[10] For such samples, the resistance of the dielectric cell is very high and its capacitance is independent of frequency.

Heterodyne-beat method The heterodyne-beat method is capable of extremely high precision, and it is the method most commonly used for dipole moment measurements. However, this method is limited to the investigation of gases or of solutions and liquids with very low conductance (cell resistance greater than 10^4 Ω). If a substance with appreciable conductance is placed in the dielectric cell, which is part of the LC tank circuit of a high-frequency oscillator tube, too much energy will be dissipated in the tank circuit and there will not be enough feedback to maintain proper oscillation. The heterodyne-beat method is fully described in Exp. 36, and considerable details on the circuitry are available in the literature.[10]

Resonance method While the resonance method is not capable of the very high accuracy of the heterodyne-beat method, it is quite satisfactory for many investigations and has the advantages of simple and low-cost circuitry. This method can best be explained in terms of the schematic diagram shown in Fig. 13. The primary circuit consists of an oscillator with the inductance coil L_1 in the plate circuit of the oscillator tube. The secondary circuit is loosely coupled via the inductances to the primary circuit. When the oscillator is in operation, a constant high-frequency alternating current flows through L_1 and a small amount of current also flows in the secondary circuit. The amount of current flowing in the secondary will depend both on the value of L_2 and on the total capacity $C = C_T + C_P + C_X$. By varying C one can achieve the *resonance* condition: $f = 1/2\pi\sqrt{L_2 C}$, where f is the frequency of the oscillator. When resonance occurs, the current flow in the secondary will be at its maximum value. This resonance point is detected by observing either a minimum in the dc plate current of the oscillator tube or a maximum in the ac voltage across the secondary. (The latter method requires the use of a vacuum-tube voltmeter.) A variable dielectric cell can be used (as in Exp. 36), or a fixed-plate cell can be used by employing the substitution method. In either case, C_X is obtained from the difference in the settings of C_P necessary to achieve resonance. The coarse tuning condenser is used only to adjust the total capacity C to an initial value near the resonance value and should *not* be disturbed during the actual measurements.

In the above discussion, the secondary circuit was treated as a pure LC network, and that is usually not true in practice. For dielectric-constant measurements this is of no concern, since there is always a unique value of C which will make the secondary circuit resonant. We can also see that the resonance method has one great advantage over the heterodyne-beat method: solutions or liquids with appreciable conductance can be studied. This is due to the fact that the dielectric cell is in the secondary circuit, which is only loosely coupled to the oscillator, and cannot affect the feedback necessary for proper oscillation. Indeed, it is possible to determine dielectric loss with this method, since the sharpness of resonance (Q of the secondary circuit) is related to the loss.[8] The lower the loss in the dielectric, the sharper the resonance and the larger the Q value.

A modified resonance apparatus for routine student use has been described in detail by Bender.[11] In this circuit a 6E5 electron-ray ("tuning-eye") tube serves both as an oscillator tube and as the resonance detector. A piezoelectric quartz crystal (1 or 2 MHz) in the grid circuit controls the frequency f of oscillation. The LC

FIGURE 13
Schematic circuit for a resonance apparatus: L_1 and L_2 are fixed inductances; C_T is a coarse tuning capacitor; C_P is a precision air capacitor; C_X is the dielectric cell. A vacuum-tube voltmeter V can be used as the detector (see text).

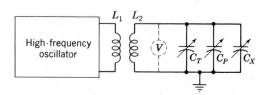

circuit (shown as the secondary circuit in Fig. 13) is part of the plate circuit and is coupled to the quartz crystal via the internal capacitance of the 6E5 tube. As the capacity in the plate tuning circuit is increased, oscillation begins as $1/2\pi\sqrt{L_2C}$ approaches f and the plate current drops. This continues up to the point of resonance where the plate current is at a minimum and therefore the shadow angle of the tuning eye is also at a minimum. With a further increase of C, oscillation stops, the plate current rises rapidly, and the shadow angle increases abruptly. This point where the eye "opens" is reproducible and can be used as the balance point for capacitance settings.

Bridge method An ac impedance bridge used for dielectric measurements is very similar in principle to the ac Wheatstone bridge described previously; however, the reactance X is now large, since the dielectric cell has a large capacitance. Both the theory and the proper experimental techniques of ac bridge measurements are extensively discussed by Hartshorn[12] and by Hague.[13] Only a very brief treatment of the "capacity bridge" will be given here.

The type of bridge often used for dipole moment work is a simple capacity bridge with two resistance arms (the "ratio arms") and two capacitance arms. Figure 14 shows a schematic diagram of a bridge with equal ratio arms of fixed resistance R and with one fixed capacitance arm C. The measuring capacitance arm consists of a parallel combination of a precision air capacitor C_P and the dielectric cell C_X. Large ($\sim 10^4$-Ω), noninductively wound resistors R_1 and R_2 are attached in parallel with the

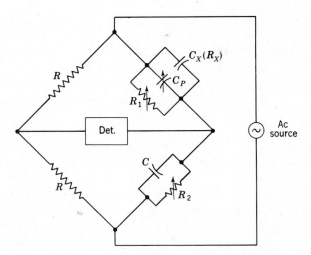

FIGURE 14
A simple capacity bridge. An oscilloscope is most useful as a detector.

capacitances. These resistors help to balance out any small differences between the resistances of the capacitance arms and are very necessary to achieve balance if the sample has appreciable conductance (i.e., if cell C_X has a finite resistance R_X).[10] Measurements can be made with a fixed-plate dielectric cell by using the substitution method, where a balance is first achieved with C_X in the bridge and then a second balance is achieved with C_X removed. Alternatively, a variable-capacity cell can be used as in Exp. 36; this has the advantage of eliminating any effects due to the capacity or resistance of the leads. If the sample has negligible conductance only an adjustment of the C_P setting is required to achieve balance. For samples having appreciable conductance (finite R_X), it is necessary to adjust both C_P and the resistors R_1 and R_2, alternately, until the balance point is reached. (The cell resistance R_X is then given by $1/R_X = 1/R_2 - 1/R_1$, and the dielectric loss can also be calculated.[8]) In either case, C_X is equal to the difference in the two settings of C_P required to balance the bridge.

For reasonably precise work it is advisable to use a commercial bridge in order to avoid the many difficult problems of design and construction involved. Several suitable types of bridge are manufactured by the General Radio Co., among them a modified Schering bridge (Model 716-C) which is capable of very high precision. In addition to a good bridge, a *stable* oscillator and suitable *high-impedance* detector are required.

REFERENCES

1. W. C. Michels, "Electrical Measurements and Their Applications," Van Nostrand, Princeton, N.J. (1957); A. W. Smith and M. L. Wiedenbeck, "Electrical Measurements," 5th ed., McGraw-Hill, New York (1948).
2. L. Page, "Principles of Electricity," 3d ed., Van Nostrand, Princeton, N.J. (1958).
3. Leeds and Northrup, "Technical Publication A12.1110: Moving-coil Galvanometers" is available from Leeds and Northrup Co., North Wales, Pa. 19454.
4. W. P. White, Potentiometers for Thermoelectric Measurements, in "Temperature: Its Measurement and Control in Science and Industry," pp. 265–278, Reinhold, New York (1941).
5. G. Jones and R. C. Josephs, *J. Amer. Chem. Soc.*, **50**, 1049 (1928); T. Shedlovsky, *J. Amer. Chem. Soc.*, **52**, 1793 (1930).
6. D. Edelson and R. M. Fuoso, *J. Chem. Educ.*, **27**, 610 (1950); G. Jones, K. T. Mysels, and W. Juda, *J. Amer. Chem. Soc.*, **62**, 2919 (1940).
7. R. A. Robinson and R. H. Stokes, "Electrolyte Solutions," 2d ed., pp. 88–95, Butterworth, London (1959); J. Braunstein and G. D. Robbins, *J. Chem. Educ.*, **48**, 52 (1971).
8. J. G. Powles and C. P. Smyth, Measurement of Dielectric Constant and Loss, in A. Weissberger (ed.), "Technique of Organic Chemistry," 3d ed., vol. I, part III, chap. XXXVIII, Interscience, New York (1959).
9. C. P. Smyth, "Dielectric Behavior and Structure," McGraw-Hill, New York (1955).

10. C. P. Smyth, Determination of Dipole Moments, in A. Weissberger (ed.), *op. cit.,* chap. XXXIX.
11. P. Bender, *J. Chem. Educ.*, **23,** 179 (1946).
12. L. Hartshorn, "Radio-frequency Measurements by Bridge and Resonance Methods," Wiley, New York (1941).
13. B. Hague, "Alternating-current Bridge Methods," 5th ed., Sir Isaac Pitman & Sons, London (1957).

GENERAL READING ON ELECTRONICS

H. V. Malmstadt, C. E. Enke, and S. R. Crouch, "Electronic Measurements for Scientists," Benjamin, Menlo Park, Calif. (1974).
A. J. Diefenderfer, "Principles of Electronic Instrumentation," Saunders, Philadelphia (1972).
S. A. Hoenig and F. L. Payne, "How to Build and Use Electronic Devices without Frustration, Panic, Mountains of Money, or an Engineering Degree," Little, Brown; Boston (1973).

XVI

TEMPERATURE

Temperature is one of the most important variables in thermodynamics and physical chemistry. In this chapter we are concerned with the methods and instruments that are used in measuring and controlling temperature.

TEMPERATURE SCALES

The nature of temperature scales has been briefly discussed in Exp. 1. The *thermodynamic temperature scale*, based on the second law of thermodynamics, embraces the Kelvin (absolute) scale and the Celsius scale, the latter being defined by the equation

$$t_{\text{Celsius}} \equiv T_{\text{Kelvin}} - 273.15 \tag{1}$$

The size of the Kelvin degree is defined by the statement that the triple point of pure water is exactly $273.16°K$. The practical usefulness of the thermodynamic scale suffers from the lack of convenient instruments with which to measure absolute temperatures routinely to high precision. Absolute temperatures can be measured over a wide range with the helium gas thermometer (appropriate corrections being

570

made for gas imperfections), but the apparatus is much too complex and the procedure much too cumbersome to be practical for routine use.

The *International Practical Temperature Scale* of 1968 (IPTS-68)[1,2] is basically arbitrary in its definition but is intended to approximate closely the thermodynamic temperature scale. It is based on assigned values of the temperatures of a number of defining fixed points and on interpolation formulas for standard instruments (practical thermometers) which have been calibrated at those fixed points. The fixed points of IPTS-68 are given in Table 1.

Between the triple point of hydrogen ($-259.34°C$) and $630.74°C$, the standard instrument is the platinum resistance thermometer. Unfortunately, the interpolation formulas are rather complicated. In the range below $0°C$, the basic formula is of the form

$$W = W_t^*(T_{68}) + \Delta W(T_{68}) \tag{2}$$

where W represents the ratio of the resistance of the platinum sensing element at a given temperature to its resistance at $0°C$, W_t^* is a single-valued reference function given numerically by a 20-term series,[1] and ΔW is a deviation function involving up to four constants which are determined by the measured deviations at various fixed points. Four different deviation functions are used in four distinct temperature ranges below $0°C$.[1] In the range from 0 to $630.74°C$, the situation is simpler. The value of t_{68} is defined by

$$t_{68} = t' + 0.045 \left(\frac{t'}{100}\right)\left(\frac{t'}{100} - 1\right)\left(\frac{t'}{419.58} - 1\right)\left(\frac{t'}{630.74} - 1\right) \tag{3}$$

Table 1 FIXED POINTS OF THE INTERNATIONAL PRACTICAL TEMPERATURE SCALE OF 1968[a]

Fixed point	Phases in equilibrium[b]	Assigned value	
		$T_{68}(°K)$	$t_{68}(°C)$
Hydrogen triple point[c]	Solid, liquid, vapor	13.81	-259.34
Hydrogen at $\frac{25}{76}$ atm[c]	Liquid, vapor	17.042	-256.108
Hydrogen boiling point[c]	Liquid, vapor	20.28	-252.87
Neon boiling point	Liquid, vapor	27.102	-246.048
Oxygen triple point	Solid, liquid, vapor	54.361	-218.789
Oxygen boiling point	Liquid, vapor	90.188	-182.962
Water triple point	Solid, liquid, vapor	273.16	0.01
Water boiling point	Liquid, vapor	373.15	100
Zinc freezing point	Solid, liquid	692.73	419.5
Silver freezing point	Solid, liquid	1235.08	961.938
Gold freezing point	Solid, liquid	1337.58	1064.43

[a] *Metrologia*, **5**, 35 (1969).
[b] Except for the triple points and the hydrogen point at $17.042°K$, all temperature values are for equilibrium states at one standard atmosphere (1,013,250 dynes cm^{-2}). Corrections can be made for small pressure deviations.
[c] Equilibrium mixture of ortho- and para-hydrogen.

where t' is determined from

$$W(t') = 1 + At' + Bt'^2 \tag{4}$$

As before, $W(t') = R(t')/R(0°C)$. The constants $R(0°C)$, A, and B are determined by calibration at the triple point of water, the boiling point of water, and the freezing point of zinc.

Between $630.74°C$ and the gold point $(1064.43°C)$, the standard instrument is a thermocouple consisting of a platinum wire and a wire which is a 10% rhodium-90% platinum alloy. The emf of this thermocouple defines the temperature t_{68} according to the formula

$$E(t_{68}) = a + bt_{68} + ct_{68}^2 \tag{5}$$

where $E(t_{68})$ is the electromotive force obtained when one junction is at t_{68} and the other is at $0°C$. The three constants are evaluated from the values of E at $630.74°C$ (as determined with a platinum resistance thermometer) and the freezing points of silver and gold.

Above the gold point, an optical pyrometer is used and IPTS-68 is defined by the Planck radiation law in the form

$$\frac{J_\lambda(T_{68})}{J_\lambda(T_{Au})} = \frac{\exp(c_2/\lambda T_{Au}) - 1}{\exp(c_2/\lambda T_{68}) - 1} \tag{6}$$

Table 2 PARTIAL LIST OF SECONDARY REFERENCE POINTS[a]

Fixed point	Phases in equilibrium[b]	Temperature $T_{68}(°K)$	$t_{68}(°C)$
Hydrogen[c]	Liquid, vapor	20.397	-252.753
Nitrogen triple point	Solid, liquid, vapor	63.148	-210.002
Nitrogen	Liquid, vapor	77.348	-195.802
Carbon dioxide	Solid, vapor	194.674	-78.476
Mercury	Solid, liquid	234.288	-38.862
Ice point	Ice, air-saturated water	273.15	0
Phenoxybenzene triple point	Solid, liquid, vapor	300.02	26.87
Indium	Solid, liquid	429.784	156.634
Bismuth	Solid, liquid	544.592	271.442
Lead	Solid, liquid	600.652	327.502
Mercury	Liquid, vapor	629.81	356.66
Sulfur	Liquid, vapor	717.824	444.674
Antimony	Solid, liquid	903.89	630.74
Copper	Solid, liquid	1357.6	1084.5
Platinum	Solid, liquid	2045	1772
Tungsten	Melting	3660	3387

[a] *Metrologia*, **5**, 35 (1969).
[b] See footnote b to Table 1.
[c] Normal hydrogen.

where J_λ is the emissivity per unit wavelength interval at wavelength λ and $c_2 = 1.4388$ cm deg.

Although the IPTS-68 conforms to the thermodynamic temperature scale much more closely than its predecessor (IPTS-48), the distinction between the two scales is not of great importance except for precise work.† In the earlier scale the simpler Callendar-van Dusen interpolation formula was used between the oxygen point and 0°C, and this may still suffice for many practical purposes. A detailed comparison of IPTS-68 with IPTS-48 has been given by Douglas,[2] who has also considered the conversion of older calorimetric data to new values based on IPTS-68.

In addition to the defining fixed points of the International Practical Temperature Scale (Table 1), a number of secondary reference points have been recommended for use (see Table 2).

TRIPLE-POINT CELL

Since the triple point of water is defined to be exactly 273.16°K on the thermodynamic temperature scale, this is an especially important fixed point. It is also a point which can be reproduced with exceptionally high accuracy. If the procedure of inner melting (described below) is used, the temperature of the triple point is reproducible within the accuracy of current techniques (about ± 0.00008°C). This precision is achieved by using the NBS-designed *triple-point cell* shown in Fig. 1. This cell, which is about 7.5 cm in outer diameter and 40 cm in overall length, has a well of sufficient size to hold all thermometers which are likely to be calibrated. Such cells can be purchased commercially or can be constructed if one pays careful attention to several details. After the glassblowing is completed, the inside of the cell should be cleaned thoroughly with chromic acid, rinsed copiously with distilled water, and steam-cleaned until the condensed water runs down the wall in a continuous film. The cell is then attached to a vacuum line, pumped down, and filled to the desired level by distillation. The water used in this vacuum distillation should be previously distilled water from which as much dissolved gas as possible has been removed. The cell is then sealed off above the side tube which serves as a support when the apparatus is hung inside a Dewar flask.

The procedure for using a triple-point cell is as follows. Cool the cell by placing it in a bath of crushed ice and distilled water. Fill the thermometer well with powdered Dry Ice in order to form a clear mantle of ice around the well. When this mantle is about 0.5 cm thick, remove the Dry Ice and place a warm tube in the well just long

---

† The difference ($T_{68} - T_{48}$) varies between 5 and 75 mdeg for the range 90–700°K.[2]

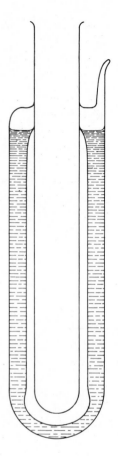

FIGURE 1
Triple-point cell for water (NBS design).
Note the clear mantle of ice formed
around the inner well.

enough to free the mantle and provide a thin layer of liquid next to the wall. If the
mantle is free, it will rotate about the well when the cell is given a quick twist about the
vertical axis. This "inner melting" provides an additional step of purification. As
the water freezes to form the mantle, impurities are left behind in the liquid near the
outer wall. When the mantle is partly melted, the water near the inner well is exceptionally pure and this water does not mix rapidly with the water outside the mantle.
The equilibrium between the inner water layer, the ice mantle, and the water vapor in
the top of the cell occurs at the defined triple-point temperature. In order to ensure
good thermal contact between the wall of the well and the thermometer being calibrated, the well is filled with ice water. It is also wise to precool the thermometer in
an ice bath before it is placed in the well.

THERMOMETERS

Mercury thermometers[3]   By far the most common type of laboratory thermometer is the mercury thermometer, based on the differential volume thermal expansion of liquid mercury (about 1.8×10^{-4} deg^{-1}) and glass (about 0.2×10^{-4} deg^{-1}). In the manufacture of such thermometers, the stem, a capillary of uniform bore, is usually marked at two points (say 0 and 100°C) and then graduated uniformly in between, on the tacit assumption that the volume of a fixed mass of mercury in glass is a linear function of the temperature. The error resulting from this assumption (about $+0.12°$ at 50°C for a 0 to 100° mercury thermometer made with Corning normal thermometer glass) is ordinarily smaller than that due to variations in the bore of the capillary.

Laboratory thermometers are commonly available in two types: solid stem and enclosed scale. The former has a stem of solid glass, with a scale engraved on the outside surface; the latter has a slender capillary and a separate engraved scale, both enclosed in an outer glass shell. The enclosed-scale type is preferable for thermometers with extremely fine threads largely because it suffers less from parallax in reading.

Since the mercury in the thread, as well as that in the bulb, is susceptible to thermal expansion, it is important in precise work to take account of the temperature of the thermometer stem. Most thermometer calibrations, especially those for enclosed-stem types, are for *total immersion*—it is assumed that the thread is at the same temperature as the bulb. Other thermometers are meant to be used with partial immersion, often to a ring engraved on the stem, and the remainder of the stem is assumed to be at room temperature (say 25°C). For precise work stem corrections should be made if the stem temperatures differ significantly from those assumed in the calibration. The correction that should be *added* to the thermometer reading is given by the equation

$$\Delta t_{corr} = -0.00016(\Delta t_{stem})(\Delta L_{stem}) \tag{7}$$

where Δt_{stem} is the amount by which the temperature of the stem *exceeds* that for which the calibration applies (or the amount by which it exceeds the thermometer reading itself, if the calibration is for total immersion) and ΔL_{stem} is the length of mercury thread, expressed in degrees Celsius, for which the temperature is different from that assumed in the calibration. To obtain Δt_{stem} a second thermometer may be positioned near the first, with its bulb near the midpoint of ΔL_{stem}.

For most purposes a partial-immersion thermometer need not be stem-corrected because of a few degrees variation in room temperature or few degrees error in the immersion level. On the other hand, it is usually worthwhile to apply stem corrections to readings of a total-immersion thermometer when used in partial immersion, particularly when reading temperatures well removed from room temperature.

Other important sources of error in mercury thermometers are parallax and sticking of the mercury meniscus. The first can be largely avoided by careful positioning of the eye when reading or use of a properly designed attached magnifier. It can be eliminated entirely by use of a cathetometer (see Chap. XVIII). Sticking of the mercury meniscus is due to the fact that the contact angle of mercury to glass (see Fig. 30-5) varies depending on whether the mercury surface is advancing or receding, and thus the capillarity pressure due to surface tension is variable. This combines with the small but finite compressibility of the mercury and the elasticity of the glass to yield a small variability in meniscus position, especially for very sensitive thermometers. Gentle tapping of the stem before taking readings usually leads to reproducible results. Readings of sensitive mercury thermometers are also slightly pressure dependent; pressure coefficients may be as high as 0.1 deg atm^{-1}.

When precision of better than about 1 percent of full scale is required, it is necessary to use a calibration chart which gives corrections to be added to or subtracted from the readings. Detailed procedures for calibrating liquid-in-glass thermometers are given by Swindells.[4] The simplest method involves comparison of the thermometer to a standard thermometer† which is itself well calibrated. The two thermometers are mounted next to each other in a well-stirred thermostat bath. The standard thermometer is immersed as specified in its calibration (usually to the top of the thread) and the test thermometer is immersed to the depth at which it will be used subsequently. The readings should be compared at several temperatures within the range of interest. Prior to this calibration procedure, the standard thermometer should be checked. Although it is best to use a triple-point cell (see p. 573), an ice bath made from finely divided ice and distilled water (see p. 588) can also be used. If necessary, a constant additive correction can be made in the calibration table for the standard thermometer. The best calibration method is to calibrate the test thermometer against a platinum resistance thermometer, since such thermometers have exceptional reproducibility and long-term stability. The calibrations of mercury thermometers should be checked at a single point from time to time, as significant irreversible changes may take place in the glass, particularly if the thermometer is used above 150°C.

Special thermometers are made for calorimetric work, where it is desired to measure very accurately (to 0.01 or even 0.001°) a temperature *difference* of the order of a few degrees. For these thermometers the fineness of scale graduation has little to do with the accuracy with which the thermometer measures a single temperature. The scale may be in error by several tenths of a degree, but this error cancels out in taking differences. A typical thermometer for bomb calorimetry has a range of 19 to

† In the past, thermometers could be submitted to the National Bureau of Standards for calibration or could be purchased with a calibration certified by the NBS. This is no longer possible, but well-calibrated thermometers can be obtained commerically.

35°, with graduations of 0.02°. For measuring freezing-point depressions with water or benzene as solvent, a range of -2 to $+6°$ with graduations of 0.01° is convenient. Such thermometers require careful handling. Not only are they relatively fragile, but they are susceptible to certain malfunctions (separation of the mercury column, bubbles in the bulb) arising principally from the extreme fineness of the thread. Whenever possible keep these thermometers upright; *avoid overly rapid heating or cooling.* If the mercury thread separates (which often happens when thermometers are shipped), cool the bulb in an ice–salt mixture to bring the mercury entirely into the bulb and tap if necessary to bring any bubbles to the top of the bulb. Then allow the thermometer to warm to room temperature in an upright position.

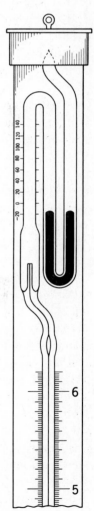

FIGURE 2
Upper end of Beckmann thermometer.

Beckmann thermometer   The Beckmann thermometer is a *differential* thermometer with a range of 5 or 6° and graduations of 0.01 or 0.02°. It differs from an ordinary thermometer principally in having a provision for changing the position of this 5 or 6° interval on the temperature scale by adjusting the amount of mercury in the bulb and thread. For this purpose a special reservoir is provided at the top of the thermometer (see Fig. 2). To raise the setting, the bulb is warmed so as to discharge droplets of excess mercury into this reservoir. To lower the setting, the bulb is warmed to bring the mercury thread into the bottom of the reservoir, where it is then united with a column of excess mercury which is brought into position by inverting the thermometer. The bulb is cooled to draw mercury into the thread. At a few degrees above the desired range the excess mercury is broken away from the thread by a sharp tap with the fingers. The thermometer is then brought to an upright position in such a way as to bring the excess mercury to its normal position in the reservoir.

Since the cubical expansion coefficient of mercury is not exactly constant, the magnitude of the true temperature difference represented by a scale division of the Beckmann thermometer varies with the temperature at which the thermometer is set to operate; accordingly, corrections are required in precise work. For a Beckmann thermometer made of Jena 16 glass (which may be taken as representative of German-made Beckmann thermometers), having a calibration which is valid when the midpoint of the scale is at 20°C, a temperature difference measured at temperature t can be corrected by multiplication with the factor $1 + 3 \times 10^{-4}(t - 20) - 10^{-6}(t - 20)^2$.

The Beckmann thermometer is very fragile and often subject to difficulties in use. Its use is best reserved for special purposes not served by available fixed-scale thermometers. Fixed-scale thermometers are ordinarily preferable for calorimetric and cryoscopic work in the usual temperature ranges.

Special liquid thermometers   For temperatures below the freezing point of mercury (−39°C), one can use several kinds of liquid-in-glass thermometers. An amalgam of 8.5 percent thallium in mercury remains liquid down to −60°C. Toluene thermometers may be used down to −95°C, and pentane thermometers will operate as low as −130°C. Mercury thermometers constructed from special glasses may be used far above the normal boiling point of mercury (357°C). However, outside the ordinary mercury range it is usually more convenient, as well as more accurate, to use thermometric devices of other types, especially thermocouples or resistance thermometers.

Gas thermometers   Although the attainment of high precision with gas thermometers entails highly complex apparatus and procedures,[5] simple gas thermometers resembling that of Exp. 1 may be used conveniently for moderately precise measurements, especially at low temperatures (see, for example, Exp. 49).

Bimetallic-strip thermometers   The direct linear expansion of a metal bar could be used as the basis of a thermometer, but very accurate measurements with a cathetometer would be required to measure small extensions of 10^{-4} to 10^{-3} cm. A bimetallic strip (two dissimilar metal strips bonded together) undergoes differential expansion. At some arbitrary temperature, the two metals are of equal length and the strip is flat. As the temperature changes one metal will expand or contract more than the other, causing the strip to bend or curl up. The temperature is indicated by a pointer activated by this motion of the strip. Although such thermometers are only accurate to about 1 percent, they are cheap, mechanically rugged, and capable of operating over a wide range. Bimetallic strips are also useful as on-off thermoswitches.

Platinum resistance thermometers[6]   The platinum resistance thermometer is capable of extremely high precision, owing to the high purity attainable for platinum and the high reproducibility of its temperature coefficient of resistivity. The resistance element is a coil of pure platinum wire, carefully annealed both before and after winding and enclosed in a tube (usually of glass) containing dry air or helium as a heat-transfer gas. Two leads are ordinarily attached to *each* end of the coil, in order to permit the resistance of the coil to be measured independently of the resistance of the leads. Platinum resistance thermometers in glass housings containing dry air may be used from -183 to $500°C$ or with a special glass housing to $630.5°C$.

The coil resistance is determined by means of a Wheatstone bridge of appropriate design, such as the Mueller type, which is shown schematically in Fig. 3. It will be observed that the thermometric resistance R_t is in the BD arm of the bridge irrespective of the choice of the two positions of the commutator switch. However, the effective position of the junction D with respect to the leads can be switched to either end of the coil (D_e or D_f). For the switch setting shown in Fig. 3

$$R + R_E = R_t + R_F \tag{8}$$

and for the other setting

$$R' + R_F = R_t + R_E \tag{9}$$

where R and R' are the settings of the precision decade resistance at galvanometer balance and R_E and R_F are the resistances of the respective leads. Clearly the thermometric coil resistance is given by

$$R_t = \frac{R + R'}{2} \tag{10}$$

The temperature coefficient of resistivity of platinum is about 0.00392 deg^{-1} at $0°C$. A thermometer with a coil of 25.5 Ω will show an increase of about 0.1 Ω deg^{-1} at that temperature. To determine the temperature to within $\pm 0.001°C$, it is

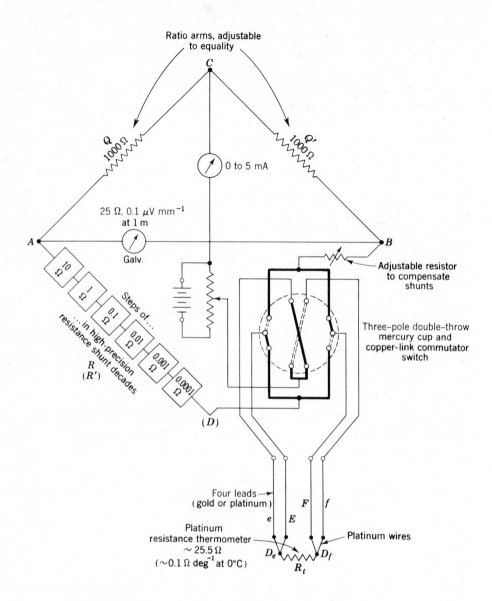

Ratio arms, adjustable
to equality

C

Q 1000 Ω

Q' 1000 Ω

0 to 5 mA

25 Ω, 0.1 μV mm^{-1}
at 1 m

A

Galv.

B

10 Ω

1 Ω

Steps of ...

0.1 Ω

0.001 Ω

... in high precision
resistance shunt decades

0.001 Ω

R
(R')

0.0001 Ω

(D)

Adjustable resistor
to compensate
shunts

Three-pole double-throw
mercury cup and
copper-link commutator
switch

Four leads
(gold or platinum)

F f

e E

Platinum
resistance thermometer
$\sim$ 25.5 Ω
($\sim$0.1 Ω deg^{-1} at 0°C)

Platinum wires

D_e

D_f

R_t

FIGURE 3
Simplified schematic diagram of platinum resistance thermometer and Mueller-
type bridge. Heavy lines in BD arm represent leads of negligible or balanced
resistance.

necessary to measure the resistance R_t to within $\pm 10^{-4}\ \Omega$. With a bridge current of 2 mA (1 mA per arm) a galvanometer with a coil resistance of about 25 Ω must be sensitive to 0.033 μV, or 0.0013 μA. At a sensitivity of 0.1 μV mm^{-1} at 1 m, which is attainable with commercially available galvanometers with this resistance, a deflection of 0.33 mm at 1 m must be observable; this is close to the limit of practical detectability. The deflection can be increased by increasing the bridge current, but there is then the disadvantage of increased ohmic heating of the thermometer coil. Under the conditions cited, about 5 μcal sec^{-1} are dissipated by that coil. Error from this source can be corrected by making measurements at different bridge currents; it is common practice to make measurements at 3 and 5 mA and to extrapolate to zero current.

Although a 25-Ω platinum resistance thermometer with a commercially available Mueller bridge (e.g., the L & N type G-2, Catalog No. 8069) and galvanometer (e.g., L & N type HS, Catalog No. 2284-d, 25-Ω coil, 0.1 μV mm^{-1} at 1 m) is capable of measuring temperatures with a reproducibility of 0.001 or 0.002°, its uncertainty in measuring on the International Practical Temperature Scale is usually somewhat larger than that in practice (about 0.01°), being determined largely by the limits within which the calibration temperatures can be established under ordinary laboratory conditions.

Thermistors Thermistors are semiconductor devices consisting of sintered mixtures of metallic oxides (such as NiO, Mn_2O_3, Co_2O_3). These ceramic-like resistors have large negative temperature coefficients of resistivity; the variation of the resistance R can be roughly approximated by

$$R = R_\infty e^{\Delta E/kT} \tag{11}$$

where ΔE is the electron energy gap. For a typical thermistor, $\Delta E/k \approx 3500$ deg which means a change in resistance of 4 percent deg^{-1} at room temperature. Thus, the magnitude of $R^{-1}(dR/dT)$ is about 10 times that of metallic resistors and high precision as a resistance thermometer can be attained with a less sensitive bridge than is required for a platinum thermometer. In addition, the resistivity is so high that thermistor elements can be very small without introducing serious complications due to lead and contact resistances. An important advantage of the small size of thermistors is their low heat capacity and rapid temperature response. Two disadvantages of thermistors in comparison with a platinum thermometer are their lower reproducibility and their poorer long-term stability. The resistance of most thermistors will change measurably with age and with temperature cycling. The type consisting of a small bead sealed in a glass probe (about 0.25 × 1.3 cm) is the most stable.[7] Such devices, which can be obtained commercially† in a variety of resistances from

† Fenwal Electronics, Inc., Framingham, Mass., and Thermistor Corp., Metuchen, N.J.

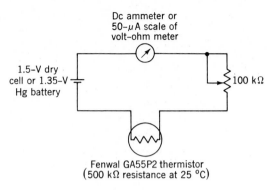

FIGURE 4
Simple circuit for measuring temperature with a thermistor. The thermistor shown here is suitable for measurements in the 75–100°C range. For other temperature ranges, thermistors with different room-temperature resistances would be selected in order to obtain an operating current of ~50 μA.

Dc ammeter or
50-μA scale of
volt–ohm meter

1.5-V dry
cell or 1.35-V
Hg battery

100 kΩ

Fenwal GA55P2 thermistor
(500 kΩ resistance at 25 °C)

10^3 to 10^4 Ω, have a useful temperature range from 0 to 300°C. Thermistors are useful for both temperature measurement and control.

When using thermistors for temperature measurement, the resistance can be determined with a simple Wheatstone bridge circuit (see Chap. XV), with a special bridge circuit for which the reading is conveniently linear in temperature,[7] or by applying a constant potential across the device and measuring the current which flows through the circuit. The latter method is less accurate but very convenient, as shown by the circuit diagram in Fig. 4. In this type of circuit, the thermistor will be heated by the current flow to the extent of approximately 1°C/mW of power. To avoid serious error, this heating effect must be negligible compared with the desired accuracy of the temperature measurement. For the circuit illustrated in Fig. 4, the *maximum* power dissipation in the thermistor is about 0.075 mW, corresponding to a temperature rise of less than 0.1°. The operating procedure is as follows: The thermistor probe is heated to the highest temperature to be measured, and the variable resistance is set to obtain a full-scale deflection on the 50-μA meter. Thereafter, this resistance setting must not be changed since any change requires recalibration of the circuit. The thermistor circuit is then calibrated against a standard thermometer in a water or oil bath to obtain a plot of temperature versus current. This calibration curve should be the average of two sets of readings, one taken before measurement of the experimental temperature and a second taken after the measurements are complete.

Thermocouples[8] Thermocouples provide one of the most convenient means of measuring temperatures over a wide range from very low ($-250°C$) to very high (1700°C) values. All that is required is a pair of wires made from two suitable metals or alloys, a good potentiometer, and a constant-temperature bath (usually an ice-water bath) for the reference junction.

When two dissimilar metals are placed in contact, a transfer of electrons from one to the other takes place and a double charge layer forms at the junction surface. As in the case of a junction between a metal electrode and an electrolyte solution, the resulting electric potential must be measured in the presence of a reference junction, which in the present instance is a junction of the same two metals at a known temperature. The thermoelectric potential is measured between the ends of two wires of the same metal—one wire leading to the junction at the reference temperature and the other to the junction at the unknown temperature; the two junctions are connected directly by a wire of the second metal completing the circuit. This potential is a measure of the unknown temperature, and in cases where the temperature difference is not large, it is roughly proportional to the difference between the unknown and reference temperatures.

The most useful thermocouple for general work is *copper-constantan*, the latter being an alloy of 60 percent copper and 40 percent nickel which is often sold under the trade name Advance. An important advantage of the use of this couple is that one of the metals is copper, which is the metal generally used for the binding posts on galvanometers and potentiometers. Thus, any stray thermoelectric potential caused by temperature differences between the contacts of the thermocouple wires with the binding posts is eliminated. The thermoelectric potential of a copper-constantan couple with the reference junction at 0°C has a temperature coefficient which varies from about 20 μV deg^{-1} at -200°C to 60 at 300°C. At room temperature, it is 43 μV deg^{-1}. This couple cannot be used for long above 500°C.

Another commonly used couple is Chromel P and Alumel, these two being iron-nickel alloys containing chromium and aluminum, respectively. The coefficient is about 40 μV deg^{-1} over a very wide range, and this thermocouple can be used at temperatures as high as 1300°C. *Iron-constantan*, with a somewhat higher coefficient (53 μV deg^{-1} at room temperature), can be used to nearly 1000°C. Platinum together with an alloy of 10 percent rhodium in platinum has a small coefficient (6 μV deg^{-1} at room temperature) but can be used as high as 1700°C.

Although extensive tables are given in various handbooks for converting measured emfs to temperatures, the best tables are those published by the National Bureau of Standards.[9] In any case, thermocouple wires should be carefully selected, and one or more specimens of each lot should be calibrated at a number of temperatures. For very precise work each thermocouple should be individually calibrated.

Junctions which are to be used only at low temperatures may be joined with soft solder (be sure to use a noncorrosive flux such as rosin), but it is better to weld the two metals together with an electric arc or an oxygen-gas flame. The two wires are stripped of any insulating material and then twisted together tightly over a distance of ~0.5 cm. If a glass-blowing torch is used, the end of the junction should be placed in the reducing region at the tip of the blue part of the flame. Remove the wires as soon as the metals melt and form a small bead at the end. This bead should be pinched with a pair of pliers and examined carefully to make sure it is hard and metallic rather than a bead of oxide (which crumbles easily). The two wires must be electrically insulated from each other, and some consideration must be given to the temperature characteristics of the insulating materials. In particular, at high temperatures nothing should come into contact with the wires (particularly Chromel P and Alumel) which will form a liquid flux (low-melting eutectic) with the protective oxide film. Wires with Fiberglas insulation in place of the usual cloth or enamel can be obtained from the Hoskins Manufacturing Company of Detroit and are usable up to 450°C. Fiberglas "spaghetti" (sleeving) can also be obtained. Bare wires and Alundum tubes or spacers are necessary above 450°C. The junctions and wires should always be protected carefully from corrosion or mechanical damage. For use in liquids or solutions, a thermocouple is usually placed in a closed glass tube (6 to 8 mm in diameter) with oil at the bottom to improve the thermal contact.

A satisfactory environment for a 0°C reference junction is provided by a slushy mixture of ice and distilled water in a Dewar flask, with a ring stirrer and a monitoring mercury thermometer.

In very rough work with large-gauge thermocouple wires the thermoelectric emf can be measured directly with a millivoltmeter or galvanometer possessing an internal resistance large in comparison with the thermocouple wires. For ordinary work a Student Potentiometer and an enclosed-scale or pointer-type galvanometer will suffice. (The Student Potentiometer should be set on the 0.01 scale; see Chap. XV.) In order to obtain a continuous record of the temperature, a thermocouple can be connected to a potentiometric strip-chart recorder. For very precise work a potentiometer designed for use in the fractional microvolt range is essential. The Wenner potentiometer (L & N Catalog No. 7559) is such an instrument; it has a low range of 0 to 0.01 V in steps of 0.1 μV and a limit of error of 0.01 percent plus 0.5 μV. Thus, with a copper-constantan thermocouple (with one measuring and one reference junction) it permits measurement with a limit of error of about 0.01° near room temperature; however, temperature differences can be measured to 0.002°.

An increase in sensitivity can be obtained by the use of a multijunction thermocouple, employing several measuring and reference junctions in alternating sequence (see Fig. 5). This gain is purchased at the price of increased complexity, particularly in regard to problems of thermal contact and electrical insulation. These problems can be solved fairly easily in the neighborhood of room temperature, and multijunction thermocouples are sometimes employed in calorimetric work.

Optical pyrometers[10] The optical pyrometer can be used for the measurement of temperature above 600°C, where blackbody radiation in the visible part of the spectrum is of sufficient intensity. The blackbody emissivity at a given wavelength in equilibrium with matter at a given temperature is given by the Planck radiation law

$$J_\lambda = \frac{c_1}{\lambda^5} \frac{1}{e^{c_2/\lambda T} - 1} \tag{12}$$

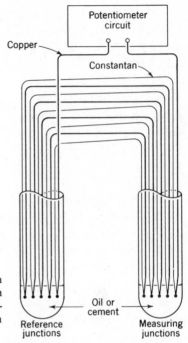

FIGURE 5
Schematic diagram of multijunction copper-constantan thermocouple. (In practice, the junctions must be positioned so that they do not touch each other and cause a short circuit.)

where J_λ is the blackbody emissivity per unit wavelength interval and c_1 and c_2 are numerical constants ($c_2 = hc/k = 1.4388$ cm deg). The optical pyrometer is essentially a photometer which measures the emissivity of the source in a given wavelength interval. We shall here consider only the most common type of optical pyrometer, the "disappearing-filament" type. The object whose temperature is to be measured is viewed through a telescope containing, at an image plane, a lamp with a carbon or tungsten filament, the temperature of which can be varied by adjustment of the current through it. A red filter in the eyepiece selects a narrow-wavelength range for visual observation. When the current has been adjusted so that the filament becomes invisible against the background, the temperature is obtained by referring the filament current to a calibration curve or chart.

Fixed points for the calibration of the optical pyrometers are the silver and gold points and higher secondary fixed points such as those given in Table 2. Calibration at other temperatures can be accomplished by use of a rotating sector or a filter of accurately known transmission factor between the fixed-point source and the pyrometer, in order to simulate a source of lower temperature in accordance with the Planck equation. Such sectors or filters are used also to permit the optical pyrometer to be used for the measurement of temperatures much above 2000°.

Optical pyrometer measurements are most reliable when the object being examined is the interior of a furnace or cavity of uniform temperature viewed through a small opening. Readings for an exposed surface are somewhat dependent upon the emissivity coefficient of the substance concerned, which for an ideal "blackbody" is unity and for actual materials is less than unity. The emissivity coefficient in the visible range is near unity for carbon and oxidized metals and considerably less for platinum and other unoxidized metals, especially when polished. Under the best conditions the optical pyrometer is accurate to about 0.2 percent of the absolute temperature. It is also a convenient instrument for less precise measurements at high temperatures, such as the routine measurement of furnace temperatures, etc.

Other thermometric devices Any physical property that depends sensitively on temperature may, in appropriate circumstances, be used to measure temperature. The vapor pressure of a liquid or a solid is such a property; the use of liquid nitrogen is suggested in Exp. 49. At very low temperatures, the vapor pressures of liquid helium (1 to 4.2°K) and of solid and liquid hydrogen (9 to 20°K) are used. Between 1 and 20°K, carbon resistance thermometers are widely used.

Excellent precision can be achieved over quite a wide range with so-called "quartz thermometers." The temperature dependence of the resonance frequency of a quartz crystal depends on the orientation of the crystal. The frequency of a temperature-sensitive crystal is beat against that of a crystal which is essentially independent of temperature. By counting the beat frequency with an electronic counter, a resolution of 10^{-4} deg is possible.

TEMPERATURE CONTROL

In addition to the measurement of temperature, it is often necessary to maintain a constant temperature. The importance of this type of control in experimental physical chemistry is illustrated by the fact that 28 of the 50 experiments described in this book require temperature control of some kind. Many physical quantities (such as rate constants, equilibrium constants, emfs, osmotic pressures) are sensitive functions of temperature and must be measured at a known temperature which is held constant to within $\pm 0.1°C$ or better. Certain physical techniques are even more demanding; for example, the use of a dilatometer in Exp. 24 requires that the temperature be controlled to within $\pm 0.002°C$.

The simplest method of achieving a constant temperature is to maintain an equilibrium (at constant pressure) between two phases of a pure substance or among three phases in a two-component system. The greatest disadvantage of this method is that one cannot attain a desired *arbitrary* temperature unless a liquid-vapor system is used with a complicated manostat to maintain an arbitrary boiling pressure. In addition, it is often difficult to maintain temperature control for very long periods of time with this method. There are, however, several advantages: economy and simplicity in operation, excellent temperature stability, and potentially high precision in the absolute temperature of the bath. Several commonly used systems of this type are listed below, together with brief comments on their use.

Liquid-nitrogen bath   Liquid nitrogen at its normal boiling point (77.3°K, −195.8°C) provides a very convenient low-temperature bath. Since O_2 dissolved in liquid N_2 will raise the temperature of the bath, the mouth of the Dewar flask should be plugged loosely with glass wool or cotton to retard the slow condensation of atmospheric oxygen. The temperature of a liquid-nitrogen bath can be calculated on the assumption that the nitrogen vapor pressure is equal to the atmospheric pressure. Clearly, temperature stability will depend on the absence of large changes in pressure; fortunately, the boiling point of nitrogen changes only 0.013°/Torr change in pressure. This bath is frequently used for freezing out vapors in a trap, especially in high-vacuum applications.

Dry Ice bath   Solid carbon dioxide in equilibrium with CO_2 vapor at 1 atm will provide a temperature of −78.5°C. Thus, Dry Ice, which is inexpensive and readily available, would seem to be very suitable for a constant-temperature bath at moderately low temperatures. Unfortunately, the use of Dry Ice alone is complicated by two difficulties: the problem of obtaining and maintaining the proper pressure of CO_2 gas and the problem of achieving good thermal contact between the Dry Ice and the object to be cooled. Although these difficulties can be overcome by careful bath design and the use of a heater to cause a constant evolution of CO_2 gas, it is much easier to use a bath consisting of Dry Ice and a liquid which does not freeze at −78.5°C. This liquid provides good thermal contact throughout the bath and prevents air from diluting the CO_2 gas at the surface of the Dry Ice as it would at an exposed Dry Ice surface. The traditional choice of liquid has been acetone, but acetone is flammable and care must be taken in the vicinity of any open flames. A recommended nonflammable liquid is trichloroethylene. In making up a Dry Ice bath, it is necessary to minimize the foaming which occurs owing to rapid evolution of gas when Dry Ice is placed in contact with liquid initially at room temperature. The Dry Ice should be pulverized. If a special grinder is not available, one can wrap chunks of Dry Ice in a towel and pound them with a mallet. Be careful in handling pieces of Dry Ice; it can cause painful "burns" if held in the bare hand for more than a few seconds. The use of tongs or insulated gloves is strongly recommended. A Dewar flask is first filled about two-thirds full with trichloroethylene or acetone, and then small quantities of very finely powdered Dry Ice are added **slowly**

with a spatula. This Dry Ice will evaporate almost immediately, and there will be considerable foaming at the surface. Add more Dry Ice only after the foaming has subsided. After a while the liquid will have cooled to the point where the evaporation of Dry Ice is much slower, and some Dry Ice will begin to accumulate on the bottom of the Dewar. At this point, small lumps of Dry Ice can be added without causing serious foaming. Good temperature control with this bath is ensured only if it is well stirred and there is a slow but steady stream of CO_2 bubbles rising from the bottom. **Caution:** Acetone is flammable and should not be used in the vicinity of any open flames. Trichloroethylene is reputedly a weak carcinogen.

Ice bath   The ice-water equilibrium at 0°C provides an excellent constant-temperature bath. The ice should be washed, and distilled water must be used. To avoid thermal gradients between water at 4°C (maximum density) at the bottom of the Dewar and a 0°C liquid surface in which ice is floating, it is usually necessary to stir this bath. As a bath for the reference junction of a thermocouple, gradients can be eliminated by completely filling the Dewar with ice and adding only a small amount of cold distilled water; the weight of ice above the liquid will force ice down to the very bottom of the Dewar.

Sodium sulfate bath   Pure sodium sulfate decahydrate can be decomposed to a mixture of the monohydrate and a saturated aqueous solution by gentle heating in a warm water bath until the temperature (as shown by a mercury thermometer) begins to rise above 32.4°C. Additional decahydrate can then be stirred in, and the container insulated (or the mixture transferred to a Dewar flask). The mixture will maintain a temperature of 32.38°C as long as the two solid phases and one liquid phase are present in equilibrium.

Vapor baths   Boiling acetone (56.5°C), water (100°C), naphthalene (218.0°C), and possibly sulfur (444.6°C) can be used as vapor baths. In each case, the object whose temperature is to be controlled is immersed in the refluxing or condensing vapor. See Exp. 1 for details of a steam bath.

The second important method of achieving temperature control is to use a thermosensing element with a feedback system to control the input of heat (or of refrigeration) to a bath so that the temperature is maintained close to any desired arbitrary value. The thermosensing device may be any thermometric instrument that provides an electrical signal, such as a thermocouple or resistance thermometer. Or it may be a mercury thermometer equipped with an auxiliary device for converting its reading into an electrical signal; commonly a wire contact is fixed above the mercury meniscus to provide a simple "off" or "on" electrical signal depending on whether the temperature is below the desired level or above it. The signal is received by a thermoregulating circuit which switches or varies the power to the heating (or refrigerating) element so as to correct the temperature deviation detected by the thermosensing element.

As in any system employing feedback to maintain a steady-state condition, certain design criteria must be met in order to obtain reasonably rapid response to environmental changes while avoiding excessive "hunting" or even uncontrolled oscillations. The performance of the system will depend upon such factors as the sensitivity and speed of response of the thermosensing element, the circulation of heat in the system (stirring, convection), and the fraction of the total energy input (heat input plus stirring work, etc.) that is being controlled by the thermoregulator. This fraction should be no larger than is required to accommodate the expected variation in heat loss to the surroundings.

To provide damping of oscillations, a "proportionating" circuit is often employed; in the vicinity of the desired temperature the current or power to the heater is made roughly proportional to the difference between the actual temperature and a temperature setting which is slightly above the desired value. Such a circuit usually has an adjustment for providing an optimum range of proportionation, since a proportionating system provides stability at the expense of some precision of temperature control. Many on-off systems provide some accidental proportionation in the form of rapid cycling of the thermosensing element due to mechanical vibration, etc.

For automatic temperature measurement thermocouples are often used with recording potentiometers; to achieve temperature control a proportionating circuit, controlling a variable transformer with a servo system, can be connected directly to the automatic potentiometer. Analogous systems are available for platinum resistance thermometers. It is also possible to use a resistance thermometer in an ac bridge and to use the magnitude and phase of the unbalance signal to control a relay or Thyratron.[11]

Water baths The water bath, equipped with stirrer, thermoregulator, control circuit, and heater, provides the most commonly required means of temperature regulation. A good water bath for general use consists of a large rectangular tub perhaps 18 by 36 in. in horizontal area and 18 in. deep, with a water capacity of about 170 liters, constructed of welded stainless steel; glass windows in two or more sides are convenient. Thermoregulators and stirrers are best mounted in the middle, leaving the ends free for experimental work. There should be adequate provision for mounting rods and clamps to support flasks, electrochemical cells, etc. Depending on the experiment, two or four pairs of students can work in a single bath of this size. The bath should be provided with the following:

Stirrer: A centrifugal water circulator, driven by a $\frac{1}{20}$-hp motor, provides adequate stirring. It should be positioned carefully so that the effluent stream will cause efficient circulation through the entire tank.

Heater: A single-blade type heater of 250-W capacity is adequate for temperature regulation up to about 30°. At higher temperatures a second one may be added. It is advisable to position the heater in the effluent stream from the stirrer. If two heaters are employed, one may be intermittent under the control of the thermoregulator and the other under power constantly or both may be intermittent, depending on their power ratings, the temperature to be maintained, and other factors.

Thermosensing element: Thermocouples, thermistors, and platinum resistance thermometers, apart from the required control circuitry, need no further discussion. Most thermosensing devices for laboratory temperature control are of the off-on thermoswitch type. Thermoswitches based on the making and breaking of contacts by bimetallic strips are useful where long-term reliability and regulation to better than about 0.5° are not required. A commonly used type of thermosensing element is a thermometer-like device in which an electrical contact is made by a platinum or other wire to a mercury meniscus in a capillary when the temperature has attained the desired value. The liquid in the bulb may be entirely mercury, or it may be in larger part a liquid such as toluene with a higher thermal expansion coefficient, and the wire contact may be fixed or adjustable. Completely sealed devices for one or several fixed temperatures or with provision for adjustment of the amount of mercury in the thread are available commercially. Unfortunately they are often unsatisfactory; their bulbs are usually too small to give the required sensitivity, their capillaries are small enough to give trouble because of breaking of the thread, and there is eventual trouble from accumulation of "dirt" at the contact due to sparking. The most reliable devices are probably homemade ones, which can be cleaned and adjusted from time to time. A suitable design is shown in Fig. 6. It contains about 200 ml of mercury in a U-shaped glass tube no larger than 20 mm o.d., with a movable platinum contact accurately centered in a capillary of 2-mm bore. The stopcock and associated storage bulb permit large changes in the temperature setting and facilitate cleaning. To clean the capillary tube and mercury meniscus, mercury is sucked out of the capillary through a long hypodermic needle attached to a suction flask, the capillary is wiped clean with a pipe cleaner, and fresh mercury is admitted from the storage reservoir.

Thermoregulating circuit: We shall consider only the regulation of power to a heater with a signal from an off-on thermoswitch, without proportionation. This can be done by electromechanical or electronic relays. A simple circuit for this purpose is shown in Fig. 7. Other circuits, including ones in which the heating load is controlled by one or more Thyratrons, can be found in the literature.[12,13]

Water-level control: When a thermostat bath is operated for long periods, it is

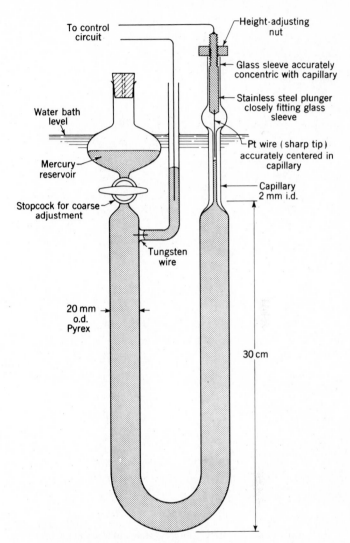

FIGURE 6
Adjustable, mercury thermoswitch-type thermosensing device.

subject to loss of water by evaporation to a point which will interfere with its normal operation unless the water is replaced. Therefore the tank should have an overflow drain and should be provided with a continuous supply of cold water that can be controlled from a few drops per minute to a small, steady stream. This supply of water also performs an important control function. A $\frac{1}{20}$-hp motor delivers con-

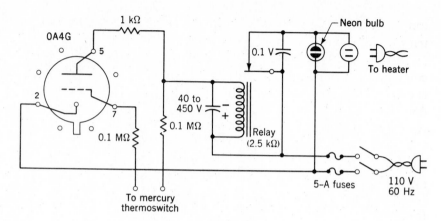

FIGURE 7

Inexpensive thermoregulator control circuit.

stantly about 37 W of mechanical energy to the bath, and, on days when the room temperature is not far below the desired bath temperature, this amount of energy is itself sufficient to maintain the bath temperature above the desired level. A stream of cold water may compensate largely or entirely for the stirring heat. The adjustment of the water flow is often critical and must be changed frequently. If the thermoregulating relay is one with double throw, the contacts opposite the heater contacts can be used to operate a solenoid valve to turn the water off and on and thereby permit a larger and less critical water flow, possibly at the expense of some fineness of temperature regulation.

Miscellaneous: To reduce rusting of iron hardware in the bath, a zinc or aluminum electrode with an applied anodic potential of a few volts with respect to the tank as ground may be provided. This should be protected from accidental contact by a perforated tube of Bakelite or other plastic.

Oil baths   Above 50°C water baths are subject to severe evaporation. Oil baths employing heavy cylinder oil can be used as high as 300°C (in a closed thermostat, because of fire hazard); with silicone oil higher temperatures may be reached. A eutectic mixture of sodium, potassium, and lithium nitrates (14, 56, and 30 percent by weight, respectively) is liquid at 120°C and with care can be used as high as 400 or even 450°C.

Circulating liquid baths   These devices consist of a fluid reservoir, a heater, a thermoregulator, and a circulating pump. The most commonly used general-purpose unit is made by Haake and will be described below. A large variety of excellent units can also be obtained from Lauda. Some of these are designed for operation at low temperatures (down to $-70°C$) as well as above room temperature.

Since the Haake constant-temperature circulating bath makes use of plastic components, the maximum temperature of these pumps is 150°C and the useful range is 25 to 150°C. Distilled water may be used as the circulating liquid in the 25 to 90° range, and Nujol, glycerine, or ethylene glycol may be used in the 90 to 150° range. Approximately 1.5 liters of liquid is required to keep the fluid in the operating range $\frac{1}{2}$ in. to 2 in. below the top of the fluid reservoir. The fluid may be pumped through a closed external system or pumped into an open bath located *above* the circulating pump with an overflow line to return fluid to the thermostat reservoir.

The electrical connections on top of the Haake unit provide circuits to the pump, the heater, and the thermoregulator. A switch allows both the pump and temperature regulating portions of the device to be turned on or off. A neon light indicates when the heater is on. The temperature of the thermostat is set by turning a rotating magnet after loosening the locking screw on the side of the thermoregulator. Rotation of the knurled screw on the top of the magnet housing rotates the magnet which moves the temperature indicator bar up or down. The temperature *at the top* of the temperature indicator bar is the approximate temperature which the reservoir will maintain. The exact temperature of the fluid is read on a separate thermometer immersed in the reservoir and held in place with a rubber stopper. The temperature of the circulating liquid will remain constant to within $\pm 0.02°C$. With the addition of special cooling accessories, this unit may also be used to provide constant-temperature liquid at temperatures below 0° (e.g., methanol in the temperature range -70 to $+30°C$). The manufacturers instructions should be consulted for operating details at low temperatures.

Air thermostats   Complex gas-handling systems of large size are often not well accommodated by water or oil baths. A double-walled air thermostat can be constructed with suitable insulation (air, fiberglass, vermiculite, foamed plastic, etc.) and with double glass windows and doors where needed. Air is circulated by a blower or fan which has its motor mounted externally (to reduce uncontrolled heating). Electrical heating elements with large surface areas, such as helical wire coils, should be used instead of the high-wattage blade-type heaters used in liquid baths. Thermoregulation is accomplished in much the same way as in a water bath.

Ovens and furnaces   Commonly available laboratory ovens can be used to temperatures as high as 200 or 300°C. They are usually regulated to 1 or 2° with a bimetallic thermoswitch. Air circulation is by convection, and the temperature is usually nonuniform. For obtaining better temperature uniformity and improved regulation, a lining made of thick sheet aluminum or copper can be installed.

For higher temperatures electric furnaces are used. These are obtainable commercially, but satisfactory ones can be made in the laboratory (see Fig. 28-1). A pregrooved Alundum core is wound with suitable heating wire (Nichrome, Chromel A, Kanthal), 12 to 20 gauge, and covered with a thick coating of Alundum cement. This is surrounded by several inches of powdered magnesia for thermal insulation, in a large metal container. After assembly the furnace is slowly heated to slightly above 1000°C to set the cement. A cover plug may be cut from firebrick. Internal metal radiation shields (platinum is best; molybdenum may be used if a neutral or reducing atmosphere is maintained) are needed if a high degree of temperature uniformity is required. Temperature is controlled with a Variac or similar transformer, operated manually or by a proportionating thermoregulator circuit with a thermocouple potentiometer or resistance thermometer bridge.

For temperatures above 1000 to 1200°C a platinum winding on a silica core or a combustion tube furnace heated by Globar or carbon rods may be used. These and other special high-temperature furnaces, including induction furnaces, vacuum furnaces, and arc furnaces, are beyond the scope of this book.

REFERENCES

1. The International Practical Temperature Scale of 1968, *Metrologia*, **5,** 35 (1969).
2. T. B. Douglas, *U.S. Natl. Bur. Stand. J. Res.*, **73A,** 451 (1969).
3. J. Busse, in "Temperature, Its Measurement and Control" (see General Reading below), vol. I, p. 228, Reinhold, New York (1941).
4. J. F. Swindells, "Calibration of Liquid-in-Glass Thermometers," *Natl. Bur. Stand. Monogr.* 90, U.S. Govt. Printing Office, Washington, D.C. (1965).
5. J. A. Beattie, in "Temperature, Its Measurement and Control" (see General Reading below), vol. II, p. 63, Reinhold, New York (1955).
6. E. F. Mueller, *op. cit.*, vol. I, p. 162 (1941).
7. C. R. Droms, *op. cit.*, vol. III, p. 339 (1962).
8. W. F. Roeser, *op. cit.*, vol. I, p. 180 (1941); D. I. Finch, *op. cit.*, vol. III (1962); F. R. Caldwell, "Thermocouple Materials," *Natl. Bur. Stand. Monogr.* 40, U.S. Govt. Printing Office, Washington, D.C. (1962).
9. R. J. Coruccini and H. Shenker, *U.S. Natl. Bur. Stand. J. Res.*, **50,** 229 (1953); *U.S. Natl. Bur. Stand. Circ.* 561 (1955).
10. W. E. Forsyth, in "Temperature, Its Measurement and Control" (see General Reading below), vol. I, p. 1115, Reinhold, New York (1941).
11. C. T. Tomizuka and D. Zimmerman, *Rev. Sci. Instrum.*, **30,** 40 (1959); A. A. Brooks, *ibid.*, **27,** 746 (1956); **28,** 297 (1957).
12. C. A. Proctor, *Rev. Sci. Instrum.*, **22,** 1023 (1951).
13. J. M. Sturtevant, Automatic Control, in A. Weissberger (ed.), "Technique of Organic Chemistry," 3d ed., vol. I, part I, pp. 17–32, Interscience, New York (1959).

GENERAL READING

R. P. Benedict, "Fundamentals of Temperature, Pressure and Flow Measurements," Wiley, New York (1969).

J. O'M. Bockris, J. L. White, and J. D. Mackenzie (eds.), "Physicochemical Measurements at High Temperatures," chaps. 2 and 3, Butterworth, London (1959).

J. M. Sturtevant, in A. Weissberger (ed.), "Technique of Organic Chemistry," 3d ed., vol. I, part I, chaps. 1, 6, Interscience, New York (1959).

"Temperature, Its Measurement and Control," vol. I (1941), vol. II (1955, H. C. Wolfe, ed.), vol. III (1962, C. M. Herzfeld, ed.), Reinhold, New York. (These volumes contain papers presented at A.I.P. symposia in November 1939, October 1954, and March 1961.)

XVII

VACUUM TECHNIQUES

This chapter contains a discussion of various topics of importance in the theory, design, and practice of handling gases at low pressures. Major emphasis is given to high-vacuum techniques, since they play a vital role in many physical chemistry research problems. For vacuum applications in this book, see Exps. 5, 6, 28, 32, and 49. Some of the material presented is also pertinent to the problem of pumping on refrigerant baths; for an application of this technique, see Exp. 49. This chapter is not intended as a comprehensive treatment; more detailed information is available in the works cited in the General Reading list.

INTRODUCTION

The term vacuum refers to the condition of an enclosed space that is devoid of all gases or other material content. It is not experimentally feasible in a terrestrial environment to achieve a "perfect" vacuum, although one can approach this condition extremely closely. It is possible routinely to obtain a vacuum of 10^{-6} Torr and with more sophisticated techniques 10^{-10} Torr (1.3×10^{-13} atm); it is even possible by special techniques to obtain a vacuum of 10^{-15} Torr or about 30 molecules/ cm^3. One *Torr*, the conventional unit of pressure in vacuum work, is the pressure equivalent to one millimeter of liquid mercury, the temperature of which is 0°C, at standard gravity of 980.665 cm sec^{-2}; 1 Torr $= \frac{1}{760}$ standard atmosphere.

For Dewar flasks, metal evaporation apparatus, and most research apparatus, a vacuum of 10^{-5} to 10^{-6} Torr is sufficient; this is in the "high vacuum" range, while 10^{-10} Torr would be termed "ultrahigh vacuum." However, for many routine purposes a "utility vacuum" or "forepump vacuum" of about 10^{-3} Torr (1 μm of mercury) will suffice, and for vacuum distillations only a "partial vacuum" of the order of 1 to 50 Torr is needed.

THEORETICAL BACKGROUND

The basic theory needed here is the kinetic theory of gases, which is discussed in Chap. IV. It will be assumed that gases always obey the perfect-gas law and have an essentially Maxwellian distribution of molecular velocities. The most important properties for our present purpose are the mean free path $\bar{l}$, the coefficient of viscosity η, and the coefficient of thermal conductivity K; in particular, we are concerned with the dependence of these three properties on pressure at constant temperature. According to Eqs. (IV-3), (IV-14), and (IV-18), $\bar{l}$ varies inversely with p while η and K are independent of p. The expression given by Eq. (IV-3) for $\bar{l}$ is correct over the entire pressure range, but Eq. (IV-14) is valid only at pressures high enough so that $\bar{l}$ is small compared with the dimensions of the container. At lower pressures, where $\bar{l}$ becomes comparable to the apparatus dimensions, both η and K decrease with decreasing pressure. When $\bar{l}$ is considerably larger than the apparatus dimensions (i.e., at pressures less than 0.01 Torr for most apparatus), η and K will vary linearly with the pressure. This limiting behavior is a consequence of molecular-flow conditions where the mechanism of momentum or kinetic-energy transport depends on collisions between gas molecules and the walls rather than on intermolecular collisions.[1] In this limiting region, the pressure dependence of viscosity or thermal conductivity serves as the basis for the operation of several types of vacuum gauges.

PUMPING SPEED FOR A SYSTEM

Detailed calculations of pumping speed are seldom necessary for designing a system to obtain a static vacuum in a small, closed line. However, certain qualitative principles of vacuum-line design are important. When one must pump out a large volume or pump on a system where there is a continuous evolution of gas inside the system (as when pumping on a refrigerant bath, for example), a careful quantitative design is essential. Overall pumping speed will depend on the characteristics of the pump (or pumps) used and also on the impedance to gas flow through the connecting tubing.

Speed of the pump   The pumping speed S, usually expressed in liter per sec, of a given pump operating at a pressure p is defined as the volume of gas measured at a constant temperature T and the given pressure p which is removed from the system per unit time. This quantity is closely related to the flux Q, which is the quantity of gas (in pressure-volume units) flowing through a given plane in unit time. For an ideal gas, the flux is given by

$$Q = \frac{\delta(pV)}{\delta t} = RT\frac{\delta N}{\delta t} = p\frac{\delta V}{\delta t} \tag{1}$$

where $\delta V/\delta t$ is the volume of gas at temperature T flowing in unit time past a plane in the system where the steady-state pressure is p, $\delta N/\delta t$ is the number of moles flowing through that plane in unit time, and the gas constant R has either the familiar value of 0.08206 liter atm deg^{-1} mol^{-1} or the equivalent value 62.37 Torr-liter deg^{-1} mol^{-1}. Thus we see that $S = Q/p$.

The pumping speed is approximately constant over a wide pressure range since it depends on the length of time required for exhausting the gas contained at a given instant within the pumping element itself. This time, the "cycle" of the pump, is the time for a shaft rotation in a rotary pump, and, for a diffusion pump or an ion pump, it is the time required for certain molecular processes to take place that are independent of pressure. The pumping speed of a given pump begins to decrease when the pressure is sufficiently low (owing to backflow of a portion of the exhaust, to dissolution of gas dissolved in the pump oil, to vaporization of the oil, or to desorption, etc.) and becomes zero when the pressure reaches some ultimate attainable pressure p^*. The variation of the pumping speed S with the pressure p is given approximately by

$$S = S^*\left(1 - \frac{p^*}{p}\right) \tag{2}$$

where S^* is the intrinsic pumping speed. It is not possible to make an accurate calculation of pump speeds, but there are reliable experimental methods for measuring S^*. Both dynamic and static methods have been used: In the dynamic method, the change in the steady-state pressure is measured when a metered leak is closed; in the static method, the rate of pressure change is measured for the evacuation of a bulb of fixed volume. Details of these methods are given by Dushman.[2] Once S^* is known, Eq. (2) permits one to calculate S, which is the quantity of practical interest.

Conductance   In pumping down a system, gas must be conducted through orifices and tubes, and frequently through valves and cold traps as well. At ordinary pressures the speed of flow is limited by the viscosity of the gas, and at very low pressures it is limited by the finite speed of molecular motion and by diffusive effects of molecular collisions with the walls. In either case, a given segment i of the apparatus conducting the gas may be said to have conductance C_i which is the flux of gas through it divided by the pressure difference:

$$C_i = \frac{Q}{(p_1{}^i - p_2{}^i)} \tag{3}$$

where $p_1{}^i$ and $p_2{}^i$ are taken as fixed pressures at the inlet and outlet of the segment, respectively. Since conductance has the same dimensions as pumping speed, the same units (liter sec^{-1}) will be used for both.

The overall conductance of two or more segments is given by

$$C = \sum_i C_i \tag{4}$$

if they are connected in parallel, and by

$$Z \equiv \frac{1}{C} = \sum_i \frac{1}{C_i} \equiv \sum_i Z_i \tag{5}$$

if they are connected in series. Here Z, the reciprocal conductance, is often called the impedance. There is an obvious analogy in behavior between Q, p, and Z on the one hand and the corresponding direct-current electrical quantities i, E, and R on the other. A consequence of Eq. (5) is that *the conductance of a series sequence of segments is always less than the conductance of any one segment.*

The conductance of a given segment is a function of its geometry and a function of the pressure, temperature, and properties of the gas being pumped. At pressures of about 10^{-2} Torr and above, gas flow in most vacuum apparatus takes place by viscous flow.† At pressures below 10^{-3} Torr, molecular flow is the principal mechanism for gas transport. We shall now cite appropriate formulas for calculating C in a variety of simple cases.

(*a*) *Viscous flow in cylindrical tubing.*   Combining Eq. (1) with Eq. (4-8) for the laminar flow of a gas (and noting that $\delta N/\delta t$ has the same meaning as ϕ_N) we obtain

$$Q = \frac{\pi d^4}{128\eta L} \frac{p_1 + p_2}{2} (p_1 - p_2) \qquad \text{dyne cm } sec^{-1} \tag{6}$$

---

† Turbulent flow will only occur if the Reynolds number (see Exp. 4) is very high. This is uncommon in vacuum work, and we shall neglect turbulence completely.

where d is the diameter of the tube in centimeters, L is its length in centimeters, η is the viscosity expressed in poise (P), and p_1 and p_2 are in dynes per square centimeter. Comparing Eqs. (3) and (6) we see that

$$C_\eta = \frac{\pi d^4}{128\eta L}\bar{p} \qquad cm^3\ sec^{-1} \tag{7}$$

where $\bar{p} = (p_1 + p_2)/2$ is the average pressure in the tube. The subscript η on C_η is a reminder that this expression is valid only for the viscous flow that occurs at high pressures. For air at 298°K, Eq. (7) reduces to

$$C_\eta(air) = \frac{180d^4}{L}\bar{p} \qquad liter\ sec^{-1} \tag{8}$$

where $\bar{p}$ is now expressed in Torr. When the tube is short enough that end effects perturb the steady-state laminar velocity distribution over an appreciable fraction of the length of the tube, Eq. (8) must be modified:[2]

$$C(air) = 180\frac{d^4}{L}\frac{\bar{p}}{[1 + 0.38(Q/L)]} \tag{9}$$

where Q is the flux in Torr-liter per second.

For viscous flow in a tube 100 cm long and 1 cm in diameter, the conductance decreases from 1350 liter sec^{-1} at 760 Torr (1 atm) to 1.78 liter sec^{-1} at 1 Torr. As we shall see below, the conductance for molecular flow in the same tube has the constant value 0.122 liter sec^{-1}. In partial-vacuum applications where viscous flow is dominant, conductance limitations are seldom a problem. Accordingly, most of our attention will be focused on the case of molecular flow, which is important in high-vacuum work.

(b) *Molecular flow.* At very low pressures the mean free path $\bar{l}$ of a gas molecule is larger than the critical dimensions of the apparatus, and collisions of a molecule with the walls are more frequent than collisions with other molecules. [At 25°C, $\bar{l} = (a/1000p)$ where $\bar{l}$ is expressed in cm and p in Torr; $a = 5.1$ for air, 3.4 for water, and 2.7 for mercury.] When the pressure is sufficiently low that intermolecular collisions are negligible, the conductance of a long uniform tube of length L, cross-sectional area A, and perimeter H was shown by Knudsen[3] to be

$$C = \frac{4}{3}\frac{A^2}{HL}\bar{c} \tag{10}$$

where $\bar{c} = (8RT/\pi M)^{1/2}$ is the mean molecular speed in cm sec^{-1} and M is the molecular weight in grams. For a cylindrical tube with $L \gg d$,

$$C = \frac{\pi d^3}{12L}\bar{c} \qquad cm^3\ sec^{-1} \tag{11}$$

Note that this expression for the conductance is independent of pressure. For air at 298°K, Eq. (11) becomes

$$C(\text{air}) = \frac{12.2d^3}{L} \qquad \text{liter sec}^{-1} \tag{12}$$

Equations (10) to (12) do not include the effect of the conductance of the inlet aperture where the tube is connected to another region of larger cross section (to a bulb, for example). Unless L is very much larger than d, this aperture conductance will have a significant effect on the total conductance of the tube. From Eqs. (3) and (IV-7) we see that for molecular flow through an orifice of area A

$$C = \tfrac{1}{4}A\bar{c} \tag{13}$$

(For air at 298°K and a circular opening of diameter d, $C = 9.15d^2$ liter sec^{-1}.) The ratio of the aperture conductance to that of a long cylindrical tube of the same diameter is $3L/4d$. A short tube may be regarded as two circuit elements—the entrance of the tube regarded as an orifice and the tube proper—in series, with their impedances additive. The entrance of the tube is then equivalent to an addition of $\Delta L = 4d/3$ to the length of the tube. The exit is not regarded as having any impedance per se, although it may have impedance by virtue of being the entrance to a much smaller tube in the sequence. This approximate treatment gives for air in a cylindrical tube at 298°K,

$$C(\text{air}) = \frac{12.2d^3}{L}\,\frac{1}{1 + 4d/3L} \qquad \text{liter sec}^{-1} \tag{14}$$

Equation (14) might be applied, for example, to the bore of a vacuum stopcock placed in a line of considerably larger diameter. A somewhat more exact treatment[2] gives, for a wide-bore stopcock with $d = 1$ cm, $L = 3.2$ cm, a conductance for air at 25°C of 2.6 liter sec^{-1}; a narrower bore stopcock with $d = 0.40$ cm, $L = 2.0$ cm will have a conductance of only 0.29 liter sec^{-1}.

A tube with a bend in it will have a lower conductance than a straight tube. A crude way to take this into account is to consider each bend as an orifice having a diameter equal to the tube diameter. This will add $4d/3$ to the length (measured along the centerline of the tube) for each bend. To make more accurate calculations would require taking into account the shape of the bend.

The conductance of a *cold trap*, chilled with liquid nitrogen to prevent mercury or pump-oil vapors from entering the system or volatile substances in the system from contaminating the pump, can also be of considerable practical importance. Such a trap is diagrammed schematically in Fig. 1. Lafferty[2] has investigated the effects of varying the trap dimensions, and has shown that the optimum conductance

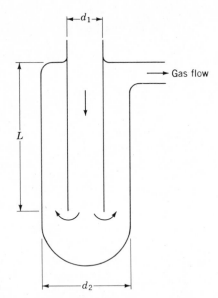

FIGURE 1
Schematic diagram of cold trap.

for a trap of outer diameter d_2 is obtained when $d_1/d_2 = 0.62$. With this condition maintained, the conductance for air at 25°C is approximately given by

$$C(\text{air}) \simeq \frac{1.4d_2^{\,3}}{L} \tag{15}$$

However, if the entire trap is cooled with liquid nitrogen the conductance will be reduced by a factor of $(\frac{77}{298})^{1/2} \simeq 0.5$. More detailed calculations[2] on a liquid-nitrogen trap with $L = 21.6$ cm and $d_2 = 4.76$ cm indicate that the conductance should be in the range 3 to 3.5 liter sec^{-1}, which is in fair agreement with the above approximation.

(c) *Intermediate pressures.* When the flow is neither purely viscous nor purely molecular, one can use an empirical expression of Knudsen[3]

$$C = C_\eta + FC_{\text{molecular}} \tag{16}$$

The factor F is a complicated function of M, T, d, η, and $\bar{p}$; it has values for air which vary between ~ 0.8 at high pressures and 1.0 at low pressures.

Overall pumping speed Consider, as an example, a large bulb connected via a tube of conductance C to a pump of speed S. We wish to find S_T, the overall pumping speed for this composite system. The rate of gas removal from the bulb Q equals $S_T p_1$ where p_1 is the pressure in the bulb. Also, Q must equal $C(p_1 - p_2)$ for the flow through the tube where p_2 is the outlet pressure at the pump end of the tube. Finally, Q equals Sp_2 at the pump itself. Since

$$\frac{Q}{C} = p_1 - p_2 = \frac{Q}{S_T} - \frac{Q}{S}$$

we can write

$$\frac{1}{S_T} = \frac{1}{C} + \frac{1}{S} \tag{17}$$

Although the speed of a pump differs from the conductance of a tube in being defined in terms of a single pressure rather than a pressure difference, Eq. (17) shows that, for the kind of system being considered, S can be treated in the same way as C in calculations involving series or parallel combinations of pumps and pumping lines.

Thus the reciprocal of the overall pumping speed S_T of a pump *plus* vacuum line is the sum of the impedances of all of the segments from the orifice representing the entrance to the vacuum line to the pump itself; typically,

$$\frac{1}{S_T} = \frac{1}{C_{\text{entrance}}} + \frac{1}{C_{\text{tube}}} + \frac{1}{C_{\text{stopcock}}} + \frac{1}{C_{\text{tube}}} + \frac{1}{C_{\text{bend}}} + \frac{1}{C_{\text{tube}}}$$

$$+ \frac{1}{C_{\text{trap}}} + \frac{1}{C_{\text{tube}}} + \frac{1}{S} \tag{18}$$

Up to this point, pumping speeds have been discussed for the case of fixed pressures. For the case of evacuating a fixed volume V, not including the space in the pump or vacuum line itself, the flux Q becomes $-V\,dp/dt$. (The negative sign is required since the pressure p in the volume V is decreasing.) Therefore, the pressure will obey the differential equation

$$S_T = \frac{Q}{p} = -\frac{V}{p}\frac{dp}{dt} \tag{19}$$

A general solution of Eq. (19) is complicated since S_T varies with the pressure.[2] For the simple case where S_T is considered constant over a range of pressures, one finds

$$p = p_0 e^{-(S_T/V)t} \tag{20}$$

where p_0 is the pressure at $t = 0$. The length of time for the pressure in volume V to fall by one decade (power of 10) is then

$$\Delta t_{0.1} = \frac{2.303}{S_T} V = \sum_i (2.303 \, Z_i) V \tag{21}$$

One frequently sees in books and catalogs the quantity $2.303Z_i \equiv 2.303/S_i$ (in sec/liter) quoted for a vacuum component rather than the impedance or conductance. The quantity $2.303Z_i$ represents the amount of additional time, per liter of space to be evacuated, that insertion of this component will add to the time for a decade fall in pressure. Thus, for the large-bore stopcock already described, $2.303Z_{stopcock} = 0.88$ sec liter^{-1} for air; including it in a line evacuating a 50-liter enclosure for which $\Delta t_{0.1}$ was previously 100 sec will increase the decade time to 144 sec.

In practice, Eq. (21) is valid only during the preliminary stages of pumping, i.e., until the pressure drops to a point where the pumping speed S begins to decrease and/or an appreciable fraction of the gas being pumped out is entering the system through leaks, desorption from the walls, or backflow from the pump, etc. Very long times are often required in the later stages to attain a vacuum close to the ultimate vacuum for the system.

DESORPTION AND VIRTUAL LEAKS

Under high or ultrahigh vacuum conditions, much or virtually all of the gas being pumped is being desorbed from the walls of the system. If we have a surface of area A which is covered with a layer of adsorbed molecules each of which occupies an average area σ, the number of moles of gas that will be produced by complete desorption is $A/\sigma N_0$, where N_0 is Avogadro's number. For A in cm^2, σ in Å^2 (1 Å $= 10^{-8}$ cm), the quantity of gas in Torr-liter is, at room temperature,

$$\int Q \, dt = 3.1 \times 10^{-4} \frac{A}{\sigma} \tag{22}$$

A 1-liter glass bulb has an inside surface area of 485 cm^2; if each adsorbed molecule occupies 10 Å^2 on the surface, the content of adsorbed gas is about 0.015 Torr-liter. At 10^{-6} Torr, 15,000 liter would have to be pumped; at 10^{-10} Torr, 150,000,000 liter. Fortunately the greater part of this gas can be pumped off at higher pressures where the volumes to be pumped are smaller.

Adsorption of gases on surfaces is of two kinds: physical and chemical. Physical adsorption is generally reversible and is characterized by heats of adsorption of only a few kilocalorie per mole. Physical adsorption is ordinarily negligible more than 50 or 100° above the boiling point of the gas. Chemical adsorption (called chemisorption) is often irreversible, comes to equilibrium very slowly except at elevated temperature (often 1000°C or higher), and is characterized by heats of adsorption of the order of 20 to 200 kcal mol^{-1}. Examples are oxygen or carbon monoxide chemisorbed on metals. Water is of special importance as it is usually present at room temperature on glass or metal surfaces in both physically and chemically adsorbed forms at the same time, with its strength of attachment to the substrate varying over a wide range. Other substances that may be adsorbed on the walls of a vacuum system are N_2, CO_2, NH_3, oxides of nitrogen, hydrocarbons, and other organic substances.

Most physically adsorbed gases are readily desorbed in the early stages of pumping, when the pressure is above 10^{-3} Torr. The approach to 10^{-6} or 10^{-7} Torr is slow because of the slow desorption of water at room temperature. For practical purposes 10^{-6} or 10^{-7} Torr may be the lowest pressure which can be achieved because the time required to desorb all of the water is so long at this temperature.

A remedy for this situation is to "bakeout" the system, in order to hasten the desorption of water and other sorbates and to increase the amount desorbed, so that these substances can be pumped off quickly at higher pressures. For most purposes 250°C is a high enough temperature, and one which is not high enough to damage most high-vacuum components; bakeout up to 400°C is occasionally practiced. A bakeout to 150° is better than none at all, as is a selective bakeout of most components of a system, leaving others (such as greased stopcocks) at room temperature. A glass system is often selectively baked out by careful flaming with a hand torch while the system is being pumped, or baked out with heating tapes or infrared lamps.

The most common source of gas in a vacuum system, excluding the gases (air) present originally and those produced by desorption, is *leaks*. The prevention of leaks—pinholes in blown glass, faulty welded seams, faulty gaskets, scratched flanges, channels in greased stopcocks and over O rings, diffusion through slightly porous materials—is a matter of careful choice of materials, apparatus design, construction, and operation; the detection and repair of leaks is a technology in itself which will be discussed later. A leak which admits only 1 mm^3 of room air per minute is putting 1.3×10^{-5} Torr-liter sec^{-1} into the pumping system; to keep up with this at 10^{-6} Torr would require a system with a pumping speed of 13 liter sec^{-1}.

The *virtual leak* is also a common problem. This is leakage into the vacuum, not of air from outside, but rather of air or vapor from some source inside the system: a drop of a slightly volatile substance slowly vaporizing, or a small pocket of air trapped in a screw thread and slowly leaking out. A cubic millimeter of room air trapped under a screw (7.6×10^{-4} Torr-liter) is 7,600,000 liter at 10^{-10} Torr; if that amount of air took a week to leak out at that pressure, an average of 12 liter sec^{-1} of the pumping capacity would be taken up for that time; the same would be true of 0.01 mg of a slightly volatile substance (M.W. 250) that required a week to volatilize completely.

A very small amount of a thermally decomposable organic substance in the system during bakeout, or in a part of the system that is operated above 200 to 300°C, can also give vacuum contamination that will take a long time to get rid of. The higher the vacuum, the more important it is that the system be clean. The presence of a greasy film on inside metal surfaces should be precluded by a "degreasing" with trichloroethylene, particularly if better than 10^{-7} Torr is desired.

Use of stopcock grease cannot easily be avoided in a glass vacuum system for lubricating stopcocks, ground joints, O rings, seals for rotating shafts, etc; vacuum pump oil is also sometimes used for lubricating purposes. These materials are specially prepared and refined to be as free as possible from volatile components. Stopcock greases may have stated "vapor pressures" as low at 10^{-6} Torr; the values may be even lower for diffusion pump oils. However, they can pick up air, water, and especially organic vapors and later release them slowly from solution into the vacuum. This nuisance should be minimized by keeping the area of these materials that is exposed to the vacuum as small as possible, by use of careful technique in greasing stopcocks, etc. Vacuum pump oil has a tendency to "creep" on surfaces; one of the functions of a cold trap is to prevent pump oil from creeping into the clean part of the vacuum system.

PUMPS

The principal types of vacuum pumps will be discussed in this section. The very simple Toepler pump is much too slow for practical use in evacuating a large system, but it is quite useful for the quantitative transfer of small amounts of gas from one part of a system to another. Rotary oil pumps are used for pumping on refrigerant baths and as the forepump for "backing" low-pressure pumps. For most high-vacuum work, diffusion pumps are used to achieve pressures of about 10^{-6} Torr; however, there are now special types of pumps which can be used to achieve ultra-high vacuums of 10^{-9} Torr or less. The water aspirator is a very useful pump for many routine operations, and the Langdon pump is well suited to vacuum distillations where vapor contamination may be a problem.

Water aspirator The water aspirator is probably the vacuum "pump" most frequently used by chemists. It produces only a partial vacuum but one that is entirely adequate for a large number of chemical operations such as filtration by suction, distillation under reduced pressure, evacuation of desiccators, and "pulling down" McLeod gauges.

This device operates through Bernoulli's principle. When a flowing fluid encounters a constriction that forces it to flow at a higher velocity, the hydrostatic pressure exerted by the fluid decreases by an amount corresponding to the increase in kinetic energy density plus any energy losses due to turbulence or viscous resistance. If the diameter of the tube increases following the constriction, the pressure may rise again if the difference in kinetic energy density is not completely lost to turbulence and viscosity. In the aspirator (see detail view in Fig. 2) water enters from a pipe at mains pressure (about 4 atm) and leaves against 1 atm external pressure from a nozzle of lesser diameter. In an internal constriction of still smaller diameter the pressure drops nearly to zero. The air from the system being evacuated enters at this constriction and is carried along with the water.

The ultimate vacuum achievable with a water aspirator is limited by the vapor pressure of the water itself; any significant drop in the pressure much below this will be reversed by the vaporization of water. The vapor pressure of water is 12.8 Torr at 15°C, 17.5 Torr at 20°C, and 23.8 Torr at 25°C. A lesser vacuum (higher pressure) can be achieved by reducing the flow of water into the aspirator and/or bleeding air into the system through a controlled leak.

When the water supply to an aspirator is shut off or greatly reduced while the system is under vacuum, the water in the aspirator can be sucked into the system. Modern aspirators often contain a check valve to prevent this, but this should not be relied upon; a trap bottle should always be provided between the aspirator and the system.

A suitable setup for the use of an aspirator is shown in Fig. 2. The aspirator is connected by rubber pressure tubing to the bottom of a trap bottle (which also serves as a "vacuum reservoir" to damp out rapid pressure fluctuations). At the top of the bottle is a glass "tee" in a rubber stopper. On one arm is attached a short piece of pressure tubing which is closed off with a pinch clamp except when air is bled in to increase the pressure or vent the system. The other arm is connected to a mercury manometer and the system. The manometer, which can be connected to and disconnected from the line by turning the stopcock at the top, is seated in a plastic cup to protect the bottom of the manometer mechanically and to retain the mercury in the event of breakage. The bottle and manometer are clamped to a short ringstand so that the setup can be stored and transported as a unit. Before using this setup check the manometer to make sure there is no significant air bubble at the top. If there is one, the manometer should be evacuated by a mechanical pump and tipped carefully so as to allow the air to escape.

In use, this setup is connected to the aspirator and system as shown, and all connections including those in the system itself are checked for tightness. Before turning the aspirator on, the vent should be fully open and the manometer closed off. With a moderate stream of water flowing through the aspirator, slowly close off the vent with a pinch clamp. *Cautiously* open the manometer; the pressure in Torr is the difference in millimeters between the heights of the two mercury menisci. Always close the manometer after each reading; it is important to keep organic vapors or water from condensing on the mercury surface. The pressure can be varied by changing the flow of water through the aspirator or the flow of air through the vent; the manometer can be left open during this process to monitor the change in pressure.

Detail view

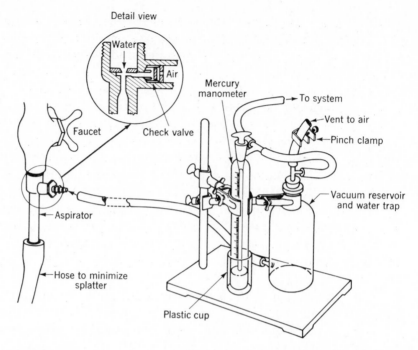

FIGURE 2

Aspirator setup with trap bottle and manometer. All connections are made with pressure tubing. The detail view shows the inside construction of a plastic aspirator.

When you are through with the vacuum, be sure the manometer is closed off, carefully open the vent fully, and then turn off the water (never turn off the water with the vent closed). Then very carefully open the manometer to vent it, and close it again.

A word of **warning**: The flow rate of water through an aspirator is subject to change, often sudden change, as the water main pressure fluctuates. This may produce changes in the pressure in the system being evacuated, or even cause "suckback" of water into the trap bottle. If a constant pressure (especially for reduced-pressure distillation) is required, one must be alert for such changes, and be ready to make required adjustments.

Toepler pump[2] The Toepler pump permits one to transfer gas from a bulb A into another bulb B without loss or contamination. This is accomplished by alternately lowering and raising the mercury level in a reservoir connected with both bulbs. When the mercury level is lowered, gas expands from A into the reservoir; when it is raised, this gas is forced into B. Many cycles of operation are necessary to effect an almost complete transfer of the gas. Obviously, this transfer could be accomplished much more easily for a condensable gas by merely cooling bulb B with liquid nitrogen. The Toepler pump is most useful in handling those few substances (such as He, H_2, CO, N_2, Ar, CH_4) which either condense below $77°K$ or have an appreciable vapor pressure at that temperature. One of the most common applications of this device is to remove gas from a reaction bulb for quantitative analysis at the end of a kinetics run; for example, in Exp. 28, a Toepler pump would be necessary to remove quantitatively the hydrogen gas from the furnace.

Rotary oil pumps   Perhaps the commonest form of vacuum pump is the mechanical pump that operates with some sort of rotary action, with moving parts immersed in oil to seal them against back streaming of exhaust as well as to provide lubrication. These pumps are used as forepumps for diffusion pumps. Other common laboratory applications are the evacuation of desiccators and transfer lines, and distillation under reduced pressure (see earlier section on Water Aspirator). These pumps have ultimate pressures ranging from 10^{-4} Torr (0.1μ) to 0.05 Torr, and pumping speeds from 0.16 to 150 liter sec^{-1} or more, depending on type and intended application. The Cenco Hyvac 2 or the Welch Duo-Seal 1400B are relatively inexpensive laboratory pumps capable of good ultimate vacuum (10^{-4} Torr) but with relatively low pumping speed (21 liter min^{-1} = 0.35 liter sec^{-1} at 1 atm; about half that at 1μ). The Cenco Pressovac or the Welch Duo-Seal 1399B are inexpensive all-purpose pumps giving a somewhat higher pumping capacity (35 liter min^{-1} = 0.58 liter sec^{-1} at 1 atm) in exchange for a less good ultimate vacuum (0.015 Torr). The Pressovac can also be used as a source of positive pressure above 1 atm. Other pumps with different capabilities are manufactured by Cenco, Welch, Edwards, and Kinney.

Most of these pumps have one or the other of two principal designs. The simplest design, as exemplified by Cenco pumps, is illustrated in Fig. 3. An eccentric cam is rotated against the cylindrical walls of the stator by means of a motor-driven pulley (not shown). For all positions of the rotor a sliding vane is pressed firmly against the top of the rotor by a spring-loaded rocker arm. Gas which expands into region A through the intake is isolated when the contact point C passes the intake opening. This gas is then compressed (in region B) by the motion of the rotor. When the cam reaches the point in its rotation corresponding to maximum compression, this gas is expelled through the exhaust valve. The surfaces of the rotor, stator, vane, and also the seat of the exhaust valve are precision ground to minimize leakage. In addition, the entire pump is immersed in a low-vapor-pressure oil. Small amounts of oil on the inside of the pump act as an additional seal as well as a lubricant; the oil check valve on the exhaust prevents appreciable flow of oil into the pump. The second design, as exemplified by Welch Duo-Seal pumps, is shown in Fig. 4. In this type of pump, a rotating drum tangent to the inside of the cylinder (with a clearance of only

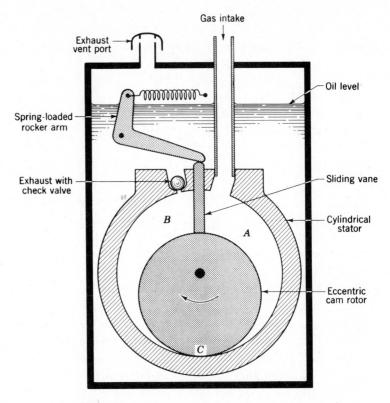

FIGURE 3
Schematic diagram of a rotary oil pump with an eccentric cam.

0.0001 in.) has deep slots which hold two (or more) vanes that slide along the cylinder wall, compressing the gas and carrying it to the exhaust port. The Cenco pumps tend to be noisy (hammering sound) when pumping a good vacuum; the hammer noises tend to be intermittent or muffled when the vacuum is less good. By contrast the Duo-Seal and Edwards pumps are most quiet when the vacuum is best; this is a definite advantage for continuous operation in a room where people are working.

There are considerable variations on these basic designs. Many commercial types are compound pumps in which two single stages are mounted on the same shaft and are connected in series as a means of increasing the pumping speed and improving the ultimate vacuum. The pumping speed at high pressures depends on both the size and design of the pump.

An important limitation on the ultimate vacuum achievable with a rotary oil pump is the characteristics of the oil used. It must of course have a negligible vapor

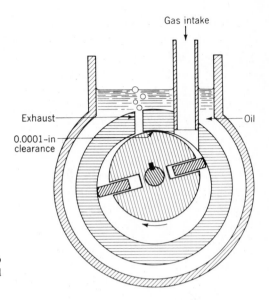

Gas intake

Exhaust

0.0001-in
clearance

Oil

FIGURE 4
Schematic diagram of a rotary oil pump
similar to the Welch Duo-Seal; based
on the Gaede oil pump design.

pressure and contain no significant quantity of dissolved volatile components; accordingly, *care should be taken to prevent vapors, especially those of organic solvents, from entering the pump.* A Dry Ice–trichloroethylene trap is adequate to prevent this. The oil should be viscous enough to serve as a lubricant, but fluid enough to permit air bubbles from the exhaust to rise rapidly to the oil surface so as not to re-enter the pump with the oil. *The oil must be clean;* dirt or metal particles suspended in the oil will cause abrasion and wear at points where tolerances are critical, and chemical fumes (e.g., hydrogen halides, sulfur dioxide, oxides of nitrogen) will cause corrosion. The oil level and cleanliness should be inspected at regular intervals, and the oil should be replaced whenever it is suspect. In the event of any serious contamination (e.g., corrosive fumes, water, or mercury) the pump must be disassembled by a competent mechanic, cleaned with solvents, dried, carefully inspected, reassembled (with due care for the extremely critical mechanical tolerances), and filled with fresh oil.

The rotary oil pump is usually belt-driven by an electric motor. The belt requires periodic inspection as a part of routine pump maintenance. It should be tight enough to prevent slippage but not so tight that it will produce wear on the motor and pump bearings and on the belt itself. A frayed or worn belt should be replaced. The belt drive of a pump presents a mechanical hazard for fingers, loose sleeves, and long hair (although the pump is normally on the floor, where the main hazard will be to skirts and trouser cuffs, especially those with flared bottoms), and should always be covered by a guard.

When a rotary oil pump must be operated under conditions where significant amounts of condensable vapors—generally water—pass into it, a pump with a *vented exhaust* should be used. In this pump a quantity of room air is introduced into the exhaust as a diluent to prevent condensation and resultant contamination of the oil. It is inadvisable to permit organic vapors to enter even a vented exhaust pump because of their high solubility in the oil; it is better to freeze them out with a cold trap.

Most rotary oil pumps are not designed to operate under conditions where large volumes of air or other gases are pumped continuously at pressures between a few Torr and atmospheric. A pump operating under these conditions is effectively compressing large quantities of gas, which results in the evolution of much heat. The initial pumpdown of a reasonable volume of air from one atmosphere to the ultimate vacuum presents no serious problem for most pumps since the quantity of heat is limited; caution should be used, however, when a system is continuously vented in order to maintain a pressure considerably higher than the normal ultimate pressure.

Langdon pump A pump specially designed for applications such as vacuum distillation, where organic vapors or other contaminants may be present and where large amounts of air may be pumped against an adjustable vent, is the Langdon pump. This pump will operate down to about 0.05 Torr (50 μ) with a capacity of 0.58 liter sec^{-1}. We shall devote special attention to this pump because its operation differs from that of other rotary oil pumps in two very important respects.

1 The pump is water-cooled (to take away the heat developed by compression of the relatively large amounts of air being pumped) and *must never be used without water cooling*.

2 The pump contains only a small amount of oil (70 ml of No. 30 SAE motor oil) *which must be replaced after each use* in order to prevent accumulation of volatile contaminants and forestall action of corrosive contaminants. This is *not* normal procedure for other rotary oil pumps.

Another important feature of the Langdon pump is that it can be more readily disassembled than most pumps and has Teflon vanes that can be replaced relatively easily.

A convenient setup for vacuum distillation and similar purposes, incorporating the Langdon pump, is shown in Fig. 5. In addition to the pump itself, this setup includes a cold trap, a needle valve as a controllable leak for adjustment of steady-state pressure, and a small swivel-type McLeod gauge. This setup must be operated with very careful attention to the following directions:

Before use inspect the cold trap; if not empty and dry remove outer member, empty and dry it and the inner tube, and reassemble tightly. Replace the Dewar flask and add trichloroethylene to fill it about half way. Add Dry Ice a few lumps at a

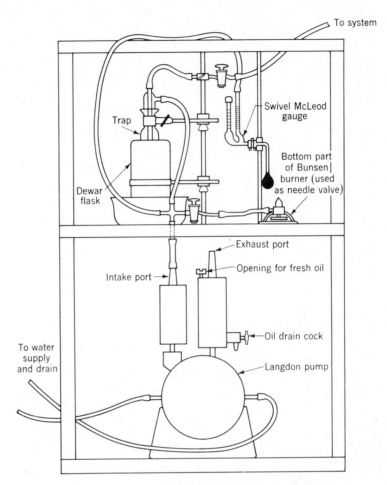

FIGURE 5
Langdon pump setup for vacuum distillation.

time, waiting for CO_2 gas evolution to cease before adding more; when no more evolution is produced add more Dry Ice to bring the liquid level within an inch or two of the ground joint. More may be added as needed during the run.

Connect the water lines; but do not yet turn the water on. Rotate the pump several revolutions by hand to ensure that it is not "frozen." Never turn on the motor if the pump does not turn easily. Close the stopcocks, turn on the pump, and *immediately afterwards turn on the cooling water* (not too large a flow). Observe the order of these last two operations; if the water is turned on first, particularly in the winter when the water is very cold, the oil viscosity may be high enough to prevent the pump from turning and the motor may burn out.

Open the stopcock to the system. If there are no leaks the vacuum may get down to 0.3 Torr or below in 3 or 4 min. Open the stopcock to the needle valve and adjust it to obtain the desired steady-state pressure. To operate the swivel McLeod gauge, rotate the arm containing the mercury supply bulb slowly upward until the mercury level in the inlet tube is at the top graduation; read the pressure at the mercury level in the closed tube. Always remember to lower the swivel arm after every reading. A general discussion of the McLeod gauge is given in the section on vacuum gauges.

When need for the vacuum has ended, open the needle valve and the stopcock all the way to admit air; turn off the pump and cooling water. Empty out any material in the glass trap; clean and reassemble the trap. The trichloroethylene may be left in the Dewar flask if there is reasonable prospect of using the setup again within a short time; otherwise pour it into a designated waste bottle.

Drain and discard the used oil. Add 70 ml fresh SAE 30 motor oil from a polyethylene graduate or syringe through the hole in the exhaust chamber that is covered by a slotted screw; replace the screw and wipe up any spilled oil. Remember: *Always* change the oil after each use to prevent corrosion from dissolved contaminants.

Diffusion pumps The diffusion pump is the standard means of achieving high vacuum (about 10^{-6} Torr); with special care, ultrahigh vacuum (10^{-10} Torr) can be achieved. The operation of the diffusion pump is superficially similar to that of an aspirator, although the principles are quite different. The molecules of the gas being pumped diffuse into a jet of hot vapor which carries them along by essentially viscous flow inside a condenser in which the vapor is gradually condensed to a liquid of low vapor pressure. The liquid falls back into a boiler where it is again vaporized and brought to the jet nozzle. The exhaust is essentially isolated from the intake by a sort of "plug" of viscous-flowing vapor, which contains the gas being pumped as a minority component; the very slight pressure difference against which the pump must operate (normally less than 10^{-2} Torr) is overcome by the momentum possessed by the vapor as it emerges from the jet nozzle. The gas being pumped is carried away by a *forepump*, which is usually a rotary oil pump.

Two very different kinds of pump fluids are used in diffusion pumps. The *mercury diffusion pump* is the most common type used in small laboratory-bench vacuum systems. The *oil diffusion pump* is the preferred choice for larger systems, particularly metal systems.

The essential features of an all-glass mercury diffusion pump are shown in Fig. 6. In operation, the mercury boils at $\sim 185°C$ under an Hg vapor-pressure head of about 10 Torr (note the difference in meniscus levels in the small U tube through which liquid mercury is returned to the boiler). One disadvantage of mercury is its high vapor pressure at room temperature ($\sim 1.5 \times 10^{-3}$ Torr). To obtain a high vacuum, a cold trap chilled with Dry Ice and trichloroethylene ($-79°C$) or liquid

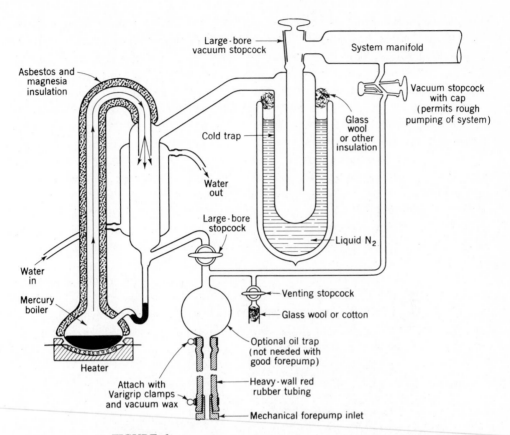

Large‑bore vacuum stopcock

System manifold

Asbestos and magnesia insulation

Vacuum stopcock with cap (permits rough pumping of system)

Glass wool or other insulation

Cold trap

Water out

Water in

Mercury boiler

Large‑bore stopcock

Liquid N₂

Venting stopcock

Glass wool or cotton

Optional oil trap (not needed with good forepump)

Heater

Attach with Varigrip clamps and vacuum wax

Heavy‑wall red rubber tubing

Mechanical forepump inlet

FIGURE 6
Typical all-glass mercury diffusion pump, shown with appropriate connections
to the vacuum manifold and to the mechanical forepump.

nitrogen ($77°K = -196°C$) is needed to condense out mercury vapor diffusing
toward the system from the pump. If not trapped out, the mercury vapor may impede
molecular flow of the gas being pumped (since its mean free path at its room-temperature vapor pressure is of the order of 1.5 cm), may tend to condense on colder
parts of the apparatus elsewhere, may amalgamate with metal components in the
system, or may contaminate a chemical product being handled in the system. Before
the heater of a mercury diffusion pump is turned on it is important to make sure that
the forepump has reduced the pressure below the limiting head pressure (about 10^{-2}
Torr), and that a *reliable* flow of cooling water is flowing through the condenser. If
water is turned on after the condenser has become hot, the glass will very likely crack,
even if it is Pyrex. A reliable (although expensive) precaution against damage is to

power the heater through the contacts of a holding relay, with the current through the hold coil passing through a flow-operated switch in the outlet side of the condenser, so that cessation of flow through the condenser—whether by water-main failure, blockage of the line, or a detached hose—will cut off the current to the heater. To turn the heater on again requires the positive act of resetting the relay. For protecting against floods from a detached hose, a solenoid valve can be placed in the water pipe line. A switch operated by a thermocouple gauge in the vacuum foreline can also be placed in the holding circuit to cut off the diffusion pump in the event of forepump failure. A suggested plumbing and electrical layout is shown in Fig. 7 (needless to say, a similar arrangement can protect a distillation setup from water failures, and the occupants of the floor below from floods). The cost of this setup must be balanced against the nuisance and health hazard of an accident in which mercury is widely dispersed throughout the laboratory; its use should be seriously considered when a system with a mercury diffusion pump is to be left unattended for prolonged periods of time.

The fluids used in the oil diffusion pump are usually high-boiling esters (e.g., di-*n*-octyl phthalate) or silicones. The latter are preferable if there is any possibility that the system may be accidentally vented to air while the pump is hot; this will usually result in destructive oxidation of the pump oil, but silicone oils are more resistant than most. **Warning** *Never vent an oil diffusion pump when it is hot.*

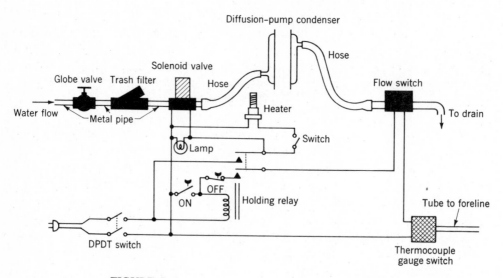

FIGURE 7
Suggested plumbing and electrical circuitry for protection from water-supply failure and floods, for use with diffusion pumps (and other things!).

An oil diffusion pump can often be used without a trap if it is sufficiently well provided with baffles to stop back-streaming molecules, for the oils used usually have room-temperature vapor pressures less than 10^{-7} Torr when pure. However, pump oils generally undergo some decomposition in service, and some decomposition products may find their way into the system. Some oil pumps (principally ones made of glass) have built-in fractionation systems to help maintain the purity of the oil. Another

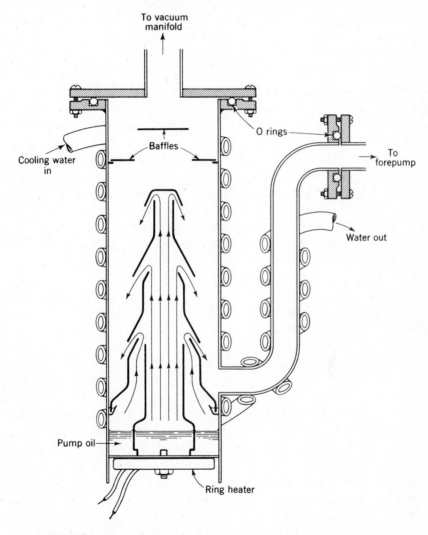

FIGURE 8
Schematic drawing of a multistage, all-metal oil diffusion pump. The arrows indicate the flow of oil vapor.

difficulty with oil diffusion pumps is the creep of a film of oil on the walls of the system; this can be prevented by use of baffles consisting of shallow trays containing a molecular sieve. Both back streaming and creep can be effectively controlled by use of a cold trap, provided the trap is opened and cleaned after each use.

All-metal oil diffusion pumps (see Fig. 8) are commonly used for applications which require high pumping speed and a low ultimate pressure. For such pumps, a cold trap is often replaced by a baffle which prevents both the back diffusion of oil and the creep of an oil film along the walls. The limiting forepressure at which a single jet diffusion pump will operate increases with decreasing annular spacing between the exit rim of the nozzle and the pump wall. Therefore, high-speed diffusion pumps usually have two or more stages of jet nozzles and are referred to as multistage pumps. By the proper design of the size and spacing of these jets, the pump acts like several separate diffusion pumps connected in series. The smallest, fastest stage is positioned nearest to the system, and the largest, slowest stage (which operates at the highest forepressure) is nearest to the forepump, as shown in Fig. 8. Oil diffusion pumps have higher intrinsic speeds than corresponding mercury pumps, since the oils have molecular weights between 300 and 500 compared with 200 for mercury. Thus an oil jet has a greater downward momentum component and will provide more efficient pumping.

To summarize: the mercury vapor pump is advantageous for small glass systems because of its cleanliness and consistent performance; good vacuums can be obtained year in and year out without replacement of the pump fluid. However, use of this pump entails the hazard of accidental exposure to mercury vapor. For large systems requiring high pumping speeds the oil pump is preferable, although contamination of the system with oil is more insidious than in the case of mercury and periodic changes of oil may be necessary.

Turbomolecular pump   This pump makes use of a turbine-like rotor, the periphery of which rotates at a speed that exceeds the mean molecular speeds of all ordinary components of air. The Welch 3102D turbomolecular pump has an ultimate vacuum of 10^{-9} Torr, and a pumping speed of 260 liter sec^{-1} from 10^{-2} to 10^{-8} Torr. Its principal limitation is that hydrogen and helium (which have high molecular speeds) are not effectively pumped, so the pump should not be used in a laboratory environment where either of these gases is involved in any way inside the system or is a significant component of the ambient air (e.g., from evaporation of liquid hydrogen or liquid helium refrigerants, or use of helium in leak-testing). This pump is not often used in laboratory applications, as it is expensive and demands careful maintenance. However, it is a very clean pump, requires no trap between itself and the system, and is not sensitive to contamination (although in some cases the forepump should be protected by a cold trap).

Ion pump A pump now much used in ultrahigh vacuum systems is the ion pump, of which the Varian Vac-Ion pump is an example. In this pump a "gas discharge" such as is commonly observed in air and other gases at about 10^{-2} Torr (e.g., neon signs, mercury and sodium lamps) is generated by a strong electric field and maintained all the way down to 10^{-10} Torr by application of a strong magnetic field from an Alnico magnet. This field constrains the ions to move in circular or helical paths thereby greatly increasing their path lengths between wall collisions and giving them the opportunity of colliding with neutral gas molecules and ionizing them so as to continue the discharge. The ions eventually strike certain parts of the specially designed titanium electrodes, eroding them by a process known as sputtering. The titanium atoms come to rest on other titanium pump surfaces, to maintain a fresh and very active titanium surface where the real pumping action takes place by surface adsorption. As the pumped molecules are adsorbed they are continually buried by additional titanium atoms which refresh the surface.

The ion pump needs no forepump, but it is necessary to reduce the pressure initially to 10^{-2} or 10^{-3} Torr before turning the pump on. With care, this may be accomplished with a rotary oil pump, but it is often done instead with a "sorption pump" such as the Varian Vac-Sorb pump in which the air in the system is adsorbed on a molecular sieve (synthetic zeolite, type 5A) chilled with liquid nitrogen. This type of pump has the advantage of presenting no danger of contaminating the system with oil.

The ion pump has an ultimate vacuum capability of 10^{-10} Torr or better, and may have a pumping speed for air of several hundred liters per second depending on its size. The components of air are chemisorbed on the fresh titanium and are thus permanently fixed by chemical bonds. Noble gases (helium, argon, etc.) are pumped at a much slower rate, as they are only physically adsorbed on the surface and are held principally by being "plastered over" by titanium atoms. An ion pump that has pumped a considerable quantity of argon is subject to a condition called argon instability in which bursts of argon are released at intervals into the system; other noble gases show a similar effect.

The ion pump is widely used in all-metal systems, usually fabricated of stainless steel (with seams heliarc welded on the inside), with copper gaskets between machined stainless-steel flanges. These systems are *baked out* for several hours at 250°C (or as high as 400°C) to desorb surface gases (mainly water). The ion pump is baked out at the same time (no higher than 250°C, to protect the magnet) while in operation. The pressure in such a system is usually measured with a Bayard-Alpert ionization gauge or a Redhead magnetron cold-cathode gauge.

An important application of systems of this kind is to fundamental studies of ultraclean surfaces, and of chemisorbed molecular monolayers on such surfaces, by such techniques as low-energy electron diffraction (LEED). The ultrahigh vacuum is necessary to provide ample time for the preparation and study of such surfaces. Assuming a molecular cross section for air of 10 Å^2 and a sticking probability of unity, it would take only about 5 sec for the surface to be covered with a monolayer at 10^{-6} Torr; this time is extended to many hours at 10^{-10} Torr.

Cryopump   As the etymology of the term cryopump or cryogenic pump implies, this pump operates at intense cold; in its most effective form it is a liquid helium-chilled "cold trap." The pumping surface is at 4.2°K if the liquid helium is at one atmosphere, or lower if the pressure over the liquid helium is reduced by pumping. The liquid helium must be protected from too rapid evaporation by cold walls and baffles chilled with liquid nitrogen (77°K) which greatly reduce the input of energy in the form of blackbody radiation. The cryopump would normally be used in a system already pumped to a high or ultrahigh vacuum by some other pump. It would be useful, for example, for rapidly pumping down argon following argon-ion bombardment of a surface in a LEED apparatus having an ion pump. The liquid helium cryopump is effective under the best conditions down to 10^{-13} to 10^{-15} Torr for all gases except helium, hydrogen, and possibly neon.

VACUUM GAUGES

A large variety of gauges is available for the measurement of low pressures, and the range of useful operation depends a great deal on the type used (see Table 1). This section contains a discussion of the principles of operation for the most commonly used vacuum gauges. More specific details of construction and operation are given

Table 1   OPERATIONAL RANGE OF
VARIOUS VACUUM GAUGES
(given in Torr)

Mercury manometer	1–1000
Oil manometer	0.03–10
Thermocouple gauge	10^{-3}–3
Pirani gauge	10^{-4}–0.3
McLeod gauge	10^{-5}–1
Ionization gauges	
Alphatron	10^{-4}–100
Philips	10^{-6}–10^{-2}
Thermionic	10^{-8}–10^{-3}
Bayard-Alpert	10^{-11}–10^{-3}
Redhead (magnetron)	10^{-12}–10^{-3}

in the references listed under General Reading and in the appropriate manufacturers' pamphlets.

Manometers   A U-tube manometer filled with mercury is simple to construct, requires no calibration, and operates over a wide pressure range. With each arm connected to a separate region, the manometer can be used to obtain differential pressures or can be used as a null device. Most commonly, one arm is evacuated and the manometer directly indicates the total pressure. To obtain the absolute pressure, a temperature correction is necessary. This correction allows one to convert the observed reading p (mm) in millimeters of mercury to Torr (1 Torr $\equiv$ 1 mm Hg at 0°C and standard gravity):

$$p \text{ (Torr)} = p \text{ (mm)} \left[1 - \frac{3\alpha t - \beta(t - t_s)}{1 + 3\alpha t} \right]$$

$$\approx p \text{ (mm)}(1 - 3\alpha t) = p \text{ (mm)}(1 - 1.8 \times 10^{-4}t) \qquad (23)$$

where 3α is the volume coefficient of expansion for mercury (average value of 3α between 0 and 40°C is 18.2×10^{-5} deg^{-1}), β is the linear expansion coefficient of the scale, t is the centigrade temperature of the manometer, and t_s is the temperature at which the scale was graduated. Equation (23) does not contain any correction for the small effect due to the difference between local and standard values of gravity. For precise work, large-bore tubing ($\sim$25 mm o.d.) is used to avoid distortion of the meniscus, and the mercury levels are measured with a cathetometer (see Chap. XVIII). A suitable manometer design for medium-precision ($\pm$0.5 mm) work is shown in Fig. 9. In this design the scale is constructed by carefully covering a smooth hardwood board with good-quality millimeter paper and the U tube is rigidly attached to this board. Before being mounted and filled with vacuum-distilled mercury, the manometer should be thoroughly cleaned inside (with nitric acid) and dried, evacuated with a good forepump, and preferably flamed with a torch to release adsorbed water.

For pressures between 0.03 and about 10 Torr, oil manometers are more accurate than mercury manometers, since oil has a much lower density. For example, dibutyl phthalate, which is often used, has a density of 1.046 g ml^{-1} at 20°C. The absolute pressure is given by

$$p_0 \text{ (in Torr)} = \frac{\rho_{\text{oil}}}{13.59} h \qquad (24)$$

where h is the reading of the oil manometer in millimeters. However, oil-filled manometers have two drawbacks. Most gases are quite soluble in oil, and this may cause frothing or erratic behavior when the pressure changes suddenly. Also, the wetting of the walls of the manometer by a viscous oil causes sluggish response.

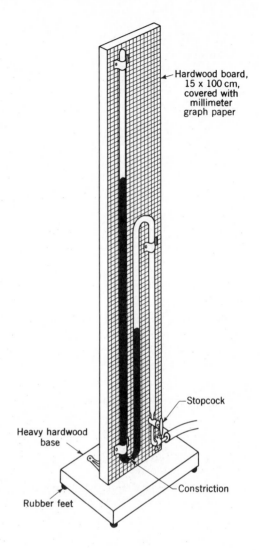

Hardwood board,
15 x 100 cm,
covered with
millimeter
graph paper

Stopcock

Heavy hardwood
base

Constriction

Rubber feet

FIGURE 9
Design for a closed-tube laboratory
manometer. The capillary constriction
acts to retard the mercury flow and thus
prevents damage in case of sudden
changes in pressure.

McLeod gauge This gauge is very widely used because of its simplicity of operation
and its wide pressure range. In addition, the readings depend only on the geometry
of the gauge and not on the properties of the gas whose pressure is being measured.
Its operation involves the compression of a known large volume V of gas at an unknown pressure p into a known small volume v where the final pressure p' can be
measured. Both p' and p are quite low, and the perfect-gas law can be used in the
form $pV = p'v$, since the compression is carried out at constant temperature.

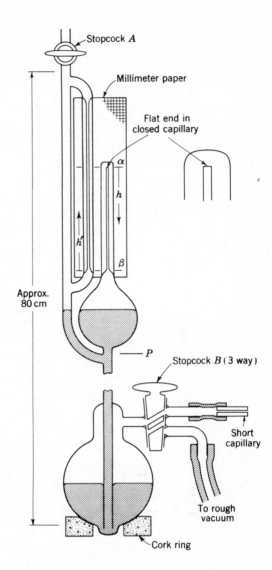

FIGURE 10
A typical McLeod gauge.

A typical McLeod gauge design is shown in Fig. 10. A closed capillary (0.5 to 2 mm i.d., length 12 to 20 cm) of uniform cross section is sealed to the top of a large bulb (volume 200 to 700 ml). A reference capillary of the same bore is mounted parallel to the closed capillary to minimize the effects of capillary depression on the readings of the differences in mercury level. The cross-sectional area A of the closed capillary is determined before the gauge is assembled by measuring the length and weight of a slug of mercury placed in the capillary. The volume V is measured from the cutoff plane P and includes the volume of the bulb and that of the capillary tubing. This is determined by measuring the weight of water required to fill the gauge prior to sealing off the closed capillary but after the side arm is attached.

To make a reading with the McLeod gauge, open stopcock A *slowly* and, if necessary, reduce the pressure on the mercury reservoir to prevent mercury from rising above the cutoff plane P. After waiting about a minute for pressure equilibrium, *carefully* open stopcock B to the atmosphere to allow the mercury level to rise slowly in the gauge. As the mercury rises above the plane P, the gas contained in the bulb is compressed into the closed capillary where its pressure can be measured.

There are two methods of reading a McLeod gauge. For low-pressure measurements the mercury is raised until the level in the reference capillary is at point α (even with the top of the closed capillary) and the level in the closed capillary h is then read. Applying the perfect-gas law we find that $pV = (h + p)(Ah)$; neglecting p in comparison with h, one obtains

$$p = \frac{A}{V} h^2 = kh^2 \qquad (25)$$

where the apparatus constant k is completely determined by the calibration data. For a typical gauge with $A = 1$ mm^2 and $V = 200$ cm^3, $k = 5 \times 10^{-6}$ Torr/(mm Hg)^2. With such a gauge, pressures down to 10^{-5} Torr can be measured easily and 10^{-6} Torr (within a factor of 2 or 3) can be measured with difficulty. Especially when measuring such low pressures, be sure to tap the capillary tube gently with finger or pencil while adjusting the meniscus position.

A "stick vacuum" is one in which the gas bubble at the top of the closed capillary is so small that when the mercury level in the comparison tube is lowered a few centimeters the column in the closed capillary "sticks" as a result of surface tension. Further lowering causes the column to break loose from the top. Usually, in determining whether there is a "stick vacuum," the level in the comparison capillary is initially raised one or two centimeters above the top of the closed capillary before it is lowered. With a gauge of the dimensions given above a stick vacuum would be less than 10^{-6} Torr.

The second method of reading a McLeod gauge is useful at higher pressures (above ~ 0.05 Torr). In this technique the mercury is raised until the level in the closed capillary is at a fixed point β (near the bottom of the capillary) and the level in the reference capillary h' is then read. Again, by using the gas law and neglecting p in comparison with h', we obtain

$$p = \frac{v}{V} h' \qquad (26)$$

where v, the volume between α and β, equals $A(\alpha - \beta)$. The range of the gauge when used in this way can be extended to higher pressures by having a small bulb just above β so that v is larger.

The procedure for operating a swivel-type McLeod gauge is briefly described on page 614. These utility gauges are useful in the 0.01- to 5-Torr range.

There are several disadvantages to the McLeod gauge which somewhat offset its attractive features. Since it contains mercury, a cold trap (containing Dry Ice or liquid nitrogen) may be necessary between the gauge and the system to avoid unwanted mercury vapor. It is slow in operation, which is a handicap when used for leak detection. Finally, it cannot measure the pressure of a vapor which will condense when compressed into the capillary. Such condensation will occur as soon as h (or h') exceeds the room-temperature vapor pressure of the substance.

Pirani gauge   The operation of this gauge depends on the pressure variation of the thermal conductivity of a gas at low pressures. A fine wire or ribbon filament, often in the form of a coil for greater sensitivity, is sealed into a glass envelope which is connected to the vacuum system. The filament is made of a metal with a high-temperature coefficient of resistance (such as platinum, nickel, or tungsten) and is heated electrically. For a fixed energy input the temperature of the filament, and therefore its resistance, will depend on the gas pressure (the resistance will decrease as the pressure increases). The filament resistance can be measured by making the gauge one arm of a Wheatstone bridge. Since the performance of the gauge depends on its geometry and design and on the nature of the gas, it is necessary to calibrate a Pirani gauge against a McLeod gauge. This can be done by maintaining a constant filament current and determining the resistance as a function of pressure or by determining the current necessary to maintain a constant resistance at each pressure. Ambient-temperature fluctuations can cause erratic readings, since thermal conduction depends on the temperature of the walls of the envelope as well as that of the filament. For accurate measurements, one can thermostat the gauge or use a compensating dummy gauge in an opposite arm of the Wheatstone bridge. This dummy gauge is made as nearly like the measuring gauge as possible except that it is evacuated and sealed off.

Thermocouple gauge   This type of gauge, which also depends on the pressure variation of gaseous thermal conductivity, is almost identical with the Pirani gauge. It differs from the Pirani gauge in that the temperature of the filament (rather than the resistance) is measured by a very fine wire thermocouple which is welded to the midpoint of the filament. The thermocouple output is usually measured with a low-resistance microammeter (~ 70 Ω) connected in series with the thermocouple. The thermocouple gauge must also be calibrated against a McLeod gauge; as with the Pirani gauge there are two ways to perform this calibration. However, the standard method of calibration and operation is to maintain constant filament current and obtain the thermocouple emf as a function of pressure. The thermocouple gauge is a rugged, inexpensive instrument which is well suited for leak detection, since it is direct reading and has a rapid response to pressure changes.

Thermocouple and Pirani gauges may be electrically coupled to a switch which will shut down the system in the event that a bad leak or pump failure causes the pressure to rise above a preset level. Alternatively, such a switch may be used to control a solenoid-operated throttle valve between system and pump so as to maintain the vacuum at a specified level (see Manostats).

Ionization gauges Ionization gauges are of several types, having in common the production of positive ions from the molecules present and the measurement of the ion current to a cathode. In the *Alphatron gauge* the ions are produced by the action of alpha rays emitted by a small amount of radioactive material. In the *Philips gauge* a cold-cathode discharge produces the ions; the *Redhead* or *magnetron gauge* uses a magnetic field to constrain the ions to long spiral paths so that the cold-cathode "discharge" can be maintained as low as 10^{-13} Torr.

However, most ionization gauges are *thermionic gauges* which obtain positive ions through bombardment of the gas molecules with electrons thermionically produced from a heated filament and accelerated by a grid which is biased 100 to 200 V positive with respect to the filament. The ions are collected by a cathode typically about 20 to 50 V negative with respect to the filament, and the ion current in the cathode circuit is measured with a sensitive galvanometer or a vacuum-tube electrometer.

Thermionic ionization gauges are usually constructed like electron tubes (with a central filament, outside cathode plate, and anode grid in between), but with their glass envelopes provided with a stem of large-diameter tubing for connection to the system. In large metal systems, however, it is common to use the gauge in the form of a so-called "nude gauge"; that is, the gauge structure is supported from electrical feedthroughs in its mounting flange so that it protrudes, without any envelope, into the vacuum chamber itself. This has the advantage of substantially eliminating errors due to the limited conductance of the connecting tube, in view of the fact that the gauge itself often acts either as an ion pump or as a virtual leak (owing to outgassing of the filament). The filament of the gauge is usually tungsten, either bare or coated with a thoria activator which permits it to be operated at a lower temperature. Tungsten is notorious for emitting large amounts of CO until it has been thoroughly outgassed by many hours of incandescent heating under high vacuum. Iridium wire is sometimes used because it gives less gas. One of the most satisfactory filaments is rhenium wire activated with a coating of lanthanum boride; this operates at a much lower temperature than a tungsten filament and can readily be prepared in the laboratory,[4] although one must have a system in which the initial sintering can be done at a vacuum better than 10^{-7} Torr. In any case, when a thermionic ionization gauge is employed, attention must be given to thoroughly outgassing the filament under vacuum before pressure readings are taken.

Thermionic ionization gauges cannot measure pressures lower than $\sim 10^{-8}$ Torr since the minimum cathode current is produced by a phenomenon altogether independent of the presence of gas molecules, namely the photoelectric ejection of electrons from the cathode plate by soft x-rays produced by the impact of the electrons on the anode grid. This led Bayard and Alpert to invent an "inverted" ionization gauge in which the cathode attracting the positive ions is a slender wire *inside* the anode grid, and the thermionic filament is outside. The *Bayard-Alpert gauge* has an "x-ray limit" which is of the order of 10^{-10} or 10^{-11} Torr, and is widely used in ultrahigh vacuum work.

The ionization gauge has to be calibrated (e.g., against a McLeod gauge at relatively high pressures) and the calibration is in general different for every gas. The ion current for a given gas is, however, linear with the gas pressure over a very wide range, extending almost down to the x-ray limit.

The ionization gauge has a number of disadvantages, besides those implicit in the foregoing paragraphs. Considerable auxiliary electrical equipment is needed. The hot filament may decompose organic vapors that may be present, it may be "poisoned" by some gases and thereby lose its emissivity, it can easily burn out if exposed to air while hot, and it gradually erodes in use through evaporation, slow oxidation, and ion bombardment. Many gauges (especially nude gauges) are so constructed that a burned-out or poisoned filament can easily be replaced in the laboratory.

The pumping or "clean-up" characteristics of the ionization gauge, which give rise to some uncertainties in measuring the system pressure, can be reduced to tolerable proportions by restricting application of high voltages to grid and collector to the very short interval needed to measure the electron and ion currents.

Further details concerning ionization gauges are given by Dushman.[2]

Partial-pressure analyzer   The partial-pressure analyzer is a form of ionization gauge in which the positive ions are mass-analyzed either by a conventional sector-mass-spectrometer arrangement (employing a permanent magnet) or by a quadrupole-mass-spectrometer arrangement (which requires no magnet). The ion current is amplified either inside by an electron multiplier arrangement or outside by a vacuum-tube electrometer, or both. Devices of this kind are sold by Varian and by CEC (now du Pont Instruments) and typically are capable of unit mass resolution for singly ionized molecules up to masses of 50 or 100 atomic units. The sensitivity is of the order of 10^{-10} Torr over most of the mass range. The partial-pressure analyzer is very useful in determining the actual composition of the residual gases inside a vacuum system, and in hunting for leaks (with a helium "torch" outside the vacuum);

it is virtually indispensible for surface adsorption and desorption studies under ultra-high vacuum.

Other gauges   The *capacitance manometer*, although expensive, possesses nearly all of the advantages of a McLeod gauge without the disadvantageous presence of mercury vapor. The pressure is sensed by the deflection of a membrane which is one element of a capacitor attached to a capacitance bridge. This gauge is an absolute instrument, is independent of the kind of gas present, is available in many ranges and sensitivities, and in its most sensitive form can measure pressures in the neighborhood of 10^{-4} Torr to within a few percent. It can also be used to measure differences in pressure between two vessels. The *Knudsen gauge* works on the radiometer principle, using an optical system to measure the angular deflection of a fiber-suspended vane which is heated on one side by a lamp so that molecules striking it rebound with increased momentum. As usually constructed, the gauge can be used in the 10^{-2} to 10^{-5} Torr range. The *vibrating fiber gauge*, which depends on the rate of damping of a vibrating fiber, is operable at 10^{-1} to 10^{-4} Torr; some vibrating quartz membrane gauges and torsion-pendulum manometers operating on the same principle can be operated to one or two decades lower pressure. The *spiral quartz manometer* (employing a helix of flattened fused silica tubing with a mirror at one end for measuring angular deflection optically) can be operated from many Torr down to about 10^{-3} Torr, and is particularly useful where the chemical inertness of fused silica is needed.

MISCELLANEOUS VACUUM COMPONENTS

Cold traps   Some discussion of cold traps has already been given; see Figs. 1, 5, 6, and the accompanying text. A trap should be designed so that a molecule must make two or more collisions with a cold surface in order to get through, but must be designed so that its conductance does not unduly limit the overall pumping speed of the system. The trap should also be designed with enough space below the inner tube to accommodate all condensate, once it has liquefied and run down the sides, without blocking the passage for air or other gases.

For a mechanical-pump vacuum, a mixture of Dry Ice and trichloroethylene ($-79°C$) is satisfactory; the latter serves simply as a medium for heat transfer from the wall of the trap to the Dry Ice. Acetone is often used instead but is not recommended because of its high volatility and flammability. Add Dry Ice slowly to the liquid, waiting for gas evolution to cease before adding the next portion; when further addition causes no evolution, as much may be added as is needed to maintain the low temperature for some time.

For high vacuum, the trap should be chilled with liquid nitrogen ($77°K$) or liquid air (90°K) in order to reduce sufficiently the vapor pressure of volatile substances that may be present, especially water. **Caution:** Avoid all contact between liquid air and any organic or other oxidizable substance. In order to avoid condensation of liquid air inside the trap when liquid nitrogen is the refrigerant, the trap should not be chilled until the pressure in the system has been reduced below about 1 Torr.

The efficiency of a trap is greatly increased by the addition of an adsorbent, such as activated charcoal or a molecular sieve (e.g., Linde 5A). However, it is very important in such a case to allow for desorption of gases as the trap warms up, especially if any significant quantity of gas may have leaked into the system and become adsorbed.

Joints and seals For many purposes it is convenient to fabricate a vacuum system in one piece: In an all-glass system the parts are fused together with a glass-blower's torch; in an all-metal system parts are soft-soldered, silver-soldered, brazed, or welded. It should be noted that soft-soldered joints, especially large ones, are relatively unreliable for vacuum purposes. (Never hard-solder or braze a joint which has been soft-soldered, unless every trace of soft solder has been removed by machining.) Metal and glass parts also can be permanently sealed together in a vacuum-tight way. This is most easily done with commercially available Kovar seals.[5] Kovar can be soldered or brazed to other metals, and the glass part, usually Pyrex with a graded seal to the special glass that is sealed to the Kovar, is easily glass blown to other Pyrex glass. Other vacuum-tight glass-to-metal seals are Housekeeper seals (glass to thin OFHC copper, where the mismatch of linear expansion coefficient is compensated by the softness of the copper), platinum wire in soft glass, and tungsten wire in Pyrex glass. These last two seals, useful for electrical feed-throughs, are fairly easy to make, although the tungsten requires a sodium nitrite flux. With skill it is possible to make a vacuum-tight seal of "five nines" (99.999 percent pure) platinum wire or ribbon in Pyrex, despite the disparity in linear expansion coefficients. For bringing electrical contacts into metal vacuum systems it is convenient to use the commercially available Stupakoff glass or ceramic-insulated lead-throughs, which may be soft-soldered into holes in the metal.

Often it is desirable to make a system demountable. For an all-glass system, standard-taper ground glass joints or ground ball joints lubricated with a good stopcock grease (Apiezon type L or N, Dow-Corning Silicone High-Vacuum) may be used. Recently glass joints have become available with O rings of neoprene (usually lubricated with a trace of stopcock grease) or of Viton (a plastic with extremely low vapor pressure, usually needing no grease to make a good seal). The two similar glass parts are clamped together with the O ring between them. This joint is more flexible than the tapered ground joint although much less flexible than the ball joint.

Metal systems are frequently assembled with flanges and O rings (circular gaskets with a circular cross section) or other gaskets. Metal gaskets (OFHC copper,

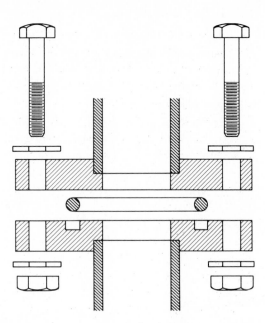

FIGURE 11
Metal O-ring flange joint.

gold, aluminum) between specially designed flanges are commonly used in all-metal ultrahigh vacuum systems which must be baked out. Neoprene O rings are usually adequate for high vacuum (10^{-6} Torr), and Viton O rings are sometimes used at very much lower pressures. When an O ring is used with metal flanges, a smooth groove (rectangular or trapezoidal in cross section) is machined into one flange and the facing surface is machined flat and as smooth as possible (see Fig. 11). Scratches across an O-ring groove or its mating surface are potential sources of leaks and must be carefully avoided. The O-ring groove should have a depth equal to about four-fifths of the cross-sectional diameter of the O ring, and a cross-sectional area only a little larger than that of the O ring. Excellent suggestions for the proper design of gasket grooves are given by O-ring manufacturers[6] and should be followed in detail. In use, the O ring is very lightly greased with stopcock grease and placed in the groove; the mating flange is then bolted on so that the metal surfaces are everywhere in direct and firm contact. O rings that are left for long periods bolted between flanges tend to lose their elasticity and may leak, especially if they have been exposed to heat.

For permanently closing off an evacuated glass system, a constriction in the line (with previously thickened walls) is first flamed for several minutes to outgas it as much as possible and is then heated with the torch until it melts, collapses, and can be pulled off. For a metal system, a tubulation of OFHC copper may be pinched off with a special tool. This is usually sufficient by itself, but a coat of soft solder carefully applied over the pinch-off will render the vacuum more secure.

Stopcocks and valves It is almost always essential that various parts of a vacuum system can be isolated from each other. This is accomplished by stopcocks and valves. In an all-glass system, glass stopcocks lubricated with a low-vapor-pressure stopcock grease are generally used. The bore of a stopcock should be large enough that its conductance, calculated with Eq. (14), is in considerable excess of that of other more expensive components in series with it. However, large-bore stopcocks are themselves expensive, and compromises may be necessary.

Stopcocks for high-vacuum use are ground to very good fit with small clearances,† so that little grease is necessary and air streaks do not easily arise on turning the stopcock. A diagonal bore stopcock is less susceptible to leaks due to streaking than a straight-bore stopcock, since the "off" position of the former is 180° from the "on" position, rather than 90° as in the latter.

The technique of greasing a stopcock with enough, but not too much, grease and no air bubbles or streaks is important in vacuum work. Both parts of the stopcock are first thoroughly cleaned with solvents (a pipe-cleaner is useful). With a spatula or matchstick the grease is applied to the stopcock plug in four streaks running the length of the plug along each side, preferably avoiding the bore orifices. The optimum amount of grease in each streak has to be learned by experience; it depends on the size of the stopcock, but as a rough guide one may say that each streak should be 2 to 3 mm wide and about $\frac{1}{4}$ mm thick; bubbles should be avoided. The plug is then inserted into the barrel in the open position, keeping the plug centered as well as possible. The plug is pressed firmly into the barrel so that the grease spreads; to facilitate this, wiggle the plug back and forth a few degrees. *After* the streaks have joined and the grease has spread over the entire surface, rotate the plug a few times. Inspect the stopcock for striations from air bubbles, which may indicate that too little grease was used, or accumulations of excess grease in the bore, which indicates that too much was used. If either is excessive, clean and regrease the stopcock.

Good-quality vacuum greases have a high viscosity and a very low vapor pressure ($\sim 10^{-6}$ Torr). It is important to choose a grease of the proper viscosity for the ambient temperature of the stopcock. If the grease is too firm, the stopcock will be difficult to turn and striations may appear; if too soft, the grease may flow out and leave too thin a film to make a vacuum seal. Apiezon (type L or N) and Dow-Corning Silicone High-Vacuum greases are recommended.

Recently glass and Teflon needle valves, sealed with O rings, have appeared as substitutes for stopcocks. These may be advantageous where fine control of gas flow is important. However, careful use and maintenance is necessary to prevent leaks

---

† Excellent vacuum stopcocks are made from Pyrex by Corning Glass Co. The less-expensive types, marked with a V, have interchangeable plugs and barrels; the better grades are individually ground and are numbered on both the plug handle and on the barrel. Certified stopcocks, for which the exact leakage rate is specified, are also available.

past the O-ring seals and seats; ordinary vacuum stopcocks are more reliable for most purposes.

For metal systems a great variety of high-conductance valves have been developed, many of which are operated through flexible metal bellows to obviate the possibility of atmospheric leaks. Some of these are sealed with neoprene or Viton O rings. For ultrahigh vacuum purposes valves have been developed in which stainless steel seats against a copper or gold gasket. Such valves may be baked out in the open position; a few can be baked out in the closed position without damage.

MANOSTATS

It is not always desirable to pump the system down to the lowest pressure possible with the means available; sometimes, as in a reduced-pressure distillation or pumping on a refrigerant bath, it is important to maintain the pressure at a fixed value. This calls for some sort of *manostat*, an analog of the thermostat which holds temperature constant.

A manostat can be constructed with any kind of pressure gauge the reading of which can be converted to an electrical signal, even if the signal is only an "off-on" signal depending on whether the pressure is less than or greater than a previously designated quantity. In the case of a mercury manometer this can be accomplished with a needle making electrical contact with the mercury meniscus. The electrical signal is used to open or close either a valve connecting the system to the pump, or a leak admitting air to the system. The latter can be constructed from a simple electrical relay in which the armature, padded with rubber or neoprene, covers or uncovers the end of a capillary tube; see Fig. 12a.

Less complicated and more ingenious devices can be made to control the pressure. Some of these depend on the fact that surface tension prevents liquid mercury from penetrating a fritted-glass disk without considerable applied pressure, while gas can penetrate it easily. The gas penetrates such a disk against a given hydrostatic head of mercury on the other side; see Fig. 12b.

A popular form of manostat makes use of the "Cartesian diver" principle;[7] see Fig. 12c. The "diver" is a float consisting of an inverted cup in a fluid such as mercury. The position of the diver depends on the difference between the system pressure and the reference pressure inside the cup (which can be adjusted to any desired value). When the system pressure is less than the reference pressure, the diver will rise until its top surface bears against the orifice to the vacuum pumping line. This cuts off the gas flow from the system until the pressure builds up again. When the system pressure becomes greater than the reference value, the diver sinks and the orifice to the pump is open. Thus the system pressure cycles back and forth over a

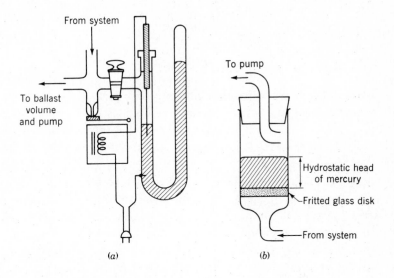

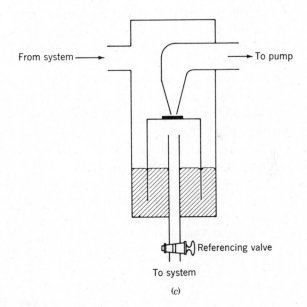

FIGURE 12

Manostats (schematic.) (*a*) Feedback manostat employing mercury manometer with adjustable wire contact and leak operated by a converted relay; (*b*) device using fritted glass disk and adjustable head of mercury; (*c*) Cartesian diver manostat.

narrow range centered at the preset value. These devices are available commercially from Greiner Scientific Corp.

DESIGN OF VACUUM LINES

Most vacuum problems encountered in physical chemistry require small "static" vacuum lines (i.e., systems of low pumping speed) capable of achieving ultimate pressures of 10^{-6} Torr or below. For such purposes an all-glass system is generally used. It has the advantage of ease of construction, even for complex designs, without the use of couplings or gaskets; it can be easily degassed after assembly; and the detection and repair of leaks are usually straightforward.

It may be helpful to make some qualitative generalizations about design parameters on the basis of the equations presented at the beginning of this chapter. In the high-vacuum region the conductance is determined by molecular flow and the overall pumping speed is given by Eq. (18). To achieve a desired pumping speed the conductances of all series components must be in excess of this value. For expensive components, such as the pump and vacuum stopcocks or valves, the speed or conductances should not be greatly in excess of those needed; for inexpensive components, such as glass tubes, the conductance should be made high enough that their impedances make up only a small part of the total. Thus the tubes in high-vacuum lines (and especially in ultrahigh-vacuum lines) should be as large in diameter as practicable, should be as short as possible, and should have as few bends as possible. Careful thought should be given to the reduction of valves and stopcocks to the minimum number really needed, as valves and stopcocks are expensive when of high conductance, often cannot be baked out, and often provide unwanted sources of gas (dissolved in stopcock grease) and opportunities for leaks. In the pressure region above 10^{-1} Torr, viscous flow becomes important and the conductance for a given tube becomes considerably larger than for pure molecular flow. Thus, for pumping at higher pressures (as in the line between a diffusion pump and forepump) smaller diameter tubes can be used.

In order to build and repair a glass vacuum system one must be familiar with certain techniques of glass blowing. An adequate description of the method for making even simple seals is beyond the scope of this chapter; the student should refer to one of the several detailed books on this subject.[8] Indeed, it is our opinion that even extensive reading about the proper techniques is not sufficient for the beginning glass blower. The advice and demonstration of a skilled worker and considerable personal practice are required to achieve the proper manual dexterity. Certain general principles can, however, be mentioned briefly. Fortunately, glass has a quite high mechanical strength for compression; to take advantage of this, vacuum lines are usually constructed with cylindrical or spherical sections. In addition to strength, a low coefficient of thermal expansion is needed to avoid cracking on sudden or local temperature change while blowing the glass. Although fused silica is excellent in this regard ($\alpha = 6 \times 10^{-7}$ deg^{-1}), it is impractical for general use owing to very high cost and a very high softening point of $\sim 1200°C$. (However, silica bulbs are used for high-temperature applications.) Pyrex glass is the proper type to use for general-purpose vacuum work. It has a linear thermal expansion coefficient of 3×10^{-6} deg^{-1}, which is about one-third that of soda-lime glass ("soft" glass).

Since the annealing temperature[†] for Pyrex is 560°C and the proper working temperature (point where glass is fluid enough to flow readily) is about 800°C, an oxygen-gas torch is required to blow this glass. In general, it is easiest to work with tubing between 8 and 15 mm in diameter. The danger with small tubing is that the walls will collapse and form a solid plug; the difficulty with large tubing is that uneven heating over the surface will cause strains and subsequent cracking. Standard practice involves slowly heating a large area of the glass with a large, gas-rich flame until it begins to soften. Then a smaller, hotter flame is used to work the glass. An inexperienced glass blower may have better success by working with the hot flame on only part of the joint at a time, but it is vital to avoid appreciable cooling anywhere until the joint has been completed and annealed. After the seal is completed, the entire area must be heated enough to relieve any internal strains and then cooled slowly through the annealing temperature range. This is commonly done by lowering the temperature of the flame until a deposit of carbon soot appears on the glass surface.

Mercury diffusion pumps made of glass are inexpensive and convenient for a general-purpose vacuum system. The diffusion pump is connected to the forepump by a short length of *heavy-wall*, large-diameter, red-rubber tubing. It is often wise to place a ballast bulb or oil trap in this section to avoid contamination if oil accidentally backs up from the forepump. Provision for isolating the diffusion pump and for rough pumping on the system is also advantageous (see Fig. 6). The Hg diffusion pump is connected to the manifold via a cold trap and large stopcock. This manifold is a length of large-diameter (~ 35 mm) tubing mounted horizontally at a height of at least 1 m from the base of the vacuum bench. A vacuum gauge (McLeod gauge or thermocouple and ionization gauge) and the desired experimental apparatus are attached directly to this manifold via stopcocks. Special vacuum stopcocks (preferably with an evacuatable cap on one end of the barrel, as shown in Fig. 6 on p. 615) must be used. The vacuum system should be carefully designed so as to minimize the number of stopcocks required, and the procedure for operating a vacuum system should be thought out so as to minimize the number of stopcock

---

[†] At the annealing temperature, strains in the glass will disappear in a few minutes and an evacuated bulb will collapse.

operations. In some cases it is necessary to make part of the apparatus demountable from the vacuum line. The use of long standard-taper joints is best for this purpose. Ball-and-socket joints or O-ring joints (p. 629) can be used if a certain amount of flexibility is needed, but they do not make as reliable a vacuum seal as tapered joints. The entire vacuum line should be clamped securely to $\frac{1}{2}$-in. rods which are mounted on a sturdy vacuum bench. The base of this bench should be at least a foot off the floor in order to provide space for locating the forepump.

In order to obtain a high vacuum of 10^{-6} Torr or less, it is necessary to eliminate "virtual leaks" due to the desorption of gases and vapors from the glass walls and the slow evolution of dissolved gas from the stopcock grease. These can be reduced by thorough cleaning of the glass prior to assembling the line and by the use of scrupulously clean grease. However, the main cause of these virtual leaks is adsorbed water vapor on the glass. This water can be driven off by degassing (heating the walls to about 200°C under vacuum). After the vacuum line is completely assembled and pumped down, the walls are carefully heated with a soft flame from a hand torch while pumping is continued. Do not heat the stopcocks or any greased joints; instead, slowly rotate each of these several times to speed up the removal of dissolved gas. It will normally require protracted pumping to achieve a good vacuum on a new line. After this initial degassing, avoid venting the line to the atmosphere if possible.

For "kinetic" vacuum lines where there is a steady influx of gas from some unavoidable source or when pumping on refrigerant baths, high pumping speed is needed. This requires large oil diffusion pumps and large-diameter tubing; therefore, all-metal systems are normally used. The construction of such lines requires considerable knowledge of "vacuum plumbing" and of special equipment such as bellows devices, Wilson seals, and vacuum gaskets.

A general-purpose vacuum system employing an oil diffusion pump and a conventional mechanical forepump is shown in Fig. 13. This system is designed to provide a high vacuum in the left-hand manifold (HMV) and a forepump vacuum in the right-hand one (LVM). The forepump manifold pressure is monitored with a mercury manometer (M) and a swivel McLeod gauge (McL 2) or a thermocouple gauge (TC). The high-vacuum manifold is monitored with a conventional McLeod gauge (McL 1) or an optional Bayard-Alpert ionization gauge (IG). Aside from the thermocouple and ionization gauges (which may be attached to the manifolds wherever desired), all pumps, traps, and gauges are fixed in position, while the system manifolds and their appendages are detachable (with O-ring joints) to provide easy adaptability to the needs of any particular experiment. The oil diffusion pump (ODP) can be isolated from the system by valves (butterfly valve BV and 3-way stopcock) and bypassed so that the high-vacuum manifold can be operated with the forepump (FP) alone.

After all joints have been checked for tightness, the traps checked for cleanliness,

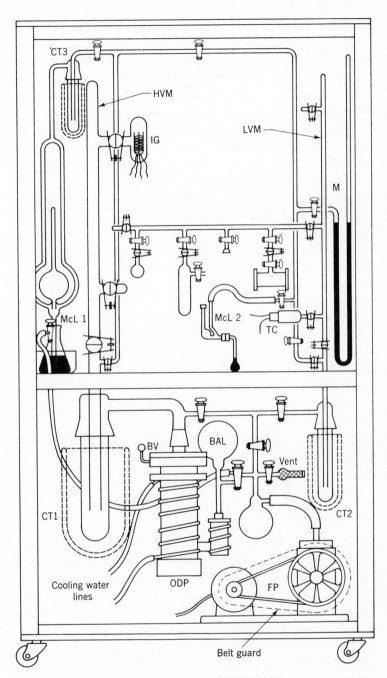

FIGURE 13
General-purpose vacuum system.

and the vent stopcock closed, the forepump may be turned on. If only a forepump vacuum is needed, all that remains is to chill the right-hand trap (CT 2) with Dry Ice and trichloroethylene. The manometer (M) and swivel McLeod gauge (McL 2) should be closed off with their stopcocks except when readings are being taken, as a precaution against contamination. If the oil diffusion pump is to be used, first wait until the forepump has reduced the pressure of the entire system (including the McLeod gauge) to less than 1 Torr; then chill the left-hand trap (CT 1) with liquid nitrogen. Do not do this earlier, or liquid oxygen may condense out of the air. When the pressure is down to 10^{-2} Torr, the water cooling of the oil pump and then the electric heater of the pump can be turned on.

The McLeod gauge (McL 1) is operated by carefully venting the mercury supply bulb to the air so that the mercury rises (not too fast!) into the bulb. Slow the mercury down to a crawl as the mercury in the reference capillary approaches the level of the top of the closed one, and tap gently with a finger or pencil. Cut off the air when that level is reached, and read the depression of the mercury in the closed capillary. To "pull down" the mercury requires that the forepump be isolated temporarily from the diffusion pump and manifolds, and opened to the mercury supply bulb long enough to reduce the pressure there to a few Torr. A ballast bulb (BAL) stores gas from the diffusion pump in the meantime. After closing the stopcock to the mercury supply bulb, wait one or two minutes before reopening the stopcock connecting the forepump to the oil diffusion pump.

When the diffusion pump is no longer needed, it is isolated at both inlet and outlet from the rest of the system, the power is turned off, and the refrigerant is taken away from the trap(s). The system may now be vented to air (preferably through a tube of drying agent). When the pump is cold it may be vented to air and the water cooling turned off. *Never vent an oil diffusion pump when it is hot.*

LEAK DETECTION

The design and construction of a vacuum system is only half the battle. To make it work properly, one must know how to detect and repair leaks. Leak trouble can be best avoided by very careful work in building the system; extra effort in constructing the line will be more than repaid in saving time in leak hunting. The most important general rule is to keep the system as simple as possible—every extra stopcock, joint, or appendage is a potential source of trouble. The entire design should be carefully thought out before any construction is started. Wherever possible use permanent joints and avoid hose connections, tapered joints, and flanges. It is best to build complicated systems in sections and vacuum-test each section on an auxiliary line before attaching it to the manifold.

Sufficiently large leaks can be detected by filling the section with air at a pressure slightly over 1 atm and locating the sound of escaping gas. This large a leak is very rare in all-glass systems but may occur in a solder joint on a metal line. Smaller leaks can be located by using a few pounds per square inch overpressure of air or some other gas (up to 1 atm overpressure can be used in all-metal systems), applying a dilute soap solution to the suspected areas, and watching for bubbles produced by the escaping gas.

In a glass system, very small pinholes (usually at a seal) can be detected by a spark discharge if the internal pressure is reduced to below about 0.1 Torr. Leak detectors utilizing a Tesla coil are available commercially; it is convenient to solder a short length of copper wire to the electrode tip of these leak testers in order to obtain a small and flexible probe. When this probe is passed slowly over the surface of the glass, a wide, low-intensity discharge occurs (which will excite a pale violet glow in the gas inside the system) unless the probe is close to a pinhole leak. Near a pinhole a concentrated, intense arc will form through the hole and excite a much brighter glow inside. The voltage on the coil should be no higher than necessary, or else the spark may puncture the glass at a weak point and create a leak. Once leaks have been detected, the system should be vented and the leaks fixed by a glass blower. Obviously, this method cannot be used near a metal electrode or clamp; a leak in such a region can often be detected by applying acetone and watching for a change in the color of the discharge produced by a Tesla coil at some point between this region and the pump. The same trick can be used for an all-metal system by connecting the system to a cold trap and mechanical pump through a length of glass tubing.

When a McLeod gauge is used to hunt for leaks in an assembled system, the best procedure is to pump down the entire system, then close off as many sections as possible (including the manifold), and let the system stand for several hours. The pressure is then measured in the guage itself, gauge plus manifold, and so on, until a section with a high pressure is located. This section can then be carefully rechecked with a Tesla coil, and the stopcocks in it can be regreased. Direct-reading gauges, such as the thermocouple or ionization gauge, are more suitable for leak hunting, since they can be read continuously. They also show a rapid response to pressure changes and a selective response to certain gases or vapors. One standard technique is to swab or flood the suspected area with liquid acetone or methyl alcohol. When a small leak is covered with the liquid, there is a rapid decrease in the leak rate probably caused by the increase in viscosity. Thus, a gauge will indicate an abrupt drop in pressure. Another technique involves covering the suspected area with a plastic bag and filling this bag with hydrogen or helium gas. Since both the thermal conductivity and the ionizability of H_2 and He differ greatly from those of air, a change in gauge reading should occur if the covered area contains a leak. Once a leak is definitely established, a small jet of helium gas from a glass-blowing torch can be used as a probe to locate the exact position. Naturally it will be difficult to find a leak by this method if there are several of them of about the same size.

It is often helpful to make a temporary repair of known leaks in order to hunt for others in the same leak-hunting operation. This can often be done with Apiezon Q, or "Q putty" as it is commonly called. This is a black puttylike material with very low vapor pressure. To be effective, however, this must be worked very patiently against the surface with the fingers to ensure good contact.

In order to find those very small and frustrating leaks which defy all other methods, one can use a mass-spectrometer leak detector. The so-called *helium leak detector* is a very useful device, incorporating a vacuum pump for evacuating the system to be tested and a mass spectrometer, adjusted to He^4, for detecting helium entering through leaks. The helium gas is sprayed over the various parts of the system, and changes in the mass spectrometer response are indicated on a meter and often audibly by changes in a whistle tone. Although helium leak detectors are complex and expensive instruments, they have the great advantages of speed in operation, extreme sensitivity, and an unambiguous response. They are of special value for metal systems.

When leaks are found, they should be repaired in a permanent way. For glass systems, this involves venting the line and reblowing the glass at the site of the leak. If necessary, emergency repairs can be made by warming the glass with a cool flame and covering the leak with de Khotinsky wax, but such repairs should always be considered as temporary. Leaks in metal systems should be repaired by mechanical means: resoldering, rewelding, or replacement of a component. This is often impractical, and recourse must be had to sealants such as Apiezon W or picein (black wax) or Glyptal (an enamel which is most satisfactory when baked on). A leaky silver-soldered joint may often be fixed by cleaning and tinning with soft solder.

When a poor vacuum persists though no leak can be found, one must suspect the possibility of virtual leaks. The system should be thoroughly cleaned and degreased inside, and retested. It may also be necessary to look for gas pockets trapped by screw threads, etc., or to replace decomposable plastic or porous ceramic components with ones giving less gas evolution.

SAFETY CONSIDERATIONS

Vacuum apparatus probably presents fewer accident hazards than almost any other kind of laboratory apparatus of comparable complexity. However, these hazards are by no means entirely negligible, and we summarize the principal ones.

Implosion This hazard is most important with glass apparatus, and is ever-present when large glass bulbs (over 1 liter in size) or flat-bottomed vessels (of any size) are evacuated. The force of atmospheric pressure makes dangerous missiles of glass fragments from imploding vessels. Where possible, avoid evacuating Erlenmeyer flasks or other flat-bottomed vessels. Reduce the hazard of flying glass by placing strips of plastic electrician's tape on all large glass vessels that are to be evacuated. Wear safety glasses.

Explosion If significant quantities of a gas have been liquefied or taken up by an adsorbent at low temperature, an explosion can result when the system warms up if adequate vents or safety valves have not been provided. An explosion of a different kind can take place if an oil diffusion pump (particularly a glass one) is vented to air while hot.

Liquid air Do not refrigerate a trap with liquid nitrogen until the system pressure is reduced below 1 Torr. (Liquid air boils at a higher temperature than liquid nitrogen and can easily be condensed at 77°K.) Liquid air present in a trap may react explosively with any organic substances that condense there. If liquid air is used as a refrigerant, keep organic materials (e.g., towels or absorbent cotton used for thermal insulation) at a distance.

Mercury poisoning Mercury diffusion pumps, McLeod gauges, and mercury manometers all pose hazards of mercury contamination in the event of fracture of glass apparatus. See Appendix D.

SOURCES OF VACUUM EQUIPMENT

Given below is a short list of some of the most important sources of vacuum equipment. The principal types of equipment which are available from each manufacturer are indicated by the following symbols: FP, rotary oil forepump; DP, diffusion pump; IP, ion pump; MV, metal valves; TG, thermocouple gauge; IG, ionization gauge; and LD, helium leak detector.

Central Scientific Co., Chicago, Ill.: FP, DP
Consolidated Vacuum Engineering Corp., Palo Alto, Calif.: DP, MV, IG, LD, IP
Edwards High Vacuum Ltd., Grand Island, N.Y.: FP, DP, IG
Granville-Phillips Co., Boulder, Colo.: MV
Hoke Inc., Cresskill, N.J.: MV
Kinney Manufacturing Co., Boston, Mass.: FP
National Research Corp., Newton Highlands, Mass.: DP, TG, IG
Varian Associates, Palo Alto, Calif.: IP
Veeco Vacuum Corp., New Hyde Park, N.Y.: DP, MV, IG, LD
W. M. Welch Scientific Co., Chicago, Ill.: FP, DP

REFERENCES

1. E. H. Kennard, "Kinetic Theory of Gases," chap. VIII (especially pp. 302, 305, 318), McGraw-Hill, New York (1938).
2. S. Dushman, "Scientific Foundations of Vacuum Technique," 2d ed. (J. M. Lafferty, ed.), Wiley, New York (1962).
3. M. Knudsen, *Ann. Physik*, **28**, 75 (1909).
4. L. J. Favreau, *Rev. Sci. Instrum.*, **36**, 856 (1965).
5. Available from the Carborundum Co., Refractories Division (formerly the Stupakoff Ceramic and Manufacturing Co.), Latrobe, Pa.
6. Goshen Rubber and Mfg. Co., Goshen, Ind.; Linear, Inc., Philadelphia, Pa.; Plastic and Rubber Products Co., Los Angeles, Calif., and Chicago, Ill.
7. R. Gilmont, *Ind. Eng. Chem., Anal. Ed.*, **18**, 633 (1946).
8. "Laboratory Glass Blowing with Pyrex Brand Glasses," Corning Glass Works, Corning, N.Y. (1957); J. D. Heldman, "Techniques of Glass Manipulation in Scientific Research," Prentice-Hall, Englewood Cliffs, N.J. (1946); E. L. Wheeler, "Scientific Glassblowing," Interscience, New York (1958).

GENERAL READING

K. Diels and R. Jaeckel, "Leybold Vacuum Handbook," Pergamon, New York (1966).
S. Dushman, "Scientific Foundations of Vacuum Technique," 2d ed. (J. M. Lafferty, ed.), Wiley, New York (1962).
A. Guthrie, "Vacuum Technology," Wiley, New York (1963).
L. H. Martin and R. D. Hill, "A Manual of Vacuum Practice," Melbourne University Press, Melbourne, Australia (1947).
P. A. Redhead, J. P. Hobson, and E. V. Kornalsen, "The Physical Basis of Ultrahigh Vacuum," Chapman & Hall, London (1968).
J. D. Strong, "Procedures in Experimental Physics," chap. III, Prentice-Hall, Englewood Cliffs, N.J. (1938).
J. Yarwood, "High Vacuum Technique," 4th ed., Chapman & Hall, London (1967).

XVIII

INSTRUMENTS

This chapter consists of brief descriptions and discussions of certain devices and instruments which are commonly used in experimental physical chemistry. Considerably greater detail can be found in the references cited. It should also be noted that most commercial scientific instruments are furnished with a detailed instruction manual. This should be reviewed and thoroughly understood *before* the instrument is used.

BALANCES

The determination of mass is one of the most common and important measurements in experimental chemistry. It is common practice to use the terms *mass* and *weight* as interchangeable, but of course they have quite different meanings. Whereas the mass m of an object in grams is a measure of the amount of matter in that object, the weight w represents the gravitational force exerted on the object by the earth and should properly be expressed in dynes. Since $w = mg$ and g varies with geographical location, the weight of an object of a given mass will depend on where it is measured. The usage of such expressions as "a 10-gram weight" to mean a mass whose weight equals that of a 10-g mass arises naturally from the common method of comparison weighing. Unless otherwise specified (as in Exp. 37), the term weight as used in this book actually means the mass in grams; this should be clear from the context and from dimensional analysis.

We shall assume that the reader has some prior experience with the use of an equal-arm analytical balance, with which an object is weighed by determining the weights which must be added to the right-hand side of the beam in order to make the rest point of the loaded balance the same as the zero point (rest point of unloaded balance).

The design, construction, and operation of such two-pan balances are described in detail by standard textbooks on quantitative chemical analysis,[1-4] and the essential features of a magnetically damped "chain" balance of this type are given in the second edition of this book. In recent years, the use of single-pan "automatic" balances has grown rapidly because of their convenience and speed of operation. Given below is a brief description of the operation of a balance of this type.

Single-pan analytical balances   Single-pan, constant-load "automatic" balances are made by over a dozen manufacturers. The principles of operation for all of them are essentially the same. The weighing pan hangs on a knife edge on one end of the beam. The beam is supported at the center of gravity on another knife edge and a counterpoise is permanently fixed at the other end of the beam. The fixed weights are hung above the weighing pan and may be removed or replaced by operation of the weight-control knobs. The balance is damped by the motion of a piston in a cylinder (air damping) to prevent the beam from acting as a pendulum and oscillating for an inconvenient length of time. A well-adjusted balance usually swings just to equilibrium and stops or it reaches equilibrium after one or two brief excursions. The rest position of the beam is displaced by putting an object on the pan. Sufficient weights are then removed with the weight-control knobs to bring the beam approximately back to its original position, and the residual displacement from the normal rest point is read on an optical scale, that is directly calibrated in milligrams. Because of the delicacy of the knife edges and the sensitivity of the balance, the beam is always arrested and the knife edges lifted out of contact with the bearing surfaces whenever anything is being put on or removed from the pan, or, for that matter, whenever a weighing is not actually in progress. Failure to observe this precaution results in rapid wear of the knife edges with resulting decrease in sensitivity of the balance and erratic behavior.

To carry out a weighing, one first checks that the balance is level and steady and that the pan is empty and clean. Chemicals must never be weighed directly on the balance pan—corrosion or contamination of the balance pan can contribute serious errors! With the balance case closed and the pan empty the beam arrest is

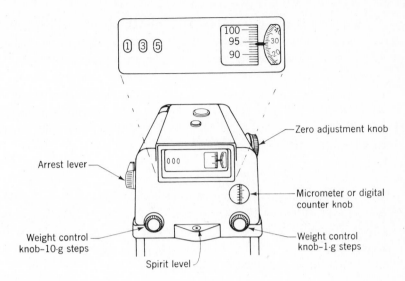

FIGURE 1
Mettler type H-5 single-pan analytical balance showing a reading of 135.9428 g.

gently released and the rest point of the balance observed. If the empty pan does not give a reading of precisely zero on the optical scale, the optical scale should be adjusted to read zero. If a zero reading cannot be achieved with the adjustment knob, a more serious adjustment is required and the instructor should be notified. When the zero has been adjusted, the object to be weighed is picked up with tongs or a paper loop and placed carefully and gently in the center of the pan. The balance case is closed and the *partial beam release* is operated. The weight knobs are then manipulated until the approximate weight of the object is determined as shown by the optical scale indicating insufficient weight with one setting and overweight with the next individual weight added. With this last weight removed, the beam is then completely released, and after equilibrium has been established the fractional weight is read on the optical scale. The beam is then arrested, the sample removed, the weights replaced, and the zero on the balance checked.

The external appearance of the controls on a typical single-pan analytical balance and the manner of reading the weight from the display panel are shown in Fig. 1.

Operation of balances   Any type of precision analytical balance should be mounted level on a sturdy bench, free from vibrations, in a room with a fairly stable temperature (no drafts or direct sunlight near the balance). Detailed instructions for the operation of the particular type to be used should be made available by the instructor. Indeed, if the student is not already familiar with the use of that particular design, demonstration by an experienced user and a practice weighing are strongly recommended. The most important general principle is the need for a sense of personal responsibility. Analytical balances are delicate instruments which are capable of excellent precision if they are used with respect and care. It is especially important to use the beam and pan arrest controls properly, since they protect the components of the balance (especially the knife edges) from damage. If any difficulties arise, consult an instructor; **do not attempt repairs or adjustments**. After use, leave the balance clean and restored to the zero settings as a courtesy to other users.

Corrections and errors   Buoyancy will affect the results of a weighing, since air will exert a buoyant effect both on the object and on the weights; in general, these two effects will not cancel. The weight in vacuo W_v of an object can be obtained from the weight in air W_a by adding the weight of air displaced by the object and subtracting the weight of air displaced by the weights. Thus,

$$W_v = W_a + V\rho - v\rho \tag{1}$$

where V is the volume of the object, v is the volume of the weights, and ρ is the density of air. Since analytical weights are almost always made of brass (lacquered or plated with gold, nickel, or chromium), we can replace v by $W_a/8.4$, where 8.4 g cm^{-3} is the density of brass, to obtain

$$W_v = W_a \left(1 - \frac{\rho}{8.4}\right) + V\rho \tag{2}$$

Equation (2) is useful if V is known, but often it is not. However, if d_0 (the density of the object) is known, V can be replaced by W_v/d_0 to give

$$W_v = W_a \frac{1 - (\rho/8.4)}{1 - (\rho/d_0)} \cong W_a \left[1 + \left(\frac{1}{d_0} - \frac{1}{8.4}\right)\rho\right] \tag{3}$$

where the final approximation is excellent as long as $\rho/d_0 \ll 1$. Although the density of air varies with temperature, pressure, and moisture content, ρ can be taken to be about 0.0012 g cm^{-3} at any relative humidity over the range 15 to 30°C and 730 to 780 Torr.[1] Buoyancy corrections are often neglected in the weighing of solids, but they are essential in weighing gases and are quite important in weighing large volumes of liquids (as when calibrating volumetric apparatus). For example, W_v for water is 0.1 percent higher than W_a, and the buoyancy correction for 100 g of water would be about 100 mg (much greater than any other source of weighing error).

The most common source of error in weighings, with conventional two-pan balances particularly, is due to inaccurate weights. Even in a good set, the weights are often in error by as much as 1 mg. It is recommended that a given set of weights be used only with a single balance and that these weights be calibrated on that balance against a good secondary standard set certified by the National Bureau of Standards. The calibration procedure is given in detail elsewhere.[2,3,5] Corrections for each weight and each rider position should be posted on the balance and should always be used. For Chainomatic balances, it is necessary to calibrate about 10 chain positions, since a linked chain is not perfectly homogeneous (i.e., the weight per unit length is not quite constant) and the corrections may vary rather erratically over the range. With single-pan balances the weights are inside the case, effectively protected from handling and dust. When delivered new or reconditioned, or after periodic servicing (which should include a check of the weights), the weights should be within the manufacturer's specifications. For the most precise work it may be advisable for the user to make a prior check of the balance with a good secondary standard set of weights, and to look carefully for any evidence of dulled knife edges (hysteresis, poor reproducibility).

Finally, there are several other weighing errors which may occur as a result of poor technique but which can usually be avoided. Volatile, hygroscopic, or efflorescent samples and samples which adsorb gases (e.g., CO_2 or O_2) should be kept in closed weighing bottles. An object should never be weighed while warm, since convective air currents will occur, causing the weighing to be in error. Weighings may also be in error because of the condensation of moisture on dry glass walls or the force produced by static charge caused by vigorous wiping of a glass surface.

BAROMETER

The Fortin barometer is simply a single-arm, closed-tube mercury manometer equipped with a precise metal scale (usually brass). The bottom of the measuring arm of the barometer dips into a mercury reservoir which is in contact with the atmosphere. The mercury level in this reservoir can be adjusted by means of a knurled screw which presses against a movable plate (see Fig. 2). When the meniscus in the reservoir just touches the tip of a pointed indicator, the zero level is properly established and the pressure can be determined from the position of the meniscus in the measuring arm. Both the front and back reference levels on a sliding vernier are simultaneously lined up with the top of this meniscus in order to eliminate parallax error, and the height of the arm can be read to the nearest tenth of a millimeter using the vernier scale. A thermometer should be mounted on or near the barometer, since the temperature must be known in order to make a correction for thermal expansion. This correction is discussed in Chap. XVII, and the appropriate formula is given by Eq. (XVII-23). The metal scale on most barometers is made of brass (linear coefficient

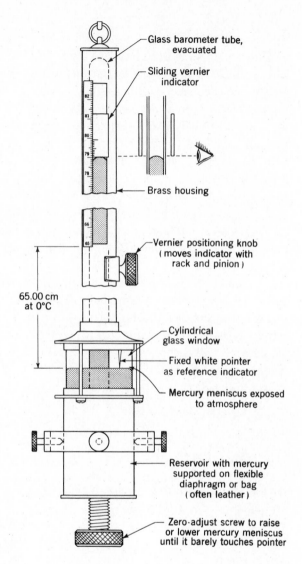

FIGURE 2
Detailed sketch of a Fortin barometer.
Uncorrected reading shown is 79.02 cm.

of thermal expansion 1.84×10^{-5} deg^{-1}) and is usually graduated so as to read correctly at 0°C. A table of barometer corrections over the range 16 to 30°C and 720 to 800 mm is given in Appendix B. High-precision work also requires corrections for the effect of gravity, residual gas pressure in the closed arm, and errors in the zero position of the scale.[6,7] Since these usually amount to only a few tenths of a millimeter, they will not be discussed here.

CATHETOMETER

A cathetometer is used for the accurate measurement of vertical distances, such as the height of the menisci in a wide-bore manometer. It consists of a heavy steel bar (or rod) mounted on a sturdy tripod stand. This steel bar is graduated in millimeters and supports a traveling telescope, which can be moved vertically through about 100 cm and can also be rotated in a horizontal plane. Leveling screws in the base of the stand are used to obtain an accurate vertical alignment of this steel scale, and the telescope mounting is equipped with a fine-adjustment screw and a spirit level to ensure accurate horizontal positioning of the telescope. The meniscus to be measured is brought into focus and aligned with respect to a cross hair in the eyepiece; the position of the telescope on the scale is then read to the nearest 0.1 or 0.05 mm with a vernier scale. A cathetometer is especially convenient for reading levels on an apparatus which must be immersed in a constant-temperature bath (e.g., an osmometer).

VOLTAGE-MEASURING INSTRUMENTS

The basic principles of dc and ac voltage measurements are discussed in Chap. XV. In many applications these measurements are carried out by more complex electronic instruments designed to produce a visual record of the detected signal, either as a trace on a luminescent screen, an inked record on chart paper, or a numerical (digital) readout of the voltage measured. It may also be necessary to discriminate a weak voltage signal from an obscuring background of electrical noise. Instruments for accomplishing these functions are described in this section. More detailed information may be found in standard textbooks on electronics.[8]

Oscilloscopes In a cathode-ray tube, electrons are emitted from a cathode and are then accelerated and focused by a series of special anodes to form a beam which impinges on the face of the tube. This face is coated with fluorescent material, and the beam produces a sharp visible spot. Displacement of this spot from the center of the screen can be achieved by passing the electron beam through the electrostatic field between a pair of charged plates. There are two independent sets of these deflection plates which control, respectively, the vertical and the horizontal position of the spot. Since high voltages are required across these plates to produce a suitable displacement of the beam, oscilloscopes have wide-band amplifiers to provide voltage amplification for the input signals.

The primary use of an oscilloscope is to display the shape of a voltage wave form (i.e., to plot out voltage vertically against a horizontal time scale). To accomplish this, there must be a sweep voltage applied to the horizontal deflection plates which will cause the beam to move from left to right at a uniform rate and then return very rapidly to the starting point. All oscilloscopes have an internal sawtooth generator to produce this linear time base sweep.

The operation of an oscilloscope can best be described by reference to Fig. 3, which shows the typical layout of the controls of a commercial instrument. The face of the cathode-ray tube, in the upper left-hand corner, is covered by a removable plastic *graticule* on which is ruled a pair of centimeter scales at right angles. The graticule can be illuminated from the side for visibility; the level of illumination is controlled by the Scale illumination knob, generally coupled to the on/off switch.

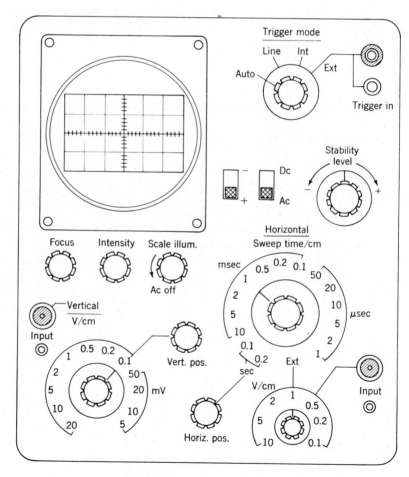

FIGURE 3
Simplified controls of a typical cathode-ray oscilloscope.

A Polaroid camera can be mounted over the tube face if it is desired to make permanent photographic records of the oscilloscope display. The sharpness of the electron-beam image is optimized by the Focus and Intensity controls. It is good practice to turn the beam intensity down to quite a low level; in particular, when the beam is resting at one spot on the tube face, the intensity *must* be lowered until the spot is barely visible. Failure to do so can result in literally burning a hole in the tube phosphor at that point.

The signal to be measured is applied to the Input terminals of the vertical deflection amplifier, and centered on the tube face by means of the Vertical position control. Voltages in excess of the deflection factor at the least sensitive position of the V/cm control must be attentuated by a suitable calibrated probe before being applied to the input terminals.

The horizontal display is controlled by the horizontal amplifier and triggering sections of the oscilloscope. The sweep is initiated whenever a preset voltage level, furnished by the triggering section, is reached. If the Trigger mode control is set to External, then this voltage must be supplied from some external source to the Trigger in terminals. At the Internal setting, the triggering signal is the same as the vertical input signal, so that the wave form to be observed triggers itself. At the Line setting, this voltage is taken from the 60-Hz line current which powers the oscilloscope. This mode is very useful in detecting sources of unwanted electronic noise which arise from the power-line voltage itself. In the Auto mode, the linear portion of the saw-tooth simply repeats itself at the end of each sweep, so that the horizontal deflection is not synchronized with any external source.

In all but the last triggering mode, the voltage at which the sweep is initiated is set by the Stability level control. Provision is also made for selecting this voltage with a positive or negative slope, and from the ac or dc portion of the input signal.

The horizontal sweep rate is determined by the Sweep time/cm control. The leading edge of the wave form should be set to the left-hand end of the graticule ruling by use of the Horizontal position control. The $y(x)$ relationship between two voltages, $x(t)$ and $y(t)$, can be obtained by setting the Horizontal sweep control to External and applying $x(t)$ to the horizontal input terminal and $y(t)$ to the vertical input ter-minal. Thus the sawtooth generator is bypassed, and the amplified $x(t)$ signal is applied directly to the horizontal deflection plates. In this mode, the oscilloscope can be used as a very sensitive device for comparing the frequencies of two different sinusoidal wave forms. When the ratio of the two frequencies is a rational fraction, a symmetric closed pattern (called a Lissajous figure) will appear on the screen. The frequency ratio can be obtained from the form of the pattern by using the formula

$$\frac{f_H}{f_V} = \frac{n_V}{n_H} \tag{4}$$

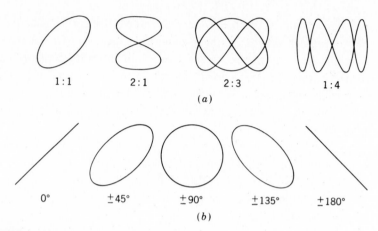

FIGURE 4

Lissajous figures: (*a*) several simple figures, with the ratios $f_H : f_V$ indicated; (*b*) the 1:1 figure for various values of the phase angle between signals of equal amplitude.

where f_H and f_V are the frequencies applied at the horizontal and vertical inputs, n_H is the number of points at which the figure is tangent to a horizontal line, and n_V is the number of points of tangency between the figure and a vertical line. Several simple types of Lissajous figures are shown in Fig. 4*a*.

Shown in Fig. 4*b* is the Lissajous figure for several different values of the phase angle between two signals of the same frequency and amplitude; note that the pattern is a circle when the two sine waves are 90° out of phase and is a straight line tilted at 45° when the phase angle is 0 or 180°. This fact provides a convenient means of determining the phase shift in a circuit. For balancing an ac Wheatstone bridge, a signal from the ac power source is applied to the horizontal input, and the unbalance signal across the bridge is applied to the vertical input. The bridge is first balanced capacitively by obtaining a straight-line pattern. After this is done, the resistive balance is indicated by obtaining a horizontal line (zero vertical amplitude).

Chart recorder A conceptually simple modification of the basic potentiometer circuit described in Chap. XV permits one to make a permanent recording of voltage variations as a function of time or another parameter. The movable tap which contacts the slide-wire is controlled by a servomotor, and the galvanometer detector is replaced by a sensitive electronic null amplifier. The output of the amplifier is connected to the servomotor, which drives the tap up or down the slide-wire according to the sign of the off-balance voltage. A pen is attached to the sliding tap, which writes a trace on a paper chart. Strip-chart recorders have a roll of chart paper which is driven (unwound) by a low-speed motor, so that the pen produces a record of voltage variations with time.

The chart can also be driven in other ways, depending on the desired application. In recording spectrophotometers (see p. 422), the chart is mechanically linked to the rotation of the prism or grating, to produce a record of voltage (related to light intensity) against wavelength. The position of the pen on a stationary chart paper can be controlled by two independent null detectors and servomotor mechanisms to record the simultaneous variation of two external voltages; such a device is called an *X,Y* recorder. *X,Y* and *X,t* recorders are manufactured by a variety of instrument makers, including Leeds and Northrup, Hewlett-Packard, and Heathkit.

Digital voltage instruments   A quite different approach to the measurement of voltage is that provided by a digital voltmeter (DVM). In these instruments, voltage discriminator and binary logic circuits are used to provide a direct numerical display of an input voltage. Since good-quality DVMs often have four-place readout, their precision is comparable to that of medium-precision slide-wire potentiometers. Further details are given in the section on Digital Voltmeters in Chap. XV.

Signal averaging devices   An important class of electronic measuring instruments is that designed to retrieve weak voltage signals from accompanying noise. Since the frequency spectrum of "white" noise is very wide, typically from 0.1 Hz to several MHz with a $1/f$ intensity distribution, much of it can be eliminated with the use of a *frequency-selective amplifier* which passes only a narrow bandwidth of the input at a specified frequency f_0. By modulating the dc signal to be measured at this same frequency, the signal-to-noise ratio is greatly improved. This modulation can be achieved in a number of ways: mechanically chopping a light beam; applying a modulated electric field to the sample in Stark-modulated microwave spectroscopy; or applying a modulated magnetic field in nuclear magnetic or electron paramagnetic resonance spectroscopy are a few examples. The combination of a frequency-selective amplifier with a *phase-sensitive detector*, which locks the amplifier input to a reference voltage, is known as a *lock-in detector*. The schematic operation of one such detector is shown in Fig. 5.

If time resolution of a signal is required, a *boxcar integrator* can be used. In such an amplifier, a narrow amplifying gate is swept across a repetitive wave form, and the averaged wave form is read out as a slowly varying dc voltage. In a more expensive version of this instrument, known as a *wave-form eductor*, the entire wave form is sampled at each repetition by a series of closely spaced gates. If the averaging is carried out by means of digital electronics, the device is known as a *computer of average transients*, or CAT.

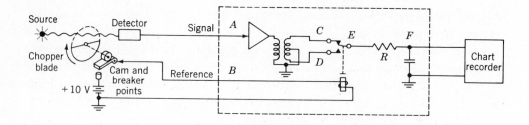

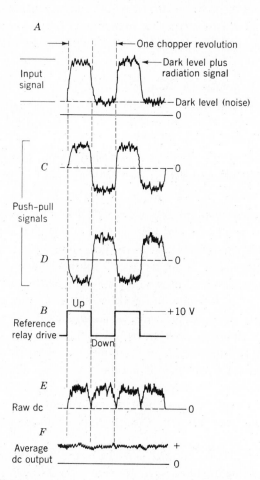

FIGURE 5

Schematic diagram of a very simple lock-in amplifier used to measure low-intensity light signals. Block diagram of the components and voltage levels at various points indicated in the diagram. The signal-plus-noise coming in at A is amplified at C and D, and referenced to the signal coming in at B. The resulting dc signal at E is smoothed by the output capacitor to the averaged signal at F.

pH METER

A pH meter is a special type of millivolt potentiometer designed to measure the emf of a cell in which the electrolyte contains hydrogen ions. A "glass electrode" is employed as the measuring electrode, and a calomel electrode is used as the reference electrode.

Calomel electrode[9] In addition to pH measurements there are many other emf cell measurements for which it is convenient to use the calomel electrode as a reference electrode against which a measuring electrode is compared. Actually, this "electrode" is really a half-cell which is connected via a KCl salt bridge to another half-cell containing the solution of interest. The *saturated* calomel electrode can be written as

$$Hg(l) + Hg_2Cl_2(s), K^+Cl^-(aq, \text{sat.}), \text{aq. electrolyte} \tag{5}$$

There are two other common versions of this half-cell: the *normal* and *tenth-normal* calomel electrodes, in which the KCl concentration is either 1.0 or 0.1 N. The saturated electrode is the easiest to prepare and the most convenient to use but has the largest temperature coefficient. The half-cell potential for each of the calomel electrodes has a different value relative to the standard hydrogen electrode; these emf values are given in Table 1. Calomel electrodes can be easily prepared in the laboratory and are also available commercially. Two typical calomel electrode designs are shown in Fig. 6.

Table 1 HALF-CELL POTENTIALS OF CALOMEL
REFERENCE ELECTRODES[a]

KCl conc.	Potential at 25°C, V
0.1N	−0.3338
1.0N	−0.2800
Saturated	−0.2415

[a] W. J. Hamer, *Trans. Electrochem. Soc.*, **72**, 45 (1937).

Glass electrode[10] This electrode is usually a silver–silver chloride electrode, surrounded by a thin membrane of a special glass which is permeable to hydrogen ions. The glass membrane is essentially a special type of salt bridge—one in which the anions are immobile (have zero transference number), since they are part of the porous glass framework through which the H^+ cations can move. The glass electrode

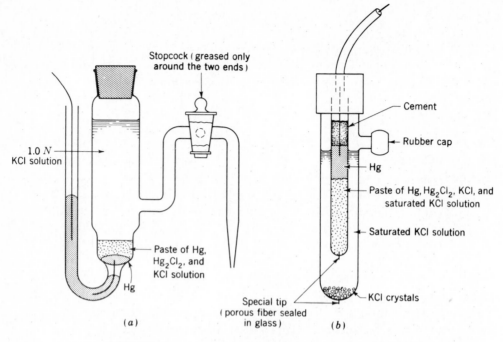

FIGURE 6

Two typical calomel electrode designs: (a) laboratory type, shown unsaturated; (b) commercial type, shown saturated.

may be formulated as

$$\text{Ag}(s) + \text{AgCl}(s), \text{H}^+\text{Cl}^-(aq, a_0), \text{H}^+(\text{glass})^-, \text{aq. electrolyte} \qquad (6)$$
$$\text{containing H}^+ \text{ at}$$
$$\text{activity } a_{\text{H}^+}$$

The change in state per faraday for this half-cell is thus

$$\text{Ag}(s) + \text{Cl}^-(a_0) + \text{H}^+(a_0) = \text{AgCl}(s) + \text{H}^+(a_{\text{H}^+}) + e^- \qquad (7)$$

The activity a_0 has a definite constant value, usually obtained by using 0.1 M HCl as the solution inside the membrane. An important advantage of the glass electrode is that it can be used under many conditions for which the hydrogen electrode is subject to serious error.[11]

pH measurement[11]   The overall emf for a cell with a calomel and a glass electrode dipping into an aqueous electrolyte solution is

$$\mathscr{E} = \mathscr{E}' - \frac{RT}{\mathscr{F}} \ln a_{\text{H}^+} = \mathscr{E}' + \frac{2.303RT}{\mathscr{F}} (\text{pH}) \qquad (8)$$

where pH $\equiv -\log a_{H^+}$ and $\mathscr{E}'$ is the difference between the half-cell potential of the calomel reference electrode and the "standard potential" of the glass electrode. Obviously, $\mathscr{E}'$ will depend on the type of calomel electrode used and on the activity a_0 of hydrogen chloride in the inner solution of the glass electrode. Since these factors are kept constant, changes in emf are a direct indication of variations in pH.

Because of the high resistance of the glass membrane (10 to 100 meg) it is not practical to measure the emf directly. Instead pH meters either use a direct-reading electronic voltmeter or amplify electronically the small current which flows through the cell and detect potentiometrically the voltage drop across a standard resistor. Both battery-operated and ac line-operated pH meters are available commercially from such firms as Leeds and Northrup Co., Beckman Instruments, Inc., Coleman Instruments, Inc., and Central Scientific Co. Such pH meters are calibrated to read directly in pH units, have internal compensation for the temperature coefficient of emf, and have provision for scale adjustments.

Since the operation of a pH meter is very simple but slightly different for each model, no detailed operational procedure will be given here. However, a few general remarks are necessary. If a glass electrode and a silver–silver chloride electrode were placed in an HCl solution for which a_{H^+} equals a_0, the emf of this cell should ideally be zero (i.e., there should be no potential difference across the glass membrane). However, there is always some small emf (1 or 2 mv) across the membrane under these conditions. This so-called *asymmetry potential* is presumably due to strains in the membrane and may change slowly with time or be temporarily changed by exposure of the electrode to very strong acid or base. Therefore, it is necessary to compensate for this asymmetry potential by calibrating the pH meter frequently against a buffer solution of known pH. Also, pH readings on solutions of pH greater than 10 are usually in error owing to a significant contribution from sodium-ion transference in the glass at these low hydrogen-ion concentrations. This difficulty can be avoided by the use of special lithium glass membranes.

POLARIMETER

The polarimeter (Fig. 7) is an instrument for measuring the optical rotation produced by a liquid or solution.[12] The *specific rotation* $[\alpha]_\lambda^t$ of a solute in solution at a given wavelength λ and Celsius temperature t is given by

$$[\alpha]_\lambda^t = \frac{100\alpha}{Lc} = \frac{100\alpha}{Lp\rho} \tag{9}$$

where α is the angle in degrees through which the electric vector is rotated, L is the path length in *decimeters*, c is the concentration of solute in grams/100 ml solution,

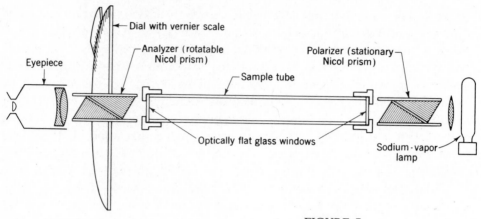

FIGURE 7
Schematic drawing of a polarimeter.

p is the weight percent of solute in the solution, and ρ is the density of the solution. The angle α is considered positive if the rotation of the electric vector as the light proceeds through the solution is in the sense of a left-hand screw or negative if in the sense of a right-hand screw. The optical-rotatory power of a compound is also expressed as a *mole rotation*, which is the specific rotation multiplied by the molecular weight. (In the older literature this product was divided by 100.)

Usually the optical rotation is measured with the sodium D yellow line (a doublet, 5890 and 5896 Å). For more precise work the 5461 Å green mercury line may be used.

Light from the source (a sodium-vapor arc lamp or a type H-4 mercury-vapor lamp with appropriate filter) is polarized by a Nicol prism, termed the polarizer, which consists of two prisms of calcite cemented together with canada balsam so that one of the two rays produced in double refraction (the "ordinary ray") is totally reflected at the interface and lost while the other (the "extraordinary ray") is transmitted. The polarized light passes through the solution and then through a second Nicol prism, termed the analyzer, which can be rotated around the instrument axis. The normal position (zero rotation) is one in which the two Nicol prisms are at 90° to each other, and no light passes through. When an optically rotating medium is introduced between the two Nicol prisms, light is transmitted; the observed rotation α is the angle in degrees through which the dial must be turned (clockwise with respect to the observer if α is positive, counterclockwise if negative) in order to restore the field to complete darkness.

As the dial is turned, the intensity of emergent light is proportional to $\sin^2(\alpha - \alpha_d)$, where α_d is the setting for complete darkness. Since this behaves approximately quadratically near the dark position, the setting is of limited sensitivity if only a single polarizer and a single analyzer are used. In many instruments the field of view is divided into two equal parts with polarization angles differing by a few degrees. This is done either by use of a composite polarizer (two Nicol prisms cemented together side by side with their planes of polarization at a small angle) or by use of an added Nicol prism covering half of the field of the polarizer. In use, the analyzer is adjusted so that the two fields appear equally bright.

Although laboratory polarimeters generally use Nicol prisms as polarizers and analyzers, dichroic crystals (such as tourmaline) or dichroic sheet polarizers (such as Polaroid) may be used in the construction of special apparatus.†

The polarimeter is commonly used in organic and analytical chemistry as an aid in identification of optically active compounds (especially natural products) and in estimation of their purity and freedom from contamination by their optical antipodes. The polarimeter has occasional application to chemical kinetics as a means of following the course of a chemical reaction in which optically active species are involved. Since the rotation α is a linear function of concentration, the polarimeter can be used (in the same way that a dilatometer might be used) in studying the acid-catalyzed hydrolysis of an optically active ester, acetal, glycocide, etc.

In modern organic chemistry optical-rotatory dispersion,[13] or the variation of optical-rotatory power (and certain related properties) with wavelength, is used in molecular-structure investigation as a means of identifying and characterizing chromophore groups. Automatic polarimetric spectrophotometers of high complexity have been developed for this purpose.

RADIATION DETECTORS

Many devices have been developed for detecting and measuring radiant energy. We shall be concerned here only with direct-reading devices. We shall give no discussion of photography, as most of the techniques required for routine work are well known or described adequately in instructions supplied with commercial photographic materials. The principles of the photographic method are adequately described elsewhere.[14]

† In the phenomenon known as dichroism, the optical absorption depends strongly on the orientation of the plane of polarization with respect to the crystallographic axes (or axis of preferred orientation). Commerical sheet polarizers are made from acicular dichroic crystals, herapathite (iodoquinine) in the case of Polaroid, suspended in a viscous or plastic medium and aligned by extrusion or stretching. Dichroic polarizers and analyzers are inferior to Nicol prisms for use in polarimeters because the transmission is considerably less than unity when the planes of polarization are parallel and the absorption is not quite complete when they are perpendicular.

Photoelectric devices[15]   Light of wavelength less than 1 μm is best detected by means of a photoelectric cell. Photocells are of three main types: the *photoconductive cell*, exemplified by the selenium cell; the *photovoltaic cell*, exemplified by the copper–cuprous oxide cell; and the *photoemission cell*, exemplified by the sodium or cesium phototubes.

The selenium cell consists of a thin film of selenium on a grid of electrodes. The operation depends on the fact that selenium is a semiconductor with a narrow energy gap. Incident light quanta of sufficient energy will generate free carriers, and this results in an increased electrical conductivity during the illumination. A cell of this kind will respond to daylight or direct illumination by an incandescent lamp; it is not useful at low illumination levels, owing to a residual conductivity which results in a "dark current." Powered by a few volts from a battery, the cell will operate a milliammeter or a sensitive electromechanical relay. These cells are useful mainly in the visible range, but selenium–tellurium cells can be used down to 1.5 μm in the near infrared.

The photovoltaic cell or barrier-layer cell, as exemplified by the copper–cuprous oxide cell, is used widely in exposure meters for photography. It is made by carefully oxidizing a clean copper surface to obtain a thin film of cuprous oxide, to which electrical contact is made by means of an exceedingly thin, optically transparent film of silver metal or other conducting material. Its operation depends upon photoexcitation and migration of electrons in the semiconducting oxide and the resulting formation of a charge double layer at the copper–cuprous oxide interface. This cell requires no battery and with sufficient illumination will operate a millivoltmeter. In principle it could be used at low illumination with a sensitive galvanometer, but ordinary vacuum-tube photocells are more satisfactory.

The photoemission cell, or phototube, depends on the photoemission of electrons from a surface having an electronic "work function" no greater than the energy of the photons which are to be detected. The simplest such cells are evacuated tubes containing a plate (or part of the tube wall) coated with an alkali metal, and a collector anode consisting of a wire, plate, or grid. This anode is operated at a few volts positive potential in order to collect the photoelectrons efficiently. The photocurrent at a given wavelength is proportional to the incident radiation intensity (with a quantum efficiency less than unity) and is relatively insensitive to applied voltage over a wide range. The photocurrent resulting from moderately strong illumination is sufficient to operate a sensitive electromechanical relay, but nearly all photocells of this kind are used with electronic amplification.

The wavelength sensitivity of a phototube depends on the composition of the photocathode material employed. Sensitivity can be stated in terms of the quantum efficiency Q (i.e., electron/photon ratio) or the radiant cathode sensitivity E in either of two units, milliampere/watt or microampere/lumen. Sensitivity values at the peak of the response curve are given in Table 2 for several typical photocathodes, and their wavelengths dependences are shown in Fig. 8.

At very low light intensities, the photoemission cell is ineffective owing to small leakage or "dark currents" which limit the sensitivity, and a modification known as a *photomultiplier tube* is used. In this tube the photoelectric current is amplified by a cascade process: the photoelectrons strike the first of several successive anodes, producing secondary electrons which are accelerated to the next anode to produce more secondary electrons, and so on. By this means the photocurrent is greatly amplified while the leakage current is unaffected. But even when a photomultiplier is operated in complete darkness, electrons are still emitted from the cathode due to other processes. The resulting dark current is amplified by the multiplier system, and sets a limit on the lowest intensity of light which can be detected directly. It is thus desirable that the dark current should be minimized. Thermionic emission appears to be responsible for the largest component of the dark current. At ambient temperatures the thermal dark current (i_t) per unit area of photocathode surface obeys Richardson's law approximately, i.e.:

$$i_t = 1.20 \times 10^2 T^2 \exp\left[\frac{-1.16 \times 10^4}{T} \phi_t\right] A \text{ cm}^{-2} \tag{10}$$

where T is the absolute temperature and ϕ_t is the thermal work function (in electron volts) for the cathode material.

It is clear that cooling the photomultiplier will have the effect of reducing the thermal component of the dark current. A reduction by a factor of 10 in the total background count rate may be obtained typically by cooling a tube with an antimony-

Table 2   SENSITIVITY VALUES AT THE PEAK OF THE RESPONSE CURVE FOR TYPICAL PHOTOCATHODES[a]

Cathode type	Stoichiometry	Peak (A)	Q (peak)	E (peak) mA/W	E (peak) μA/lumen
S20	Na_2KSb–Cs	3800	0.23	67.5	150
S11	Cs_3Sb–O	3900	0.19	59.8	70
"S"	Cs_3Sb	3800	0.16	49.0	50
"Super" S11	Cs_3Sb–O	4100	0.23	76.0	95
S10	BiAgOCs	4200	0.068	23.0	55
S1	AgOCs	8000	0.004	2.58	25
Bialkali (ambient temp)	K_2CsSb	3800	0.27	82.8	80
Bialkali (high temp)	Na_2KSb	3600	0.21	60.9	40

[a] EMI Electronics, Ltd., Hayes, Middlesex, England.

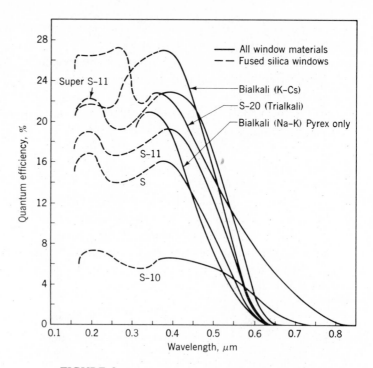

FIGURE 8

Quantum efficiencies 100 Q of various photocathode surfaces.

cesium cathode from room temperature to 0°C; a corresponding reduction by a factor of about 16 may be obtained under similar conditions for a trialkali cathode. For all cathodes except the AgOCs (S1) type, it is found that the thermal component of the dark current may be virtually eliminated by cooling to −40°C, and no significant improvement is obtained in cooling to temperatures below this. The S1 cathode has an intrinsically high thermal dark current, and in this case it is profitable to cool to liquid-nitrogen temperatures. When cooling to such extreme temperatures great care should be taken to cool the whole tube and to avoid strain due to differential contraction of the tube base and socket.

Although photomultipliers are normally stored in the dark, they may be exposed to daylight providing, of course, that the high voltage is not applied! However, such exposure to daylight results in the trapping of energy in the cathode. When the tube is subsequently placed in the dark it takes some time for this energy to be dissipated in the emission of electrons. This effect will manifest itself as an excessive dark current. After a period of 24 to 48 hr in the dark (during which time the high voltage need not be applied), the dark current should have fallen to its equilibrium value.

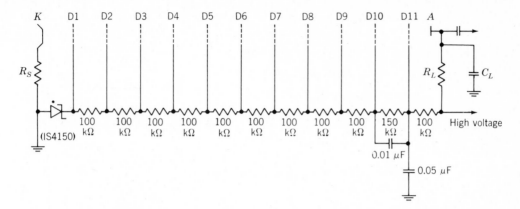

FIGURE 9

Dynode chain circuit for an EMI type 6097 photomultiplier. R_L is the anode load; R_S is a protective resistor of about 100 kΩ; C_L is the capacitance of the anode circuit to earth (including parasitic anode capacitance).

The typical output circuit for a photomultiplier is shown in Fig. 9. A high voltage (> 1000 V) is applied between the cathode and the anode; the string of resistors, called a dynode chain, serves to distribute the voltage uniformly across each amplifying stage. The anode output should be fed into a high-gain preamplifier with a high input impedance.

The short-wavelength limit for a photomultiplier is determined by the transmission of the window material covering the photocathode. Transmission curves for typical window materials are shown in Fig. 10. The region below the "quartz-cutoff" at 1700 Å down to the soft x-ray region of 100 Å or so is known as the *vacuum ultraviolet*. Photomultipliers may be used in this region, provided they are made an intrinsic part of the vacuum system; alternatively, the window may be coated with sodium salicylate, which fluoresces in the near ultraviolet when struck by vacuum uv radiation.

Thermal detectors[16]   Unlike light, which can be detected by a photoelectric effect, infrared photons are too low in energy to trigger individual counting events. Instead, a detection system which measures the thermal energy delivered by the radiation must be employed. The radiation detector in this case is basically an approximation to an isolated blackbody absorber of small heat capacity combined with a sensitive device for detecting changes in temperature.

The *bolometer* is essentially a resistance thermometer, usually with a platinum element, although a thermistor may also be used. In the platinum bolometer two arms of the bridge are thin (~1 μm), narrow (0.5 mm) strips; one of these, which

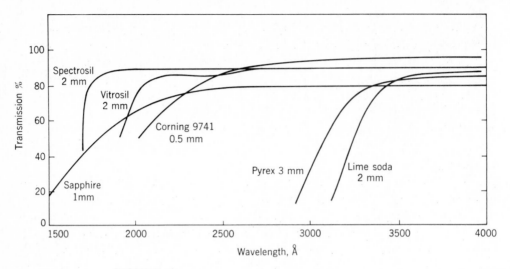

FIGURE 10

Ultraviolet transmission of typical photomultiplier windows. Spectrosil and Vitrosil are different types of fused quartz.

serves as the receiver of the radiation, is blackened with an evaporated metal. The strips are usually mounted in an evacuated chamber. Under optimum conditions the bolometer is capable of detecting as little as 10^{-3} μW.

More often used is the *thermopile* (a multijunction thermocouple). The receiver is usually a thin metal foil, blackened with a metal black customarily deposited onto the metal surface by evaporation. The thermocouple leads are narrow strips cut from thin foils of copper and constantan or bismuth and silver; the junctions are soldered with pure tin and cemented to the receiver. The reference junctions are cemented to a similar "compensation" foil, which is not exposed to the radiation. The thermopile is protected from air currents by means of an enclosure and is frequently operated in vacuum to prevent heat loss by gas conduction.

An alternative method of construction is to deposit the receiver foil and the thermocouple metals by successive vacuum evaporation, through masks, onto a very thin plastic film such as Formvar. The thermocouple metals are in this case usually antimony and bismuth.

The thermopile may be calibrated by use of a special carbon-filament lamp obtainable from the NBS.

A thermal detector which is sometimes used is the *Golay cell*. This consists of a small chamber filled with gas and having a flexible wall or diaphragm. Radiation impinging on the blackened outer surface of this chamber causes thermal expansion of the gas, and the resulting distention of the wall causes the deflection of light by an attached mirror. This deflection is usually detected by a photocell. The Golay cell is particularly useful in the far infrared.

Bolometers, thermopiles, and Golay cells are much too slow and insensitive for such modern applications as laser measurements and remote sensing of surface temperatures by orbital satellites. What is employed instead is a photovoltaic or photoconductive semiconductor diode, which has a high resistance when cold (77°K or 4°K) and a low resistance when warm, i.e., when struck by infrared radiation. The wavelength response of such detectors, expressed in terms of their normalized spectral

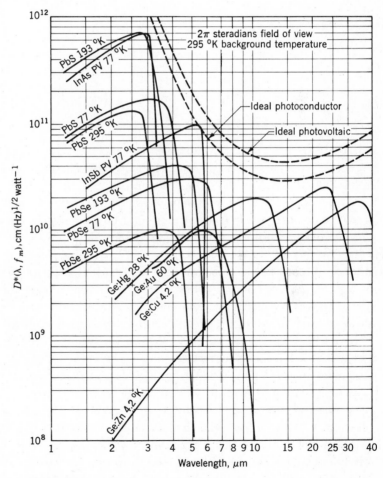

FIGURE 11
Detectivity of photovoltaic and photoconductive semiconductor diodes.

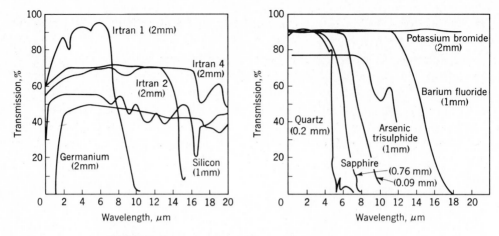

FIGURE 12

Spectral transmission of commonly used window materials. Irtran is the proprietary name for several polycrystalline alkaline-earth sulfides produced by the Eastman Kodak Co.

detectivity $D^*(\lambda, f_m) = (A\,\Delta f)^{1/2}/NEP_\lambda$, depends on their composition, as shown in Fig. 11. The quantity NEP_λ is the incident power at λ which gives a signal-to-noise ratio of unity, A is the area of the detector, and Δf is the electrical bandwidth.

The long-wavelength limit of infrared detectors depends on the window material used, just as does the short-wavelength limit of photoelectric devices. Transmission curves for typical infrared window materials are shown in Fig. 12.

Radiation counters[17] Electromagnetic radiations with wavelengths below about 10 Å are called x-rays or gamma rays. These have a penetrating power which increases strongly as the wavelength decreases and is dependent on the atomic numbers of the atoms present rather than their state of chemical combination.

A 1-Å photon has about 8000 eV of energy, while ultraviolet radiation has only 3 or 4. This much energy, absorbed in matter, may result in the virtually simultaneous production of many ions—enough so that under optimum conditions the absorption of a single x-ray or gamma-ray photon can be detected with an efficiency approaching 100 percent. The measurement of x-ray and gamma-ray intensities thus amounts to a *counting* of discrete events whereby individual photons are detected.

The best known detector is the *Geiger-Müller counter*. This usually consists of a cylindrical container (the cathode) filled with an absorbing gas such as argon or krypton, with an insulated central wire (the anode) to serve as a collector for electrons. When an x-ray photon is absorbed, producing ions and electrons, acceleration of the ions and electrons to the electrodes results in collisions resulting in more ionization; thus a cascade process develops which results in a general gas discharge. This continues until the fall of potential between the electrodes is sufficient to quench the discharge and allow the potential to be restored. The Geiger-Müller counter with its associated circuitry is relatively simple but for many radiation-counting purposes suffers from a long "dead time" (0.1 to 1 msec) resulting from the complete discharge and required recharge. This dead time results in coincident counts and nonlinear response at counting rates higher than about 100 counts per sec. Another limitation of the Geiger-Müller counter is that the size of the pulse generated is independent of the energy of the incoming particle.

To overcome these limitations the self-quenching *proportional counter* has been developed. This counter is very much like the Geiger counter in construction. The electrons and positive ions from the primary ionizing event go to their respective electrodes, but the production of additional electrons through positive-ion bombardment of the wall is prevented by molecules of some organic compound (i.e., ethanol) which are present as a "quench gas." Thus the tube does not discharge completely. The pulse is of very short duration, of the order of 1 μsec. Therefore, counting rates of up to 10,000 counts per sec are essentially linear with intensity. Since the pulse is very small, the proportional counter requires an exceedingly sensitive (high-gain) preamplifier, well shielded from electrical disturbances. Most important for many purposes is the fact that the pulse height depends upon the energy of the incident photon or other particle. The output of the preamplifier may be fed to an electronic pulse-height discriminator circuit, connected to two or more scaling and counting circuits, among which the pulses are distributed according to the height ranges in which they fall. The self-quenching proportional counter does not last indefinitely; after about 10^{11} counts the quench gas is entirely consumed, and the tube thereafter behaves like a Geiger-Müller counter.

Another commonly used detector is the *scintillation detector*. This makes use of a crystal which produces a scintillation (pulse of visible light) on absorption of an x-ray photon. The visible light is detected by a photomultiplier tube and associated amplifier circuit, which is sensitive enough to detect nearly every scintillation. The scintillating crystal is usually sodium iodide doped with an activator such as thallous iodide.

The counters described above are widely used in counting nuclear radiations: gamma rays (electromagnetic radiations, usually of higher energy than x-rays), alpha rays (helium nuclei), beta rays (electrons and positrons), neutrons, etc. For beta rays and especially for alpha rays the counter windows must be very thin. For thermal-neutron counting (e.g., in neutron diffraction) $B^{10}F_3$ is added to the counter gas.

In most counting applications the counting rate is too high to permit the direct use of a mechanical register. An electronic scaling circuit with a binary scale going to 32 or 64 is sufficient for use with a mechanical register if the counting rate is within the linear range of a Geiger-Müller counter. Much higher scaling ranges are required in order to make best use of a proportional or scintillation counter; these usually operate on a scale of 10 and comprise several decades, completely obviating the use of a mechanical register. For many purposes (such as x-ray powder spectrogoniometry) the counting rate, which is proportional to the "intensity," is obtained from the rapid stream of pulses by an electrical circuit and recorded directly on a strip-chart recorder.

REFRACTOMETERS[18]

The term *refractometer* is principally applied to instruments for determining the index of refraction of a liquid, although instruments also exist for determining the indexes of refraction of a solid. The index of refraction n for a liquid or an isotropic solid is the ratio of the phase velocity of light in a vacuum to that in the medium. It can be defined relative to a plane surface of the medium exposed to vacuum as shown in Fig. 13a; it is the ratio of the sine of the angle ϕ_v which a ray of light makes with a normal to the surface in vacuum to the sine of the corresponding angle ϕ_m in the medium:

$$n = \frac{c_v}{c_m} = \frac{\sin \phi_v}{\sin \phi_m} \tag{11}$$

It is common practice to refer the index of refraction to air (at 1 atm) rather than to vacuum, for reasons of convenience; the index referred to vacuum can be obtained from that referred to air by multiplying the latter by the index of refraction of air referred to vacuum, which is 1.00027.

The index of refraction is a function of both wavelength and temperature. Usually the temperature is specified to be 20 or 25°C. The former is more in accord with past practice, but the latter is somewhat easier to maintain with a constant-temperature bath under ordinary laboratory conditions. The wavelength is usually specified to be that of the yellow sodium D line (a doublet, 5890–5896 Å), and the index is given the symbol n_D.

Most refractometers operate on the concept of the *critical angle* ϕ_{crit}; this is the angle ϕ_m for which ϕ_v (or ϕ_{air}) is exactly 90° (see Fig. 13b). A ray in the medium with any greater angle ϕ_{m_1} will be totally reflected at an equal angle ϕ_{m_2} as shown in Fig. 13c. The index of refraction is given in terms of the critical angle by

$$n = \frac{c_v}{c_l} = \frac{1}{\sin \phi_{crit}} \tag{12}$$

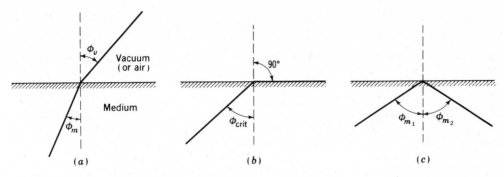

FIGURE 13
Reflection and refraction at an interface: (a) $\phi_m < \phi_{crit}$, (b) $\phi_m = \phi_{crit}$, (c) $\phi_{m_1} > \phi_{crit}$.

In a refractometer the critical angle to be measured is that inside a glass prism in contact with the liquid, since the index of refraction of the glass is higher than that of the liquid. Therefore

$$\frac{c_l}{c_g} = \frac{c_l}{c_v}\frac{c_v}{c_g} = \frac{n_g}{n} = \frac{1}{\sin \phi_g}$$

where n_g is the index of refraction of the prism glass and ϕ_g is the critical angle in the glass. By trigonometry it can be shown that the index of refraction of the liquid is given by

$$n = \sin \delta \cos \gamma + \sin \gamma \sqrt{n_g^2 - \sin^2 \delta} \tag{13}$$

where γ is the prism angle (angle between the two transmitting faces) and δ is the angle of the critical ray in air with respect to the normal to the glass-air prism face (see Fig. 14).

The most precise type of refractometer is the *immersion refractometer*. It contains a prism fixed at the end of an optical tube containing an objective lens, an engraved scale reticule, and an eyepiece. It also contains an Amici compensating prism (see below). In use, the instrument is dipped into a beaker of the liquid clamped in a water bath for temperature control. A mirror in the bath or below it reflects light into the bottom of the beaker at the requisite angle and with some angular divergence. The field of view is divided into an illuminated area and a dark area, as shown in Fig. 14; the scale reading which corresponds to the boundary-line (critical-ray) position is read and referred to a table to obtain the refractive index. This instrument is capable of measuring the refractive index to ± 0.00003. Its scale normally covers only a small range; a set containing several refractometers or detachable prisms is required to cover the ordinary range of refractive indexes for liquids (1.3 to 1.8).

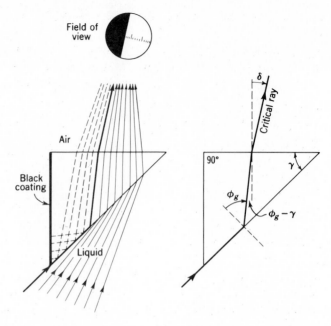

FIGURE 14

Essential features of an immersion refractometer. The behavior of the critical ray is shown in detail, since this represents the basic principle of almost all refractometers.

The most commonly used form of refractometer is the *Abbe refractometer*, shown schematically in Fig. 15. This differs from the immersion refractometer in two important respects. First, instead of dipping into the liquid, the refractometer contains only a few drops of the liquid held by capillary action in a thin space between the refracting prism and an illuminating prism. Second, instead of reading the position of the critical-ray boundary on a scale, one adjusts this boundary so that it is at the intersection of a pair of cross hairs by rotating the refracting prism until the telescope axis makes the required angle δ with the normal to the air interface of the prism. The index of refraction is then read directly from a scale associated with the prism rotation.

The Abbe refractometer commonly contains two Amici compensating prisms, geared so as to rotate in opposite directions. An Amici prism is a composite prism of two different kinds of glass, designed to produce a considerable amount of dispersion but to produce no angular deviation of light corresponding to the sodium D line. By use of two counterrotating Amici prisms the net dispersion can be varied from zero to some maximum value in either direction. The purpose of incorporating the Amici

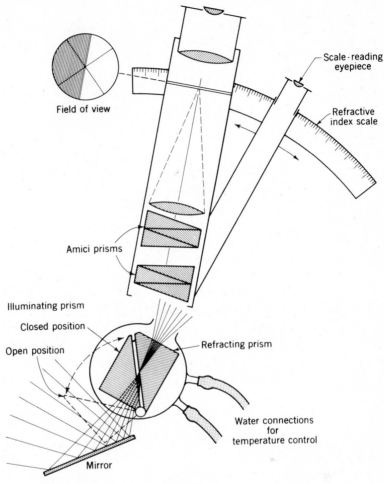

Field of view

Scale-reading eyepiece

Refractive index scale

Amici prisms

Illuminating prism

Closed position

Open position

Refracting prism

Water connections for temperature control

Mirror

FIGURE 15
Schematic diagram of an Abbe refractometer.

prisms is to compensate for the dispersion of the sample so as to produce the same result with white light that would be obtained if a sodium arc were used for illumination. This is achieved by rotating the prisms until the colored fringe disappears from the field of view and the boundary between light and dark fields becomes sharp. It should be borne in mind that the dispersion of a sample is not always exactly compensated, for dispersion is not exactly defined by a single parameter for all substances. The most precise results are obtained with illumination from a sodium arc, the Amici prisms being set at zero dispersion.

The Abbe refractometer is less precise (± 0.0001) than the immersion refractometer and requires somewhat less exact temperature control ($\pm 0.2°C$). For this purpose water from a thermostat bath is circulated through the prism housings by means of a circulating pump. Alternatively, tap water is brought to the temperature of a thermostat bath by flow through a long coil of copper tubing immersed in the bath, and is then passed once through the refractometer and down the drain.

The procedure for the use of an Abbe refractometer is as follows:

1   If a sodium arc is being used, check to see that it is operating properly. The sodium arc should be treated carefully and should be turned on and off as infrequently as possible. It should be turned on at least $\frac{1}{2}$ hr before use.

2   Check to see that the temperature is at the required value by reading the thermometer attached to the prism housing.

3   Open the prism (rotate the illuminating prism with respect to the stationary refracting prism). Wipe both prism surfaces *gently* with a *fresh* swab of cotton wool dampened with acetone or benzene. **Caution:** The prisms must be treated with great care since scratches will decrease the sharpness of the boundary and permanently reduce the accuracy of the instrument.

4   When the prism surfaces are clean and dry, introduce the sample. In some models the refracting prism surface is horizontal and faces upward. In this case place a few drops of the liquid sample on the refracting prism and move the illuminating prism to the closed position. Other models have a prism configuration like that shown in Fig. 15. In that case bring the prisms into the closed position first and then squirt a small amount of sample into the filling hole.

5   Rotate the prism until the boundary between light and dark fields appears in the field of view. If necessary, adjust the light source or the mirror to obtain the best illumination.

6   If necessary, rotate the Amici prisms to eliminate the color fringe and sharpen the boundary.

7   Make any necessary fine adjustment to bring the boundary between light and dark fields into coincidence with the intersection of the cross hairs.

8   Turn on the lamp (if any) that illuminates the scale, and read off the value of the refractive index.

9   Open the prism and wipe it *gently* with a clean swab of cotton wool, dampened with acetone or benzene. When dry, close the prism.

If the sample is very volatile, it may evaporate before the procedure is completed. In this case or in the event of drift, add more sample.

One of the worst enemies of the refractometer is *dust*. A gritty particle may scratch the prisms badly enough to require their replacement. The cotton wool used for wiping the prisms should be kept in a covered jar. Each swab of cotton wool should be used only once and then discarded. *Do not rub* the prisms with cotton wool, and do not attempt to wipe them dry: if streaks are left when the acetone or benzene evaporates, wipe again with a fresh swab dampened with fresh solvent. Do not use lens tissue on the prism surfaces. Finally, the instrument should be protected with its dust cover when not in use, and the table on which the instrument is used should be kept scrupulously clean.

For adjustment of the scale a small "test piece" (rectangular block of glass of accurately known index of refraction) is usually provided with the refractometer. The illuminating prism is swung out and the surface of both the refracting prism and the test piece are carefully cleaned. They are then carefully brushed with a clean camel's-hair brush (which is normally kept in a stoppered container) and inspected at grazing incidence to detect particles of dust or grit. A very small drop (ca. 1 mm^3) of a liquid (such as 1-bromonaphthalene or methylene iodide) which has a higher refractive index than the refracting prism is placed on the test piece, and the latter is then carefully pressed against the refracting prism and carefully moved around to spread the liquid. The reading of refractive index is made in the usual way. If it is not in agreement with the true value of the test piece, an adjustment of the instrument scale is made or a correction is calculated.

The procedure for determining the index of refraction of an isotropic solid sample is similar; like the test piece it must have at least one highly polished plane face.

The refractometer is essentially an analytical instrument, used to determine the composition of binary mixtures (as in Exp. 17) or to check the purity of compounds. Its most common industrial application is in the food and confectionary industries, where it is used in "saccharimetry"—the determination of the concentration of sugar in syrup. Many commercially available refractometers have two scales: one calibrated directly in refractive index, the other in percent sucrose at 20°C.

The refractive index of a compound is a property of some significance in regard to molecular constitution. The *mole refraction,* defined by Eq. (36-13), is a constitutive and additive property; for a given compound it may be approximated by the sum of contributions of individual atoms, double bonds, aromatic rings, and other structural features.[18]

TIMING DEVICES

For measuring long time intervals (10 min or longer), a sweep-second-hand watch or electric clock is often adequate. The accuracy of a good mechanical watch is usually between ± 0.05 and 0.2 percent, depending on the quality of the watch and the precision of its adjustment. However, this adjustment will change with time and cannot be relied on over a period of months. An electric clock operating on the 60-cycle ac power line has the advantage of excellent long-term stability. The time accuracy of an electric clock depends directly on the frequency stability of the power source, and the *average* line frequency over a 24-hr period is maintained very close to 60 cycles. However, the line frequency may differ from 60 cycles by as much as ± 0.1 percent over a period as long as several hours.

For timing short intervals (less than 10 min), there is an appreciable problem in accurately reading a moving sweep-second hand. The uncertainty in a time interval caused by this difficulty can easily be as large as ± 1 sec. This error can be greatly reduced by using a stopwatch or electric interval timer. Although the reading error per se is then eliminated, one must recognize the error due to the reaction time of the experimenter who is manually operating the start-and-stop mechanism. Reaction times vary greatly from one individual to another, but a reasonable estimate of the error in a time interval from this source would be ± 0.2 sec.

A considerable improvement in timing accuracy can be achieved by the use of a precision electric timer[19] driven by a synchronous motor which is operated from a constant-frequency ac power supply. Power supplies controlled by a tuning-fork oscillator are subject to frequency variations of less than 0.01 percent if the input voltage and ambient temperature are reasonably stable. Such precision timers have an electrically activated mechanical clutch which allows the motor to run continuously but which permits the timer hands to move only while this clutch is engaged. The clutch action will introduce an uncertainty of about ± 0.01 sec for an ac-operated clutch or about ± 0.005 sec for a dc clutch. The error due to human reaction time is still present unless the timer is operated automatically.† The performance of an interval timer can be checked against a secondary time standard (such as a frequency counter) or against the time signals broadcast by the NBS over radio station WWV. Such a calibration will considerably reduce the systematic error due to an operating frequency which differs from the nominal value (often by about 0.1 percent).

For the most precise timing, a high-speed electronic frequency counter can be used to count the oscillations of an ultrastable crystal-controlled oscillator. The Hewlett-Packard Co., Palo Alto, Calif., offers several instruments capable of making time measurements accurate to within a few parts in 10^8. The international standard of time at present is the cesium atomic-beam magnetic-resonance "atomic clock," which is stable to a few parts in 10^{12}.

---

† For example, the heating period in a calorimetry experiment can be timed automatically by using a fast double-pole switch to control both the timer and heater circuit simultaneously.

WESTPHAL BALANCE

The Westphal balance[20] is an instrument for measuring the density or specific gravity of a liquid by application of the principle of Archimedes. Although it is not usually capable of the very high accuracy obtainable with a pycnometer (Exp. 11), it is easier and more rapid to use. It is far more accurate than a hydrometer.

The Westphal balance measures gravimetrically the buoyancy exerted on a glass-enclosed body of definite volume immersed in the liquid. This body is an elongated glass bulb weighted with mercury and containing a thermometer. It is suspended by means of a slender platinum wire from one arm of a special balance. The volume of the test body is carefully adjusted to some definite value, say 5 ml, by grinding the glass at the bottom. A detailed description of the operation of this device is given on pp. 450–451 of the second edition of this book.

REFERENCES

1. H. A. Fales and F. Kenny, "Inorganic Quantitative Analysis," chap. IV, Appleton-Century-Crofts, New York (1939).
2. D. A. Skoog and D. M. West, "Fundamentals of Analytical Chemistry," 2d ed., Holt, New York (1969).
3. I. M. Kolthoff, E. B. Sandell, E. J. Meehan, and S. Bruckenstein, "Quantitative Chemical Analysis," 4th ed., chap. 19, Macmillan, New York (1969).
4. W. C. Pierce, E. L. Haenisch, and D. T. Sawyer, "Quantitative Analysis," 4th ed., chap. 3, Wiley, New York (1948).
5. Design and Test of Standards of Mass, *Natl. Bur. Stand. Circ.* 3, 3d ed., Washington (1918).
6. G. W. Thomson, Determination of Vapor Pressure, in A. Weissberger (ed.), "Technique of Organic Chemistry," 3d ed., vol. I, part I, chap. IX, Interscience, New York (1959).
7. W. G. Brombacher, D. P. Johnson, and J. L. Cross, "Mercury Barometers and Manometers," *Natl. Bur. Stand. Monogr.* 8, Washington, D.C. (1960).
8. H. V. Malmstadt, C. G. Enke, and E. C. Toren, Jr., "Electronics for Scientists," Benjamin, New York (1963).
9. D. J. G. Ives and G. J. Janz (eds.), "Reference Electrodes," Academic, New York (1961).
10. M. Dole, "The Glass Electrode," Wiley, New York (1941).
11. H. H. Willard, L. L. Merritt, Jr., and J. A. Dean, "Instrumental Methods of Analysis," 3d ed., pp. 448–460, Van Nostrand, Princeton, N.J. (1958).
12. T. R. P. Gibb, "Optical Methods of Chemical Analysis," chap. VIII, McGraw-Hill, New York (1942).
13. C. Djerassi, "Optical Rotatory Dispersion," McGraw-Hill, New York (1960).
14. C. E. K. Mees, "The Theory of the Photographic Process," Macmillan, New York (1952).
15. K. S. Lion, "Instrumentation for Scientific Research," chap. 5, McGraw-Hill, New York (1959).

16. G. R. Harrison, R. C. Lord, and J. R. Loofbourow, "Practical Spectroscopy," chap. 12, Prentice-Hall, Englewood Cliffs, N.J. (1948); R. A. Smith, F. E. Jones, and R. P. Chasmar, "The Detection and Measurement of Infra-Red Radiation," 2d ed., Clarendon, Oxford (1968); W. L. Wolfe (ed.), "Handbook of Military Infrared Technology," U.S. Govt. Printing Office, Washington, D.C. (1965).

17. S. A. Korff, "Electron and Nuclear Counters," 2d ed., Van Nostrand, Princeton, N.J. (1955).

18. T. R. P. Gibb, *op. cit.*, chap. VII.

19. See, for example, "Precision Timers," Publication 198-A of the Standard Electric Time Co., Springfield, Mass.

20. N. Bauer and S. Z. Lewin, Determination of Density, in A. Weissberger (ed.), *op. cit.*, chap. IV.

XIX

MISCELLANEOUS PROCEDURES

Physical chemistry laboratory work involves many manual arts, techniques, and procedures in addition to those described in earlier chapters. In this chapter we shall deal with some of the more important of these.

VOLUMETRIC PROCEDURES

Several experiments in this book require either titration of solutions to determine concentrations of a given chemical species or successive dilutions of a solution to obtain a series of solutions with known concentration ratios. For the benefit of those without previous experience with the necessary volumetric techniques, we shall present a summary of the more important aspects of those volumetric methods that are commonly encountered in physical chemistry laboratory work. Considerably greater detail is available in many standard textbooks of quantitative chemical analysis.[1]

Volumetric apparatus Volumetric glassware (Fig. 1) of importance to us is of three principal kinds: (1) volumetric flasks, (2) pipettes, and (3) burettes.

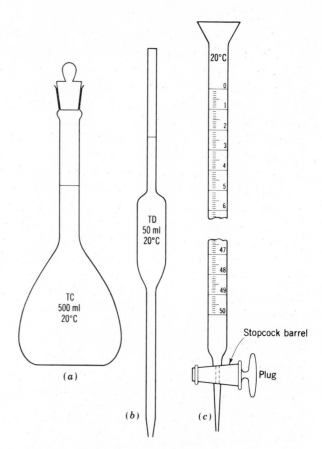

FIGURE 1
Typical volumetric glassware: (*a*) volumetric flask, (*b*) transfer pipette, (*c*) burette (Geissler type).

The *volumetric flask* has a ring engraved around its neck; when the bottom of the liquid meniscus is level with this ring, at the designated temperature, the volume of liquid contained is that indicated by the engraved label. Frequently the letters "TC," meaning "to contain," are present on the label. When such a flask is used "to deliver," it should be allowed to drain for at least 1 min; even so, its accuracy is not so high when used to deliver as when used to contain.

The technique for using a volumetric flask will be illustrated by a description of the proper way to make up a standard solution from a weighed amount of a solid. Once the solid has been introduced into the clean flask with the aid of a funnel, solvent is added to rinse out the funnel and rinse down the neck of the flask. A sufficient amount of the solvent is then added to dissolve the solid; shaking and swirling is most efficient when the flask is only about half full. When the solid has dissolved, additional solvent is added until the liquid level is about 2 cm below the calibration mark. The flask is stoppered, and the solution is mixed well by repeated inversion and shaking. The flask is then placed upright, the stopper removed and rinsed down with a small amount of solvent. The flask is allowed to drain for a minute or two, and the neck is inspected to make sure that there are no clinging drops above the calibration mark. (If necessary, such drops may be removed by careful blotting with a piece of filter paper.) Finally the liquid level is raised until the bottom of the meniscus is even with the calibration mark by addition of solvent dropwise from a pipette or a dropper. The flask is then stoppered, and the solution is mixed thoroughly.

It should also be noted that graduated cylinders are often used for rough measurements of liquid volume. When used to deliver a given volume, the smaller sizes are not at all accurate. When used to contain, the larger graduated cylinders may be accurate to ~ 3 percent. Graduated cylinders are much more convenient than volumetric flasks when solutions of approximate concentration are made up by dilution of concentrated reagents.

The *transfer pipette* has a single ring engraved on its upper stem, and its label frequently contains the letters "TD," meaning "to deliver." The pipette is filled by drawing liquid above the ring (use a rubber bulb if there is any possibility that the liquid is toxic!), and the top of the upper stem is closed with a finger tip. By careful manipulation of the finger tip the meniscus is allowed to fall to the ring. On delivering the contents into a vessel, the pipette is held at an angle of about $35°$ from the vertical and the tip is held against the wall of the vessel to prevent spattering. When the meniscus stops falling, the pipette should be held for about 10 sec to allow drainage and then withdrawn after again touching the tip to the wall of the receiving vessel but *without* blowing out the liquid held in the tip by capillary action.

Measuring pipettes are essentially small burettes without stopcocks. As with burettes, the need for adequate drainage time cannot be ignored. Because of the double reading involved in the use of a measuring pipette and the additional manipulation required they are neither as precise nor as accurate as are transfer pipettes, but are quite satisfactory for semiquantitative operations.

Micropipettes are made in sizes from 0.1 μl to 500 μl (0.5 ml) in several designs. Some are simply capillaries which fill completely by capillary action and are blown or rinsed out. Others are capillaries with a single calibration mark which are filled beyond the mark by capillary action and then drawn down to the mark by touching the tip carefully to absorbent paper. Very often micropipettes are essentially miniature measuring pipettes fitted with a "pipette control," usually an uncalibrated hypodermic syringe or a threaded piston in a cylinder which permits suction or pressure to be applied to adjust the level of the liquid with high precision. Micropipettes can be used with considerable precision if sufficient care is exercised, but the importance of droplets left on the outside of the tip, internal cleanliness, and uniformity of technique is even greater than with pipettes of ordinary size.

The *burette* is a cylindrical tube of uniform cross section which is graduated in volume units along its length. A stopcock and drawn-out tip are attached at the bottom. A burette with bad chips in the tip should be replaced or repaired. Burette stopcocks are made in a variety of styles; those having a smooth glass barrel and a fitted Teflon plug are the most desirable. The set nut which holds the Teflon stopcock in position must be tightened sufficiently to avoid leakage but not overtightened lest the plug be deformed or the threads damaged. The traditional ground-glass stopcock requires careful lubrication with an appropriate stopcock grease. (Silicone grease should never be used since it tends to form a water repellant film on glass surfaces causing the burette to drain poorly.) If a greased stopcock does not turn smoothly or shows streaks or grease plugs, the stopcock should be cleaned (see below) and re-greased. Small particles of stopcock grease which clog the burette tip may be removed with a fine wire.

In use, the burette should be mounted vertically and firmly by means of a clamp that allows the entire burette scale to be visible from the front. With the burette graduations facing the operator, the stopcock handle is generally at his right. A right-handed individual should learn to operate the stopcock with his left hand in order to leave his right hand free for swirling, manipulating a wash bottle, etc. The crook between the thumb and forefinger fits around the burette at the top of the barrel, and the handle is grasped with the thumb and first two fingers. With practice very fine control can be obtained in this way, and a fraction of a drop can be delivered if necessary.

In reading the burette, avoid parallax error. The line of sight should be accurately perpendicular to the burette axis; the rings engraved all the way around the burette will help. Read the *bottom* of the meniscus to about a tenth of the smallest division. Some help in seeing the meniscus can be obtained from a white card containing a very broad horizontal black line, held behind the burette so that the line is reflected in the meniscus. Move the card upward until the top edge of the line and the bottom edge of its reflection become tangent in your field of view; read the position of the point of tangency.

The burette is filled with the required solution (after two or three rinsings with portions of the same solution; see below) to a point above the zero graduation. Solution is run out into a beaker until the meniscus has dropped to zero or some position on scale. Be sure that no air bubbles are present, especially in the tip, and that no drop is hanging when the initial reading is made. The burette is now ready to deliver solution. After running out any significant volume, allow at least 30 sec for drainage from the burette walls before reading. Also, before reading, remove any hanging drop by touching the tip against the wall of the receiving vessel. The volume delivered is the difference between initial and final readings, subject to calibration corrections and temperature corrections if needed.

For the delivery of small volumes, particularly of highly volatile materials or materials susceptible to the atmosphere, the use of a *hypodermic syringe* is recommended. Good quality syringes may be read with reasonable accuracy or the amount of liquid discharged may be found by weighing. Pyrex syringes are reasonably resistant to most reagents but some care should be exercised with the use of syringes made from soft glass. The use of ordinary stainless-steel hypodermic needles as delivery tips with syringe burettes has a number of hazards. Although stainless steel tends to be quite resistant to many reagents, the needles are brazed into the shanks with an acid-susceptible alloy. Teflon needles are available commercially and are to be preferred.

Accuracy and calibration The National Bureau of Standards has established specifications[2] concerning the shape, dimensions, graduations, and capacity tolerances of precision volumetric glassware. Various capacity tolerances are given in Table 1. Volumetric flasks which meet these specifications are designated "Class A" and are usually so marked. Precision pipettes, with a serial number and certificate, can also be obtained. The tolerances of most commercially available glassware are greater than those listed in Table 1 by approximately a factor of 2 or 3.

Whenever very reliable volumetric measurements are required, the glassware

Table 1 NATIONAL BUREAU OF STANDARDS SPECIFICATIONS FOR CAPACITY TOLERANCES OF PRECISION VOLUMETRIC GLASSWARE. (THE TOLERANCES FOR THE "ANALYTICAL" QUALITY OF GLASSWARE MADE BY MOST MANUFACTURERS ARE ABOUT TWICE THOSE LISTED.)

| Capacity in ml less than and including | Volumetric flasks | | Pipettes | |
	Calibrated to contain (TC) $\pm$ ml	Calibrated to deliver (TD) $\pm$ ml	Transfer (TD) $\pm$ ml	Measuring (TD) $\pm$ ml
1	0.01	—	0.006	0.01
2	0.015	—	0.006	0.01
5	0.02	—	0.01	0.02
10	0.02	0.04	0.02	0.03
25	0.03	0.05	0.03	0.05
50	0.05	0.10	0.05	
100	0.08	0.15	0.08	
200	0.10	0.20	0.10	
300	0.12	0.25		
500	0.20	0.40		
1000	0.30	0.60		
2000	0.50	1.00		
Above 2000	1 part in 4000	1 part in 2000		

should be calibrated, preferably at 20°C. Volumetric flasks (TC) are best calibrated by weighing them empty and then filled with water. Pipettes are calibrated by filling with water, delivering their contents into a weighed glass-stoppered weighing bottle, and then reweighing. A burette is filled with water, and 10 to 20 percent at a time is run out into a weighed glass-stoppered weighing bottle or flask, which is stoppered and weighed after each addition (without being emptied in between); a calibration curve (required correction plotted against burette reading) is then prepared, analogous to a thermometer calibration curve. In each case the correct density of water at that temperature (0.99823 g ml^{-1} at 20°C) must be used in the calculations, and air-buoyancy corrections (see Chap. XVIII) are applied. Further details on calibration procedures and calculations are given elsewhere.[1,2]

Thermal expansion   The NBS has specified 20°C as the *normal temperature* for volumetric work. The cubical coefficient of expansion of Pyrex is about 0.9×10^{-5} deg^{-1}; that of water at 20°C is about 2.1×10^{-4} deg^{-1}. The expansion of glass will be of importance only in very precise work; that of water, however, will affect molar concentrations and will be of significance if the actual temperature is more than about 5° removed from 20°C.

Cleaning of glassware   Volumetric work of any quality depends upon glassware with a surface clean enough to be wet uniformly by water and aqueous solutions; if the meniscus pulls away from areas of the glass leaving dry spots, the glass requires cleaning. If cleaning with an ordinary laboratory detergent (such as trisodium phosphate or an organic sulfonate) is not sufficient, a chromic acid cleaning solution may be needed. About 10 g of $Na_2Cr_2O_7$ is dissolved in the minimum quantity of hot water, and after cooling, about 200 ml of concentrated sulfuric acid is slowly added with stirring. The cleaning solution must be kept in a glass-stoppered bottle. If after much use it appears greenish, it should be replaced. Contact with any organic material (e.g., wood, cloth) or with the skin must be carefully avoided. Glassware to be cleaned is ordinarily filled with this solution, which may be moderately warm but should not be hot. Stopcocks should be dismantled and degreased with solvent beforehand; glassware should be reasonably dry to avoid dilution of the solution with water. The solution will attack most fillers and pigments used to fill graduations of burettes and other volumetric glassware; confine the solution to the inside surfaces as much as possible. After 15- to 30-min contact with this solution the glassware should be emptied and thoroughly rinsed with distilled water. The clean glassware usually is allowed to drain if it is not to be used immediately; burettes, however, are often filled with distilled water and covered by an inverted beaker pending use.

Rinsing   Volumetric glassware need not be dry before filling with the appropriate aqueous solution if the vessel is first rinsed two or three times with the solution.

Several *small* portions are more effective than the same total volume of solution in a single portion.

Titration   In titration with a standard solution, the amount of solution required to react quantitatively with a given sample is measured with a burette. A description of a burette and its proper manipulation has been given on p. 680. In order to determine the equivalence point (the point at which exactly stoichiometric quantities of sample and titrant have been brought together), it is necessary to find a chemical or physical property which changes very rapidly at this point. Many properties have been used successfully,[1] but the most common method is the visual observation of a color change in a chemical indicator present in very small concentration. This observable change takes place at the so-called "end point," which must lie very close to the equivalence point. The technique of titration is principally concerned with approaching the end point with reasonable speed without "running over"; it is better learned by practice than by reading a description, but a few words here may help.

The burette is mounted on a stand with a white glass or porcelain base to facilitate observation of indicator color changes. The titration vessel is typically a 125-ml Erlenmeyer flask. Into it the reactant to be titrated is introduced, by pipette (typically 20 or 25 ml) if it is in solution or from a weighing bottle if it is a solid. In the latter case, the transfer is made quantitative by washing the weighing bottle with a stream of distilled water from a wash bottle (a polyethylene squeeze-type bottle is recommended). Other reagents are added as required; the indicator may be added at this point or in certain cases (e.g., starch solution in iodimetry) at a later time.

Initially the solution is allowed to run out of the burette into the titration vessel at full speed, the vessel being held so that the stream runs against a wall to avoid spattering. The flask is continuously swirled to mix the reagents, so that only a portion of the mixture shows the indicator color change. As the color change becomes more general, the stream is slowed down, and when the color change almost pervades the entire mixture, the stream is stopped. The walls of the flask are then rinsed down with distilled water from the wash bottle, and the flask is swirled until the color change disappears. Allow the solution to run from the burette at the rate of a drop every few seconds with continuous swirling, until a significant change in behavior after successive drops is observed. Thereafter add only one drop or a fraction of a drop at a time; in each case touch the tip to the wall, rinse down, and swirl. When the desired color change persists on swirling, the end point has been reached.

If you "overrun" the end point, you may back-titrate with another burette containing an additional quantity of the same solution originally pipetted into the flask. With the accumulation of some experience the necessity for back-titration will become a rare occurrence.

For acid-base titrations, 0.1 M NaOH and HCl standard solutions are useful. A NaOH solution made up to approximately this concentration from carbonate-free NaOH (prepared as described in the next section) is standardized by titrating, with phenolphthalein as indicator, a weighed quantity of potassium acid phthalate, $KH(C_8H_4O_4)$. This salt should be kept dry in a desiccator pending use. The NaOH solution can then be used in the standardization of an HCl solution. When a strong base is used to titrate a weak acid, phenolphthalein is the preferred indicator; for titrating a weak base with a strong acid, methyl red or methyl orange should be used. Any of these indicators or certain others (e.g., brom-thymol blue) can be used in titrating a strong acid with a strong base.

Redox, or oxidation-reduction, titrations are well exemplified by the titration of iodine with thiosulfate in Exp. 15. Commonly used standard solutions for redox titrimetry include iodine (in excess potassium iodide) and potassium permanganate as oxidizing reagents and sodium thiosulfate as reducing agent. Sodium thiosulfate solutions can be standardized by titration of the iodine liberated when a precisely weighed quantity of potassium iodate is dissolved in water and potassium iodide and sulfuric acid are added in excess.

When iodine is titrated with sodium thiosulfate, a commonly used indicator is starch. A satisfactory starch indicator solution is obtained by grinding about 1 g of soluble starch powder in a small amount of hot water to form a smooth paste and adding this to about 200 ml of boiling water. About 0.5 mg of mercuric iodide may be ground with the starch to serve as a preservative. Alternatively, Thyodene, a proprietary indicator based on starch, may be used instead of this preparation.

Preparation and storage of solutions in large quantities   In a physical chemistry laboratory experiment that is performed by many students many liters of each solution are normally required. For the preparation and storage of such solutions, 18-liter carboys (such as those used for commercial handling of acids) are convenient. Reagents are weighed out on a triple-beam balance or measured out with a graduated cylinder and introduced into the carboy, water is added to fill the carboy, the contents are well mixed, and samples are withdrawn by pipette for standardization by titration. The carboy is clearly labeled and placed in the laboratory for student use. A convenient arrangement is shown in Fig. 2.

If the solution must be made up to a precisely specified concentration, a 2000-ml volumetric flask can be used one or more times. If a very large quantity is required, the concentration of the solution in a carboy can be adjusted by two or three progressively smaller additions of reagent or of water each followed by a titration.

A special word should be said about *mixing*, particularly in the case of large volumes. It will not suffice merely to swirl the carboy for a few minutes. When the carboy is filled, enough air space (1 or 2 liter) should be left to permit effective

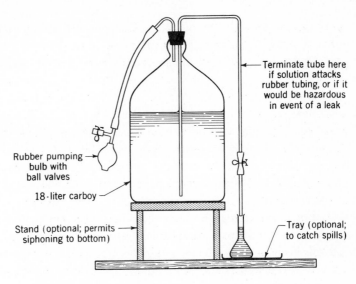

Terminate tube here
if solution attacks
rubber tubing, or if it
would be hazardous
in event of a leak

Rubber pumping
bulb with
ball valves

18-liter carboy

Stand (optional; permits
siphoning to bottom)

Tray (optional;
to catch spills)

FIGURE 2
Method of using a carboy for dispensing large quantities of solution.

mixing. Most convenient for thorough mixing is an electric stirrer, comprising a motor with a long shaft and a swivel-mounted propeller which will go through the mouth of the carboy. This will produce adequate mixing in 15 to 20 min. If such a device is not available, the carboy should be tightly stoppered and turned onto its side on a table covered with a towel. It should then be *vigorously shaken* by a rapid back-and-forth rolling motion sufficient to distribute air bubbles throughout the volume. This should be continued for at least 1 or 2 min. After the carboy has been allowed to stand upright for a few minutes, samples should be withdrawn both from the bottom and from the top for titration.

PURIFICATION METHODS

Mercury   Reagent or triple-distilled mercury is available commercially and can be used without further treatment for virtually all purposes in a physical chemistry laboratory. Through ordinary use mercury becomes dirty and should be either returned in exchange for new reagent mercury or else cleaned in the laboratory. If the contamination is merely a surface accumulation of dust and oxides, sufficient cleaning can be accomplished by "filtering" the mercury through a piece of filter paper in which a small pinhole has been punched; the small quantity of mercury which does not flow through the pinhole will support all the scum and can be poured off into a flask for further purification at a later time.

When mercury is contaminated with dissolved metals, more rigorous purification methods are required. A thin stream of mercury can be allowed to fall through a tall column containing a dilute solution of nitric acid and mercurous nitrate; several repetitions of this procedure may be required before all metal contamination has been removed. Alternatively, the solution is placed in a flask with the mercury, and clean air is bubbled through the mercury so as to agitate it and put it into good contact with the solution. Wet mercury may be first dried over filter paper and then "filtered" through a small hole in a dry filter paper as described above in order to remove the last droplets of water.

For highest purity, mercury (especially when contaminated with noble metals, copper, and lead) should be redistilled repeatedly until a satisfactory spectrographic analysis is obtained. It is now seldom necessary for a laboratory to maintain its own mercury still unless mercury with purity higher than that of commercially obtainable mercury is required for special purposes.

Mercury vapor is toxic. Mercury spills should be meticulously cleaned up. Where inaccessible, mercury droplets may be covered with a fine dusting of sulfur to retard evaporation.

Water Ordinary distilled water is pure enough for most purposes. For some purposes dissolved air is objectionable and can be removed by boiling for a short period. For conductance work, ions other than those resulting from the ionization of water itself must be reduced to the minimum possible concentrations.

Neutral organic substances are usually not objectionable in trace amounts; therefore deionized water, obtainable from columns containing ion exchange resins, is suitable for almost all purposes. The best conductivity water is triple-distilled water (ordinary once-distilled water is distilled a second time from dilute acidified permanganate to oxidize organic impurities, and a third time with a block-tin condenser from dilute barium hydroxide to remove volatile acids and CO_2). Conductivity water can be stored in polyethylene bottles or in glass bottles that have been washed for long periods or otherwise treated to remove the more soluble constituents from the glass surface. Exposure to air should be minimized to prevent contamination by carbon dioxide.

Organic liquids The principal contaminant of *benzene* is thiophene, C_4H_4S, which cannot be removed from benzene by distillation but can be removed by treatment with concentrated sulfuric acid. Water can be removed by fractional distillation, a binary azeotrope coming over first. *Ethanol* cannot be freed of water by a simple fractionation, since an azeotrope with 5 percent water forms. Benzene can be added, permitting the water to be removed in a ternary azeotrope by fractionation. Alternatively, the 95 percent ethanol can be digested with calcium oxide to remove water and then distilled. For highest purity, this should be preceded by treatment with silver oxide to remove aldehydes. Absolute alcohol should be protected from exposure to air in order to prevent water contamination. *Ether* should be dried over sodium or lithium wire and then distilled to remove peroxide; since these peroxides are explosive, the distillation must not be continued to dryness. *Hydrocarbons* can be purified by extensive fractionation combined with treatments with molecular sieves.

Sodium hydroxide   For use as a reagent in acidimetric titration, sodium hydroxide must be freed from contamination by sodium carbonate, which rapidly forms on exposure to air. Commercially obtainable reagent sodium hydroxide needs no further purification if available in stick form, since any carbonate that forms on exposure to air can be quickly washed off with distilled water before dissolving the sticks in CO_2-free water (distilled water, reboiled if necessary) to make up the desired solution. This probably cannot be done effectively with the usual pellets. However, sodium carbonate is virtually insoluble in a saturated solution (15 to 18 M) of sodium hydroxide. Such a solution can be made up by stirring the solid hydroxide with cracked ice and allowing the resulting hot solution to cool. This syrupy liquid can be stored in a polyethylene bottle. When the insoluble carbonate has settled, the clear hydroxide solution can be drawn off as needed by pipette or siphon and diluted with CO_2-free water to the desired concentration. Sodium hydroxide solutions should not be stored for long in untreated glass flasks or bottles. Polyethylene bottles, or glass bottles coated inside with paraffin, are satisfactory.

Handling of sodium hydroxide solutions of any concentration, but especially the high concentrations involved here, entails extreme hazard of the destruction of the cornea if any solution gets into the eye. **Wear a face shield.** Carefully avoid skin contact

Other procedures   It is seldom necessary in physical chemistry laboratory work to purify chemical compounds beyond the purity attainable commercially. For special purposes such techniques as crystallization, fractional distillation, chromatographic separation, and zone refining may be used.

GAS-HANDLING PROCEDURES

Many of the experiments in this book involve the use of one or more gases such as oxygen, nitrogen, hydrogen, helium, argon, and carbon dioxide. We shall be concerned here with procedures for handling these gases.

Cylinders, reducing valves, regulators   Although some gases (e.g., hydrogen) can be prepared in chemical generators, it is far more convenient to obtain gases commercially in steel cylinders. These cylinders are available in several types and sizes. The large size ordinarily used in the laboratory is illustrated in Fig. 3. A typical cylinder is 51 in. in height and 9 in. in diameter; at 2200 psi and 70°F it contains 244 ft^3 of gas (1 atm). In addition, smaller cylinders and very small "lecture bottles" are useful for supplying occasionally needed gases such as hydrogen chloride, ethylene, etc.

Each cylinder is delivered with a protective cap which should be removed only when the cylinder has been chained against a laboratory table or a wall. **Cylinders should always be chained** to prevent upset, which has been known to cause violent release of the gas or even bursting of the cylinder, with serious consequences.

At the top of the cylinder are a needle valve and a threaded outlet. To clear the outlet of dust the needle valve should be barely opened for an instant and reclosed. In most laboratory work a *regulator* is attached to the cylinder, as shown in Fig. 3. This usually comprises a Bourdon gauge to indicate the cylinder pressure (up to 4000 psi), another gauge to indicate outlet pressure (ordinarily up to 60 psi), and an adjust-

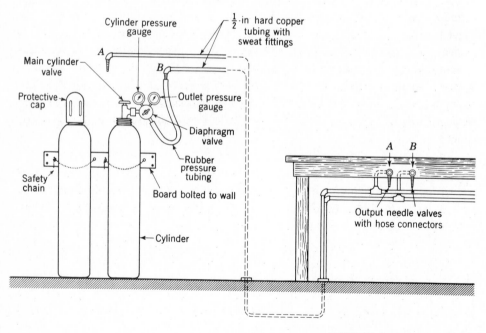

FIGURE 3
Design for laboratory gas lines.

able diaphragm valve to regulate the outlet pressure. The diaphragm valve is a simple form of manostat (see Fig. 3) in which the needle of a needle valve is attached to a flexible diaphragm separating the low-pressure gas from the outlet. When the outlet pressure exceeds the desired value (controlled by a spring between the diaphragm and an adjusting screw), the motion of the diaphragm is such as to close the needle valve, shutting off the flow of gas from the high-pressure side; when the outlet pressure is too low, the diaphragm opens the needle valve.

The fittings and threads on the cylinder outlet are of several types, depending on the kind of gas, in order to prevent the wrong regulator or other fitting from being used with a given cylinder.

The regulator fastens to the cylinder with a metal-to-metal contact, no gasket ordinarily being required except in the case of CO_2 cylinders. The connection is made tight with a large wrench. To verify that the seal is gastight, the cylinder valve is opened and then closed and the pressure gauge is watched for a few minutes.

When it is desired to use the gas, the cylinder valve is opened *all the way*. It is good practice to close this valve when gas is no longer needed (e.g., at the end of a laboratory period). Do not allow a cylinder to be emptied down to 1 atm; there should be some pressure left when the cylinder is returned. Unneeded cylinders should be returned to the supplier to avoid needless demurrage charges.

A simple *reducing valve*, consisting of a needle valve with fittings to match the cylinder and a hose connector for the low-pressure side, can be used in place of a regulator for some purposes where gas flow through an open system is desired. This device is less expensive than a regulator, but its use requires more care because if the system should accidentally be closed, the pressure will build up to whatever is required to burst it at its weakest point if less than the full cylinder pressure. It is advisable to connect the system to the reducing valve with a lightweight rubber tube that will easily blow off the hose connector if the pressure becomes too high, or provide some other safety valve near the inlet end of the system.

Warning: Never connect a gas cylinder directly to a closed system which is not specially designed to withstand high pressures (such as a combustion bomb). Even when a regulator is used, remember that the diaphragm valve may slowly leak; arrange for some kind of safety pressure release, if only a rubber tubing connection that can be easily blown off.

Most gases undergo a Joule-Thomson cooling when they expand in a regulator or reducing valve. In the case of CO_2 the cooling at high flow rates is often large enough to be troublesome, causing frosting of apparatus or even clogging of the regulator or the reducing valve with solid. In addition to this effect, there may be a compressional heating of the gas if a significant pressure is built up quickly in some part of the system. If a gas must be maintained at a constant temperature, it should be passed through a long coil of copper tubing immersed in a constant-temperature bath. One hundred feet of $\frac{1}{4}$-in. copper tubing is adequate for flow rates up to about 5 liter min^{-1}.

Needle valves   For control of gas flow at ordinary pressures, needle valves give much better control than stopcocks. The hard-steel tapered needle, at the end of a screw-threaded shaft, seats in a cylindrical hole so that the area of open space for gas flow is gradually increased or decreased on rotating the shaft. Persons whose acquaintance with valves is limited to water faucets often damage needle valves by needlessly overtightening them when shutting off the flow. This results in a decrease in the sensitivity of control of the gas flow. It is important not to exert any more force than necessary.

Gas-distribution lines   For an individual experiment, the cylinder may be chained at the laboratory table and the gas carried from the regulator to the experimental apparatus by a length of rubber tubing. If the same gas is needed simultaneously in several experiments, a gas-distribution line (see Fig. 3) is a great convenience. The cylinders are chained against a wall, and the gas is conducted through $\frac{1}{2}$-in. copper tubes to the laboratory tables, where they service as many outlets as are needed. Each outlet consists of a needle valve with a convenient knob and a nipple to which a length of rubber tubing may be attached. If two or more gas lines are available, they should be clearly distinguished by color coding. When a line is changed from one gas to another, it should be flushed out thoroughly before use.

Hoses   Gases for open systems may be carried by ordinary $\frac{1}{4}$- or $\frac{5}{16}$-in. gum-rubber tubing. Closed systems may require rubber or plastic pressure tubing, which can safely be used with pressures up to several atmospheres. **Warning:** Do not subject glass apparatus containing bulbs more than 2 or 3 in. in diameter to internal pressures of more than 1 atm above the outside pressure.

Gas purification   Although in many cases the gas from the cylinder is sufficiently pure for direct use, for certain purposes it should be subjected to one or more purification procedures. Hydrogen is frequently contaminated with small amounts of oxygen, which should be removed if the gas is to be used in a hydrogen electrode. The most convenient procedure is to use a catalytic purifier such as Deoxo, which contains palladium; the oxygen combines with hydrogen to form water, which is subsequently removed with a drying tube if objectionable. Oxygen can also be removed by bubbling the gas through an alkaline solution of pyrogallol.

Another frequent contaminant is water vapor. This can be removed by passing the gas through a U tube filled with a suitable drying agent. A tube of this type is shown in Fig. 4-4. Gas flow should not be too fast, or the drying will be incomplete. Use of two or more drying tubes in series may provide better drying. Suitable adsorbents are anhydrous magnesium perchlorate (sold commercially as Anhydrone), anhydrous calcium sulfate (Drierite, often colored blue with added cobalt chloride which turns pink when hydrated), and activated molecular sieves (such as Linde type 4A). Phosphorus pentoxide and calcium chloride are effective drying agents but eventually liquefy after absorbing sufficient water. Drierite and molecular sieves can be reactivated by heating in a laboratory oven at 250°C in the presence of air. **Caution:** Cloth and other organic materials impregnated with magnesium perchlorate can be dangerously flammable.

Soda-lime is commonly used for removing carbon dioxide; alternatively, sodium hydroxide–impregnated asbestos (Ascarite) followed by a drying agent such as magnesium perchlorate can be used.

Water saturation   Gases to be used in systems containing water or aqueous solutions should be saturated with water before they are admitted. For this purpose a bubbler containing a fritted disk which disperses the gas in the form of very small bubbles is far superior to the ordinary laboratory bubbler. The temperature of the bubbler should be the same as that of the system, and the connection to the system should be as short as possible to avoid condensation of water from the gas before it enters the system.

Flowmeters   Figure 4 shows a simple flow manometer[3] which can be easily constructed in the laboratory. In this device a pressure difference, caused by viscous flow in a capillary tube (see Exp. 4), is measured by a simple manometer containing mercury or a colored organic liquid (such as dibutyl phthalate with added eosin). The response of this flowmeter is very nearly linear over the range in which the gas flow in the capillary is laminar and may be roughly linear over a useful range even when the flow is turbulent. By variation of the capillary length and diameter and of the liquid density a wide range of flow rates can be measured.

A steady gas flow for calibrating a flowmeter of this kind can be obtained from a needle valve attached to a regulator set to 5–10 psi or more (to avoid perturbations due to small variations in outlet pressure). At small flow rates the volume of gas flow over an interval of time measured with a stopwatch can be determined with a gas burette; a three-way stopcock can be used to switch the gas burette in and out of the system. For larger flow rates a water-filled inverted graduated cylinder or volumetric flask, its mouth held under the surface of a water bath, can be used to collect gas from a rubber tube held underneath it for a time interval measured with a stopwatch. For precise work a correction should be made for the partial pressure of water vapor in the gas collected.

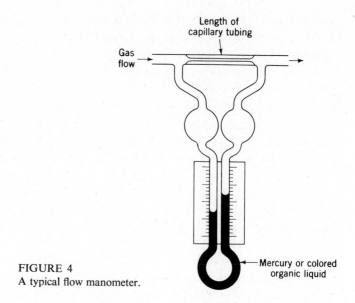

FIGURE 4
A typical flow manometer.

Another type of flowmeter is the Rotameter manufactured by the Brooks Rotameter Co., Lansdale, Pa. A small "float," contained in a vertical glass tube with an inside taper, is supported by the flowing gas and has a steady-state position which depends upon the rate of gas flow. This device can be made closely linear and is available in a wide variety of ranges of flow rate.

Pressure gauges The Bourdon gauge is a dial gauge which can be used for measuring high pressures approximately. It makes use of a curved metal tube of oval or flattened cross section; high pressure tends to straighten out this tube slightly, producing a motion that is converted to a rotation of the indicating needle by a rack and pinion. A direct-reading gauge for lower pressures makes use of a flexible metal diaphragm, usually corrugated, with appropriate mechanical linkages.

For precise work other devices may be used. The most accurate is the manometer (see Chap. XVII). For very high precision a closed-tube mercury manometer is used and the meniscus positions are determined with the aid of a cathetometer (see Chap. XVIII). Clearly the mercury manometer is convenient only for pressures that do not exceed a few atmospheres. For higher pressures a *deadweight gauge* may be used.[4] This gauge measures the force exerted by the gas on a highly polished piston with a close sliding fit in a highly polished vertical cylinder, the mean diameter being known accurately. The effect of friction is largely eliminated by the presence of a lubricant and by a reciprocating mechanism that rotates the piston or the cylinder back and forth a few degrees around the vertical axis. The force due to gas pressure is balanced by weights placed on a pan supported by the floating piston. Pressures up to thousands of atmospheres can be measured with high precision.

Manostats   A manostat is a device for maintaining the pressure in a system at a constant value. A regulator like that described above, operating in conjunction with an outlet leak, constitutes a simple manostat. More precise manostats, also operating on the feedback principle, make use of mercury or oil manometers. An electrical contact, made or broken when a mercury meniscus reaches a certain level, can be made to cause the opening or closing of a solenoid-operated valve. Somewhat better control, which may include a proportionating feature, can be obtained by using a light source and photocell to sense the position of the manometer meniscus. Manostats of this kind are analogous to thermoregulating systems. Several simple manostat designs are discussed in Chap. XVII.

Precautions   In addition to several precautions suggested above for gas handling, certain particular precautions should be followed in handling hydrogen and oxygen.

Hydrogen is very flammable and also forms explosive mixtures with air over a wide range of compositions. If used in any quantity the effluent hydrogen should be vented out of doors by a tube through a window. This is not essential for the slow rates of flow necessary with a hydrogen electrode, but good ventilation is important; there should be no open flames, and no smoking permitted.

Oxygen is hazardous when in contact with flammable substances, particularly oil. *No oil or other organic substance should be allowed to come into contact with oxygen under pressure.* Oxygen lines, valves, and regulators must be kept scrupulously free of oil. Organic packing materials should not be used in needle valves employed with oxygen; graphite-impregnated asbestos is satisfactory to moderately high pressures.

ELECTRODES FOR ELECTROCHEMICAL CELLS

We shall describe below the electrodeposition procedures required in the preparation of the platinum and silver–silver chloride electrodes used in the electrochemical experiments described in this book. We shall not attempt to give a general treatment of electroplating or of electrode preparation. Further details can be found in various monographs.[5]

Platinum electrodes   These are used in conductivity cells and in hydrogen electrodes. For these purposes they are usually covered with a deposit of platinum black to increase the surface area. For certain other purposes, such as use in redox electrodes, this is usually not necessary.

Platinum can be used in the form of thin sheet or screen. Platinum wire can be welded to a small square of sheet or screen by placing it in position on a metal surface, heating to red heat with a torch, and striking lightly with a small ball peen hammer. The platinum wire can then be sealed into the end of a piece of soda-glass tubing, and electrical contact can be made through mercury placed in this tube.

Platinum can be plated onto other metals by use of a solution containing 1 g platinum diammino nitrate, 1 g sodium nitrate, 10 g ammonium nitrate, and 5 ml concentrated ammonium hydroxide in enough water to make 100 ml of solution. The cleaned metal is made the cathode, and a strip of platinum metal the anode; three dry cells (4.5 V), a rheostat, and a milliammeter are connected between the electrodes, and the plating is carried out at a current density of 50 to 100 mA/cm^2 of cathode surface.

For depositing platinum black a solution of 3 g platinic chloride and 0.2 g lead acetate in 100 ml of distilled water is prepared. This can be kept in a glass-stoppered bottle and used repeatedly. The platinum electrode should be treated with warm aqua regia (one part concentrated HNO_3 to three parts concentrated HCl) to clean the surface and, if necessary, to remove old platinum black. It is then rinsed thoroughly with distilled water. While still wet the electrode is immersed in (or the cell filled with) the platinizing solution. If there is only one electrode to be treated, platinum wire will serve as the anode. Two dry cells and a rheostat in series are connected between the electrodes, the electrode to be platinized being the cathode (negative). The rheostat is adjusted so that gas is produced only slowly. If both electrodes are to be platinized, the polarity is reversed every 30 sec. The electrolysis should be stopped as soon as the electrodes are sooty black; an excessive deposit should be avoided. The platinic chloride solution is then returned to its stock bottle; the electrodes are rinsed thoroughly in distilled water, and electrolysis is continued with a very dilute solution of sulfuric acid in order to remove traces of chlorine. After a final washing with distilled water the electrodes are ready for use. Pending use they should be stored in contact with water; they should never be allowed to dry out.

Silver–silver chloride electrodes   These electrodes may be made from thin sheet silver of high purity, but it is probably better to plate silver onto a clean square of platinum sheet or screen. This is made the cathode and a strip of very pure sheet silver (at least 99.95 percent pure) the anode in a plating bath containing 41 g silver cyanide, 40 g potassium cyanide, 11 g potassium hydroxide, and 62 g potassium carbonate per liter. Use three dry cells connected in series with a rheostat and milliammeter. Set the rheostat to about 1000 Ω and make electrical connections to the electrodes *before* immersion; then adjust the rheostat so as to obtain a current density of about 5 $mA\ cm^{-2}$. Plate for a few hours. Remove the electrodes, and wash very thoroughly with distilled water to remove all trace of cyanides.

The silver-plated electrode is "aged" in an acidified solution of silver nitrate and is then made the anode (and a platinum wire the cathode) in a 1 M HCl solution at a current density of 5 to 10 mA cm^{-2}. In a few minutes a brownish coat of silver chloride will appear and the electrolysis can be stopped. The electrode should be aged for a few days in distilled water before use and should never be allowed to dry out. After long periods the potential of the electrode changes, probably owing to crystal growth. The old silver chloride coating can be removed with ammonia or cyanide, and after washing, a fresh coating can be prepared.

In use, care should be taken to avoid exposure of a silver–silver chloride electrode to bromides. Best results are obtained when oxygen is removed from the solution by a stream of nitrogen.

MATERIALS FOR CONSTRUCTION

In this section we shall comment on a wide variety of materials that can be used in the construction and repair of laboratory apparatus.

Glass Glasses[6] and glass blowing are discussed briefly in Chap. XVII in connection with vacuum systems. The transparency, low thermal and electrical conductivity, and chemical inertness of glass make it a valuable material for constructing many pieces of apparatus. Because of its low thermal expansion coefficient, Pyrex glass can be blown into many complex shapes without too great a danger of cracking due to internal strains. Pyrex is also preferable to soft glass, since it can be used at temperatures up to 500°C without undergoing plastic deformation.

The thermal-expansion coefficient of Pyrex is lower than that of most metals; platinum will not make a satisfactory glass-to-metal seal in Pyrex. Tungsten can be sealed in Pyrex if cleaned with sodium nitrite in a flame beforehand. For vacuum-tight glass-to-metal seals commercially available Kovar seals are recommended.

Soft glass (soda-lime glass) has a lower softening point and a higher expansion coefficient than Pyrex and is much more subject to thermal stress. It is mainly useful in the laboratory as a glass that will produce a vacuumtight seal with platinum. It will not seal directly to Pyrex, but "graded seals" (tubes of Pyrex and soft glass joined together with several intermediate glasses in between) are commercially available.

Fused silica or "fused quartz" has high tensile strength and very low mechanical hysteresis; these properties make quartz fibers useful for certain instrument suspensions. Fused silica is very useful for apparatus that must be used above 500°C. Vycor glass, which contains a few percent of other oxides and has a working temperature intermediate between fused silica and Pyrex, can also be used. Vycor-Pyrex graded seals are available.

Metals Hot-rolled steel is the least expensive metal for construction purposes and in sheet, strip, rod, girder, and angle form is useful in the laboratory for constructing stands and frameworks. To prevent rusting, it should be galvanized (zinc dipped), cadmium plated, or at least painted. Several commercial primers are effective for the retardation of rusting; they can be covered with additional coats of ordinary paint if desired. Cold-rolled (low-carbon) steel is also used for heavy-duty structural purposes, but owing to its excellent machining properties it also has many uses in apparatus construction. It can be galvanized or plated with cadmium, copper, nickel, chromium, or any of several other metals. Copper should be plated on steel before nickel and both before chromium. Steel parts can be joined by welding, silver soldering, soft soldering, or copper brazing (if previously copper- or nickel-plated). High-carbon steels, such as tool steels (e.g., drill rod), are less easily machinable but have the advantage that they can be hardened by heat-treatment.

Stainless steels are in a special class, differing from other steels in being non-magnetic and essentially free from rusting and corrosion. So-called "18-8" stainless steel contains 18 percent chromium and 8 percent nickel. The chemical resistance of stainless steel makes it very attractive for many purposes, but its high cost and the difficulty of machining it limit its use somewhat. It can be silver-soldered or welded; soft soldering is difficult.

Aluminum and its alloys are excellent structural metals, with good machinability, fair corrosion resistance, good electrical and thermal conductivity. It is readily sand-cast or die-cast. Aluminum is seldom plated; it can be anodized (to produce a thick oxide layer) and then dyed or painted. Aluminum is difficult to weld owing to its flammability. Soldering is also difficult but can be aided by tinning with indium metal. The low melting point of aluminum (660°C) somewhat restricts its application. Aluminum should not be allowed to come into contact with mercury.

Aluminum and its alloys are often used in the laboratory in sheet form for making electrical chassis and panels and in rod form for constructing frames for apparatus support. Aluminum foil is useful for heat-reflecting shields in low-temperature work. A common structural alloy is Duralumin (Dural), which contains about 4 percent copper and traces of manganese and magnesium. Duralumin that is heated to 530°C and quenched in water is ductile for $\frac{1}{2}$ hr or more and can be readily cold-worked; thereafter the alloy hardens and attains considerable strength. The hardening can be delayed for long periods (for rivets, etc.) by storage at Dry Ice temperatures.

Copper is used where its high electrical and thermal conductivity, its malleability and ductility, and its ease of soft soldering confer advantages. OFHC (oxygen-free, high-conductivity) copper should be used where highest conductivity is required or when employed in a vacuum system, particularly when soldering or welding is to be done in a hydrogen atmosphere. Copper can be made very soft by heating and is useful for making vacuum gaskets; the oxide which forms on heating can be removed with ammonia solution. Soft copper becomes hard on cold-working; accordingly, a soft copper gasket should be used only once or else reannealed before each reuse. Soft copper is exceedingly difficult to machine. It is subject to oxidation and should not be heated above 100 to 200°C for any length of time except in a reducing atmosphere or unless adequately plated. Copper can be joined by brazing, silver soldering, or soft soldering. Copper amalgamates readily with mercury.

Brass is basically a copper-zinc alloy; bronze a copper-tin alloy. In practice both often contain many other metals. Their high machinability, resistance to corrosion, and ease of soft-soldering make them very useful in apparatus construction. Owing to the volatility of zinc, brass should not be used in high-vacuum components that must be baked out or operated hot. Certain bronzes such as phosphor bronze are useful for springs and diaphragms; beryllium copper is also useful in these applications.

Monel and Inconel are basically Ni–Cu–Co alloys containing small amounts of iron and manganese. These alloys have fair machinability and high corrosion resistance even at elevated temperatures; some are magnetic, others not. They can be brazed, silver-soldered, and soft-soldered. Monel and cupronickel are very useful in the construction of apparatus for low-temperature work owing to their low thermal conductivities. Inconel can be used for heating elements; Nichrome (Ni–Cr or Ni–Cr–Fe–Mn) and Chromel (Ni–Cr–Fe) are also useful for this purpose.

Invar (64 percent iron, 36 percent nickel) has a very low coefficient of thermal expansion (1×10^{-6} deg^{-1}). It is magnetic and only moderately corrosion resistant. Kovar (53.7 percent iron, 29 percent nickel, 17 percent cobalt, 0.3 percent manganese) has been mentioned in Chap. XVII as a glass-sealing metal.

Silver is an excellent conductor of heat and electricity and a good reflector of light. It is relatively immune to oxidation but becomes tarnished by exposure to sulfur compounds in exceedingly small concentrations. It is an excellent electroplating metal and can also be deposited in thin films by evaporation. In Dewar flasks and other vacuum glassware it is deposited from an aqueous medium by the Brashear process.[7] Silver is an excellent brazing material and an important constituent of "silver solder." The term silver is often applied to alloys of silver with copper; for example, "Sterling" silver contains 7.5 percent copper. "Fine" silver is 99.9 + percent silver.

Platinum and palladium are useful because of their chemical inertness, electrical conductivity, high reflectivity, and high melting point; their high cost restricts their use to applications in which only small amounts are employed: electrical contacts, suspension wires, heating elements, radiation shields, etc. They absorb hydrogen;

Table 2 MECHANICAL PROPERTIES OF METALS

Metal	Hardness (Brinell)	Tensile strength (units of 10^3 lb/in.^{2})	Young's modulus of elasticity (units of 10^6 lb/in.^{2})	Melting point (°C)	Density (g/cm^3)
Aluminum (pure)	16	8.5	8–11	660	2.7
Duralumin	125	88.2	10	~640	3.0
Brass (67 Cu–33 Zn)	145	66.8	13	940	8.4
Copper		33	18	1082	8.5
Wrought iron		50	26–29	1510	
Cast iron	77	16	12–14	1200	7.7
Mild steel	16	50–60			to
Carbon steel	400	200	28	1430	7.9
Stainless steel	~500	250–300		~1500	
Gold		36	11.4	1063	19.3
Silver		40	11.2	961	10.5
Platinum	64	48	24	1773	21.5
Monel	130–300	75–170	24–26	1350	8.9

palladium is very permeable to it and may be mechanically damaged by exposure to it. These metals easily spot-weld to themselves and to each other. Gold is also very useful because of its inertness. It is an excellent plating material and can be deposited by vacuum evaporation or chemical deposition.

Tungsten has the highest melting point of any known metal (3380°C) and is useful for heating elements and various vacuum-cell components. It is difficult to work and is usually handled in the form of wire or ribbon. Tungsten wires tend to have a fibrous structure; lead-through tungsten-Pyrex seals may not be vacuum-tight unless one end of the wire is welded over with nickel. Tungsten spot-welds to itself and to nickel and tantalum. Tungsten wires will oxidize if heated in air. Molybdenum and tantalum are much more easily machinable and workable, are chemically resistant, and also have high melting points; they are much used in vacuum tubes.

The mechanical properties of a number of commonly used metals are summarized in Table 2.

High-polymeric materials There now exists a vast array of high-polymeric materials, both natural and synthetic. We shall concern ourselves here only with a few which are of particular usefulness in laboratory apparatus construction.

Rubber is the most commonly encountered of such materials. In its vulcanized form it is used in rubber tubing and rubber stoppers. A sulfur coating may appear on the surface of such stoppers in the course of time. For tubing (other than pressure tubing) pure gum rubber is preferable, although it must be replaced more frequently.

Neoprene (du Pont) is a rubber-like material which is a polymer of 2-chlorobutadiene-1,3. Somewhat less flexible than natural rubber, it has much greater resistance to oils, greases, and chemicals. Neoprene is useful in vacuum work in the form of gaskets, O rings, and tubing.

A convenient substitute for rubber tubing in the laboratory is transparent vinyl-plastic tubing such as Tygon B44-3, a compounding of polyvinyl chloride with certain liquid plasticizers. This tubing is tough and flexible and makes a very good seal with glass fittings. It tends to become yellow with age, and after long exposure to water it may become somewhat milky. It is attacked by many organic solvents but has fairly good resistance to most other ordinary chemicals.

Cellulose is a material of construction which is used in the laboratory in the natural forms represented by cork, wood, and cotton and in the reconstituted forms represented by rayon fiber and cellophane film. These have been supplanted for most uses by synthetic fibers and films. However, unlacquered cellophane (available as sheet and as sausage casing) is useful for semipermeable membranes to be used in dialysis and osmotic pressure work.

Bakelite, a phenolic resin, is often used by itself or in combination with paper, textiles, or natural fibers to provide inexpensive, readily machinable, electrically insulating materials for panels and various small parts. Vulcanite (hard rubber) is also useful for this purpose.

Lucite and Plexiglas (polymethylmethacrylate as marketed by du Pont and by Rohm and Haas, respectively) and polystyrene are transparent thermoplastic materials. Their machinability is fairly good, but somewhat limited by their thermoplasticity. They are strongly attacked by solvents such as acetone. They can be cemented with solvents alone or with such cements as Duco. Over long periods cracks may develop at points of strain, and discoloration may result from prolonged exposure to strong light.

Nylon (du Pont polyhexamethylenediamine adipic polyamide) is available in solid form as well as fiber and sheet. It has high strength and mechanical stability, excellent machinability, and low surface friction; it is excellent for bearings, small gears and cams, etc., where it can be used with minimal lubrication or none at all. Nylon fibers and threads are useful in apparatus construction.

Teflon (du Pont polytetrafluoroethylene) is a somewhat more flexible solid which is virtually unsurpassed in chemical inertness, electrical insulating properties, and self-lubricating qualities. It is available in the form of rod, tube, tape, and sheet and is readily machinable. It is useful for gaskets and bushings, unlubricated vacuum seals for rotating shafts, etc. It can be used in dynamic high-vacuum systems if these are not baked out much above 130°C. **Caution:** When Teflon is heated to decomposition, it reportedly gives off fumes which are extremely toxic. After machining it, clean up all chips and scraps at once.

Saran (Dow vinylidene dichloride) is a tough, horny, chemically resistant

Table 3 PROPERTIES OF COMMON PLASTICS

	Plexiglas or Lucite	Bakelite	Saran	Polystyrene	Polyethylene	Teflon
Composition	Polymethyl-methacrylate	Phenol-formaldehyde	Polyvinylidene chloride	Polystyrene	Polyethylene	Polytetra-fluoroethylene
High-temperature limit, °C	60	115	66	72	60–120	132
Chemically resistent to:	Aqueous acids and alkalis, most oils	Weak acids, all organic solvents	Aqueous acids and alkalis, most oils, some organic solvents	Aqueous acids and alkalis, alcohols	Aqueous acids and alkalis, most oils, hydroxylic solvents	Almost everything
Chemically sensitive to:	Alcohols, ketones, esters, aromatic hydrocarbons, halogenated hydrocarbons	Alkalis, strong acids	Ketones, aromatic hydrocarbons, oxidizing acids	Most organic solvents	Hydrocarbon solvents, some ketones and esters	Almost nothing
Machinability	Good	Good	Poor	Fair	Poor	Good
Tensile strength (units of 10^3 lb/in.^{2})	4–10	4–18	0.85–9	3–10	1.3	1.8
Young's modulus (units of 10^6 lb/in.^{2})	0.5	0.7–4.5	0.2–2.0	0.2–0.6	0.02	0.06
Other properties	High optical quality, high electrical resistivity, good mechanical strength	Fireproof, high electrical resistivity, excellent mechanical strength	Fireproof (but gives off toxic HCl fumes), poor mechanical strength		Poor mechanical strength, flexible	Fireproof (but gives off toxic fumes), low friction coefficient

plastic available in a variety of forms useful in the laboratory. Saran pipe or tubing can easily be welded to itself or sealed to glass and is useful for handling corrosive solutions. Thin Saran film, available as a packaging material, is useful for windows, support films, etc. Mylar and other polyester films are also useful for these purposes and for electrical insulation, dielectrics in capacitors, etc. Much thinner than these are films that can be made in the laboratory by allowing a dilute ethylene dichloride solution of Formvar (polyvinyl acetal) to spread on a water surface and dry. Such films are commonly used in electron microscopy for specimen supports but have many other potential uses in research.

Polyethylene and polypropylene are somewhat similar to Teflon but are inferior in chemical resistance and many other respects. They are useful in the form of bottles, flasks, and beakers for containing such reagents as hydrofluoric acid, strong bases, etc., which attack glass. Polyethylene tubing is much less flexible than rubber or Tygon but more flexible than Saran; it can be used for handling caustics, corrosive gases, etc. Polyethylene film has better chemical resistance than Saran and Mylar but lower strength and poorer optical properties.

Mechanical and chemical properties of a number of commonly used plastics are summarized in Table 3.

Miscellaneous construction materials Transite, a product of the Johns Manville Co., consists of asbestos and portland cement; it is available in various forms, including sheets and tubes, and is resistant to flames and moderate heat. Because of its asbestos content (see Appendix D) its use should be attended by appropriate precautions especially when cutting or drilling.

Mica (muscovite or phlogopite) is useful for electrical insulation, thin windows, etc. It can be cleaved to very thin sheets by scraping an edge and inserting a razor blade between the laminae.

Lava is the tradename of various natural stone materials (talc, soapstone) marketed by the American Lava Corporation; they can be machined with ordinary machine tools operated very carefully at low speeds. On firing at a temperature of about 1050°C, Lava No. 1137 loses its natural water, undergoes a slight shrinkage (2 percent), and assumes the character of a hard ceramic. The firing should be done gradually; the rate of heating should be only 150°C hr^{-1}, and the final temperature should be held for 45 min. The furnace is then shut off and allowed to cool before opening. These materials are useful in fabricating structural elements and insulators in vacuum cells.

Thermal insulating materials Vermiculite (expanded mica) is useful for insulating ovens and furnaces where the temperature is not too high. At high temperatures (~ 1000°C) powdered magnesia, diatomaceous earth, or firebrick can be used.

At room temperature and below, foamed plastics are very effective. Polystyrene foam is easily cut into any shape desired. Certain commercially available urethane foams can be "foamed in place" to fill awkwardly shaped insulating spaces.

Cements and adhesives   Little need be said here regarding general-purpose cements, such as household glue and Duco, which dry by evaporation of solvent. We shall mention several special-purpose cements, adhesives, and sealing agents with which the student is less likely to be familiar.

For making semipermanent seals in vacuum systems, *de Khotinsky cement* (shellac compounded with wood tar), *picein*, and *Apiezon W* (black waxes of petroleum origin) are useful. These soften or melt on warming to 50–150°C, flow readily on warm surfaces, and stick well to clean glass or metal surfaces. They have fairly low vapor pressures, but their exposed surfaces in vacuum systems should be kept to a minimum.

For a wide variety of vacuum-sealing applications, *Glyptal* (General Electric glycol phthalate), a lacquer with or without added pigment and with a solvent such as xylene, is useful. It adheres well to clean glass or metal and has a relatively low vapor pressure after baking. It should not be heated above 150°C. Glyptal hardens very slowly because it forms a surface film which retards evaporation of solvent from beneath the surface. It should be allowed to dry for several days at room temperature or for 12 hr under an infrared lamp.

Temporary seals in vacuum systems can be made with a mixture of *beeswax* and *rosin*, melted together in equal parts; this mixture is particularly useful in sealing around bell jars. It is applied smoking hot and smoothed with a heated soldering iron. It is not very strong and can easily be cut loose with a knife. *Apiezon Q*, a compounding of petroleum residues and graphite, has a low vapor pressure and a putty-like consistency; it can be used for sealing bell jars, temporary sealing of leaks, etc.

For miscellaneous mounting and positioning jobs, such as seating the mercury supply bulb for a McLeod gauge, *plaster of paris* (dehydrated calcium sulfate) may be useful. This is mixed with water to form a thick paste, which gradually hardens. A mortar of *portland cement* and sand can be used for the same purpose.

An irreversible cement useful for many purposes is *litharge-glycerin*. Pulverized litharge (PbO) is first heated to 400°C and after cooling is mixed with glycerin to form a thick paste. It sets to form a tough, adherent solid, which will withstand temperatures as high as 250°C. It is useful in plumbing but should not be used in vacuum work.

An irreversible ceramic cement which is not vacuumtight but which can be used for anchoring structures in vacuum systems is *sauereisen cement* (Central Scientific Co.). This is made by suspending ceramic powders in sodium silicate solution (water-glass). The cement sets very hard and withstands temperatures up to 590°C.

Zinc oxychloride (dental cement), made by mixing calcined zinc oxide powder with concentrated zinc chloride solution, can also be used at high temperatures.

An increasing number of cementing problems are now being solved with cements based on *epoxy resins* and related substances. Among these are Araldite (Ciba Co.), A-6 (Armstrong Products Co.), F-88 (American Consolidated Dental Co.), and Tygoweld (U.S. Stoneware Co.). Each consists of a resin liquid plus an activator which is mixed in before applying. These cements set irreversibly to form tough, adherent solids which bond well to metals (usually including aluminum) and most other materials. They are relatively inert to chemicals and solvents. For joints that must be exposed in high-vacuum systems, a special grade of vacuum epoxy cement must be used; no other epoxies should be used in such systems.

Epoxy resin cements available in stationery stores often contain fillers and are not necessarily as good for laboratory purposes as those mentioned above. **Caution:** Epoxy resins are considered to be toxic until thoroughly set; skin contact should be avoided.

A special-purpose adhesive which forms an exceedingly strong bond to metals and most other materials, including careless experimenters, is Eastman 910 Adhesive (Eastman Kodak Co.). It is expensive, but only small amounts are needed. (Do not keep more than you will need in a few months, as it slowly deteriorates; store in the refrigerator.) Another useful line of flexible adhesives is the silicones, such as General Electric RTV. They can be clear or opaque, are fairly inert chemically, and some are good to 300°C or above. These are just poured or extruded from a tube, and cured by several hours exposure to the atmosphere.

TUBING CONNECTIONS

It is often necessary to seal joints between dissimilar materials against leakage, especially in vacuum systems. A number of materials are available for doing this.

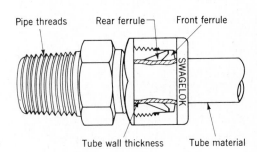

FIGURE 5
Typical Swagelok compression fitting.

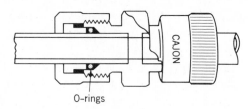

FIGURE 6
Cajon Ultra-Torr compression fitting. O–rings

Metal tubing is often terminated in a pipe thread. This is an accurately tapered and threaded unit, and an assembly must be well tightened with a *pair* of wrenches in order to seat the joint. Small leaks can be sealed by wrapping the inner member with a thin film of Teflon tape (available as Strip-eeze or Tape Dope) before assembling the joint. An older method uses a penetrating cement called Loctite applied to the joint.

Various types of compression fittings are used as an alternative to pipe threads. The Swagelok fittings (Fig. 5) are more expensive than the standard types but can be disassembled more easily. They can also be used to couple wide varieties of tubing by using the following combinations of materials:

Tubing	Threaded body	Front ferrule	Rear ferrule	Nut
Glass	Brass or stainless	Teflon	Nylon	Brass or stainless
Teflon	Teflon	Teflon	Stainless	Brass
Polyethylene	Brass	Nylon	Nylon	Brass
Tygon	(Either metal, Nylon, or polyethylene fittings can be used, but a special tube insert must be included.)			

The Ultra-Torr fittings (Fig. 6) seal by means of an O ring, usually made of Viton. This is a rubber-like material with a very low vapor pressure which can be baked to ~200°C. Teflon is also occasionally used as a material for O rings. It should be emphasized that male and female elements of the different types of fittings are *not* interchangeable.

SHOPWORK

Experimental work in physical chemistry is not limited to the assembly of standard pieces of laboratory apparatus and the making of measurements with them. New or modified apparatus is frequently needed. This must be designed and often constructed by the experimenter himself. Therefore he must be acquainted not only with the appropriate materials for constructing various apparatus but also with many of the techniques of machine work, soldering, wiring, etc. We shall here limit ourselves to a very brief comment on machine tools and a short discussion of various soldering and welding techniques.

The most important machine tools are the drill press, the band saw, the grinding wheel, the lathe, and the milling machine. The first three of these are very simple basic tools that every experimenter should learn how to use. The lathe and the milling machine are more complex and more expensive tools of great value in constructing special apparatus; if possible, anyone interested in physical chemistry research work should learn how to operate these tools also. At the very least, it is necessary to understand the principles of their operation and to appreciate what they can and cannot do. Indeed, a familiarity with machine tools is a vital part of the ability to design complex apparatus properly. No attempt will be made here to describe the operation of any machine tools. Written descriptions are available elsewhere,[8] but it is of great importance to obtain basic instruction in the use of these tools from a qualified machinist in order to avoid risk of personal injury to the operator and damage to the machines.

Soldering is the technique of joining two metals with a fused metal of lower melting point which wets both surfaces.[8] *Soft solder* is a low-melting alloy of lead, tin, and sometimes other metals (e.g., bismuth or antimony). It is commonly used in making electrical connections, making connections between copper or brass tubes or fixtures by means of "sweat fittings" (such as sleeves, elbows, tees, etc.), assembling small parts of an apparatus, and occasionally assembling large parts when reliance is not placed on it for much mechanical strength. It can be used for ordinary vacuum work but is not satisfactory for high-vacuum work, since its low melting point ($\sim 200°C$) is lower than most bakeout temperatures.

Silver solder is a common form of *hard solder*; it is an alloy of silver and copper often with tin, zinc, or cadmium added. Silver solder melts at 600 to 700°C and has high mechanical strength. Gold solders are hard solders consisting mainly of gold and copper, melting generally above 800°C.

In electrical wiring, soft solder is commonly used with noncorrosive rosin fluxes and soldering pastes. Soldering of electrical connections with rosin-core solder is easiest when both parts to be joined have been "tinned," i.e., wet with a coating of solder. Hookup wire is often pretinned, as are pigtails of resistors and capacitors, lugs of tube sockets and terminal strips, etc. Copper wire usually requires no tinning but may require scraping to remove Formvar or other lacquer (which is often invisible) or oxide and dirt. Before a connection is soldered, it should first be made in such a manner that the solder is not relied on for mechanical strength. A tinned soldering iron carrying a drop of solder is applied to the connection and held there until the solder spreads over the metal by capillary action; more solder can be applied if necessary, but the minimum required amount should be used.

Soft-soldering of copper, iron, steel, and brass objects of larger size is usually accomplished by the use of a burner or hand torch, with an acid flux—frequently an aqueous solution of zinc chloride which is brushed onto the hot metal concurrently with the addition of solder. The surfaces to be joined should be pre-tinned, if convenient, and excess solder is shaken off or wiped off with a cloth. The two surfaces to be joined are then placed in contact and heated with the torch until the solder begins to flow; more solder is then added as required, and the pieces are allowed to cool undisturbed. The finished work should be washed thoroughly with water to remove the flux.

Hard-soldering, or silver-soldering, requires the use of a hand torch and a soldering paste of borax and boracic acid. The pieces to be joined are placed in contact, and the surfaces brushed with the flux during heating. Solder is applied as soon as the work is hot enough to melt it and encouraged to spread if necessary by further application of flux. Silver solders containing zinc or cadmium should not be used in vacuum systems that must be baked out or operated hot.

The cleanest and most reliable joints in vacuum systems are made by *hydrogen furnace brazing*. The parts to be assembled are clamped together with a thin gasket or sheet of the brazing alloy (silver or gold solder) between them. They are heated in a hydrogen atmosphere until the brazing alloy melts, runs, and wets both metal surfaces. No flux is required.

Welding techniques are many and varied and will not be discussed in detail. The laboratory worker will usually prefer to have any needed welding done by a qualified welder. Thin sheet metals and wires can be spot-welded; this is the most usual means of assembling the internal components of vacuum tubes. Stainless-steel housings of vacuum systems should be heliarc-welded on the *inside* of the joint.

REFERENCES

1. D. A. Skoog and D. M. West, "Fundamentals of Analytical Chemistry," 2d ed., Holt, New York (1969); I. M. Kolthoff, et al., "Quantitative Chemical Analysis," 4th ed., Macmillan, New York (1969); I. M. Kolthoff and P. J. Elving (eds.), "Treatise on Analytical Chemistry," Interscience, New York (1959 *et seq.*, a multivolume, continuing work).
2. E. L. Peffer and G. C. Mulligan, Testing of Glass Volumetric Apparatus, *Natl. Bur. Stand. Circ.* 602, Washington, D.C. (1959).
3. A. F. Benton, *Ind. Eng. Chem.*, **11**, 623 (1919).
4. J. A. Beattie and W. L. Edel, *Ann. Physik*, 5F, **11**, 633 (1931).

5. C. A. Hampel (ed.), "The Encyclopedia of Electrochemistry," Reinhold, New York (1964); D. J. G. Ives and G. J. Janz, "Reference Electrodes," Academic, New York (1961).

6. Corning Glass Works, Bulletin B-88 (1951).

7. J. Strong, "Procedures in Experimental Physics," pp. 154–157, Prentice-Hall, Englewood Cliffs, N.J. (1943).

8. *Ibid.*, pp. 569–582.

XX

LEAST-SQUARES FITTING PROCEDURES

Among the most common computational operations in experimental science is the fitting of experimental data to a theoretical function or model. Traditionally, this has often been done graphically when there is only one independent variable, as in fitting a straight line to a number of experimental points (x,y), with reliance on visual judgement for maximizing the quality of the fit. This is still widely done, particularly when the experimental results are of only modest quality. When the number of variables or the number of fitting parameters is large, or when it is important to make the best possible use of the data, the fitting must be done numerically. The computational method most commonly used is the method of least squares.

INTRODUCTION

The method of least squares has been used for many years, but the computations were once slow and onerous. The availability of modern computers has made least-squares fitting a routine operation. Indeed, *least squares has become so easy to do that it is possible to be seduced into a false sense of security concerning the validity of the results obtained.* Least-squares fitting is not magic and is subject to the maxim of computer science, "garbage in, garbage out." It is important to realize that a set of experimental points can often be fit quite well with an analytical expression bearing no relation whatsoever to any valid physical theory. There is nothing particularly wrong with fitting a data set with an artificial function for purely empirical purposes, such as to obtain an analytical representation of the data convenient for computational purposes (interpolation, differentiation, integration). For example, one might use a least-squares fit to the gaussian error function, Eq. (II-2), in various contexts having little or nothing to do with the gaussian error function in principle; atomic electron densities, for instance. However, it is important to keep in mind which features of the model being tested are physically meaningful and which are not. One should resist cluttering up a theoretical model with features of doubtful physical significance merely to improve the fit. Furthermore, in trying to decide between two theoretical models on the basis of the "best" least-squares fit, one must be very careful in assessing the magnitude and the character of the experimental errors.

We will not undertake to present here a comprehensive and rigorous treatment of least squares, since this would require the scope of an entire book. Several such books are listed under General Reading at the end of this chapter. Our aim is to give a succinct and practical description of least squares as used in physical chemistry.

It must be kept in mind that the entire treatment given in this chapter is concerned with uncertainty in the data caused by random errors and not with the possible presence of systematic errors.

FOUNDATIONS OF LEAST SQUARES

Let it be assumed that for m different values of an independent variable x (x_i; $i = 1$, ..., m) there are corresponding measured values of a dependent variable y (y_i; $i = 1$, ..., m). Let us further assume that there is a theoretical model predicting the way in which y is expected to depend on x, and that this model may be represented by an analytical function f:

$$y = f(\alpha_1, \ldots, \alpha_n; x) \tag{1}$$

The quantities α_j ($j = 1, \ldots, n$)† are independently adjustable parameters which are initially unknown or only approximately known. The initial trial values of these

---

† In appropriate circumstances the index j can be chosen to run instead from 0 through $n - 1$, for example when the model function is a polynomial

$$y = \alpha_0 + \alpha_1 x + \alpha_2 x^2 + \cdots + \alpha_{n-1} x^{n-1}$$

as in Eq. (31) and in the Sample Least Squares Calculation at the end of this chapter.

parameters will be denoted as $\alpha_1{}^0, \ldots, \alpha_n{}^0$. It should be pointed out that we might just as easily have more than one independent variable; let there be p of them $(x^k; k = 1, \ldots, p)$. The model is now represented by

$$y = f(\alpha_1, \ldots, \alpha_n; x^1, \ldots, x^p) \qquad (2)$$

There are virtually no additional complexities or difficulties of any consequence resulting from having more than one independent variable, but for the sake of simplicity of expression we will assume only one in the treatment presented below.

Least-squares fitting deals with the problem of determining the "best" values of the adjustable parameters $\alpha_1, \ldots, \alpha_n$ so as to maximize the agreement between the observed y values and the values of y calculated with Eq. (1). The number m of observations exceeds (usually by an order of magnitude or more) the number n of adjustable parameters. Thus, the mathematical problem is overdetermined. [For a situation in which $m = n$, the corresponding set of simultaneous equations

$$y_i = f(\alpha_1, \ldots, \alpha_n; x_i) \qquad (3)$$

could in principle be solved uniquely for the parameters α_j. But this "exact" fit to the data would provide no test at all of the validity of the model.] In least squares as properly applied, the number of observations is made large compared to the number of parameters in order (1) to sample adequately a domain of respectable size for testing the validity of the model, (2) to increase the accuracy and precision of the parameter determinations, and (3) to obtain statistical information as to the quality of the parameter determination and the applicability of the model.

The least-squares criterion of best fit depends upon the concept known as *maximum likelihood*: the best set of parameters α_j is one which maximizes the *probability* function for the full set of measurements y_i. The probability P_i for making at $x = x_i$ a single measurement of y equal to y_i is [see Eq. (II-2)]

$$P_i = \frac{1}{\sqrt{2\pi}\,\sigma_i} \exp\left\{-[y_i - f(\alpha_1, \ldots, \alpha_n; x_i)]^2/2\sigma_i{}^2\right\}$$

assuming that the measurements y_i are normally distributed (i.e., are samples of a gaussian distribution) with standard deviations σ_i. For making the entire set of measurements y_i the probability is

$$P = \prod_i P_i = \prod_i \frac{1}{\sqrt{2\pi}\,\sigma_i} \exp\left\{-\sum_i [y_i - f(\alpha_1, \ldots, \alpha_n; x_i)]^2/2\sigma_i{}^2\right\}$$

Consider now the variation of P with respect to the α_j values. This affects only the sum in the exponent; maximizing P is the same thing as minimizing this sum. Thus, maximum likelihood for P becomes the *least-squares principle*

$$\delta_{\alpha_1} X^2 = \delta_{\alpha_2} X^2 = \cdots = \delta_{\alpha_n} X^2 = 0 \qquad (4)$$

where δ signifies variation with respect to infinitesimal and independent variations of the α;† and X^2 is defined by

$$X^2 \equiv \sum_i \left[\frac{y_i - f(\alpha_1, \ldots, \alpha_n; x_i)}{\sigma_i} \right]^2 \tag{5}$$

Actually, least squares is often applied in cases where it is not known with any certainty that measurements of y_i conform to a normal distribution, or even when it is in fact known that they do *not* conform to a normal distribution. Does this destroy the applicability of the maximum-likelihood criterion? The answer is: not necessarily. The central-limit theorem,[1] simply stated, says that the sum (or average) of a large number of measurements conforms very nearly to a normal distribution irrespective of the distributions of the individual measurements *provided* that no one measurement contributes more than a small fraction to the sum (or average) and that the variations in the widths of the individual distributions are within reasonable bounds. (As we shall see, the average of a group of numbers is a special case of a least-squares determination.)

The factor $1/\sigma_i^2$ in Eq. (5) has the significance of the weight of the ith observation; i.e., it determines how heavily that observation contributes to the sum. However, it is not always possible to know even approximately the values of σ_i at the time the measurements are made. In this case, it is important and usually possible to assign *relative* weights w_i^R which are estimates related to the *true* weights $w_i^T \equiv \sigma_i^{-2}$ by

$$w_i^R \approx k w_i^T = k \sigma_i^{-2} \tag{6}$$

where k is an unknown scale factor. Thus we will rewrite Eq. (5) as

$$X^2 = \sum_i w_i [y_i - f(\alpha_1, \ldots, \alpha_n; x_i)]^2 \tag{7}$$

and use this form below on the assumption that true weights w_i^T are available. If relative weights w_i^R are used, this will only introduce an unknown constant multiplicative factor into X^2 and will have no effect on minimizing this quantity.

If by reason of difficulty or lack of time problems of weighting are to be ignored, all factors w_i can be set equal to one in the equations which follow. In that case, however, subsequent sections dealing with evaluation of uncertainties and goodness of fit lose much of their validity unless applied to a case in which all the observations just happen to be of equal weight.

---

† The validity of Eq. (4) depends on the function f being well behaved, i.e., possessing no discontinuities in the function itself or in the first derivative with respect to any α.

NORMAL EQUATIONS

For X^2 to be a minimum, we require that

$$\frac{\partial X^2}{\partial \alpha_1} = \frac{\partial X^2}{\partial \alpha_2} = \cdots = \frac{\partial X^2}{\partial \alpha_n} = 0$$

For the derivative with respect to α_1 we have

$$-2 \sum_i w_i[y_i - f(\alpha_1, \ldots, \alpha_n; x_i)] \frac{\partial f}{\partial \alpha_1} = 0 \tag{8}$$

For the general case that f is not a linear function of the α_j, let us expand it in a Taylor series around the values

$$y_i^0 \equiv f(\alpha_1^0, \ldots, \alpha_n^0; x_i)$$

calculated with the trial values α_i^0 of the parameters. *Keeping only terms to first order*

$$f(\alpha_1, \ldots, \alpha_n; x_i) = y_i^0 + \frac{\partial f_i}{\partial \alpha_1} \Delta\alpha_1 + \frac{\partial f_i}{\partial \alpha_2} \Delta\alpha_2 + \cdots + \frac{\partial f_i}{\partial \alpha_n} \Delta\alpha_n \tag{9}$$

we obtain after rearranging Eq. (8)

$$\sum_i w_i \left(\frac{\partial f_i}{\partial \alpha_1}\right)^2 \Delta\alpha_1 + \sum_i w_i \frac{\partial f_i}{\partial \alpha_1} \frac{\partial f_i}{\partial \alpha_2} \Delta\alpha_2 + \cdots$$

$$+ \sum_i w_i \frac{\partial f_i}{\partial \alpha_1} \frac{\partial f_i}{\partial \alpha_n} \Delta\alpha_n = \sum_i w_i \frac{\partial f_i}{\partial \alpha_1} (y_i - y_i^0) \tag{10}$$

where by $\Delta\alpha_j$ we mean $\alpha_j - \alpha_j^0$ and by $\partial f_i/\partial \alpha_j$ we mean $\partial f/\partial \alpha_j$ evaluated at $\alpha_1 = \alpha_1^0, \ldots, \alpha_n = \alpha_n^0, x = x_i$. To make the notation more compact, we define

$$A_{jk} \equiv \sum_i w_i \frac{\partial f_i}{\partial \alpha_j} \frac{\partial f_i}{\partial \alpha_k} \tag{11a}$$

$$h_j \equiv \sum_i w_i \left(\frac{\partial f_i}{\partial \alpha_j}\right)(y - y_i^0) \tag{11b}$$

Then Eq. (10) takes the form

$$A_{11} \Delta\alpha_1 + A_{12} \Delta\alpha_2 + \cdots + A_{1n} \Delta\alpha_n = h_1 \tag{12a}$$

Similarly
$$A_{21} \Delta\alpha_1 + A_{22} \Delta\alpha_2 + \cdots + A_{2n} \Delta\alpha_n = h_2$$

$$\cdots$$

$$\cdots \tag{12b}$$

$$A_{n1} \Delta\alpha_1 + A_{n2} \Delta\alpha_2 + \cdots + A_{nn} \Delta\alpha_n = h_n$$

These n equations in the n unknowns $\Delta\alpha_j$ are called the *normal equations*. If the determinant of the coefficients $|A_{ik}|$ does not vanish, these equations can be solved to obtain the values of $\Delta\alpha_j$. Then a new and improved set of α's can be obtained:

$$\alpha_1{}^1 = \alpha_1{}^0 + \Delta\alpha_1$$
$$\alpha_2{}^1 = \alpha_2{}^0 + \Delta\alpha_2$$
$$\vdots \tag{13}$$
$$\alpha_n{}^1 = \alpha_n{}^0 + \Delta\alpha_n$$

If the function f is linear in the parameters α_j the fitting procedure is now complete and the values $\alpha_j{}^1$ are the best values in a least-squares sense. If f is nonlinear in terms of any of the α_j values, it is usually necessary to improve the α_j values by carrying out another "cycle" of minimization where $\alpha_j{}^1$ plays the role previously played by $\alpha_j{}^0$. The iteration process is repeated as many times as necessary to obtain convergence of the α_j values to some predetermined level of accuracy. This procedure for solving overdetermined nonlinear equations will only converge if the initial trial parameters $\alpha_j{}^0$ are sufficiently close to the correct final values, so that errors resulting from the omission of second- and higher-order terms from the Taylor expansion are not too severe. In some cases, this means that $\alpha_j{}^0$ must be very close to the true values, which is an awkward limitation.† It should be stressed that the use of linearized normal equations provides a clear exposition of the general principles; however, good *nonlinear* least-squares algorithms for many applications do not rely on the use of the normal equations described above.†

It is useful to derive the normal equations [Eq. (12)] in another way that will lead directly to a matrix formulation. Let us approximate the experimental value y_i by the first-order Taylor expansion of $f(\alpha_1, \ldots, \alpha_n; x_i)$ given in Eq. (9). Thus we obtain the *equation of condition* or *observational equation* for y_i:

$$\frac{\partial f_i}{\partial\alpha_1}\Delta\alpha_1 + \frac{\partial f_i}{\partial\alpha_2}\Delta\alpha_2 + \cdots + \frac{\partial f_i}{\partial\alpha_n}\Delta\alpha_n = (y_i - y_i{}^0) \tag{14}$$

or

$$a_{i1}\Delta\alpha_1 + a_{i2}\Delta\alpha_2 + \cdots + a_{in}\Delta\alpha_n = \Delta y_i \tag{15}$$

where

$$a_{ij} \equiv \frac{\partial f_i}{\partial\alpha_j} \tag{16}$$

---

† It should be stressed that while the use of linearized normal equations clearly exemplifies the general principles, good *nonlinear* least-squares algorithms for many applications do not rely on the use of the normal equations described above. There is no single best method for carrying out nonlinear least-squares fitting—the most advantageous procedure depends on the particular problem. Other methods include grid searches, gradient searches, and mixed algorithms such as Marquardt's. See Refs. 2 and 3 for more details.

To reduce the m observational equations [Eq. (14)] to the smaller number n of normal equations we proceed as follows. Multiply each of equations [Eq. (15)] by the respective $w_i a_{i1}$ and add. This gives the first normal equation, Eq. (10) or (12a). Multiply each of equations [Eq. (15)] through by $w_i a_{i2}$ and add; this gives the second normal equation; continue until all normal equations [Eqs. (12a) and (12b)] have been generated.

We may also write the observational and normal equations in matrix form. The *observational matrix* $\mathbf{a}$ is a rectangular matrix with m rows (indexed by i) and n columns (indexed by j), the elements of which are the a_{ij} given in Eq. (16):

$$
\mathbf{a} = \begin{pmatrix}
a_{11} & a_{12} & \cdots & a_{1n} \\
a_{21} & a_{22} & \cdots & a_{2n} \\
a_{31} & a_{32} & \cdots & a_{3n} \\
\vdots & \vdots & & \vdots \\
a_{m1} & a_{m2} & \cdots & a_{mn}
\end{pmatrix}
$$

We shall also require the transpose $\mathbf{a}^T$ of the matrix (i.e., the matrix with rows and columns interchanged). We will employ a square $m \times m$ weight matrix, with diagonal elements $w_{ii} = w_i$ and (in our case) off-diagonal elements that are zero:

$$
\mathbf{w} = \begin{pmatrix}
w_1 & 0 & 0 & \cdots & 0 \\
0 & w_2 & 0 & \cdots & 0 \\
0 & 0 & w_3 & \cdots & 0 \\
\vdots & \vdots & \vdots & & \vdots \\
0 & 0 & 0 & \cdots & w_m
\end{pmatrix}
\tag{18}
$$

More generally, $\mathbf{w}$ is not required to be diagonal, and indeed there must be nonzero off-diagonal elements if the different y_i have been determined in such a manner that their errors are correlated. Finally, $\Delta\boldsymbol{\alpha}$ and $\Delta\mathbf{y}$ are column vectors of n and m elements, respectively:

$$\Delta\boldsymbol{\alpha} = \begin{pmatrix} \Delta\alpha_1 \\ \Delta\alpha_2 \\ \vdots \\ \Delta\alpha_n \end{pmatrix} ; \qquad \Delta\mathbf{y} = \begin{pmatrix} \Delta y_1 \\ \Delta y_2 \\ \Delta y_3 \\ \vdots \\ \Delta y_m \end{pmatrix} \qquad (19)$$

The observational equation (15) may be written in matrix form as

$$\mathbf{a} \cdot \Delta\boldsymbol{\alpha} = \Delta\mathbf{y} \qquad (20)$$

To obtain the normal equations multiply both sides of Eq. (20) on the left by $\mathbf{a}^T \cdot \mathbf{w}$:

$$\mathbf{a}^T \cdot \mathbf{w} \cdot \mathbf{a} \cdot \Delta\boldsymbol{\alpha} = \mathbf{a}^T \cdot \mathbf{w} \cdot \Delta\mathbf{y}$$

or
$$\mathbf{A} \cdot \Delta\boldsymbol{\alpha} = \mathbf{h} \qquad (21)$$

where

$$\mathbf{A} = \begin{pmatrix} A_{11} & A_{12} & \cdots & A_{1n} \\ A_{21} & A_{22} & \cdots & A_{2n} \\ \vdots & \vdots & & \vdots \\ A_{n1} & A_{n2} & \cdots & A_{nn} \end{pmatrix} = \mathbf{a}^T \cdot \mathbf{w} \cdot \mathbf{a} \qquad (22)$$

is the normal equations matrix (sometimes called the design matrix), and

$$\mathbf{h} = \begin{pmatrix} h_1 \\ h_2 \\ \vdots \\ h_n \end{pmatrix} = \mathbf{a}^T \cdot \mathbf{w} \cdot \Delta\mathbf{y} \qquad (23)$$

is an n-element column vector. The elements of $\mathbf{A}$ and $\mathbf{h}$ are given by Eqs. (11a) and (11b), respectively.

To solve Eq. (21) for $\Delta\boldsymbol{\alpha}$, multiply both sides by $\mathbf{B} \equiv \mathbf{A}^{-1}$, the inverse of the matrix $\mathbf{A}$. We obtain

$$\Delta\boldsymbol{\alpha} = \mathbf{B} \cdot \mathbf{h} \tag{24}$$

It is an important condition on solving the normal equations that the determinant of $\mathbf{A}$ not vanish:

$$\det A \equiv |\mathbf{A}| \neq 0 \tag{25}$$

The equations in matrix form are particularly convenient for computation; matrix multiplication and matrix inversion routines are present in virtually all computer-center library systems and in software packages provided by computer manufacturers.

WEIGHTS

In general, a weight w_i must be assigned for each measurement y_i. If estimates of the standard deviations σ_i are available, the true weights $w_i^T = \sigma_i^{-2}$ can be used. If σ_i are not known but it is manifest that all measurements should have the same uncertainty, the weights are all equal and may be set equal to unity for convenience. There may be other circumstances in which the standard deviations are not known, but in which *relative weights* w_i^R can be assigned on an arbitrary scale by judgement based on experience or by common-sense criteria. (For example, a value of y_i, which is the mean of two or three measurements of y at the same x_i, has a weight twice or three times that of a single measurement at that x_i.) Such relative weights are related to the true weights by Eq. (6).

The best basis for assigning weights is to make several measurements y_{ik} ($k = 1, \ldots, N$) at the same x_i, and determine the experimental variance S_i^2 of each measurement with Eq. (II-6). Since the mean of those N measurements will enter the least-squares calculations as y_i, that variance divided by N will constitute the first approximation to σ_i^2. Since N is likely to be small (less than 6), σ_i^2 will be rather uncertain and consequently a certain amount of pooling or smoothing of the σ_i^2 values may be appropriate. This may be done by plotting the σ_i^2 against x_i and drawing a smooth curve to obtain the values to be used in the least-squares calculation. Or, the σ_i^2 may be averaged in blocks of data having in each block uniform conditions of measurement. We may summarize by writing

$$\frac{1}{w_i} = \sigma_i^2(\text{est}) = \left[\frac{1}{N(N-1)} \sum_{k=1}^{N} (y_{ik} - y_i)^2 \right]_{\text{smoothed}} \tag{26}$$

where y_i is the mean of the y_{ik}.

Since y is functionally dependent on x, the uncertainty in y must contain a contribution from any uncertainty in x. If in the above-described procedure the variable x is set separately and independently to its assigned valued value x_i for each of the N measurements y_i, then presumably the contribution of the uncertainty in x_i will automatically be reflected in the uncertainty of y_i and Eq. (26) will apply directly. However, if x is set only once to x_i for the N measurements y_i, so that the error contributed by x_i is the same for all N measurements, then the uncertainty in both must be reflected in the weight:

$$w_i = \frac{1}{\sigma_i{}^2} = \frac{1}{\sigma_{iy}{}^2 + \left(\dfrac{\partial f}{\partial x}\right)_{x_i}^2 \sigma_{ix}{}^2} \tag{27}$$

where σ_{iy} and σ_{ix} are the independent standard deviations in y and x, respectively, and $(\partial f/\partial x)_{x_i}$ is evaluated with $\alpha_j = \alpha_j{}^0$ (the trial values of the parameters). Equation (27) can also be used in other cases where the uncertainties in y and x are independent and have been estimated separately. In principle, the least-squares method assumes that all the uncertainty is in the y_i value at some exactly known x_i value, but Eq. (27) is a good approximation for adjusting the weights to reflect the experimental fact that both x and y are usually subject to experimental error.

It should be stressed that in any experiment for which there is to be a least-squares refinement of parameters, intended to yield not only the best possible parameter values but also respectable estimates of their uncertainties *and* a test of the validity of the model, it is vital to take the trouble to analyze the methods and the circumstances of the experiment carefully in order to get the best possible values of the a priori weights.

REJECTION OF DISCORDANT DATA

There may be one or more y_i values that deviate markedly from the trend of the others. Not only may y_i values be in error because of the kinds of occurrences discussed in Chap. II (reading errors, transposition of digits, power line transients), there may also be deviations occurring at discrete x_i values arising from physical effects not taken into account in the model. (Examples: in spectroscopy, an unexpected resonance happening at a certain frequency ν_i, enhancing or reducing the spectral response relative to that envisaged by the model; in X-ray crystallography, multiple reflection or thermal diffuse scattering producing abnormal intensity for a particular Bragg reflection.)

Sometimes the deviation is so gross that the offending measurement can be discarded at once. More often, it is necessary to carry out one or more cycles of least squares and to obtain weighted residuals Δ_i based on properly scaled weights [defined below by Eqs. (37a) and (37b)]. It must then be decided by inspection of the Δ_i values whether the apparently discordant y_i should be rejected and one or more additional cycles of least squares carried out. This question is seldom discussed adequately, and there seem to be no well-defined and generally accepted criteria. We present some suggestions based on the magnitudes of the weighted residuals.

For values of m no larger than 10, one can use a Q test criterion (see Chap. II) for considering the rejection of *one* outlying datum y_i. In the present context, the quantity Q is defined by

$$Q = (\Delta_{max} - \Delta_{near})/(\Delta_{max} - \Delta_{far}) \tag{28}$$

The quantities appearing in this equation may be best understood with the aid of an example. Let the residuals for a set of 10 measurements be arranged from the largest negative to the largest positive:

$$-1.0 \quad -.9 \quad -.7 \quad -.7 \quad -.3 \quad .0 \quad .3 \quad .4 \quad .4 \quad 2.0$$

Here Δ_{max} (the Δ with the largest absolute magnitude) is 2.0, Δ_{near} (the Δ nearest to Δ_{max}) is .4, and Δ_{far} (the Δ farthest from Δ_{max}) is -1.0. Thus $Q = (2.0 - .4)/[2.0 - (-1.0)] = 0.53$. From Table II-2 the critical value Q_c is 0.41. Therefore the y_i point that yields $\Delta_i = 2.0$ should be rejected.

For values of m ranging from 11 to 50 we suggest that a seemingly discordant datum for which $\Delta_i > 2.6$ be rejected. (A value of 2.58 corresponds to an a priori estimate that a given measurement has about a 1 percent chance of being valid.) For larger values of m it is difficult or impossible to distinguish data having large weighted residuals because of faulty measurement from those having large weighted residuals because they happen to be in the tail of the normal distribution. However, data with "wild" values of Δ_i (i.e., exceeding 4) may always be rejected.

SIMPLE APPLICATIONS

A trivial example of a least-squares calculation is the calculation of the arithmetic mean. In this case the independent variable x does not appear and there is only one parameter α, which is the desired average value of y. The observational equations are of the form

$$\Delta\alpha = \alpha = y_i - 0$$

since we may take $\alpha^0 = 0$. The single normal equation is

$$\sum_i w_i(1)^2 \alpha = \sum_i (1) w_i y_i$$

whence

$$\alpha = \bar{y} = \frac{\sum w_i y_i}{\sum w_i} \tag{29}$$

When the weights are all taken as unity this becomes

$$\alpha = \bar{y} = \frac{1}{m} \sum_i y_i \tag{30}$$

A more significant example is that of a linear relationship

$$y = f(\alpha_0, \alpha_1; x) = \alpha_0 + \alpha_1 x \tag{31}$$

This is the equation of a straight line, where α_1 is the slope and α_0 is the y intercept. Here again, because of linearity, we may take $\alpha_0{}^0 = \alpha_1{}^0 = 0$. The m observational equations are of the form

$$\alpha_0 + \alpha_1 x_i = y_i$$

and the two normal equations are

$$\sum_i w_i(1)^2 \alpha_0 + \sum_i w_i(1) x_i \alpha_1 = \sum_i w_i(1) y_i$$

$$\sum_i w_i x_i(1) \alpha_0 + \sum_i w_i x_i{}^2 \alpha_1 = \sum_i w_i x_i y_i \tag{32}$$

Solving, we obtain

$$\alpha_0 = \frac{1}{D} \left(\sum_i w_i x_i{}^2 \sum_i w_i y_i - \sum_i w_i x_i \sum_i w_i x_i y_i \right) \tag{33a}$$

$$\alpha_1 = \frac{1}{D} \left(-\sum_i w_i x_i \sum_i w_i y_i + \sum_i w_i \sum_i w_i x_i y_i \right) \tag{33b}$$

where

$$D = \sum_i w_i \sum_i w_i x_i{}^2 - \left(\sum_i w_i x_i \right)^2 \tag{34}$$

is the determinant $|A|$ of the matrix $\mathbf{A}$. In matrix terms we have for this case

$$\mathbf{A} = \begin{pmatrix} \sum w_i & \sum w_i x_i \\ \sum w_i x_i & \sum w_i x_i{}^2 \end{pmatrix}, \quad \mathbf{h} = \begin{pmatrix} \sum w_i y_i \\ \sum w_i x_i y_i \end{pmatrix},$$

$$\mathbf{B} = \begin{pmatrix} \dfrac{1}{D} \sum w_i x_i{}^2 & -\dfrac{1}{D} \sum w_i x_i \\ -\dfrac{1}{D} \sum w_i x_i & \dfrac{1}{D} \sum w_i \end{pmatrix} \tag{35}$$

If unit weights are employed, all w_i are deleted and $\sum w_i$ is replaced by m.

The student may find it illuminating to derive expressions for α_0, α_{1x}, and α_{1y} for the three-parameter least-squares case with two independent variables:

$$z = \alpha_0 + \alpha_{1x}x + \alpha_{1y}y$$

This is the equation of a plane in three-dimensional space.

CALCULATIONAL PRECAUTIONS

Whether done with a computer or with a hand calculator, considerable attention must be given to retaining in the calculations sufficient numerical precision. Parameter values often appear as small differences among large quantities. Errors in the various quantities generated in the calculations tend to be highly correlated. Therefore it is advisable to treat the values taken for the weights w_i as being exact, and to ascribe to the observations y_i as much precision as possible—at least two significant figures more than those in which they are given. This precision should be retained throughout the calculation. In inverting the matrix on a hand calculator, it is well to use the full precision of the calculator, in floating point mode where possible. If a computer is used, this will naturally be done, but in some cases the program may need to be modified for double precision. Single precision ordinarily gives 6 to 10 decimal digits in the mantissa and 2 in the exponent, with the algebraic signs for both.

Just as important as retaining the full precision of your calculator or computer is the matter of properly "conditioning" your normal equation matrix. This can be illustrated by a simple example. Suppose one wants to fit a low-order polynomial such as $y = \alpha_0 + \alpha_1 x + \alpha_2 x^2$ to the following data:

x (year)	1969	1970	1971	1972	1973
y (units sold)	500	501	504	506	504

If one computes $\sum x_i$, $\sum x_i{}^2$, $\sum x_i{}^3$, $\sum x_i y_i$, etc. directly, the resulting matrix will have very large elements and the parameters α_j are given as small differences between large numbers. However, if one first forms an auxiliary data set $\tilde{y}_i$, $\tilde{x}_i$, where $\tilde{y}_i = y_i - \bar{y}$ and $\tilde{x}_i = x_i - \bar{x}$,

$\tilde{x}$	−2	−1	0	1	2
$\tilde{y}$	−3	−2	1	3	1

and if one then fits these numbers, this problem is avoided and one can achieve two or three more significant digits in the α_j values. Such a procedure is especially important if one is working on a calculator with only 6- to 8-digit precision.

When doing calculations with a hand calculator it is important to realize that *one* small arithmetic error can spoil the entire calculation. Since the overall calculation may involve many hundreds of steps, it is advisable to do every computational step twice. When polynomial fits are carried out with models having a large number of adjustable parameters (greater than 5 or 6), there are often precision problems and ill-conditioned matrices can occur. In such cases it may be best to avoid the use of normal equations and do the fitting with orthogonal polynomials.[6] When doing nonlinear fits with a computer, one must specify a desired accuracy for the parameters to avoid needless iterations before "convergence" is reached.

GOODNESS OF FIT

The χ^2 test   Having completed the least-squares computations (with enough cycles to obtain convergence if the function is nonlinear in the α_j) and having determined the best values α_j^* of the parameters, it remains to determine how good they are and how good the model is. In this and the following sections a number of equations are presented without proof; detailed developments can be found in the books referred to at the end of this chapter.

At this point the observed y_i are not in perfect agreement with the calculated y_i^* given by the model with the refined parameters

$$y_i^* = f(\alpha_1^*, \ldots, \alpha_n^*; x_i) \tag{36}$$

There remain residuals $y_i - y_i^*$, which we may subject to statistical analysis. To reduce these to the same statistical population we define a *weighted residual* Δ_i:

$$\Delta_i \equiv \frac{y_i - y_i^*}{\sigma_i} \tag{37a}$$

where the σ_i are the a priori standard deviations. In practice, the weighted residuals are approximated by

$$\Delta_{i0} = \sqrt{w_i}(y_i - y_i^*) \tag{37b}$$

where w_i are the a priori estimated weights. If the true weights were used, these two quantities would be identical; however, in the case where the weights are estimated on an arbitrary relative basis [see Eq. (6)] it is necessary to make a distinction between them.

Analogous to Eq. (II-6) there is the following expression for the *variance of an observation of unit weight*:

$$S_{(1)}^2 = \frac{\sum \Delta_{i0}^2}{m - n} = \frac{X_{\min}^2}{m - n} \tag{38}$$

where m is the number of observations and n is the number of adjustable parameters. The square root of this quantity $S_{(1)}$ is an *estimate* of the standard deviation of an observation of unit weight. If $S_{(1)}$ were indeed equal to the standard deviation of an observation of unit weight, it would have the value unity. Presumably such a value would be obtained if there were an infinite number of observations, the model were rigorously correct, and the correct values $w_i = \sigma_i^{-2}$ were used in Eq. (37b). In general, $S_{(1)}$ differs from unity. For an error-free model, $S_{(1)}$ may be greater than or less than unity depending on whether the weights are overestimated or underestimated.† If the weights are correct, errors in the model tend to make $S_{(1)}$ differ from unity. Thus we see the reason for the emphasis on assigning a priori weights realistically and accurately. If one has confidence in the weights, the value of $S_{(1)}$ can provide a real test of the validity of the model. On the other hand, if weights have been assigned only on a relative basis, $S_{(1)}$ does not afford an effective test of a single model but may be of value in comparing two models (see below).

The test of a fit can be put into the form of the following question: "At a specified level of confidence, is the value of $S_{(1)}^2$ consistent with the assumption that the residuals Δ_{i0} are representative of a normal distribution as they would be if the model were correct and the weights properly assigned?" Here we make use of the fact that in principle the minimized quantity X^2 should conform to the so-called *chi-square distribution*. The probability distribution function for χ^2, with $v = (m - n)$ degrees of freedom, is[2,4]

$$P(\chi^2, v) = \frac{(\chi^2)^{(v-2)/2} e^{-\chi^2/2}}{2^{v/2} \Gamma(v/2)} \tag{39}$$

where $\Gamma(x)$ is the gamma function. The probability that χ^2 will exceed a certain limiting value is

$$P_{int}(\chi^2, v) = \int_{\chi^2}^{\infty} P(x^2, v) \, dx^2 \tag{40}$$

One can refer to published tables[5,6] of reduced chi-square $\chi_v^2 \equiv \chi^2/v$ for given probability P_{int} and a number of degrees of freedom v. An abridged table for $P_{int} = 0.95$ and $P_{int} = 0.05$ is given in Table 1.

If the model is known in advance to be above reproach, the value of $S_{(1)}^2$ can be used to answer the question: "At a specified level of confidence is the value of $S_{(1)}^2$ consistent with the assumption that the weighted residuals Δ_{i0} represent a normal distribution with mean zero and standard deviation unity?" This could be crudely

---

† One can rescale the weights a posteriori on the approximate basis that $w_i(\text{new}) = w_i(\text{old})/S_{(1)}^2$ if one is confident the model is correct.

Table 1 LIMITING VALUES OF $\chi_v^2 = \chi^2(v)/v$ AND $F(1,v)$, FOR STATED PROBABILITY OF EXCEEDING THESE VALUES[a]

	Limiting χ_v^2		Limiting $F(1, v)$
v	$P_{int}=0.95$	$P_{int}=0.05$	$P_{int}=0.05$
1	.0039	3.84	161
2	.0515	3.00	18.5
3	.117	2.60	10.1
4	.178	2.37	7.7
5	.229	2.21	6.6
6	.273	2.10	6.0
8	.342	1.94	5.3
10	.394	1.83	5.0
12	.436	1.75	4.8
15	.484	1.67	4.5
20	.543	1.57	4.4
30	.616	1.46	4.2
50	.695	1.35	4.0

[a] Taken from Refs. 5 and 6

translated into the question: "Is the value of $S_{(1)}^2$ consistent with the assumption that the weights have been properly assigned?" Here $S_{(1)}^2$ can be either less than or greater than unity; for 90 percent confidence limits, $S_{(1)}^2$ should lie between the limiting χ_v^2 values corresponding to $P_{int} = 0.05$ and $P_{int} = 0.95$. For example, when $v = m - n = 30$, we find from Table 1 that $S_{(1)}^2$ should lie between 0.616 and 1.46 if the above question is to be answered in the affirmative. An affirmative answer is *not* a guarantee that the weights have been assigned correctly, but only an assertion that they cannot be criticized on the basis of the statistics available.

If the assignment of weights is known in advance to be above reproach, the value of $S_{(1)}^2$ can be used to answer the more interesting question: "At a specified level of confidence is the value of $S_{(1)}^2$ consistent with the assumption that the data set conform to the assumed model function?" This could be rephrased as: "Do the residuals conform to a normal distribution?" For the above example of $v = 30$ and 90 percent confidence limits, $S_{(1)}^2$ should lie between 0.616 and 1.46 if the question is to be answered positively. That is, if the model is valid, there is only a 5 percent a priori statistical probability that $S_{(1)}^2$ would lie below 0.616 and the same probability that it might lie above 1.46.

At this point it is convenient to make a change in notation. It has become common practice to cite the quantity "reduced chi square" χ_v^2:

$$\chi_v^2 \equiv \frac{1}{v} \sum_i \frac{(y_i - y_i^*)^2}{\sigma_i^2} \simeq \frac{1}{v} \sum_i w_i(y_i - y_i^*)^2 \tag{41}$$

defined by the first expression and, in practice, calculated with the second expression. Thus, in the literature, χ_v^2 is the symbol often given to exactly what we have been calling $S_{(1)}^2$. Good fits require χ_v^2 values that lie between the limiting values specified (in Table 1, for example) for a given confidence level and v value. To continue with our example of a fit to a data set with $m = 32$ and $n = 2$, let us assume that χ_v^2 calculated with Eq. (41) has the value 1.7. At the 90 percent confidence level, the validity of the model, the assignment of weights, or both are suspect. However, one must be quite certain about the weights before this χ_v^2 value be used to reject the model (within the specified confidence limits). Let us now assume that the fit to our data set had given $\chi_v^2 = 1.2$. For this χ_v^2 value, the validity of the model and the assignment of weights cannot be challenged on a statistical basis at this confidence level. However, this does not guarantee that the model is sound unless one can show that artificially low weights have not been used.

If the model function is a nonlinear function of the parameters α_j, and the trial values α_j^0 are not quite close to the true values, the least-squares treatment may converge on a false minimum for χ_v^2. Usually when this happens, $\chi_v^2 \simeq S_{(1)}^2 \gg 1$. However, in problems where m and n are both very large (as in X-ray crystallography), there are a great many false minima, some of which are not far from the true minimum in n-dimensional parameter space and yield χ_v^2 values not greatly different from unity. This is one additional argument for making careful a priori estimates of σ_i. When the weights for the observations are properly estimated, it is unlikely (although occasionally possible) that a false minimum will satisfy the χ^2 test within the appropriate confidence limits.

Another way to avoid false minima when using simple nonlinear fitting functions such as

$$y = \alpha_0 + \alpha_1 x^{\alpha_2}$$

is to fix the value of α_2 and carry out a linear least-squares fit; then step α_2 through a range of values and plot χ_v^2 versus the assigned α_2 values.

Distribution of residuals   When the number of measurements is large (preferably more than 100) one can carry out a χ^2 test of the frequency distribution of the Δ_i values.[2] This frequency test can be more instructive than the χ^2 test of $S_{(1)}^2$ since it may allow a diagnosis of defects in the model or defects in the weight distribution apart from a mere scaling error in the weights.

Some workers have introduced a more detailed statistical test of the least-squares fit and the weighting scheme by using *normal probability plots*.[7,8] This test compares the actual distribution of the observed weighted residuals Δ^{obs} to the ideal values Δ^{ideal} expected for a normal distribution of mean zero and standard deviation unity. A plot of Δ^{obs} versus Δ^{ideal} should yield points close to a straight line with slope unity which passes through the origin, and most of the scatter should occur at the ends. If the plotted points look linear but with a slope considerably different from unity, the weights may be relatively correct but either over-estimated or underestimated. If the plotted points exhibit a pronounced deviation from linearity, either the model is defective or erroneous relative weights have been assigned.

In the spirit of the above, it is helpful to make a qualitative inspection for trends in the residuals even if a formal analysis is not undertaken. In some fitting situations, one can see that dropping a few points will allow a change in the adjustable fitting parameters that will appreciably decrease all the remaining Δ_i values and thus significantly lower χ_v^2. Such a procedure may or may not be defensible on purely statistical grounds, but it can at least lead one to carefully inspect the experimental validity of possibly errant points. Even when there are no experimentally suspect points with peculiar residuals, it is important to be aware of the trend in the residuals across the data set. Models that give a best fit with systematic trends in the residuals (say a block of negative residuals at each end of the data set with a central block of positive residuals) are suspect even if the χ_v^2 values appear to be satisfactory.

STANDARD DEVIATIONS IN THE PARAMETERS

We assume now that the least-squares refinement has converged satisfactorily, that any necessary rejection of discordant data has taken place before the final cycles were carried out, and that the statistical tests on the weighted residuals have given reassuring results. We can now proceed to estimate the standard deviations in the adjustable parameters. If the results of statistical tests were *not* reassuring, only qualified confidence can be given to the estimated standard deviations; it may be appropriate in such a case, if they are to be quoted at all, to multiply them by a factor of 2 or 3.

If the weights were originally assigned correctly on the basis of realistic values of σ_i, the standard deviation of an observation of unit weight should indeed be unity. In that case a propagation-of-error treatment of Eq. (24) yields for the estimated standard deviation in parameter α_j

$$\sigma(\alpha_j) = B_{jj}^{1/2} \tag{42}$$

where B_{jj} is the jth diagonal element in the inverse normal equations matrix $\mathbf{B} = \mathbf{A}^{-1}$. This equation would be used only in those infrequent cases where the number of observations is small (less than about 10) and where the reliance that can be placed on the σ_i values is uncommonly high. In the more typical case, where the weights are not that highly reliable and m is fairly large, an estimate of the standard deviation of unit weight $S_{(1)}$ given by Eq. (38) is needed. One then obtains

$$\sigma(\alpha_j) = B_{jj}{}^{1/2}S_{(1)} \tag{43}$$

If m is still small, in the context of Chap. II, it may be better to quote confidence limits λ rather than estimated standard deviations. In this case,

$$\lambda(\alpha_j) = tB_{jj}{}^{1/2}S_{(1)} \tag{44}$$

where t of Student's distribution is found in Table II-1 with DF $= m - n$.

For the trivial one-parameter case where the parameter to be determined is the arithmetic mean,

$$B_{11} = \frac{1}{\sum\limits_i w_i}$$

Making use of Eqs. (38) and (43) we obtain

$$\sigma(\bar{y}) = \left[\frac{\sum \Delta_i{}^2}{(N-1)\sum w_i} \right]^{1/2}$$

If all weights are taken as unity, $\sum_i w_i = m = N$ and

$$\sigma(\bar{y}) = \left[\frac{\sum (y_i - y_i^*)^2}{N(N-1)} \right]^{1/2} \tag{45}$$

which corresponds to the result cited in Chap. II.

For the two-parameter case of the linear relationship we may apply Eqs. (38) and (43) to the estimation of the standard deviations in the intercept α_0 and the slope α_1 of the corresponding straight line. From Eq. (35) we have

$$B_{00} = \frac{1}{D} \sum_i w_i x_i{}^2, \qquad B_{11} = \frac{1}{D} \sum_i w_i$$

where D is given by Eq. (34). Thus

$$\sigma(\alpha_0) = \left(\frac{1}{D} \sum_i w_i x_i{}^2 \right)^{1/2} \left(\frac{\sum \Delta_i{}^2}{m-n} \right)^{1/2} \tag{46a}$$

$$\sigma(\alpha_1) = \left(\frac{1}{D} \sum_i w_i \right)^{1/2} \left(\frac{\sum \Delta_i{}^2}{m-n} \right)^{1/2} \tag{46b}$$

For unit weights

$$\sigma(\alpha_0) = \left(\frac{1}{D} \sum_i x_i{}^2 \right)^{1/2} \left[\frac{\sum (y_i - y_i^*)^2}{m-n} \right]^{1/2} \tag{47a}$$

$$\sigma(\alpha_1) = \left(\frac{m}{D} \right)^{1/2} \left[\frac{\sum (y_i - y_i^*)^2}{m-n} \right]^{1/2} \tag{47b}$$

It frequently happens that one wishes to obtain an estimated standard deviation for a quantity F which is calculated from two or more of the α_j according to a functional relationship of some sort, say $F(\alpha_1, \ldots, \alpha_n)$. It would *not* be correct to estimate separately the $\sigma(\alpha_j)$ with Eqs. (38) and (43) and apply Eq. (II-50). The reason is that the errors in the α_j are usually not independent; in general they are *correlated*. A simple example will show how errors can be correlated. Imagine three points a, b, c *along a straight line*. Let the distance ab be α_1, and let the distance bc be α_2. If the estimated standard deviation (e.s.d.) in the position of any of the points is *independently* σ_p, the e.s.d. of α_1 is $\sqrt{2}\sigma_p$ and the e.s.d. of α_2 is also $\sqrt{2}\sigma_p$. What is the estimated standard deviation of $F = \alpha_1 + \alpha_2 =$ the distance ac? If we use the propagation-of-errors treatment without allowing for correlation, we get the answer that $\sigma(F) = [\sigma^2(\alpha_1) + \sigma^2(\alpha_2)]^{1/2} = 2\sigma_p$, while on going back to first principles we see that the true answer is $\sqrt{2}\sigma_p$. The two distances α_1 and α_2 are correlated; when point b moves toward a, it is moving away from c. For a general discussion of this problem and methods for handling it with the correlation matrix see Refs. 2 and 4. The correlation matrix may also provide useful information for the design of an experiment.

COMPARISON OF MODELS

It often happens that a decision has to be made between two somewhat different models for describing the data. Both models have physical significance but they are based on different physical hypotheses. If both appear to provide reasonably good fits to the data, can preference be given for one model over the other?

The judgement may be based on the reduced chi-square values χ_v^2 for the fits with the two respective models. However, even if fitting with the two models involves the same number of degrees of freedom v, it should not be said that model 2 is significantly preferable to model 1 just because χ_v^2 for model 2 is smaller than χ_v^2 for model 1. One must answer the question: "Is the difference *significant?*" Here we need to look at the probability distribution for the ratio of reduced chi squares. Such ratios, χ_{v1}^2/χ_{v2}^2, should conform to a distribution known as the F distribution, for which the probability distribution function is[5]

$$P_F(F, v_1, v_2) = \frac{\Gamma\left(\dfrac{v_1 + v_2}{2}\right)}{\Gamma\left(\dfrac{v_1}{2}\right)\Gamma\left(\dfrac{v_2}{2}\right)}\left(\frac{v_1}{v_2}\right)^{v_1/2}\frac{F^{(v_1-1)/2}}{\left(1 + F\dfrac{v_1}{v_2}\right)^{(v_1+v_2)/2}} \tag{48}$$

The test is made with the integral probability

$$P_{\text{int}}(F, v_1, v_2) = \int_F^\infty P_F(f, v_1, v_2)\, df \tag{49}$$

which expresses the probability that F obtained from a random data set exceeds a certain value. We make use of the fact that

$$F = \frac{\chi_{v_1}{}^2}{\chi_{v_2}{}^2} = \frac{\chi_1{}^2/(m - n_1)}{\chi_2{}^2/(m - n_2)} = \frac{S_{(1)1}^2}{S_{(1)2}^2} \tag{50}$$

and look up in published tables[5,6] the value of

$$P_{\text{int}}(F, v_1, v_2) = P_{\text{int}}\left[\frac{S_{(1)1}^2}{S_{(1)2}^2}, m - n_1, m - n_2\right]$$

or else look up the limiting value of $F(v_1, v_2)$ for a given probability level, say 5 percent.

For example, suppose that there are two models both with the same number of adjustable parameters. Each fit will involve the same number of degrees of freedom, say 20. Reference to statistical tables gives an $F(20, 20)$ value of 2.12 for a probability level of .05. That means that if $\chi_{v_1}{}^2/\chi_{v_2}{}^2 > 2.12$ it can be said at the 95 percent confidence level that model 1 may be rejected in favor of model 2. Note the important feature than an accurate value for the ratio $\chi_{v_1}{}^2/\chi_{v_2}{}^2$ does *not* require a priori knowledge of the σ_i values and true weights. If good relative weights are used [see Eq. (6)], the reduced chi-square ratio will be correct.

This F test, in the form indicated, can also be used when the number of degrees of freedom is different for the two models. The most frequently encountered circumstance of this kind results from least-squares fitting with two models, one having n parameters and the other having the same n parameters plus p additional parameters associated with an elaboration of the first model. As a simple example, consider a unimolecular gas-phase kinetics experiment like Exp. 28. Strict first-order kinetics predicts that the partial pressure of the decaying molecular species will vary as

$$\ln p = \ln p_0 - kt \qquad \text{(model 1)} \tag{51}$$

However, at low pressures, owing to decreased probability of collisional deactivation of the activated molecule, the behavior may be better represented by

$$\ln p = \ln p_0 - kt + \frac{b}{p_0}(e^{kt} - 1) \qquad \text{(model 2)} \tag{52}$$

Equation (51) has two adjustable parameters: p_0 and k; Eq. (52) has three: p_0, k, and b. The experimental data consist of measurements of p as a function of t over a range in which p decreases by a factor of about e^2. Least-squares parameter determinations are carried out separately with each of the two models. They give slightly different sets of values of p_0 and k. It is found that $S_{(1)}^2$ for model 2 is less than $S_{(1)}^2$ for model 1. Is model 2 to be preferred? That is: does a nonzero coefficient b really arise from a physically better model, or does it simply provide a cosmetic improvement in the fit by adding an extra and nonphysical adjustable parameter?

The F test in this case can be cast in a somewhat different form:

$$\frac{\chi^2(m-n) - \chi^2(m-n-p)}{p(m-n-p)^{-1}\chi^2(m-n-p)} = F_\chi(p, m-n-p) \tag{53}$$

or, substituting in the $S_{(1)}^2$ values

$$\frac{(m-n)S_{(1)}^2(m-n) - (m-n-p)S_{(1)}^2(m-n-p)}{pS_{(1)}^2(m-n-p)} = \frac{v_1}{p}\frac{\chi_{v_1}^2}{\chi_{v_2}^2} - \frac{v_2}{p} = F_\chi(p, v_2) \tag{54}$$

In the present case $n = 2$, $p = 1$; let us suppose that $m = 20$. For model 1, $v_1 = m - n = 18$; for model 2, $v_2 = m - n - p = 17$. Therefore

$$\frac{18S_{(1)1}^2 - 17S_{(1)2}^2}{1 \cdot S_{(1)2}^2} = 18\frac{\chi_{v_1}^2}{\chi_{v_2}^2} - 17 = F_\chi(1, 17) \tag{55}$$

At $P = .05$, the limiting value of $F(1, 17)$ is found from Table 1 to be 4.5. The corresponding limiting ratio of reduced chi squares is $\chi_{v_1}^2/\chi_{v_2}^2 = 1.19$. Suppose in our hypothetical example that the actual ratio is 1.24. Then at the 95 percent confidence level model 2 is to be preferred, and the hypothesis that decay of the activated species is competing significantly with deactivation is confirmed. If, on the other hand, this ratio is found to be only 1.12, the "better" fit with model 2 is not statistically significant. We might have strong theoretical reasons for preferring model 2, but this set of data on this particular reacting system does not allow one to reject model 1.

More detailed comparison of two models could involve the use of normal probability plots[7,8] or at least inspection of residuals for systematic trends as discussed earlier. If two models produce fits of equivalent quality in the sense of an F test, one might still prefer the model which gave the more random sequence of residuals (or, put differently, be suspicious of a model that gave blocks of residuals with alternating signs). Another way to compare models that yield almost equivalent fits is *range shrinking*, in which the data set is systematically truncated from either end of the range in the variable x and the stability of the parameter value is inspected.

Finally, one should pay considerable attention to the physical reasonableness of the adjustable parameters α_j before choosing one model over another. To embrace model 2 and reject model 1 on a purely statistical analysis of one set of data is dangerous. If possible, one should vary the experimental conditions, analyze several sets of data, and look at the behavior of the α_j values. In our gas kinetics example, one could vary the temperature at which the rate of decomposition is studied and test the temperature dependence of the k values obtained from least-squares fits at each temperature. What if model 2 were statistically better for fitting the data at each T but gave k (and probably b) values that were erratic functions of T, while model 1 gave k values that smoothly varied with T in a way that was theoretically pleasing? One might then prefer model 1 but report the statistical problems associated with the data. In particular, one should then look carefully for a posteriori evidence that the relative weights might be reassigned. Unsuspected systematic errors present over part of the range could have influenced the quality of the overall fit.

SUMMARY OF PROCEDURES

This section will provide a brief "road map" of the steps recommended in carrying out a least-squares fit.

1 Establish the weights w_i to be assigned to each observed value y_i. It is vital to have good relative weights, and preferable to have true weights based on the standard deviations σ_i. Think carefully about your measurements before taking the easy way out and adopting equal weights (see Weights). It may be necessary to smooth the observed variances over the entire data set (see Sample Least-Squares Calculation).

2 Reject any obviously bad points (see Rejection of Discordant Data).

3 Decide on the model function to be used. This choice may be guided by a theoretical prediction or by a rough empirical assessment of the type of dependence likely.

4 Choose a reasonable set of trial values $\alpha_j{}^0$ for the adjustable parameters of the model. In nonlinear fitting, a good choice of $\alpha_j{}^0$ is very important to reduce computational time and to avoid false minima (see Goodness of Fit, towards the end of the section on χ^2 tests).

5 Carry out the least-squares minimization of the quantity X^2 in Eq. (7) according to any one of a number of algorithms (see Normal Equations for one basic approach). During this process one should be careful about accuracy and precision (see Calculational Precautions).

6 After convergence has been achieved (or several cycles are completed with a hand calculator), check the residuals $y_i - y_i{}^*$ to see if any data points can and should be deleted from the data set (see end of the section on Goodness of Fit). If so, rerun the least-squares fit on the new, edited data set.

7 Obtain the reduced chi square value $\chi_v{}^2$ given by Eq. (41) and carry out the appropriate statistical tests for goodness of fit, including inspection of the weighted residuals for systematic trends.

8 If either theory or clear systematic trends in the residuals suggest alternative models, carry out a new least-squares fit to the same data set with the new model. Make a statistical analysis to help decide whether one of the models can be rejected in favor of the other. Be careful to consider the physical reality of the adjustable parameters in the model that you finally choose (see Comparison of Models).

9 Evaluate the statistical uncertainties in the adjustable parameters obtained from the best fit (see Standard Deviations in the Parameters).

10 *Think about what you are doing. Do not treat least-squares fitting as a magical mathematical game.*

SAMPLE LEAST-SQUARES CALCULATION

We will illustrate here as many as possible of the principles and techniques discussed in this chapter with a sample "curve fitting" by least squares.

For values of x_i ranging from 2.1 to 5.0 ($i = 1$ through 30) in intervals of 0.1, measurements of y were made, four y_{ik} for each x_i value. The measured values are listed in Table 2 and plotted in Fig. 1. For each x_i, the four y_{ik} were averaged to

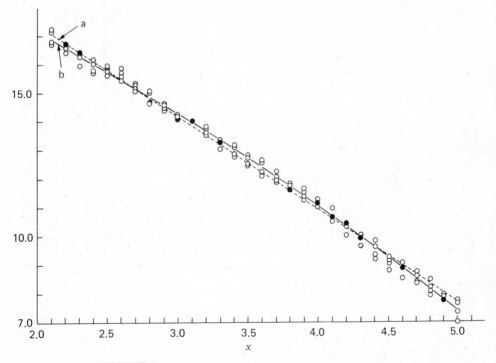

FIGURE 1

Plot of experimental points and least-squares lines: (a) straight line fit (2 parameters), (b) quadratic fit (3 parameters). A filled-in circle represents two or more nearly concident points.

give y_i, and the variance S_i^2 was calculated. Since the measurements were made under uniform conditions and since the variances, when plotted against x, showed (apart from their natural statistical scatter) a smooth variation suggestive of quadratic behavior, the variances were smoothed by least-squares fitting

Table 2 SAMPLE LEAST SQUARES CALCULATION: DATA AND LEAST SQUARES FITS WITH TWO AND THREE PARAMETERS

		THE DATA SET								2 Parameters[a]			3 Parameters[b]		
i	x_i	y_{ik}, $k=1,\dots,4$				$y_i^{(mean)}$	S_i^2	$S_{sm_i}^2$	$S_{sm_i}^{2(mean)}$	w_i	y_i^{cal}	Δ_i^2	Δ_i	y_i^{cal}	Δ_i^2	Δ_i
1	2.1	17.15	16.73	16.80	17.24	16.98	.0638	.0372	.0093	107	17.13	2.41	-1.55	16.85	1.81	1.34
2	2.2	16.62	16.46	16.62	16.60	16.57	.0060	.0342	.0086	116	16.81	6.68	-2.58	16.58	.01	-.11
3	2.3	16.33	15.94	16.35	16.26	16.22	.0363	.0315	.0079	127	16.48	7.32	-2.70	16.30	.81	-.90
4	2.4	16.14	15.70	15.77	15.97	15.89	.0396	.0291	.0073	137	16.16	9.99	-3.16	16.03	2.69	-1.64
5	2.5	15.64	15.87	15.90	15.61	15.75	.0228	.0270	.0068	147	15.83	.94	-.97	15.74	.01	.12
6	2.6	15.66	15.84	15.55	15.62	15.62	.0303	.0251	.0063	159	15.50	2.29	1.51	15.46	4.07	2.02
7	2.7	15.10	15.28	15.20	15.32	15.22	.0094	.0235	.0059	169	15.18	.27	.52	15.16	.61	.78
8	2.8	14.70	14.97	14.67	15.06	14.85	.0355	.0222	.0056	179	14.85	.00	.00	14.87	.07	-.27
9	2.9	14.64	14.40	14.44	14.58	14.52	.0129	.0212	.0053	189	14.53	.02	-.14	14.57	.47	-.69
10	3.0	14.23	14.11	14.12	14.18	14.16	.0031	.0204	.0051	196	14.20	.31	-.56	14.27	2.37	-1.54
11	3.1	14.03	14.06	14.06	14.04	14.05	.0002	.0199	.0050	200	13.88	5.78	2.40	13.96	1.62	1.27
12	3.2	13.66	13.75	13.52	13.57	13.63	.0103	.0197	.0049	204	13.55	1.31	1.14	13.65	.08	-.29
13	3.3	13.33	13.28	13.05	13.28	13.24	.0158	.0197	.0049	204	13.23	.02	.14	13.34	2.04	-1.43
14	3.4	12.81	13.11	12.88	13.16	12.99	.0293	.0200	.0050	200	12.90	1.62	1.27	13.02	.18	-.42
15	3.5	12.79	12.48	12.55	12.74	12.64	.0221	.0206	.0052	192	12.58	.69	.83	12.69	.48	-.69
16	3.6	12.13	12.62	12.67	12.25	12.42	.0718	.0215	.0054	185	12.25	6.35	2.31	12.37	.46	.68
17	3.7	12.27	12.08	11.91	11.97	12.06	.0250	.0226	.0056	179	11.93	3.03	1.74	12.04	.07	.27
18	3.8	11.67	11.66	11.81	11.75	11.72	.0050	.0240	.0060	167	11.60	2.40	1.55	11.70	.07	.26
19	3.9	11.34	11.42	11.66	11.47	11.47	.0185	.0257	.0064	156	11.28	5.63	2.37	11.36	1.89	1.37
20	4.0	11.26	11.24	11.03	11.28	11.20	.0135	.0276	.0069	145	10.95	9.06	3.01	11.02	4.70	2.17
21	4.1	10.67	10.53	10.64	10.98	10.70	.0372	.0298	.0074	135	10.63	.66	.81	10.67	.12	.35
22	4.2	10.39	10.40	10.08	10.33	10.30	.0225	.0323	.0081	123	10.30	.00	.00	10.32	.05	-.22
23	4.3	9.96	10.00	9.65	9.93	9.88	.0254	.0357	.0088	114	9.98	1.14	-1.07	9.96	.73	-.85
24	4.4	9.31	9.68	9.21	9.84	9.51	.0893	.0381	.0095	105	9.65	2.06	-1.43	9.61	1.05	-1.02
25	4.5	9.09	8.95	9.16	9.25	9.09	.0294	.0414	.0104	96	9.33	5.53	-2.35	9.24	2.16	-1.47
26	4.6	9.07	8.86	8.86	8.53	8.83	.0498	.0450	.0112	89	9.00	2.57	-1.60	8.88	.22	-.47
27	4.7	8.71	8.39	8.58	8.34	8.50	.0294	.0488	.0122	82	8.68	2.66	-1.63	8.50	.00	.00
28	4.8	7.94	8.52	8.17	8.04	8.17	.0641	.0529	.0132	76	8.35	2.46	-1.57	8.13	.12	.35
29	4.9	7.74	7.88	7.71	7.84	7.79	.0065	.0573	.0143	70	8.03	4.03	-2.01	7.75	.11	.33
30	5.0	7.01	7.37	7.70	7.62	7.42	.0963	.0620	.0155	65	7.70	5.10	-2.26	7.37	.16	.40

[a] $y_i^{cal} = 23.957 - 3.251\,x_i$

[b] $y_i^{cal} = 21.618 - 1.850\,x_i - .2000\,x_i^2$

$\sum \Delta_i^2 = 91.33$
$\div 28$
$S_{(i)}^2 = 3.262$
$S_{(i)} = 1.806$

$\sum \Delta_i^2 = 29.23$
$\div 27$
$S_{(i)}^2 = 1.083$
$S_{(i)} = 1.040$

with a quadratic using unit weights. The variance function obtained was $S_{sm}^2 = 0.162 - .0879x + .01357x^2$. The values of $S_{sm_i}^2$ calculated with this equation are also listed in Table 2.

No measurements were rejected as being discordant. For example, let us look at the measurements at $x_i = 2.2$, where the observed y_{ik} values were

$$16.62 \qquad 16.46 \qquad 16.62 \qquad 16.60$$

The value of Q is $24/26 = 0.923$, which exceeds $Q_c(90\%)$ for $N = 4$. Should the value 16.46 be rejected? If these four measurements were the only ones, one should certainly reject that value. Here, however, we have the benefit of much more information, and a glance at the rest of the data show that the range of y_{ik} values at $x_i = 2.2$ is reasonably representative. In applying rejection criteria or other statistical criteria one must always take into account the context in which the criteria are given, and be careful to make the best possible use of *all* the data at hand.

There being four measurements averaged, the value of the smoothed variance for y_i, the mean of the y_{ik}, was taken as $S_{sm_i}^2/N$ where $N = 4$. This now is our best estimate of σ_i^2. Its reciprocal, rounded to an integer for computational convenience, was taken as the weight w_i. In many other circumstances, there will not be as many as four measurements of each y_i to form a basis for estimating w_i values. Often only one y_i value is available at each x_i. If this is the case, it is important to carry out multiple measurements at least for some x_i values to provide a feel for "typical" σ_i values. Also, one should use all one's general knowledge and pay careful attention to the range of y_i, x_i values in guessing the best weights.

Since the plot in Fig. 1 appears to confirm roughly a theoretical expectation of linear behavior, a two-parameter least-squares fit to the function

$$y = \alpha_0 + \alpha_1 x$$

was undertaken. For this, using $\alpha_0{}^0 = \alpha_1{}^0 = 0$ we find

$$A = \begin{pmatrix} 4313.0 & 14706.5 \\ 14706.5 & 52600.8 \end{pmatrix}$$

$$h = \begin{pmatrix} 55011.9 \\ 181330.4 \end{pmatrix}$$

Inversion of the matrix A and multiplication with h gives

$$B = \begin{pmatrix} 4.9688515\text{-}03 & -1.3892262\text{-}03 \\ -1.3892262\text{-}03 & 4.0742072\text{-}04 \end{pmatrix}$$

$$\Delta\alpha = B \cdot h = \begin{pmatrix} 23.9567 \\ -3.2507 \end{pmatrix} = \begin{pmatrix} \Delta\alpha_0 \\ \Delta\alpha_1 \end{pmatrix}$$

or $\alpha_0 = 23.9567$, $\alpha_1 = -3.2507$

In **B**, floating point notation is used, -03 denoting $\times 10^{-3}$. Values of y_i^{cal} calculated with these parameters are given in Table 2. The value $\chi_v^2 \simeq S_{(1)}^2 = 3.262$ is quite different from unity; for $P_{int} = 0.05$ and $v = m - n = 30 - 2 = 28$, the limiting χ_v^2 is 1.476. Therefore, at the 95 percent confidence level, our χ_v^2 value is *not* consistent with the assumption that the residuals Δ_i are representative of a normal distribution. Either our weighting scheme is wrong or the data are not well represented by a linear fit.

Examination of the fit of the straight line to the data points (as characterized by the residuals) shows that the points tend to be a little below the line near the ends and a little above near the middle, as if the correct fitting function should have a small amount of curvature. Let us now consider a quadratic model, which may be justified by a more refined theory or may be purely empirical. Accordingly, a new fit was made with the function

$$y = \alpha_0 + \alpha_1 x + \alpha_2 x^2$$

For this, when $\alpha_0^0 = \alpha_1^0 = \alpha_2^0 = 0$ we find

$$\mathbf{A} = \begin{pmatrix} 4313.0 & 14706.5 & 52600.8 \\ 14706.5 & 52600.8 & 196491.5 \\ 52600.8 & 196491.5 & 762679.0 \end{pmatrix} \; ; \quad \mathbf{h} = \begin{pmatrix} 66032.1 \\ 220628.7 \\ 773646.0 \end{pmatrix}$$

$$\Delta\alpha = \mathbf{B} \cdot \mathbf{h} = \begin{pmatrix} 9.0559662\text{-}02 & -5.2868645\text{-}02 & 7.3749622\text{-}03 \\ -5.2868645\text{-}02 & 3.1370214\text{-}02 & -4.4357422\text{-}03 \\ 7.3749622\text{-}03 & -4.4357422\text{-}03 & 6.3546621\text{-}04 \end{pmatrix} \begin{pmatrix} 66032.1 \\ 220628.7 \\ 773646.0 \end{pmatrix}$$

$$\begin{pmatrix} \Delta\alpha_0 \\ \Delta\alpha_1 \\ \Delta\alpha_2 \end{pmatrix} = \begin{pmatrix} 5027.1389 & -9586.6925 & +4580.6627 \\ -2934.8389 & +5688.3734 & -2755.0838 \\ 409.39816 & -804.33490 & +394.69441 \end{pmatrix} = \begin{pmatrix} 21.1091 \\ -1.5493 \\ -0.24232 \end{pmatrix}$$

where the intermediate terms of the matrix-vector product are displayed in order to emphasize the point that the parameter values obtained are often small differences among large quantities! Because of concern about limitations of arithmetic precision when using a hand calculator with only eight significant figures in the above calculation, a second cycle of least squares was carried out based on $\Delta y_i = y_i^{obs} - (21.1091 - 1.5493x_i - .24232x_i^2)$. This gave

$$\Delta\alpha = \mathbf{B} \cdot \Delta\mathbf{h} = \mathbf{B} \cdot \begin{pmatrix} -0.0865 \\ -0.4423 \\ -1.4181 \end{pmatrix} = \begin{pmatrix} 0.5092 \\ -0.3012 \\ 0.0423 \end{pmatrix}$$

or $\alpha_0 = 21.6183$, $\alpha_1 = -1.8505$, $\alpha_2 = -0.2000$

The variance of an observation of unit weight, $S_{(1)}^2$, was 1.083, while that from the first cycle was 1.193. (An approach that might have eliminated the need for a second cycle with three parameters when using a hand calculator would have been to choose $\alpha_0{}^0 = 23.957$, $\alpha_1{}^0 = -3.251$, $\alpha_2{}^0 = 0$ as the initial trial values since these are the results from the two-parameter fit.)

Having found the best parameter values for our three-parameter model, we can now use Eq. (43) to estimate the standard deviations in these parameters. The final results of this least-squares fit are

$$\alpha_0 = 21.62(31), \quad \alpha_1 = -1.85(18), \quad \alpha_2 = -0.200(26)$$

where the digits in parentheses are the estimated standard deviations in the corresponding final digits of the parameter values. It is seen that the parameters from the first cycle of three-parameter least squares differ from those of the second by more than one but less than two standard deviations, while the parameters from the linear least-squares calculation differ from those of the quadratic by about nine standard deviations!

It now remains to apply some of the statistical tests previously described. We first examine our value $\chi_v^2 \simeq S_{(1)}^2 = 1.083$. For $P_{\text{int}} = 0.05$ and $v = 27$, the limiting value of χ_v^2 is 1.486, from which it appears that our χ_v^2 value is eminently satisfactory at the 95 percent confidence level; in fact, the probability of exceeding 1.083 with 27 degrees of freedom is about 65 percent.

While it is already clear from other evidence that the three-parameter fit is satisfactory while the two-parameter fit is not, let us apply the F test to see if the addition of the quadratic parameter produces an improvement in the model that is significant at the 95 percent confidence level. From Eq. (54) we obtain

$$F(1, 27) = \frac{28 S_{(1)2}^2 - 27 S_{(1)3}^2}{1 \cdot S_{(1)3}^2} = \frac{28 \times 3.262 - 27 \times 1.083}{1.083} = 57.34$$

at $P = .05$, the limiting value of $F(1, 27)$ is 4.22; thus, the improvement is significant at the 95 percent confidence level; indeed, it is still significant at about the 99.99 percent confidence level.

Coda. Now it can be told: The "experimental" y_i values in Table 2 were actually generated from *known* "true" values of the parameters and *known* standard deviations! Thus we have here a demonstration of how well the method of least squares works.

We started out with

$$y_i^{\text{true}} = 21.33 - 1.697 x_i - 0.2232 x_i^2$$

and the y_i values were assigned a standard deviation of

$$\sigma_i^{\text{true}} = 0.8 - 0.4 x_i + 0.06 x_i^2$$

Numbers from a random number table were used to generate normal deviates. These were multiplied by σ_i^{true} as given above and the resulting synthetic "errors" were added to the y_i^{true} to produce the "experimental" data.

Let's see how well the least squares worked. The smoothed variance came out to be $S_i^2 = 0.162 - 0.088x_i + 0.136x_i^2$. This cannot correspond exactly with $(\sigma_i^{true})^2$, which is a quartic. The least-squares smoothing does not look particularly good, especially near the ends of the range. In the following comparison, results of a completely independent visual smoothing (not used) are also included:

x_i	$(\sigma_i^{true})^2$	S_i^2(l.s.)	S_i^2(visual)
2.1	.0504	.0372	.041
3.0	.0196	.0204	.017
4.0	.0256	.0276	.024
5.0	.0900	.0620	.061

The least-squares smoothing in this case cannot be said to be really superior to visual smoothing; although neither the least squares nor the visual variances are particularly satisfying compared to the true variances, they are the best we have. Certaintly either of them is better than equal variances, and very much better than the individual variances, for the purpose of assigning weights. (For example, the variance determined for the four measurements at $x_i = 3.1$ is 0.0002, corresponding to a weight of 5000, about 25 times the maximum weight eventually assigned.) However, the least-squares method is not very sensitive to modest errors in the weights. If such errors are quite large, statistical tests should indicate the existence of problems with the assigned weights.

Now let us look at the parameter values with their estimated standard deviations:

	2-parameter	3-parameter	true
α_0	23.96(13)	21.62(31)	21.33
α_1	$-3.251(36)$	$-1.85(18)$	-1.697
α_2	—	$-0.200(26)$	-0.2232

The values from the three-parameter determination are all within one standard deviation of the true values. However, the values from the two-parameter determination are 20 and 43 of their own estimated standard deviations away from the true values! This illustrates the danger of relying on estimated standard deviations when the model may be defective.

Finally, it should be noted that all the calculations for this example were first done by hand on a $30 pocket calculator. They were checked on a computer, but this example illustrates that one does not need always to be dependent on the availability of high-speed, high-cost computers.

REFERENCES

1. E. Whittaker and G. Robinson, "The Calculus of Observations," 4th ed., pp. 167–173, Blackie, Glasgow (1944).
2. P. R. Bevington, "Data Reduction and Error Analysis for the Physical Sciences," pp. 81–88, 92–118, 134–186, 204–245, McGraw-Hill, New York (1969).
3. F. S. Acton, "Numerical Methods that Work," chap. 17, Harper & Row, New York (1970).
4. W. C. Hamilton, "Statistics in Physical Science," pp. 49–60, 124–144, Ronald, New York (1964).
5. P. R. Bevington, *op. cit.*, pp. 313–318.
6. "CRC Handbook of Tables for Probability and Statistics," 2d ed., pp. 125–181, 295–298, 304–329, 504–517, The Chemical Rubber Co., Cleveland (1968); "CRC Standard Mathematical Tables," 25th ed., pp. 524, 537, 538–543, CRC Press, Inc., Boca Raton, Florida (1973).
7. S. C. Abrahams and E. T. Keve, *Acta Cryst.* **A27,** 157 (1971).
8. W. C. Hamilton and S. C. Abrahams, *Acta Cryst.* **A28,** 215 (1972).

GENERAL READING

F. S. Acton, "Analysis of Straight-Line Data," Dover, New York (1966).

P. R. Bevington, *op. cit.*

W. C. Hamilton, *op. cit.*

XXI

USE OF COMPUTERS

In many fields of physical chemistry and chemical physics, the use of a computer is considered absolutely indispensible. Many things are done today which would be impossible without modern computers. These include, for example, Hartree-Fock ab initio quantum mechanical calculations, least-squares refinements of X-ray crystal structures with hundreds of adjustable parameters and thousands of observational equations, and Monte Carlo calculations of molecular dynamics, to name only a few. Moreover, an important contemporary application of computers is to control instruments, such as mass spectrometers, X-ray diffractometers, and NMR spectrometers.

Accordingly, acquaintance with computers and with the basics of programming is essential to a sound foundation in physical chemistry. The student interested in either experimental or theoretical physical chemistry is well advised to obtain some experience with computers. A laboratory course in physical chemistry is an excellent place to start,† if a start has not already been made. The purpose of this chapter is to provide a brief introduction to computers.

† The arithmetical calculations required for nearly all of the experiments in this book can be done with nothing more than a good pocket calculator. In a number of cases, however, the use of a digital computer can convey great benefits in convenience, speed, and reliability. For example, Exp. 11, on partial molal volume, is one in which the calculations are somewhat tedious and susceptible to errors when done with a pocket calculator. Experiments 2, 5, 23–29, 42, 45, and 49 are others in which the use of a computer may be helpful.

The field of computer science is large and is growing rapidly. No one book, much less a single chapter like this one, can provide all the information that a beginner might need in the course of setting up calculations for experiments such as those in this book. Much of the necessary information, which includes not only the principles and techniques of programming but also many practical aspects of dealing with a specific machine, is best gained by experience with the local computer system. (Computer systems are highly individual and idiosyncratic.) The student is urged to consult general programming manuals, such as those listed at the end of this chapter, specific manuals describing the computer intended to be used, and any available literature describing the local system. The present chapter is intended as a prelude to such detailed study.

Before proceeding further, it would be well to sound a warning. Computers may be essential to modern science, but they can also be a tremendous time sink. The feeling of power that one may get from programming a complicated calculation can be intoxicating, but one should not lose sight of the physical problem that is being solved. Serious physical scientists would do well to hang a sign over their desks saying something like

WARNING:
EXCESSIVE COMPUTER PROGRAMMING MAY BE
HAZARDOUS TO YOUR SCIENTIFIC HEALTH

They should then resolve to avoid programming overkill: make it a practice to do the least amount of programming necessary to carry out a desired calculation, and then STOP.

For the purpose of this chapter, the word computer will mean a machine (or set of machine components) which can store a large number of words of digital information and can operate on these words in accordance with stored instructions to perform the required operations of arithmetic and logic.† The flexibility of a digital computer arises in part from the fact that instructions are stored like data and may be modified by arithmetical and logical commands in the instruction set. Much more important is the versatility of the available instructions. There are instructions for the arithmetic operations (add, subtract, multiply, divide), certain logical instructions, word modification instructions (shift left or right, truncate, round off, split words), and input/output instructions (read, write). In addition, there are the very important control instructions that enable the machine to depart from a strict sequential reading of the stored instructions, often subject to a stated

---

† We shall not deal here with analog computers, which solve problems by modeling the physical or mathematical situations with analogous electronic circuits. In such computers the variables are continuous voltages and currents, the measurements of which are subject to random errors.

condition. This last feature confers upon the machine the power to make decisions, i.e., gives rise to program branching.

The computer stores all of its information in the form of binary numbers, for which the only digits are 0 and 1. A single binary digit is known as a *bit*; a sequence of eight binary digits is known as a *byte*. The information to be stored is organized into *words*, and computer instructions normally operate on one word at a time. A given computer usually operates with a standard word length, typically 32 bits (but sometimes as much as 64 bits) for large computers and 16 bits (or as few as 12 or 8) for small computers called mini-computers and microcomputers. A single number is usually stored in one word, but it may be stored in two or more words (or a fraction of a word) depending on the precision needed and the capabilities of the instruction set. An instruction typically occupies one 32-bit word, the first part of which contains the *instruction code* and the second part the *address* of a word in storage being referred to; in 16-bit computers an instruction typically occupies two (or more) words.

HARDWARE

The elements of a computing system can be divided into two major categories, colloquially known as *hardware* and *software*. The hardware consists of the electronic and mechanical equipment. The software consists of the information that tells the computer what to do.

The organization of the hardware of a large central computer is illustrated in Fig. 1. The essential elements are the central processing unit (CPU), the random-access memory (RAM), and the input-output (I/O). The CPU is the "brain" of the computer which operates on the information stored in the RAM, and one obviously needs I/O devices to get information into and out of the machine. Other hardware components which may or may not be used are a read-only memory (ROM), an auxiliary memory, and an I/O buffer. Each of these elements will be described briefly.

Central processing unit   This unit reads an instruction from a location in memory determined by the sequential position of that instruction in the program, or, in certain cases, by a previously executed transfer/control instruction. Then, in accord with the instruction read, the CPU performs some arithmetic, logical, or control function, referring in most cases to the contents of a register in memory identified by its address. Built into the CPU is the "hard-wired" circuitry which represents the

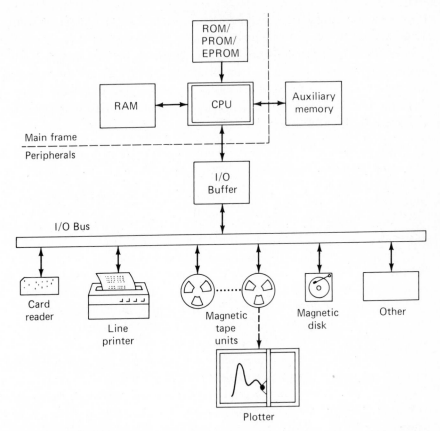

FIGURE 1
Schematic diagram of a hypothetical large computer system. "Other" includes type-writer, video monitor, card punch, paper tape reader and punch, buffer-converter for instrument control, buffer for telephonic communication with remote terminals and/or other computers, etc.

instruction set—the translation of the binary instruction code into the elementary step or sequence of steps that must be performed in order to execute the instruction.

Random access memory   The RAM contains a large number of registers for storing individual words. Usually the number of registers (words of storage) is expressed in units of K (1 K = 2^{10} = 1024), and more often than not it is 2^n K. With 16 bits for the address of an instruction, any of 2^{16} = 65,536 registers (64 K) may be addressed. All registers are accessible with equal rapidity, usually in 1 or 2 μ-sec. At the start of a calculation the RAM generally contains only the operating system program, which will read into the RAM the user's program with needed constants and necessary subroutines. The program (with the aid of the operating system) reads data into the RAM, carries out the intended calculations, and reads the results out of the RAM. Information stored in the RAM may be "volatile" or "nonvolatile," i.e., may disappear or remain intact when power to the machine is turned off. RAM memory on LSI (large-scale integrated) silicon chips is volatile; memory in a network of tiny ferrite magnetic doughnuts called "cores" is nonvolatile. (RAM storage is often called core storage.)

Read-only memory The ROM is a nonvolatile memory in which the contents are stored permanently; information can be read by the CPU from a ROM but cannot be written into the ROM by the CPU. The ROM, like the RAM, is randomly addressable. The information stored in the ROM is incorporated during the process of manufacturing it. ROMs are useful for storing control programs and sub-programs such as loaders, I/O controllers, decimal-binary interconverters, pro-gramming language assemblers, compilers and interpreters, operating systems, and subroutines—in short, any standard routines that will be used again and again without modification. A variant of ROM, the programmable read-only memory (PROM) can be written into *once* by a special procedure; what is written into it is permanently stored. Another variant, the erasable programmable read-only memory (EPROM) can be written into in the normal way but the contents can only be erased, so that new information can be read in, by exposure of the surface of the silicon chip to ultraviolet light through a small window over the chip. The EPROM is particularly valuable for storing routines used by a microprocessor to control an instrument (see below), as it is nonvolatile and not inadvertently erasable.

Auxiliary memory Usually the capacity of the RAM is no more than 64 K storage registers (words), but a large computer may have access to many millions of words stored in auxiliary devices. These devices are not random access, and time of access is measured at best in milliseconds rather than microseconds, and at worst in seconds or minutes. The magnetic bubble memory and the rapidly spinning magnetic disk are important high-capacity devices in the millisecond category. In the case of the disk, it is necessary to wait for any desired word to rotate around to the position of the read-write head. Thus, this kind of device is most efficiently used in reading or writing a large block or file of sequential words into or out of the RAM. Use of magnetic tape vastly increases the memory capacity, but seconds or minutes may be required to locate any given word on a reel of tape.

I/O buffer Input-output devices are slow by comparison with the operations of the CPU. Therefore, in large computers, a specialized small computer called an I/O buffer operates the I/O devices and controls traffic on the I/O bus on command of the CPU, leaving the CPU free to go on with other work while the I/O devices are running. The I/O buffer may also handle transfer of information from one I/O device to another, e.g., from card reader to magnetic tape or disk, or magnetic tape or disk to line printer.

I/O devices The most rudimentary input-output devices are a set of toggle switches and lights on the computer console. Such devices for introducing and reading information one binary digit at a time are not yet entirely obsolete. To start operations on some minicomputers, it is necessary to "toggle in" a short program of instructions to enable the computer to operate more appropriate I/O devices. Many minicomputers and microcomputers use a typewriter-style *keyboard* for input and a *video monitor* for output. The monitor may be used to display the computer response either in alphanumeric characters or in the form of a plot.

In large installations, the primary input is the standard *punch card*, shown in Fig. 2. This card (introduced by IBM and often called an IBM card) has 80 columns, one for each alphanumeric character to be punched, and 12 rows numbered (from the top edge) 12, 11, 0, 1, 2, ..., 9. A single punch in row 0 through 9 of a given column denotes a decimal digit; a combination of a punch in row 1 through 9 with a punch in row 12 through 0 denotes an alphabetic letter. Special symbols (punctuation marks, arithmetic symbols) use other combinations. A high-speed *card reader* may deliver the information from a deck of cards directly into the CPU or I/O buffer; in large installations the information generally flows from the card reader to a magnetic tape unit or magnetic disk unit for temporary storage, to be read into the computer later on command from the CPU. Likewise the CPU or I/O

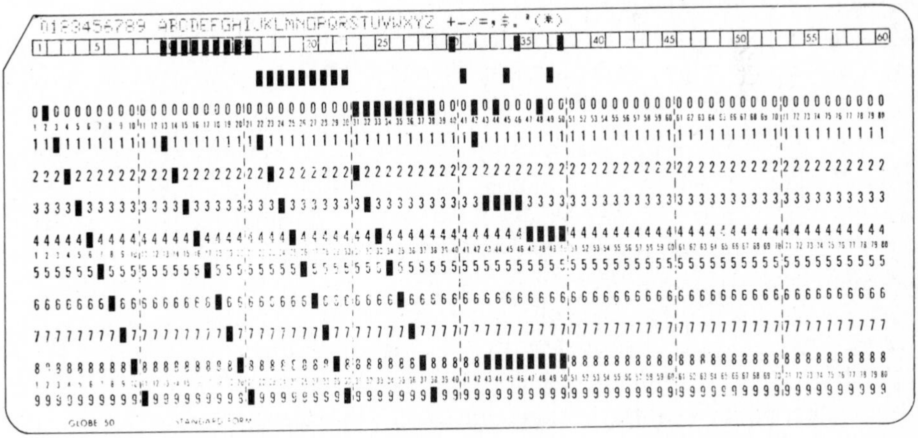

FIGURE 2
IBM punch card in IBM 026 BCD code.

buffer can deliver output information directly to a line printer or a card punch, or else store the information on magnetic tape or disk for later printing, punching, or plotting.

The most important output device in a large installation is the *line printer*. Usually this prints at high speed on a continuous roll of paper 15 in. wide which is easily separated into sheets 11 in. high. Up to 137 characters and spaces can be printed in a single line, and there are typically 60 lines per page.

Magnetic tape units are important workhorses of a large computer; a typical installation might have five to ten of them. Modern computer tapes usually contain nine tracks to accommodate at each position along the tape eight binary bits (one byte) and a parity check bit. A *magnetic disk unit* usually has several disks, each with many tracks, spinning typically at 3000 rpm. There is one read-write head for each disk surface, which may be positioned within a fraction of a second to any track on the surface. Disk units are important especially for time-sharing interactive computing with remote terminals because they enable programs and data to be rapidly read in and out of the RAM as the system rotates from one user to the next. A unit much used with small computers is the so-called "floppy disk," which is handled much like a phonograph record. Although its capacity is much smaller than that of the other types of disks, it is definitely superior in speed and capacity to the cassette tape units used on some small computers.

SOFTWARE

The software comprises all the information written on paper and encoded on various computer media (cards, tape, disk, core storage, etc.). The formal aspects are based on information theory, logic, algorithmic principles, and other areas of mathematics. Of major concern to the typical user is the *program*, a sequence of instructions in a well-defined language that defines the operation of the computer. Such a computer program is a detailed statement of one or more algorithms: step-by-step logical procedures for solving problems. The entire program used in carrying out a given calculation is almost never limited to that written by the programmer for a particular calculation (see Fig. 3). In addition to the user's "source" program, large numbers of instructions are needed to load the program and data into the machine, compile or interpret or assemble the language in which the program is written into binary machine language, provide diagnostic information about programming errors, convert the data from decimal to binary, compute needed functions (logarithms, exponentials, trigonometric functions, etc.), invert matrices, perform sorting operations, convert numerical results to decimal, and unload the results into the output device. These various manipulations are given by

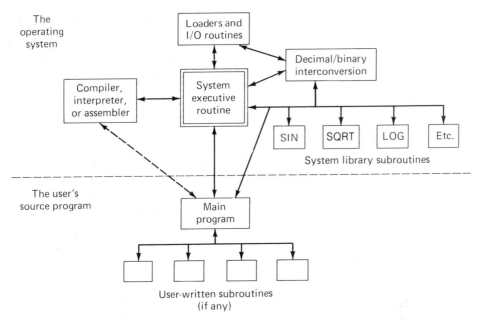

FIGURE 3
Schematic diagram of the entire computer program and system software involved in a given calculation.

programmatic blocks called *subroutines*, the logic and structure of which need not be known in detail by the casual user. A subroutine is a sequence of instructions intended to perform a specific function or set of functions; control passes to it from the *executive* routine (or from another subroutine) which "calls" the subroutine, and when the job is done control is returned to the calling program. These subroutines, together with an executive routine, constitute the *operating system.*

Figure 3 gives a schematic diagram of the organization of the software used in any given calculation. Usually the part of the material written by the user is a rather small proportion of the total; the major part is in the operating system. However, an extensive user program may require the greater part of the RAM for the source program, data, and storage of intermediate results in the course of a calculation, and may require use of auxiliary memory besides.

TYPES OF COMPUTERS

The spectrum of computer capability is virtually continuous from huge computers with extremely sophisticated and fast central processors to minicomputers and small microcomputers for personal and even recreational use. Table 1 lists a variety of computers classified roughly according to size and cost. Within each classification there is a very wide range of capabilities, generally determined by the presence or absence of certain advanced CPU capabilities (e.g., hardware floating-point arithmetic), the amount of random-access memory (4 K to 64 K words or more), and by the number and kind of peripherals.

The speed of a computer depends upon much more than the possible speed at which the elemental electronic steps involved in the computer logic take place (typically 0.1 to 1 μsec). It depends upon the number of bits that can be processed simultaneously (ranging from only 8 in the microcomputer to 32 or even 64 in the large computers), the number of operations related to the execution of a single instruction that can be handled simultaneously, the extent to which upcoming operations can be anticipated and the required words fetched ahead of time for quicker processing (the "look-ahead" feature), the extent to which the CPU can be relieved from housekeeping chores (like I/O), and other factors, which also influence greatly the complexity and cost of the machine.

Not only the hardware, but also the software requirements, vary widely over this spectrum. With a minicomputer, the operating system may be fairly rudimentary and casual, as there is no great need to keep the machine operating every

Table 1 REPRESENTATIVE COMPUTERS

Type	Examples	Typical cost	Typical applications
Very large and ultra-fast computers	CDC 7600￼CRAY 1	$5M	Institutional and regional computing; local and wide area batch and time-sharing
Large and fast computers	CDC 6600￼CDC Cyber￼IBM 360, IBM 370￼UNIVAC 1108￼PDP 10	$1M–$5M	Institutional computing; local batch and time-sharing; commercial and banking
Mini-maxi computers	PDP VAX	$100K–$500K	Institutional, departmental, or business computing
Minicomputers	PDP 8, PDP 11￼NOVA 4000	$5K–$100K	Departmental, laboratory, or business computing; instrument automation
Microcomputers	North Star Horizon￼Heath H8, H11￼Apple II, III￼Commadore PET￼Radio Shack TRS-80￼Rockwell AIM￼HP 85￼Atari 400, 800	$500–$3000	Laboratory computing; instrument automation; small business, home, or personal computing

minute. In a multimillion dollar installation it is important to keep the machine operating continuously and efficiently with the aid of a carefully designed operating system. Although instrument-control microcomputers are often coded directly in assembly language or machine language, other applications of these microcomputers depend upon typewriter input and video output and on use of a simple algorithmic language such as BASIC; all of these require appropriate software.

MODES OF OPERATION

Small computers generally have a single console (keyboard and video monitor or printer) and are intended for the exclusive use of one person at a given time. They are commonly used in the interactive mode to be described below. It is also possible to use a large computer in this manner through timesharing on a remote terminal. However, the simplest and most direct mode of operation on large machines is "batch processing."

In the batch-processing mode, the various "jobs" (in the form of decks of punched cards) are loaded sequentially onto magnetic tape or disk. Each job deck is preceded by one or more control cards which give the system some necessary information such as the user's name and account number (and/or secret password), expected maximum CPU time, maximum number of lines of output expected, magnetic tape units to be used, etc. The operating system program reads each deck from the tape or disk, compiles or assembles the program into binary machine language (if it is not already in binary form from previous compilation or assembly), and executes the program, loading from magnetic disk any library subroutines requested by the user's program. The system then records the results on magnetic tape or disk, to be printed and/or punched later, and goes on to the next job. The printout delivered to the user commonly concludes with a statement of the time used and perhaps the charge. If in the compilation or assembly of the program the system detects a programming error, the output will typically consist only of diagnostic information. Input-output for batch processing need not take place at the computer installation itself; a remote terminal, perhaps in a distant city, can submit jobs and receive output over telephone wires. Sometimes the remote terminal is a smaller computer which has already partially processed the data and/or will further process the output received.

In the interactive mode, the user sits at a terminal and establishes a direct communications link with the computer. For a large installation operating many remote timesharing terminals linked by telephone lines to the central computer, the computer serves these terminals on a rapidly rotating basis. It visits each console in sequence, loading from disk any information already provided for that console and proceeding with additional operations for a limited period of time determined by a system algorithm. At the end of that time, if the job is still not finished, it is put onto disk again, and the computer goes on to the next console. In the course of a calculation at any given console, the central computer may visit that console many times. Systems for handling timeshared interactive computing have been under development for many years and are highly sophisticated.

The interactive mode is particularly attractive when one wants to vary certain input parameters in the course of the calculation on the basis of an inspection of provisional or partial outputs. A number of different parameter sets or a number of different computational conditions may be tried without having to submit several successive jobs in the batch mode and wait out the "turn-around time" (which may be several hours or even days). Interactive computing also provides important conveniences in writing programs. Having roughed out the program ahead of time, one can sit at the terminal and type the developing program into the keyboard in one of the algorithmic languages to be described. The programmer may initiate compilation of this program from the terminal. If program errors prevent successful compilation, diagnostic information comes back. The programmer may correct and edit the program with appropriate commands until it has been successfully compiled. The data can then be typed in or read in from disk or tape on which they have already been stored. The results will come back to the remote monitor or printer, or can be stored on disk or tape at the computer, or can be printed there off-line to be delivered later. Interactive computing by timesharing on a central computer has the advantage of access to the full capabilities of a large machine, but there are also some disadvantages not encountered on a minicomputer. The principal problem with timesharing is saturation, when too many users are at their terminals at the same time, causing the response of the computer to be so slow as to be virtually useless.

BINARY NOTATION

The positional arithmetic notation of common usage deals with numbers written in the form

$$N \equiv d_m d_{m-1} \cdots d_2 d_1 d_0 . d_{-1} d_{-2} \cdots d_{-n} \tag{1}$$

where in the decimal case each of the d_i represents a digit 0 through 9. In this notation, the given array of digits represents the numerical value given by

$$N = \sum_{i=m}^{-n} d_i B^i \tag{2}$$

where B is the base of the number system; for decimal numbers, B is 10. There is nothing particularly special about the number 10 for the base of a number system; its choice presumably arose historically from the fact that the human being is endowed with 10 fingers. The basic memory devices of computers have, in effect, only two fingers, represented by the two states of a bistable electronic circuit or the two magnetic polarities of a ferrite core. Therefore, computers are built to do arithmetic in the binary system where B is 2 and where 0 and 1 are the only digits. The binary computer is also well-equipped to do logical operations, because formal mathematical logic recognizes only two values: TRUE and FALSE, represented in the computer by the respective binary digits 1 and 0.

Long strings of binary numbers are tedious to write and to read. Since 8 and 16 are powers of 2, they represent bases that can be converted easily to and from binary. In the octal system (base 8), each octal digit from 0 through 7 represents a binary number of up to three digits; one can simply partition the sequence of binary digits into groups of three to obtain the conversion to octal. In the hexadecimal system (base 16) each digit from 0 through 9 and A through F represents a binary number of up to four digits; hence a byte is equivalent to two hexadecimal digits. Examples:

Decimal	Binary	Octal	Hexa-decimal	Binary
425	$= 110\ 101\ 001_2$	$= 651_8$	$= 1A9_{16}$	$= 1\ 1010\ 1001_2$
125	$=\quad 1\ 111\ 101_2$	$= 175_8$	$= 7D_{16}$	$=\quad 111\ 1101_2$
11.4375	$= 1\ 011.011\ 1_2$	$= 13.34_8$	$= B.7_{16}$	$=\quad 1011.0111_2$

The binary numbers are presented twice to show the different partitioning. Note that by convention, when it is necessary to distinguish among bases, the bases are indicated by subscripts. Since octal and hexadecimal are so easy to convert to and from binary, instruction codes and addresses of words in storage are commonly expressed in one of these bases, usually octal, in computer literature. The contents of a 16-bit word can be expressed with four hexadecimal digits, but it is more common to use 6 octal digits of which the sixth (left end) is limited to the two values 0 and 1.

The binary expressions for the digits in the octal, decimal, and hexadecimal systems are given in Table 2. Arithmetic operations in these various bases are intrinsically identical to those in base 10 except that the addition and multiplication tables are different.

There are two types of numerical data—integers, and real numbers (which represent continuous quantities)†; these are represented differently in the computer.

† The treatment of imaginary and complex quantities will not be discussed in this chapter.

Table 2 NUMBER SYSTEMS USED WITH COMPUTERS

Number systems

Binary B = 2	Octal B = 8	Decimal B = 10	Hexadecimal B = 16	Expression in binary form
0	0	0	0	0
1	1	1	1	1
	2	2	2	10
	3	3	3	11
	4	4	4	100
	5	5	5	101
	6	6	6	110
	7	7	7	111
		8	8	1000
		9	9	1001
			A	1010
			B	1011
			C	1100
			D	1101
			E	1110
			F	1111

An integer is stored directly in its binary form with the left-hand digit (e.g., bit 16 of a 16-bit word) representing the algebraic sign $(0 = +, \ 1 = -)$. Thus $-425_{10} = -651_8$ would be stored as

$$1\ 0\ 0\ 0\ 0\ 0\ 0\ 1\ 1\ 0\ 1\ 0\ 1\ 0\ 0\ 1$$

in a 16-bit-word machine. (Some computers, such as the PDP 11, use a different method for dealing with negative numbers.[1]) A real number is stored in the floating-point mode. In this case, a single 32-bit word (or two 16-bit words) contains a pair of numbers that correspond to the exponent and the mantissa (fractional part) of the number. Typically one bit is used for the sign, a string of 7 bits represents the exponent, and the remaining 24 bits represent the mantissa. By convention, the decimal point in a floating-point number is usually placed at the far left; for example, 73.34×10^4 is written as 0.7334×10^6. Thus the exponent tells one how far to the right (positive exponent) or left (negative exponent) the decimal point must be "floated" if the number were written out in positional notation.

In addition to the use of numbers for input data, the FORTRAN or BASIC programming languages to be described later will also use alphabetic characters and special characters (punctuation, arithmetical characters). The computer can only accept alphanumeric information of these kinds in some kind of binary coded form. For this it is becoming increasingly standard practice to use an 8-binary-digit (1-byte) code to represent an alphanumeric character. One byte can accommodate

$2^8 = 256$ combinations of 0s and 1s. This is far more than enough to cover all characters normally used for computation in English (decimal digits, capital Roman letters, ordinary punctuation, arithmetical characters, and other mathematical symbols). There is room enough left for lowercase Roman letters, Greek letters (and other alphabets if desired), as well as control characters for Teletype and related equipment. A standard 8-bit code for frequently used characters known as ASCII (American Standard Code for Information Interchange) is now generally accepted, particularly in working with small computers and components.[2]

PROGRAM LANGUAGES

Instructions are seldom written in the binary (or the equivalent octal or hexadecimal) form in which they will reside in the computer. Generally the *source program* is written in some kind of programming language,[3] and computer processing is required before the program assumes its final binary form as the *object program*. A source program which is closely related to the object program can be written in what is called *symbolic language*, in which the binary instruction codes are represented by alphabetical mnemonic codes and the addresses are represented by symbolic alphanumeric names. Most programmers seldom get down to the level of symbolic language.

Nearly all user-written programs are written in one or another of several *algorithmic languages*, of which BASIC and FORTRAN are the best-known examples. The common characteristics of these algorithmic languages are: capability of expressing, in recognizable English words (or abbreviations) and phrases, the component parts of an algorithm or flow diagram; minimally essential rules of grammar and syntax; algebraic expression of arithmetical commands; convenient specification of I/O requirements; and maximum flexibility for program formulation and modification.

FORTRAN (from the phrase FORmula TRANslator) was developed in 1954 and has evolved over the years into several versions, the best known of which are FORTRAN II and FORTRAN IV. Although regarded as somewhat inelegant by purists of algorithmic theory, FORTRAN has achieved a commanding position throughout the world for general scientific and engineering computation.† In recent years, the IBM company has introduced PL/1 (Programming Language 1), an extremely powerful and general algorithmic language of which subsets could replace FORTRAN. However, it is very formal and has not achieved much acceptance outside the professional computer-science fraternity.

† Other languages used for scientific programming include ALGOL, APL, and PASCAL.

In the other direction, the language BASIC (Beginners' All-purpose Symbolic Instruction Code) was developed in 1965 at Dartmouth College to respond to the needs of students and users of smaller and simpler computers. BASIC has a somewhat simpler structure than FORTRAN and is not as powerful in the hands of an experienced programmer, but it is very convenient for users of minicomputers. It is also the most popular language for the small and inexpensive microcomputers and "personal" computers and exists in many versions, often with restrictions based on hardware limitations. However, it is usually available at installations with large computers as well.

Typically, once a program is written in FORTRAN, a system subprogram known as a *compiler* converts it into an object program in machine language. The object program can be stored and used again and again if desired. In the case of BASIC, a system subprogram known as an *interpreter* typically converts each command statement into a block of machine-language instructions *at execution time*. Thus, in the case of FORTRAN, the compiler has finished its work before the object program starts execution; in the case of BASIC, the object program is being continuously generated by the interpreter during execution, and the interpreter is similarly involved each time the program is run. (It is also possible in some systems for a BASIC source program to be compiled, or for a FORTRAN source program to be interpreted.) The compiler-type operation produces the more efficient execution, particularly if the object program is to be run repeatedly; the interpreter-type operation is in some respects more convenient and flexible especially with smaller computers.

Even with a given machine and a given language, there may be different compilers to choose from. Certain compilers are designed to produce a very efficient machine-language program, at the expense of more computer time devoted to compilation. Others, more attuned to one-shot calculations, are designed for "load-and-go" programming, in which the programmer submits the data together with a FORTRAN source program. If the program successfully compiles, execution immediately follows. Still other compilers are oriented strongly toward diagnostic help against programmer errors. WATFOR and WATFIV, which are FORTRAN compilers equipped with fairly elaborate diagnostic structure, are popular for student instruction.

PROGRAMMING

Computers are often characterized in the popular press as wonderful gigantic brains. Wonderful they may be, but in truth the computer is a literal-minded idiot. It will do anything within its power that you tell it to do, no matter how stupid. It makes no allowance for human frailties: any detection of errors or diagnostics that it gives were programmed in detail by some human being. However, the computer has two vastly important saving graces. Whatever you tell it to do will be done quickly and accurately. It never quits out of boredom or makes errors because it is tired (although it does suffer occasional breakdowns).

Since "to err is human," no one writing a program of any length should expect it to work properly the first time. Something as trivial as a missing or misplaced punctuation mark will prevent successful compilation of a program. Fortunately, something this trivial is easy to correct with the aid of compiler diagnostics. The hardest mistakes to avoid, and ones that the computer will usually be unable to point out to you, are *mistakes in the logical structure of the program.* The logical structuring of the program is the aspect of programming that requires the greatest care and thought; the rest is merely mechanical.

A program is an algorithm, i.e., a rigorously defined computational procedure. As such it can be diagrammed on paper, before the actual coding begins. When written out, the program is basically a linear rendering of this algorithmic *flow diagram.* In such a rendering, branching (represented two dimensionally in the flow diagram much as a highway is branched on a road map) is accomplished with conditional transfer statements. The construction of a flow diagram for a new program, although often neglected, is strongly recommended to be done before the program is reduced to formal expression in some computer language, as it permits a careful check to be made on the logic of the program. The ability to generate such flow diagrams is a basic skill needed by every physical scientist desiring to perform complex calculations on a computer. Once the flow diagram has been constructed and carefully checked, the writing of the program in whatever language the available computer center will accept involves not much more than meticulous observance of the rules and syntax of that particular language.

We will give here, as an example of a computation to be programmed for a computer, the problem of finding the greatest common divisor of two positive integers. We will first describe the algorithm in words, then present a flow diagram, then give programs in FORTRAN and BASIC which exemplify many features of these languages, in particular the use of conditional transfer statements to implement the necessary logical decisions of the algorithm.

We will employ Euclid's algorithm. Let the two integers be A and B. Euclid tells us to divide the larger by the smaller; for clarity we will interchange the two magnitudes *if* necessary so that the smaller is A and the larger is B. We examine the remainder R in B/A. *If* that remainder is 0, the answer is almost trivial since A is the greatest common divisor. *If* R is not zero, we rename the quantity A as B and rename the quantity R as A, repeat the division, and again examine the remainder. This procedure is repeated until we find a R value that is zero. Then the last value of A is the greatest common divisor. (Note the occurrences of the italicized word "*if*"; each of them will be found in the FORTRAN and BASIC programs that follow.)

A flow diagram for this algorithm is presented in Fig. 4. We have added an extra feature permitting this algorithm to be applied repeatedly to a large number of A, B integer pairs. A and B are tested to be sure that they are both positive integers; if one is found to be 0 or negative, the program stops. If the A, B pairs are punched on successive cards, a blank card at the end (A, $B = 0, 0$) signals the end of the computation.

This flow diagram uses some conventional styles of boxes for the various computational steps: circles for the beginning and end of the program, facsimile punch-card for input, scroll for printed output, diamonds for branch points, rectangular boxes for computational operations, and small circles for junction points.

Figure 5 shows the program in FORTRAN, as submitted for load-and-go batch processing, together with the output obtained with a small data deck. We explain here a few features of FORTRAN as applied to this problem; for a fuller understanding, the reader is referred to FORTRAN programming manuals.[4-6] A letter C in card column 1 indicates that the card contains only a comment, which is ignored by the compiler. Card columns 2 to 5 may contain a statement number, which is obligatory if the statement is referred to by another statement, otherwise optional. Statement numbers need not follow any particular numerical order. The first operation of the program (PRINT 5) was to print the heading of the output table. Next the program read in a data card with A and B values; in FORTRAN these are coded IA1 and IB1. The I is used because in FORTRAN any variable that is to be treated as an integer must have as its first character one of the six letters I, J, K, L, M, or N; the 1 is placed afterward to distinguish the numbers as read in from those (IA, IB) used later in the program, the values of which may change. The abbreviation .LE. stands for "less than or equal to"; .EQ. stands for "equal to." MOD (argument 1, argument 2) is one of the system library functions; it gives (argument 1) *modulo* (argument 2), i.e., the remainder when argument 1 is divided by argument 2. The various control cards are not shown since every installation has its particular requirements. It is important not only to know the local rules, but to know them down to the last comma and period. This is at least equally important in writing the FORTRAN program. (As a matter of fact, this particular program did not compile successfully the first time because in statement 11 the comma was inadvertently omitted after "READ 20.")

Figure 6 shows the same problem programmed in BASIC. This program was run on a PDP 11, but it might have been run with no essential difference in procedure on a large central computer using a timesharing terminal. On the system command RUN the computer began execution, and asked successively for the numerical values needed. For example, the computer typed INPUT A1?. The user

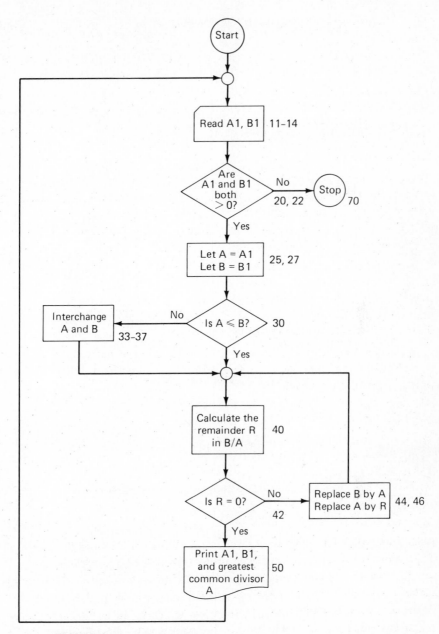

FIGURE 4
Flow diagram for Euclid's algorithm to find the greatest common divisors of pairs of positive integers A1 and B1. Numbers outside of boxes correspond to BASIC statement numbers in Fig. 6.

```
      PROGRAM EUCLID(INPUT,OUTPUT)
C     TO FIND GREATEST COMMON DIVISORS OF PAIRS OF POSITIVE
C     INTEGERS A AND B
      PRINT 5
    5 FORMAT(1H ,8X,1HA,9X,1HB,3X,13HGTST COM DVSR)
   11 READ 20,IA1,IB1
   20 FORMAT (2I10)
C        IF A AND B DO NOT BOTH EXCEED 0 END OF DATA,STOP
      IF (IA1.LE.0) GO TO 70
      IF (IB1.LE.0) GO TO 70
      IA=IA1
      IB=IB1
C        IF B IS LESS THAN A INTERCHANGE A,B
      IF (IA.LE.IB) GO TO 40
      ITEMP=IA
      IA = IB
      IB = ITEMP
C        FIND REMAINDER R IN B/A
   40 IR=MOD(IB,IA)
C        IF R IS ZERO, CURRENT A IS GREATEST COMMON DIVISOR.
C        OTHERWISE REPLACE IB,IA BY IA,IR AND REPEAT
      IF (IR.EQ.0) GO TO 50
      IB=IA
      IA =IR
      GO TO 40
   50 PRINT 60,IA1,IB1,IA
   60 FORMAT (3I10)
      GO TO 11
   70 STOP
      END
```

A	B	GTST COM DVSR
64	28	4
39	16	1
50	26	2
11	13	1

FIGURE 5
Euclid program in FORTRAN, and output of sample calculations.

replied by typing 64. Next the computer asked INPUT B1? and the user replied 28. Then the computer typed out the result for this pair of numbers.

Again we mention some features of BASIC as applied to this problem; the reader is referred to programming manuals for a fuller understanding.[7-9] Some differences from FORTRAN will be noted. Remarks to be ignored by the interpreter are indicated by REM. *All* statements in BASIC must be numbered, and the numbers must monotonically increase. Arithmetic statements are introduced by the word LET. Variable names must be only a single letter, or a single letter followed by a single digit, while in FORTRAN a variable name may contain up to six letters and digits, the first of which must be a letter. The distinction between

```
EUCLID                  MU BASIC/RT-11 V01-01C

10 REM FIND GREASTEST COMMON DIVISOR OF TWO INTEGERS A,B
11 PRINT "INPUT A1";
12 INPUT A1
13 PRINT "INPUT B1";
14 INPUT B1
20 IF A1<=0 THEN  GO TO 70
22 IF B1<=0 THEN  GO TO 70
25 LET A=A1
27 LET B=B1
30 IF A<B THEN  GO TO 40
33 LET T=A
35 LET A=B
37 LET B=T
38 REM FIND REMAINDER R IN B/A
40 LET R=B-(INT(B/A))*A
42 IF R=0 THEN  GO TO 50
44 LET B=A
46 LET A=R
48 GO TO 40
50 PRINT "FOR A=";A1,"AND B=";B1,"GTST COM DVSR=";A
52 PRINT
55 GO TO 11
70 STOP
80 END

READY
RUN

EUCLID                  MU BASIC/RT-11 V01-01C

INPUT A1? 64
INPUT B1? 28
FOR A= 64       AND B= 28       GTST COM DVSR= 4

INPUT A1? 39
INPUT B1? 16
FOR A= 39       AND B= 16       GTST COM DVSR= 1

INPUT A1? 50
INPUT B1? 26
FOR A= 50       AND B= 26       GTST COM DVSR= 2

INPUT A1? 11
INPUT B1? 13
FOR A= 11       AND B= 13       GTST COM DVSR= 1

INPUT A1? 0
INPUT B1? 0

STOP AT LINE 70

READY
```

FIGURE 6
Euclid program in BASIC, and output of sample calculations.

integers and reals does not exist in BASIC. The functions available are somewhat different; here we use the BASIC function INT (argument), which gives the largest integer that is not greater in magnitude than the argument. In some versions of BASIC, liberties may be taken with the standard language: the word LET may be abbreviated to L or even omitted altogether; in an IF statement, the word THEN and the words GO TO may be optional.

Despite the differences between FORTRAN and BASIC, a great deal of similarity will be noted between the programs given in Figs. 5 and 6. This should be expected, because (1) both programs express the same algorithm, and (2) both are written in codes designed to be as close to English language and to conventional arithmetic expression as reasonably possible. Nevertheless, the rigor required in accuracy of syntax is equally severe in both languages.

In interactive computing with BASIC, there is a vocabulary of system control commands which may depend somewhat on the machine or the system. These commands typically include: NEW (program name) = user desires to type in a new BASIC program of that name (e.g., NEW EUCLID); OLD (program name) = find the program of that name in auxiliary memory and load it into RAM; LIST = list the previously referenced program on the video monitor, Teletype, or printer; RUN = commence execution of the program; SAVE = place the program in storage for possible later use or modification; BYE = log out the user. Each command is transmitted to the system on carriage return. When the system has fully responded to the command it types READY as a signal to the user that he may type in the next system command.

The above programming example does not demonstrate the features of FORTRAN and BASIC that are most important in arithmetical calculation. To illustrate the capabilities of these languages for coding the evaluation of arithmetic expressions, we will use a few of the equations in this book.

Joule-Thomson coefficient of a van der Waals gas (Eq. 2-11):

$$\mu = \frac{(2a/RT) - b}{C_p}$$

FORTRAN:

$$EMU = (2.*A/R*T - B)/CP$$

BASIC:

$$100\ LET\ M = (2*A/R*T - B)/C2$$

In both languages spaces can be introduced at will for readability, as in these examples. In both, the asterisk (*) is used as a multiplication sign. In both languages, all variables on the right of the equal sign must have been previously defined in the program. In FORTRAN a decimal point follows the 2 because the expression is one involving reals as opposed to integers; reals and integers may not be mixed in the same FORTRAN expression, except that exponents may be integers in an expression of reals and subscripts are *always* positive integers. As already mentioned, the distinction between reals and integers does not exist in BASIC; this gives an advantage of simplicity and convenience at some expense in efficiency.

Apparent molal volume (Eq. 11-17):

$$\phi = \frac{1}{d}\left(M_2 - \frac{1000}{m}\frac{W - W_0}{W_0 - W_e}\right)$$

FORTRAN:

$$\text{PHI} = (\text{EM2} - 1000.*(W - W0)/\text{EM}*(W0 - WE))/D$$

BASIC:

$$150 \text{ LET } F = (M2 - 1000*(W - W0)/M*(W0 - W1))/D$$

Root-mean-square velocity of a gas molecule (Eq. IV-1):

$$u = \left(\frac{3RT}{M}\right)^{1/2}$$

FORTRAN:

$$U = \text{SQRT}(3.*R*T/\text{EM}) \qquad \text{or} \qquad U = (3.*R*T/\text{EM})**0.5$$

BASIC:

$$200 \text{ LET } U = \text{SQR}(3*R*T/M) \qquad \text{or} \qquad 200 \text{ LET } U = (3*R*T/M)^{\wedge}0.5$$

Note the different exponentiation signs: ** for FORTRAN and ^ (or ↑) for BASIC.

A powerful feature of both languages is the ability to code repetitive operations. The most common application is in summing. Repetitive operations that do not depend upon a conditional transfer for exit, particularly those which deal with subscripted variables such as the elements of an array, are handled in FORTRAN with a "DO loop" and in BASIC with "FOR ... NEXT" statements. For example,

$$S = \sum_{i=1}^{N} T_i$$

FORTRAN:

$$\text{SUM} = 0.$$

$$\text{DO } 150 \text{ I} = 1,N$$

$$150 \text{ SUM} = \text{SUM} + \text{TERM(I)}$$

BASIC:

> 120 S = 0
>
> 130 FOR I = 1 TO N
>
> 140 LET S = S + T(I)
>
> 150 NEXT I

As an example of coding repetitive calculations we will code Eq. (XX-24) in which an $n \times n$ inverted normal equations matrix **B** is multiplied by an n-element column vector **h** to give the n-element column vector $\Delta\alpha$ representing the least-squares shifts in the parameters. This equation may be written in summation form as

$$\Delta\alpha_j = \sum_{k=1}^{n} B_{jk} h_k$$

FORTRAN:

> DIMENSION DEL(20), B(20,20), H(20)
>
> $\vdots$
>
> DO 100 J = 1, N
>
> DEL(J) = 0.
>
> DO 100 K = 1, N
>
> 100 DEL(J) = DEL(J) + B(J,K)*H(K)

BASIC:

> 10 DIM D(20), B(20,20), H(20)
>
> $\vdots$
>
> 50 FOR J = 1 TO N
>
> 60 LET D(J) = 0
>
> 70 FOR K = 1 TO N
>
> 80 LET D(J) = D(J) + B(J,K)*H(K)
>
> 90 NEXT K
>
> 100 NEXT J

The DIMENSION (DIM) statement is needed to inform the system of the storage requirements for the arrays used. This statement must occur in the program ahead of any statement in which the listed arrays are referenced. For this example, we assume that the order n or N is no greater than 20, and may be less. (The storage reserved by the DIMENSION statement may be larger than that needed, subject of course to the capacity of the RAM.) To adapt the program to a problem in which, say, $n = 30$, we need only change the DIMENSION statement accordingly.

The above is an example of two "nested" DO loops or FOR ... NEXT loops. More than two loops may be nested, the limiting number depending on the compiler-interpreter and the system. However, the number of subscripts permitted for an array element is strictly limited to three in FORTRAN and to two in BASIC, in the standard versions of these languages.

Instruction in programming is beyond the scope of this chapter. Programming is best learned by experience with an actual calculation on a computer, with the aid of appropriate programming manuals.[4-9] There are two levels of concern. First, the logic and syntax must be correct, or the program will not run. Second, even if the program gives a correct result, it may be grossly inefficient and costly. We will content ourselves with offering one hint for programming efficiency: if there are operations in the calculation that need be done only once, keep them out of loops. As a simple example, consider

F = 0.

DO 100 I = 1,50

100 F = F + COS(6.28318*XJ*X)*SIN(6.28318*YJ(I)*Y)

Since the cosine term does not depend on the index I, it needs to be calculated only once. A more efficient version would be

$F = 0.$

A = COS(6.28318*XJ*X)

DO 100 I = 1,50

100 F = F + A*SIN(6.28318*YJ(I)*Y)

and a still more efficient version would be

F = 0.

A = COS(6.28318*XJ*X)

B = 6.28318*Y

DO 100 I = 1,50

100 F = F + A*SIN(B*YJ(I))

Examples of FORTRAN programs for data analysis and least-squares and statistical calculations are given by Bevington.[9] A variety of programs useful in physical chemistry are given in the references listed under General Reading. Programs for a few of the experiments in this book (Exp. 5, 11, 26, 42, and 49) were prepared as a supplement to the third edition, but locally generated programs have the advantage of making optimum use of available computers and providing some programming experience.

MICROCOMPUTERS AND INSTRUMENT CONTROL

For well over a decade, minicomputers (PDP 8, PDP 11, NOVA 4000, etc.) have been used to control scientific instruments. All that is needed in addition to the instrument and the computer is an "interface" to communicate between the two, usually translating the digital output of the computer into analog input to the instrument and analog (sometimes digital) output of the instrument into digital input to the computer. The automation of an X-ray diffractometer provides a good example. The orientation of the crystal specimen is determined by the rotations of up to four shafts; these are continuous physical (analog) quantities. On each shaft is a motor which can be turned on and off, in either direction, on command from the computer. On each shaft also (in some cases integral with the motor) is a shaft encoder, which reads out its angular position in binary form for the computer. The computer can turn on the motor and then turn it off when the encoder signals back the required angular position of the shaft. The radiation detector of the diffractometer gives its output in discrete "counts" which can be processed directly by the computer. Other instruments give outputs as voltages, and appropriate circuitry is used to convert these to digital signals, as in the familiar digital voltmeters (see Chap. XV).

The development of single-chip circuitry is rapidly bringing instrument automation within reach of much more modest applications. The VLSI (very large scale integrated) technology permits incorporating thousands of transistors, diodes, and other circuit elements into the surface of a single silicon wafer or chip a few square centimeters in size. Most significant of these VLSI devices is the *microprocessor*. This device is a completely self-contained, although somewhat primitive, computer CPU with an instruction set adequate for a fairly wide range of intended applications. Among important microprocessors are the Intel 4040 and 8080 and the Motorola 6502 and MC 6800. Generally these microprocessors are designed with an 8-bit (1-byte) word length, but some newer ones are based on a 16-bit word. The "personal" microcomputers, such as those listed in Table 1, are based on these and similar microprocessors. Such a microcomputer consists basically of a microprocessor as CPU plus one or more RAMs on chips, and very typically, ROMs and/or PROMS to handle operating systems, loaders, compiler-interpreters, and other permanent software (often called "firmware"), and EPROMS to store the more or less permanent working control programs. In addition to the microprocessor and memory chips, digital-to-analog and analog-to-digital circuits are needed as already described.

Microprocessors have already received wide application to the control of instruments, e.g., automatic cameras and electronic automobile ignition systems. In a physical chemistry laboratory, a microprocessor system could be used for as simple an application as a temperature controller or programmer-controller for a water or air thermostat, a distilling column, or a thermogravimetric analysis apparatus. It could serve as an integrator for a gas chromatograph, or it might be used to optimize the settings of a spectrophotometer. At Oregon State University a Rockwell AIM microcomputer has been used with an ADC converter to measure and record transmittances on a millisecond time scale in a flash photolysis experiment.

The programming of a microprocessor is often done in assembler language, and usually needs to be done only once. In that case, the program can be stored in a PROM or EPROM (described earlier). However, even relatively simple systems are amenable to BASIC programming with a ROM-stored compiler-interpreter.

The details involved in the use of microprocessors are described elsewhere.[11] We give references here to two published examples of microprocessor control, one for temperature programming the direct inlet probe of a high-resolution mass spectrometer,[12] the other for control of a differential titrator.[13] The descriptions in these papers are highly technical and not likely to be understood in detail unless the reader has studied the subject of microprocessors in considerable depth, but they will serve to illustrate many of the factors involved in applying microprocessors to practical instrument control.

REFERENCES

1. J. W. Cooper, "The Mini-Computer in the Laboratory, with Examples Using the PDP 11," pp. 14–16, Wiley, New York (1977).
2. The Diebold Group (eds.), "Automatic Data Processing Handbook," pp. 2–308, 2–309, McGraw-Hill, New York (1977).
3. J. A. Feldman, "Programming Languages," Scientific American, **241**, 94 (December 1979).
4. D. D. McCracken, "A Guide to FORTRAN IV Programming," 2d ed., Wiley, New York (1972).
5. E. I. Organick and P. Meissner, "FORTRAN IV," 2d ed., Addison-Wesley, Reading, Mass. (1974).
6. V. Y. Dock, "FORTRAN IV Programming," 2d ed., Reston Publishing Co., Reston, Va. (1976).
7. T. E. Hull and D. D. F. Day, "An Introduction to Programming and Applications with BASIC," Addison-Wesley, Reading, Mass. (1979).

8. W. N. Hubin, "BASIC Programming for Scientists and Engineers," Prentice-Hall, Engle-wood Cliffs, N.J. (1978).

9. J. G. Kemeny and T. E. Kurtz, "BASIC Programming," 2d ed., Wiley, New York (1971).

10. P. R. Bevington, "Data Reduction and Error Analysis for the Physical Sciences," McGraw-Hill, New York (1969).

11. L. A. Leventhal, "Introduction to Microprocessors: Software, Hardware, Programming," Prentice-Hall, Englewood Cliffs, N.J. (1978).

12. C. L. Pomernacki, "Microprocessor-Based Controller For Temperature Programming the Direct Inlet Probe of a High-Resolution Mass Spectrometer," *Rev. Sci. Instrum.*, **48**, 1420 (1977).

13. N. Busch, P. Freyer and H. Szameit, "Microprocessor-Controlled Differential Titrator," *Anal. Chem.*, **14**, 2167 (1978).

GENERAL READING

J. A. Feldman, "Programming Languages," Scientific American, **241**, 94 December (1979).

P. A. D. deMaine and R. D. Seawright, "Digital Computer Programs for Physical Chem-istry," 2 vols., Macmillan, New York (1963).

D. F. DeTar (ed.), "Computer Programs for Chemistry," 4 vols., Benjamin, New York (1968-72).

T. R. Dickson, "The Computer and Chemistry," Freeman, San Francisco (1968).

T. L. Isenhour and P. C. Jurs, "Introduction to Computer Programming for Chemists," Allyn-Bacon, Boston (1972).

K. B. Wiberg, "Computer Programming for Chemists," Benjamin, New York (1965).

C. R. Bauer and A. P. Peluso, "Basic FORTRAN IV with WATFOR and WATFIV," Addison-Wesley, Reading, Mass. (1974).

P. Henrici, "Computational Analysis with the HP-25 Pocket Calculator," Wiley, New York (1977).

C. B. Kreitzberg and B. Shneiderman, "The Elements of FORTRAN Style," Harcourt Brace Jovanovich, New York (1972).

D. D. McCracken, "A Simplified Guide to FORTRAN Programming," Wiley, New York (1974).

C. J. Sass, "FORTRAN IV Programming and Applications," Holden-Day, San Francisco (1974).

Listed below are the most common meanings of those symbols which occur frequently in this book; special usages of these symbols and the meanings of any unlisted symbols are defined in the text wherever they occur.

Symbol	Meaning
a	Activity
c	Concentration, speed of light
a, b, c	Crystal unit cell dimensions
d	Molecular diameter, density, Bragg lattice-plane spacing
e	Electronic charge, base of natural logarithms
f	Force, function
g	Acceleration due to gravity, gas, gyromagnetic ratio
h	Planck's constant, height
i	Electric current
k	Boltzmann's constant, specific rate constant
i, j, k	Indices of array elements (in vectors, matrices)
l	Liquid, length
h, k, l	Crystal Miller indices
$\bar{l}$	Mean free path
m	Mass, mass of molecule, molality
n	Number of molecules, index of refraction
$\bar{n}$	Concentration in molecules per cubic centimeter
p	Pressure

Symbol	Meaning
q	Heat absorbed by the system
r	Radius
s	Solid
t	Celsius (centigrade) temperature, time
u	Root-mean-square velocity
v	Velocity, vibrational quantum number
w	Work done by the system, statistical weight
x, y, z	Linear coordinates
z	Valence of an ion, molecular collision frequency, molecular partition function
A	Helmholtz free energy, area, absorbancy
B	Second virial coefficient
C	Heat capacity, capacitance, number of components
C_p	Heat capacity at constant pressure
C_v	Heat capacity at constant volume
D	Diffusion constant
D_0, D_e	Dissociation energy, referenced to ground state and potential minimum respectively
E	Internal energy, potential difference, scalar electric field intensity
F	Distribution of χ^2 ratios
G	Gibbs free energy
H	Enthalpy (heat content), scalar magnetic-field intensity
I	Intensity of radiation, moment of inertia
J	Rotational quantum number
K	Equilibrium constant, coefficient of thermal conductivity
K_f	Molal freezing constant
L	Specific conductance
M	Molecular weight, molarity
N	Number of moles, number of equivalents, normality
N_0	Avogadro's number
P	Number of phases
P_M	Molar polarization
Q	Electric charge, generalized thermodynamic quantity, canonical partition function, vibrational coordinate
R	Gas constant, resistance
S	Entropy
S^2	Statistical variance
$S_{(1)}^2$	Variance of an observation of unit weight ("goodness of fit")
T	Absolute temperature
U	Potential energy, ionic mobility
V	Volume
W	Weight
X, Y	Mole fraction
Z	Molar collision frequency, atomic number, number of formula units per unit cell

Symbol	Meaning	
E	Electric field intensity	
F	Local (internal) electric field intensity	
H	Magnetic field intensity	vector quantities
I	Magnetization	
P	Polarization	

$\mathscr{E}$	Electromotive force (emf)
$\mathscr{F}$	Faraday constant
$\mathscr{R}$	Rydberg constant

α	Thermal-expansion coefficient, degree of dissociation, polarizability, optical rotation, adjustable parameter (least squares)
α_0	Distortion polarizability
β	Coefficient of compressibility, Bohr magneton
γ	Activity cofficient, surface tension
δ	Deviation, chemical shift (NMR)
ε	Molecular energy, dielectric constant, error, extinction coefficient
η	Coefficient of viscosity
θ	Surface coverage, angle (e.g., Bragg angle)
λ	Wavelength, equivalent ionic conductance, limit of error
μ	Ionic strength, chemical potential, Joule-Thomson coefficient, dipole moment, reduced mass
ν	Frequency
$\tilde{\nu}$	Wavenumber (in cm^{-1})
ρ	Density
σ	Molecular area or dimension, standard error, order parameter
τ	Relaxation time
ϕ	Apparent molal volume, angle
χ	Magnetic susceptibility
χ^2	Distribution of goodness of fit

Θ	Debye characteristic temperature
Λ	Equivalent conductance
Π	Osmotic pressure

aq	Aqueous solution
ln	Natural logarithm
log	Logarithm to the base 10
pH	$-\log(a_{H^+})$
°C	Degree Celsius (centigrade)
°K	Degree Kelvin
Q^0	Any thermodynamic property Q of a substance in its standard state
$\tilde{Q}$	Molal quantity Q
$\bar{Q}_A$	Partial molal quantity Q for component A

APPENDIX B
BAROMETER CORRECTIONS

The entries in the table below are calculated from Eq. (XVII-23) on the assumption that the barometer has a *brass* scale graduated to be accurate at 0°C (see Chap. XVIII). These corrections should be **subtracted** from the observed barometer readings. (If the brass scale is accurate at 20°C, the appropriate corrections are approximately 0.3 mm greater than those given.) Once the barometer reading has been corrected to 0°C, the pressure is referred to as Torr rather than mm Hg.

t, °C	720 mm	740 mm	760 mm	780 mm	800 mm
16	1.88	1.93	1.98	2.03	2.09
17	1.99	2.05	2.10	2.16	2.22
18	2.11	2.17	2.23	2.29	2.35
19	2.23	2.29	2.35	2.41	2.48
20	2.34	2.41	2.47	2.54	2.60
21	2.46	2.53	2.60	2.67	2.73
22	2.58	2.65	2.72	2.79	2.86
23	2.69	2.77	2.84	2.92	2.99
24	2.81	2.89	2.97	3.05	3.12
25	2.93	3.01	3.09	3.17	3.25
26	3.04	3.13	3.21	3.30	3.38
27	3.16	3.25	3.34	3.42	3.51
28	3.28	3.37	3.46	3.55	3.64
29	3.39	3.49	3.58	3.68	3.77
30	3.51	3.61	3.71	3.80	3.90

APPENDIX C

CONCENTRATION UNITS FOR SOLUTIONS

Name	Symbol	Definition
Weight percent	%	(Grams of solute per grams of solution) $\times$ 100
Mole fraction[a]	X_A	Moles of A per total number of moles
Molarity	M	Moles of solute per liter of solution
Normality	N	Equivalents of solute per liter of solution
Formality[b]	F	Formula weights of solute per liter of solution
Molality	m	Moles of solute per 1000 g of solvent
Weight formality[b]	f	Formula weights of solute per 1000 g of solvent

[a] The symbol Y_A is often used for the mole fraction of A in a gas phase which is in equilibrium with a liquid solution.

[b] These units are infrequently used but are of great convenience in expressing the overall composition of a solution when the solute is partially associated or dissociated.

APPENDIX D
SAFETY

Electrical hazards Several experiments make use of 110-V ac or dc electrical power and employ apparatus in which exposed metal parts are "live." These may include innocent-looking potentiometer connections when a potentiometer is employed to measure potentials or current in such a circuit. If the laboratory table has a metal surface, cover it with an insulating sheet of plywood or other material before assembling an electrical circuit. Remember that metal fixtures of all kinds and pipes or tubes of any kind that carry water are usually grounded. Turn off all electrical apparatus before altering circuits, if possible; if apparatus must be left on, use properly insulated test prods and leads. Be on the lookout for charged condensers, which may not be discharged owing to a broken circuit or a defective bleeder resistor. Naturally, 220 V provides a greater hazard than 110 V. It should be kept in mind that the laboratory is often served with 220 V in a three-wire system, with 110 V each side of ground. If 110-V outlets are supplied with a ground wire and one side of the 220-V line, as much as a 220-V difference can be obtained in accidental contact between circuits plugged into outlets serviced by opposite sides of the 220-V line. Shock, if it does occur, can be a serious matter; medical help should be summoned at once. Keep the victim quiet and comfortable; administer no stimulants of any kind.

Chemical hazards These are many and varied. It should be taken for granted that any chemical substance taken by mouth or inhaled is toxic until and unless definite assurance has been given to the contrary. Reactions that evolve toxic fumes or vapors or entail risk of fire should always take place in a fume hood. Poisonous solutions (such as cyanides) and even doubtful ones must not be pipetted by mouth; use a rubber bulb. Mercury vapor can attain a hazardous concentration in the laboratory atmosphere. Mercury should be kept in covered vessels at all times. Spills should be carefully cleaned up (a capillary tube attached to a suction flask is convenient for this), and inaccessible droplets in floor cracks and hard-to-reach places should be covered with a light dusting of powdered sulfur. Another insidious hazard is that of vapors from organic solvents. Such solvents should not be used indiscriminately for cleaning purposes, and spills should be avoided. Good ventilation is important.

Environmental exposure to chemical hazards is currently a subject of increasing concern and awareness. Zero exposure or zero risk of exposure is impossible in practice either in the chemical laboratory or elsewhere. Part of the professional role of chemists is to acquire knowledge and to develop judgement as to which precautions are necessary to limit these risks. One should not be blindly afraid of every chemical in the laboratory, nor should one be foolishly fearless. Many chemical hazards can be avoided by simply not eating or drinking or breathing large quantities of the chemical. Most chemical poisons are eliminated from the body, so that the effects of exposure gradually diminish. However, some poisons are not eliminated completely, and they accumulate, usually in particular tissues. (For example, carbon tetrachloride accumulates in the liver.)

Recently, chronic exposure to low levels of certain chemicals has been shown to increase significantly the incidence of cancer. Such chemicals are referred to as carcinogens. The following list gives those chemicals classified as strong carcinogens by OSHA (Occupational Safety and Health Administration):

1	2-Acetylaminofluorene	11	β-Naphthylamine
2	4-Aminobiphenyl	12	4-Nitrobiphenyl
3	Benzidine (salts implied)	13	N-Nitrosodimethylamine
4	Bis(chloromethyl) ether	14	β-Propiolactone
5	Chloromethyl methyl ether	15	Vinyl chloride
6	3,3'-Dichlorobenzidine	16	Coal tar pitch volatiles
7	4-Dimethylaminoazobenzene	17	Asbestos
8	Ethylenimine	18	Benzene
9	4,4'-Methylene-bis(2-chloroaniline)	19	Acrylonitrile
10	α-Naphthylamine		

None of these chemicals is used in the experiments described in this book. There are other chemicals thought to be weak carcinogens on the basis of statistical inferences involving data from long-term studies on large numbers of subjects. In these cases, there may be a hazard in industrial settings from chronic exposure to vapors or direct contact with the skin, but the risk from brief use in a research or educational setting is not serious.

Chemical burns Strong acids (particularly oxidizing acids) and bases may cause severe burns to the skin. If skin contact is made, wash copiously with water. If the exposure is to a strong acid, washing with a very dilute weak base (ammonia) is helpful; for a strong base use a very dilute weak acid (acetic acid). Particular attention should be directed to eye protection (use of *safety glasses*, goggles, or a face shield) and to prompt and effective action should any chemical agent get into the eyes; a strong base such as sodium hydroxide can permanently destroy the cornea in a few seconds. *Speed is all-important* in getting the exposed individual to an eyewash fountain or other source of copiously flowing (but low stream-pressure) water and *thoroughly* bathing the eyeball. The eyelids should be lifted away from the eyeball to facilitate effective washing. Use nothing but water. *Get medical help promptly.*

Fire and explosion Any flammable substance provides a potential fire hazard. In experiments which make use of hydrogen gas or other flammable gases, not only open flames but also cigarettes and sparking electrical contacts provide the possibility of explosion. The distillation of flammable liquids must be carried out in the absence of open flames; use a steam bath or electrical heating mantle. If an experiment involves an irreducible risk of fire or explosion, arrange for an adequate barrier. *Safety glasses* are strongly recommended in all circumstances in which fire or explosion is a possible eventuality. Know the location of the nearest water and fire extinguishers (use water only on paper or cloth fires). In the event of serious burns, do not apply ointments or medications; summon medical help.

Radiation hazards Ultraviolet light from a mercury lamp or carbon arc is highly damaging to the eyes. Ordinary glasses give some protection, but the experimental arrangement should be well shielded so as to decrease the possibility of accidental exposure to a minimum. Prolonged exposure of the skin to such radiation can produce a severe "sunburn." An optical laser beam, even from a laser of very low power, which enters directly into the eye or is accidentally reflected into the eye by a surface can cause irreparable damage to the retina as a result of focusing by the lens of the eye. Eyeglasses provide no protection from this hazard. Exposure to strong radio-frequency or microwave fields can "cook" tissue and produce deep internal burns. Exposure to x-rays and to the radiation from radioactive materials must be carefully guarded against in experiments dealing with them. Any such experiments should be done under the direct supervision of an experienced research worker who will assume personal responsibility for all required safety measures, and under an appropriate license if radioactive materials are involved.

Mechanical and other hazards Most mechanical hazards are too clearly apparent to warrant mention here. Vacuum systems often carry a hazard of collapse or implosion; bulbs more than 1 liter in volume should be surrounded by a metal screen or else wrapped with electrical tape to reduce hazard from flying glass particles in the event of implosion. The bursting of a container due to overpressure is a frequent cause of accident or injury. A compressed-air line (usual pressure: of the order of 50 psi) should never be connected to a closed system containing rubber tubing or glass bulbs. No closed system, except a properly designed combustion bomb, should be attached to a cylinder of compressed gas (usual maximum pressure: about 3000 psi) unless a suitable reducing valve is attached; even then, a relief valve should be provided to guard against accidental overpressure. Gas cylinders must be chained or strapped to prevent their falling over. The protective cap must be in place whenever a gas cylinder is being moved; cylinders should be moved with an appropriate hand truck, not dragged across the floor. Mechanical pumps must have belt guards.

Safety equipment   *Safety glasses* have been mentioned in this appendix and in other places in this book in connection with specific hazards. However, use of safety glasses (plain or ground to prescription) equipped with side shields, or other approved means of eye protection (plastic goggles alone or over ordinary prescription glasses, plastic face shields) *at all times in the laboratory* is increasingly becoming mandatory in instructional laboratories as it has already generally become in industrial laboratories, and is strongly recommended as standard laboratory policy. In any case, omission of this precaution in circumstances where there exists the possibility of fire, explosion, implosion, spattering of caustic chemical agents, or flying fragments from machine-shop operations, is nothing short of foolhardy. The laboratory should be equipped with a conveniently accessible safety shower and an eye-wash fountain; more than one of each in a large laboratory. Increasingly, the fixed type of eye-wash fountain is being superseded by a spray nozzle at the end of an extensible hose; there should be one of these on each laboratory bench. In lieu of such devices—or in addition to them—2- or 3-ft lengths of rubber hose (*not* small-bore pressure tubing) attached with wire or clamps to water faucets are certainly better than nothing.

The laboratory should have convenient access to one or more fume hoods (with a face velocity of at least 100 ft/min) for any operations involving more than insignificant quantities of volatile chemicals in open containers.

An approved fire extinguisher [the "dry chemical" (bicarbonate) type is preferred, but the CO_2 type is satisfactory] should be mounted near at least one exit and refilled after every use, no matter how small. The laboratory should be arranged so as to provide two or more avenues of escape from any experimental setup in case of emergency. A first-aid kit containing Band-Aids, sterile gauze, adhesive tape, petroleum jelly, a mild antiseptic, sterile cotton swabs, tweezers, a set of sewing needles, a packet of razor blades, and a quick-reference first-aid manual will provide adequately for most emergencies.

The location of an inhalator, a stretcher, and other rescue equipment, if not in the laboratory itself, should be known. The telephone number of the nearest medical emergency room and the local ambulance service should be posted conspicuously. Instructions for emergency evacuation, including special procedures for evacuating physically handicapped persons, should also be posted. An evacuation drill held near the beginning of each academic term is recommended.

Under no circumstances should a person be allowed to work in the laboratory alone.

Finally, safety depends on habits that must be gained outside the laboratory as well as inside. Thus, on your way to and from the laboratory, look both ways before crossing the street; after finishing the writing of that laboratory report, don't smoke in bed!

APPENDIX E
RESEARCH JOURNALS

Out of the hundreds of different research periodicals currently being published, the 43 listed below describe research of special interest to physical chemists. A very complete tabulation of periodicals, the correct abbreviation of their titles, and their distribution in American and Canadian libraries can be found in "ACCESS: Key to the Source Literature of the Chemical Sciences," American Chemical Society (1969). In 1970 this volume was renamed "Chemical Abstracts Service Source Index," and a *CAS Source Index Quarterly* has been published since then to update the listings. Many libraries will still have the volume under its original title.

Abbreviated Title	*Full Title*
Accounts Chem. Res.	Accounts of Chemical Research
Acta Crystallogr. Sect. A,B	Acta Crystallographica
Appl. Spectrosc.	Applied Spectroscopy
Bull. Chem. Soc. Jap.	Bulletin of the Chemical Society of Japan (no papers in Japanese!)
Can. J. Chem.	Canadian Journal of Chemistry
Can. J. Phys.	Canadian Journal of Physics
Chem. Phys. Lett.	Chemical Physics Letters
C. R. Acad. Sci.	Comptes Rendus Hebdomadaires des Seances de l'Academie des Sciences, serie C
Discuss. Faraday Soc.	Discussions of the Faraday Society

Abbreviated Title	*Full Title*
Dokl. Phys. Chem.	Doklady Physical Chemistry (English translation of Doklady Akademii Nauk SSSR: Seriya Fiz. Khim.)
Helv. Chim. Acta	Helvetica Chimica Acta
Infrared Phys.	Infrared Physics
J. Amer. Chem. Soc.	Journal of the American Chemical Society
J. Appl. Phys.	Journal of Applied Physics
J. Chem. Educ.	Journal of Chemical Education
J. Chem. Phys.	Journal of Chemical Physics
J. Chem. Soc., Faraday I, II	Journal of the Chemical Society, Faraday transactions I and II
J. Chem. Thermodyn.	Journal of Chemical Thermodynamics
J. Geophys. Res.	Journal of Geophysical Research
J. Inorg. Nucl. Chem.	Journal of Inorganic and Nuclear Chemistry
J. Mol. Spectrosc.	Journal of Molecular Spectroscopy
J. Phys. A,B,C,D	Journal of Physics, four sections
J. Phys. Chem.	Journal of Physical Chemistry
J. Phys. Chem. Solids	Journal of Physics and Chemistry of Solids
J. Phys. Soc. Jap.	Journal of Physical Society of Japan
J. Polym. Sci.	Journal of Polymer Science
J. Quant. Spectrosc. Radiat. Transfer	Journal of Quantitative Spectroscopy and Radiative Transfer
Opt. Spectrosc.	Optics and Spectroscopy (English translation of Optika i Spektroskopiya)
Phil. Mag.	Philosophical Magazine
Physica	Physica
Phys. Lett.	Physics Letters
Phys. Rev. A,B	Physical Review, Section A and B
Phys. Rev. Lett.	Physical Review Letters
Proc. Roy. Soc. (London), Ser. A.	Proceedings of the Royal Society of London, Series A, Mathematical and Physical Sciences
Pure Appl. Chem.	Pure and Applied Chemistry
Rev. Sci. Instrum.	Review of Scientific Instruments
Russ. J. Phys. Chem.	Russian Journal of Physical Chemistry (English translation of Zhurnal Fizicheskoi Khimii)
Sov. Phys.-Crystallogr.	Soviet Physics, Crystallography (English translation of Kristallografiya)
Sov. Phys.-JETP	Soviets Physics-JETP (English translation of Zh. Eksp. Teor. Fiz.)
Spectrochim. Acta	Spectrochimica Acta
Surface Sci.	Surface Science
Trans. Faraday Soc.	Transactions of the Faraday Society (after 1971 see J. Chem. Soc.)
Z. Phys. Chem.	Zeitschrift für Physikalische Chemie (Frankfurt am Main)

INDEX

1. Values of Defined Constants

Standard gravity, 980.665 cm sec^{-2}
Standard atmosphere, 1,013,250 dyne cm^{-2} = 101325 pascal
Torr, 1/760 standard atm
Thermochemical calorie, 4.184 abs joule

Temperature of the triple point of water $\begin{cases} 273.16° \text{ Kelvin} \\ 0.01° \text{ Celsius} \end{cases}$

2. Values of Basic and Derived Constants

Velocity of light (c), 2.997925 $\times$ 10^{10} cm sec^{-1}
Planck's constant (h), 6.62559 $\times$ 10^{-27} erg sec
Faraday constant ($\mathscr{F}$), 96,487.0 coulomb equiv^{-1}
Avogadro's constant (N_0), 6.02252 $\times$ 10^{23} mol^{-1}
Absolute temperature of ice point (T_0), 273.150°K
Gas constant (R), 82.055 cm^3 atm deg^{-1} mol^{-1}
             8.31434 joule deg^{-1} mol^{-1}
             1.9872 cal deg^{-1} mol^{-1}
Electronic charge (e), 1.60210 $\times$ 10^{-19} coulomb
Boltzmann constant (k), 1.38054 $\times$ 10^{-16} erg deg^{-1}
Atomic mass unit (amu), 1.66043 $\times$ 10^{-24} g

3. Conversion Factors

1 inch = 2.54 cm

1 Ångstrom = 10^{-8} cm = 0.1 nm

1 liter = 1000.028 cm^3

1 bar = 10^6 dyne cm^{-2} = 0.1 MPa $\simeq$ 0.9869 atm

1 pascal = 10^{-5} bar = 9.8692 $\times$ 10^{-6} atm = 7.501 mTorr

1 joule = 10^7 erg
      = 9.8691 cm^3 atm
      = 0.23901 cal

1 l-atm = 24.219 cal = 101.323 J

1 coulomb = 0.1 emu = 2.99793 $\times$ 10^9 esu
       = 0.23901 cal volt^{-1}

1 electron volt = 1.60210 $\times$ 10^{-12} erg molecule^{-1}
          = 23.061 kcal mol^{-1} = 96.487 kJ mol^{-1}
          = 8065.7 cm^{-1}

ln x = 2.302585 log x

† These values are taken from E. R. Cohen and J. W. M. Du Mond, *Rev. Mod. Phys.*, 37, 537 (1965).